Fortschritte der Chemie organischer Naturstoffe

Progress in the Chemistry of Organic Natural Products

65

Founded by L. Zechmeister
Edited by W. Herz, G. W. Kirby, R. E. Moore,
W. Steglich, and Ch. Tamm

Author:
Y. Asakawa

Springer-Verlag
Wien New York 1995

Prof. W. HERZ, Department of Chemistry,
The Florida State University, Tallahassee, Florida, U.S.A.

Prof. G. W. KIRBY, Chemistry Department,
The University, Glasgow, Scotland

Prof. R. E. MOORE, Department of Chemistry,
University of Hawaii at Manoa, Honolulu, Hawaii, U.S.A.

Prof. Dr. W. STEGLICH, Institut für Organische Chemie der Universität
München, München, Federal Republic of Germany

Prof. Dr. CH. TAMM, Institut für Organische Chemie der Universität Basel,
Basel, Switzerland

© 1995 by Springer-Verlag/Wien
Softcover reprint of the hardcover 1st edition 1995

Library of Congress Catalog Card Number AC 39-1015

Typesetting: Macmillan India Ltd., Bangalore-25

Printed on acid-free and chlorine-free bleached paper

With 2 Figures

ISSN 0071-7886
ISBN-13:978-3-7091-7427-2   e-ISBN-13:978-3-7091-6896-7
DOI: 10.1007/978-3-7091-6896-7

Contents

List of Contributors IX

Chemical Constituents of the Bryophytes. By Y. ASAKAWA. 1

I. Introduction 5
II. Chemical Constituents of Hepaticae (Liverworts) 12
    1. Monoterpenoids 12
    2. Tris-normonoterpenoids 15
    3. Homomonoterpenoids 15
    4. Tris-norsesquiterpenoids 24
    5. Sesquiterpenoids 25
       5.1 Acoranes and Noracoranes 26
       5.2 Africanes, Norafricanes and Secoafricanes 27
       5.3 Aristolanes 32
       5.4 Aromadendranes, Secoaromadendranes and
          Norsecoaromadendranes 33
       5.5 Azulenes and Indenes 42
       5.6 Barbatanes (Gymnomitranes) 47
       5.7 Bazzananes 49
       5.8. Bergamotanes, Bicycloelemanes and Elemanes 51
       5.9 Bicyclogermacranes, Isobicyclogermacranes and Lepidozanes 54
       5.10 Bisabolanes, Bourbonanes and Brasilanes 57
       5.11 Cadinanes, Amorphanes and Muurolanes 60
       5.12 Calamenanes, Calacoranes and Cadalenes 61
       5.13 Caryophyllanes, Cedranes and Chamigranes 65
       5.14 Chiloscyphanes, Oppositanes, Copaanes and Cubebanes 136
       5.15 Cuparanes and Herbertanes (Isocuparanes) 140
       5.16 Daucanes (Carotanes) and Drimanes 152
       5.17 Eremophilanes and Eudesmanes 157
       5.18 Farnesanes and Germacranes 171
       5.19 Gorgonanes and Guaianes 176
       5.20 Himachalanes and Humulanes 177
       5.21 Longifolanes, Longibornanes, Longipinanes and Longicyclanes 179
       5.22 Maalianes and Monocyclofarnesanes 182
       5.23 Myltaylanes and Cyclomyltaylanes 186
       5.24 Pacifigorgianes and Patchoulanes 188
       5.25 Pinguisanes and Norpinguisanes 190

5.26 Santalanes, Spirovetivanes, Thujopsanes and Valencanes 202
5.27 Vitranes, Widdranes and Zieranes 205
5.28 Miscellaneous Sesquiterpenoids. 208
6. Diterpenoids 210
    6.1 Abietanes and Cembranes 211
    6.2 Clerodanes, Secoclerodanes and Spiroclerodanes 212
    6.3 Dolabellanes 219
    6.4 Fusicoccanes 226
    6.5 Kauranes 229
    6.6 Labdanes and Chettaphanins 236
    6.7 Phytanes, Pimaranes and Rearranged Pimaranes 246
    6.8 Sacculatanes 268
    6.9 Sphenolobanes 272
    6.10 Trachylobanes and Verrucosanes 273
    6.11 Verticillanes 278
7. Steroids and Triterpenoids 280
8. Aromatic Compounds. 296
    8.1 Benzoic Acid and Cinnamic Acid Derivatives 296
    8.2 Bibenzyls 296
    8.3 Bis-Bibenzyls 335
    8.4 Long Chain Alkyl Phenols 351
    8.5 Naphthalenes and Isocoumarins 352
    8.6 Neolignan, Phenanthrenes and Phthalides 355
    8.7 Miscellaneous Aromatic Compounds 358
    8.8 Flavonoids 361
9. Lipids 380
    9.1 n-Alkanes and Related Compounds 380
    9.2 Fatty Acids 381
10. Sulfur-Containing Compounds 384
11. Carbohydrates 384
III. Chemical Constituents of Musci (Mosses) 391
    1. Mono- and Diterpenoids 391
    2. Steroids, Triterpenoids and Carotenoids 392
    3. Aromatic Compounds. 403
        3.1 Benzoic and Cinnamic Acid Derivatives 403
        3.2 Flavonoids 408
            3.2.1 Flavones 408
            3.2.2 Isoflavones 422
            3.2.3 Biflavones 422
            3.2.4 Aurones and Anthocyanins 428
        3.3 Benzonaphthoxanthenones 432
    4. Lipids 432
        4.1 n-Alkanes and Related Compounds 432
        4.2 Fatty Acids 433
    5. Miscellaneous 456
IV. Chemical Constituents of Anthocerotae (Hornworts) 460
    1. Sesquiterpenoids 460
    2. Aromatic Compounds. 460
        2.1 Cinnamic Acid Derivatives 460
        2.2 Lignans. 461

V. Biologically Active Substances of Bryophytes 464
   1. Characteristic Scents 465
   2. Pungency, Bitterness and Sweetness 468
   3. Allergenic Contact Dermatitis 469
   4. Antitumor Activity 470
   5. Antimicrobial and Antifungal Activity 473
   6. Insect Antifeedant and Molluscicidal Activity 475
   7. Plant Growth Regulatory Activity 477
   8. Superoxide Release Inhibitory Activity 479
   9. 5-Lipoxygenase, Calmodulin and Thromboxane Synthetase
      Inhibitory Activity 479
  10. Vasopressin (VP) Antagonist and Cardiotonic Activity 480
  11. Piscicidal Activity 480
  12. Neuritic Sprouting Activity 481
  13. Muscle Relaxing Activity 481
  14. Miscellaneous 482
VI. Chemosystematics of Hepaticae 483
   1. Jungermanniidae 486
     1.1 Metzgeriales 486
       1.1.1 Metzgeriaceae 486
       1.1.2 Aneuraceae (Riccardiaceae) 486
       1.1.3 Pallaviciniaceae 487
       1.1.4 Blasiaceae 487
     1.2 Calobryales 488
       1.2.1 Haplomitriaceae 488
     1.3 Jungermanniales 488
       1.3.1 Jungermanniaceae 488
       1.3.2 Lophoziaceae 489
       1.3.3 Gymnomitriaceae (Marsupellaceae) 490
       1.3.4 Arnelliaceae 490
       1.3.5 Plagiochilaceae 490
       1.3.6 Lophocoleaceae 491
       1.3.7 Scapaniaceae 492
       1.3.8 Balantiopsidaceae 492
       1.3.9 Adelanthaceae 493
       1.3.10 Schistochilaceae 493
       1.3.11 Antheliaceae 494
       1.3.12 Lepidoziaceae 494
       1.3.13 Calypogeiaceae 495
       1.3.14 Isotachidaceae 495
       1.3.15 Trichocoleaceae 495
       1.3.16 Ptilidiaceae 495
       1.3.17 Herbertaceae 496
       1.3.18 Radulaceae 496
       1.3.19 Pleuroziaceae 497
       1.3.20 Porellaceae 498
       1.3.21 Frullaniaceae 499
       1.3.22 Lejeuneaceae 499
   2. Marchantiidae 501
     2.1 Sphaerocarpales 501

2.2 Monocleales 501
2.3 Marchantiales. 502
   2.3.1 Targioniaceae 502
   2.3.2 Aytoniaceae (Grimaldiaceae) 502
   2.3.3 Conocephalaceae 504
   2.3.4 Lunulariaceae 505
   2.3.5 Marchantiaceae 505
   2.3.6 Ricciaceae 507
VII. Chemosystematics of Musci and Anthocerotae 507
VIII. Chemical Relationships Between Algae, Bryophytes and
Pteridophytes–Evolution of Bryophytes 509
   1. Similarities and Differences in Terpenoid and Steriod Content. 511
   2. Similarities and Differences in Content of Aromatic Compounds 520
   3. Similarities and Differences in Alkane and Fatty Acid Content 522
Acknowledgements 525
References 525

Author Index 563

Subject Index 575

List of Contributors

ASAKAWA, Prof. Dr. Y., Faculty of Pharmaceutical Sciences, Tokushima Bunri University, Yamashiro-cho, Tokushima 770, Japan.

Chemical Constituents of the Bryophytes

Y. ASAKAWA, Faculty of Pharmaceutical Sciences,
Tokushima Bunri University, Tokushima, Japan

Contents

I. Introduction	5
II. Chemical Constituents of Hepaticae (Liverworts)	12
1. Monoterpenoids	12
2. Tris-normonoterpenoids	15
3. Homomonoterpenoids	15
4. Tris-norsesquiterpenoids	24
5. Sesquiterpenoids	25
5.1 Acoranes and Noracoranes	26
5.2 Africanes, Norafricanes and Secoafricanes	27
5.3 Aristolanes	32
5.4 Aromadendranes, Secoaromadendranes and Norsecoaromadendranes	33
5.5 Azulenes and Indenes	42
5.6 Barbatanes (Gymnomitranes)	47
5.7 Bazzananes	49
5.8 Bergamotanes, Bicycloelemanes and Elemanes	51
5.9 Bicyclogermacranes, Isobicyclogermacranes and Lepidozanes	54
5.10 Bisabolanes, Bourbonanes and Brasilanes	57
5.11 Cadinanes, Amorphanes and Muurolanes	60
5.12 Calamenanes, Calacoranes and Cadalanes	61
5.13 Caryophyllanes, Cedranes and Chamigranes	65
5.14 Chiloscyphanes, Oppositanes, Copaanes and Cubebanes	136
5.15 Cuparanes and Herbertanes (Isocuparanes)	140
5.16 Daucanes (Carotanes) and Drimanes	152
5.17 Eremophilanes and Eudesmanes	157
5.18 Farnesanes and Germacranes	171
5.19 Gorgonanes and Guaianes	176
5.20 Himachalanes and Humulanes	177
5.21 Longifolanes, Longibornanes, Longipinanes and Longicyclanes	179
5.22 Maalianes and Monocyclofarnesanes	182
5.23 Myltaylanes and Cyclomyltaylanes	186
5.24 Pacifigorgianes and Patchoulanes	188
5.25 Pinguisanes and Norpinguisanes	190
5.26 Santalanes, Spirovetivanes, Thujopsanes and Valencanes	202

5.27 Vitranes, Widdranes and Zieranes. 205
5.28 Miscellaneous Sesquiterpenoids. 208
6. Diterpenoids 210
    6.1 Abietanes and Cembranes 211
    6.2 Clerodanes, Secoclerodanes and Spiroclerodanes 212
    6.3 Dolabellanes. 219
    6.4 Fusicoccanes 226
    6.5 Kauranes. 229
    6.6 Labdanes and Chettaphanins 236
    6.7 Phytanes, Pimaranes and Rearranged Pimaranes 246
    6.8 Sacculatanes. 268
    6.9 Sphenolobanes 272
    6.10 Trachylobanes and Verrucosanes 273
    6.11 Verticillanes 278
  7. Steroids and Triterpenoids 280
  8. Aromatic Compounds. 296
    8.1 Benzoic Acid and Cinnamic Acid Derivatives 296
    8.2 Bibenzyls 296
    8.3 Bis-Bibenzyls 335
    8.4 Long Chain Alkyl Phenols. 351
    8.5 Naphthalenes and Isocoumarins. 352
    8.6 Neolignans, Phenanthrenes and Phthalides 355
    8.7 Miscellaneous Aromatic Compounds 358
    8.8 Flavonoids 361
  9. Lipids 380
    9.1 n-Alkanes and Related Compounds 380
    9.2 Fatty Acids 381
  10. Sulfur-Containing Compounds 384
  11. Carbohydrates 384
III. Chemical Constituents of Musci (Mosses) 391
  1. Mono- and Diterpenoids. 391
  2. Steroids, Triterpenoids and Carotenoids 392
  3. Aromatic Compounds. 403
    3.1 Benzoic and Cinnamic Acid Derivatives 403
    3.2 Flavonoids 408
        3.2.1 Flavones 408
        3.2.2 Isoflavones 422
        3.2.3 Biflavones 422
        3.2.4 Aurones and Anthocyanins 428
    3.3 Benzonaphthoxanthenones. 432
  4. Lipids 432
    4.1 n-Alkanes and Related Compounds 432
    4.2 Fatty Acids 433
  5. Miscellaneous. 456
IV. Chemical Constituents of Anthocerotae (Hornworts) 460
  1. Sesquiterpenoids 460
  2. Aromatic Compounds 460
    2.1 Cinnamic Acid Derivatives. 460
    2.2 Lignans 461
 V. Biologically Active Substances of Bryophytes. 464
  1. Characteristic Scents 465
  2. Pungency, Bitterness and Sweetness 468

3. Allergenic Contact Dermatitis 469
4. Antitumor Activity 470
5. Antimicrobial and Antifungal Activity. 473
6. Insect Antifeedant and Molluscicidal Activity 475
7. Plant Growth Regulatory Activity 477
8. Superoxide Release Inhibitory Activity 479
9. 5-Lipoxygenase, Calmodulin and Thromboxane Synthetase
   Inhibitory Activity 479
10. Vasopressin (VP) Antagonist and Cardiotonic Activity 480
11. Piscicidal Activity 480
12. Neuritic Sprouting Activity 481
13. Muscle Relaxing Activity 481
14. Miscellaneous 482
VI. Chemosystematics of Hepaticae 483
1. Jungermanniidae 486
   1.1 Metzgeriales. 486
       1.1.1 Metzgeriaceae 486
       1.1.2 Aneuraceae (Riccardiaceae) 486
       1.1.3 Pallaviciniaceae 487
       1.1.4 Blasiaceae 487
   1.2 Calobryales 488
       1.2.1 Haplomitriaceae 488
   1.3 Jungermanniales 488
       1.3.1 Jungermanniaceae. 488
       1.3.2 Lophoziaceae 489
       1.3.3 Gymnomitriaceae (Marsupellaceae). 490
       1.3.4 Arnelliaceae 490
       1.3.5 Plagiochilaceae. 490
       1.3.6 Lophocoleaceae 491
       1.3.7 Scapaniaceae 492
       1.3.8 Balantiopsidaceae 492
       1.3.9 Adelanthaceae 493
       1.3.10 Schistochilaceae 493
       1.3.11 Antheliaceae 494
       1.3.12 Lepidoziaceae. 494
       1.3.13 Calypogeiaceae 495
       1.3.14 Isotachidaceae. 495
       1.3.15 Trichocoleaceae 495
       1.3.16 Ptilidiaceae. 495
       1.3.17 Herbertaceae 496
       1.3.18 Radulaceae. 496
       1.3.19 Pleuroziaceae 497
       1.3.20 Porellaceae. 498
       1.3.21 Frullaniaceae 499
       1.3.22 Lejeuneaceae 499
2. Marchantiidae 501
   2.1 Sphaerocarpales 501
   2.2 Monocleales. 501
   2.3 Marchantiales 502
       2.3.1 Targioniaceae 502
       2.3.2 Aytoniaceae (Grimaldiaceae) 502
       2.3.3 Conocephalaceae 504

2.3.4 Lunulariaceae 505
2.3.5 Marchantiaceae 505
2.3.6 Ricciaceae 507
VII. Chemosystematics of Musci and Anthocerotae 507
VIII. Chemical Relationships Between Algae, Bryophytes and
Pteridophytes – Evolution of Bryophytes 509
1. Similarities and Differences in Terpenoid and Steroid Content 511
2. Similarities and Differences in Content of Aromatic Compounds 520
3. Similarities and Differences in Alkane and Fatty Acid Content 522
Acknowledgements 525
References 525

Table Ia. Classification of Hepaticae (Liverworts).   6
Table Ib. Classification of Musci (Mosses).   8
Table Ic. Classification of Anthocerotae (Hornworts).   11
Table IIa. Mono-, Tris-normono- and Homomonoterpenoids Found in the
          Hepaticae.   16
Table IIb. Tris-norsesqui- and Sesquiterpenoids Found in the Hepaticae   66
Table IIc. Diterpenoids Found in the Hepaticae.   247
Table IId. Steroids and Triterpenoids Found in the Hepaticae.   281
Table IIe. Aromatic Compounds Found in the Hepaticae   298
Table IIf. Flavonoids Found in the Hepaticae   363
Table IIg. Chemical Constituents (Lipids and Miscellaneous) Found in the
          Hepaticae.   385
Table IIh. Sulfur-Containing Compounds Found in the Hepaticae   390
Table IIIa. Mono-, Di- and Triterpenoids, Steroids and Carotenoids Found
           in the Musci   393
Table IIIb. Aromatic Compounds Found in the Musci   404
Table IIIc. Aromatic Compounds (Flavones, Isoflavones, Biflavones, Aurones,
           Anthocyanins and Benzonaphthoxanthenones) Found
           in the Musci   410
Table IIId. Chemical Constituents (Lipids and Miscellaneous) Found in the
           Musci   434
Table IV. Terpenoids and Aromatic Compounds Found in the Anthocerotae . .   462
Table Va. Distribution of Terpenoids in Algae, Bryophytes and Pteridophytes . .   513
Table Vb. Distribution of Steroids in Algae, Bryophytes and Pteridophytes . . .   516
Table Vc. Distribution of Aromatic Compounds in Algae, Bryophytes and
          Pteridophytes   517
Table VI. Evolutionary Level of Algae, Bryophytes and Higher Plants by Fatty
          Acids and Sterols   520

The following tables (approximately 152 pages) are available only on disk as supplementary material and can be ordered directly from the author:

Table VII. Hepaticae Containing Terpenoids, Steroids, Aromatic Compounds,
           Lipids and Miscellaneous Compounds
Table VIII. Musci Containing Terpenoids, Aromatic Compounds, Lipids and
            Miscellaneous Compounds
Table IX. Anthocerotae Containing Terpenoids and Aromatic Compounds

I. Introduction

In the twelve years since the first review article dealing with chemical constituents of the Hepaticae appeared in this series as Volume 42 (*19*), several short reviews concerned with chemical constituents of bryophytes have been published (*22, 96, 144, 265, 271, 647, 649, 650*). In 1988, a Symposium on Chemistry and Chemical Taxonomy of Bryophytes was organised on the behalf of the Phytochemical Society of Europe; the proceedings of this meeting appeared as a book entitled Bryophytes: Their Chemistry and Chemical Taxonomy (*651*). The symposium concerned itself with phytochemical, biochemical, botanical, chemotaxonomical, pharmaceutical, biotechnological and environmental aspects of bryophytes as well as with the synthesis of the terpenoids and aromatic compounds bryophytes elaborate. The physiological and biochemical aspects of bryophytes have also been described in a recent book Bryophytes Development: Physiology and Biochemistry (*139*).

Almost all chemical structures presented in Volume 42 were established by 60, 90, or 100 MHz NMR spectroscopy. Recent development of NMR spectroscopy at high fields such as 400, 500 and 600 MHz and two-dimensional (2D) NMR techniques have resulted in a dramatic increase of papers concerned with structure determination of chemical constituents isolated not only from higher plants but also from bryophytes. By use of such techniques the structures of a few sesquiterpenoids and aromatic compounds described in Volume 42 have been revised. In the present review, isolation, structure determination and total synthesis of naturally occurring terpenoids, aromatic compounds and lipids of Hepaticae encountered after 1982, as well as their biological activity and the chemosystematics of Hepaticae, will be discussed. The organization follows that adopted in the previous review. In addition, however, the present review will also include the known terpenoids, aromatic compounds and lipids of Musci (mosses) and Anthocerotae (hornworts). It will also deal with the chemosystematics of both of these classes and the chemical relationships between algae, bryophytes and pteridophytes. Because of this, it is necessary to describe the classification of bryophytes in some detail.

The bryophytes are taxonomically placed between the algae and the pteridophytes and 25000 species are now known world-wide. They are divided into three classes, Musci (mosses, 14000 species), Hepaticae (liverworts, 6000 species) and Anthocerotae (hornworts, 300 species). In the modern classification of the bryophytes (*232, 301, 493–495*), the Hepaticae are divided into two subclasses and 7 orders (Table 1a). The Musci are divided into 7 subclasses and 15 orders (Table 1b) while the Anthocerotae comprises only three families (Table 1c).

Table 1a. *Classification of Hepaticae (Liverworts)*

Subclass: Jungermanniidae
  Order: Metzgeriales
    Suborder: Metzgeriineae
      Family: Metzgeriaceae: *Metzgeria, Apometzgeria*
              Phyllothalliaceae: *Phillothallia*
              Aneuraceae (= Riccardiaceae): *Aneura, Cryptothallus,*
                                              Riccardia
              Pelliaceae (= Dilaenaceae): *Pellia*
              Pallaviciniaceae: *Jensenia, Pallavicinia, Moerckia*
              Blasiaceae: *Blasia, Cavicularia*
    Suborder: Codoniaceae (= Fossombroniaceae)
      Family: Treubiaceae: *Treubia*
              Codoniaceae: *Fossombronia, Petalophyllum, Neteroclada*
              Hymenophytaceae: *Verdoonia, Symphyogyna, Allisonia,*
                                Hymenophyton Podomitrium,
                                Xenothullus
  Order: Takakiales
    Family: Takakiaceae: *Takakia*
  Order: Calobryales
    Family: Haplomitriaceae: *Haplomitrium*
  Order: Jungermanniales
    Suborder: Jungermanniineae
      Family: Jungermanniaceae
        Subfamily: Lophozioideae: *Chandonanthus, Barbilophozia,*
                                   Anastrepta, Lophozia, Gymnocolea,
                                   Sphenolobopsis, Sphenobus,
                                   Anastrophyllum, Tritomaria
                   Jamesonielloideae: *Jamesoniella*
                   Myliioideae: *Mylia*
                   Jungermannioideae: *Jungermannia, Nardia*
      Family: Gymnomitriaceae: Marsupella, *Gymnomitrion, Prasanthus*
              Acrobolbaceae: *Acrobolbus, Tylimanthus*
              Arnelliaceae: *Arenellia, Southbya, Gongylanthus*
              Plagiochilaceae: *Pedinophyllum, Plagiochila*
              Lophocoleaceae: *Heteroscyphus, Leptoscyphus, Lophocolea,*
              Chiloscyphus, Clasmatocolea
              Geocalycaceae: *Geocalyx, Harpanthus, Saccogyna*
              Scapaniaceae: *Douinia, Diplophyllum, Scapania*
      Family: Balantiopsidaceae: *Balantiopsis*
              Schistochilaceae: *Schistochila*
    Suborder: Lepidoziineae
      Family: Adelanthaceae
        Subfamily: Adelanthoideae: *Adelanthus*
                   Odontoschismatoideae: *Odontoschisma, Jackiella*
      Family: Cephaloziellaceae: *Cephaloziella*
              Cephaloziaceae: *Cephalozia, Nowellia, Cladopodiella,*
        Subfamily: Cephalozioideae: *Cephalozia, Nowellia, Cladopodiella,*
                                    Pleuroclada

Table 1a (*continued*)

Hygrobielloideae: *Hygrobiella*
Family: Antheliaceae: *Anthelia*
Lepidoziaceae
Subfamily: Lepidoziodeae: *Telaranea, Kurzia, Lepidozia*
Acromastigoideae (= Bazzaioideae): *Bazzania*
Zoopsidoideae: *Zoopsis, Bonneria, Paracromastigum*
Family: Calypogeiaceae: *Calypogeia*
Suborder: Ptilidiineae
Family: Isotachidaceae: *Isotachis, Neesioscyphus, Eoisotachis*
Pseudolepicoleaceae
Subfamily: Pseudolepicoleideae: *Pseudolepicolea*
Blepharostomatoideae: *Blepharostoma*
Family: Trichocoleaceae: *Trichocolea, Trichocoleopsis,*
Neotrichocolea
Ptilidiaceae: *Ptilidium*
Lepicoleaceae
Subfamily: Mastigophoroideae: *Mastigophora*
Family: Herbertaceae: *Herbertus, Triandrophyllum*
Suborder: Radulineae
Family: *Radula*
Suborder: Pleuroziineae
Family: Pleuroziaceae: *Pleurozia*
Suborder: Porellineae
Family: Porellaceae: *Porella*
Suborder: Jubulineae
Family: Frullaniaceae (= Jubulaceae): *Frullania, Jubula*
Lejeuneaceae
Subfamily: Ptychanthoideae: *Marchesinia, Dicranolejeunea*
Lejeuneodieae: *Cheilolejeunea, Harpalejeunea,*
Drepanolejeunea, Lejeunea
Cololejeuneoideae: *Colura, Cololejeunea*
Subclass: Marchantiidae
Order: Sphaerocarpales
Family: Riellaceae: *Riella*
Sphaerocarpaceae: *Shaerocarpos*
Order: Monocleales
Family: Monocleaceae: *Monoclea*
Order: Marchantiales
Suborder: Marchantiineae
Family: Targioniaceae: *Targionia*
Aytoniaceae (= Grimaldiaceae)
Subfamily: Aytonioideae: *Plagiochasma*
Reboulioideae: *Reboulia, Mannia, Asterella*
Family: Conocephalaceae: *Conocephalum*
Lunulariaceae: *Lunularia*
Cleveaceae: *Peltolepsis, Sauteria, Athalamia*
Marchantiaceae: *Bucegia, Preissia, Marchantia, Dumortiera,*
Neohodgsonia

Table 1a (*continued*)

Family: Exormothecaceae: *Exormotheca*
            Corsiniaceae: *Corsinia*
  Suborder: Ricciineae
    Family: Oxymitraceae: *Oxymitria*
            Ricciaceae: *Ricciocarpos, Riccia*

Table 1b. *Classification of Musci (Mosses)*

Subclass: Andreaeidae
    Family: Andreaeaceae: *Andreaea*
Subclass: Sphagnidae
    Family: Sphagnaceae: *Sphagnum*
Subclass: Tetraphidae
  Order: Tetraphidales
    Family: Tetraphidaceae: *Tetrodontium, Tetraphis*
Subclass: Polytrichidae
  Order: Polytrichales
    Family: Polytrichaceae: *Atrichum, Bartramiopsis, Oligotrichum,*
                            Pogonatum, Polytrichastrum, Polytrichum
Subclass: Buxbaumiidae
  Order: Buxbaumiales
    Family: Buxbaumiaceae: *Buxbaumia, Diphyscium, Theriotia*
Subclass: Archidiidae
  Order: Archidiales
    Family: Archidiaceae: *Archidium*
Subclass: Bryidae
  Order: Fissiidentales
    Family: Fissidentaceae: *Fissidens*
  Order: Dicranales
    Family: Ditrichaceae: *Ceratodon, Ditrichium, Eccremidium,*
                          Pleuridium, Pseudephemerum, Saelania
                          Trichodon
            Bryoxiphiaceae: *Bryoxiphium*
            Seligeriaceae: *Blinidia, Brachydontium, Seligeria*
            Dicranaceae: *Aongstroemia, Arctoa, Brothera, Bruchia,*
                         Bryohumbertia, Campylopodium,
                         Campylopus, Cynodontium, Dichodontium,
                         Dicranella, Dicranodontium, Dicranoloma,
                         Dicranoweisia, Dicranum, Garckea,
                         Holomitrium, Kiaeria, Leucoloma,
                         Oncophorus, Oreas, Oreoweisia,
                         Paraleucobryum, Rhabdoweisia, Trematodon
    Family: Leucobryaceae: *Leucobryum, Leucophanes*
  Order: Pottiales
    Family: Calymperaceae: *Calymperes, Exostratum, Syrrhopodon*

Table 1b (*continued*)

Pottiaceae: *Anoectangium, Barbula, Bryoerythrophyllum,*
Chenia, Desmatodon, Dialytrichia, Didymadon,
Eucladium, Gymnostronum, Hydrogonium
Hymenostylium, Hyophila, Leptodontium,
Luisierella, Molendoa, Oxystegus, Phascum,
Pottia, Pseudosymblepharis, Scopelophila,
Streblotrichum, Timmilella, Tortella, Tortula,
Trichostomum, Tuerckheimia, Weissia, Weisiopsis

Encalyptaceae: *Encalypta*

Order: Grimmiales

Family: Grimmiaceae: *Campylostelium, Coscinodon, Grimmia,*
Ptycomitrium, Racomitrium, Schistidium

Erpodiaceae: *Aulacopilum, Glyphomitrium, Venturiella*

Order: Funariales

Family: Disceliaceae: *Discelium*

Ephemeraceae: *Ephemerum, Micromitrium*

Funariaceae: *Entosthodon, Funaria, Physcomitrella,*
Physcomitrium

Splachnaceae: *Gymnostomiella, Oedipodium, Splachnum,*
Tayloria, Tetraplodon

Order: Schistostegales

Family: Schistostegaceae: *Schistostega*

Order: Eubryales

Family: Bryaceae: *Anomobryum, Brachymenium, Bryum,*
Epipterygium, Leptobryum, Mielichhoferia,
Plagiobryum, Pohlia, Rhodobryum

Mniaceae: *Cinclidum, Cyrtomnium, Mnium, Orthomnium,*
Plagiomnium, Pseudobryum, Rhizomnium,
Trachycystis

Aulacomniaceae: *Aulacomnium*

Rhizogoniaceae: *Pyrrhobryum*

Hypnodendraceae: *Hypnodendron*

Meesiaceae: *Paludella*

Timmiaceae: *Timmia*

Bartramiaceae: *Bartramia, Bartramidula, Breutelia,*
Conostomum, Fleischerobryum, Philonotis,
Plagiopus

Order: Orthotrichales

Family: Rhachitheciaceae: *Hypnodontopsis, Rhachithecium*

Orthotrichaceae: *Amphidium, Drummondia, Macrocoma,*
Macromitrium, Orthotrichum, Schlotheimia,
Ulota, Zygodon

Order: Isobryales

Family: Rhacopilaceae: *Rhacopilum*

Fontinalaceae: *Dichelyma, Fontinalis*

Climaciaceae: *Climacium*

Pleuroziopsidaceae: *Pleuroziopsis*

Hedwigiaceae: *Hedwigia*

Table 1b *(continued)*

Cryphaeaceae: *Cryphaeae, Cryptodontopsis,*
        Forsstroemia, Pilotrichopsis.
Leucodontaceae: *Dozya, Felipponea, Leucodon*
Prionodontaceae: *Taiwanobryum*
Trachypodaceae: *Duthiella, Pseudospiridentopsis, Trachypus*
Pterobryaceae: *Calyptothecium, Eumyurium, Garovaglia,*
        Oedicladium, Palisadula, Pterobryum
Meteoriaceae: *Aerobryopsis, Aerobryum, Barbella,*
        Floribundaria, Meteoriella, Meteoriopsis,
        Meteorium, Pseudobarbella
Neckeraceae: *Bissetia, Himantocladium, Homalia,*
        Homaliadelphus, Homaliodendron, Neckera,
        Neckeropsis, Pinnatella, Thamnobryum
Lembophyllaceae: *Dolichomitra, Dolichomitriopsis,*
        Isothecium, Neobarbella

Order: Hookeriales
  Family: Hookeriaceae: *Callicostella, Calyptrochaeta,*
        Chaetomitrium, Distichophyllum, Hookeria,
        Thamniopsis
  Symphyodontaceae: *Symphyodon*
  Hypopterygiaceae: *Cyathophorella, Dendrocyathophorum,*
        Hypopterygium, Lopidium

Order: Hypnobryales
  Family: Theliaceae: *Fauriella, Myurella*
  Fabroniaceae: *Anacamptodon, Fabronia, Habrodon,*
        Helicodontium, Schwetschkea,
        Schwetschkeopsis
  Leskeaceae: *Iwatsukiella, Lescuraea, Leskea, Leskeella,*
        Lindbergia, Okamuraea, Orthoamblystegium,
        Pseudoleskea, Psudoleskeella,
        Pseudoleskeopsis, Rigodiadelphus
  Thuidiaceae: *Abietinella, Anomodon, Boulaya, Bryonoguchia,*
        Claopodium, Haplocladium, Haplohymenium,
        Helodium, Herpetineuron, Heterocladium,
        Hylocomiopsis, Miyabea, Rauiella, Thuidium
  Amblystegiaceae: *Amblystegium, Calliergon, Calliergonella,*
        Campyliadelphus, Campylium,
        Campylophyllum, Cratoneuron,
        Drepanocladus, Hamatocaulis,
        Hygroamblystegium, Hygrohypnum,
        Leptodictyum, Limprichtia, Loeskypnum,
        Palustriella, Platydictya, Pleurozium,
        Pseudohygrophypnum, Sanionia,
        Sarmenthypnum, Sasaokaea
  Brachytheciaceae: *Brachythecium, Bryhnia, Camptothecium,*
        Cirriphyllum, Cratoneurella, Eurhynchium,
        Homalothecium, Kindbergia,
        Kurohimephypnum, Myuroclada,

Table 1b (*continued*)

Palamocladium, Platyphypnidium,
Rhynchostegiella, Rhynchostegium,
Scorpiurium
Entodontaceae: *Entodon, Orthothecium, Pseudoscleropodium,*
Pterigynandrum, Sakuraia
Plagiotheciaceae: *Isopterygiopsis, Plagiothecium*
Sematophyllaceae: *Acroporium, Aptychella, Brotherella,*
Clastobryella, Heterophyllium, Meiothecium,
Neacroporium, Pylaisiadelpha, Radulina,
Rhaphidorrhynchium, Rhaphidostichum,
Taxithelium, Trichosteleum, Tristischella,
Wijkia
Hypnaceae: *Callicladium, Ctenidium, Ectropothecium,*
Eurohypnum, Glossadelphus, Gollania,
Herzogiella, Homomallium, Hondaella, Hypnum,
Isopterygium, Phyllodon, Platygyrium, Podperaea,
Pseudotaxiphyllum, Ptilium, Pylaisiella,
Rhytidiadelphus, Stereodontopsis, Vesicularia
Rhytidiaceae: *Rhytidium*
Hylocomiaceae: *Hylocomiastrum, Hylocomium,*
Loeskeobryum, Macrothamnium

Table 1c. *Classification of Anthocerotae (Hornworts)*

Order: Anthocerotae
  Family: Anthocerotaceae: *Anthoceros, Megaceros, Phaeoceros*
  Dendrocerotaceae: *Dendroceros*
  Notothyladaceae: *Notothylas*

Among the bryophytes, the chemical constituents of the Hepaticae
have been studied in more detail, because liverworts possess cellular oil
bodies, while the other two classes lack complex oil bodies. MUES (*416*)
estimated that only 6% of all liverwort species have been investigated
chemically. MARKHAM (*362*) estimated that the figure for mosses is
probably less than 2%.

II. Chemical Constituents of Hepaticae (Liverworts)

1. Monoterpenoids

Previously twenty-nine monoterpenoids were reported as having been isolated from or detected in Hepaticae (*19*). The absolute configurations of the monoterpenoids found in liverworts have not been established since almost all of them were detected by GC-MS analysis.

A miniature thalloid liverwort, *Targionia hypophylla* (Targioniaceae), emits an intense fragrance. The methanol extract was chromatographed on silica gel-Lobor to give *cis*-(**29**) and *trans*-pinocarveyl acetates (**30**), together with a sesquiterpene alcohol, drimenol (**328**) (*75*). Identification of the acetates was carried out as follows. *Trans*-pinocarveol (**31a**) was oxidized by pyridinium chlorochromate (PCC), followed by LiAlH$_4$ reduction to give a mixture of *cis*- and *trans*-pinocarveols which were isolated by chromatography on a silica gel-Lobor column and then acetylated to furnish *cis*- (**29**) and *trans*-pinocarveyl acetates (**30**). This species contains not only the above acetates but also the related monoterpene hydrocarbons, limonene (**9**) and β-phellandrene (**12b**), α- (**27**) and β-pinene (**28**), as minor components. α-Terpineol (**17**) has been isolated from stem-leafy liverwort *Jungermannia vulcanicola* (*437*).

European *Conocephalum conicum* elaborates (−)-thujanol (**21**) and its epimer (**22**) (*146*). The ^{13}C NMR data of the former compound correspond closely with those (+)-thujanol. The chirality of the monoterpene hydrocarbons of *C. conicum* has been studied by two dimensional gas chromatography using permethylated β-cyclodextrin or dipentyl butyryl γ-cyclodextrin (*608*). Almost all of the monoterpenes showed high optical purity, the (−)-enantiomer prevailing by more than 96% (*i.e.* in 92% e.e.) except in the case of (+)-α-thujene (**20a**).

The enantiomeric composition [%(+)/(−)] of the monoterpene fraction has been estimated as limonene (**9**) (3/97), β-phellandrene (**12b**) (31/69), α-thujene (**20a**) (> 95/ < 5), β-sabinene (**23**) (0/100), α-pinene (**27**) (2/98), β-pinene (**28**) (0/100) and camphene (**37**) (4/96). In order to determine the exact enantiomeric composition of the monoterpene hydrocarbons found in *C. conicum*, (+)-β-phellandrene (**12b**) ([α]$_D$ + 48) and (−)-α-thujene (**20b**), the enantiomer of (**20a**), were synthesized. In addition to the above monoterpenoids, the non-chiral monoterpene hydrocarbons, myrcene (**1**), α-terpinene (**10**), γ-terpinene (**11**), terpinolene (**13**) and *p*-cymene (**15a**), as well as a trace amount of α-phellandrene (**12a**) have been identified in *C. conicum*. Such studies on the enantiomeric compositions of terpenoids in liverworts provide useful information on their biosynthesis in plants.

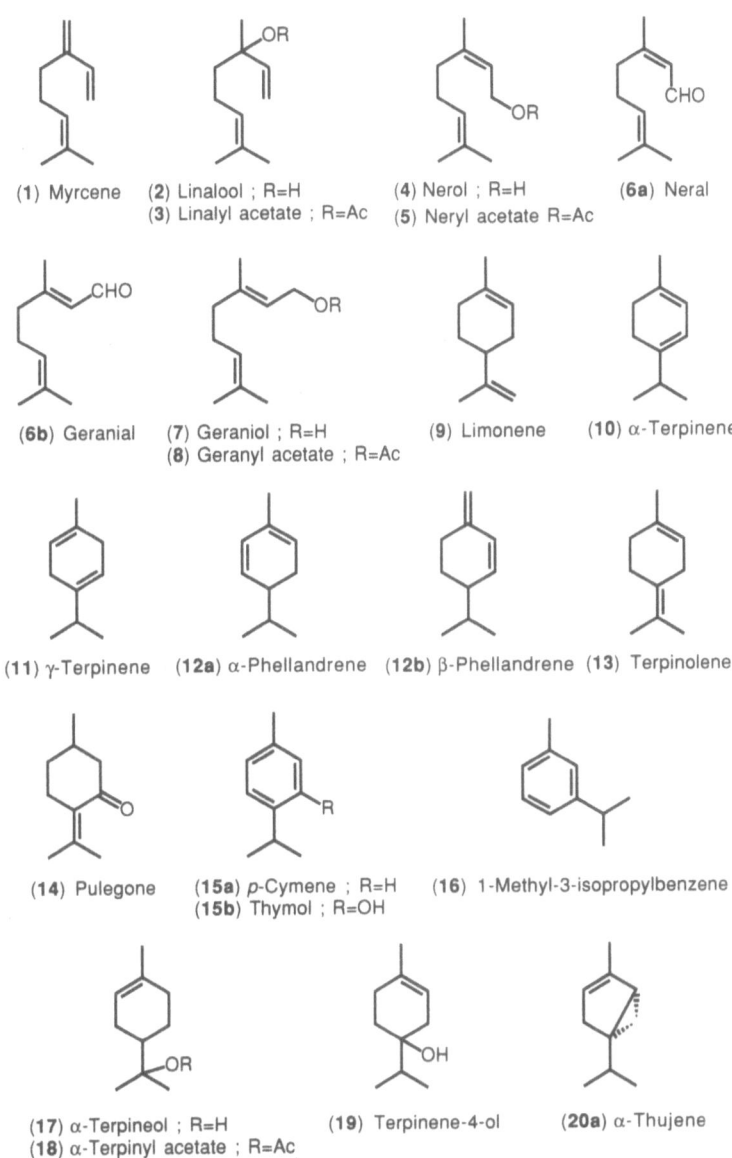

(1) Myrcene   (2) Linalool ; R=H   (4) Nerol ; R=H   (6a) Neral
              (3) Linalyl acetate ; R=Ac   (5) Neryl acetate R=Ac

(6b) Geranial   (7) Geraniol ; R=H   (9) Limonene   (10) α-Terpinene
              (8) Geranyl acetate ; R=Ac

(11) γ-Terpinene   (12a) α-Phellandrene   (12b) β-Phellandrene   (13) Terpinolene

(14) Pulegone   (15a) p-Cymene ; R=H   (16) 1-Methyl-3-isopropylbenzene
              (15b) Thymol ; R=OH

(17) α-Terpineol ; R=H   (19) Terpinene-4-ol   (20a) α-Thujene
(18) α-Terpinyl acetate ; R=Ac

Chart 1a. Monoterpenoids found in the Hepaticae

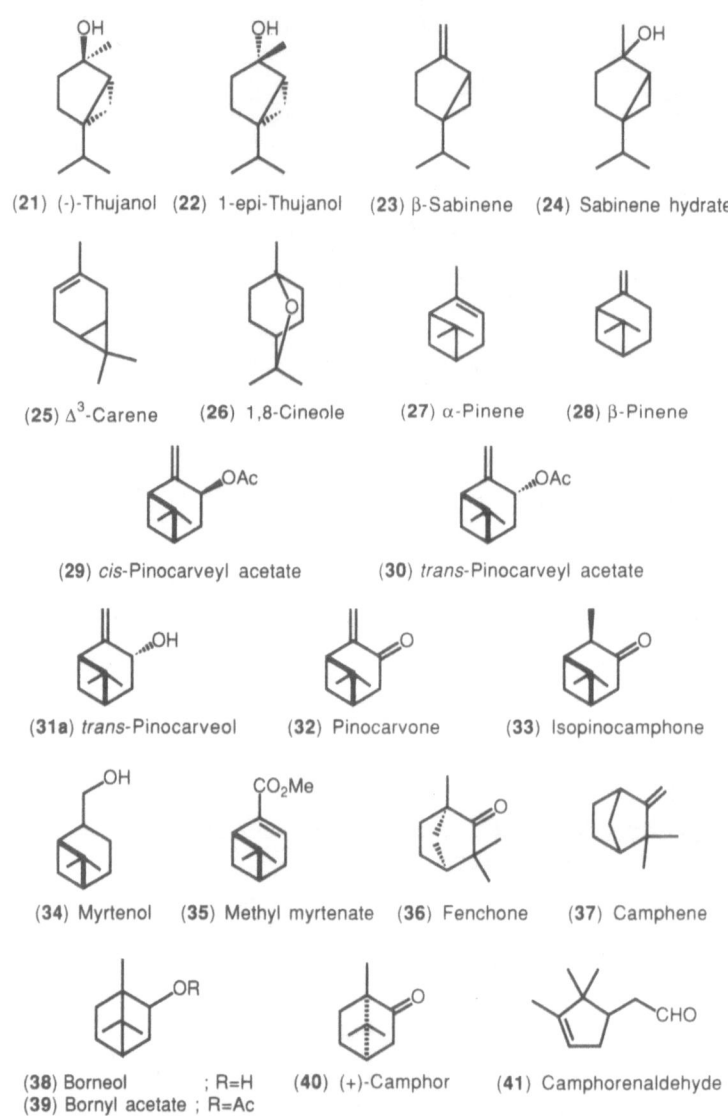

(21) (-)-Thujanol (22) 1-epi-Thujanol   (23) β-Sabinene   (24) Sabinene hydrate

(25) Δ³-Carene   (26) 1,8-Cineole   (27) α-Pinene   (28) β-Pinene

(29) cis-Pinocarveyl acetate   (30) trans-Pinocarveyl acetate

(31a) trans-Pinocarveol   (32) Pinocarvone   (33) Isopinocamphone

(34) Myrtenol   (35) Methyl myrtenate   (36) Fenchone   (37) Camphene

(38) Borneol   ; R=H   (40) (+)-Camphor   (41) Camphorenaldehyde
(39) Bornyl acetate ; R=Ac

Chart 1b. Monoterpenoids found in the Hepaticae

Nerol (**4**) and its acetate (**5**), neral (**6a**), geranyl acetate (**8**), pulegone
(**14**), 1-methyl-3-isopropyl benzene (**16**), α-terpinyl acetate (**18**), sabinene
hydrate (**24**), Δ³-carene (**25**), 1,8-cineole (**26**), *trans*-pinocarveol (**31a**),
pinocarvone (**32**), isopinocamphone (**33**), myrtenol (**34**), methyl myrtenate
(**35**), fenchone (**36**), camphorenaldehyde (**41**) and cyclocitral (**42**) have

been newly found in liverworts together with previously known monoterpenoids as shown in Chart 1 and Table IIa. The monoterpenoids found most frequently in liverworts are α-pinene (**27**), β-pinene (**28**) and limonene (**9**).

2. Tris-normonoterpenoids

Jungermannia obovata contains the tris-normonoterpene ketone (**44**) which possesses a strong carrot-like odour (*143, 146*). It has been synthesized by thermolysis of ascaridole (**45**) (*444*).

(**45**)

3. Homomonoterpenoids

It was known that a camphoraceous compound with a very mossy aroma of unknown structure was present in *Lophocolea heterophylla* (*268*). From the *n*-hexane extract, a homomonoterpene alcohol (**48**) has been isolated. Its spectral data and physical constants identified it as (−)-2-methylisoborneol (**48**) prepared from D-(+)-camphor (**40**) (*586*).

This is the first report of the isolation of 2-methylisoborneol from plants, although it has been isolated from several species of microorganisms in garden soil (*203*). GERBER (*203*) has proposed that it is biosynthesized from a typical sesquiterpene precursor, eudesmane, as shown in Scheme 1. Further investigation of the essential oil of *Lophocolea heterophylla* resulted in the identification of 2-methyl-2-bornene (**46**), 2-methylenebornane (**47**), exo-2-methylfenchol (**49**) and exo-3-methyl-3-borneol (**50**) (*527a*). The latter two compounds (**49, 50**) have been detected in the essential oil of *L. bidentata*, along with (−)-2-methylisoborneol (**48**) (*527a*). It is interesting to note that later, α-selinene (**354**) and related sesquiterpene lactones were isolated from *L. heterophylla* (*586*).

Table IIa. *Mono-, Tris-normono- and Homomonoterpenoids Found in the Hepaticae*

Structure number	Name of compounds	Formula	m.p.°C	[α]$_D$	Plant source	Order	References	Comments
(1)	Myrcene	$C_{10}H_{16}$			*Conocephalum conicum*	M	(62, 608)	
					Lejeunea glaucescens	J	(46)	GC-MS
					Lophocolea heterophylla	J	(527a)	
					Plagiochila subdura	J	(46)	GC-MS
					Porella cordaeana	J	(593)	GC-MS
					Wiesnerella demudata	M	(62)	GC-MS
(2)	Linalool	$C_{10}H_{18}O$			*Conocephalum conicum*	M	(62)	
					Lophocolea heterophylla	J	(527a)	
(3)	Linalyl acetate	$C_{12}H_{20}O_2$			*Wiesnerella demudata*	M	(62)	GC-MS
(4)	Nerol	$C_{10}H_{18}O$			*Conocephalum conicum*	M	(62)	GC-MS
					Wiesnerella demudata	M	(62)	GC-MS
(5)	Neryl acetate	$C_{12}H_{20}O_2$			*Conocephalum conicum*	M	(62)	GC-MS
					Wiesnerella demudata	M	(62)	GC-MS
(6a)	Neral	$C_{10}H_{16}O$			*Plagiochila subdura*	J	(46)	GC-MS
(6b)	Geranial	$C_{10}H_{16}O$			*Conocephalum conicum*	M	(62)	GC-MS
(7)	Geraniol	$C_{10}H_{18}O$			*Wiesnerella demudata*	M	(62)	GC-MS
(8)	Geranyl acetate	$C_{12}H_{20}O_2$			*W. demudata*	M	(62)	GC-MS
(9)	Limonene	$C_{10}H_{16}$			*Acrolejeunea torulosa*	J	(219)	GC-MS
					Archilejeunea parviflora	J	(219)	GC-MS
					Bazzania harpago	J	(341)	GC-MS
					B. praerupta	J	(341)	GC-MS
					Chandonanthus hirtellus	J	(51)	GC-MS
					Conocephalum conicum	M	(62, 106, 608)	(+/−)3/97
					C. japonicum	M	(62)	GC-MS
					Lepidolejeunea ornata	J	(219)	GC-MS

No.	Compound	Formula	Species		Ref.	Method
			Lophocolea heterophylla	J	(527a)	GC-MS
			Lopholejeunea howei	J	(219)	GC-MS
			Marchantia paleacea var. diptera	M	(74)	GC-MS
			M. tosana	M	(74)	GC-MS
			Plagiochila bispinosa	J	(46)	GC-MS
			P. oresitropha	J	(48)	
			P. trabeculata	J	(589)	GC-MS
			Riccardia lobata var. yakushimensis	J	(42)	GC-MS
			Ricciocarpos natans	M	(639, 640)	GC-MS
			Schistochila aligera	J	(51)	GC-MS
			Stictolejeunea balfourii var. bekkei	J	(219)	GC-MS
			Targionia hypophylla	M	(75)	GC-MS
			Trichocolea pluma	J	(51)	GC-MS
			Thysananthus amazonicus	J	(219)	GC-MS
			Wiesnerella denudata	M	(62)	
(10)	α-Terpinene	$C_{10}H_{16}$	Conocephalum conicum	M	(608)	GC-MS
			Plagiochila subdura	J	(46)	
(11)	γ-Terpinene	$C_{10}H_{16}$	Conocephalum conicum	M	(62, 593, 608, 642)	GC-MS
			Mylia nuda	J	(642)	
			Wiesnerella denudata	M	(62)	
(12a)	α-Phellandrene	$C_{10}H_{16}$	Conocephalum conicum	M	(608)	
(12b)	β-Phellandrene	$C_{10}H_{16}$	Conocephalum conicum	M	(608)	
			Mylia nuda	J	(624)	
			Plagiochila geniculata	J	(513)	(+/−) 31/69
			Targionia hypophylla	M	(27)	GC-MS

Table IIa (continued)

Structure number	Name of compounds	Formula	m.p.°C	$[\alpha]_D$	Plant source	Order	References	Comments
(13)	Terpinolene	$C_{10}H_{16}$			Conocephalum conicum	M	(62, 608)	GC-MS
					Lophocolea heterophylla	J	(527a)	
					Wiesnerella demudata	M	(62)	GC-MS
(14)	Pulegone	$C_{10}H_{16}O$			Radula boryana	J	(513)	GC-MS
(15a)	p-Cymene	$C_{10}H_{14}$			Conocephalum conicum	M	(608)	GC-MS
					Stictolejeunea balfourii var. bekkei	J	(219)	GC-MS
(15b)	Thymol	$C_{10}H_{14}O$			Trichocolea pluma	J	(133)	
(16)	1-Methyl-3-isopropyl-benzene	$C_{10}H_{14}$			Stictolejeunea balfourii var. bekkei	J	(219)	GC-MS
(17)	α-Terpineol	$C_{10}H_{18}O$			Conocephalum conicum	M	(62)	GC-MS
					Jungermannia vulcanicola	J	(437)	
					Lophocolea bidentata	J	(527a)	
					L. heterophylla	J	(527a)	
(18)	α-Terpinyl acetate	$C_{12}H_{20}O_2$			Wiesnerella demudata	M	(62)	GC-MS
(19)	Terpinene-4-ol	$C_{10}H_{18}O$			Wiesnerella demudata	M	(62)	GC-MS
					Conocephalum conicum	M	(62)	GC-MS
					Lophocolea heterophylla	J	(527a)	
(20a)	(+)-α-Thujene	$C_{10}H_{16}$			Conocephalum conicum	M	(62, 580, 608)	(+/−)95/5
(21)	(−)-Thujanol	$C_{10}H_{18}O$			Porella cordaeana	J	(593)	GC-MS
(22)	1-epi-Thujanol	$C_{10}H_{18}O$			Conocephalum conicum	M	(146)	GC-MS
(23)	(−)-β-Sabinene	$C_{10}H_{16}$			Conocephalum conicum	M	(146)	GC-MS
					Conocephalum conicum	M	(62, 106)	GC-MS
							(608)	(+/−)0.1/99.9

	Compound	Formula	Species		Ref.	Method
			Plagiochila sciophila (= *P. acanthophylla* subsp. *japonica*	J	(62)	GC-MS
			P. subdura	J	(46)	GC-MS
			Porella cordaeana	J	(593)	GC-MS
(24)	Sabinene hydrate	C$_{12}$H$_{18}$O	*Conocephalum conicum*	M	(62)	GC-MS
			Wiesnerella denudata	M	(62)	GC-MS
			Lophocolea heterophylla	J	(527a)	
			Lophocolea heterophylla	J	(527a)	
(25)	Δ³-Carene	C$_{10}$H$_{16}$	*Acrolejeunea torulosa*	J	(219)	GC-MS
(26)	1,8-Cineole	C$_{10}$H$_{18}$O	*Conocephalum conicum*	M	(62, 608)	(+ / −) 2/98
(27)	(−)-α-Pinene	C$_{10}$H$_{16}$	*Frullania falciloba*	J	(64)	GC-MS
			Fossombronia pusilla	Me	(491)	GC-MS
			Gackstroemia magellanica	J	(45)	GC-MS
			Heteroscyphus planus	J	(427a)	GC-MS
			Isotachis humectata	J	(45)	GC-MS
			Lophocolea bidentata	J	(527a)	
			L. heterophylla	J	(527a)	
			Marchantia paleacea var. *diptera*	M	(74)	GC-MS
			Mylia nuda	J	(622, 624)	GC-MS
			Nardia scalaris	J	(106)	GC-MS
			Omphalanthus filiformis	J	(47)	GC-MS
			Plagiochila elata	J	(46)	GC-MS
			P. oresitropha	J	(48)	GC-MS
			P. pachyloma	J	(48)	GC-MS
			P. sciophila (= *P. acanthophylla*) subsp. *japonica*	J	(62)	GC-MS
			P. spinulosa	J	(495)	GC-MS
			P. squamurifera	J	(47)	GC-MS
			P. stephensoniana	J	(39)	GC-MS
			P. subdura	J	(46)	GC-MS

Table IIa (continued)

Structure number	Name of compounds	Formula	m.p.°C	[α]_D	Plant source	Order	References	Comments
					P. tambillensis	J	(48)	GC-MS
					P. verruculosa	J	(47)	GC-MS
					Porella cordaeana	J	(593)	
					P. densifolia subsp. appendiculata	J	(67)	GC-MS
					Symbiezidium barbiflorum	J	(219)	GC-MS
					Targionia hypophylla	M	(75)	GC-MS
					Trichocolea pluma	J	(133)	GC-MS
(28)	(−)-β-Pinene	C₁₀H₁₆			Acrolejeunea torulosa	J	(219)	GC-MS
					Archilejeunea parviflora	J	(219)	GC-MS
					Conocephalum conicum	M	(62, 608)	(+/−) 0.1/99.9
					Fossombronia pusilla	Me	(491)	GC-MS
					Lophocolea bidentata	J	(527a)	
					L. heterophylla	J	(527a)	
					Marchantia paleacea var. diptera	M	(74)	GC-MS
					Mylia nuda	J	(622, 624)	
					Plagiochila elata	J	(46)	GC-MS
					P. sciophila (= P. acanthophylla subsp. japonica)	J	(62)	GC-MS
					P. stephensoniana	J	(39)	GC-MS
					Pleurozia acinosa	J	(625)	GC-MS
					Porella cordaeana	J	(593)	GC-MS
					P. densifolia subsp. appendiculata	J	(67)	GC-MS
					Targionia hypophylla	M	(75)	GC-MS

(29)	cis-Pinocarveyl-acetate	$C_{12}H_{18}O_2$	*Targionia hypophylla*	M	(27)	
(30)	trans-Pinocarveyl-acetate	$C_{12}H_{18}O_2$	*Targionia hypophylla*	M	(27)	
(31a)	trans-Pinocarveol	$C_{10}H_{16}O$	*Lophocolea heterophylla*	J	(527a)	
(32)	Pinocarvone	$C_{10}H_{14}O$	*Lophocolea heterophylla*	J	(527a)	
(33)	Isopinocamphone	$C_{10}H_{16}O$	*Lophocolea heterophylla*	J	(527a)	
(34)	Myrtenol	$C_{10}H_{16}O$	*Lophocolea heterophylla*	J	(527a)	
(35)	Methyl myrtenate	$C_{11}H_{16}O_2$	*Lophocolea heterophylla*	J	(527a)	
(36)	Fenchone	$C_{10}H_{16}O$	*Lophocolea heterophylla*	J	(527a)	
(37)	(−)-Camphene	$C_{10}H_{16}$	*Conocephalum conicum*	M	(62, 106)	GC-MS
			Isotachis humectata	J	(45)	GC-MS
			Lepidolejeunea ornata	J	(219)	GC-MS
			Lophocolea heterophylla	J	(527a)	
			Mylia nuda	J	(622, 624)	
			Plagiochila oresitropha	J	(48)	GC-MS
			Porella cordaeana	J	(593)	GC-MS
			P. densifolia subsp. *appendiculata*	J	(67)	GC-MS
			Thysananthus amazonicus	J	(219)	GC-MS
(38)	Borneol	$C_{10}H_{18}O$	*Conocephalum conicum*	M	(62, 580)	GC-MS
			Lophocolea bidentata	J	(527a)	
			L. heterophylla	J	(527a)	
			Wiesnerella denudata	M	(51)	
(39)	Bornyl acetate	$C_{12}H_{20}O_2$	*Conocephalum conicum*	M	(62)	
			Lophocolea heterophylla	J	(343, 372a)	
			Wiesnerella denudata	M	(51)	
(40)	(+)-Camphor	$C_{10}H_{16}O$	*Lophocolea heterophylla*	J	(527a)	
(41)	Camphorenaldehyde	$C_{10}H_{16}O$	*Lophocolea heterophylla*	J	(527a)	
(42)	Cyclocitral	$C_{10}H_{16}O$	*Lophocolea heterophylla*	J	(527a)	

Y. Asakawa

Table IIa (continued)

Structure number	Name of compounds	Formula	m.p.°C	$[\alpha]_D$	Plant source	Order	References	Comments
(44)	4-Hydroxy-4-methyl-cyclohex-2-en-1-one	$C_7H_{10}O_2$			Jungermannia obovata	J	(143, 146)	
(46)	2-Methyl-2-bornene	$C_{11}H_{18}$			Lophocolea heterophylla	J	(527a)	
(47)	2-Methylenebornane	$C_{11}H_{18}$			Lophocolea heterophylla	J	(527a)	
(48)	(−)-2-Methylisoborneol	$C_{11}H_{20}O$		−4.8	Lophocolea bidentata	J	(527a)	
					L. heterophylla	J	(372a, 586)	
(49)	Exo-2-methylfenchol	$C_{11}H_{20}O$			Lophocolea bidentata	J	(527a)	
					L. heterophylla	J	(527a)	
(50)	Exo-3-methyl-3-borneol	$C_{11}H_{20}O$			Lophocolea bidentata	J	(527a)	
					L. heterophylla	J	(527a)	
(51)	Trinoranastreptene	$C_{12}H_{16}$			Calypogeia granulata	J	(321, 453, 539)	
					Lophozia ventricosa	J	(429, 567a)	
(54)	trans-1,4a-Dimethyl-(1,2,3,4,4a,5,6,7-octahydronaphthalene	$C_{12}H_{20}$		+86	Bazzania angustifolia	J	(626, 633)	
					B. fauriana	J	(626, 633)	
					Lophocolea bidentata	J	(527a)	
(55)	Geosmin	$C_{12}H_{22}O$			Lophocolea bidentata	J	(527a)	
					L. heterophylla	J	(527a)	
					Symphyogyna brongniartii	J	(513)	in vitro culture
(56)	β-Ionone	$C_{13}H_{20}O$			Fossombronia pusilla	Me	(491)	GC-MS

J: Jungermanniales; M: Marchantiales; Me: Metzgeriales

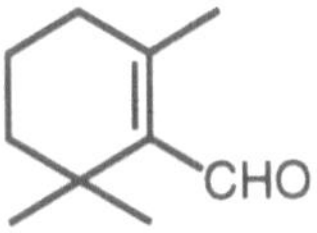

(**42**) Cyclocitral

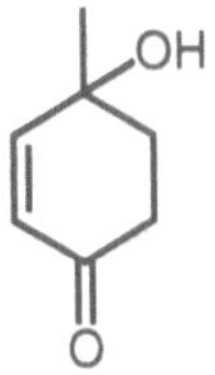

(**44**) 4-Hydroxy-4-methyl-
cyclohex-2-en-1-one

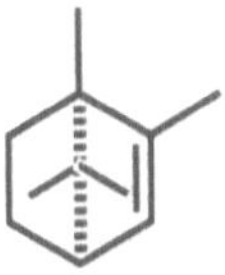

(**46**) 2-Methyl-2-bornene

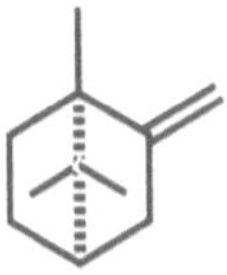

(**47**) 2-Methylenebornene

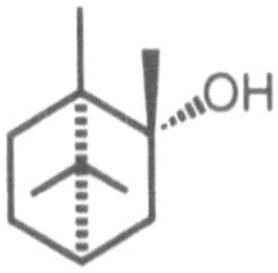

(**48**) (−)-2-Methylisoborneol

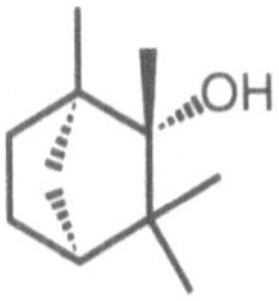

(**49**) Exo-2-methylfenchol

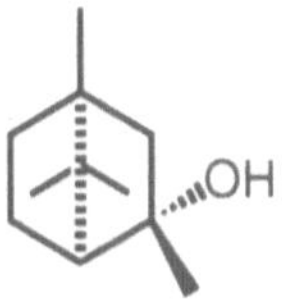

(**50**) Exo-3-methyl-3-borneol

Chart 1c. Mono-, tris-normono- and homomonoterpenoids found in the Hepaticae

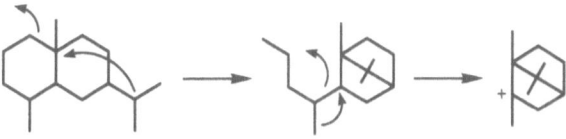

Scheme 1. Possible biogenetic pathway for 2-methylbornanes from eudesmane-type sesquiterpenoids

4. Tris-norsesquiterpenoids

Trinoranastreptene **(51)** was isolated from cultured cells of *Calypogeia granulata* along with azulenes. The structure was deduced by comparing its NMR spectrum with that of anastreptene **(93)**, spin decoupling and NOE experiments *(321, 453, 539)*. Compound **(51)** has also been isolated from European *Lophozia ventricosa (429)*. It is noteworthy that the same trinoranastreptene, then named inflatene *(303)* or clavukerin B *(332)* and the similar tris-norsesquiterpene, clavukerin A **(52)** *(331)* and isoclavukerin A **(53)** *(151, 354)* have been isolated from marine organisms. The absolute configuration of clavukerin A has been established as (8S, 8aS)-(−)-3,8-dimethyl-1,2,6,7,8,8a-hexahydroazulene *(331)*. The absolute configuration of **(51)** isolated from both the liverworts and marine organisms was suggested to be the same as that of anastreptene **(93)**. Hypothetical biogenetic pathways for trinoranastreptene **(51)** and clavukerin A **(52)** are proposed by KOBAYASHI *et al.* shown in Scheme 2 *(332)*.

A unique tris-norsesquiterpene **(54)** has been isolated from Taiwanese *Bazzania fauriana* and detected in *B. angustifolia (633)* and in *Lophocolea bidentata (527a)*. Its structure was elucidated by various NMR techniques and the usual decoupling experiments *(633)*. The spectral data were

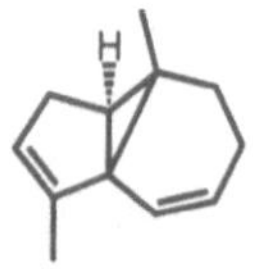

(51) Trinoranastreptene 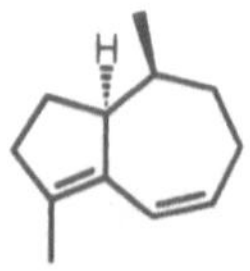 **(52)** Clavukerin A 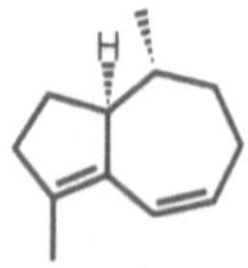   **(53)** Isoclavukerin A

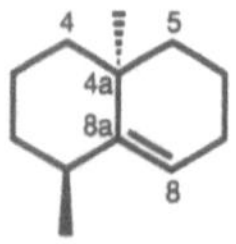

(54) *trans*-1,4a-Dimethyl-1,2,3,4,4a,5,6,7-octahydronaphthalene

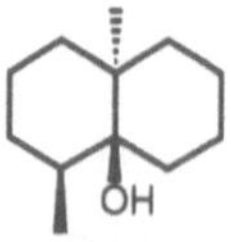

(55) Geosmin 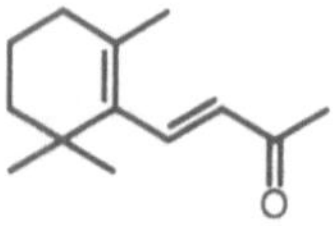   **(56)** β-Ionone

Chart 2. Tris-norsesquiterpenoids found in the Hepaticae and their related compounds

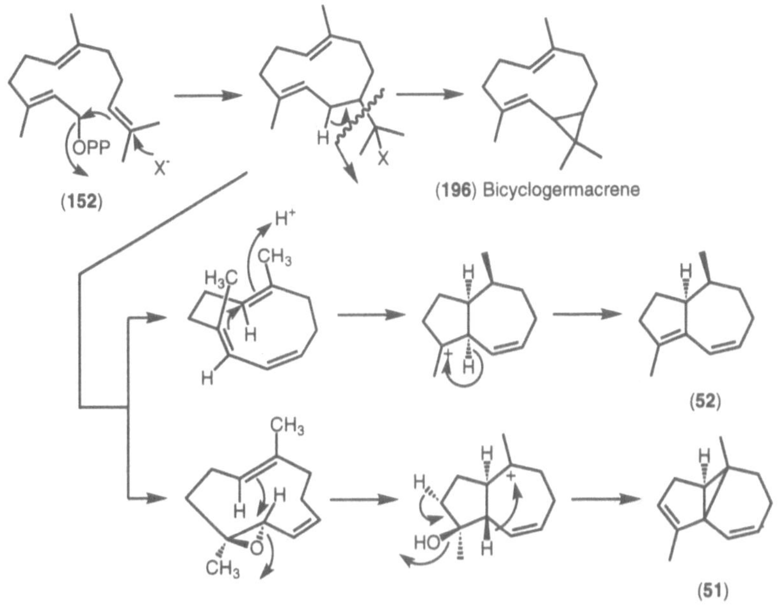

Scheme 2. Possible biogenetic pathways for trinoranastreptene and clavukerin from farnesyl pyrophosphate

identical with those of synthetic **54** (*370, 471*). The absolute configuration of (**54**) was confirmed by total synthesis of (+)-(**54**) using 2,6-dimethyl-cyclohexanone and (*R*)-(+)-methylbenzylamine (*471*). *In vitro* cultured *Symphyogyna brongniartii* produces geosmin (**55**) (*513*). The same alcohol has been detected in *Lophocolea bidentata* and *L. heterophylla* (*527a*) and isolated from the soil organism, *Actinomyces* species as an earthy-smelling metabolite (*203*). Synthetic (**54**) has a strong earthy-musty odor, but natural (**54**) has a rather light and pleasant odor (*633*). Compound (**54**) may originate biogenetically from eudesmane-type sesquiterpenes since eudesmols (**359, 360**) have also been found in the oil of *B. fauriana* (*633*).

β-Ionone (**56**) has been detected in *Fossombronia pusilla* by GC-MS (*491*).

5. Sesquiterpenoids

Liverworts are rich sources of sesquiterpenoids. Acorane-, aristolane-, aromadendrane-, azulene-, barbatane- (= gymnomitrane), bazzanane-,

bergamotane-, bicycloelemane-, bicyclogermacrane-, bisabolane-, bour-
bonane-, cadinane-, calamenane-, caryophyllane-, cedrane-, chamigrane-,
copaane-, cubebane-, cuparane-, drimane-, elemane-, eremophilane-,
eudesmane-, farnesane-, germacrane-, gorgonane-, guaiane-, pseudo-
guaiane-, himachalane-, humulane-, longibornane-, longifolane-, longi-
pinane-, maaliane-, monocyclofarnesane-, muurolane-, pinguisane-, san-
talane-, 2,3-secoaromadendrane-, thujopsane-, vitrane-, widdrane- and
ylangane-type sesquiterpenoids have been found in liverworts (*19*). Among
them, the presence of bergamotane-, cedrane-, muurolane-, santalane-,
thujopsane- and widdrane-type sesquiterpenoids in liverworts has been
confirmed by GC-MS. An important endogenous character of the
Hepaticae is that most of the sesquiterpenoids isolated from liverworts
are enantiomeric to those found in higher plants (*19*).

Noracorane-, africane-, norafricane-, secoafricane-, brasilane, dau-
cane- (= carotane-), chiloscyphane-, herbertane- (= isocuparane-), myl-
taylane-, cyclomyltaylane-, oppositane-, pacifigorgiane-, patchoulane-,
spirovetivane-, valencane-, and zierane-type sesquiterpenoids have been
newly isolated from liverworts, together with the previously known
sesquiterpenoids enumerated in the previous paragraph: Eudesmane-
type sesquiterpenoids are the most frequently encountered sesquiterpenoids in liverworts.

5.1 Acoranes and Noracoranes

Acoradiene (**57**) has been detected not only in Jungermanniales but
also in Metzgeriales. Its absolute configuration remains to be clarified.

A noracorane-type sesquiterpenoid, inflatenone (**58**) has been isolated
from *Gymnocolea inflata* (*429, 566, 567a*). Its structure and absolute
configuration were established by various NMR techniques including
HMQC (= Heteronuclear Multiple Quantum Coherence), HMBC ([1]H-
Detected Multiple Bond Heteronuclear Multiple Quantum Coherence
Spectrum Connectivity), NOESY (Nuclear Overhauser and Exchange

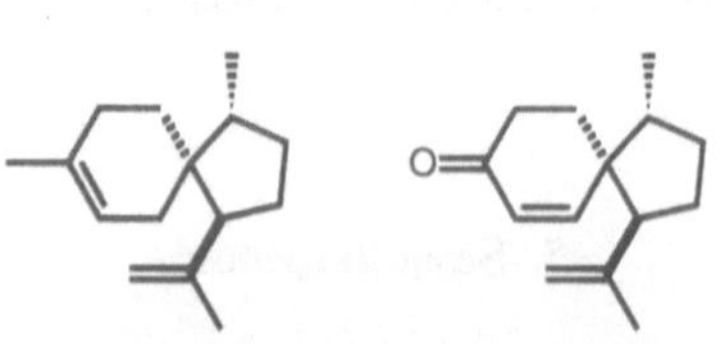

(**57**) Acoradiene   (**58**) Inflatenone

Chart 3. Acorane and noracorane found in the Hepaticae

Spectroscopy) and the CD spectrum. Compound (**58**) has been obtained as a synthetic intermediate (*154, 428*).

5.2 Africanes, Norafricanes and Secoafricanes

Porella caespitans var. *setigera* produces not only aristolane-type sesquiterpenoids (**82**) but also sacculatane-type diterpenoids (*19*). Further fractionation of the ethyl acetate extract of *P. caespitans* var. *setigera* resulted in the isolation of two africane-type sesquiterpenes (**59, 60**) (*596,* *597*) and one norafricane-type sesquiterpene, norafricanone (**66**) (*583*). Detailed analysis of the NMR data including 2D-COSY and spin decoupling experiments established the gross structure of (**59**). Confirmation of its structure was provided by chemical degradation as shown in Scheme 3. Reduction of (**59**) with LiAlH$_4$ gave a triol (**59a**) and its epimer (**59c**) which were esterified by *p*-bromobenzoyl chloride to afford mono-benzoates (**59b, 59d**). The relative stereochemistry of C-2 and C-3 in triol (**59a**) has been confirmed as *cis* since it forms an acetonide (**59e**). Dehydration of (**59a**) with *p*-TsOH gave three ring-cleaved compounds (**59f–59h**). LiAlH$_4$ reduction of (**59f**) afforded a diol (**59i**), followed by benzoylation with *p*-bromobenzoyl chloride to furnish a mono-benzoate (**59j**). The relative stereochemistry of the methyl groups on C-4 and C-10 and the acetyl group at C-2 was confirmed by an NOE difference spectrum. The absolute configuration of (**59**) was established by combination of NOE difference spectroscopy and the CD spectra of the *p*-bromobenzoates (**59b**) and (**59d**) (*135, 233*). The negative Cotton effects at 244 nm (Δε − 8.7) for (**59b**) and 253 nm (Δε − 53) for (**59d**) showed a 2*R*-configuration.

The structure of the second africane sesquiterpenoid (**60**) was deduced from the close resemblance of its spectroscopic data to those of (**59**) and was confirmed by the following chemical correlation. Acetylation of (**60**) gave a mono acetate (**59k**) which was reduced with LiAlH$_4$ to (**59a**). The ^{1}H- and ^{13}C-NMR spectra and spin decoupling of norafricanone (**66**) showed the presence of a bicyclo[5.0.1] skeleton with acetoxymethyl and acetyl groups. The location of each functional group was confirmed by 2D-COSYs (^{1}H-^{1}H, long range ^{13}C-^{1}H). Compound (**66**) may originate from co-metabolite (**59**).

From the ether extract of a Colombian liverwort, *Porella swartziana*, a new african-type sesquiterpenoid, caespitenone (**61**) for which pseudo-guaiane-type structure had been proposed previously (*19*) and a new 3,4-secoafricane-type sesquiterpenoid, secoswartzianin A (**67**), were isolated as the major components (*441a, 571, 571a, 571b, 568*). The functional

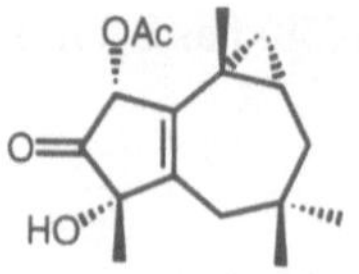

(59) 3α-Hydroxy-5α-acetoxyafrican-2(6)-en-4-one

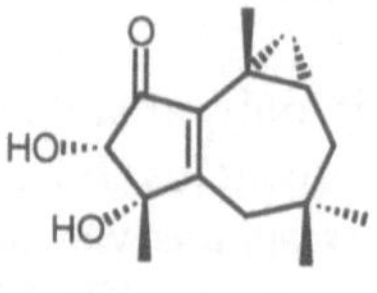

(60) 3α,4α-Dihydroxyafrican-2(6)-en-5-one

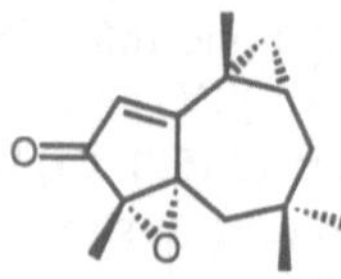

(61) Caespitenone 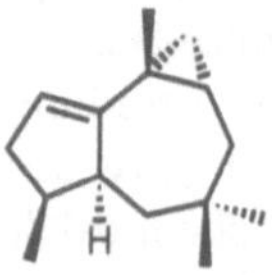   (62) Swartzianin A 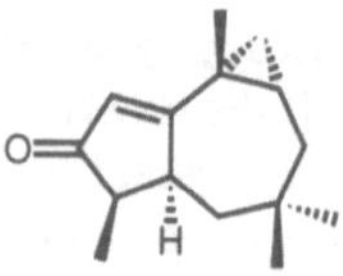   (63) Swartzianin B 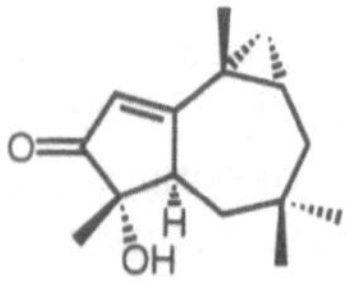   (64) Swartzianin C

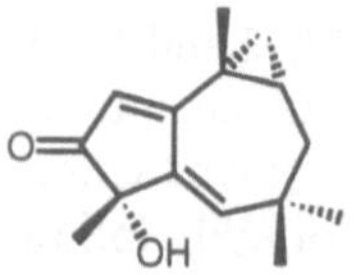

(65) Swartzianin D 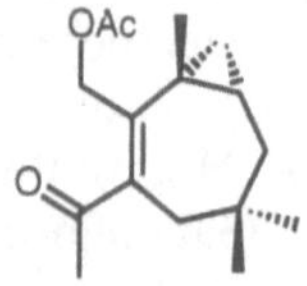   (66) Norafricanone 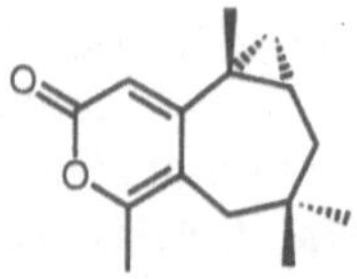   (67) Secoswartzianin A

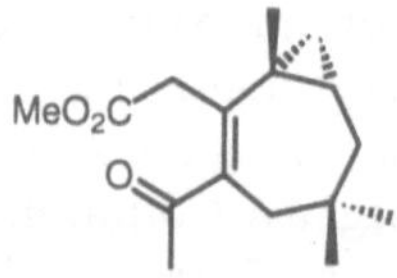

(68) Secoswartzianin B 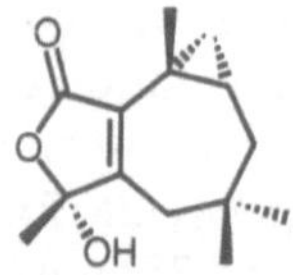   (69) Norswartzianin 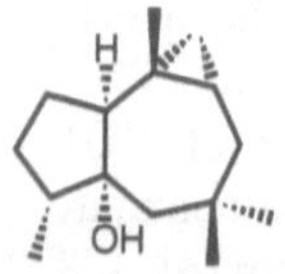   (70) Isoafricanol

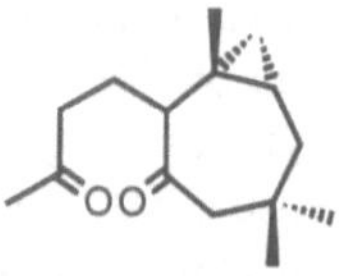

(71) 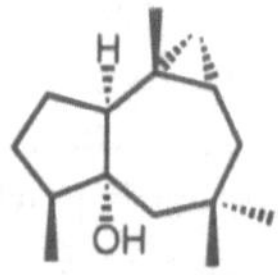   (72) 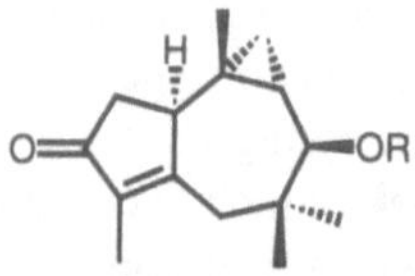   (73) R=Ang
(74) R=H   (75)

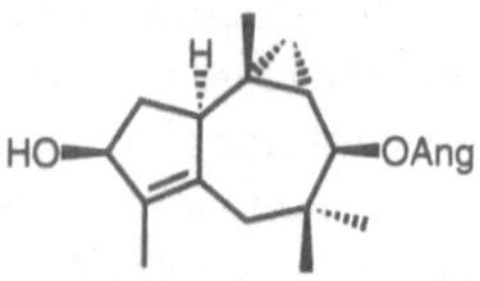

Chart 4. Africanes, norafricanes and secoafricanes found in the Hepaticae and their related compounds

groups present in (61) were deduced from the IR, UV and NMR data while the africane-type structure (61) was established by ^{1}H-^{1}H-, ^{13}C-^{1}H-, long range ^{13}C-^{1}H-2D-COSYs and HMBC experiments. The absolute configuration of (61) was settled by the 600 MHz NOESY spectrum of a

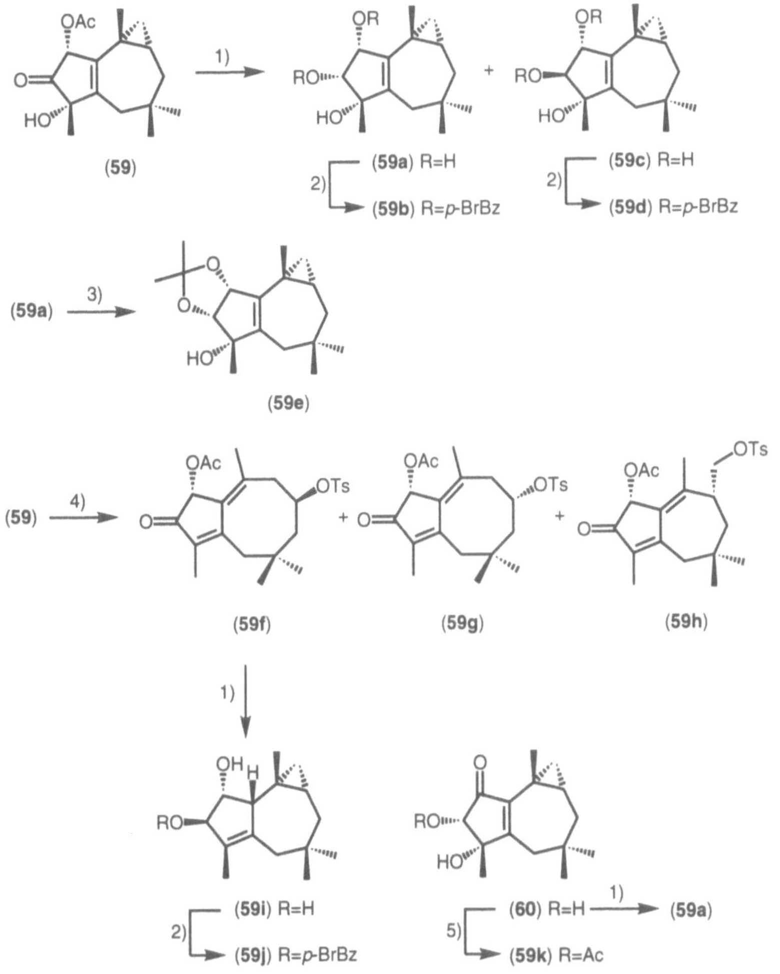

1) LiAlH$_4$/Et$_2$O 2) p-BrBzCl/Py 3) Me$_2$CO/CuSO$_4$ 4) p-TsOH/C$_6$H$_6$ 5) Ac$_2$O/Py

Scheme 3. Reactions of africane-type sesquiterpenoids

mono-alcohol (**61a**) prepared from (**61**) with (PhSe)$_2$ in NaBH$_4$ and the negative Cotton effect at 236 nm ($\Delta\varepsilon$ − 11.2) of a mono-bromobenzoate (**61b**) prepared from (**61a**) by reduction with NaBH$_4$ in CeCl$_3$ followed by benzoylation with p-bromobenzoyl chloride as indicated in Scheme 4. The structure of (**67**) was based on extensive NMR techniques, mainly 2D-COSYs, while the absolute configuration assigned to (**67**) was based

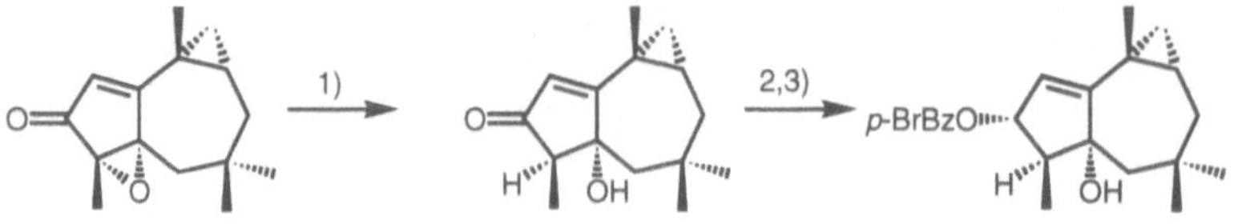

(61)   (61a)   (61b)

1) (PhSe)₂/NaBH₄/EtOH 2) NaBH₄/CeCl₃/MeOH 3) p-BrBzCl/DMAP/Py

Scheme 4. Reaction of caespitenone

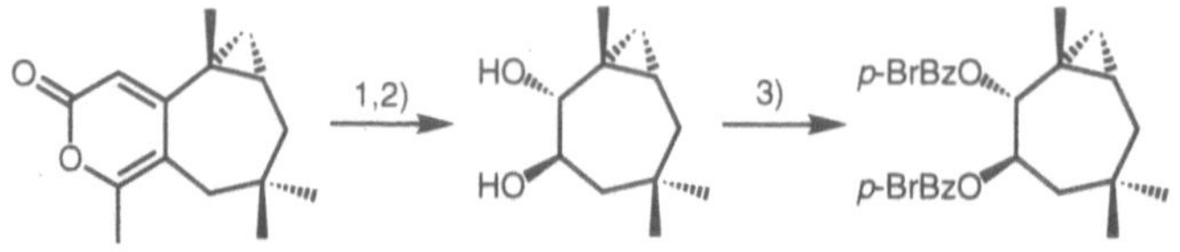

(67)   (67a)   (67b)

1) O₃/MeOH 2) NaBH₄ 3) p-BrBzCl/DMAP/Py

Scheme 5. Reaction of secoswartzianin A

on the CD Cotton effect at 249 nm ($\Delta\varepsilon - 37.7$) of a dibenzoate (67b) which was prepared from (67) by ozonolysis followed by reduction and then benzoylation as shown in Scheme 5 (*571a*).

Further fractionation of the ether extract of *P. swartziana* yielded four africane-type, swartzianins A–D (62–65), one 3,4-secoafricane-type seco-swartzianin B (68) (*441a, 571, 571a*) and one norafricane-type sesquiter-penoid, norswartzianin (69) (*574a, 574b*). The relative stereostructures of swartzianins C (64) and D (65) were established by spectroscopic compar-ison with (61) and by X-ray crystallographic analysis. The structure of (69) was also established by X-ray analysis. Structures of the other new compounds were elucidated by comparing their spectral data with those of (61), (64) and (65). From *Nardia scalaris*, isoafricanol (70), the C-4 epimer of africanol (72) which has been found in a sapwood staining ascomycete fungus (*1a*) has been isolated together with new diterpenoids (*vide infra*) (*354a*).

A similar 4,5-secoafricane derivative (71) has been isolated from the essential oil of a higher plant, *Lippia integrifolia* (*131*).

Africane-, norafricane- and secoafricane-type sesquiterpenoids are quite rare in nature, although africanes (72–75) have been found in a marine invertebrate (601) and in the roots of Senecio oxyriifolius (Compositae) (118). However the above report comprises the first isolation of the africane-, norafricane- and secoafricane-type sesquiterpenoids from Hepaticae. The africane skeleton may originate from humulene by cyclization as shown in Scheme 6 (118, 410, 601).

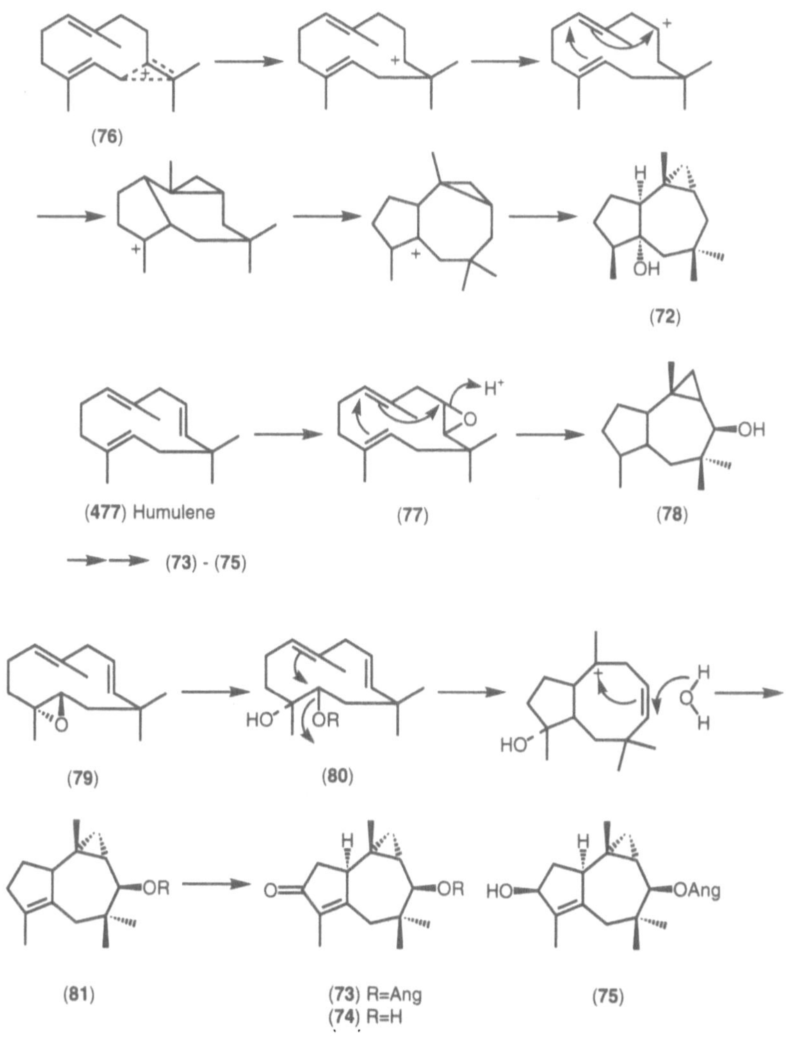

Scheme 6. Possible biogenetic pathways for africane-type sesquiterpenoids

5.3 Aristolanes

Aristolane-type sesquiterpenoids are rare in liverworts. The presence of *ent*-aristolone (**82**) has been confirmed in *Porella caespitans* var. *setigera* (*19*, *596*). The same compound has been isolated from *P. cordaeana* (*239a*) and *P. roellii* (*543*). *Reboulia hemisphaerica* also produces *ent*-aristolone (**82**) (*55*), together with the new labile *ent*-aristol-1,8-diene (**85**) (*249*) and *ent*-8β-hydroxyaristolene (**83**) (*73*). (+)-Aristol-9-ene (= α-ferulene) (**84**) is the minor component of *Pleurozia acinosa* (*625*) while *Riccardia jackii* produces *ent*-(+)-aristolan-10β-ol (**86**) (*380*). Two new aristolanes, (+)-aristol-9-en-12β-al (**87**) and (+)-aristol-9-en-12β-oic acid (**88**) have been isolated from Taiwanese *Bazzania tridens*; the structure assignment was based on analysis of the ^{1}H- and ^{13}C-NMR spectra and comparison of the spectral data with those of previously known aristolone (**82**) and aristol-9-ene (**84**) (*629*). The absolute configurations of both compounds were established by single X-ray crystallographic analysis of (**88**) and comparison of the specific optical rotations of (**87**) and (**88**) with those of (**84**) and (**82**) (*629*).

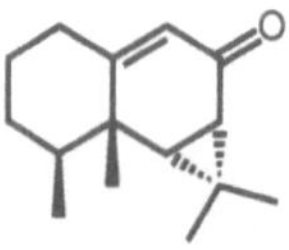

(**82**) *ent*-Aristolone 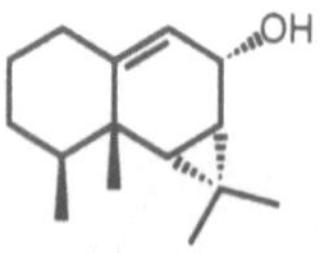   (**83**) *ent*-8β-Hydroxyaristolene 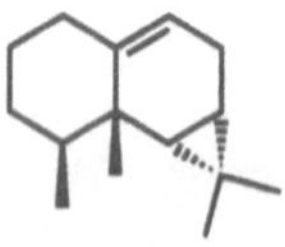   (**84**) (+)-Aristol-9-ene
                                                                    (= (+)-α-Ferulene)

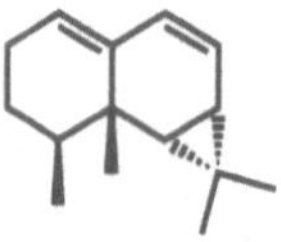

(**85**) Aristol-1(10),8-diene 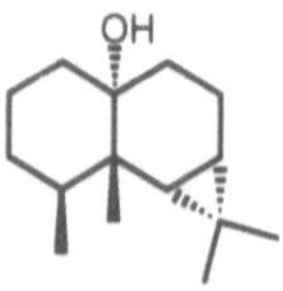   (**86**) *ent*-Aristolan-10β-ol 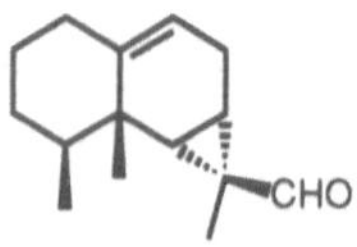   (**87**) *ent*-Aristol-9-en-12β-al

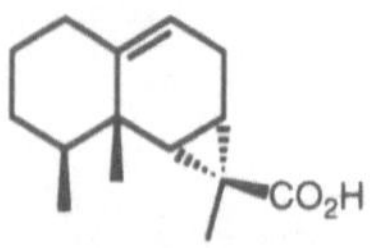

(**88**) *ent*-Aristol-9-en-12β-oic acid

Chart 5. Aristolanes found in the Hepaticae

5.4 Aromadendranes, Secoaromadendranes and Norsecoaromadendranes

Ent-aromadendrane-type sesquiterpenoids are widespread in Jungermanniales. Secoaromadendrane-type sesquiterpenoids are mainly distributed in *Plagiochila* species. The previously known aromadendrene (**89**), alloaromadendrene (**90**), α- (**91**) and β-gurjunene (**92**), anastreptene (**93**), (−)-ledene (**94**), (+)-cyclocolorenone (**96**), spathulenol (**97**), myliol (**98**), dihydromylione (**99**), and C-10-*epi*-globulol (**100**) have been detected in Jungermanniales. β-spathulene (**95**) (*622, 624*) and (−)-viridiflorol (**101**) (*324, 435*) have been newly found in Hepaticae. Anastreptene and *ent*-spathulenol are the most frequently encountered aromadendrane-type sesquiterpenoids.

Anastreptene (**93**) and β-barbatene (**162**) (a barbatane-type sesquiterpenoid) coexist in certain liverworts and cannot be separated at all on HPLC using silica gel at room temperature. Complete separation on a semipreparative scale was only possible at temperatures below − 35°, using *n*-pentane as solvent (*107*). (+)-Aromadendrene, the enantiomer of (**89**), has been converted to (+)-spathulenol, the enantiomer of (**97**), by two subsequent ozonizations the second of which is regio- and stereoselective, followed by a Wittig reaction with methylenetriphenylphosphorane (*609*). The total synthesis of anastreptene (**93**) has been approached by EICHER (*168*). So far, it has been impossible to reduce (−)-9-anastreptone to anastreptene (**93**). Two sesquiterpenoids, β-diploalbicene and diploalbicanol have been isolated from *Diplophyllum albicans* and *D. taxifolium* (*451*) and were subsequently shown to be identical with (−)-aromadendrene (**89**) and (+)-C-10-epiglobulol (**100**) (*560*) which belong to the *ent*-series. The spectral data of (**100**) and *ent*-globulol (**102**) are almost identical, but the melting points are different. The structure of what was thought to be *ent*-globulol (**102**), previously isolated from several *Plagiochila* species (*19*), should be revised to *ent*-10-epi-globulol because of the identity of the spectral data (*460*).

Mylia taylorii is a rich source of aromadendrane- and secoaromadendrane-type sesquiterpenoids (*19*). Further fractionation of the ethanol extract of *M. taylorii* resulted in the isolation of four new aromadendranes, (+)-myli-4(15)-en-9-one (**103**), (−)-3-epi-myliol (**104**), (+)-4(15)-dehydroledol (**105**) and (+)-4(15)-dehydroglobulol (**106**), together with (+)-*ent*-globulol (**102**) (*386*). The structure of (**103**) followed from ^{1}H- and ^{13}C-NMR spectroscopy including the shift reagent technique and from chemical degradation. Hydrogenation of (**103**) in the presence of Pd-C gave a dihydro derivative (**103a**) which was treated with tosylhydrazine followed by reduction with $NaBH_4$ to afford (−)-α-gurjunene (**91**)

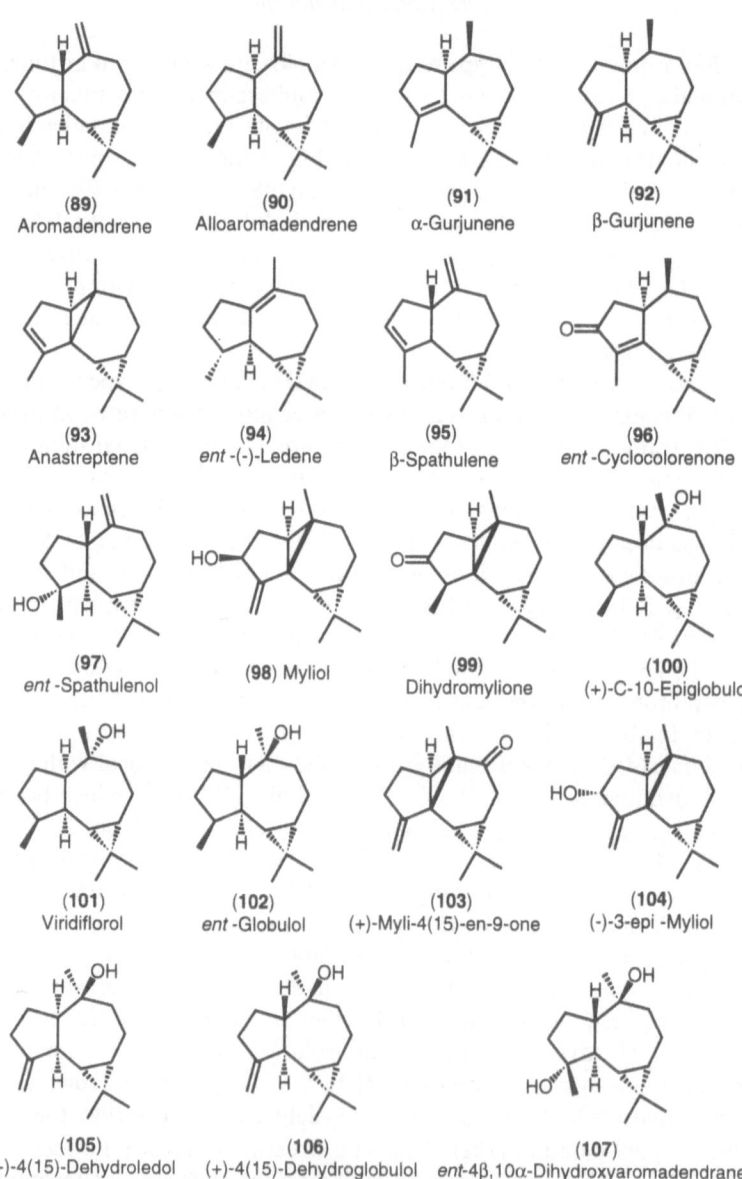

(89)
Aromadendrene

(90)
Alloaromadendrene

(91)
α-Gurjunene

(92)
β-Gurjunene

(93)
Anastreptene

(94)
ent -(-)-Ledene

(95)
β-Spathulene

(96)
ent -Cyclocolorenone

(97)
ent -Spathulenol

(98) Myliol

(99)
Dihydromylione

(100)
(+)-C-10-Epiglobulol

(101)
Viridiflorol

(102)
ent -Globulol

(103)
(+)-Myli-4(15)-en-9-one

(104)
(-)-3-epi -Myliol

(105)
(+)-4(15)-Dehydroledol

(106)
(+)-4(15)-Dehydroglobulol

(107)
ent-4β,10α-Dihydroxyaromadendrane

Chart 6a. Aromadendranes found in the Hepaticae

(*19*). The structure of 3-epi-myliol (**104**) was assigned by spectroscopic comparison of (**104**) and its monoacetate (**104a**) with the spectra of the previously known myliol (**98**) isolated from the same liverwort (*19*). The NMR spectra of (**105**) and (**106**) closely resembled those of spathulenol (**97**) suggesting that (**105**) and (**106**) might be aromadendrane-type sesquiterpenoids. Oxidation of (**105**) with OsO_4 gave a diol, which was treated with $NaIO_4$ to give a cyclopentanone (**105a**). The absolute stereochemistry assigned to (**105**) was based on the NOE spectrometry, the small coupling constant ($J = 4$ Hz) between H-1 and H-5 and the positive Cotton effect at 316 nm ($\Delta\varepsilon + 0.49$) of ketoalcohol (**105a**). The NMR spectrum of (**106**) was closely related to that of (**105**), indicating that (**106**) might be a stereoisomer of (**105**). Hydrogenation of (**106**) in the presence of 10% Pd-C gave two dihydro derivatives in the ratio 30:1. The minor product was identical with (+)-*ent*-globulol (**102**) while the major product was the C-4 epimer. The stereostructure of (**106**) also followed from NOESY and NOE experiments. *ent*-4β,10α-Dihydroxyaromadendrane (**107**) has been isolated from *Plagiochila ovalifolia* (*435a*). Its enantiomer has been found in higher plant, *Brasilia sickii* (*114a*).

(103a)   (104a)   (105a)

Three dimeric sesquiterpenoids, myltaylorione A (**108**), myltaylorione B (**109**) and bitaylorione (**110**), were isolated from an ethanol extract of *Mylia taylorii* (*531*). Structures (**108, 109**) were suggested by comparison of the ^{1}H- and ^{13}C-NMR spectra with those of (−)-dihydromylione (**99**) and (−)-dihydrotaylorione (**112**) derived from taylorione (**111**) (*385*) and analysis of the usual 2D-NMR experiments. The spiro structure at C-4 and the stereochemistry of (**108**) and (**109**) were characterized by 2D-HOHAHA (Mononuclear Hartmann-Hahn Spectrum) and NOESY, respectively. Compounds (**108**) and (**109**) are isomers formed by Diels-Alder type reactions of taylorione isomer (**113**) and mylione (**114**) although **113** and **114** have not been found in *M. taylorii* (Scheme 7). Bitaylorione (**110**) was also shown to be a dimeric sesquiterpenoid with a tricyclic (5,6,5)-ring system on the basis of 2D-^{1}H-^{1}H and COLOC (Correlation Spectroscopy *via* Long-Range Couplings) spectra. The compound is not of the normal Diels-Alder adduct (2,2'; 3,4') type.

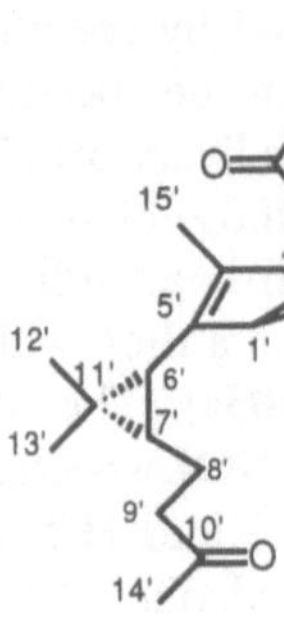

(108) Myltaylorione A

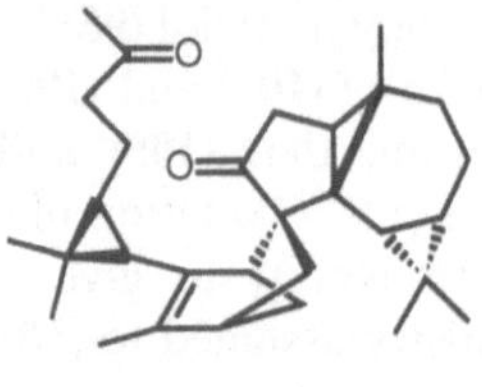

(109) Myltaylorione B

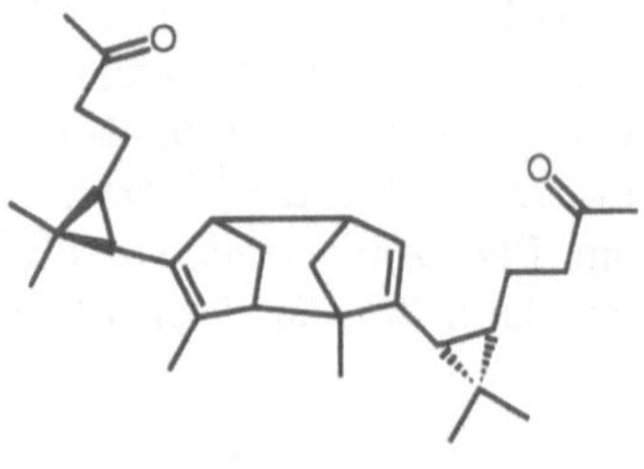

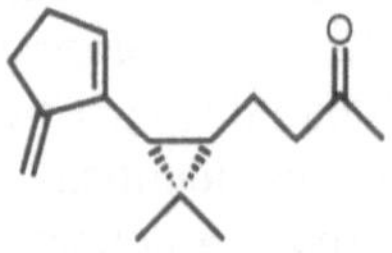

(111) Taylorione

(110) Bitaylorione

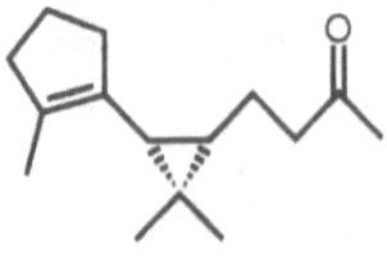

(112)

Chart 6b. Aromadendrane-secoaromadendrane and secoaromadendrane-secoaromaden-
drane dimers found in the Hepaticae

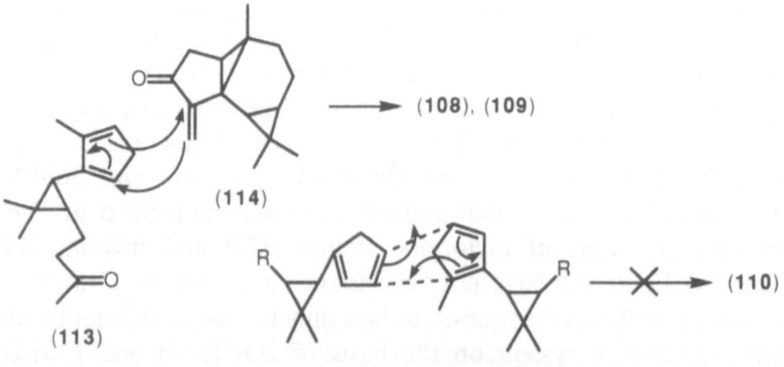

(108), (109)

(114)

(113)

Normal Diels-Alder reaction

(110)

Scheme 7. Formation of dimeric sesquiterpenoids by a Diels-Alder type reaction

The known 2,3-secoaromadendrane-type sesquiterpenoids plagio-
chilide (**121**), plagiochiline A (**115**), B (**116**), C (**117**), D (**118**), E (**119**), H
(**120**), hanegokedial (= plagiochilal A) (**123**), ovalifolienal (**124**) and 9α-
acetoxyovalifoliene (**125**) have been isolated from several South American
and Asiatic *Plagiochila* species (*46, 48*). Further fractionation of the
dichloromethane extract of *P. fruticosa* resulted in the isolation of three
novel 2,3-secoaromadendrane-type sesquiterpenoids, plagiochilal B (**126**)
possessing powerful pungency, plagiochiline J (**127**) and plagiochiline K

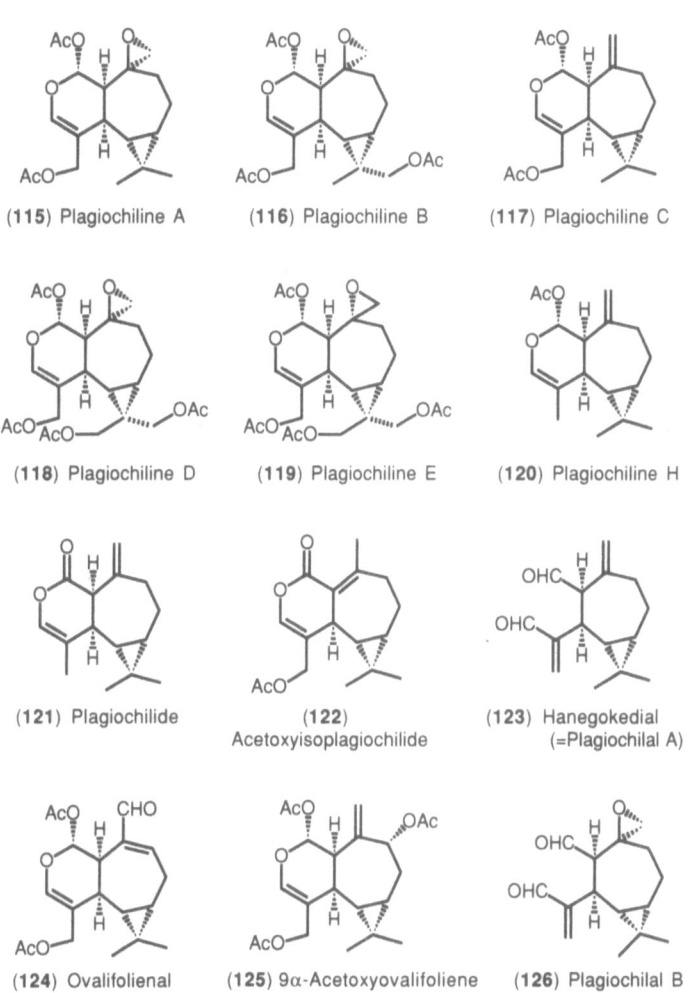

(**115**) Plagiochiline A   (**116**) Plagiochiline B   (**117**) Plagiochiline C

(**118**) Plagiochiline D   (**119**) Plagiochiline E   (**120**) Plagiochiline H

(**121**) Plagiochilide   (**122**)
                                  Acetoxyisoplagiochilide   (**123**) Hanegokedial
                                                                   (=Plagiochilal A)

(**124**) Ovalifolienal   (**125**) 9α-Acetoxyovalifoliene   (**126**) Plagiochilal B

Chart 6c. Secoaromadendranes found in the Hepaticae

(**128**), along with plagiochilide (**121**) and plagiochiline A (**115**) (*185*). Absolute stereostructures were established by extensive 2D-COSY NMR data, NOE difference spectroscopy and a chemical correlation with plagiochiline A (**115**) shown in Scheme 8 (*185*). Most of the previously known 2,3-secoaromadendrane-type sesquiterpene hemiacetals such as plagiochiline A (**115**) and plagiochilide (**121**) contain a trisubstituted end-double bond and are most likely to be biosynthesized *via ent*-hanegokedial (**123**) from a key precursor (−)-bicyclogermacrene (**195**), a cometabolite in *Plagiochila* species (*19*).

The three new hemiacetals (**126**–**128**) are presumably formed with an intact exo-double bond from (**123**) through a series of oxidation and reduction processes shown in Scheme 9. The occurrence of compounds (**126**–**128**) with an exo-double bond may suggest an alternative biosynthetic pathway *via* (**127**) or (**128**) to highly oxygenated plagiochiline A (**115**) and its related hemiacetals (*185*).

When one chews plagiochiline A (**115**), one notices a persistent pungent taste as a result of the following transformation. Enzymatic treatment of plagiochiline A (**115**) with amylase or saliva produces the intensely pungent dialdehyde, plagiochilal B (**126**) which has been isolated from some collections of *Plagiochila fruticosa* (*185*) together with furanoplagiochilal (**126d**) in good yield (*248, 252*). Treatment of (**115**) with

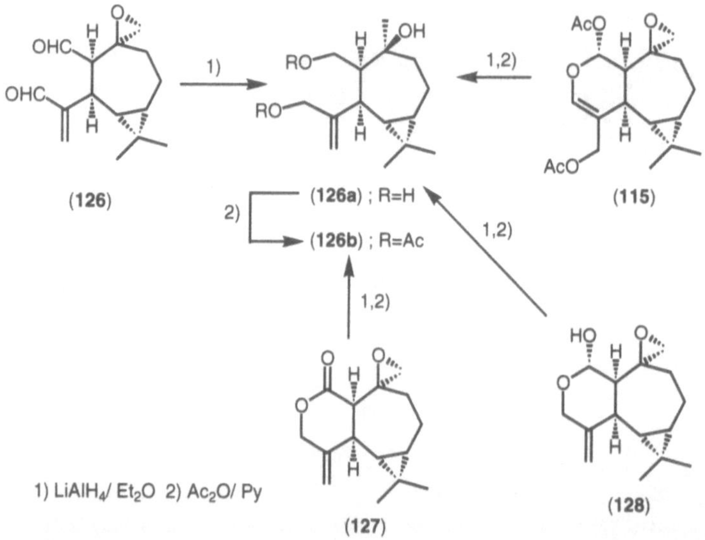

(**126**) (**115**) (**127**) (**128**)

1) $LiAlH_4$/ Et_2O 2) Ac_2O/ Py

Scheme 8. Correlation of plagiochilal B, plagiochilines J and K with plagiochiline A

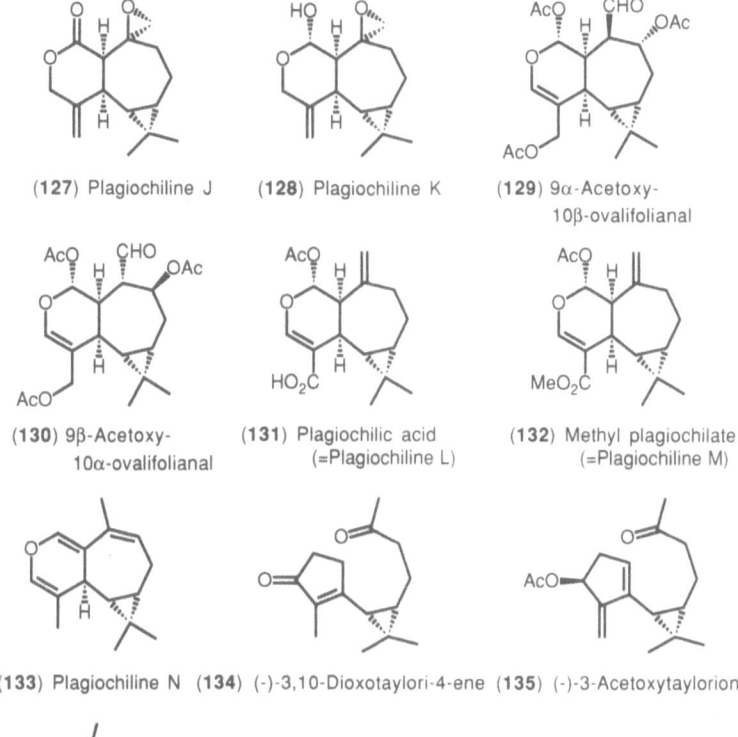

(127) Plagiochiline J   (128) Plagiochiline K   (129) 9α-Acetoxy-
                                                       10β-ovalifolianal

(130) 9β-Acetoxy-   (131) Plagiochilic acid   (132) Methyl plagiochilate
   10α-ovalifolianal   (=Plagiochiline L)   (=Plagiochiline M)

(133) Plagiochiline N   (134) (-)-3,10-Dioxotaylori-4-ene   (135) (-)-3-Acetoxytaylorione

(136) 2-Nor-1,3-epoxy-1,10-secoaromadendra-1(5),3-dien-10-one

Chart 6d. Secoaromadendranes and norsecoaromadendranes found in the Hepaticae

KHCO₃ in aqueous MeOH afforded the same aldehydes, (126) and (126d) shown in Scheme 10.

Two novel hemiacetals, 9α-acetoxy-10β-ovalifolianal (129) and 9β-acetoxy-10α-ovalifolianal (130), have been isolated from *Plagiochila peculiaris*; structures were established by comparing their spectroscopic properties with those of ovalifolienal (124) (*623*). In addition to plagiochiline C (117), two new 2,3-secoaromadendrane-type sesquiterpene hemiacetals, plagiochiline L (131) and plagiochiline M (132) have been isolated from *Heteroscyphus planus*; their structures were elucidated by spectroscopic and chemical correlation with plagiochiline C (117) (*249, 252*). P.

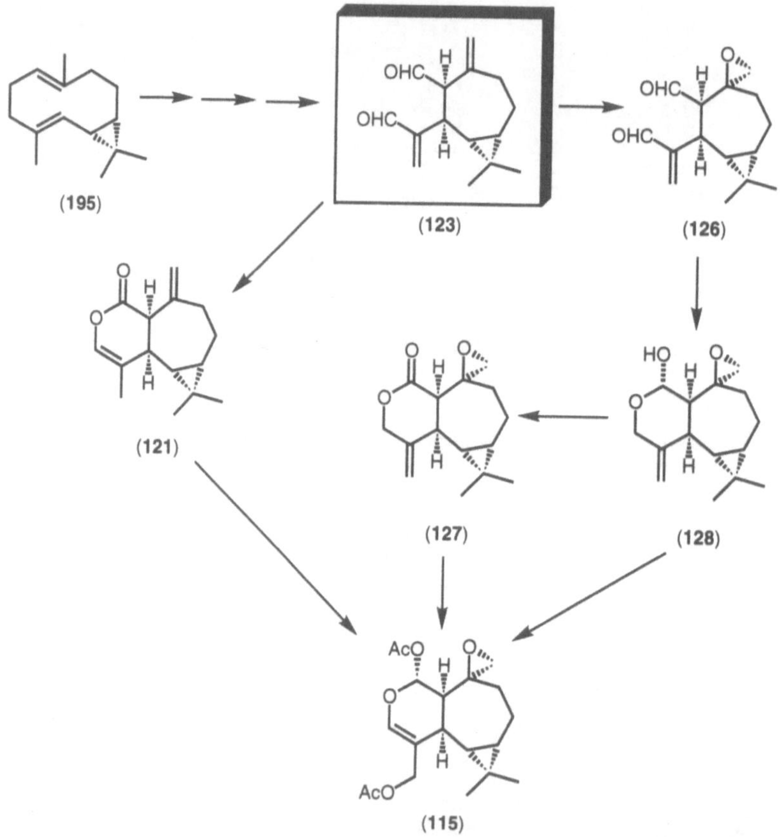

Scheme 9. Possible biogenetic pathways for plagiochilal B, plagiochilines A, J and K

ovalifolia produces various 2,3-secoaromadendrane-type sesquiterpenoids (*19*). Further fractionation of the ether extract of *P. ovalifolia* resulted in the isolation of two new 2,3-secoaromadendrane-type sesquiterpenoids, acetoxyisoplagiochilide (**122**) (*435a*) and plagiochiline N (**133**) possessing the 3H-pyron skeleton (*249*). The total synthesis of (+)-*ent*-hanegokedial (= plagiochilal A) (**123**) has been accomplished by TAYLOR *et al.* (*548*) from (−)-bicyclo[5.1.0]enone (*547*) in five steps.

Taylorione (**111**) a 1,10-secoaromadendrane-type sesquiterpene ketone, has been isolated from *Mylia taylorii*. Its absolute stereostructure was established by spectral and chemical evidence as well as total synthesis (*19, 647*).

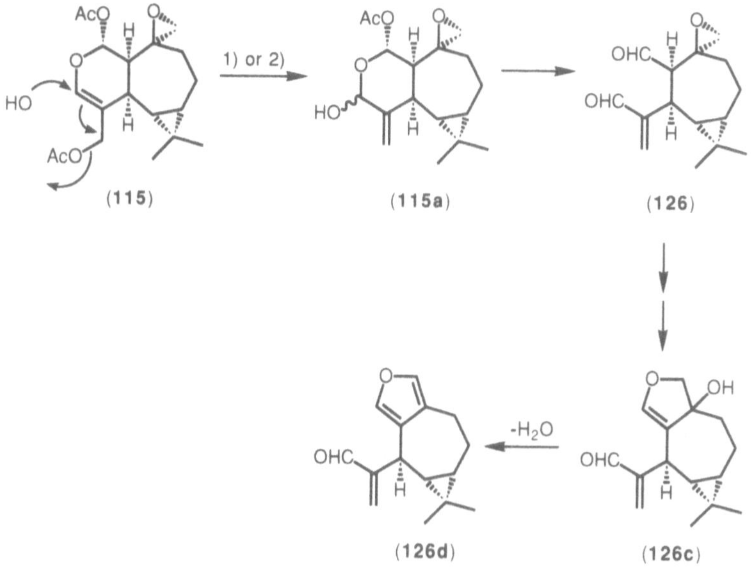

Scheme 10. Formation of plagiochilal B and furanoplagiochilal from plagiochiline A

Two further secoaromadendranes, (−)-3,10-dioxotaylori-4-ene (**134**) and (−)-3-acetoxytaylorione (**135**) have been isolated from the same liverwort (*386*). The structure of the former was arrived at by chemical evidence and by means of ^{1}H-NMR spectroscopy using shift reagent and by comparing the ^{13}C-NMR spectrum with that of taylorione (**111**). The spectral data of a hydroxyketone (**135a**) obtained from (**135**) with KOH in methanol were quite similar to those of taylorione (**111**), suggesting that (**135**) might be taylorione containing an extra secondary acetoxyl group. The position of the acetoxyl group at C-3 has been clarified by spin decoupling experiments on the ketoalcohol (**135a**).

A novel 1,10-norsecoaromadendrane-type sesquiterpenoid (**136**), has been isolated from European *Mylia taylorii*, together with the previously known *ent*-aromadendranes, myliol (**98**), dihydromylione (**99**) and taylorione (**111**) (*242*). The NMR spectrum of (**136**) was very similar to that of taylorione (**111**) except that the signals of the exomethylene cyclopentene were replaced by those of a 3,4-disubstituted furan, thus suggesting the novel structure (**136**). The presence of a small allylic coupling (J = 1.2 Hz) between H-6 and H-1 and the formation of the known carbocyclic acid (**136a**) from (**136**) by ozonolysis supported the proposed structure of the nor-secoaromadendrane-type sesquiterpenoid.

(135a)   (136a)

5.5 Azulenes and Indenes

Calypogeia species (Calypogeiaceae) are rich sources of azulenes and indenes (*19*). A sterile culture of *Calypogeia azulea* produces 1,4-dimethyl-azulene (**137**), 4-methyl-1-methoxycarbonylazulene (**138**) and 3,7-dimethyl-5-methoxycarbonylindene (**148**) (*106*). The cultured cells of *C. granulata* produce the same azulenoids as those of intact and re-differentiated plants (*321, 453, 539*). The blue oil obtained from the cultured cells of *C. granulata* was chromatographed on silica gel to give (+)-1,8a-dihydro-3,8-dimethylazulene (**150**) together with 3,7-dimethylindene-5-carboxaldehyde (**149**) and 1,4-dimethylazulene (**137**) (50% of oil) (*321, 453, 538*). Dihydroazulene (**150**) is too unstable to be collected by preparative GC. It has been isolated by rapid flash chroma-tography (neutral SiO_2, *n*-hexane elution, N_2 atmosphere) and subse-quent preparative HPLC using *n*-hexane as a solvent. When the solvent was removed under N_2, (**150**) was converted to 1,4-dimethylazulene (**137**) and a colorless polymer. Compound (**150**) had UV absorption maxima at 211 nm (sh), 288 (log ε 4.10), 234 (4.40) and 312 (3.73) indicating the presence of a conjugated tetraene system. The structure was characterized by NMR spectrometry and facile oxidation to 1,4-dimethylazulene, the very intense optical activity ([α]$_D$ + 1165°) and CD Cotton effects (235 nm, $\Delta\varepsilon$ − 47.4 and 314.0 nm, $\Delta\varepsilon$ + 19.7) suggesting a strongly distorted conjugated system (*235*). The structure of (**149**) was also established by NMR spectrometry and comparison with the NMR spectrum of 3,7-dimethyl-5-methoxycarbonylindene (**148**) isolated from *C. trichomanis* (*19*). 3,7-Dimethylindene-5-carboxaldehyde (**149**) is prob-ably derived from a possible precursor (**153**) *via* route a or b shown in Scheme 11. In order to confirm this, [2-^{13}C]-labeled acetate was fed to the culture to give (**137**) and (**149**) in which the asterisked carbons should be labeled (*321, 453, 538*). The ^{13}C enriched, the natural abundance and difference spectra of both (**137**) and (**149**) showed that carbons 2,5,7,9,10, 1-Me and the 4-Me of (**137**) (terpenoid biosynthetic route, Scheme 11) and 2,5,6,8,9,3-Me and 7-Me, but not the carbonyl carbon of (**149**), were

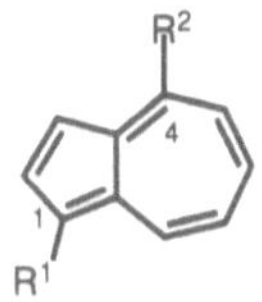

(**137**) 1,4-Dimethylazulene ; $R^1=R^2$=Me

(**138**) 4-Methyl-1-methoxycarbonylazulene ; R^1=CO$_2$Me, R^2=Me

(**139**) 4-Methylazulene-1-carbaldehyde
(=4-Methyl-1-formylazulene) ;R^1=CHO, R^2=Me

(**140**) 4-Methyl-1-carboxyazulene ; R^1=CO$_2$H, R^2=Me

(**141**) 4-Methoxymethyl-1-carboxyazulene ; R^1=CO$_2$H, R^2=CH$_2$OMe

(**142**) 1,4-Dicarboxyazulene ; $R^1=R^2$=CO$_2$H

(**143**) 4-Hydroxymethyl-1-methoxycarbonylazulene ;
R^1=CO$_2$Me, R^2=CH$_2$OH

(**144**) 4-Carboxy-1-methoxycarbonylazulene ;
R^1=CO$_2$Me, R^2=CO$_2$H

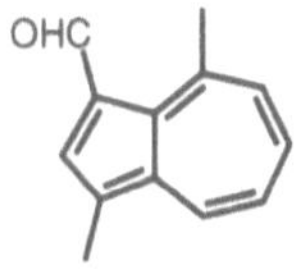

(**145**) 1,4-Dimethyl-3-formylazulene

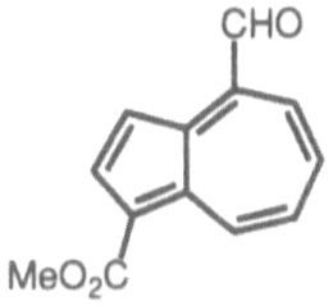

(**146**) 4-Formyl-1-methoxycarbonylazulene

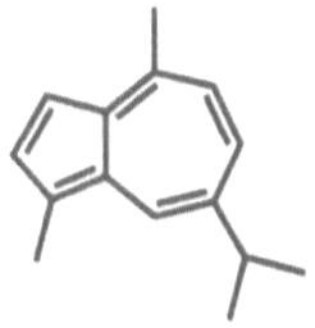

(**147**) Guaiazulene

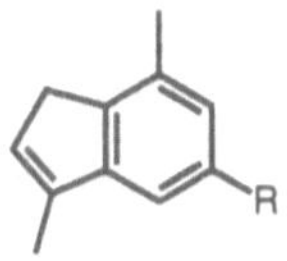

(**148**) 3,7-Dimethyl-5-methoxy-carbonylindene ; R=CO$_2$Me
(**149**) 3,7-Dimethylindene-5-carbaldehyde ; R=CHO

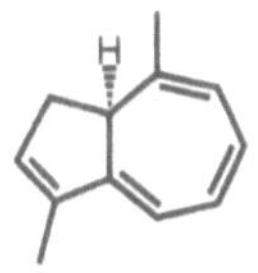

(**150**) (+)-1,8a-Dihydro-3,8-dimethylazulene
(=3,10-Dihydro-1,4-dimethylazulene)

Chart 7a. Azulenes and indenes found in the Hepaticae

enriched. These results indicate that (**137**) is formed *via* optically active 3,10-dihydro-1,4-dimethylazulene (**150**), and (**149**) *via* hypothetical trinorsesquiterpenoid intermediate by route a and not by route b. However, the possibility that azulene (**137**) might also originate from bicyclogermacrene (**196**), trinoranastreptene (**51**) and anastreptene (**93**) is not excluded

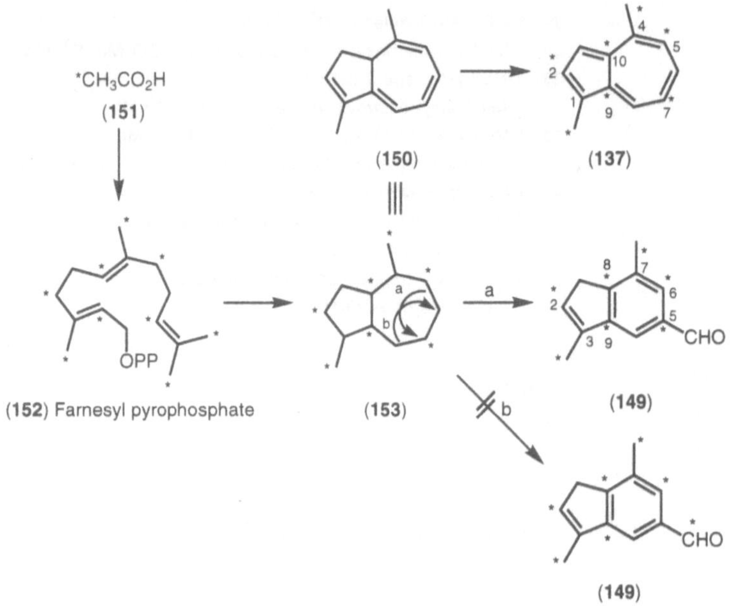

Scheme 11. Incorporation of ^{13}C from [2-^{13}C]-labeled acetate into 1,4-dimethylazulene and 3,7-dimethylindene-5-carboxaldehyde

because the yields of trinoranastreptene and anastreptene depend on the culture period. It should be noted that cell culture techniques constitute a convenient and efficient method for production of very unstable secondary metabolites of bryophytes and for studies of their biosynthesis because most bryophytes are very small plants and it is difficult to collect pure mats on a large scale.

The absolute configuration of 1,8a-dihydro-3,8-dimethylazulene (**150**) was shown to be 8a*S* by comparing the CD spectrum with that calculated for (8a*R*)-1,8a-dihydroazulene (**154**) by the SCF-CI-dipole velocity MO method (*235*). This has also been proved experimentally by synthesis of the model compounds (1*S*,8a*S*)-(+)-1,8a-dihydro-1-methoxy-8a-methylazulene (**155**) and (1*S*,8a*S*)-(+)-1,8a-dihydro-1-methoxy-6,8a-dimethylazulene (**156**) (*235*), as well as by synthesis of two more closely related model compounds, (8a*S*)-(+)-1,8a-dihydro-8a-methylazulene (**157**) and (8a*S*)-(+)-1,8a-dihydro-6,8a-dimethylazulene (**158**), and measuring their chiroptical properties (*236*). The CD Cotton effects of (**157**) were stronger and closer in intensity to those of natural dihydroazulene (**150**) than to those of dihydroazulenes (**155**) and (**156**). The above evidence and X-ray crystallographic analysis

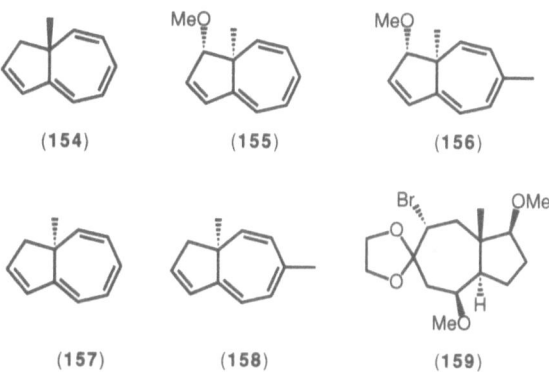

(154)   (155)   (156)

(157)   (158)   (159)

Chart 7b. Synthetic azulene derivatives related to the naturally occurring dihydroazulene
(150)

of (1*S*,3*SR*,4*S*,7*R*,8*SR*)-(+)-7-bromo-2,3,3a,4,5,7,8,8-octahydro-1,4-di-
methoxy-8a-methyl-6(1H)-azulene-6-ethylene acetal (159) completely
established the absolute configuration of (150) with its twisted π-electron
system (236).

Further investigation of the hexane extract of *in vitro* cultures of *C.
azulea* resulted in isolation of two additional new azulenoids, 4-
methylazulene-1-carbaldehyde (= 4-methyl-1-formylazulene) (139) and
4-methyl-1-carboxyazulene (140), together with the previously known
azulenes (137) and (138). Structures of the new compounds were based on
spectroscopic evidence using 2D-COSY (HMBC) and NOESY spectrometry (439). SIEGEL *et al.* (507) studied the *n*-hexane-dichloromethane-
EtOAc extract of *in vitro* cultures of *Calypogeia azulea* and isolated seven
novel azulenes (139, 140, 142 as an ester, 143–146), together with 1,4-
dimethylazulene (137) and (138). The major azulene was (137) (0.05% of
dry weight of plant). Structures of the new azulenoids were elucidated by
comparing the ¹H-NMR and mass spectral data with those of the known
azulenes (137) and (138). Presence of a sharp singlet at δ 10.35 (C*H* = O)
and the far downfield shift of the H-8 signal (doublet at δ 9.64) supported
the structure proposed for (139). The structures of (139), (140), (143) and
(144) were established by synthesis using 4-methylazulene and 1,4-
dimethylazulene (137) (507). The ester of (142) is unstable being easily
hydrolyzed to stable azulene-1,4-carboxylic acid (142), which was also
obtained by saponification of (144); hence it was suggested that an
unknown alcohol esterifies the carboxyl group at C-4. The ¹H-NMR and
mass spectral data of (145) were identical with those of 1,4-dimethyl-3-
formylazulene which was synthesized from 1,4-dimethylazulene by

formylation with DMF-POCl$_3$. The structure proposed for **(146)**, 4-formyl-1-methoxycarbonylazulene is based on the identity of the mass fragmentation pattern with that of synthetic **(146)**. Compound **(146)** has also been found in the higher plants *Helichrysum ambiguum* subsp. *ambiguum* (*305*) and *Ixiolaena leptolepis* (*356*), both in Compositae.

Compounds **(137–139)** and **(145)** but not the others, have also been isolated from *C. azulea* collected in the field. The yield of the major azulene **(137)** – 2.3 g (0.45% dried weight) – was considerably higher than the yield from the *in vitro* cultures while the yield of **(138)** was little different.

It had been thought that 1,4-dimethylazulene **(137)** and its analogues were characteristic compounds of *Calypogeia* species (Calypogeiaceae).

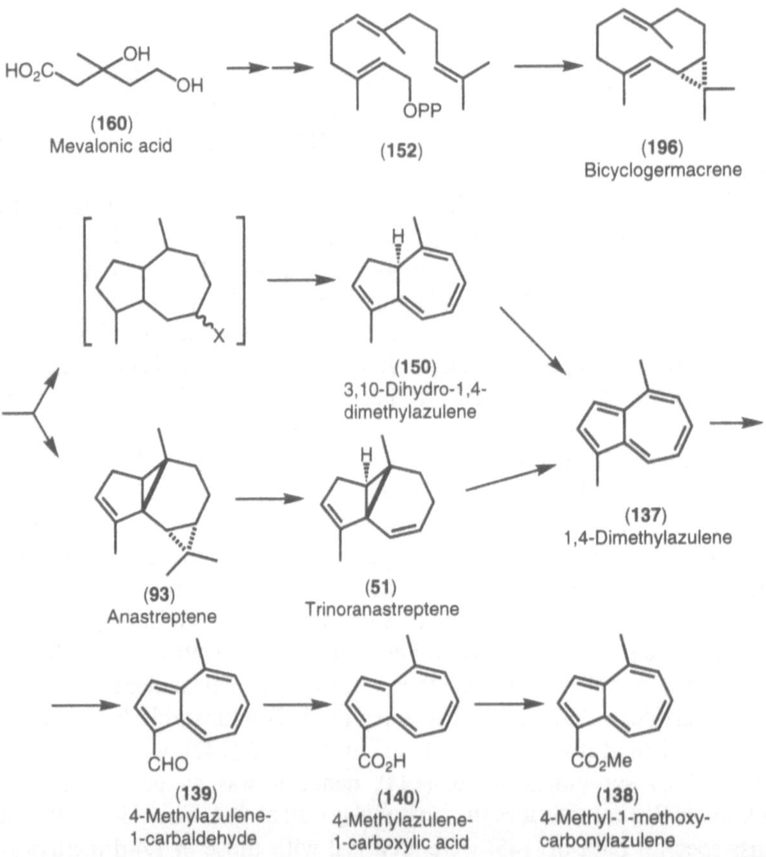

Scheme 12. Possible biogenetic pathways for naturally occurring oxygenated azulenoids

Subsequently, however 1,4-dimethylazulene was also isolated from *Macrolejeunea pallescens* (Lejeuneaceae) and *Plagiochila micropterys* (Plagiochilaceae) *(433)*. *Plagiochila longispina* contains two new azulenes, 4-methylazulene-1-carbaldehyde **(139)** and 4-methoxymethyl-1-carboxyazulene **(141)** *(507)*. Structures assigned to the new azulenes were based on a comparison of ^{1}H-NMR and mass spectral data with those of 1,4-dimethylazulene **(137)** and 4-methyl-1-methoxycarbonylazulene **(138)**. The structure of **(141)** was conclusively confirmed by synthesis using **(138)** as starting material *(507)*. Azulene **(139)** exhibited a deep purple color in non-polar solvents which changed to carmine-red in polar solvents. Possible biogenetic pathways for oxygenated azulenes have been proposed by NAKAGAWARA et al. *(439)* (Scheme 12).

Guaiazulene **(147)** has been detected in *Mastigophora diclados* *(622, 635)*.

5.6 Barbatanes (Gymnomitranes)

Barbatane-type sesquiterpenoids are common constituents of Jungermanniidae and Marchantiidae. Nine barbatanes from liverworts were already mentioned in the earlier report *(19)*. Among them, the hydrocarbon β-barbatene (= gymnomitrene) **(162)** is the most familiar compound. However, oxygenated barbatane-type sesquiterpenoids are rare. α-Barbatene **(161)** is often detected by GC-MS; however it seems to be an artifact since during preparative GC the facile isomerization of β-barbatene to the α-isomer was observed *(342)*. A total synthesis of (±)-gymnomitrol **(163)** has been reported as preliminary communication *(19)*, while a 16-step total synthesis of (163) has been described in a full paper *(141)*.

Bazzania fauriana contains α-barbatene **(161)**, β-barbatene **(162)** and gymnomitrol **(163)** *(581)*. *Bazzania trilobata* elaborates a number of sesquiterpenoids such as barbatenes, bazzanenes and calamenenes *(19)*. The composition of the essential oil from *B. trilobata* does not change with the seasons *(283)*. Three new barbatane-type sesquiterpenoids, 9-oxogymnomitryl acetate **(164)**, 9α-hydroxygymnomitryl acetate **(165)** and 9α-hydroxygymnomitryl cinnamate **(166)** have been isolated from the non-pungent *Plagiochila trabeculata*, together with β-barbatene **(162)** and gymnomitrol **(163)**; their structures were elucidated by a combination of extensive NMR studies and chemical correlations *(589)*. That compound **(166)** was 9α-hydroxygymnomitryl cinnamate and not 9α-cinnamoxygymnomitrol was established by extensive spin decoupling and 2D-COSY

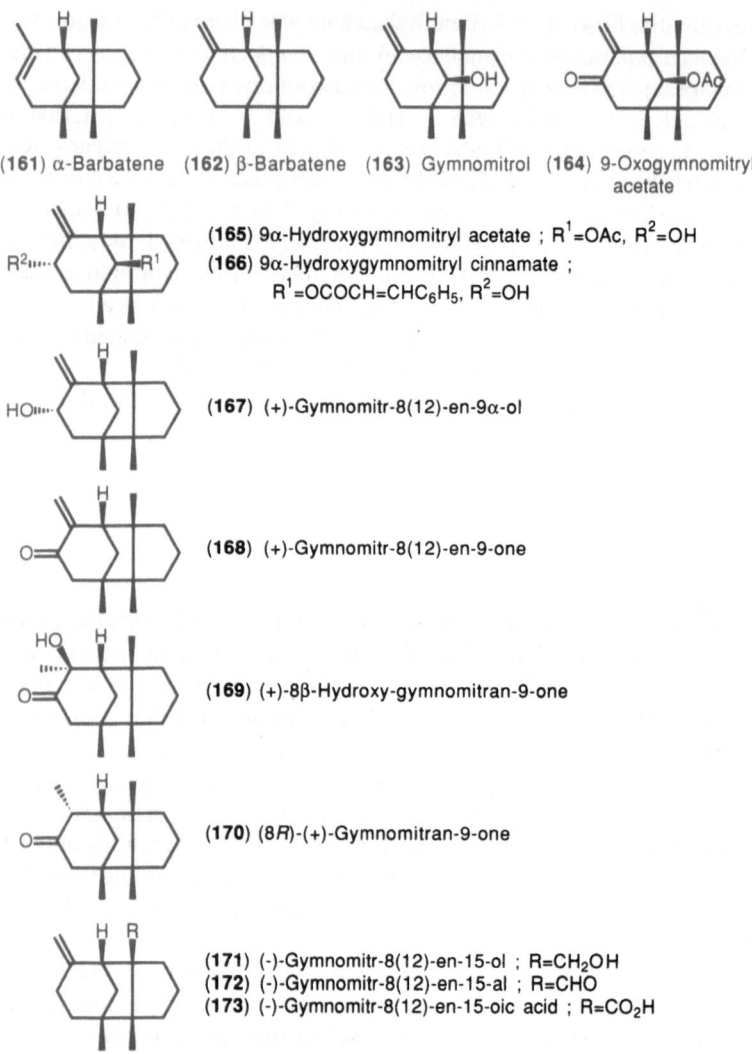

(161) α-Barbatene   (162) β-Barbatene   (163) Gymnomitrol   (164) 9-Oxogymnomitryl
                                                                        acetate

(165) 9α-Hydroxygymnomitryl acetate ; R^1=OAc, R^2=OH
(166) 9α-Hydroxygymnomitryl cinnamate ;
        R^1=OCOCH=CHC$_6$H$_5$, R^2=OH

(167) (+)-Gymnomitr-8(12)-en-9α-ol

(168) (+)-Gymnomitr-8(12)-en-9-one

(169) (+)-8β-Hydroxy-gymnomitran-9-one

(170) (8R)-(+)-Gymnomitran-9-one

(171) (-)-Gymnomitr-8(12)-en-15-ol ; R=CH$_2$OH
(172) (-)-Gymnomitr-8(12)-en-15-al ; R=CHO
(173) (-)-Gymnomitr-8(12)-en-15-oic acid ; R=CO$_2$H

Chart 8. Barbatanes (gymnomitranes) found in the Hepaticae

NMR spectrometry, while the stereochemistries of (164–166) were determined by NOE difference spectrometry.

European wild *Reboulia hemisphaerica* and its *in vitro* cultured thallus produces (+)-gymnomitr-8(12)-en-9α-ol (167) and (+)-gymnomitr-8(12)-en-9-one (168) (*411, 412*). (+)-8β-Hydroxygymnomitran-9-one

(169) has been isolated only from agar cultured *R. hemisphaerica* while (8*R*)-(+)-gymnomitran-9-one (170) has been found only in material gathered in the field. The structure assigned to (167) was based on ^{1}H- and ^{13}C-NMR spectral data, comparison with the spectra of gymnomitrol (163) and conversion (167) by CDCl$_3$ in an NMR tube to a mixture of naturally occurring (+)-gymnomitr-8(12)-en-9-one (168) and (8*R*)-(+)-gymnomitran-9-one (170), respectively. The stereochemistry of (167) at C-9 was deduced by NOE spectrometry of its monoacetate while its absolute configuration was derived from the positive Cotton effect ($\Delta\varepsilon$ 300 nm, + 0.37) of the saturated ketone (170). The structure assigned to (+)-8β-hydroxygymnomitran-9-one (169) was also based on a comparison of its NMR spectral data with those of 170.

The new gymnomitrenols (167–170) belong to the same series of barbatane-type sesquiterpenoids found in other liverworts (*19, 589*). While barbatane-type sesquiterpenoids have not been detected in Japanese *R. hemisphaerica* (*55*), three new barbatane-type sesquiterpenoids, (−)-gymnomitr-8(12)-en-15-ol (171), (−)-gymnomitr-8(12)-en-15-al (172) and (−)-gymnomitr-8(12)-en-15-oic acid (173) have been isolated from *Marsupella emarginata* var. *patens*, along with (−)-β-barbatene (162) (*384*). Their structures and absolute configurations were established by spectroscopic and chemical correlation and by X-ray crystallographic analysis of a *p*-bromobenzoate prepared from (172).

M. emarginata var. *patens* is chemically different from *M. emarginata* and *M. aquatica* because the latter two species produce longipinane-type sesquiterpenoids (*19*). The biogenetic precursor of the gymnomitranes is undoubtedly β-bazzanene (175) whose acid catalyzed rearrangement has been investigated by WU and LIU (*634*).

5.7 Bazzananes

Three bazzanane-type sesquiterpenoids, α-bazzanene (174), β-bazzanene (175) and bazzanenol were previously known in liverworts (*19*). Structure (177) was proposed for bazzanenol on the basis of the chemical and spectroscopic data (*19*). Further study of the essential oils of seven *Bazzania* species led to the isolation of (+)-bazzanenol the structure of which was revised to (176) (*264*). α-Bazzanene (174) has been obtained by isomerization of β-bazzanene (175) with formic acid at room temperature, together with α-tricodiene (182) (Scheme 13) (*19*). A new bazzanane-type hydrocarbon, isobazzanene (178) has been isolated from *Bazzania fauriana* and *B. angustifolia*; its structure was elucidated by analysis of the ^{1}H NMR and MS spectra and by its partial synthesis from β-bazzanene

(174) α-Bazzanene 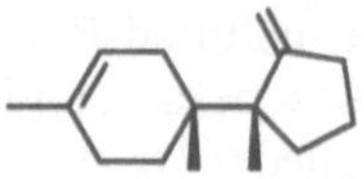   (175) β-Bazzanene

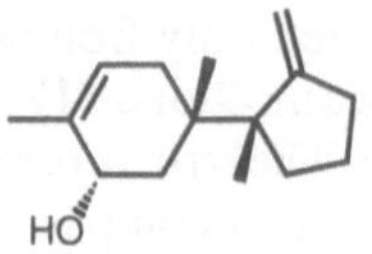

(176) Bazzanenol

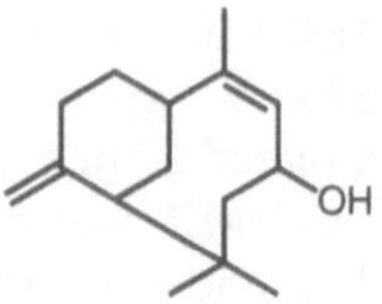

(177)

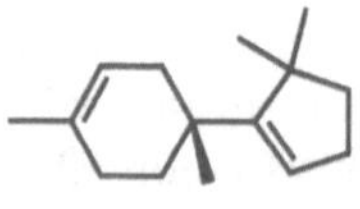

(178) Isobazzanene

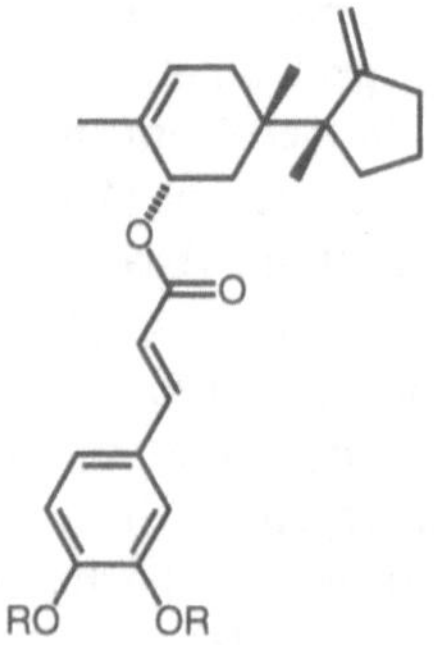

(179) Bazzanenyl caffate ; R=H
(180) R=Me

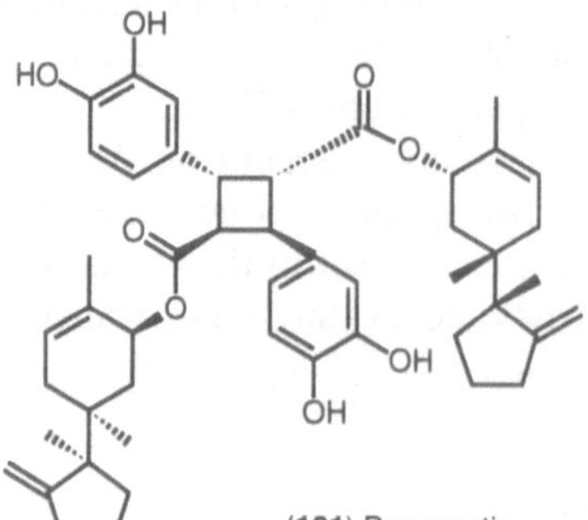

(181) Bazzanetin

Chart 9. Bazzananes found in the Hepaticae

(175) using HCOOH (Scheme 13) (*634*). Structure (175a) previously assigned to the acid rearrangement product (*19*) has to be revised. Other products produced on acid treatment of β-bazzanene were cuparene (284), α-bazzanene (174), isocyclobazzanene (178a) and cyclobazzanene (178b) (*634, 636*) (Scheme 14). The sesquiterpenes (178a, 178b) with the bicyclo[2.2.2]octane system which has less energy strain than the bicyclo[3.2.1]system have not been found in nature. Further study of the chemical constituents of *B. fauriana* resulted in the isolation of a new

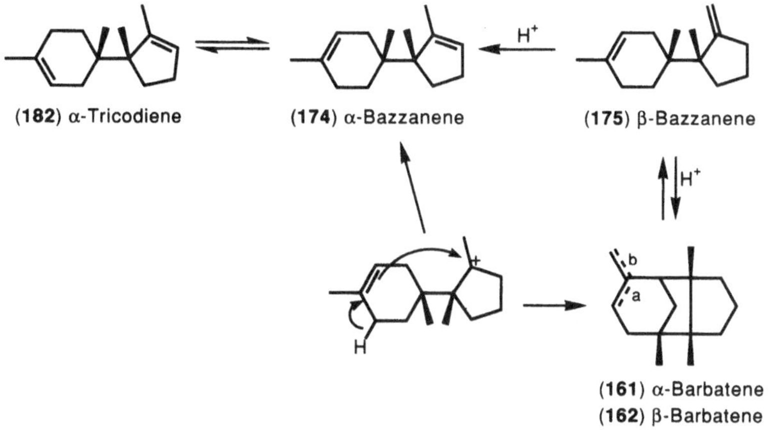

(182) α-Tricodiene   (174) α-Bazzanene   (175) β-Bazzanene

(161) α-Barbatene
(162) β-Barbatene

Scheme 13. Formation of α-bazzanene and α-tricodiene from β-bazzanene and possible biogenetic pathway for barbatenes

bazzanenyl ester (179) and β-bazzanene (175) (581). Methylation of (179) with methyl iodide gave a dimethyl ether (180) which was further reduced with LiAlH$_4$ to afford 3,4-dimethoxycinnamyl alcohol and bazzanenol (176), indicating that (179) was bazzanenyl caffeate. That the caffeate ester on C-2 was α-oriented was established by NOE difference spectrometry.

Bazzanetin (181), a new bis-bazzanenyl cyclobutyrate possessing two dihydroxyphenyl groups has been isolated from the methanol extract of *Bazzania pompeana*; its structure was elucidated by chemical degradation (hydrolysis, methylation, acetylation and reduction) (249a).

5.8 Bergamotanes, Bicycloelemanes and Elemanes

As reported earlier (19), trans-α- (183) and trans-β-bergamotene (184) have been found in two Jungermanniales species. The latter hydrocarbon has been detected in *Monoclea forsteri* and *Radula boryana* by GC-MS analysis (513).

Bicycloelemene (185) has been isolated from or detected in Marchantiales and Jungermanniales. α- (186), β- (187), γ- (188) and δ-elemenes (189) and elemol (190) have been detected in liverworts; among these β-elemene (187) is the most common as shown in Table IIb. *Clasmatocolea humilis* produces dehydrosaussurea lactone (191) and saussurea lactone (192) (45). The former compound has been detected in *Plagiochila hondurensis* (48). From the ethereal extract of French *Plagiochasma*

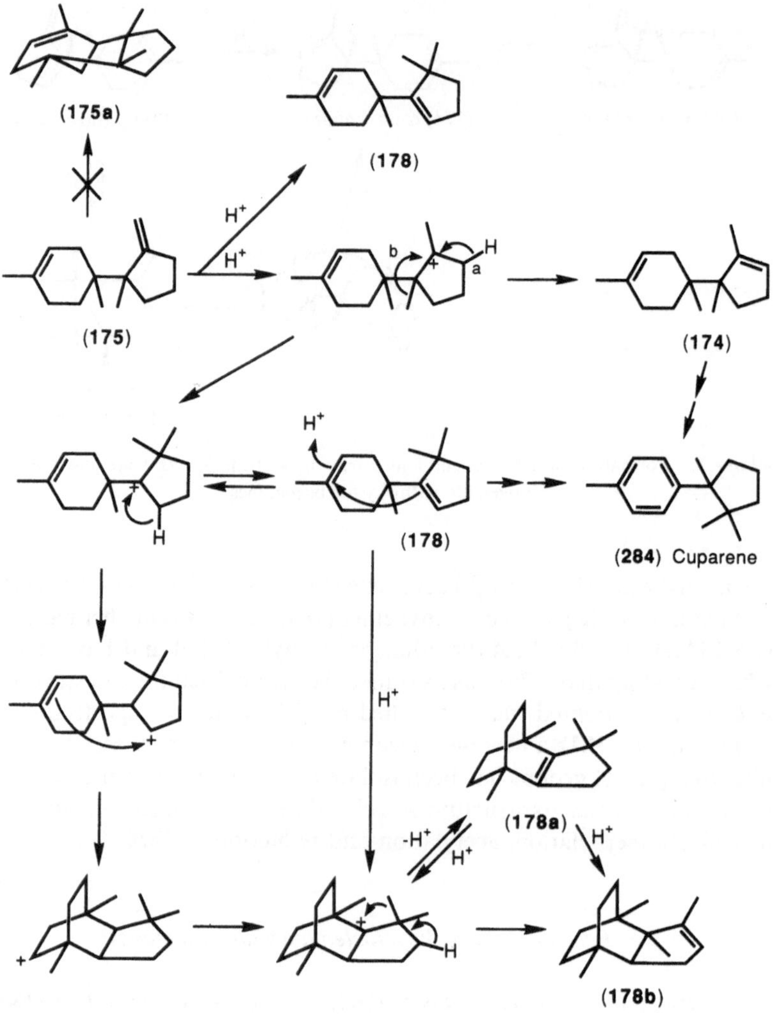

(175a)

(178)

H⁺

H⁺

(175)

(174)

H⁺

(178)

(284) Cuparene

H⁺

H⁺

(178a)

-H⁺

H⁺

H⁺

+

(178b)

Scheme 14. Acid treatment of β-bazzanene

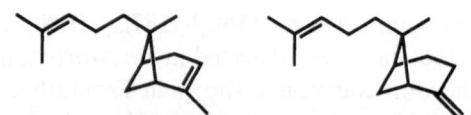

(183) *trans-α*-Bergamotene   **(184)** *trans-β*-Bergamotene

Chart 10. Bergamotanes found in the Hepaticae

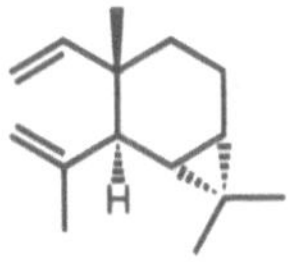

(185) Bicycloelemene 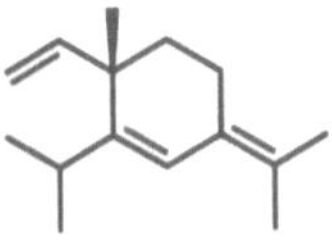   (186) α-Elemene 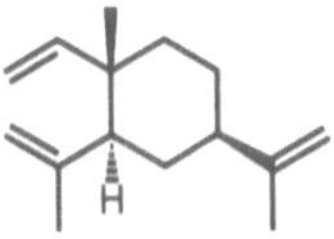   (187) β-Elemene 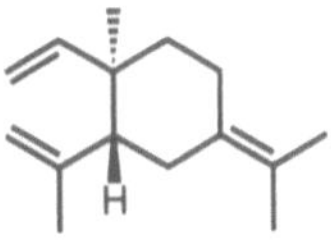   (188) γ-Elemene

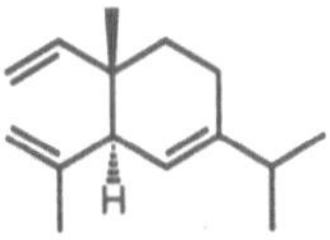

(189) δ-Elemene 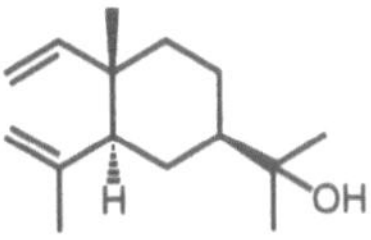   (190) Elemol 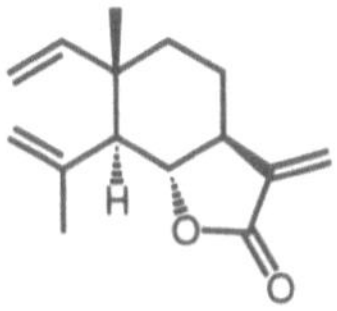   (191) Dehydrosaussurealactone

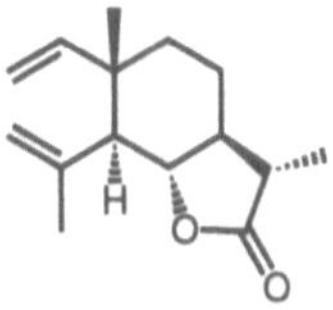

(192) Saussurealactone 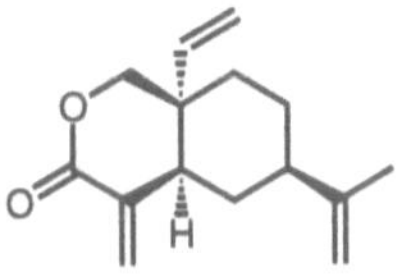   (193) Elema-1,4(15),11-trien-3,14-olide

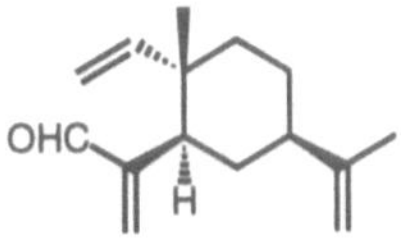

(194) Elema-1,4(15),11-trien-3-al

Chart 11. Bicycloelemanes and elemanes found in the Hepaticae

rupestre, two novel elemane-type sesquiterpenoids (193) and (194) have been isolated (*243*). The relative stereochemistry of (193) has been elucidated by application of ^{1}H-^{1}H 2D- and delayed ^{1}H-^{1}H COSY spectroscopy as well as by NOE difference spectrometry. The spectroscopic properties of (193) are in good agreement with those of 11,13-dihydrovernodalin (195) from *Vernonia* species (Compositae) (*193*). The

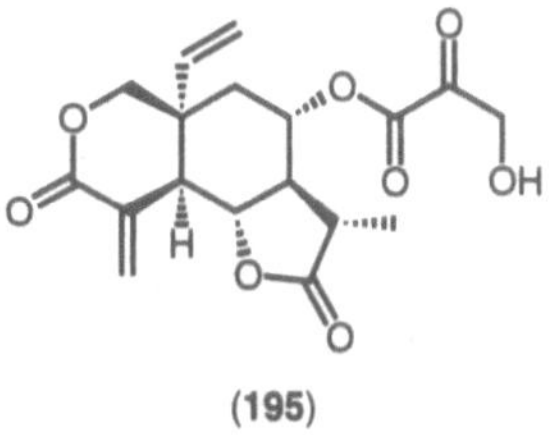

(195)

structure of (**194**) was suggested by spectroscopic comparison with (**193**) and confirmed by ^{1}H-^{1}H 2D shift-correlated and NOE difference spectrometry. The absolute configuration of both new sesquiterpenoids remains to be established.

5.9 Bicyclogermacranes, Isobicyclogermacranes and Lepidozanes

Ent-bicyclogermacrene (**196**), which is the biosynthetic precursor of bicycloelemanes, aromadendranes, alloaromadendranes, secoaromadendranes, maalianes, aristolanes and vitranes is the most widely distributed sesquiterpene hydrocarbon in Hepaticae. 3α-Acetoxybicyclogermacrene (**197**) has been isolated only from Jungermanniales, particularly from Plagiochilaceae (*19*). A new labile bicyclogermacrane-type sesquiterpene aldehyde, (−)-bicyclogermacrenal (**198**), has been isolated from the thalloid liverwort *Conocephalum conicum* (*595*). The position of the C-13 aldehyde group was suggested by NOE's between H-5 and CHO, and H-7 and C11-Me. Attempted conversion of (**198**) to the known bicyclogermacrene (**196**) was unsuccessful due to instability of (**198**) under reductive conditions. *C. conicum* also biosynthesizes bicyclogermacrene-14-al (**199**) (*146*). *Calypogeia* species elaborate not only azulenoids but also bicyclogermacrane-type sesquiterpenoids. From the essential oil of the freshly cultured cells of *C. granulata*, *C. trichomanis* (*453, 539*), *C. muelleriana*, *C. peruviana* and *C. tosana* (*321*), 2-acetoxy-3-hydroxybicyclogermacrene (**200**) and 3-acetoxy-2-hydroxybicyclogermacrene (**201**) were isolated along with 3α-acetoxybicyclogermacrene (**197**) and 3α-hydroxybicyclogermacrene (**203**) (*539*). *Plagiochila fruticosa* produces 3α, 14-diacetoxy-2-hydroxybicyclogermacrene (**202**) (*249*).

A small thalloid liverwort, *Conocephalum japonicum* (= *C. supradecompositum*) produces not only germacranolides (*19*) but also two new bicyclogermacrenes, 5-methoxybicyclogermacrene (**204**) and 14-methoxybicyclogermacrene (**205**) (*584*). These methoxylated compounds were obtained from the dichloromethane extract; however, they might be artifacts formed during the fractionation process on Sephadex LH-20 using CHCl$_3$-MeOH.

Bazzania japonica produces (−)-isobicyclogermacrenal (**206**) (*84*) which has also been isolated from *Lepidozia vitrea* belonging to the same family (*19*). The stereochemistry of isobicyclogermacrenal was conclusively established by a combination of chemical degradation (Scheme 15) and X-ray crystallographic analysis of (+)-isobicyclogermacrenol (**206a**). The chair-twist conformation for (**206**) is in agreement with that of the mother hydrocarbon, isobicyclogermacrene (**206b**) (*383*).

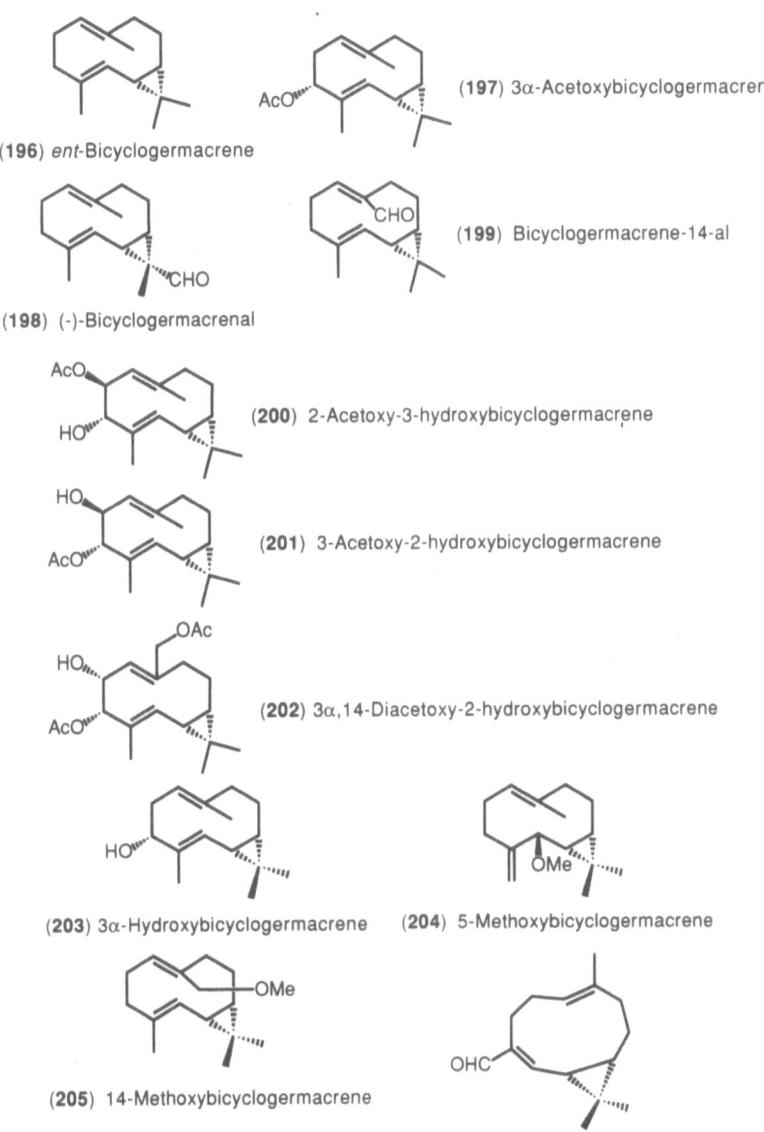

(196) *ent*-Bicyclogermacrene

(197) 3α-Acetoxybicyclogermacrene

(198) (-)-Bicyclogermacrenal

(199) Bicyclogermacrene-14-al

(200) 2-Acetoxy-3-hydroxybicyclogermacrene

(201) 3-Acetoxy-2-hydroxybicyclogermacrene

(202) 3α,14-Diacetoxy-2-hydroxybicyclogermacrene

(203) 3α-Hydroxybicyclogermacrene

(204) 5-Methoxybicyclogermacrene

(205) 14-Methoxybicyclogermacrene

(206) (-)-Isobicyclogermacrenal

Chart 12a. Bicyclogermacranes and isobicyclogermacranes found in the Hepaticae

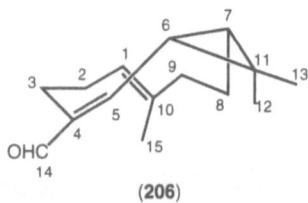

(206)   (206a)   (206d) : R=H
                                                      (206e) ; R=Me

(206b)   +   (206c)

1) LiAlH₄ 2) MnO₂ 3) Py-SO₃, LiAlH₄ 4) O₃, H₂O₂ 5) CH₂N₂

(206)

Scheme 15. Reactions of isobicyclogermacrenal

A new lepidozane-type sesquiterpene alcohol (**207**) has been isolated from *Trocholejeunea sandvicensis* (*14, 556*) and *Porella swartziana* (*441a, 569*). The presence of two protons on a cyclopropane ring was confirmed by two high field signals at δ − 0.20 and 0.24. The structure was further characterized as (4S*,5S*,6R*,7R*)-1(10)E-lepidozen-5-ol by ¹H-, ¹³C-NMR spectrometry as well as by ¹H-¹H-2D-COSY and HMBC experiments. The relative stereochemistry was deduced from the NOE spectrometry. The 4(15)-dehydro derivative (**209**) of (**207**) has been isolated from the marine Actinia *Anthopleura pacifica* (*646*).

Bicylcogermacrene (**196**) which occurs naturally and the related non-naturally occurring isobicyclogermacrene, lepidozene and isolepidozene have been synthesized by short routes starting from geranylacetone (*396*).

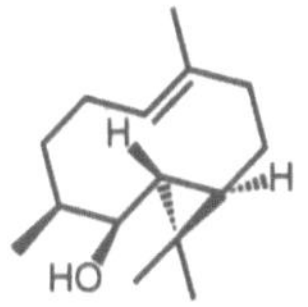

(207) (4S*,5S*,6R*,7R*)-1(10)E-Lepidozen-5-ol

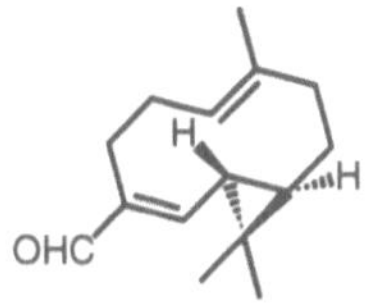

(208) Lepidozenal

Chart 12b. Lepidozanes found in the Hepaticae

(±)-Isobicyclogermacrenal (**206**) has been synthesized from piperitone (*359*). (−)-Lepidozenal (**208**) has been isolated from *Lepidozia vitrea* (*19*). The biosynthesis of *ent*-bicyclogermacrene (**196**) and its enantiomer (**196a**), (−)-isobicylcogermacrenal (**206**) and (−)-lepidozenal (**208**) may conceivably involve the stereoselective 1,3-diprotonation reactions shown in Scheme 16 (*383*). It is noteworthy that bicyclogermacrene (**196**), (−)-lepidozenal (**208**) and lepidozenol (**210**) have also been isolated from marine animals (*332, 646*).

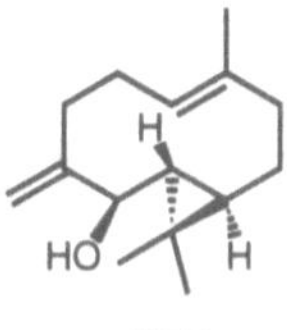

(209) 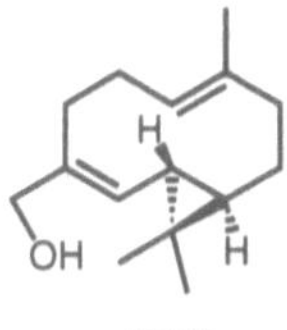   (210)

5.10 Bisabolanes, Bourbonanes and Brasilanes

Bisabolane-type sesquiterpenoids are rare in Hepaticae. Only α- (**212**) and β-bisabolene (**213**) and *ar*-curcumene (**214**) were previously detected in liverworts by GC-MS analysis (*19*). Since then, (*S*)-(+)-*Z*-α-bisabolene (**212**) has been isolated from European *Scapania crasiretis* (*435*). Two new hydroxybisabolanes, (6R,7R)-(+)-α-bisabolol (**216**) and its enantiomer (**217**) have more recently been isolated from *Jungermannia vulcanicola* (*438*) and *Frullania brasiliensis* (*433*), respectively. The latter compound has also been found in Taiwanese *Bazzania tridens* (*629*). It is

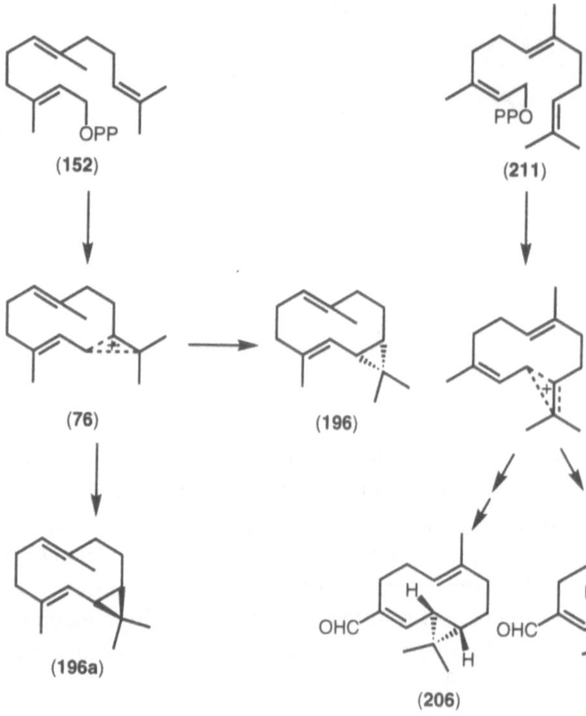

Scheme 16. Possible biogenetic pathways for bicyclogermacrane-type sesquiterpenoids

interesting that both normal and enantiomeric sesquiterpenoids have been found in different species of the same or different genera of the Hepaticae. South American *Marchantia chenopoda* produces bisabola-triene (= β-sesquiphellandrene) (215) (*345c*). An enantiomeric *ar*-cur-cumene (214) has been isolated from *Marsupella emarginata* var. *patens*, together with barbatane-type sesquiterpenoids (*384*). The same com-pound has been detected in cultured cells of *Heteroscyphus planus* (*427a*) and *Lophocolea heterophylla* (*527a*). (−)-*Ent*-nuciferal (218) has been obtained from *Gymnomitrion concinnatum* (*434*). The (+)-enantiomer has been isolated from the higher plant, *Torreya nucifera* (*480*). *Anthelia julacea* produced a new bisabolal, julaceal (219), together with (218) (*435*). The absolute and stereostructure of (219) was established by 2D-COSY NMR spectroscopy and by its conversion to (−)-*ent*-nuciferal (218) by dehydrogenation with DDQ.

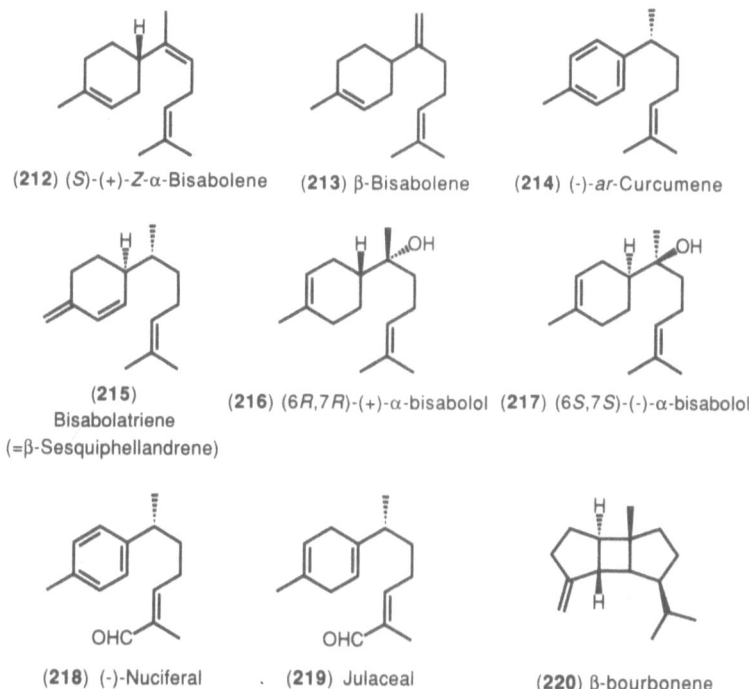

(212) (S)-(+)-Z-α-Bisabolene   **(213)** β-Bisabolene   **(214)** (-)-ar-Curcumene

(215)
Bisabolatriene
(=β-Sesquiphellandrene)

(216) (6R,7R)-(+)-α-bisabolol   **(217)** (6S,7S)-(-)-α-bisabolol

(218) (-)-Nuciferal   .   **(219)** Julaceal   **(220)** β-bourbonene

Chart 13. Bisabolanes and bourbonane found in the Hepaticae

As previously reported (*19*), β-bourbonene (**220**) has been detected in a *Dumortiera* and a *Scapania* species. *Mylia nuda* produces also β-bourbonene but its absolute configuration remains to be clarified (*624*).

A new acid-labile brasilane-type tertiary alcohol, conocephalenol (**221**) has been isolated from the European thalloid liverwort *Conocephalum conicum*, its structure being deduced by 2D-INADEQUATE (Incredible Natural Abundance Double Quantum Transfer Experiment) NMR spectroscopy (*145*, *146*). The total synthesis of (+)-conocephalenol ($[\alpha]_D$ + 5.85) has been accomplished by TORI *et al.* in 12 steps (*552, 570, 573*). For comparison conocephalenol was reisolated from German *Conocephalum conicum*. As the natural product had a negative rotation ($[\alpha]_D$ − 4.77), the synthetic compound was the enantiomer of (**221**), thus establishing the absolute stereostructure of natural conocephalenol (*441a*).

Brasilanes (**222–224**) previously known only from marine organisms (*517, 620*) have now been found in a liverwort. TORI *et al.* (*552*) have

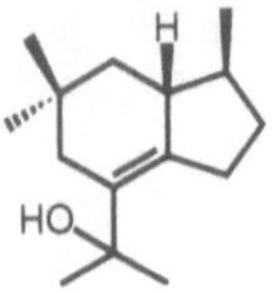

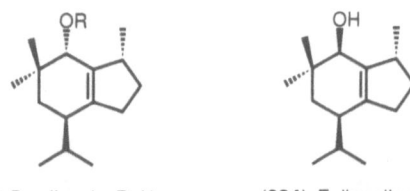

(221) Conocephalenol

Chart 14a. Brasilane found in the Hepaticae

(222) Brasilenol ; R=H   **(224)** Epibrasilenol
(223) Brasilenol acetate ; R=Ac

Chart 14b. Brasilanes found in algae

proposed a possible biogenetic route to conocephalenol by ring closure of humulene (Scheme 17).

5.11 Cadinanes, Amorphanes and Muurolanes

The earlier report mentioned that γ-cadinene **(226)** and α- **(228)** and γ-muurolenes **(229)** had been detected in several liverworts by GC-MS analysis *(19)*. The same sesquiterpene hydrocarbons and α-amorphene **(230)** have more recently been found in Jungermanniales, Marchantiales and Monocleales. The essential oil of *Wiesnerella denudata* contains α-cadinol **(231)** *(62)*. T-cadinol **(232)** and torreyol **(236)** have been detected in essential oils of *Lophocolea heterophylla* and *L. bidentata*, respectively *(527a)*. (+)-*Ent*-epicubenol **(233)** has been isolated from two *Scapania* species, *S. undulata* *(153)* and *S. uliginosa* *(289)*, and *Conocephalum conicum* *(553)*; its structure was supported by dehydration to (+)-*ent*-cubebene on treatment with SOCl₂ in pyridine *(153)*. (−)-Epicubenol **(234)**, the antipode of **(233)** has been isolated from commercial cubeb oil *(452)*. γ-Cadinene **(226)**, δ-cadinene **(227)** and *epi*-cubenol **(233)** have been detected in cultured cells of *Heteroscyphus planus* *(427a)*. The latter two

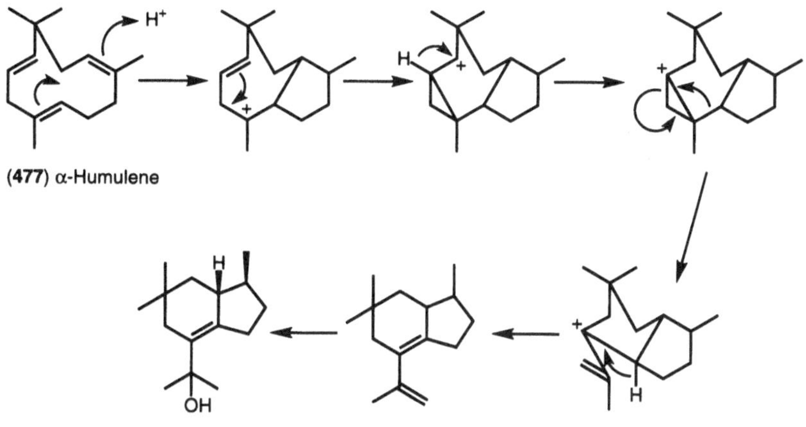

(477) α-Humulene

(221)   (225)

Scheme 17. Possible biogenetic pathway for (−)-conocephalenol

compounds have been detected in essential oils of *Lophocolea hetero-phylla* and *L. bidentata* (*527a*). Epicubenol (**234**) and cubenol (**235**) have been detected in brown algae, *Dictyopteris* species (*179*) and red tide (*310*).

Mannia subpilosa and *Reboulia hemisphaerica* belonging to Aytoni-aceae (= Grimaldiaceae) biosynthesize three new cadinane-type sesquiter-penoids, cadina-4,11-dien-14-al (**237**), cadina-4,11-dien-14-ol (**238**) and 14-acetoxycadina-4,11-diene (**240**) (*613*). Further study of the latter species resulted in the isolation of a cadinene carboxylic acid (**239**) (*505a, 623a*). *Gongylanthus ericetorum* produces a new cadinane-type ether (**241**) (*44*). *Lepidozia fauriana* biosynthesizes two cadinane lactones, (−)-lepidozenolide (**242**) and its peroxy compound (**243**) (*505a, 623a*). The latter compound has also been isolated from *L. vitrea* (*505a, 623a*).

5.12 Calamenanes, Calacoranes and Cadalenes

As reported earlier (*19*) calamenene (**244**), 5-hydroxycalamenene (**245**) and α- (**246**) and β-calacorenes (**247**) occur in liverworts. Since then 7-hydroxycalamenene (**248a**) (*342*), 8-hydroxycalamenene (**250**), 5,8-dihy-droxycalamenene (**251**) (*249*) and 7-acetoxy-8-hydroxycalamenene (**253**) (*586*) have also been found in liverworts. *Lophocolea heterophylla* elabo-rates cadalene (**258**) (*343*). *Bazzania trilobata* had a rather high content of 5-hydroxycalamenene (**245**) (*19*) but without assignment of stereochemis-try. From the coupling constants of H-1, H-4 and H-13, the relative

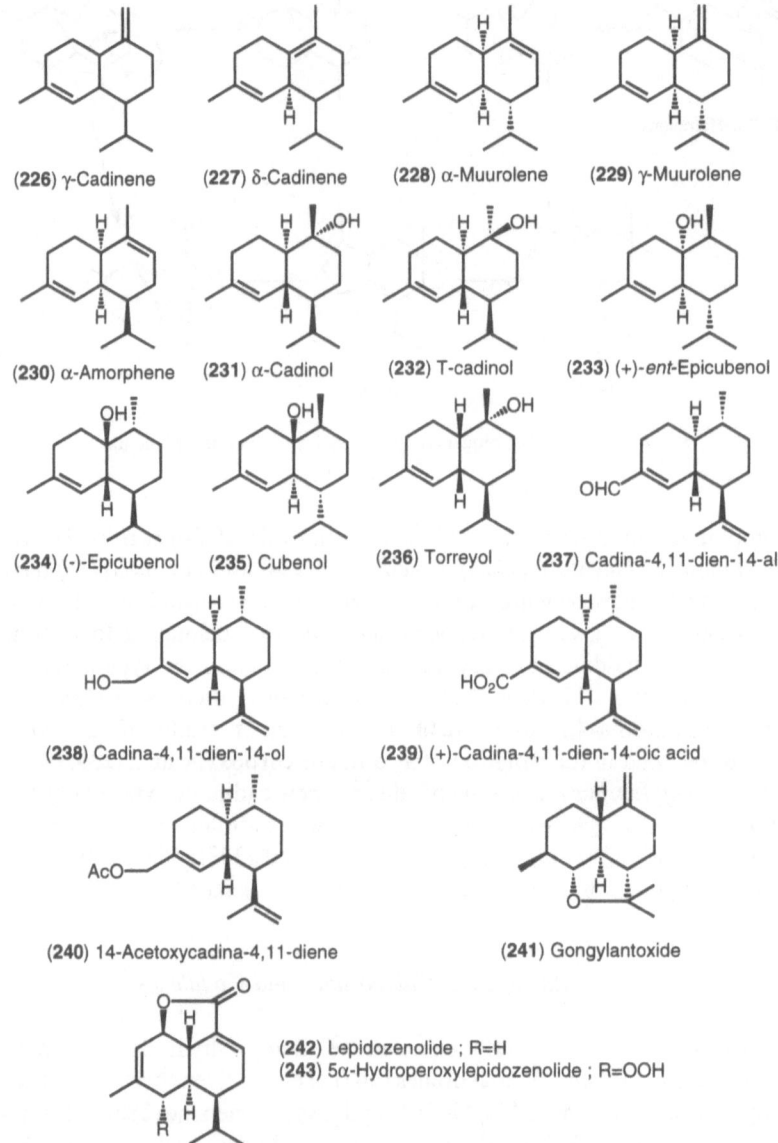

(226) γ-Cadinene   (227) δ-Cadinene   (228) α-Muurolene   (229) γ-Muurolene

(230) α-Amorphene   (231) α-Cadinol   (232) T-cadinol   (233) (+)-*ent*-Epicubenol

(234) (-)-Epicubenol   (235) Cubenol   (236) Torreyol   (237) Cadina-4,11-dien-14-al

(238) Cadina-4,11-dien-14-ol   (239) (+)-Cadina-4,11-dien-14-oic acid

(240) 14-Acetoxycadina-4,11-diene   (241) Gongylantoxide

(242) Lepidozenolide ; R=H
(243) 5α-Hydroperoxylepidozenolide ; R=OOH

Chart 15a. Cadinanes, amorphanes and muurolanes found in the Hepaticae and their related compounds

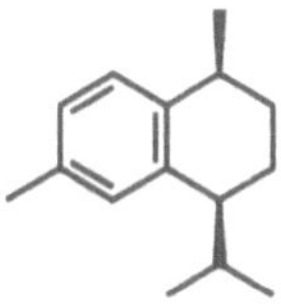

(244) 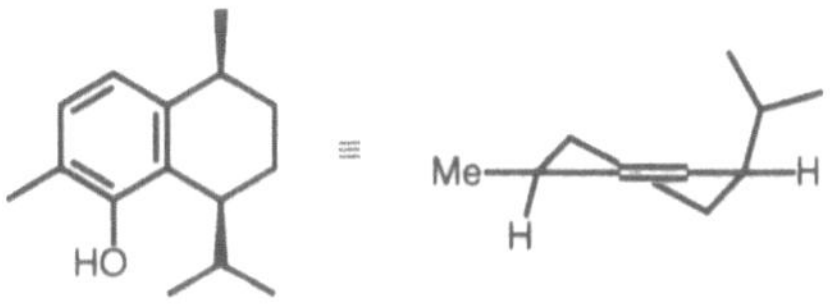   (245)

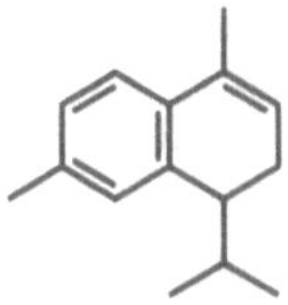

(246) 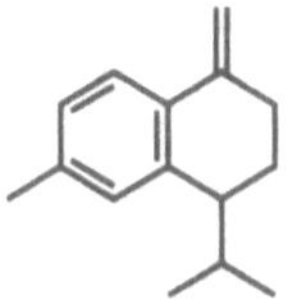   (247)

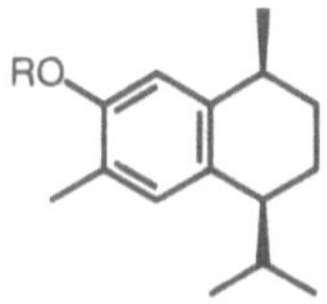

(248a) (1S,4S)-7-Hydroxycalamenene ; R=H
(248b) R=Me

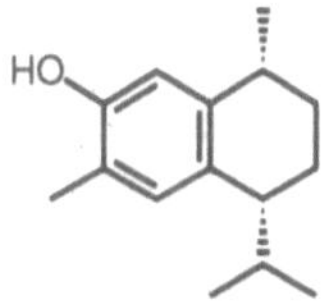

(249) (1R, 4R)-7-Hydroxycalamenene

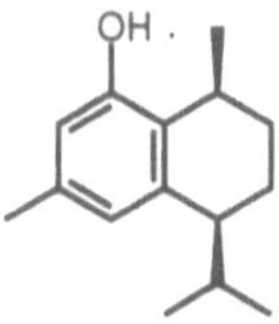

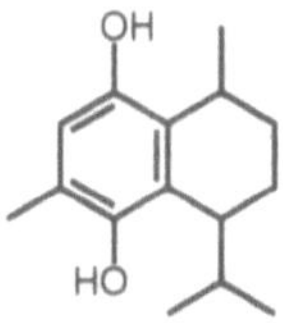

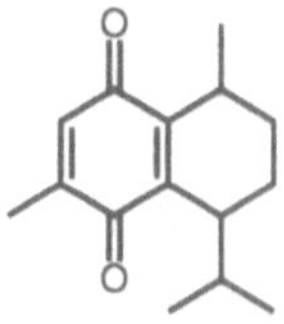

(250) 8-Hydroxycalamenene   (251) 5,8-Dihydroxycalamenene   (252)

Chart 15b. Calamenanes and calacoranes found in the Hepaticae

configuration of this substance has since been shown to be **(245)** with the
alkyl groups *cis* as in 7-hydroxycalamenene **(249)** whose structure was
established by X-ray crystallographic analysis (*157*). The absolute configuration of **(245)** is based on its CD spectrum. The sign of the Cotton effect
within the α-band is determined mainly by the helicity of the cyclohexane
ring, but depends also on the substitution in the benzene chromophore
within the α-band. As **(245)** exhibited a positive Cotton effect at 278 and
272 nm, the absolute configuration at C-1 and C-4 was 1S and 4S (*283*). A

diastereomeric 5-hydroxycalamenene and its methyl ether have been synthesized from *o*-cresol isobutyrate (*486*).

Lophocolea heterophylla produces not only calamenanes (**244, 248a, 253**) but also eudesmane-type sesquiterpenoids (*586*). The structure assigned to (**253**) was based on the ¹H NMR spectrum, the NOE difference spectrum of its methyl ether and the cooccurrence of *cis*-calamenene (**244**) and 7-hydroxycalamenene (**248a**) (*586*). *Heteroscyphus planus* elaborates 7-hydroxycalamenene (**248a**) and 5,8-dihydroxycalamenene (**251**). The structure of the latter compound is based on autooxidation of (**251**) to a 1,4-quinone (**252**) (*249*). The structure of a substance earlier thought to be 7-hydroxycalamenene (**248a**) isolated from Japanese *Bazzania trilobata* (*587*) was revised to 8-hydroxycalamenene (**250**) (*343*).

In vitro cultured gametophytes and suspension cells of *Heteroscyphus planus* produce volatile terpenoids. Chromatography of the methanol extracts of the gametophyte grown on Murashige-Skoog (MSK)-3 and modified MSK media and suspension cells resulted in the isolation of four additional new calamenene-type sesquiterpenoids, (1*S*)-7-methoxy-1,2-dihydrocadalene (**254**), (1*S*,4*R*)-7-methoxycalamenene (**255**), (1*S*,4*R*)-7-

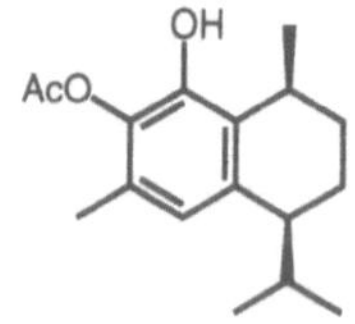

(**253**) 7-Acetoxy-8-hydroxycalamenene

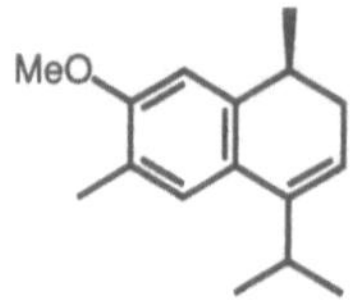

(**254**) (1*S*)-7-Methoxy-1,2-dihydrocadalene

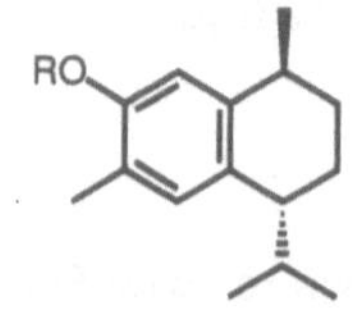

(**255**) (1*S*,4*R*)-7-Methoxycalamenene ; R=Me
(**256**) (1*S*,4*R*)-7-Hydroxycalamenene ; R=H

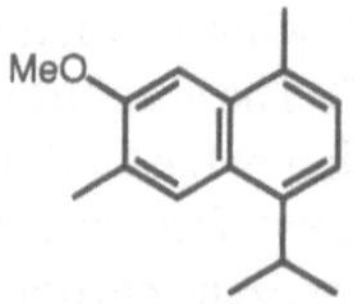

(**257**) 7-Methoxycadalene

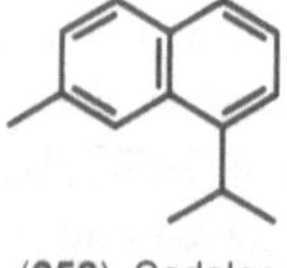

(**258**) Cadalene

Chart 15c. Calamenanes and cadalenes found in the Hepaticae and their related compounds

hydroxycalamenene (**256**) and 7-methoxycadalene (**257**), along with calamenene (**244**) the absolute configuration of which remain to be clarified (*427a*). Hydrogenation of (**254**) gave *cis*-(**248b**) and *trans*-calamenene (**255**) while methylation of (**256**) with diazomethane afforded (**255**).

5.13 Caryophyllanes, Cedranes and Chamigranes

As reported earlier (*19*) β-caryophyllene (**259**) has been found not only in Jungermanniales and Marchantiales but also in Monocleales; its absolute configuration is the same as that isolated from higher plants. (−)-β-Caryophyllene oxide (**260**) has been newly isolated from *Marchantia paleacea* var. *diptera* together with β-caryophyllene (**259**) (*74*). The absolute configuration of (**260**) is the same as that found in tracheophytes.

While cedrane-type sesquiterpenoids are very rare in Hepaticae, α- (**261**) and β-cedrenes (**262**) have been detected in several liverworts by GC-MS analysis as shown in Table IIb.

As reported earlier, the enantiomeric α- (**263**) and β-chamigrenes (**264**) have been isolated from *Scapania* species (*19*). β-Chamigrene is widely distributed in Hepaticae and has since been found in 24 other liverworts. *Ent*-chamigrenic acid (**265**) has been isolated from the Colombian liverwort, *Omphalanthus filiformis*, together with *ent*-β-chamigrene (**264**) (*441a*, *571*). The antipode of (**265**) has been isolated from a conifer, *Juniperus squamata* (*352*). A novel chamigrane-type sesquiterpene ketone, (−)-*ent*-9-oxo-α-chamigrene (**266**) has been isolated from German *Marchantia polymorpha* (*71*). Its relative and absolute stereochemistry has

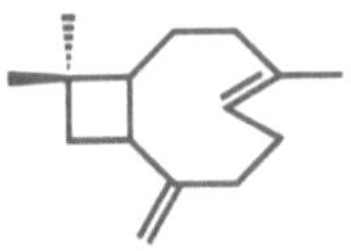

(**259**) β-Caryophyllene 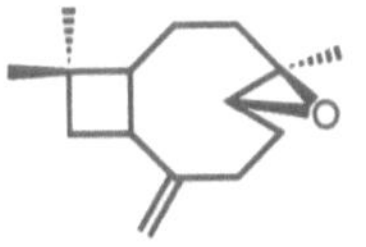   (**260**) β-Caryophyllene oxide

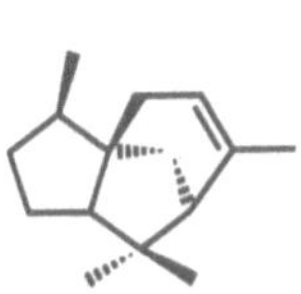

(**261**) α-Cedrene 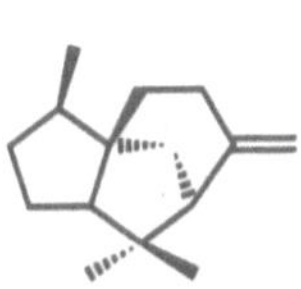   (**262**) β-Cedrene

Chart 16. Caryophyllanes and cedranes found in the Hepaticae

Table IIb. *Tris-norsesqui- and Sesquiterpenoids Found in the Hepaticae*

Structure number	Name of compounds	Formula	m.p. °C	$[\alpha]_D$	Plant source	Order	References	Comments
(57)	Acoradiene	$C_{15}H_{24}$			*Chiloscyphus hookeri*	J	(45)	
					Gymnocolea inflata	J	(429)	
					Heteroscyphus planus	J	(427a)	
					Jamesoniella colorata	J	(45)	
					Lejeunea lumbricoides	J	(219)	
					Pellia epiphylla	Me	(106)	
					Plagiochila moritziana	J	(513)	
					P. neesiana	J	(46)	
					P. parvidens	J	(46)	
					P. Subdura	J	(46)	
					Riccardia andina	Me	(513)	GC-MS
					Schistochila laminigera	J	(45)	
(58)	Inflatenone	$C_{14}H_{22}O$		− 7.94	*Gymnocolea inflata*	J	(429, 566, 567a)	CD 352 nm + 84.9
(59)	3α-Hydroxy-5α-acetoxyafrican-2(6)-en-4-one	$C_{17}H_{24}O_4$	96–8	− 141	*Porella caespitans* var. *setigera*	J	(596, 597)	
(60)	3α,4α-Dihydroxy-african-2(6)-en-5-one	$C_{15}H_{22}O_3$	194–5	− 267	*Porella caespitans* var. *setigera*	J	(596, 597)	
(61)	Caespitenone	$C_{15}H_{20}O_2$			*Porella swartziana*	J	(47, 441a, 568, 571, 571b)	CD 286 nm − 7.50, CD 354 nm − 3.35
(62)	Swartzianin A	$C_{15}H_{24}$		− 116.5	*Porella swartziana*	J	(441a, 571)	
(63)	Swartzianin B	$C_{15}H_{22}O$		− 183.9	*Porella swartziana*	J	(441, 571)	CD 247 nm − 9.57, CD 262 nm + 14.29, CD 297 nm − 2.14

No.	Compound	Formula	mp	[α]	Species		Ref.	CD
(64)	Swartzianin C	$C_{15}H_{22}O_2$	90–5	− 368.1	Porella swartziana	J	(441a, 571)	CD 213 nm + 4.96; CD 252 nm − 12.76; CD 330 nm − 2.13
(65)	Swartzianin D	$C_{15}H_{20}O_2$	106–7	− 153.0	Porella swartziana	J	(441a, 571)	CD 220 nm + 21.5; CD 252 nm − 9.76; CD 295 nm − 7.62
(66)	Norafricanone	$C_{16}H_{24}O_3$	127–8.5	− 204	Porella caespitans var. setigera	J	(583)	
(67)	Secoswartzianin A	$C_{15}H_{20}O_2$		− 39.0	Porella swartziana	J	(441a, 571, 571a)	CD 304 nm + 4.27
(68)	Secoswartzianin B	$C_{16}H_{24}O_3$		− 111	Porella swartziana	J	(441a, 571, 571a)	CD 228 nm − 5.75
(69)	Norswartzianin	$C_{14}H_{20}O_3$		− 42.1	Porella swartziana	J	(574a, 574b)	
(70)	Isoafricanol	$C_{15}H_{26}O$		+ 13.3	Nardia scalaris	J	(354a)	
(82)	ent-Aristolone	$C_{15}H_{22}O$			Porella caespitans var. setigera	J	(596)	
				+ 252.5	P. cordaeana	J	(239a)	
					P. roellii	J	(543)	
					Reboulia hemisphaerica	M	(55)	
				+ 243	Reboulia hemisphaerica	M	(73)	
(83)	ent-8β-Hydroxy-aristolene	$C_{15}H_{24}O$			Bazzania tridens	J	(629)	
(84)	(+)-Aristol-9-ene (= (+)-α-Ferulene)	$C_{15}H_{24}$		+ 45	Pleurozia acinosa	J	(625)	
(85)	Aristol-1,8-diene	$C_{15}H_{22}$			Reboulia hemisphaerica	M	(249)	
(86)	ent-Aristolan-10β-ol	$C_{15}H_{26}O$	87–8	+ 6	Riccardia jackii	Me	(380)	
(87)	ent-Aristol-9-en-12β-al	$C_{15}H_{22}O$		+ 87.5	Bazzania tridens	J	(629)	
(88)	ent-Aristol-9-en-12β-oic acid	$C_{15}H_{22}O_2$	154–8	+ 50.8	Bazzania tridens	J	(629)	
(89)	Aromadendrene	$C_{15}H_{24}$			Diplophyllum albicans	J	(451, 461)	
					D. taxifolium	J	(451, 461)	
					Plagiochila peculiaris	J	(621, 637, 638)	
					Scapania ornithopodioides	J	(622, 638)	

Table IIb (continued)

Structure number	Name of compounds	Formula	m.p. °C	$[\alpha]_D$	Plant source	Order	References	Comments
(90)	Alloaromadendrene	$C_{15}H_{24}$			*S. robusta*	J	(621)	
					Schistochila acuminata	J	(621, 630)	
					S. rigidula	J	(638)	
					Calypogeia granulata	J	(321, 539)	
					Frullania jackii	J	(420)	
					Lejeunea lumbricoides	J	(219)	
					Lopholejeunea eulopa	J	(219)	
					Mastigolejeunea humilis	J	(219)	
					Monoclea forsteri	J	(513)	
					M. gottschei subsp. *neotropica*	J	(514)	
					Plagiochila dura	J	(46)	
					P. fuegiensis	J	(46)	
					P. geniculata	J	(513)	
					P. lecheri	J	(46)	
					P. stephensoniana	J	(46)	
					P. subdura	J	(46)	
					Scapania robusta	J	(638)	
					Schiffneriolejeunea omphalanthoides	J	(219)	
					Thysananthus convolutus	J	(219)	
					T. mollis	J	(219)	
					Trichocolea pluma	J	(133)	
(91)	α-Gurjunene	$C_{15}H_{24}$			*Chiloscyphus pallido-virens*	J	(45)	
					Gymnocolea inflata	J	(429)	
					Lopholejeunea subfusca	J	(219)	

No.	Compound	Formula	Species		Ref.
			Plagiochila moritziana	J	(513, 515)
			Roivainenia jacquinotii	J	(45)
			Scapania ornithopodioides	J	(621)
			Lopholejeunea subfusca	J	(219)
			Radula boryana	J	(513)
			Scapania ornithopodiodes	J	(621)
			S. robusta	J	(621)
			Trichocolea pluma	J	(133)
(92)	β-Gurjunene	C$_{15}$H$_{24}$	Anastrophyllum minutum	J	(106, 108)
			Bazzania fauriana	J	(622)
			B. praerupta	J	(341)
			B. stolonifera	J	(513)
			B. tridens	J	(621, 627)
			Calypogeia azulea	J	(106)
			C. granulata	J	(539)
			C. muelleriana	J	(321)
			C. peruviana	J	(321)
			C. trichomanis	J	(321)
			C. tosana	J	(321)
(93)	Anastreptene	C$_{15}$H$_{22}$	Chandonanthus hirtellus	J	(51)
			Clasmatocolea vermicularis	J	(47)
			Diplophyllum serrulatum	J	(585a)
			Frullania bicornistipula	J	(513)
			Heteroscyphus aselliformis	J	(51)
			Isotachis haematodes	J	(47)
			Leptoscyphus liebmanianus	J	(47)
			Lophocolea bidentata	J	(527a)
			L. heterophylla	J	(527a)
			Mastigophora diclados	J	(621, 635)
			Monoclea forsteri	J	(513)

Table IIb (continued)

Structure number	Name of compounds	Formula	m.p. °C	$[\alpha]_D$	Plant source	Order	References	Comments
					Nardia scalaris	J	(513)	
					Pellia epiphylla	Me	(106)	
					Plagiochila amazonica	J	(48)	
					P. cipaconensis	J	(48)	
					P. cucullata	J	(48)	
					P. dichotoma	J	(48)	
					P. dilatata	J	(48)	
					P. falcata	J	(48)	
					P. friabilis	J	(48)	
					P. guayrapurinensis	J	(48)	
					P. kroneana	J	(48)	
					P. moritziana	J	(513, 515)	
					P. peculiaris	J	(621, 637)	
					P. tenerrima	J	(48)	
					P. trabeculata	J	(589)	
					Porella swartziana	J	(47)	
					Scapania ornithopodioides	J	(638)	
					S. robusta	J	(621, 631, 623)	
					Schistochila acuminata	J	(621, 630)	
					Sphenolobus minutus	J	(107)	
					Syzygiella anomala	J	(47)	
					Thysananthus amazonicus	J	(219)	
(94)	ent-(–)-Ledene	$C_{15}H_{24}$			Bazzania trilobata	J	(342)	
					Calypogeia granulata	J	(539)	
					C. muelleriana	J	(321)	
					C. peruviana	J	(321)	

No.	Compound	Formula	[α]	Species		Method	Ref.
(95)	β-Spathulene	C$_{15}$H$_{24}$	−65	C. trichomanis		J	(321)
				C. tosana		J	(321)
				Riccardia jackii		Me	(380)
(96)	ent-Cyclocolorenone	C$_{15}$H$_{22}$O		Mylia nuda		J	(622, 624)
				Schiffneria hyalina		J	(622)
				Bazzania tridens		J	(629)
				Chiloscyphus hookeri		J	(45)
				Diplasiolejeunea patelligera		J	(219)
				Frullania falciloba		J	(60)
				F. gaudichaudii		J	(60)
				Jubula japonica		J	(585b)
				Jungermannia colorata		J	(45)
				Lejeunea discreta		J	(219)
				Leucolejeunea aff. decurrens		J	(219)
				Monoclea forsteri		M	(513)
				Plagiochila oresitropha		J	(48)
				P. parvidens		J	(46)
				Schiffneriolejeunea omphalanthoides		J	(219)
				Thysananthus pterobryoides	GC-MS	J	(513)
(97)	ent-Spathulenol	C$_{15}$H$_{24}$O	−36.3	Anastrophyllum minutum		J	(106)
				Archilejeunea olivacea		J	(219)
				Balantiopsis rosea		J	(65)
				Barbilophozia floerkei		J	(565, 567a)
				B. hatcheri		J	(565, 567a)
				Bazzania praerupta		J	(341)
				B. tridens		J	(623, 629)
				Bryopteris filicina		J	(432a)

Table IIb *(continued)*

Structure number	Name of compounds	Formula	m.p. °C	$[\alpha]_D$	Plant source	Order	References	Comments
					Clasmatocolea vermicularis	J	*(47)*	
					Diplasiolejeunea patelligera	J	*(219)*	
					Diployphyllum serrulatum	J	*(585a)*	
					Heteroscyphus aselliformis	J	*(51)*	
					Isotachis haematodes	J	*(47)*	
					I. humectata	J	*(45)*	
					Jackiella javanica	J	*(247)*	
					Lejeunea albescens	J	*(217)*	
					L. discreta	J	*(219)*	
					L. glaucescens	J	*(219)*	
					L. lumbricoides	J	*(219)*	
					Lepidolejeunea ornata	J	*(219)*	
					Leptoscyphus liebmanianus	J	*(47)*	
					Lethocolea glossophylla	J	*(47)*	
					Lophocolea bidentata	J	*(527a)*	
					L. heterophylla	J	*(527a)*	
					Lopholejeunea subfusca	J	*(219)*	
					Makinoa crispata	Me	*(622)*	
					Marchantia tosana	M	*(74)*	
					Monoclea forsteri	Mo	*(513)*	

	Mo	(514)
M. gottschei subsp. neotropica	J	(514)
Plagiochila acanthoda	J	(48)
P. adiantoides	J	(48)
P. asplenioides	J	(435)
P. beskeana	J	(48)
P. bursata	J	(513)
P. cipaconensis	J	(48)
P. corniculata	J	(48)
P. cristatissima	J	(48)
P. cucullata	J	(48)
P. dichotoma	J	(48)
P. elata	J	(48)
P. falcata	J	(48)
P. friabilis	J	(48)
P. gayana	J	(46)
P. geniculata	J	(513)
P. goebeliana	J	(48)
P. hondurensis	J	(513)
P. hookeriana	J	(48)
P. kroneana	J	(48)
P. moritziana	J	(513, 515)
P. ovalifolia	J	(435a)
P. oxyphylla	J	(48)
P. pachyloma	J	(48)
P. parvidens	J	(46)
P. parvitexta	J	(48)
P. pittieri	J	(48)
P. pulcherrima	J	(190)
P. scopulosa	J	(295)
P. stephensoniana	J	(188)

Table IIb *(continued)*

Structure number	Name of compounds	Formula	m.p. °C	$[\alpha]_D$	Plant source	Order	References	Comments
					P. tambillensis	J	*(48)*	
					P. tenerrima	J	*(48)*	
					P. trabeculata	J	*(589)*	
					Pleurozia gigantea	J	*(52)*	
					Porella densifolia subsp. *appendiculata*	J	*(67)*	
					P. densifolia var. *fallax*	J	*(67)*	
					Riccardia andina	Me	*(513)*	
				– 22	*R. jackii*	Me	*(380)*	
					Scapania crasiretis	J	*(435)*	
					S. javanica	J	*(51)*	
					S. uliginosa	J	*(434a)*	
					Schistochila acuminata	J	*(136)*	
					S. aligera	J	*(51)*	
					S. laminigera	J	*(45)*	
					S. reflexa	J	*(45)*	
					Schiffneriolejeunea omphalanthoides	J	*(219)*	
					Stephaniella paraphyllina	J	*(47)*	
					Stictolejeunea balfourii var. *bekkei*	J	*(219)*	
					S. squamata	J	*(219)*	
					Syzygiella anomala	J	*(47)*	
					Tetralophozia setiformis	J	*(429, 567a)*	
					Thysananthus amazonicus	J	*(219)*	

No.	Compound	Formula	mp (°C)	$[\alpha]$	Source	Method	Ref.
(98)	Myliol	$C_{15}H_{22}O$	111–2	−20	T. convolutus; T. mollis; T. pterobryoides; Tritomaria quinquedentata; Mylia taylorii	J	(219); (219); (513); (565, 567a); (242, 246, 386)
(99)	Dihydromylione	$C_{15}H_{22}O$			Mylia taylorii	J	(242)
(100)	(+)-C-10-Epiglobulol	$C_{15}H_{26}O$	47–8		Diplophyllum albicans; D. taxifolium	J	(451, 461); (451, 461)
(101)	Viridiflorol	$C_{15}H_{26}O$	73.5–4.5		Bazzania trilobata; Lophocolea bidentata; L. heterophylla	J	(342); (527a); (527a)
(102)	ent-Globulol	$C_{15}H_{26}O$		+87	Mylia taylorii; Lophocolea bidentata; L. heterophylla; Plagiochila neesiana; Riccardia andina; Thysananthus pterobryoides	J; J; J; J; Me; J	(386, 531); (527a); (527a); (46); (513); (513)
(103)	(+)-Myli-4(15)-en-9-one	$C_{15}H_{20}O$	145–5.5	+1	Mylia taylorii	J	(386)
(104)	(−)-3-epi-Myliol	$C_{15}H_{22}O$	93–4	−126	Mylia taylorii	J	(386)
(105)	(+)-4(15)-Dehydroledol	$C_{15}H_{24}O$		+18	Mylia taylorii	J	(386)
(106)	(+)-4(15)-Dehydroglobulol	$C_{15}H_{24}O$	68–9	+12	Mylia taylorii	J	(386)
(107)	ent-4β,10α-Dihydroxyaromadendrane	$C_{15}H_{26}O$		+18.2	Plagiochila ovalifolia	J	(435a)
(108)	Myltaylorione A	$C_{30}H_{42}O_2$			Mylia taylorii	J	(531)
(109)	Myltaylorione B	$C_{30}H_{42}O_2$			Mylia taylorii	J	(531)

Table IIb (*continued*)

Structure number	Name of compounds	Formula	m.p. °C	$[\alpha]_D$	Plant source	Order	References	Comments
(110)	Bitaylorione	$C_{30}H_{44}O_2$			*Mylia taylorii*	J	*(531)*	
(111)	Taylorione	$C_{15}H_{22}O$		− 28	*Mylia taylorii*	J	*(242, 246, 386)*	
(115)	Plagiochiline A	$C_{19}H_{26}O_6$			*Plagiochila adiantoides*	J	*(48)*	
					P. asplenioides	J	*(435)*	
					P. cristatissima	J	*(48)*	
					P. dura	J	*(46)*	
					P. fruticosa	J	*(185)*	
					P. goebeliana	J	*(48)*	
					P. guayrapurinensis	J	*(48)*	
					P. guilleminiana	J	*(48)*	
					P. micropterys	J	*(433)*	
					P. pittieri	J	*(48)*	
					P. pulcherrima	J	*(190)*	
(116)	Plagiochiline B	$C_{21}H_{28}O_8$			*P. micropterys*	J	*(433)*	
					P. pulcherrima	J	*(190)*	
(117)	Plagiochiline C	$C_{19}H_{26}O_5$			*Heteroscyphus planus*	J	*(249, 252)*	
					Plagiochila adiantoides	J	*(48)*	
					P. cipaconensis	J	*(48)*	
					P. cristatissima	J	*(48)*	
					P. cucullata	J	*(48)*	
					P. dilatata	J	*(48)*	
					P. dura	J	*(46)*	
					P. falcata	J	*(48)*	
					P. gayana	J	*(48)*	
					P. goebeliana	J	*(48)*	
					P. guayrapurinensis	J	*(48)*	

No.	Compound	Formula	$[\alpha]$	mp	Species		Ref.
(118)	Plagiochiline D	$C_{23}H_{30}O_{10}$			P. guilleminiana	J	(48)
					P. hookeriana	J	(48)
					P. lecheri	J	(46)
					P. micropterys	J	(433)
					P. moritziana	J	(513, 515)
					P. ovalifolia	J	(435a)
					P. pittieri	J	(48)
					P. scopulosa	J	(48)
					P. tenerrima	J	(48)
					Plagiochila beskeana	J	(48)
					P. dura	J	(46)
					P. falcata	J	(48)
					P. lecheri	J	(46)
					P. scopulosa	J	(48)
(119)	Plagiochiline E	$C_{23}H_{30}H_{10}$			Plagiochila lecheri	J	(46)
(120)	Plagiochiline H	$C_{17}H_{24}O_{3}$			Plagiochila falcata	J	(48)
					P. lecheri	J	(46)
					P. pittieri	J	(48)
(121)	Plagiochilide	$C_{15}H_{20}O_{2}$			Plagiochila falcata	J	(48)
					P. fruticosa	J	(185)
(122)	Acetoxyisoplagiochilide	$C_{17}H_{22}O_{4}$	+ 14.3		Plagiochila ovalifolia	J	(435a)
(123)	Hanegokedial (= Plagiochilal A)	$C_{15}H_{20}O_{2}$			Plagiochila falcata	J	(48)
					P. gayana	J	(46)
(124)	Ovalifolienal	$C_{19}H_{24}O_{6}$			Plagiochila beskeana	J	(48)
					P. peculiaris	J	(623)
(125)	9α-Acetoxyovalifoliene	$C_{21}H_{28}O_{7}$			Plagiochila dura	J	(46)
					P. lecheri	J	(46)
(126)	Plagiochilal B	$C_{15}H_{20}O_{3}$	− 4.0	132–5	Plagiochila fruticosa	J	(185)
(127)	Plagiochiline J	$C_{15}H_{20}O_{3}$	+ 120		Plagiochila fruticosa	J	(185)
(128)	Plagiochiline K	$C_{15}H_{22}O_{3}$	± 0		Plagiochila fruticosa	J	(185)

Table IIb (continued)

Structure number	Name of compounds	Formula	m.p. °C	$[\alpha]_D$	Plant source	Order	References	Comments
(129)	9α-Acetoxy-10β-ovalifolianal	$C_{21}H_{28}O_8$			*Plagiochila peculiaris*	J	(623)	
(130)	9β-Acetoxy-10α-ovalifolianal	$C_{21}H_{28}O_8$			*Plagiochila peculiaris*	J	(623)	
(131)	Plagiochilic acid (= Plagiochiline L)	$C_{17}H_{22}O_5$	158–161	+ 9.6	*Heteroscyphus planus*	J	(249, 252)	CD 271 nm − 0.53
(132)	Methyl plagiochilate (= Plagiochiline M)	$C_{18}H_{24}O_5$	94–6	+ 9.7	*Herteroscyphus planus*	J	(249, 252)	CD 271 nm − 0.53
(133)	Plagiochiline N	$C_{15}H_{20}O$		+ 46.1	*Plagiochila ovalifolia*	J	(249, 435a)	
(134)	(−)-3,10-Dioxo-taylori-4-ene	$C_{15}H_{22}O_2$		− 19	*Mylia taylorii*	J	(386)	
(135)	(−)-3-Acetoxy-taylorione	$C_{17}H_{24}O_3$		− 25	*Mylia taylorii*	J	(386)	
(136)	2-Nor-1,3-epoxy-1,10-secoaromadendra-1(5),3-dien-10-one	$C_{14}H_{20}O_2$		− 25.8	*Mylia taylorii*	J	(242)	
(137)	1,4-Dimethylazulene	$C_{12}H_{12}$			*Calypogeia azulea*	J	(106, 507)	
					C. glamulata	J	(321, 453, 439, 507, 538, 539)	
					C. peruviana	J	(321)	
					C. tosana	J	(321)	
					Macrolejeunea pallescens	J	(433)	
					Plagiochila longispina	J	(507)	
					P. micropterys	J	(433)	

(138)	4-Methyl-1-methoxycarbonyl azulene	$C_{13}H_{12}O_2$	Calypogeia azulea	J	(106)
			C. glanulata	J	(321, 439, 507, 539)
			Plagiochila longispina	J	(507)
(139)	4-Methylazulene-1-carbaldehyde (= 4-Methyl-1-formylazulene)	$C_{12}H_{10}O$	Plagiochila longispina	J	(507)
(140)	4-Methyl-1-carboxyazulene	$C_{12}H_{10}O_2$	Calypogeia azulea	J	(439, 507)
(141)	4-Methoxymethyl-1-carboxyazulene	$C_{13}H_{12}O_3$	Plagiochila longispina	J	(507)
(142)	1,4-Dicarboxyazulene	$C_{12}H_8O_4$	Calypogeia azulea	J	(507)
(143)	4-Hydroxymethyl-1-methoxycarbonyl-azulene	$C_{13}H_{12}O_3$	Calypogeia azulea	J	(507)
(144)	4-Carboxy-1-methoxy-carbonylazulene	$C_{13}H_{10}O_4$	Calypogeia azulea	J	(507)
(145)	1,4-Dimethyl-3-formylazulene	$C_{13}H_{12}O$	Calypogeia azulea	J	(507)
(146)	4-Formyl-1-methoxycarbonyl-azulene	$C_{13}H_{10}O_3$	Calypogeia azulea	J	(507)
(147)	Guaiazulene	$C_{15}H_{18}$	Mastigophora diclados	J	(622, 635)
(148)	3,7-Dimethyl-5-methoxy-carbonylindene	$C_{12}H_{14}O_2$	Calypogeia azulea	J	(507)
			C. glanulata	J	(321, 453, 538, 539)
			C. peruviana	J	(321)
			C. tosana	J	(321)
			C. trichomanis	J	(321)
(149)	3,7-Dimethylindene-5-carbaldehyde	$C_{11}H_{12}O$	C. azulea	J	(106)
			C. glanulata	J	(538, 539)

Table IIb (continued)

Structure number	Name of compounds	Formula	m.p. °C	$[\alpha]_D$	Plant source	Order	References	Comments
(150)	(+)-1,8a-Dihydro-3,8-dimethylazulene (= 3,10-Dihydro-1,4-dimethyl-azulene)	$C_{12}H_{14}$		+ 1165	Calypogeia glomulata	J	(321, 453, 538)	CD 232 nm − 133.3 CD 312 nm + 53
(161)	α-Barbatene	$C_{15}H_{24}$			Anastrophyllum minutum	J	(106)	
					Archilejeunea mariana	J	(219)	
					Bazzania angustifolia	J	(621, 622)	
					B. fauriana	J	(581, 636)	
					B. tridens	J	(621, 623)	
					Lejeunea albescens	J	(219)	
					Lophocolea bidentata	J	(527a)	
					L. heterophylla	J	(343, 527a)	
					Marchantia polymorpha	M	(382)	
					Plagiochila engelii	J	(46)	
					P. neesiana	J	(46)	
					P. parvidens	J	(46)	
					P. pulcherrima	J	(190)	
					Radula frondescens	J	(47)	
					Reboulia hemisphaerica	M	(411)	
					Riccardia jackii	Me	(380)	
(162)	β-Barbatene	$C_{15}H_{24}$			Adelanthus lindenbergianus	J	(45)	
					Anastrophyllum minutum	J	(106, 108)	
					Archilejeunea mariana	J	(219)	
					Bazzania fauriana	J	(581, 621, 622, 636)	

B. harpago		J	(341)
B. tridens		J	(621, 623, 627)
Chiloscyphus hookeri		J	(45)
C. pallido-virens		J	(45)
C. polyanthos		J	(60)
Clasmatocolea vermicularis		J	(60)
Frullania clavata		J	(60)
F. falciloba		J	(60, 64)
F. gaudichaudii		J	(60)
Jamesoniella colorata		J	(45)
Jensenia erythropus		J	(47)
Jubula japonica		J	(585b)
Lejeunea albescens		J	(219)
Lejeunea sp.		J	(47)
Lepidozia reptans		J	(147)
Leptoscyphus liebmanianus		J	(47)
Lophocolea bidentata		J	(527a)
L. coadunata		J	(47)
L. heterophylla		J	(343, 527a)
Makinoa crispata		Me	(622)
Marchantia berteroana		M	(39)
M. polymorpha	− 22	M	(74, 382)
Marsupella emarginata var. patens	− 16.9	J	(384)
Mastigophora diclados		J	(621, 622, 635)
Nardia scalaris		J	(106)
N. subclavata		J	(585)
Plagiochila asplenioides		J	(435)
P. beskeana		J	(48)

Table IIb *(continued)*

Structure number	Name of compounds	Formula	m.p. °C	$[\alpha]_D$	Plant source	Order	References	Comments
					P. bursata	J	*(513)*	
					P. dilatata	J	*(48)*	
					P. dura	J	*(46)*	
					P. engelii	J	*(46)*	
					P. friabilis	J	*(48)*	
					P. guilleminiana	J	*(48)*	
					P. hondurensis	J	*(513)*	
					P. peculiaris	J	*(621, 637)*	
					P. neesiana	J	*(46)*	
					P. parvidens	J	*(46)*	
					P. rosariensis	J	*(48)*	
					P. sciophila	J	*(62)*	
					(= *P. acanthophylla* subsp. *japonica*)			
					P. scopulosa	J	*(48)*	
					P. tenerrima	J	*(48)*	
					P. trabeculata	J	*(589)*	
					Reboulia hemisphaerica	M	*(106)*	
					Riccardia andina	Me	*(513)*	
					Roivainenia jacquinotii	J	*(45)*	
					Scapania javanica	J	*(51)*	
					S. ornithopodioides	J	*(638)*	
					Schistochila acuminata	J	*(621, 630)*	
					S. laminigera	J	*(45)*	
					S. nobilis	J	*(54)*	

No.	Compound	Formula	mp (°C)	$[\alpha]$	CD	Species		Ref.
(163)	Gymnomitrol	$C_{15}H_{24}O$	114.5–6	−18.6		Sphenolobus minutus	J	(107)
						Symphyogyana brongniartii	J	(513)
						Bazzania fauriana	J	(581)
						B. trilobata	J	(283, 342)
						Chiloscyphus hookeri	J	(45)
						Plagiochila trabeculata	J	(589)
						Plagiochila trabeculata	J	(589)
(164)	9-Oxogymnomitryl acetate	$C_{17}H_{24}O_3$						
(165)	9α-Hydroxy-gymnomitryl acetate	$C_{17}H_{26}O_3$				Plagiochila trabeculata	J	(589)
(166)	9α-Hydroxy-gymnomitryl cinnamate	$C_{24}H_{30}O_3$				Plagiochila trabeculata	J	(589)
(167)	(+)-Gymnomitr-8(12)-en-9α-ol	$C_{15}H_{24}O$	63–5	+41.7		Reboulia hemisphaerica	M	(411, 412)
(168)	(+)-Gymnomitr-8(12)-en-9-one	$C_{15}H_{22}O$	52–4	+0.95		Reboulia hemisphaerica	M	(411, 412)
(169)	(+)-8β-Hydroxygymno-mitran-9-one	$C_{15}H_{24}O_2$	94–5	+60.32		Reboulia hemisphaerica	M	(411, 412)
(170)	(8R)-(+)-Gymmo-mitran-9-one	$C_{15}H_{24}O$	54–7	+0.45	CD 300 nm + 0.37	Reboulia hemisphaerica	M	(411, 412)
(171)	(−)-Gymnomitr-8(12)-en-15-ol	$C_{15}H_{24}O$	88–90	−28.1		Marsupella emarginata var. patens	J	(384)
(172)	(−)-Gymnomitr-8(12)-en-15-al	$C_{15}H_{22}O$		−10.0		Marsupella emarginata var. patens	J	(384)
(173)	(−)-Gymnomitr-8(12)-en-15-oic acid	$C_{15}H_{22}O_2$	131–2	−0.7		Marsupella emarginata var. patens	J	(384)

Table IIb (continued)

Structure number	Name of compounds	Formula	m.p. °C	$[\alpha]_D$	Plant source	Order	References	Comments
(174)	α-Bazzanene	$C_{15}H_{24}$			Bazzania angustifolia	J	(621, 622)	
					B. fauriana	J	(621, 622, 636)	
					B. tridens	J	(621, 623)	
					Plagiochila moritziana	J	(515)	
(175)	β-Bazzanene	$C_{15}H_{24}$		+45	Riccardia jackii	Me	(380)	
					Adelanthus lindenbergianus	J	(45)	
					Archilejeunea mariana	J	(219)	
					Bazzania angustifolia	J	(621, 622)	
					B. fauriana	J	(581, 621, 622, 636)	
					B. tridens	J	(621, 623, 627)	
					Frullania davurica	J	(420)	
					F. falciloba	J	(60)	
					F. jackii	J	(420)	
					Lejeunea albescens	J	(219)	
					Lejeunea sp.	J	(47)	
					Lepidozia borneensis	J	(51)	
					Mastigophora diclados	J	(621, 622, 635)	
					Mylia nuda	J	(621)	
					Omphalanthus filiformis	J	(47)	
					Plagiochila beskeana	J	(48)	
					P. excisa	J	(48)	
					P. hondurensis	J	(48)	
					P. kroneana	J	(48)	
					P. moritziana	J	(513)	

No.	Compound	Formula	$[\alpha]$	Species		Ref.
				P. neesiana	J	*(46)*
				P. peculiaris	J	*(621, 637)*
				Reboulia hemisphaerica	M	*(411)*
				Riccardia jackii	Me	*(380)*
				Scapania ornithopodioides	J	*(638)*
				S. robusta	J	*(621, 623)*
				Schistochila acuminata	J	*(621, 630)*
				S. glaucoviridis	J	*(621)*
				S. reflexa	J	*(45)*
				Thysananthus pterobryoides	J	*(513)*
(176)	Bazzanenol	$C_{15}H_{24}O$		*Bazzania fauriana*	J	*(581)*
				B. pompeana	J	*(264)*
(178)	Isobazzanene	$C_{15}H_{24}$	0	*Bazzania angustifolia*	J	*(621, 622, 634)*
				B. fauriana	J	*(621, 622, 634)*
				B. tridens	J	*(627)*
(179)	Bazzanenyl caffeate	$C_{24}H_{30}O_4$		*Bazzania fauriana*	J	*(581)*
(181)	Bazzanetin (= 1α,3β-Di(3,4-dihydroxyphenyl)-2α,4β-dibazzanenyl cyclobutylate)	$C_{48}H_{60}O_8$	− 15.9	*Bazzania pompeana*	J	*(249a)*
(184)	*trans*-β-Bergamotene	$C_{15}H_{24}$		*Monoclea forsteri*	M	*(513)*
				Radula boryana	J	*(513)*
(185)	Bicycloelemene	$C_{15}H_{24}$		*Calypogeia granulata*	J	*(539)*
				C. muelleriana	J	*(321)*
				C. peruviana	J	*(321)*
				C. trichomanis	J	*(321)*
				C. tosana	J	*(321)*
				Clasmatocolea humilis	J	*(45)*
				Marchantia polymorpha	M	*(74)*
				M. tosana	M	*(74)*

Table IIb (continued)

Structure number	Name of compounds	Formula	m.p. °C	$[\alpha]_D$	Plant source	Order	References	Comments
(186)	α-Elemene	$C_{15}H_{24}$			Plagiochila parvidens	J	(46)	
(187)	β-Elemene	$C_{15}H_{24}$			Porella densifolia subsp. appendiculata	J	(67)	
					P. densifolia var. fallax	J	(67)	
					Scapania ornithopodioides	J	(622)	
					Pleurozia acinosa	J	(622)	
					Acrolejeunea mariana	J	(219)	
					A. pycnoclada	J	(219)	
					A. torulosa	J	(219)	
					Archilejeunea parviflora	J	(219)	
					Calypogeia granulata	J	(539)	
					C. muelleriana	J	(321)	
					C. peruviana	J	(321)	
					C. trichomanis	J	(321)	
					C. tosana	J	(321)	
					Clasmatocolea humilis	J	(45)	
					Conocephalum conicum	M	(62)	
					C. japonicum	M	(62)	
					Frullania bicornistipula	J	(513)	
					F. brasiliensis	J	(47)	
					F. serrata	J	(51)	
					Heteroscyphus planus	J	(427a)	
					Lejeunea albescens	J	(219)	
					L. lumbricoides	J	(219)	
					Lejeunea sp.	J	(47)	

(188)	γ-Elemene	$C_{15}H_{24}$	*Lophocolea bidentata*	J	(527a)
			L. coadunata	J	(47)
			L. heterophylla	J	(527a)
			Marchantia plicata	M	(47)
			M. polymorpha	M	(74)
			Plagiochila adiantoides	J	(48)
			P. beskeana	J	(48)
			P. gayana	J	(46)
			P. hondurensis	J	(48)
			P. pachyloma	J	(48)
			P. panamensis	J	(513)
			Pleurozia acinosa	J	(622)
			Porella squamurifera	J	(47)
			Scapania ornithopodioides	J	(638)
			Thysananthus amazonicus	J	(219)
			T. convolutus	J	(219)
(189)	δ-Elemene	$C_{15}H_{24}$	*Conocephalum conicum*	M	(62, 106)
			Lophozia ventricosa	J	(429, 567a)
			Plagiochila panamensis	J	(513)
			Anastrophyllum minutum	J	(106)
			Bazzania stolonifera	J	(513)
			Clasmatocolea vermicularis	J	(47)
			Frullania davurica	J	(420)
			F. jackii	J	(420)
			F. sphaerocephala	J	(47)
			Lejeunea albescens	J	(219)
			Lepidozia borneensis	J	(51)
			Lophozia ventricosa	J	(429, 567a)
			Marchantia polymorpha	M	(74)

Table IIb (continued)

Structure number	Name of compounds	Formula	m.p. °C	$[\alpha]_D$	Plant source	Order	References	Comments
					Monoclea forsteri	Mo	(513)	
					Mylia nuda	J	(621)	
					Plagiochila adiantoides	J	(48)	
					P. beskeana	J	(48)	
					P. bifaria	J	(48)	
					P. cipacomensis	J	(48)	
					P. cucullata	J	(48)	
					P. dichotoma	J	(48)	
					P. dilatata	J	(48)	
					P. engelii	J	(48)	
					P. excisa	J	(48)	
					P. gayana	J	(46)	
					P. goebeliana	J	(48)	
					P. guayrapurinensis	J	(48)	
					P. guilleminiana	J	(48)	
					P. hondurensis	J	(48)	
					P. kroneana	J	(48)	
					P. moritziana	J	(513)	
					P. neesiana	J	(46)	
					P. pachyloma	J	(48)	
					P. parvidens	J	(46)	
					P. pulcherrima	J	(190)	
					P. rosariensis	J	(48)	
					P. scopulosa	J	(48)	
					P. tenerrima	J	(48)	
					Pleurozia acinosa	J	(622)	

No.	Compound	Formula	$[\alpha]$	Species		Ref.
(190)	Elemol	$C_{15}H_{26}O$		Radula boryana	J	(513)
				Riccardia sp.	Me	(48)
				Reboulia hemisphaerica	M	(106)
				Scapania glaucoviridis	J	(621)
				Schistochila aligera	J	(51)
				Symphyogyna brasiliensis	J	(513)
				Thysananthus convolutus	J	(219)
				Gackstoemia magellanica	J	(45)
				Pleurozia acinosa	J	(622)
				Clasmatocolea humilis	J	(45)
(191)	Dehydrosaussurea lactone	$C_{15}H_{20}O_2$		Clasmatocolea humilis	J	(45)
(192)	Saussurea lactone	$C_{15}H_{22}O_2$		Plagiochila hondurensis	J	(48)
(193)	Elema-1,4(15),11-trien-3,14-olide	$C_{15}H_{20}O_2$	+ 18.1	Plagiochasma rupestre	M	(243)
(194)	Elema-1,4(15),11-trien-3-al	$C_{15}H_{22}O$	+ 25.7	Plagiochasma rupestre	M	(243)
(196)	ent-Bicyclo-germacrene	$C_{15}H_{24}$		Acrolejeunea pusilla	J	(219)
				Adelanthus bisetulus	J	(45)
				A. lindenbergianus	J	(45)
				Anastrophyllum minutum	J	(106, 108)
				Balantiopsis cancellata	J	(45)
				B. rosea	J	(65)
				Bazzania fauriana	J	(622)
				B. praerupta	J	(341)
				B. tridens	J	(627)
				Calypogeia azulea	J	(106)
				C. granulata	J	(531)
				C. muelleriana	J	(321)
				C. peruviana	J	(321)

Table IIb. (continued)

Structure number	Name of compounds	Formula	m.p. °C	$[\alpha]_D$	Plant source	Order	References	Comments
					C. trichomanis	J	(321)	
					C. tosana	J	(321)	
					Chiloscyphus hookeri	J	(45)	
					Clasmatocolea humilis	J	(45)	
					C. vermicularis	J	(47)	
					Conocephalum japonicum	M	(62)	
					Frullania brasiliensis	J	(47)	
					F. clavata	J	(47)	
					F. davurica	J	(420)	
					F. falciloba	J	(60, 64)	
					F. jackii	J	(420)	
					Frullanoides densifolia	J	(47)	
					Herbertus divergens	J	(513)	
					Isotachis haematodes	J	(47)	
					I. humectata	J	(45)	
					Jackiella javanica	J	(247)	
					Jamesoniella colorata	J	(45)	
					Jensenia erythropus	J	(47)	
					Lejeunea albescens	J	(219)	
					L. discreta	J	(219)	
					Lejeunea sp.	J	(47)	
					Lepidozia reptans	J	(147)	
					Leptoscyphus liebmanianus	J	(47)	
					Lethocolea glossophylla	J	(47)	
					Lophozia ventricosa	J	(429, 567a)	

Makinoa crispata	Me	(622)
Marchantia berteroana	M	(39)
M. tosana	M	(74)
Mastigophora diclados	J	(621)
Neotrichocolea bissetii	J	(585a)
Neteroclada confluens	J	(45)
Omphalanthus filiformis	J	(47)
Plagiochila acanthoda	J	(47)
P. adiantoides	J	(48)
P. amazonica	J	(48)
P. beskeana	J	(48)
P. bifaria	J	(48)
P. bispinosa	J	(46)
P. cipaconensis	J	(48)
P. cristatissima	J	(48)
P. cucullata	J	(48)
P. dichotoma	J	(48)
P. dilatata	J	(48)
P. duricaulis	J	(46)
P. elata	J	(46)
P. engelii	J	(46)
P. excisa	J	(48)
P. falcata	J	(48)
P. friabilis	J	(48)
P. gayana	J	(46)
P. goebeliana	J	(48)
P. guayrapurinensis	J	(48)
P. guilleminiana	J	(48)
P. hondurensis	J	(48)
P. hookeriana	J	(48)
P. kroneana	J	(48)
P. moritziana	J	(513, 515)

Table IIb (continued)

Structure number	Name of compounds	Formula	m.p. °C	$[\alpha]_D$	Plant source	Order	References	Comments
					P. neesiana	J	(46)	
					P. oresitropha	J	(48)	
					P. ovalifolia	J	(435a)	
					P. oxyphylla	J	(48)	
					P. pachyloma	J	(48)	
					P. panamensis	J	(513)	
					P. parvidens	J	(46)	
					P. peculiaris	J	(622)	
					P. pittieri	J	(48)	
					P. pulcherrima	J	(190)	
					P. rosariensis	J	(48)	
					P. scopulosa	J	(48)	
					P. spinulosa	J	(295)	GC-MS
					P. stephensoniana	J	(39)	
					P. tambillensis	J	(48)	
					P. tenerrima	J	(48)	
					P. trabeculata	J	(589)	
					P. verruculosa	J	(48)	
					Pleurozia acinosa	J	(622)	
					Porella elegantula	J	(39)	
					P. densifolia subsp. appendiculata	J	(67)	
					P. densifolia var. fallax	J	(67)	
					P. squamurifera	J	(47)	
					P. swartziana	J	(47)	
					Radula boryana	J	(513)	

No.	Compound	Formula	[rotation]	Species		Ref.
(197)	3α-Acetoxy-bicyclogermacrene	$C_{17}H_{26}O_2$	− 55	Reboulia hemisphaerica	M	(106)
				Riccardia andina	Me	(513)
				R. jackii	Me	(380)
				Roivainenia jacquinotii	J	(45)
				Scapania subalpina	J	(289)
				S. ornithopodioides	J	(622)
				Schistochila acuminata	J	(622)
				S. aligera	J	(51)
				S. glaucoviridis	J	(621)
				S. laminigera	J	(45)
				Stephaniella paraphyllina	J	(47)
				Symphyogyna brongniartii	J	(513)
				Triandrophyllum subtrifidum var. trifidum	J	(45)
				Thysananthus convolutus	J	(219)
				T. pterobryoides	J	(513)
				Archilejeunea mariana	J	(219)
				Calypogeia granulata	J	(453, 539)
				C. muelleriana	J	(321)
				C. peruviana	J	(321)
				C. trichomanis	J	(453, 539)
				C. tosana	J	(321)
				Plagiochila moritziana	J	(513, 515)
				P. trabeculata	J	(589)
(198)	(−)-Bicyclogermacrenal	$C_{15}H_{22}O$	− 52.0	Conocephalum conicum	M	(595)
(199)	Bicyclogermacrene-14-al	$C_{15}H_{22}O$		Conocephalum conicum	M	(146)

Table IIb (continued)

Structure number	Name of compounds	Formula	m.p. °C	$[\alpha]_D$	Plant source	Order	References	Comments
(200)	2-Acetoxy-3-hydroxy-bicyclogermacrene	$C_{17}H_{26}O_3$		− 10.3	Calypogeia granulata	J	(453, 539)	
					C. muelleriana	J	(321)	
					C. peruviana	J	(321)	
					C. trichomanis	J	(453, 539)	
					C. tosana	J	(321)	
(201)	3-Acetoxy-2-hydroxy-bicyclogermacrene	$C_{17}H_{26}O_3$		+ 63.2	Calypogeia granulata	J	(453, 539)	
					C. muelleriana	J	(321)	
					C. peruviana	J	(321)	
					C. trichomanis	J	(453, 539)	
					C. tosana	J	(321)	
(202)	3α,14-Diacetoxy-2-hydroxybicyclo-germacrene	$C_{19}H_{28}O_5$			Plagiochila fruticosa	J	(249)	
(203)	3α-Hydroxy-bicyclogermacrene	$C_{15}H_{24}O$			Calypogeia granulata	J	(453, 539)	
					C. muelleriana	J	(321)	
					C. peruviana	J	(321)	
					C. trichomanis	J	(453, 539)	
					C. tosana	J	(321)	
(204)	5-Methoxy-bicyclogermacrene	$C_{16}H_{26}O$			Conocephalum japonicum	M	(584)	
(205)	14-Methoxybicyclo-germacrene	$C_{16}H_{26}O$			Conocephalum japonicum	M	(584)	
(206)	(−)-Isobicyclo-germacrenal	$C_{15}H_{22}O$			Bazzania japonica	J	(84)	
					Lepidozia vitrea	J	(383)	

No.	Compound	Formula	[α]	Species		Ref.	CD
(207)	(4S*, 5S*, 6R*, 7R*)-1(10)E-lepidozen-5-ol	$C_{15}H_{26}O$	−29.9	Bryopteris filicina	J	(432a)	CD 285 + 0.012
				Porella swartziana	J	(441a, 569)	
				Trocholejeunea sandvicensis	J	(14, 556)	
(208)	Lepidozenal	$C_{15}H_{22}O$		Lepidozia vitrea	J	(383)	
(212)	α-Bisabolene	$C_{15}H_{24}$		Chandonanthus hirtellus	J	(51)	
				Neteroclada confluens	J	(45)	
				Riccardia andina	Me	(513)	
				Scapania crasiretis	J	(435)	
(212)	(S)-(+)-Z-α-Bisabolene	$C_{15}H_{24}$		Clasmatocolea vermicularis	J	(47)	
(213)	β-Bisabolene	$C_{15}H_{24}$		Marchantia polymorpha	M	(74)	
				Monoclea forsteri	Mo	(513)	
				Riccardia andina	Me	(513)	
(214)	(−)-ar-Curcumene	$C_{15}H_{22}$		Bazzania fauriana	J	(621)	
				Heteroscyphus planus	J	(427a)	
				Lophocolea heterophylla	J	(527a)	
			−30	Marsupella emarginata var. patens	J	(384)	
				Scapania ornithopodioides	J	(638)	
(215)	Bisabolatriene (= β-Sesquiphellandrene)	$C_{15}H_{24}$	−7.48	Marchantia chenopoda	M	(553)	
(216)	(6R,7R)-(+)-α-Bisabolol	$C_{15}H_{26}O$	+55	Jungermannia vulcanicola	J	(437)	
(217)	(6S,7S)-(−)-α-Bisabolol	$C_{15}H_{26}O$	−59	Bazzania tridens	J	(629)	
			−64	Frullania brasiliensis	J	(433)	

Table IIb (continued)

Structure number	Name of compounds	Formula	m.p. °C	$[\alpha]_D$	Plant source	Order	References	Comments
(218)	(−)-Nuciferal	$C_{15}H_{20}O$	107.5	+60.1	Anthelia julacea	J	(435)	
			109.5		Gymnomitrion concinnatum	J	(434, 434a)	
(219)	Julaceal	$C_{15}H_{22}O$		−24.4	Anthelia julacea	J	(435)	
(220)	β-Bourbonene	$C_{15}H_{24}$			Lophocolea bidentata	J	(527a)	
					Mylia nuda	J	(622, 624)	
(221)	Conocephalenol	$C_{15}H_{26}O$		−4.77	Conocephalum conicum	M	(145, 146, 441a)	
(226)	γ-Cadinene	$C_{15}H_{24}$			Acrolejeunea torulosa	J	(219)	
					Frullania davurica	J	(420)	
					F. jackii	J	(420)	
					Gymnocolea inflata	J	(429)	
					Heteroscyphus planus	J	(427a)	
					Isotachis humectata	J	(45)	
					Marchantia foliacea	M	(39)	
					Mylia nuda	J	(622, 624)	
					Schistochila aligera	J	(51)	
(227)	δ-Cadinene	$C_{15}H_{24}$			Heteroscyphus planus	J	(427a)	
					Lophocolea bidentata	J	(527a)	
					L. heterophylla	J	(527a)	
					Mylia nuda	J	(622, 624)	
(228)	α-Muurolene	$C_{15}H_{24}$			Gackstroemia magellanica	J	(45)	
					Trichocolea pluma	J	(133)	
(229)	γ-Muurolene	$C_{15}H_{24}$			Balantiopsis erinacea	J	(45)	
					B. cancellata	J	(45)	
					Gackstroemia magellanica	J	(45)	

No.	Name	Formula		Species		Ref.
(230)	α-Amorphene	$C_{15}H_{24}$		*Marchantia foliacea*	M	(39)
(231)	α-Cadinol	$C_{15}H_{26}O$		*M. tosana*	M	(74)
				Monoclea forsteri	Mo	(513)
				Plagiochila bispinosa	J	(46)
				P. elata	J	(46)
				P. lecheri	J	(46)
				P. moritziana	J	(513)
				P. subdura	J	(46)
				Scapania robusta	J	(621, 631)
				Lophocolea heterophylla	J	(527a)
(232)	T-Cadinol	$C_{15}H_{26}O$		*Wiesnerella denudata*	M	(62)
				Lophocolea heterophylla	J	(527a)
(233)	(+)-*ent*-Epicubenol	$C_{15}H_{26}O$		*Conocephalum conicum*	M	(553)
				Heteroscyphus planus	J	(427a)
				Lophocolea bidentata	J	(527a)
				L. heterophylla	J	(527a)
				Scapania uliginosa	J	(289)
			+ 111.6	*Scapania undulata*	J	(153, 289, 434a)
(236)	Torreyol	$C_{15}H_{26}O$		*Lophocolea heterophylla*	J	(527a)
(237)	Cadina-4,11-dien-14-al	$C_{15}H_{22}O$		*Mannia subpilosa*	M	(613)
				Reboulia hemisphaerica	M	(613)
(238)	Cadina-4,11-dien-14-ol	$C_{15}H_{24}O$		*Mannia subpilosa*	M	(613)
				Reboulia hemisphaerica	M	(613)
(239)	Cadina-4,11-dien-14-oic acid	$C_{15}H_{22}O_2$		*Reboulia hemisphaerica*	M	(505a, 623a)
(240)	14-Acetoxycadina-4,11-diene	$C_{17}H_{26}O_2$		*Mannia subpilosa*	M	(613)
				Roboulia hemisphaerica	M	(613)
(241)	Gongylantoxide	$C_{15}H_{24}O_2$		*Gongylanthus ericetorum*	J	(44)
(242)	Lepidozenolide	$C_{15}H_{20}O_2$		*Lepidozia fauriana*	J	(505a, 623a)

Table IIb (continued)

Structure number	Name of compounds	Formula	m.p. °C	$[\alpha]_D$	Plant source	Order	References	Comments
(243)	5α-Hydroperoxy-lepidozenolide	$C_{15}H_{20}O_4$			Lepidozia fauriana	J	(505a, 623a)	
(244)	(1S,4S)-Calamenene	$C_{15}H_{22}$			L. vitrea	J	(505a, 623a)	
				−41.6	Bazzania fauriana	J	(621)	
					B. trilobata	J	(342)	
					Calypogeia azulea	J	(106)	
					Heteroscyphus planus	J	(427a)	
					Lophocolea heterophylla	J	(343, 586)	
					Mastigophora diclados	J	(621, 622, 635)	
					Monoclea forsteri	Mo	(513)	
					Mylia nuda	J	(621, 622, 624)	
					Plagiochila bispinosa	J	(46)	
					P. cucullata	J	(48)	
					P. duricaulis	J	(46)	
					P. falcata	J	(48)	
					P. oxyphylla	J	(48)	
					P. parvitexta	J	(48)	
					P. peculiaris	J	(621)	
					Scapania robusta	J	(631)	
					S. subalpina	J	(289)	
					Schistochila aligera	J	(51)	
					S. glaucoviridis	J	(621)	
					Triandrophyllum subtrifidum var. trifidum	J	(45)	

(245)	(1S,4S)-(−)-5-Hydroxy-calamenene	C$_{15}$H$_{22}$O	−8.4	Anastrophyllum minutum	J	(106)	
				Archilejeunea olivacea	J	(219)	
				Bazzania trilobata	J	(283, 342)	CD 272 nm CD 278 nm
				Chiloscyphus polyanthos	J	(60)	
				Clasmatocolea vermicularis	J	(47, 60)	
				Lophocolea coadunata	J	(47)	
				Lopholejeunea eulopa	J	(219)	
				L. howei	J	(219)	
				L. subfusca	J	(219)	
				Mastigolejeunea undulata	J	(219)	
				Mylia nuda	J	(624)	
				Plagiochila oresitropha	J	(48)	
				Triandrophyllum subtrifidum var. trifidum	J	(45)	
(246)	α-Calacorene	C$_{15}$H$_{20}$		Lophocolea heterophylla	J	(343, 586)	
				Mastigophora diclados	J	(621, 622, 635)	
				Mylia nuda	J	(624)	
				Plagiochila peculiaris	J	(637)	
(247)	β-Calacorene	C$_{15}$H$_{20}$		Bazzania fauriana	J	(621)	
				Mastigophora diclados	J	(621, 622, 635)	
				Mylia nuda	J	(624)	
				Plagiochila bispinosa	J	(46)	
				P. chacabucensis	J	(46)	
				P. duricaulis	J	(46)	
				P. elata	J	(46)	
				P. lecheri	J	(46)	
				Scapania ornithopodioides var. trifidum	J	(638)	

Table IIb (continued)

Structure number	Name of compounds	Formula	m.p. °C	$[\alpha]_D$	Plant source	Order	References	Comments
(248a)	(1S,4S)-7-Hydroxy-calamenene	$C_{15}H_{22}O$			Lophocolea heterophylla	J	(527a, 586)	
(250)	8-Hydroxy-calamenene	$C_{15}H_{22}O_1$		+ 31.5	Bazzania trilobata	J	(342, 587)	
(251)	5,8-Dihydroxy-calamenene	$C_{15}H_{22}O_2$			Heteroscyphus planus	J	(249)	
(253)	7-Acetoxy-8-hydroxycalamenene	$C_{17}H_{24}O_3$			Lophocolea heterophylla	J	(586)	
(254)	(1S)-7-Methoxy-1,2-dihydrocadalene	$C_{16}H_{22}O$		− 60	Heteroscyphus planus	J	(427a)	
(255)	(1S,4R)-7-Methoxy-calamenene	$C_{16}H_{24}O$		+ 42	Heteroscyphus planus	J	(427a)	
(256)	(1S,4R)-7-Hydroxy-calamenene	$C_{15}H_{22}O$		+ 59	Heteroscyphus planus	J	(427a)	
(257)	7-Methoxycadalene	$C_{16}H_{20}O$			Heteroscyphus planus	J	(427a)	
(258)	Cadalene	$C_{15}H_{18}$			Lophocolea heterophylla	J	(343)	
(259)	β-Caryophyllene	$C_{15}H_{24}$			Archilejeunea olivacea	J	(219)	
					A. parviflora	J	(219)	
					Balantiopsis rosea	J	(65)	
					Conocephalum conicum	M	(62)	
					Diplasiolejeunea patelligera	J	(219)	
					Fossombronia pusilla	J	(491)	
					Frullania bicornistipula	J	(513)	
					F. davurica	J	(420)	

F. falciloba	J	(64)
F. jackii	J	(420)
F. sphaerocephala	J	(47)
Frullanoides densifolia	J	(47)
Gackstroemia magellanica	J	(45)
Heteroscyphus planus	J	(427a)
Lopholejeunea eulopa	J	(219)
Lophozia ventricosa	J	(429, 567a)
Marchantia berteroana	M	(39)
M. paleacea var. *diptera*	M	(74)
M. polymorpha	M	(74)
M. tosana	M	(74)
Marchesinia brachiata	J	(219)
Monoclea forsteri	Mo	(513)
M. gottschei subsp. *neotropica*	Mo	(514)
Omphalanthus filiformis	J	(47)
Pellia epiphylla	Me	(106)
Plagiochila gayana	J	(46)
Porella densifolia subsp. *appendiculata*	J	(67)
P. densifolia var. *fallax*	J	(67)
P. swartziana	J	(47)
Thysananthus convolutus	J	(219)
T. fruticosus	J	(219)
T. mollis	J	(219)
Wiesnerella denudata	M	(62)
Marchantia paleacea var. *diptera*	M	(74)

(260) β-Caryophyllene oxide $C_{15}H_{24}O$

Table IIb (continued)

Structure number	Name of compounds	Formula	m.p. °C	$[\alpha]_D$	Plant source	Order	References	Comments
(261)	α-Cedrene	$C_{15}H_{24}$			Gymnocolea inflata	J	(429)	
					Heteroscyphus planus	J	(427a)	
					Marchantia polymorpha	M	(74)	
					Triandrophyllum subtrifidum var. trifidum	J	(45)	
(262)	β-Cedrene	$C_{15}H_{24}$			Bazzania stolonifera	J	(513)	
(263)	α-Chamigrene	$C_{15}H_{24}$			Bazzania fauriana	J	(621, 622)	
					Lophocolea bidentata	J	(527a)	
					L. heterophylla	J	(527a)	
(264)	β-Chamigrene	$C_{15}H_{24}$			Bazzania angustifolia	J	(621, 622)	
					B. fauriana	J	(621, 622, 636)	
					B. stolonifera	J	(513)	
					Cheilolejeunea excisula	J	(219)	
					Chiloscyphus polyanthos	J	(60)	
					Clasmatocolea vermicularis	J	(60)	
					Conocephalum conicum	M	(106)	
					Frullanoides densifolia	J	(47)	
					Heteroscypus aselliformis	J	(51)	
					H. planus	J	(427a)	
					Lepidozia borneensis	J	(51)	
					Leucolejeunea aff. decurrens	J	(219)	
					Lophozia ventricosa	J	(429)	
					Marchantia paleacea var. diptera	M	(74)	

No.	Compound	Formula	mp	$[\alpha]_D$	Species		Ref.
					M. palmata	M	(72)
					M. plicata	M	(47)
				+66	*M. polymorpha*	M	(71, 72, 74, 382, 553, 579)
					Monoclea forsteri	Mo	(513)
					M. gottschei subsp. *neotropica*	Mo	(514)
					Omphalanthus filiformis	J	(47, 441a, 571)
					Plagiochila moritziana	J	(513, 515)
					P. peculiaris	J	(621, 637)
					Pleurozia acinosa	J	(622)
					Porella densifolia subsp. *appendiculata*	J	(67)
					P. densifolia var. *fallax*	J	(67)
					Schiffneriolejeunea nymannii	J	(219)
					Scapania ornithopodioides	J	(638)
					Schistochila acuminata	J	(621, 630)
					Thysananthus pterobryoides	J	(513)
(265)	*ent*-Chamigrenic acid	$C_{15}H_{22}O_2$			*Omphalanthus filiformis*	J	(441a, 571)
(266)	*ent*-9-Oxo-α-chamigrene	$C_{15}H_{22}O$		−43	*Marchantia polymorpha*	M	(71) CD 321 nm − 0.70
(267)	Chiloscyphone	$C_{15}H_{22}O$		−24.4	*Chiloscyphus pallescens*	J	(149, 241)
(268)	Chiloscypholone	$C_{15}H_{24}O_2$	93–4	−53.9	*Chiloscyphus pallescens*	J	(149, 241)
(269)	11ζ,12-Epoxy-chiloscypholone	$C_{15}H_{24}O_3$		+37.8	*Chiloscyphus pallescens*	J	(241)
(270)	*ent*-(5R,6S,9R)-4α-Hydroxyoppositan-10-one	$C_{15}H_{26}O_2$		+76.1	*Chiloscyphus pallescens*	J	(241)

Table 11b (continued)

Structure number	Name of compounds	Formula	m.p. °C	$[\alpha]_D$	Plant source	Order	References	Comments
(280)	α-Copaene	$C_{15}H_{24}$			*Bazzania angustifolia*	J	*(622)*	
					B. fauriana	J	*(621, 622)*	
					Chiloscyphus pallido-virens	J	*(45)*	
					Frullania davurica	J	*(420)*	
					F. jackii	J	*(420)*	
					Lophocolea bidentata	J	*(527a)*	
					L. coadunata	J	*(47)*	
					L. heterophylla	J	*(527a)*	
					Marchantia foliacea	M	*(39)*	
					M. plicata	M	*(47)*	
					Monoclea forsteri	Mo	*(513)*	
					Trichocolea pluma	J	*(133)*	
(281)	α-Cubebene	$C_{15}H_{24}$			*Gymnocolea inflata*	J	*(429)*	
					Marchantia foliacea	M	*(39)*	
(282)	β-Cubebene	$C_{15}H_{24}$			*Herbertus divergens*	J	*(513)*	
					Heteroscyphus planus	J	*(427a)*	
					Lophocolea heterophylla	J	*(527a)*	
					Pellia endiviifolia	Me	*(457)*	
					Symphyogyna brongniartii	J	*(513)*	
(283)	Cubebol	$C_{15}H_{26}O$			*Conocephalum conicum*	M	*(249)*	
(284)	Cuparene	$C_{15}H_{22}$			*Bazzania angustifolia*	J	*(621, 622)*	
					B. fauriana	J	*(581, 636)*	
					B. praerupta	J	*(341)*	

	Species		Reference
	B. stolonifera	J	(531)
	B. tridens	J	(621, 623, 629)
− 28.0	*B. trilobata*	J	(342)
	Frullania gaudichaudii	J	(60)
	Gymnocolea inflata	J	(429)
	Herbertus aduncus	J	(57)
	H. divergens	J	(47)
	H. subdentatus	J	(57)
	Heteroscyphus aselliformis	J	(51)
	H. planus	J	(427a)
	Lepidozia borneensis	J	(51)
	L. reptans	J	(147)
	Lophocolea bidentata	J	(527a)
	L. heterophylla	J	(343)
	Lopholejeunea eulopa	J	(219)
	L. howei	J	(219)
	Lophozia ventricosa	J	(429)
	Marchantia berteroana	M	(39)
	M. paleacea var. *diptera*	M	(74)
− 52.5	*M. polymorpha*	M	(61, 71, 74, 553, 579)
− 63	*M. polymorpha*	M	(382)
	Mastigophora diclados	J	(134, 621, 622, 626, 635)
	Mylia nuda	J	(621, 622, 624)
	Monoclea forsteri	Mo	(513)
− 78.1	*Nardia scalaris*	J	(106, 354a)
	Plagiochila adiantoides	J	(48)
	P. bispinosa	J	(46)
	P. chacabucensis	J	(46)

Table IIb (continued)

Structure number	Name of compounds	Formula	m.p. °C	$[\alpha]_D$	Plant source	Order	References	Comments
					P. cucullata	J	(48)	
					P. dichotoma	J	(48)	
					P. dura	J	(46)	
					P. duricaulis	J	(46)	
					P. falcata	J	(48)	
					P. parvitexta	J	(48)	
					P. guayrapurinensis	J	(48)	
					P. kroneana	J	(48)	
					P. peculiaris	J	(621, 637)	
					P. pittieri	J	(48)	
					P. subdura	J	(46)	
					Radula boryana	J	(513)	
					R. buccinifera	J	(66)	
					R. complanata	J	(66)	
					R. oyamensis	J	(66)	
					Reboulia hemisphaerica	M	(411)	
					Riccardia andina	Me	(513)	
				−28	R. jackii	Me	(380)	
					Ricciocarpos natans	M	(639, 640)	
					Scapania ornithopodioides	J	(638)	
					S. robusta	J	(621, 631)	
					Schistochila acuminata	J	(621, 630)	
					S. glaucoviridis	J	(621)	
					S. laminigera	J	(45)	
					S. reflexa	J	(45)	

No.	Name	Formula	mp	[α]	Species		Ref.
(285)	2-Hydroxycuparene	$C_{15}H_{22}O$			Targionia hypophylla	M	(65)
					Thysananthus pterobryoides	J	(513)
					Bazzania fauriana	J	(581)
					B. japonica	J	(84)
					Gymnomitrion concinnatum	J	(434a)
					Herbertus aduncus	J	(57)
					H. divergens	J	(47)
					H. sakuraii	J	(57)
					H. subdentatus	J	(57)
					Makinoa crispata	Me	(622)
					Marchantia berteroana	M	(39)
					M. paleacea var. diptera	M	(74)
			−43		M. polymorpha	M	(61, 74, 382)
					Plagiochila bispinosa	J	(46)
					P. dura	J	(46)
					P. duricaulis	J	(46)
					P. subdura	J	(46)
					Radula kojana	J	(50)
					Roivainenia jacquinotii	J	(45)
					Scapania robusta	J	(631)
					S. reflexa	J	(45)
					Triandrophyllum subtrifidum var. trifidum	J	(45)
(286)	3-Hydroxycuparene	$C_{15}H_{22}O$			Herbertus subdentatus	J	(57)
(287)	2,3-Dihydroxycuparene	$C_{15}H_{22}O_2$	124–5	−20.9	Bazzania fauriana	J	(581)
					Herbertus aduncus	J	(57)
					H. divergens	J	(47)
					H. subdentatus	J	(57)

Table IIb (continued)

Structure number	Name of compounds	Formula	m.p. °C	$[\alpha]_D$	Plant source	Order	References	Comments
(288)	(R)-(−)-α-Cuparenone	$C_{15}H_{20}O$	50–2		*Plagiochila bispinosa*	J	(46)	
					P. duricaulis	J	(46)	
					P. subdura	J	(46)	
					Radula perrottetii	J	(66)	
					Herbertus aduncus	J	(57)	
(289)	α-Cuparenol	$C_{15}H_{22}O$		− 134.2	*Reboulia hemisphaerica*	M	(411, 412)	
(290)	1-Formyl-2,3-dihydroxy-cuparene	$C_{15}H_{20}O_3$			*Lepidozia concinna*	J	(37)	
					Plagiochila bispinosa	J	(46)	
					P. subdura	J	(46)	
(291)	(−)-2-(1,2,2-Trimethyl)-cyclopentyl-6-methyl-1,4-quinone	$C_{15}H_{20}O_2$			*Radula javanica* (= *R. variabilis*)	J	(49)	
(292)	α-Cuprenene	$C_{15}H_{24}$		− 73	*Marchantia polymorpha*	M	(74, 579)	
(293)	β-Cuprenene	$C_{15}H_{24}$		− 68.5	*Marchantia polymorpha*	M	(74)	
(294)	γ-Cuprenene	$C_{15}H_{24}$			*Bazzania angustifolia*	J	(621)	
					B. fauriana	J	(621)	
					Lophozia ventricosa	J	(567a)	
(295)	(−)-δ-Cuprenene (= Angustifolene)	$C_{15}H_{24}$		− 40	*Bazzania angustifolia*	J	(621, 622)	
					B. tridens	J	(621, 629)	
					Mastigophora diclados	J	(621, 622)	
					Marchantia polymorpha	M	(74, 382)	
					Mylia nuda	J	(621)	
					Plagiochila peculiaris	J	(621, 622, 637)	

No.	Compound	Formula	mp (°C)	[α]	Species	M/J	Refs.
(296)	(+)-ε-Cuprenene	$C_{15}H_{24}$			Scapania ornithopodioides	J	(621)
					S. robusta	J	(621)
					Schistochila acuminata	J	(621)
				+168	Marchantia polymorpha	M	(382)
(297)	(R)-(−)-8,11-Dihydro-α-cuparenone	$C_{15}H_{22}O$	39–41	−116.4	Reboulia hemisphaerica	M	(411, 412)
(298)	(−)-Cyclopropane-cuparenol	$C_{15}H_{26}O$			Marchantia paleacea var. diptera	M	(71, 74)
					M. polymorpha	M	(71, 74)
(299)	(−)-Epi-cyclopropane-cuparenol	$C_{15}H_{26}O$			Marchantia paleacea var. diptera	M	(74)
					M. polymorpha	M	(74)
(300)	(+)-Cuprenenol	$C_{15}H_{26}O$			Jungermannia rosulans	J	(19)
(301)	ent-Cuprenenol	$C_{15}H_{26}O$			Marchantia polymorpha	M	(74)
(302)	Demotarisiol	$C_{15}H_{26}O$		−39.6	Demotarisia linguifolia	J	(431)
(303)	Cuparenolide	$C_{15}H_{22}O_2$	59–60	+24.4	Ricciocarpos natans	M	(640, 641)
(304)	Cuparenolidol	$C_{15}H_{22}O_3$		+38.7	Ricciocarpos natans	M	(640, 641)
(305)	Grimaldone	$C_{15}H_{22}O$	91–2	−49.6	Mannia fragrans	M	(276)
(306)	3,6-Peroxocupar-1-ene	$C_{15}H_{24}O_2$			Nardia scalaris	J	(354a)
(307)	Herbertene (= (1S)-1,2,2-Trimethyl-m-tolyl-cyclopentane) (= Isocuparene)	$C_{15}H_{22}$		−48	Herbertus aduncus	J	(57, 392–394)
					H. sakuraii	J	(57)
					H. subdentatus	J	(57)
					Mastigophora diclados	J	(51, 134, 621, 622)
					M. woodsii	J	(134, 626)

ORD 400 nm − 600
ORD 312 nm − 6200
ORD 294 nm ± 0
ORD 274 nm + 7600
ORD 230 nm ± 0

Table IIb (*continued*)

Structure number	Name of compounds	Formula	m.p. °C	$[\alpha]_D$	Plant source	Order	References	Comments
(308)	α-Herbertenol	$C_{15}H_{22}O$		− 55	*Herbertus acanthelius*	J	(433)	
					H. aduncus	J	(57, 393, 395)	
					H. sakuraii	J	(57)	
					H. subdentatus	J	(57, 433)	
					Mastigophora diclados	J	(134, 622, 626)	
					M. woodsii	J	(134, 626)	
					Plagiochila bispinosa	J	(46)	
(309)	β-Herbertenol	$C_{15}H_{22}O$	80–1	− 47	*Herbertus aduncus*	J	(393, 395)	
					Herbertus divergens	J	(513)	
					H. subdentatus	J	(57, 433)	
				− 39	*Marchantia polymorpha*	M	(74, 382)	
					Mastigophora diclados	J	(51, 134, 186, 191, 622, 626)	
					M. woodsii	J	(134, 626)	
(310)	(−)-α-Formyl-herbertenol	$C_{15}H_{20}O_2$	134–5	− 66	*Herbertus aduncus*	J	(393, 395)	
					Triandrophyllum subtrifidum var. *trifidum*	J	(45)	
					Plagiochila subdura	J	(46)	
(311)	(−)-Herbertenediol	$C_{15}H_{22}O_2$	90.5–91.5	− 46.5	*Herbertus aduncus*	J	(392, 393)	
					H. subdentatus	J	(433)	
					Mastigophora diclados	J	(51, 134, 186, 191, 621, 626)	
					M. woodsii	J	(134, 626)	
					Plagiochila subdura	J	(46)	

No.	Compound	Formula	M.p.	Rotation	Species		Ref.	Notes
(312)	(−)-Herbertenolide	$C_{15}H_{18}O_2$	95.5–96.5	−86.4	Herbertus aduncus	J	(392, 393)	
					H. sakuraii	J	(623a)	
(313)	Isocuparene-3,4-diol	$C_{15}H_{22}O_2$	150–1	−73.6	Mastigophora diclados	J	(51, 186, 191)	
(321)	Mastigophorene A	$C_{30}H_{42}O_4$	258–261	−65.3	Mastigophora diclados	J	(51, 186, 191)	CD 202 nm −24.1 / CD 222 nm +4.4
(322)	Mastigophorene B	$C_{30}H_{42}O_4$	210–211	−39.1	Mastigophora diclados	J	(51, 186, 191)	CD 202 nm +24.6 / CD 215 nm −15.5
(323a)	Mastigophorene C	$C_{30}H_{42}O_4$		−46.7	Mastigophora diclados	J	(51, 186, 191)	
(324a)	Mastigophorene D	$C_{30}H_{42}O_4$		−46.1	Mastigophora diclados	J	(51, 186, 191)	
(327)	Hercynolactone	$C_{15}H_{20}O_2$			Barbilophozia barbata	J	(432)	
				−84.5	B. floerkei	J	(565, 567a)	
				−83.7	B. hatcheri	J	(565, 567a)	
				−84.2	B. lycopodioides	J	(143, 275)	ORD 400 nm −300 / ORD 300 nm −1650 / ORD 266 nm −5300 / ORD 250 nm 0
(328)	Drimenol	$C_{15}H_{26}O$	95–6	−18.8	Bazzania fauriana	J	(581)	
					B. praerupta	J	(341)	
					B. tridens	J	(629)	
					B. trilobata	J	(342)	
					Diplophyllum serrulatum	J	(585a)	
					Frullania clavata	J	(60)	
					Porella vernicosa	J	(458)	
					Targionia hypophylla	M	(75)	Suspension culture
					Trichocolea pluma	J	(133)	
(329)	Albicanol	$C_{15}H_{26}O$			Bazzania japonica	J	(84)	
					Diplophyllum serrulatum	J	(585a)	
(330)	Albicanyl acetate	$C_{17}H_{28}O_2$			Bazzania japonica	J	(84)	
(331)	Albicanal	$C_{15}H_{24}O$		−69.8	Diplophyllum serrulatum	J	(585a)	
(332)	Isoalbicanal	$C_{15}H_{24}O$			Diplophyllum serrulatum	J	(585a)	
(333)	Albicanic acid	$C_{15}H_{24}O_2$		−23.1	Dipolphyllum serrulatum	J	(585a)	

Table IIb (continued)

Structure number	Name of compounds	Formula	m.p. °C	$[\alpha]_D$	Plant source	Order	References	Comments
(334a)	Drimenyl caffeate	$C_{24}H_{32}O_4$			Bazzania fauriana	J	(581)	
(335)	Albicanyl caffeate	$C_{24}H_{32}O_4$			Bazzania fauriana	J	(581)	
					B. japonica	J	(84)	
(336)	Polygodial	$C_{15}H_{22}O_2$			Porella roellii	J	(543)	Suspension culture
					P. vernicosa	J	(458)	Suspension culture
(337)	Isopolygodial	$C_{15}H_{22}O_2$			Porella vernicosa	J	(458)	
(338)	Drimeninol	$C_{15}H_{24}O_2$			Porella roellii	J	(543)	
(339)	Isodrimeninol	$C_{15}H_{24}O_2$			Porella roellii	J	(249)	
(340)	6α-Hydroxy-isodrimeninol	$C_{15}H_{24}O_3$			Porella roellii	J	(249)	
(341)	Dehydroconfertifolin	$C_{15}H_{20}O_2$			Porella roellii	J	(543)	
(342)	Cinnamolide	$C_{15}H_{22}O_2$			Lejeunea flava	J	(219)	
					Porella roellii	J	(543)	
					P. vernicosa	J	(458)	Suspension culture
(343)	Drimenin	$C_{15}H_{22}O_2$	129	− 30.3	Porella cordaeana	J	(239a)	
(344)	7-Ketoisodrimenin	$C_{15}H_{20}O_3$		+ 40.8	Porella cordaeana	J	(239a)	
(345)	7-Ketoisodrimenin-5-ene	$C_{15}H_{18}O_3$		+ 1.5	Porella cordaeana	J	(239a)	
(346)	7α-Chloro-6β-hydroxy-confertifolin	$C_{15}H_{21}O_3Cl$	194.5–196	+ 36.0	Makinoa crispata	Me	(258)	
(347a)	6β,7α-Dihydroxy-confertifolin	$C_{15}H_{22}O_4$	212–5	+ 47.9	Makinoa crispata	Me	(258)	
(348)	6β,7β-Epoxy-confertifolin	$C_{15}H_{20}O_3$	138–140	+ 16.7	Makinoa crispata	Me	(258)	

No.	Compound	Formula	[α]	Species		Ref.
(350)	Eremophilene	$C_{15}H_{24}$		Acrolejeunea pycnoclada	J	(219)
				Calypogeia granulata	J	(539)
				C. muelleriana	J	(321)
				C. peruviana	J	(321)
				C. tosana	J	(321)
				C. trichomanis	J	(321)
				Frullania serratta	J	(51)
				Monoclea forsteri	Mo	(513)
				Plagiochila fuegiensis	J	(219)
				Thysananthus convolutus	J	(219)
				T. mollis	J	(219)
(351)	(+)-(4S*,5R*, 7S*,8R*)- Eremophila-9,11- dien-8α-ol	$C_{15}H_{24}O$	+ 205.8	Marsupella emarginata	J	(244)
(354)	ent-α-Selinene	$C_{15}H_{24}$		Bazzania praerupta	J	(341)
				B. spiralis	J	(341)
				Frullania tamarisci subsp. obscura	J	(82)
				Lophocolea bidentata	J	(527a)
				L. heterophylla	J	(343)
				Monoclea forsteri	Mo	(513)
				Pellia epiphylla	Me	(106)
				Plagiochila peculiaris	J	(621, 637)
				Riccardia jackii	Me	(380)
				Scapania ornithopodioides	J	(622, 638)
				S. robusta	J	(631)
(355)	ent-β-Selinene	$C_{15}H_{24}$	− 15	Frullania serratta	J	(51)
				Lophocolea heterophylla	J	(527a)
				Marchantia paleacea var. diptera	M	(74)

Table IIb *(continued)*

Structure number	Name of compounds	Formula	m.p. °C	$[\alpha]_D$	Plant source	Order	References	Comments
					Marchantia palmata	M	(72)	
					M. polymorpha	M	(72, 74)	
					Plagiochila peculiaris	J	(621, 637)	
					Porella cordaeana	J	(593)	
					Riccardia crassa	Me	(583)	
				− 38	*R. jackii*	Me	(380)	
					R. prehensilis	Me	(45)	
					Scapania ornithopodioides	J	(622, 638)	
					S. robusta	J	(631)	
					Thysananthus convolutus	J	(219)	
(356)	Selina-4,11-diene	$C_{15}H_{24}$			*Mylia nuda*	J	(622, 624)	
					Scapania ornithopodioides	J	(622, 638)	
					Scapania robusta	J	(621, 631)	
(357)	γ-Selinene	$C_{15}H_{24}$			*Frullania davurica*	J	(420)	
(358)	δ-Selinene	$C_{15}H_{24}$			*Monoclea gottschei* subsp. *neotropica*	Mo	(514)	
					Roivainenia jacquinotii	J	(45)	
					Symphyogyna brongniartii	J	(513)	
(359)	α-Eudesmol	$C_{15}H_{26}O$			*Fossombronia pusilla*	Me	(491)	
					Isotachis humectata	J	(45)	
(360)	β-Eudesmol	$C_{15}H_{26}O$			*Balantiopsis rosea*	J	(65)	
					Bazzania fauriana	J	(622)	
					Fossombronia pusilla	Me	(491)	
					Radula frondescens	J	(47)	
					Riccardia prehensilis	Me	(45)	
					Thysananthus convolutus	J	(219)	

No.	Compound	Formula	[α] / mp	Species		Ref.
(361)	Selina-11-en-4-ol	$C_{15}H_{26}O$		Conocephalum conicum	M	(569)
				Lophocolea heterophylla	J	(527a)
			+26	Riccardia jackii	Me	(380)
(362)	Neointermediol	$C_{15}H_{26}O$		Lophocolea heterophylla	J	(585a, 623a)
(363)	Eudesm-3-en-7α-ol	$C_{15}H_{26}O$		Bazzania tridens	J	(585a, 623a)
				Lepidozia vitrea	J	(585a, 623a)
(364)	4β-Methoxy-eudesmanal	$C_{16}H_{26}O_3$		Frullania tamarisci subsp. obscura	J	(82)
(365)	Eudesmanal	$C_{15}H_{22}O_2$		Frullania tamarisci subsp. obscura	J	(82)
(366)	(+)-5α,7β(H)-Eudesm-4α,6α-diol	$C_{15}H_{26}O_2$	+14.7 107-8	Frullania tamarisci subsp. obscura	J	(582)
(368)	ent-Eudesm-4(15)-en-6α,7α-diol	$C_{15}H_{26}O_2$		Chiloscyphus pallescens	J	(149)
				Lepidozia reptans	J	(143, 149)
(369)	Eudesm-4(15)-en-6β,7α-diol	$C_{15}H_{26}O_2$	+4.6	Lepidozia vitrea	J	(585a)
(370)	Eudesm-3-en-6α-acetoxy-7α-ol	$C_{17}H_{28}O_3$	−43.8	Lepidozia vitrea	J	(585a)
(371)	Eudesm-3-en-6β,7α-diol	$C_{15}H_{26}O$	+34.6 94-5	Chiloscyphus pallescens	J	(143, 149)
				Lepidozia reptans	J	(143, 146, 147)
				L. vitrea	J	(585a)
				Lophozia ventricosa	J	(567a)
(372)	ent-4(15),7(11)-Eudesmadien-8-one (= β-Cyclogermacrone) (= Ventricosin A)	$C_{15}H_{22}O$	−15.6	Barbilophozia floerkei	J	(429, 567a)
			−80.5	Lophozia ventricosa	J	(286, 287, 429, 565, 567a)

Table IIb (continued)

Structure number	Name of compounds	Formula	m.p. °C	$[\alpha]_D$	Plant source	Order	References	Comments
(373)	ent-4(15),11(12)-Eudesmadien-8-one	$C_{15}H_{22}O$		− 18.9	Lophozia ventricosa	J	(429, 565, 567a)	CD 250 nm − 0.20 CD 298 nm + 0.36
(374)	6β-Hydroxyeudesm-3-ene	$C_{15}H_{26}O$		− 35	Bazzania fauriana	J	(581)	
(375)	α-Cyclogermacrone	$C_{15}H_{22}O$			Bazzania fauriana	J	(581)	
					B. tridens	J	(623a)	
(376)	ent-8β-Hydroxy-eudesm-3,11-diene	$C_{15}H_{24}O$		+ 40.5	Bazzania spiralis	J	(341)	
(377)	ent-α-Cyperone	$C_{15}H_{22}O$			Marchantia polymorpha	M	(71)	
(378)	6β-Acetoxy-vitranoxide	$C_{17}H_{28}O_3$			Lepidozia fauriana	J	(505a, 623a)	
					L. vitrea	J	(505a, 623a)	
(379)	Furanoeudesma-1,3-diene	$C_{15}H_{18}O$			Lophocolea heterophylla	J	(527a)	
(380a)	(+)-Frullanolide	$C_{15}H_{20}O_2$			Frullania dilatata	J	(19)	
(380b)	(−)-Frullanolide	$C_{15}H_{20}O_2$			Frullania apiculata	J	(60)	
					F. asagrayana	J	(63)	
					F. bicornistipula	J	(513)	
					F. brasiliensis	J	(47)	
					F. ternatensis	J	(60)	
					F. Serratta	J	(60)	
					F. sphaerocephala	J	(47)	
					Clasmatocolea vermicularis	J	(60)	
(381)	Oxyfrullanolide	$C_{15}H_{20}O_3$						

No.	Compound	Formula	mp	[α]	CD	Source		Ref.
(382)	Dihydrooxy-frullanolide	$C_{15}H_{22}O_3$				Clasmatocolea vermicularis	J	(60)
(384)	Dihydrofrullanolide	$C_{15}H_{22}O_2$				Frullania asagrayana	J	(63)
						F. bicornistipula	J	(513)
						F. brasiliensis	J	(47)
						F. sphaerocephala	J	(47)
						F. tamarisci subsp. nisquallensis	J	(63)
						Plagiochila tenerrima	J	(48)
(385)	(+)-β-Frullanolide	$C_{15}H_{20}O_2$	165–7	+178	CD 254 nm −1.5	Frullania brotheri	J	(541)
(386)	(+)-Brothenolide	$C_{15}H_{20}O_2$	113–4	+153	CD 255 nm −1.6	Frullania brotheri	J	(541)
(387)	Dihydro-β-frullanolide	$C_{15}H_{22}O_2$				Frullania bicornistipula	J	(513)
(388a)	(+)-Arbusculin B	$C_{15}H_{20}O_2$		+35		Frullania brasiliensis	J	(433)
						F. hamatiloba	J	(588)
						F. serrata	J	(51)
(388b)	(−)-Arbusculin B	$C_{15}H_{20}O_2$				Frullania usamiensis	J	(56)
(389a)	β-Cyclocostunolide	$C_{15}H_{20}O_2$				Frullania bicornistipula	J	(513)
						Frullania tamarisci subsp. obscura	J	(82, 582)
(389b)	ent-β-Cyclocostunolide	$C_{15}H_{20}O_2$				Frullania usamiensis	J	(56)
(390)	4-epi-Arbusculin A	$C_{15}H_{22}O_3$				Frullania tamarsci subsp. obscura	J	(56, 582)
(391)	Arbusculin A	$C_{15}H_{22}O_3$				Conocephalum japonicum	M	(584)
(392)	(11S)-Dihydro-arbusculin A (= Colartin)	$C_{15}H_{24}O_3$				Conocephalum japonicum	M	(584)
(393)	(11S)-Dihydro-β-cyclocostunolide	$C_{15}H_{22}O_2$				Frullania bicornistipula	J	(513)

Table IIb (continued)

Structure number	Name of compounds	Formula	m.p. °C	$[\alpha]_D$	Plant source	Order	References	Comments
(394)	8α-Acetoxy-β-cyclo-costunolide	$C_{17}H_{22}O_4$			Wiesnerella denudata	M	(51)	
(395a)	Rothin A acetate	$C_{17}H_{22}O_4$			Frullania sp.	J	(441a)	
(396)	α-Santonin	$C_{15}H_{18}O_3$			Fossombronia pusilla	Me	(91)	
(397)	Nepalensolide A	$C_{15}H_{20}O_2$	99	+ 98.5	Frullania nepalensis	J	(63, 564, 575)	CD 223 nm + 0.42 CD 245 nm + 0.35
					F. serratta		(51)	
(398)	Nepalensolide B	$C_{15}H_{22}O_2$			Frullania nepalensis	J	(63, 575)	
					Frullania serratta	J	(51)	
(399)	Nepalensolide C	$C_{15}H_{22}O_2$			Frullania nepalensis	J	(63, 575)	
					F. serratta	J	(51)	
(400)	Nepalensolide D	$C_{15}H_{20}O_2$			Frullania nepalensis	J	(63, 575)	
(401)	α-Cyclocostunolide	$C_{15}H_{20}O_2$			Frullania tamarisci subsp. obscura	J	(82)	
(402)	Methoxyfrullanolide	$C_{16}H_{24}O_3$			Frullania tamarisci	J	(249)	
(403a)	Lactone dimer A	$C_{30}H_{42}O_5$	207–8		Frullania tamarsci	J	(70, 146)	
(404a)	Lactone dimer B	$C_{30}H_{40}O_4$			Frullania tamarsci	J	(70)	
(405)	5α-Hydroxyeudesm-4(15),7(11),8(9)-trien-12,8-olide	$C_{15}H_{18}O_3$		− 149.3	Lophocolea heterophylla	J	(586)	
(406)	5α,8β-Dihydroxy-eudesm-4(15),7(11)-dien-12,8-olide	$C_{15}H_{20}O_4$	216–7		Lophocolea heterophylla	J	(586)	
(407)	ent-Isoalantolactone	$C_{15}H_{20}O_2$	107–9	− 155	Lophocolea coadunata	J	(47)	
					L. heterophylla	J	(343)	

No.	Compound	Formula	mp / [α]	Species		References
(408)	ent-Dihydroiso-alantolactone	$C_{15}H_{22}O_2$		Lophocolea coadunata	J	(47)
(409)	ent-Diplophyllolide	$C_{15}H_{20}O_2$		Chiloscyphus polyanthos	J	(60, 143)
				Clasmatocolea vermicularis	J	(60)
				Diplophyllum albicans	J	(143)
				Lophocolea bidentata	J	(527a)
				Plagiochila moritziana	J	(513, 515, 516)
				Tritomaria quinquedentata	J	(429, 565)
(410)	ent-Diplophyllin	$C_{15}H_{20}O_2$		Chiloscyphus polyanthos	J	(60)
				Clasmatocolea vermicularis	J	(60)
(411)	ent-3-Oxo-diplophyllin	$C_{15}H_{18}O_3$		Diplophyllum albicans	J	(143)
				Plagiochila moritziana	J	(513, 515, 516)
				Chiloscyphus polyanthos	J	(60, 143)
(413)	ent-7α-Hydroxy-diplophyllolide	$C_{15}H_{20}O_3$		Chiloscyphus polyanthos	J	(60, 143)
				Diplophyllum albicans	J	(143)
(414)	ent-Dihydro-diplophyllolide	$C_{15}H_{22}O_2$	99.5–100.5, −3.9	Tritomaria quinquedentata	J	(429, 565, 567a)
(415)	Plagiospirolide A	$C_{35}H_{52}O_2$	197 ± 1, +411.5	Plagiochila moritziana	J	(515, 516)
(416)	Plagiospirolide B	$C_{15}H_{52}O_2$	+59.2	Plagiochila moritziana	J	(515, 516)
(417)	Plagiospirolide C	$C_{35}H_{52}O_3$	+42.9	Plagiochila moritziana	J	(515)
(418)	Plagiospirolide D	$C_{35}H_{52}O_3$	+42.0	Plagiochila moritziana	J	(515)
(419)	Plagiospirolide E	$C_{30}H_{42}O_2$	+498.8	Plagiochila moritziana	J	(515)
(420)	Dihydroagarofuran	$C_{15}H_{26}O$		Fossombronia pusilla	J	(491)
				Symphyogyna brasiliensis	J	(513)
(423)	α-Farnesene	$C_{15}H_{24}$		Scapania ornithopodioides	J	(638)

Table IIb (continued)

Structure number	Name of compounds	Formula	m.p. °C	$[\alpha]_D$	Plant source	Order	References	Comments
(424)	trans-β-Farnesene	$C_{15}H_{24}$			Acrolejeunea torulosa	J	(219)	
					Adelanthus lindenbergianus	J	(45)	
					Blepharolejeunea incongrua	J	(219)	
					Gymnocolea inflata	J	(429)	
					Jensenia erythropus	J	(47)	
					Leptoscyphus liebmanianus	J	(47)	
					Lophocolea heterophylla	J	(527a)	
					Monoclea forsteri	Mo	(513)	
					Plagiochila gayana	J	(46)	
					P. lecheri	J	(46)	
					P. parvidens	J	(48)	
					P. scopulosa	J	(48)	
					Radula boryana	J	(513)	
					R. frondescens	J	(47)	
					R. kojana	J	(50)	
					R. voluta	J	(47)	
					Roivainenia jacquinotii	J	(45)	
					Scapania ornithopodioides	J	(638)	
					S. subalpina	J	(289)	
					Syzygiella anomala	J	(47)	
					Tylimanthus urvilleanus	J	(45)	

No.	Name	Formula	mp	[α]	Species		Ref.
(425)	trans, cis-Farnesol	$C_{15}H_{26}O$			Plagiochila ovalifolia	J	(62)
(426)	(+)-trans-Nerolidol	$C_{15}H_{26}O$			Gymnocolea inflata	J	(432)
					Lophocolea heterophylla	J	(527a)
					Plagiochila ovalifolia	J	(62)
					Wiesnerella denudata	M	(62)
					Dumortiera hirsuta	M	(585)
(427)	4,5-Dehydro-nerolidol	$C_{15}H_{24}O$			Conocephalum conicum	M	(62)
(428)	Germacrene-B	$C_{15}H_{24}$			Plagiochila ovalifolia	J	(662)
					Bazzania praerupta	J	(341)
(429)	Germacrene-D	$C_{15}H_{24}$			Conocephalum conicum	M	(62)
					Gackstroemia magellanica	J	(45)
					Lejeunea albescens	J	(219)
					L. discreta	J	(219)
					L. lumbricoides	J	(219)
					Lophocolea bidentata	J	(527a)
					L. heterophylla	J	(527a)
					Lopholejeunea subfusca	J	(219)
					Marchantia foliacea	M	(39)
					Neteroclada confluens	J	(45)
					Thysananthus amazonicus	J	(219)
					T. convolutus	J	(219)
(431a)	ent-Germacra-4(15), 5,10(14)-trien-1β-ol	$C_{15}H_{24}O$		+ 146.3	Jackiella javanica	J	(437)
(432)	ent-Germacra-4(15), 5,10(14)-trien-1α-ol	$C_{15}H_{24}O$		+ 88.4	Jackiella javanica	J	(437)
(435)	1,6-Diketogerma-crene	$C_{15}H_{24}O_2$	69–74	− 92.5	Porella swartziana	J	(569, 574a, 574b)

Table IIb (continued)

Structure number	Name of compounds	Formula	m.p. °C	$[\alpha]_D$	Plant source	Order	References	Comments
(436)	4β-Hydroxy-germacra-1(10),5-diene	$C_{15}H_{26}O$			Conocephalum conicum	M	(553)	
				− 130	Marchantia plicata	M	(433)	
					Porella swartziana	J	(569)	
(437)	ent-1(10)E,5E-Germacradien-11-ol	$C_{15}H_{26}O$		− 110.2	Dumortiera hirsuta	M	(585)	
(438)	Germacra-1(10),5-dien-4,11-diol	$C_{15}H_{26}O_2$			Bryopteris filicina	J	(432a)	
					Ptychanthus striatus	J	(249a)	
(439)	2-Acetoxy-8-keto-germacrene	$C_{17}H_{26}O_3$			Conocephalum conicum	M	(553)	
(440)	Furanogermacra-1(10),4-diene (= Furanodiene)	$C_{15}H_{20}O$			Lophocolea heterophylla	J	(527a)	
(441)	Costunolide	$C_{15}H_{20}O_2$			Clasmatocolea humilis	J	(45)	
					Frullania californica	J	(63)	
					F. tamarisci subsp. nisquallensis	J	(63)	
					F. tamarisci subsp. obscura	J	(582)	
					F. serratta	J	(51)	
					Plagiochila hondurensis	J	(48)	
					Wiesnerella denudata	M	(51)	
(442)	Dihydrocostunolide	$C_{15}H_{22}O_2$			Clasmatocolea humilis	J	(45)	
					Plagiochila hondurensis	J	(48)	
(443)	Tulipinolide	$C_{17}H_{22}O_4$			Frullania serratta	J	(51)	
					Wiesnerella denudata	M	(51)	

(444)	(11 R)-Dihydro-tulipinolide	$C_{17}H_{24}O_4$		Frullania serratta	J	(51)
(445)	(11 S)-Dihydro-tulipinolide	$C_{17}H_{24}O_4$		Frullania serratta	J	(51)
				Wiesnerella demudata	M	(51)
(446)	4α,5β-Epoxy-8-epi-inunolide	$C_{15}H_{20}O_3$	117	Porella acutifolia subsp. tosana	J	(82, 600)
(447)	4α,5β-Epoxy-7α,8β,11α-H-germacra-1(10)-en-12,8α-olide (= 11(13)-Dihydro-4α,5β-epoxy-8-epi-inunolide)	$C_{15}H_{22}O_3$	+ 19.5	Porella acutifolia subsp. tosana	J	(82, 600)
(448)	5-Keto-7α,8β,11α-H-germacra-1(10)-en-12,8α-olide	$C_{15}H_{22}O_3$		Porella acutifolia subsp. tosana	J	(82, 600)
(449)	1α-Hydroperoxy-4α,5β-epoxy-germacra-10(14),11(13)-dien-12,8α-olide	$C_{15}H_{20}O_5$	− 14.3	Porella acutifolia subsp. tosana	J	(82, 600)
(450)	1β-Hydroperoxy-4α,5β-epoxy germacra-10(14),11(13)-dien-12,8α-olide	$C_{15}H_{20}O_5$		Porella acutifolia subsp. tosana	J	(82, 600)

Table IIb (*continued*)

Structure number	Name of compounds	Formula	m.p. °C	$[\alpha]_D$	Plant source	Order	References	Comments
(453)	*ent*-Maalioxide (= Ventricosin B)	$C_{15}H_{26}O$	64–5	−36.1 −41.7	*Barbilophozia floerkei* *Jubula japonica* *Lophozia ventricosa* *L. ventricosa* *Radula perrottetii*	J J J J J	(429) (585a) (287) (429, 565) (585a)	
(454)	Iso-α-gurjunene B	$C_{15}H_{24}$			*Monoclea forsteri* *M. gottschei* subsp. neotropica	Mo Mo	(513, 514) (514)	
(455)	Guai-4,11(12)-diene	$C_{15}H_{24}$			*Monoclea forsteri* *M. gottschei* subsp. neotropica	Mo Mo	(514) (514)	
(456)	(1S,10R)-Guai-4,6-diene	$C_{15}H_{24}$		+ 62	*Bryopteris filicina*	J	(432a)	
(457)	α-Gurjunene	$C_{15}H_{24}$			*Monoclea forsteri*	Mo	(513)	
(458)	β-Guaiene	$C_{15}H_{24}$			*Frullania bicornistipula* *Monoclea forsteri* *Radula boryana*	J Mo J	(513) (513) (513)	
(459)	γ-Gurjunene	$C_{15}H_{24}$			*Mylia nuda*	J	(624)	
(460)	(1R*,5S*,7S*,10R*)-Guai-4(15)-en-6-one-1-ol	$C_{15}H_{24}O_2$		+ 23.4	*Porella swartziana*	J	(249, 574a, 574b)	
(461)	(1R*,5S*,7S*,10R*)-Guai-3-en-6-one-1-ol	$C_{15}H_{24}O_2$		+ 33.2	*Porella swartziana*	J	(249, 574a, 574b)	

(462)	Isoporelladiolide	$C_{15}H_{16}O_4$		−14.9		Porella acutifolia subsp. tosana	J	(600)
(463)	Guai-1(10),3,11(13)-trien-14,2β,12,6α-diolide (= Dehydroiso-porelladiolide)	$C_{15}H_{14}O_4$				Porella acutifolia subsp. tosana	J	(600)
(464)	Porelladiolide	$C_{15}H_{16}O_4$			CD 300 nm − 0.1 CD 260 nm + 0.3	Porella acutifolia subsp. tosana	J	(600)
(465)	Dihydroestafiatin	$C_{15}H_{20}O_3$	88–9	−12.5		Frullanoides densifolia	J	(14, 554, 556)
(466)	Estafiatin	$C_{15}H_{18}O_3$		−10.4		Frullanoides densifolia	J	(14, 554, 556)
(467)	8α-Acetoxy-zaluzanin D	$C_{19}H_{22}O_6$				Wiesnerella denudata	M	(51)
(474)	α-Himachalene	$C_{15}H_{24}$				Plagiochila chacabucensis	J	(46)
						P. duricaulis	J	(46)
						P. engelii	J	(46)
						P. lecheri	J	(46)
(475)	β-Himachalene	$C_{15}H_{24}$				Lophocolea heterophylla	J	(527a)
						Plagiochila elata	J	(46)
						P. fuegiensis	J	(46)
						P. parvidens	J	(46)
						Radula boryana	J	(513)
						Triandrophyllum subtrifidum var. trifidum	J	(45)
(477)	α-Humulene	$C_{15}H_{24}$				Acrolejeunea pusilla	J	(219)
						A. pycnoclada	J	(219, 567a)
						Lophozia ventricosa	J	(429)
						Scapania robusta	J	(631)
						Stictolejeunea balfourii var. bekkei	J	(219)

Table IIb (continued)

Structure number	Name of compounds	Formula	m.p. °C	$[\alpha]_D$	Plant source	Order	References	Comments
(478)	Humulenyl acetate	$C_{17}H_{28}O_2$			Frullania sp.	J	(249)	
(479)	(+)-Bicyclo-humulenone	$C_{15}H_{24}O$			Plagiochila sciophila (= P. acanthophylla subsp. japonica)	J	(19)	
(480)	5-Hydroxyisobicyclo-humulenone	$C_{15}H_{24}O_2$		+ 163	Jubula japonica	J	(585b)	
(481)	Longifolene	$C_{15}H_{24}$			Mastigophora diclados	J	(621, 635)	
					Plagiochila moritziana	J	(513)	
					Scapania robusta	J	(621, 631)	
					S. subalpina	J	(289)	
					S. uliginosa	J	(289)	
					S. undulata	J	(289)	
(482)	Longiborneol	$C_{15}H_{26}O$			Chiloscyphus pallescens	J	(149)	
					Scapania undulata	J	(289, 434a)	
(483)	α-Longipinene	$C_{15}H_{24}$			Scapania undulata	J	(19)	
(484)	β-Longipinene	$C_{15}H_{24}$			Scapania undulata	J	(434a)	
(485)	Longipinanol	$C_{15}H_{26}O$		− 49.4	Scapania undulata	J	(289, 434a)	
(487)	Isolongifolene	$C_{15}H_{24}$			Scapania subalpina	J	(289)	
					S. uliginosa	J	(289)	
					S. undulata	J	(434a)	
(488)	Marsupellol	$C_{15}H_{24}O$			Marsupella emarginata	J	(244, 434, 434a)	
(489)	Marsupellone	$C_{15}H_{22}O$			Marsupella emarginata	J	(244, 434, 434a)	
					Plagiochasma rupestre	M	(243)	
(490)	Acetoxymarsupellone	$C_{17}H_{24}O_3$			Marsupella emarginata	J	(434, 434a)	

No.	Name	Formula	mp	$[\alpha]_D$	Organism		Ref.	CD
(491)	ent-12β-Acetoxy-longipin-2(10)-en-3-one	$C_{17}H_{24}O_3$	95–6	−17.1	Marsupella aquatica	J	(279)	CD 197 nm − 0.82 CD 202 nm − 1.22 CD 221 nm 0 CD 240 nm − 0.76 CD 275 nm 0 CD 299 nm + 0.04 CD 310 nm + 0.08 CD 322 nm + 0.15 CD 335 nm + 0.21 CD 349.5 nm + 0.27 CD 365.5 nm + 0.13 CD 380 nm + 0.04 CD 395 nm 0
(492)	9,11α,14-Triacetoxy-marsupellone (= ent-9,11α,14-Triacetoxylongipin-2(10)-en-3-one)	$C_{21}H_{28}O_7$		− 65	Marsupella emarginata	J	(434, 434a)	CD 248 nm − 0.11 CD 325 nm + 0.14
(493)	9,11β,14-Triacetoxy-marsupellone (= ent-9,11β,14-Triacetoxylongipin-2(10)-en-3-one)	$C_{21}H_{28}O_7$		+ 45.5	Marsupella emarginata	J	(434, 434a)	CD 260 nm − 0.03 CD 326 nm + 0.12
(494)	9,14-Diacetoxy-marsupellone (= ent-9,14-Diacetoxylongipin-2(10)-en-3-one)	$C_{19}H_{26}O_5$		− 37.9	Marsupella emarginata	J	(434, 434a)	
(495)	β-Maaliene	$C_{15}H_{24}$			Frullania bicornistipula Monoclea forsteri	J Mo	(513) (514)	

Table IIb (continued)

Structure number	Name of compounds	Formula	m.p. °C	$[\alpha]_D$	Plant source	Order	References	Comments
(496)	γ-Maaliene	$C_{15}H_{24}$			M. gottschei subsp. neotropica	Mo	(514)	
(497)	Maaliol	$C_{15}H_{26}O$			Plagiochila geniculata	J	(513)	
					Radula boryana	J	(513)	
					Scapania ornithopodioides	J	(638)	
					Riccardia chamedryfolia	Me	(435)	
					Plagiochila dura	J	(46)	
					P. lecheri	J	(46)	
					P. moritziana	J	(513, 515)	
(498)	Maalian-5-ol	$C_{15}H_{26}O$			Lepidozia vitrea	J	(585a)	
(499)	(+)-ent-Maali-4(15)-en-1β-ol	$C_{15}H_{24}O$			Plagiochila ovalifolia	J	(249, 438)	
					Mylia taylorii	J	(386, 531)	
(500)	Striatene	$C_{15}H_{24}$			Archilejeunea mariana	J	(219)	
					Bryopteris diffusa	J	(219)	
					Cheilolejeunea excisula	J	(219)	
					C. imbricata	J	(219)	
					Diplasiolejeunea patelligera	J	(219)	
					Lejeunea lumbricoides	J	(219)	
					Leucolejeunea aff. decurrens	J	(219)	
				+72.7	Ptychanthus striatus	J	(540)	
					Mastigolejeunea humilis	J	(219)	

No.	Compound	Formula	mp	$[\alpha]$	Species		Ref.
					Porella densifolia subsp. appendiculata	J	(67)
					P. densifolia var. fallax	J	(67)
					Schiffneriolejeunea nymannii	J	(219)
(501a)	Striatenic acid	$C_{15}H_{22}O_2$			S. omphalanthoides	J	(219)
(502)	Striatol	$C_{15}H_{26}O$			Thysananthus amazonicus	J	(219)
					T. fruticosus	J	(219)
					Cheilolejeunea trifaria	J	(249)
					Lejeunea discreta	J	(219)
					Leucolejeunea aff. decurrens	J	(219)
				$+49.5$	Porella densifolia subsp. appendiculata	J	(67)
					P. densifolia var. fallax	J	(67)
					Ptychanthus striatus	J	(540)
					Schiffneriolejeunea nymannii	J	(219)
					S. omphalanthoides	J	(219)
					Thysananthus amazonicus	J	(219)
(503)	β-Monocyclo-nerolidol	$C_{15}H_{26}O$		$+3.2$	Cheilolejeunea excisula	J	(219)
					Ptychanthus striatus	J	(540)
					Spruceanthus polymorphus	J	(438)
(504)	trans-γ-Monocyclo-farnesol	$C_{15}H_{26}O$		$+18.2$	Diplophyllum serrulatum	J	(585a)
(505)	Striatenone	$C_{15}H_{24}O$		-9.1	Porella cordaeana	J	(593)
					Porella navicularis	J	(591)
(506)	Ricciocarpin A	$C_{15}H_{20}O_3$	110–111	$+17.8$	Ricciocarpos natans	M	(639, 640)
(507)	Ricciocarpin B	$C_{15}H_{20}O_4$	160–1	$+6.3$	Ricciocarpos natans	M	(639, 640)
(508)	Ricciofuranol	$C_{15}H_{22}O_3$		$+16.2$ CD 319 nm $+1.6$	Ricciocarpos natans	M	(639, 640)

Table IIb (continued)

Structure number	Name of compounds	Formula	m.p. °C	$[\alpha]_D$	Plant source	Order	References	Comments
(511)	(−)-Tridensenal	$C_{15}H_{26}O$		−35.8	Bazzania tridens	J	(623, 628)	
(512)	(−)-Tridensone	$C_{15}H_{26}O$		−15.1	Bazzania tridens	J	(629)	
(515)	Abscisic acid	$C_{15}H_{20}O_4$			Conocephalum conicum	M	(442a)	
					Marchantia polymorpha	M	(442a)	
(516)	Myltaylenol (= Myltayl-4(12)-en-15-ol)	$C_{15}H_{24}O$	69–70.5	−59	Mylia taylorii	J	(386, 533)	
(517)	Cyclomyltaylenol	$C_{15}H_{24}O$		−1.9	Mylia taylorii	J	(386, 535)	
(518a)	Cyclomyltaylane-3-ol	$C_{15}H_{24}O$	78–9	−19.1	Bazzania japonica	J	(84, 599)	
(519)	Cyclomyltaylane-5-ol	$C_{15}H_{24}O$			Mannia subpilosa	M	(614)	
					Reboulia hemisphaerica	M	(614)	
(520)	Cyclomyltaylyl-3-caffeate	$C_{24}H_{30}O_4$			Bazzania japonica	J	(84, 599)	
(521b)	Cyclomyltaylane (= Tridensene)	$C_{15}H_{24}$			Bazzania tridens	J	(627)	
(523)	Tamariscol	$C_{15}H_{26}O$		−21.8	Frullania tamarisci subsp. asagrayana	J	(63)	
				−22.6	F. nepalensis	J	(63)	
				−21.8	F. tamarisci subsp. obscura	J	(63)	
				−21.5	F. tamarisci subsp. tamarisci	J	(63, 150)	
(527)	α-Patchoulene	$C_{15}H_{24}$			Mastigophora diclados	J	(621, 635)	
					Plagiochila panamensis	J	(513)	
					Scapania ornithopodioides	J	(622, 638)	

No.	Name	Formula	$[\alpha]_D$	Species		Suspension culture	References
(528)	β-Patchoulene	$C_{15}H_{24}$		Bazzania fauriana	J		(622)
(529)	α-Pinguisene	$C_{15}H_{24}$		Porella elegantula	J		(189)
(530)	Pinguisenol	$C_{15}H_{26}O$		Porella acutifolia subsp. tosana	J		(600)
(531)	7-Keto-8-carbomethoxy-pinguisenol	$C_{16}H_{24}O_4$	+ 15.7	Porella acutifolia subsp. tosana	J		(600)
(532)	Naviculol	$C_{15}H_{26}O$	+ 48.5	Frullanoides densifolia	J		(14, 554–556)
				Porella navicularis	J		(591)
(533)	Isonaviculol	$C_{15}H_{26}O$	− 17.9	Frullanoides densifolia	J		(14, 554–556)
				Porella cordaeana	J		(593)
(535)	Porellapinguisenone	$C_{15}H_{22}O_3$		Plagiochila alternans	J		(433)
(536)	Deoxopinguisone	$C_{15}H_{22}O$		P. rosariensis	J		(48)
				Porella densifolia subsp. appendiculata	J		(67)
(538)	Dehydropinguisone	$C_{15}H_{18}O_2$	− 209.2	P. vernicosa	J		(458)
				Ptychanthus striatus	J		(540)
(539)	Dehydrodeoxo-pinguisone (= Pinguisenene)	$C_{15}H_{20}O$		Plagiochila retrospectans	J		(435a)
				Plagiochila rosariensis	J		(48)
				Ptychanthus striatus	J		(540)
				Thysananthus fruticosus	J		(219)
(540)	Furanopinguisanol (= 7α-Hydroxydeoxo-pinguisone)	$C_{15}H_{22}O_2$	− 1.96	Trocholejeunea sandvicensis	J		(44, 554, 556)
(541)	Dehydropinguisenol	$C_{15}H_{20}O_2$		Acrolejeunea pusilla	J		(219)
				Trocholejeunea sandvicensis	J		(44, 88, 554, 556)
(542)	Dehydropinguisenol methyl ether (= 7-Methoxy-dehydro-pinguisenene)	$C_{16}H_{22}O_2$	+ 82.3	Trocholejeunea sandvicensis	J		(44, 554, 556)

Table IIb *(continued)*

Structure number	Name of compounds	Formula	m.p. °C	$[\alpha]_D$	Plant source	Order	References	Comments
(543)	2-Hydroxy-pinguisenene	$C_{15}H_{20}O_2$			*Porella platyphylla*	J	*(146)*	
(544)	2-Hydroxy-7-methoxy-deoxopinguisone	$C_{16}H_{24}O_3$			*Porella platyphylla*	J	*(146)*	
(545)	2-Hydroxy-11ζ-methoxypinguis-5(10),6-diene	$C_{16}H_{24}O_3$			*Porella platyphylla*	J	*(146)*	
(546)	Pinguisenene methyl ester (= Bryopterin A)	$C_{16}H_{20}O_3$		− 13	*Bryopteris filicina*	J	*(432a)*	
(547)	Deoxopinguisone methyl ester	$C_{16}H_{22}O_3$			*Porella vernicosa*	J	*(458)*	Suspension culture
(548)	4-Hydroxydeoxo-pinguisone-12,15-dimethyl ester (= Bryopterin C)	$C_{17}H_{22}O_6$		− 44	*Bryopteris filicina*	J	*(432a)*	
(549)	Deoxopinguisone-12,15-dimethyl ester (= Bryopterin B)	$C_{17}H_{22}O_5$		− 19	*Bryopteris filicina*	J	*(432a)*	
(550)	6α,11α-Dimethoxy-pinguis-5(10)-ene	$C_{17}H_{28}O_3$			*Trocholejeunea sandvicensis*	J	*(44, 554, 556)*	
(551)	6α,11β-Dimethoxy-pinguis-5(10)-ene	$C_{17}H_{28}O_3$			*Trocholejeunea sandvicensis*	J	*(14, 554, 556)*	

No.	Name	Formula	mp	$[\alpha]$	Species		Ref.
(552)	Ptychanolactone	$C_{15}H_{22}O_3$			Ptychanthus striatus	J	(249)
(554)	Pinguisanin	$C_{15}H_{20}O_2$			Acrolejeunea pusilla	J	(219)
					A. pycnoclada	J	(219)
					Frullanoides densifolia	J	(47)
					Neotrichocolea bissetii	J	(585a)
					Plagiochila rosariensis	J	(48)
					Porella cordaeana	J	(593)
					P. platyphylla	J	(40, 146)
					Trocholejeunea sandvicensis	J	(14, 82, 554, 556)
(555)	Dehydropinguisanin	$C_{15}H_{18}O_2$			Acrolejeunea pusilla	J	(219)
					A. pycnoclada	J	(219)
					A. torulosa	J	(219)
					Frullanoides densifolia	J	(47)
					Trocholejeuna sandvicensis	J	(14, 82, 554, 556)
(556)	Pinguisenal	$C_{15}H_{20}O_3$			Acrolejeuena pusilla	J	(219)
					A. pycnoclada	J	(219)
					Frullanoides densifolia	J	(47)
(557)	Porellapinguisanolide	$C_{15}H_{22}O_5$		+55.5	Porella cordaeana	J	(593)
(558)	Ptychanolide	$C_{15}H_{22}O_3$	143	+13.3	Frullanoides densifolia	J	(44, 554–556)
			143–4	+23.2	Ptychanthus striatus	J	(540)
				+18.9	Trocholejeunea sandvicensis	J	(14, 554–556)
(560)	Pinguisanolide	$C_{15}H_{20}O_4$			Acrolejeunea pycnoclada	J	(219)
					A. torulosa	J	(219)
					Frullanoides densifolia	J	(47)
					Ptychanthus striatus	J	(540)
					Neotrichocolea bissetii	J	(585a)

Table IIb (continued)

Structure number	Name of compounds	Formula	m.p. °C	$[\alpha]_D$	Plant source	Order	References	Comments
(562)	Isopinguisanolide	$C_{15}H_{20}O_4$			Acrolejeunea pusilla	J	(219)	
					A. pycnoclada	J	(219)	
					A. torulosa	J	(219)	
					Frullanoides densifolia	J	(47)	
					Porella platyphylla	J	(146)	
(563)	Spirodensifoliin A	$C_{17}H_{22}O_6$	193–5	−40.6	Frullanoides densifolia	J	(44, 554, 556)	
(564)	Spirodensifoliin B	$C_{15}H_{18}O_4$		−36.8	Frullanoides densifolia	J	(44, 554, 556)	
(565a)	Spiropinguisanin	$C_{15}H_{20}O_5$			Porella cordaeana	J	(593)	
(566)	Norpinguisone	$C_{14}H_{18}O_2$			Porella densifolia subsp. appendiculata	J	(67)	
					P. navicularis	J	(591)	
					P. vernicosa	J	(458)	Suspension culture
(567a)	Norpinguisone methyl ester	$C_{15}H_{18}O_4$			Bryopteris filicina	J	(432a)	
					Lejeunea discreta	J	(219)	
					Porella cordaeana	J	(593)	
					P. densifolia subsp. appendiculata	J	(67)	
				−60.2	P. elegantula	J	(189)	
					P. navicularis	J	(591)	
					P. vernicosa	J	(458)	
(568)	Bryopterin D	$C_{15}H_{16}O_5$			Bryopteris filicina	J	(432a)	
(569)	Norpinguisanolide	$C_{14}H_{14}O_4$			Porella cordaeana	J	(239a)	
				−125	P. elegantula	J	(189)	
(580)	β-Santalene	$C_{15}H_{24}$			Gackstroemia magellanica	J	(45)	
(581)	α-Santalan-12(R), 13-diol	$C_{15}H_{26}O_2$	89–90	−27	Porella caespitans var. setigera	J	(597, 598)	

No.	Compound	Formula	mp	$[\alpha]$	Species		Ref.	CD
(582)	α-Spirovetivene	$C_{15}H_{24}$			*Scapania maxima*	J	(623)	CD 208 nm − 1.1
					S. robusta	J	(621, 623, 632)	
(583)	β-Spirovetivene	$C_{15}H_{24}$		− 12	*Scapania maxima*	J	(623)	
					S. robusta	J	(623)	
(588)	*ent*-Thujopsene	$C_{15}H_{24}$		+ 74.6	*Bazzania angustifolia*	J	(621, 622)	
					B. stolonifera	J	(513)	
					Heteroscyphus planus	J	(427a)	
(589)	*ent*-Thujopsenone	$C_{15}H_{22}O$		− 80	*Marchantia polymorpha*	M	(74, 382, 579)	
					Mastigophora diclados	J	(621, 635)	
					Plagiochila peculiaris	J	(621, 637)	
					Schistochila acuminata	J	(621, 630)	
(590)	*ent*-Thujopsan-7β-ol	$C_{15}H_{26}O$		± 0	*Marchantia polymorpha*	M	(74, 382)	
(592)	7α-Hydroxyvalenc-1(10)-ene	$C_{15}H_{26}O$		+ 75	*Marchantia polymorpha*	M	(71)	
					Bazzania fauriana	J	(581)	
(594)	Vitrenal	$C_{15}H_{22}O$		+ 107	*Lepidozia vitrea*	J	(389)	
(596)	(−)-Widdrol	$C_{15}H_{26}O$		− 91	*Marchantia polymorpha*	M	(74, 382)	
(597)	Saccogynol	$C_{15}H_{22}O$			*Saccogyna viticulosa*	J	(146)	
(600)	Peculiaroxide	$C_{15}H_{26}O$		+ 1.5	*Plagiochila peculiaris*	J	(269, 623, 631a)	
(601)	Conicumol	$C_{15}H_{26}O$		− 13.2	*Conocephalum conicum*	M	(553)	
(602a)	Omphalic acid	$C_{15}H_{22}O_2$			*Omphalanthus filiformis*	J	(441a, 571)	
(603a)	Riccardiphenol A	$C_{21}H_{28}O_2$		+ 143	*Riccardia crassa*	Me	(583)	
(604)	Riccardiphenol B	$C_{22}H_{30}O_2$		− 72	*Riccardia crassa*	Me	(583)	
(607a)	Serpentiphenol	$C_{16}H_{22}O_2$		± 0	*Cheilolejeunea serpentina*	J	(249)	
(609)	Trifarienol A	$C_{15}H_{26}O_2$	59–60	+ 10.2	*Cheilolejeunea trifaria*	J	(249)	
(610)	Trifarienol B	$C_{15}H_{26}O_2$	105–105.5	− 3.6	*Cheilolejeunea trifaria*	J	(249)	
(611)	Trifarienol C	$C_{17}H_{28}O_3$	99–100	+ 13.0	*Cheilolejeunea trifaria*	J	(249)	
(612)	Trifarienol D	$C_{17}H_{28}O_3$	83–5	− 0.2	*Cheilolejeunea trifaria*	J	(249)	
(613)	Trifarienol E	$C_{17}H_{24}O_2$		+ 2.4	*Cheilolejeunea trifaria*	J	(249)	
(614)	Rebouliadienol	$C_{15}H_{24}O$			*Reboulia hemisphaerica*	M	(249)	
(615)	Chenopodene	$C_{15}H_{24}$		+ 32.7	*Marchantia chenopoda*	M	(553)	CD 235 nm + 0.07

J: Jungermanniales; M: Marchantiales; Me: Metzgeriales; Mo: Monocleales

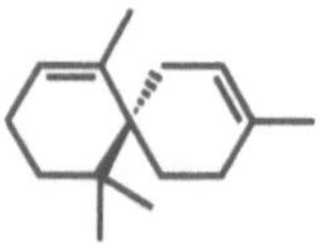

(263) α–Chamigrene 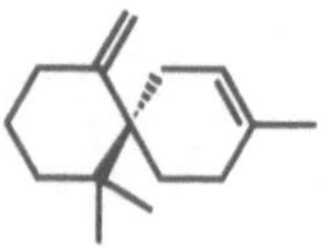   (264) β–Chamigrene 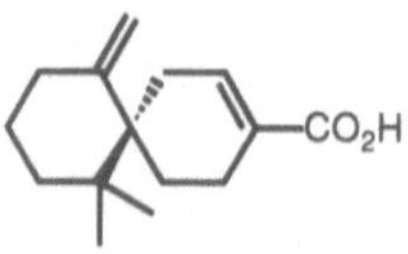   (265) *ent*-Chamigrenic acid

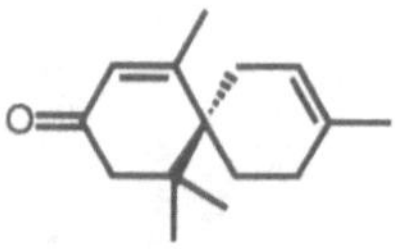

(266) *ent*-9-Oxo-α-chamigrene

Chart 17. Chamigranes found in the Hepaticae

been determined by a combination of NMR spectroscopy and the identity of its CD spectrum with that of synthetic (−)-*ent*-9-oxo-α-chamigrene (*213*). This is the first report of the isolation of an *ent*-oxo-α-chamigrene from natural sources.

5.14 Chiloscyphanes, Oppositanes, Copaanes and Cubebanes

Chiloscyphone (**267**) isolated from Japanese *Chiloscyphus polyanthos* was originally presumed to be a cadinane derivative (**272**) (*19*). However, because the properties of synthetic (**272**) were not identical with those of chiloscyphone, the structure originally assigned to chiloscyphone had to be in error (*222*). This was confirmed by further work (*149*). Extraction of Scottish *Chiloscyphus pallescens* gave an alcohol, chiloscypholone (**268**) as well as an eudesmene diol (**368**) and longiborneol (**482**) (*149*). Dehydration of chiloscypholone furnished chiloscyphone (**267**) and isochiloscyphone (**268a**) (Scheme 18). The structures assigned to (**267, 268, 268a**) which embody a new carbon skeleton were based mainly on ^{1}H- and ^{13}C-NMR spectrometry. Biogenetically eudesmene diol (**368**) accompanying the chiloscyphanes is an attractive candidate for ring contraction, methyl migration (1,2-shift) and dehydration processes which would lead to the chiloscyphane carbon skeleton with the correct stereochemistry indicated in Scheme 19 (*146, 149*).

Fractionation of the ether extract of Taiwanese *C. pallescens* resulted not only in the isolation of chiloscyphone (**267**) and chiloscypholone (**268**), but also in 11ζ,12-epoxychiloscypholone (**269**) and *ent*-(5R,6S,9R)-4α-hydroxyoppositan-10-one (**270**) (*241*). Structure (**269**) was confirmed

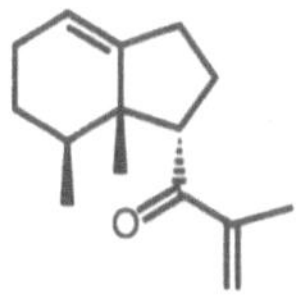

(267) Chiloscyphone

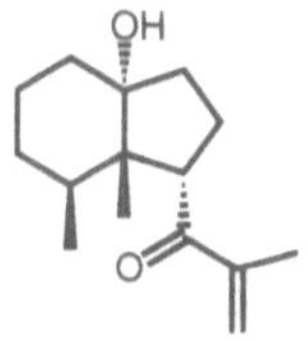

(268) Chiloscypholone

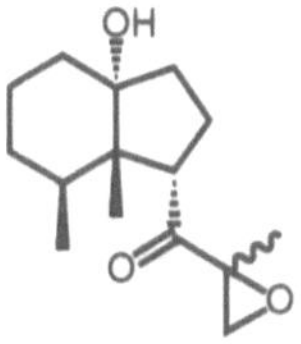

(269) 11ζ,12-Epoxychiloscypholone

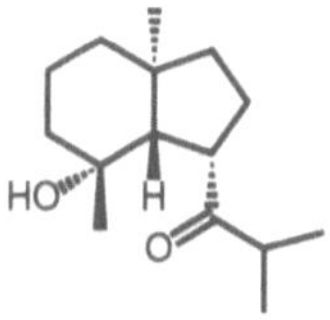

(270) ent-(5R,6S,9R)-4α-Hydroxyoppositan-10-one

Chart 18. Chiloscyphanes and oppositane found in the Hepaticae

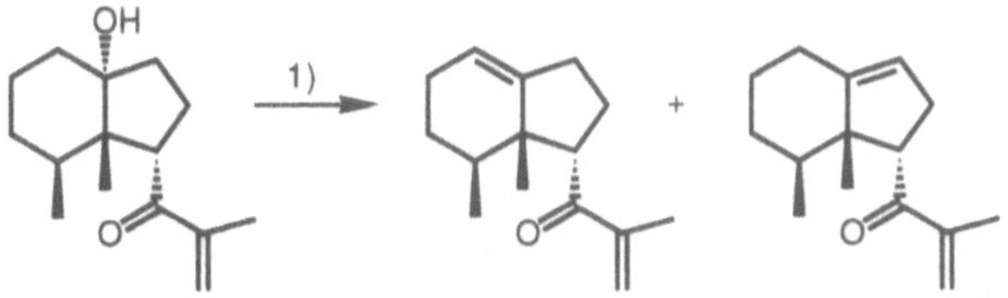

(268)

(267) Chiloscyphone (268a) Isochiloscyphone

1) SOCl₂/Py, -20°

Scheme 18. Formation of chiloscyphone and isochiloscyphone from chiloscypholone

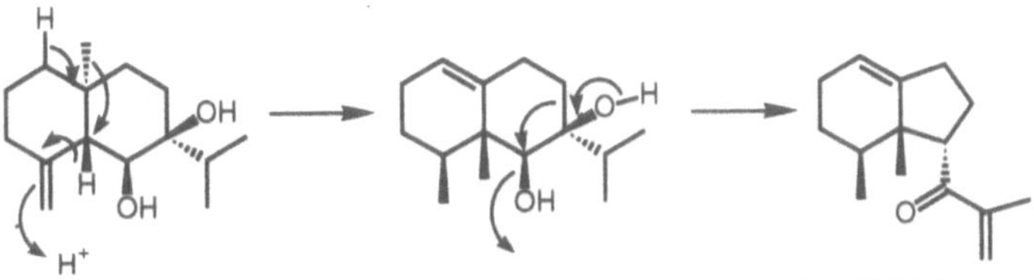

(368)

(273) Eremophilane

(267) (-)-Chiloscyphone

Scheme 19. Possible biogenetic pathway for chiloscyphone

by epoxidation of (268) the major epoxide being identical in all respects
with the natural product. The oppositane structure for (270) was proved
by analysis of the ^{1}H-, ^{13}C-NMR and ^{1}H-^{1}H-2D-COSY spectra and by
NOE spectrometry. The co-occurrence of oppositane- and chiloscy-
phane-type sesquiterpenoids supports the biogenetic scheme mentioned
in the previous paragraph. A very similar *ent*-oppositane-type brominated sesquiterpene alcohol (271) has been isolated from the red algae,
Laurencia subopposita (151).

(271) Oppositol   (272)

The revised structure of chiloscyphone (267) has been confirmed by
two total syntheses of (±)-chiloscyphone (204, 220, 221 and 551, 557).
    The absolute configurations of (−)-chiloscyphone (267) and (+)-
chiloscypholone (268) have also been determined as a result of total
syntheses of the optically active compounds by the route outlined in
Scheme 20, the key intermediate alcohol having been resolved by use of
chiroptical (1S)-(−)-camphanic chloride (558, 559). Racemic ketone
(276a + 276b) was reduced with NaBH$_4$ to give isomeric alcohols (274,
277) which were separated by silica gel chromatography, followed by
esterification with (1S)-(−)-camphanic chloride to furnish a mixture of
diastereoisomers (275a, 275b, 278a, 278b). Each isomer was separated by
HPLC, hydrolyzed and finally oxidized to give two ketones, (+)-(276a)
and (−)-(276b), respectively. The CD spectrum of (+)-(276a) exhibited a
positive Cotton effect. Alcohol (+)-(276a) was dehydrated to give only
(+)-(279) which subjected to the same reactions as before (551, 557)
afforded (−)-chiloscyphone (267). The absolute configuration of (+)-
chiloscypholone (268) then followed from its conversion to (−)-
chiloscyphone (267) (149).
    However, it turned out that the absolute configuration of (+)-(276a)
was misassigned, leading to the incorrect absolute stereochemistry for
(−)-chiloscyphone and (+)-chiloscyphone. An X-ray crystallographic
analysis of the ω-camphanic ester (275a) (558, 559) clearly showed that
(275a) has the 1S,2R,5R,6S,7R configuration, since the absolute configurations of the camphanic ester part are known. Since the results of the X-
ray analysis of (275a) are unambiguous and (−)-chiloscyphone has been

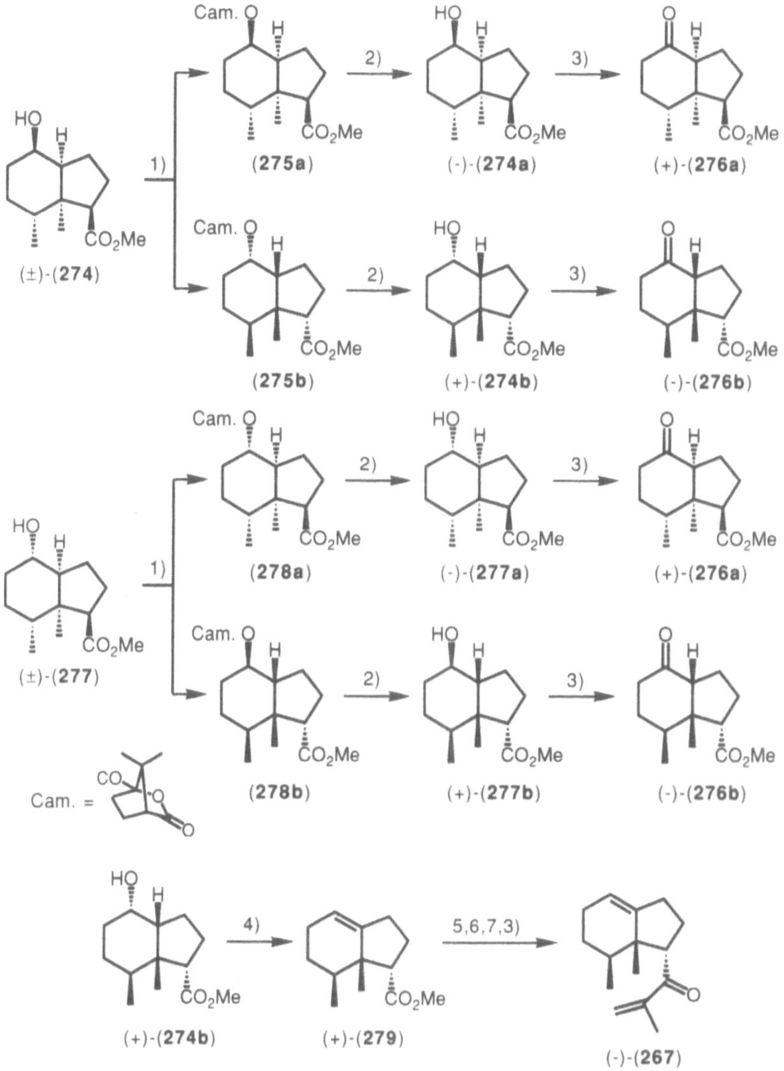

1) (1*S*)-(-)-Camphanic chloride/CH₂Cl₂-Py/DMAP 2) KOH/MeOH 3) Jones oxd.
4) POCl₃/Py 5) LiAlH₄ 6) Swern oxd. 7) CH₂=CH(Me)MgBr

Scheme 20. Total synthesis of (−)-chiloscyphone (*558, 559*)

synthesized from the diastereoisomeric (275b), the absolute configuration
of (267) and (268) should be revised as shown in Chart 18 and Scheme 20
(552, 560).

As noted earlier (19) α-copaene (280) is widespread in *Frullania*
species. *Bazzania, Lophocolea, Chiloscyphus, Marchantia, Monoclea* and
Trichocolea species also elaborate α-copaene as shown in Table IIb. α-
Cubebene (281) has been detected in *Gymnocolea* and *Marchantia* and β-
cubebene (282) in *Herbertus, Lophocolea* and *Symphyogyna* species. β-
Cubebene has also been found in cell culture of *Heteroscyphus planus*
(427a) and *Jungermannia infusca* (457). (−)-α-Cubebene (281) and (−)-
β-cubebene (282) have been isolated from the brown algae, *Dictyopteris*
species (179). *Conocephalum conicum* contains cubebol (283) as a minor
component (249).

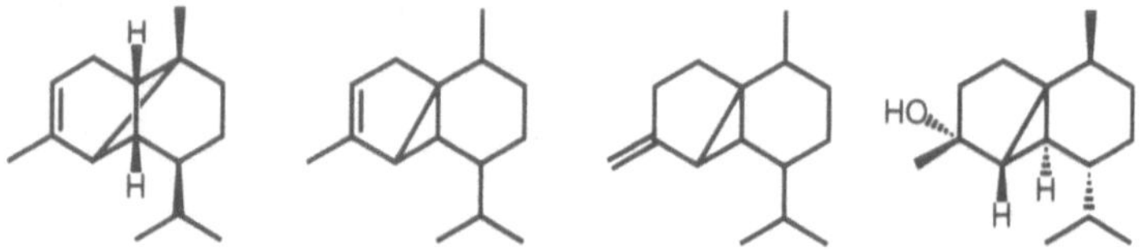

(280) α–Copaene   (281) α-Cubebene   (282) β-Cubebene   (283) Cubebol

Chart 19. Copaanes and cubebanes found in the Hepaticae

5.15 Cuparanes and Herbertanes (Isocuparanes)

The most common cuparane-type sesquiterpenoid is (−)-cuparene
(284). 2-Hydroxycuparene (285) is also widespread in stem-leafy and
thalloid liverworts as shown in Table IIb. *Herbertus* species are rich
sources of cuparane-type sesquiterpenoids. *H. aduncus* produces cuparene (284), 2-hydroxycuparene (285), 2,3-dihydroxycuparene (287) and α-
cuparenone (288). *H. subdentatus* contains not only (284–287) but also 3-
hydroxycuparene (286) (57). The ether extract of *Lepidozia concinna*
contains α-cuparenone (288) and the previously unknown α-cuparenol
(289) (37). Some *Bazzania, Plagiochila* and *Radula* species contain 2,3-
dihydroxycuparene (287). 1-Formyl-2,3-dihydroxycuparene (290) has
been detected in *Plagiochila bispinosa* and *P. subdura*, along with
isocuparane-type sesquiterpenoids (46).

(−)-2-(1,2,2-Trimethyl)-cyclopentyl-6-methyl-1,4-quinone   (= cupareno-quinone) (291) has been newly isolated from *Radula javanica*
(= *R. variabilis*) (49); it was prepared previously from 2-hydroxycuparene

(285) by chromic oxide (19) or m-chloroperbenzoic acid oxidation (59). Ent-α- (292), β- (293), γ- (294), δ-cuprenenes (= angustifolene) (295) and ε-cuprenene (296) have been isolated from liverworts. (−)-δ-Cuprenene (295) on treatment with 2,3-dichloro-5,6-dicyano-1,4-benzoquinone gives (−)-cuparene (284) (382). Both Japanese *Bazzania tricrenata* and *B. trilobata* produce an unknown sesquiterpene hydrocarbon whose mass spectrum had the base peak at m/z 69 and/or 111. Wu (621) suggested that this compound might be δ-cuprenene (295).

A new cuparene-type sesquiterpenoid, (R)-(−)-8,11-dihydro-α-cuparenone (297) has been isolated from the dichloromethane extract of southern French *Reboulia hemisphaerica*, together with the previously known (R)-(−)-α-cuparenone (288) (412). The ¹H-NMR spectrum suggested that it was the 8,11-dihydro derivative of (288), a suggestion which was confirmed by preparation of (288) from (297) by dehydrogenation with DDQ in benzene.

(−)-Cyclopropanecuparenol (298) and (−)-epicyclopropanecuparenol (299), two novel cuparane-type sesquiterpenoids, have been isolated from *Marchantia polymorpha* and *M. paleacea* var. *diptera* (74, 579). The gross structure of (298) was elucidated using 2D-COSY techniques as well as by its treatment with p-toluenesulfonic acid which gave three ent-cuparene-type sesquiterpene hydrocarbons, α-(292) and β-cuprenenes (293) and cuparene (284), indicating that the stereochemistry at C-1 was S. The relative stereochemistry of the tertiary hydroxyl group and the cyclopropane ring of (298) is based on NOE spectrometry. The absolute configuration of the six-membered ring of (298) and (299) remains to be established.

(+)-Cuprenenol (300) has been isolated from *Jungermannia rosulans* (19) while (−)-ent-cuprenenol (301) has been isolated from *Marchantia polymorpha* (74). A new rearranged cuparene-type sesauiterpenoid (302) has been obtained from *Demotarisia linguifolia* (431). The spectral data closely resembled those of the cuprenenols (300, 301), suggesting that it was a cuparenenol-like sesquiterpenoid. Dehydration of (302) with POCl₃ gave an unsaturated hydrocarbon (302a) (Scheme 21). ¹H- and ¹³C-NMR spectrometry suggested that the gross structure of (302) was that of a rearranged cuparene-type sesquiterpene alcohol. The mass fragment ions shown in Scheme 21 support this assumption. No useful information was obtained from NOE difference spectrometry.

Ricciocarpos natans is the only species in Ricciaceae containing oil bodies. It produces not only cuparene (284), but also two cuparene-type lactones, cuparenolide (303) and cuparenolidol (304) (639, 640). The relative configurations were established by NOE spectrometry. The δ-lactone and the cyclohexane ring were found to be trans-fused while the

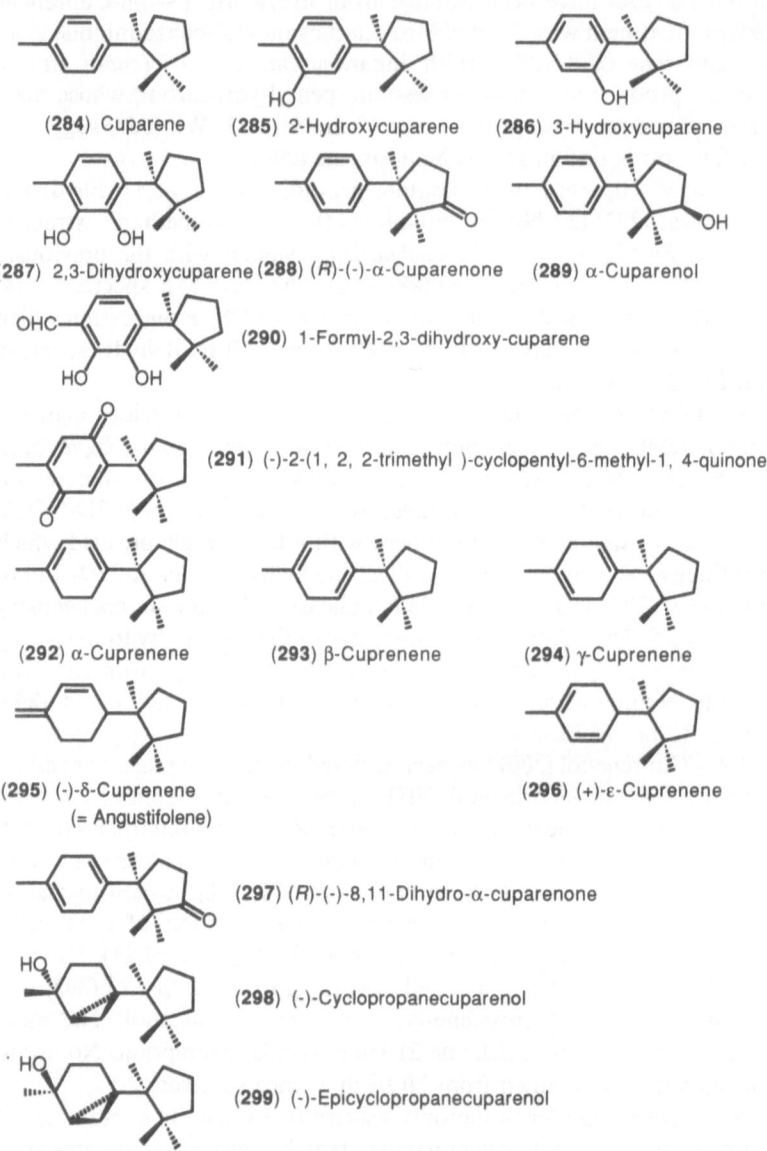

(**284**) Cuparene   (**285**) 2-Hydroxycuparene   (**286**) 3-Hydroxycuparene

(**287**) 2,3-Dihydroxycuparene (**288**) (*R*)-(-)-α-Cuparenone   (**289**) α-Cuparenol

(**290**) 1-Formyl-2,3-dihydroxy-cuparene

(**291**) (-)-2-(1, 2, 2-trimethyl)-cyclopentyl-6-methyl-1, 4-quinone

(**292**) α-Cuprenene   (**293**) β-Cuprenene   (**294**) γ-Cuprenene

(**295**) (-)-δ-Cuprenene   (**296**) (+)-ε-Cuprenene
   (= Angustifolene)

(**297**) (*R*)-(-)-8,11-Dihydro-α-cuparenone

(**298**) (-)-Cyclopropanecuparenol

(**299**) (-)-Epicyclopropanecuparenol

Chart 20a. Cuparanes found in the Hepaticae

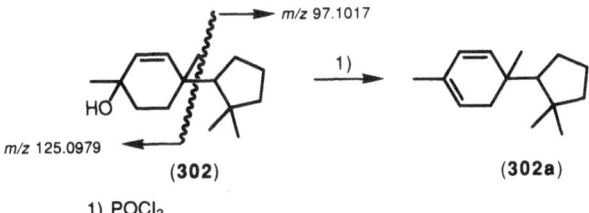

Scheme 21. Mass fragmentation of rearranged cuparene-type sesquiterpenoid

cyclopentane and the lactone ring are *cis*-fused. The absolute configuration remains to be clarified. This is the first report of terpenoids in the whole family of the Ricciaceae.

In addition to the above cuparenes, a new odoriferous cuparene-type ketone, grimaldone (305), has been isolated from *Mannia fragrans*; its relative and absolute stereochemistry is based on an X-ray crystallographic analysis and the CD spectrum (276). *Nardia scalaris* produces two diastereomeric 3,6-peroxy-cupar-1-enes (306), together with 2-hydroxycuparene (285) (354a).

Total syntheses of (±)-cuparene (284) and the non-natural (±)-β-cuparenone have been achieved by ISHIBASHI et al. (297) in 2 and 7 steps, respectively. (+)-β-Cuparenone has also been synthesized by TAKANO et al. (530).

Herbertus (Herbertaceae) and *Mastigophora* (Lepicoleaceae) are rich sources of herbertane (= isocuparane)-type sesquiterpenoids. South American *Plagiochila bispinosa* and *P. subdura* (46), and *Marchantia polymorpha* also contains isocuparane-type sesquiterpenoids (74, 382).

(−)-Herbertene (= isocuparene) [(1S)-1,2,2-trimethyl-*m*-tolyl-cyclopentane] (307), (−)-α-herbertenol (= 5-hydroxyisocuparene) (308), (−)-β-herbertenol (= 3-hydroxyisocuparene) (309), (−)-α-formyl-herbertenol (= 2-formyl-5-hydroxyisocuparene) (310), (−)-herbertenediol (= 4,5-dihydroxyisocuparene) (311) and (−)-herbertenolide (312) have been isolated from *Herbertus aduncus* (57, 392–395). *H. sakuraii* produces the latter lactone (623a).

The mass and NMR spectra of (307) containing a 1,3-disubstituted benzene nucleus are very similar to those of cuparene (284), formation of (dimethyl isophthalate) by oxidation of (307) with dilute nitric acid and methylation supported the relative locations of the methyl and cyclopentyl groups on the aromatic ring. While the structure and absolute configuration of the cyclopentyl group was established by ozonolysis of

(300) (+)-Cuprenenol 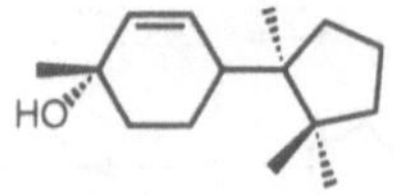   (301) ent-Cuprenenol 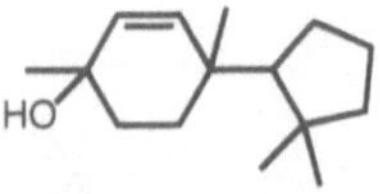   (302) Demotarisiol

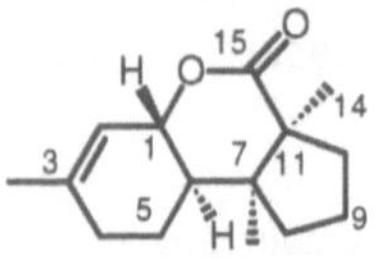

(303) Cuparenolide 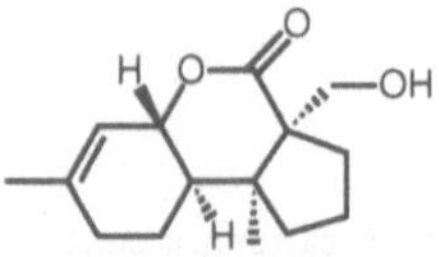   (304) Cuparenolidol 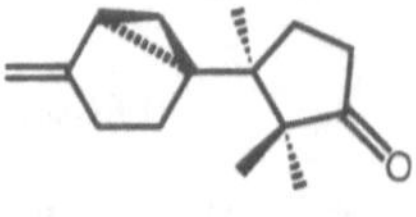   (305) Grimaldone

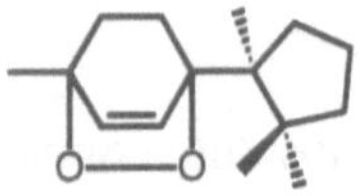

(306) 3,6-Peroxocupar-1-ene

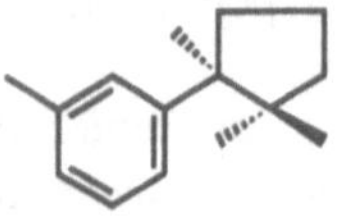

(307) Herbertene (= (1S)-1,2,2-Trimethyl-m-tolylcyclopentane)

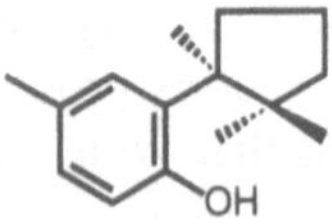

(308) α-Herbertenol 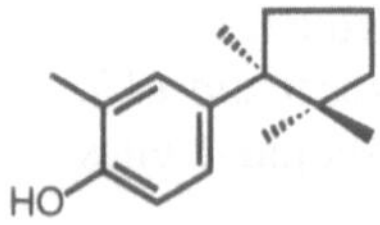   (309) (-)-β-Herbertenol 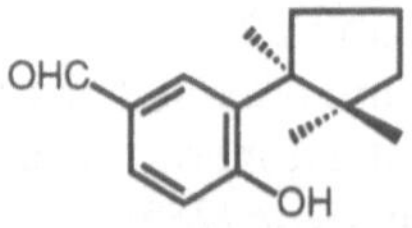   (310) (-)-α-Formylherbertenol

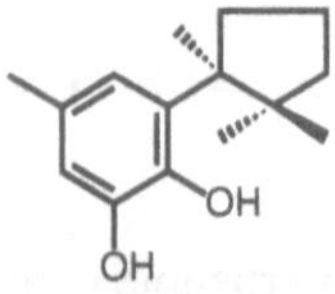

(311) (-)-Herbertenediol 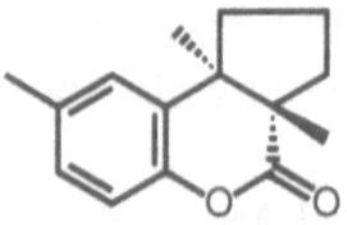   (312) (-)-Herbertenolide 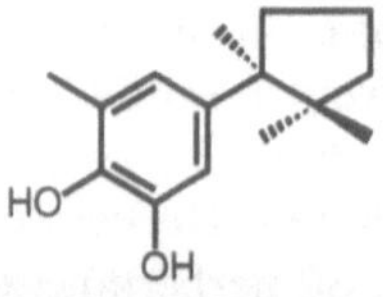   (313) Isocuparene-3,4-diol

Chart 20b. Cuparanes and herbertanes (isocuparanes) found in the Hepaticae

(307) which gave (−)-camphanic acid. Structure and absolute configuration of (308) were similarly settled by combination of ^{1}H- and ^{13}C-NMR spectrometry and chemical degradation to 1,2,4-trisubstituted benzene derivatives, and (−)-camphanic acid as well as by comparison of the spectral data with those of 2-hydroxycuparene (285). Structure (309) was deduced by analysis of the coupling pattern of the aromatic protons and its production from (−)-camphonyl bromide by reaction with butyllithium-nitrobenzene.

The structure of a third compound (310) is based on a chemical correlation with (308). The methylation of (−)-α-herbertenol with MeI followed by MnO$_2$ oxidation produced an aromatic aldehyde which was

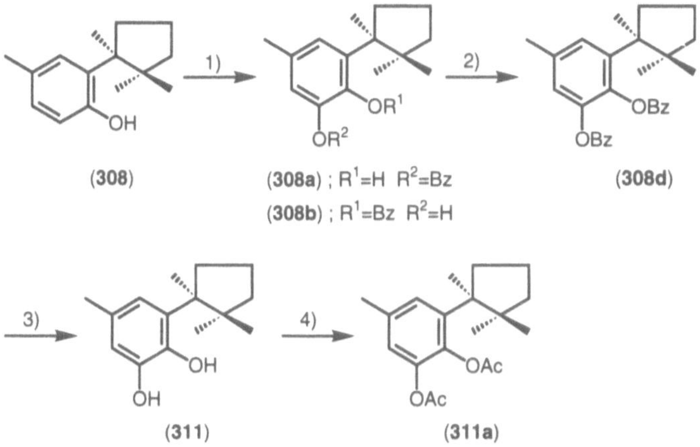

(308)   (308a) ; R¹=H R²=Bz   (308d)

(308b) ; R¹=Bz R²=H

(311)   (311a)

1) (PhCOO)₂/C₆H₆ 2) BzCl/Py 3) LiAlH₄/Et₂O 4) Ac₂O/Py

Scheme 22a. Reactions of herbertenediol

demethylated with boron tribromide to give a substance identical with (310) while the structure of diol (311) was solved by the chemical correlation outlined in Scheme 22a. (−)-α-Herbertenol (308) was benzoyloxylated with benzoyl peroxide to afford a mixture of benzoyloxyphenols (1:1) which was esterified with benzoyl chloride in pyridine to give a single dibenzoate (308d). Further reduction with LiAlH₄ furnished a diol whose optical rotation and spectral data were in good agreement with those of natural (−)-herbertenediol (311).

The structure of lactone (312) was also solved by correlation with α-herbertenol (308) (Scheme 22b). Reduction of (312) with LiAlH₄ gave the hydroxyphenol (312a), which was converted to a monomethyl ether (312b). Reduction of (312a) and (312b) with triphenylphosphite-methiodide [(PhO)₃PMeI] in hexamethylphosphoric acid (HMPA) and reaction of the resulting mixtures with sodium cyanoborohydride afforded a cyclic ether (312c) which was more easily obtained from (312a) with (PhO)₃PMeI in HMPA or p-toluenesulfonic acid in benzene under reflux. When (312c) was treated with ethanethiol and AlCl₃, a hydroxythioether (312d) was obtained in good yield whose reduction with Raney-Ni gave α-herbertenol (308). The stereochemistry of the lactone ring is based on the chemical shifts of the tertiary methyl signals in (312a) and (312b). In each case one of the signals resonates at high field and exhibits a shift similar to one of the methyl signals of (308) and (308c) (Scheme 22b). The methyl groups responsible for these signals are presumably β-oriented and shielded by the anisotropic effect of the benzene ring.

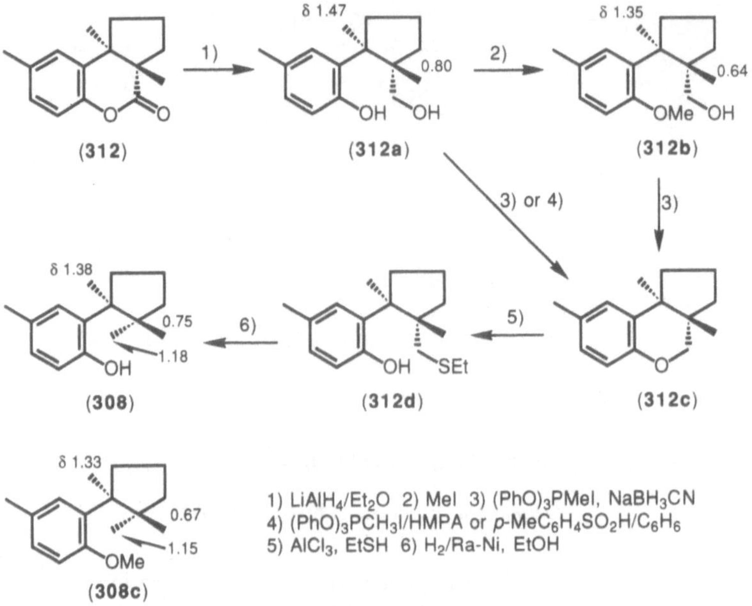

Scheme 22b. Correlation of cuparenolide with α-herbertenol

H. aduncus produces not only herbertane-type but also cuparane-type sesquiterpenoids such as (−)-cuparene (**284**), 2-hydroxycuparene (**285**), 2,3-dihydroxycuparene (**287**) and α-cuparenone (**288**). *Herbertus subdentatus* (*57, 433*), *H. sakuraii* (*57*), *H. divergens* (*513*) and *H. acanthelius* (*433*) also elaborate cuparane- and herbertane-type sesquiterpenoids. (−)-β-Herbertenol (**309**) has been isolated from *Marchantia polymorpha* (*382*). *Plagiochila bispinosa* and *P. subdura* elaborate α-herbertenol (**308**) and herbertenediol (**311**), respectively (*46*). α-Formylherbertenol (**310**) has been found in *Triandrophyllum subtrifidum* var. *trifidum* (*45*).

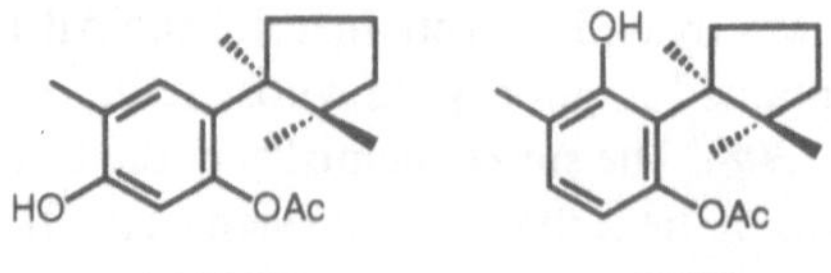

Similar herbertane-type sesquiterpenoids (**314, 315**) have been found in the pathogenic fungus *Helicobasidium mompa* (*309*).

Herbertene (**307**) was unexpectedly obtained by reaction of a dri-mane-type alcohol, *trans*-didehydrobicyclofarnesol (**317**), with formic acid in the presence of perchloric acid (Scheme 23) (*182*). The mechanistic interpretation of the rearrangement of (**317**) to (**307**) is uncertain. Three cyclopropylcarbinyl ions (Scheme 23) are possible intermediates for the formation of herbertene (**307**) and two other such species may be responsible for cuparanes as shown in the scheme. Among these isomers, (**317b**) from which (**307**) would be formed, seems to be the most stable. In order to clarify the mechanism for the rearrangement, the following experiments were performed (*183*). Optically active [1,9- ^{13}C]- and [4a-^{13}C]-*trans*-didehydrobicyclofarnesol (**317**) were synthesized. Acid treatment of 1,9-labeled (**317**) (Scheme 24) gave 2,4-labeled (**307**) and (**318**) in which the labels were randomized over C-5, C-6, C-7 and C-8. Acid treatment of 4a-labeled (**317**) gave 1-labeled (**307**) and (**318**) in which the label was randomized over C-1 and C-10. Solvolysis of the very labile tosylate (**319**) with AcOH-AcONa gave an acetate (**320**) which on treatment with formic acid in the presence of a catalytic amount of HClO$_4$ furnished (**307**) and (**318**) in the ratio 3.5:1. These results suggest that cation (**317d**) corresponding to (**320**) might be a common intermediate for both (**307**) and (**318**). The proposed mechanism for formation of herbertene (**307**) from (**317**) is shown in Scheme 25.

The formal total synthesis of ($\pm$)-herbertene (**307**) was accomplished by BANERJEE *et al.* (*90*) while the first enantio-controlled synthesis of ($-$)-herbertene (**307**) has been accomplished by TAKANO *et al.* (*530*).

The ethyl acetate extract of the rather primitive liverwort *Mastigophora diclados* contains cuparene (**284**), herbertene (**307**), α- (**308**) and β-herbertenols (**309**) as well as herbertenediol (**311**) (*134, 622, 626*). The same compounds have also been found in *Mastigophora woodsii* (*134, 626*). Further fractionation of the ether extract of Malaysian *Mastigophora diclados* resulted in the isolation of four novel isocuparane-type sesquiterpene dimers, mastigophorenes A (**321**), B (**322**), C (**323a**) and D (**324a**), along with a new isocuparane-type monomer, isocuparene-3,4-diol (**313**), and the known β-herbertenenol (**309**) and ($-$)-herbertenediol (**311**) (*51, 186, 191*).

The dimers A (**321**) and B (**322**) have the same molecular formula C$_{20}$H$_{42}$O$_4$ while the IR and UV spectra showed the presence of a hydroxyl group and a benzene ring. The ^{1}H- and ^{13}C-NMR spectra were almost identical. The NMR spectra were closely related to those of herbertenediol (**311**) except that one of the *meta*-coupled aromatic proton signals of (**311**) was missing. Also the aromatic doublet at δ 113.4 of (**311**) was replaced by an aromatic singlet at δ 117.1 and 117.0 in (**321**) and (**322**), respectively. These spectral data were compatible with symmetrical

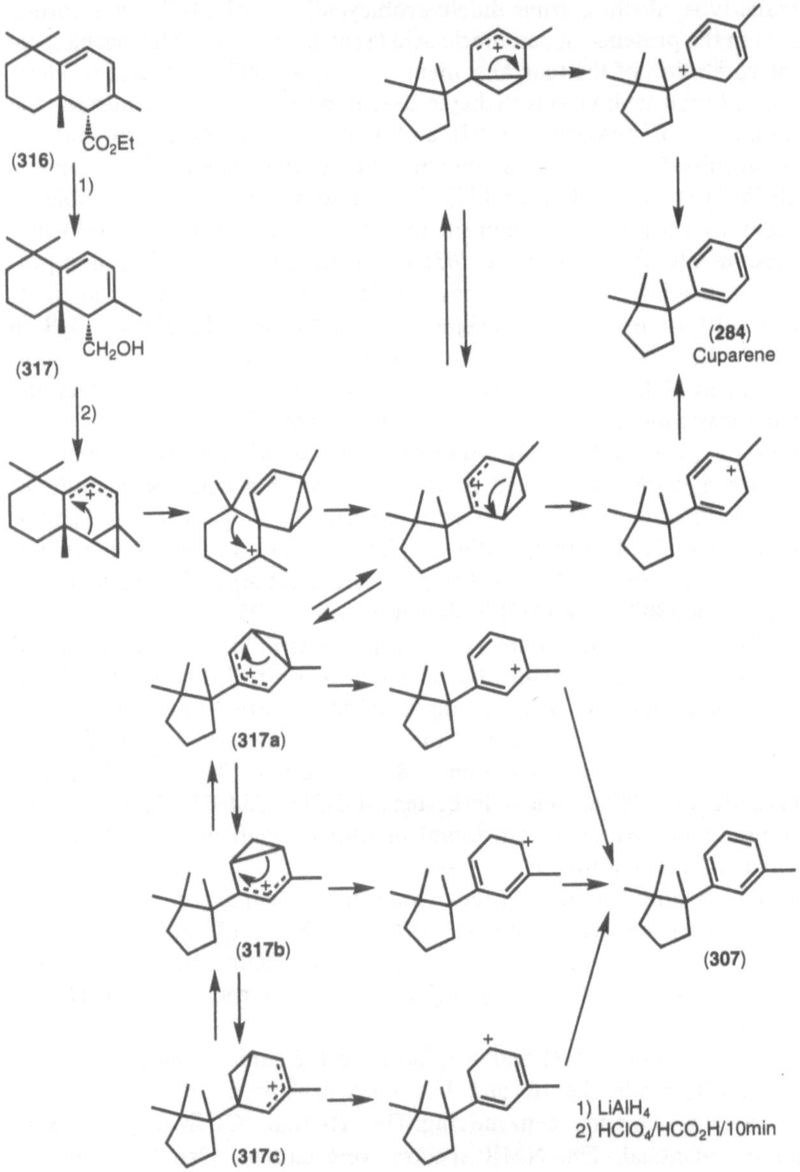
CO2Et
(316)
1)
CH2OH
(317)
2)
(284)
Cuparene
(317a)
(317b)
(307)
(317c)
1) LiAlH4
2) HClO4/HCO2H/10min

Scheme 23. Formation of herbertene from *trans*-didehydrobicyclofarnesol

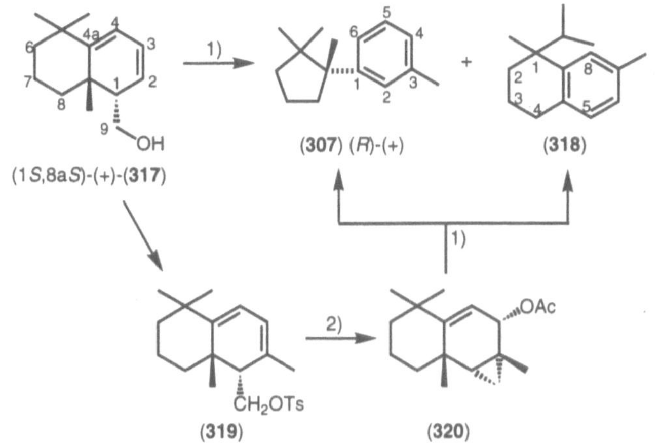

1) HCO$_2$H/cat. HClO$_4$, reflux 2) AcOH/AcONa

Scheme 24. Acid-catalyzed rearrangement of *trans*-didehydrobicyclofarnesol

• or • = ^{13}C

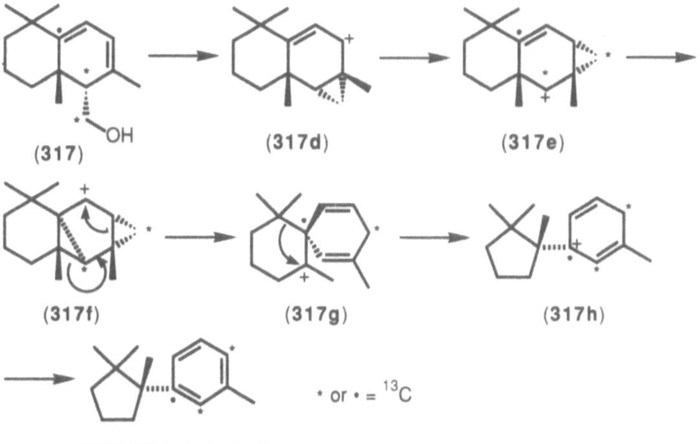

(**307**) (*R*)-(+)-Herbertene

Scheme 25. Proposed mechanism for formation of herbertene from *trans*-didehydrobicyclofarnesol

dimers of herbertenediol (**311**), presumably linked through an aryl-aryl bond at the C-1 or C-3 position of (**311**). The slight spectral differences between (**321**) and (**322**) could be rationalized as being due to the diastereomeric environment caused by the existence of atropisomers with respect to the biaryl bond. The aryl-aryl bonds in (**321**) and (**322**) were

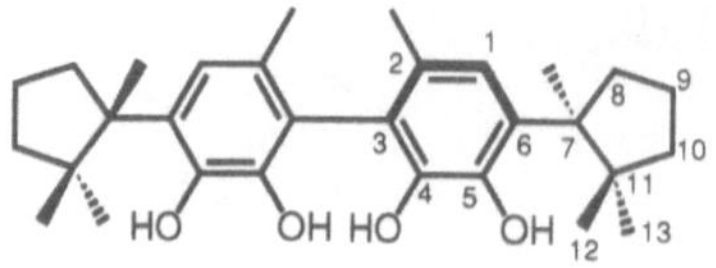

(321) Mastigophorene A

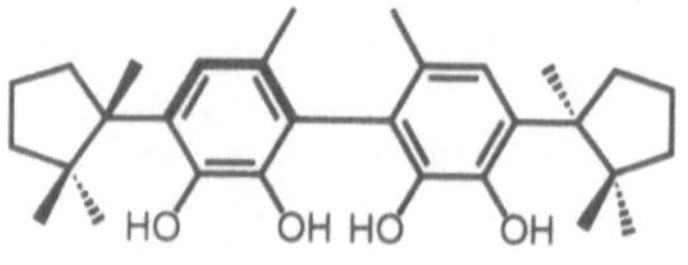

(322) Mastigophorene B

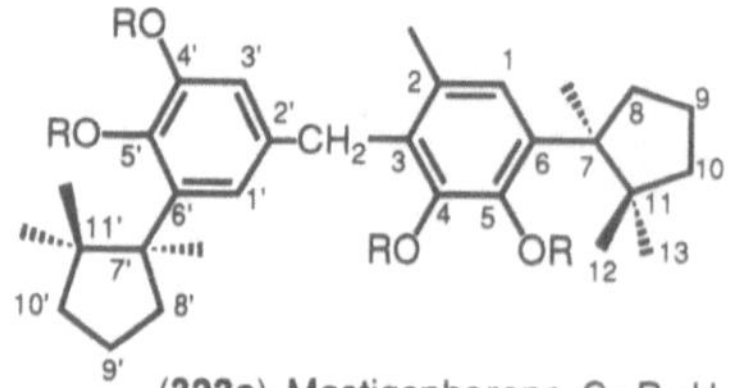

(323a) Mastigophorene C ; R=H
(323b) R=Ac

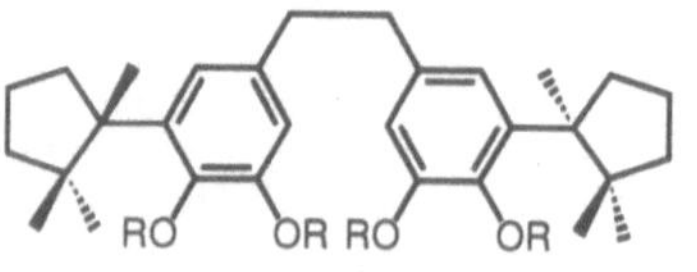

(324a) Mastigophorene D ; R=H
(324b) R=Ac

Chart 20c. Herbertanes (isocuparanes) found in the Hepaticae and their derivatives

presumably formed between the C-3 positions of the two molecules of (311) because the chemical shifts of the quaternary carbons involved in the biphenyl bond were observed at relatively high field, near δ 117. In fact, the sole aromatic proton at δ 6.86 resp. 6.85 showed an NOE interaction with the H-14 methyl signal. The tentative structures (321) and (322) were also supported by long range ^{13}C-^{1}H-2D-COSY experiments. On the basis of the spectral data, the mastigophorenes A and B are therefore atropisomers at the biphenyl axis linked at the C-3 positions of the two molecules of (311).

The absolute configurations at the aryl-aryl axes of (321) and (322) were established by the CD exciton chirality rule (*237*). The CD spectrum of (321) showed the first positive Cotton effect at 222nm and the second negative Cotton effect at 202 nm, indicating (*S*)-configuration at the biaryl axis, whereas (322) had the (*R*)-configuration because of the first negative and second positive Cotton effects at 215 and 202 nm. Accordingly, the structures of the mastigophorenes A and B are established to be (*S*)-3,3'-biherbertenediol (321) and (*R*)-3,3'-biherbertenediol (322), respectively.

Structures of the mastigophorenes C (323a) and D (324a) were established by a combination of chemical reaction (acetylation) and extensive NMR spectroscopy including 2D-COSY and NOESY techniques. The molecular formula of isocuparene-3,4-diol (313) was identical with that of herbertenediol (311) and the ^{1}H NMR sepctrum contained the same spin systems as (311). The arrangement of substituents in the

benzene ring of (313) was established by the NOE difference spectrum
(186).

Biosynthesis of the mastigophorenes might involve phenoxy radicals
produced by one electron oxidation of (−)-herbertenediol (311) which is

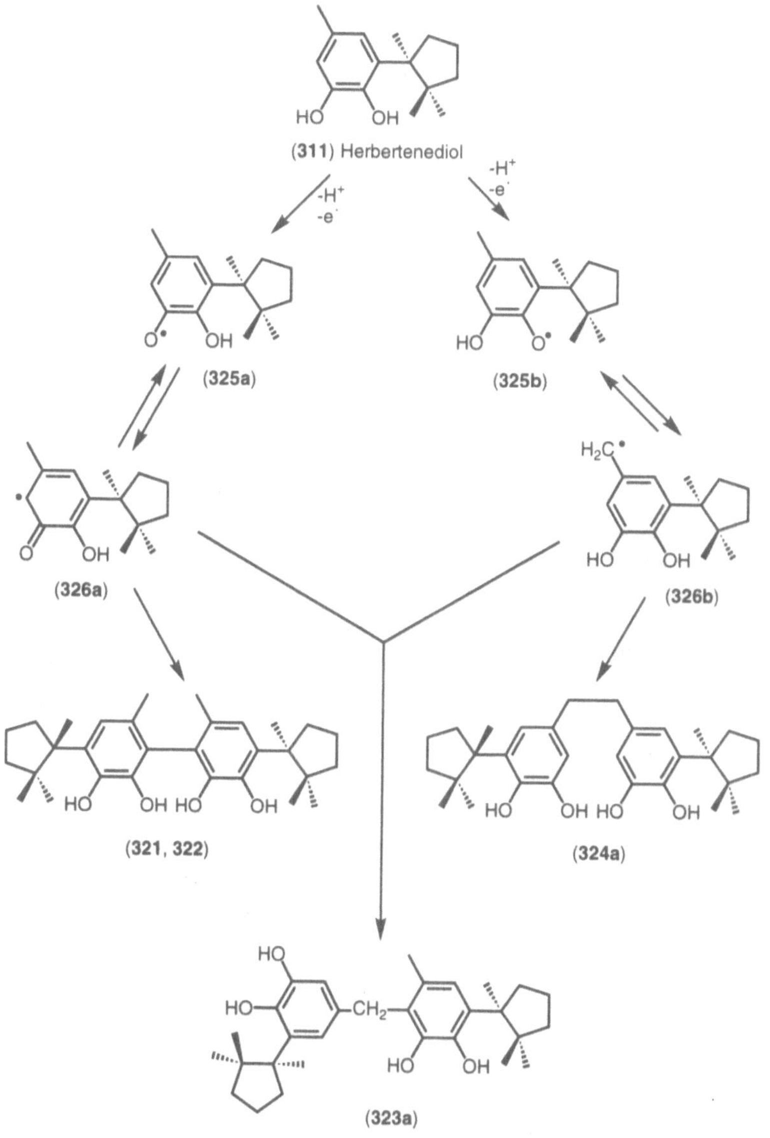

Scheme 26. Possible biogenetic pathways for mastigophorenes A–D

a co-metabolite in *M. diclados*. The phenoxy radicals (**325a, 325b**) subsequently form radical A (**326a**) or an unstable benzyl radical B (**326b**) which might evolve into a quinonemethide by further oxidation or loss of ·H. Homocoupling between two radicals (**326a**) should lead to (**321**) and (**322**) followed by aromatization. On the other hand, mastigophorene D (**324a**) is most likely to be derived by the direct coupling of the two benzyl radicals (**326b**). Another hetero-coupling between radicals (**326a**) and (**326b**) should give mastigophorene C (**323a**) as shown in Scheme 26.

5.16 Daucanes (Carotanes) and Drimanes

Only one carotane-type sesquiterpene lactone, hercynolactone (**327**) has so far been isolated from three *Barbilophozia* species, *B. lycopodioides* (*143, 275*), *B. hatcheri* (*275, 429, 565*), *B. floerkei* (*429, 565*) and *B. barbata* (*432*). The biogenesis of hercynolactone can be represented by cyclization of all-*trans* farnesyl pyrophosphate (**152**) as shown in Scheme 27 (*143*).

The earlier report (*19*) noted that eleven drimane-type sesquiterpenoids has been isolated from liverworts. Drimenol (**328**) has since been found also in *Bazzania, Diplophyllum, Frullania, Targionia* and *Trichocolea* species. *Lejeunea flava* elaborates cinnamolide (**342**) (*219*). Six drimane sesquiterpenoids, polygodial (**336**), drimeninol (**338**), isodrimeninol (**339**), 6α-hydroxyisodrimeninol (**340**), cinnamolide (**342**), and a new drimanolide, dehydroconfertifolin (**341**) have been isolated from the pungent American species *Porella roellii* (*249, 543*). The pungency is due to polygodial (**336**). The stereochemistry of (**341**) has been determined by chemical transformation of the known hemiacetal (**338**) to (**341**). *P. roellii* is chemically similar to *P. fauriei* and belongs to the *Porella vernicosa* complex (*19*).

Cell suspension cultures of *Porella vernicosa* produce drimenol (**328**), polygodial (**336**), isopolygodial (**337**) and cinnamolide (**342**) (*458*). It is interesting that *in vitro* cultures of *P. vernicosa* produce isopolygodial (**337**) which has not been found in living material. American *Porella cordaeana* produces drimenin (**343**), 7-ketoisodrimenin (**344**), and 7-ketoisodrimenin-5-ene (**345**) (*239a*) which have not been found in the same species collected in Europe (*593*).

Bazzania species belonging to the Lepidoziaceae are rich sources of many kinds of sesquiterpenoids. A new drimane-type sesquiterpene ester (**334a**) has been isolated from *Bazzania fauriana*, together with the previously known albicanyl caffeate (**335**) (*581*). Methylation of (**334a**) with methyl iodide gave a dimethyl ether (**334b**) which on reduction with LiAlH$_4$ afforded 3,4-dimethoxycinnamyl alcohol and drimenol (**328**).

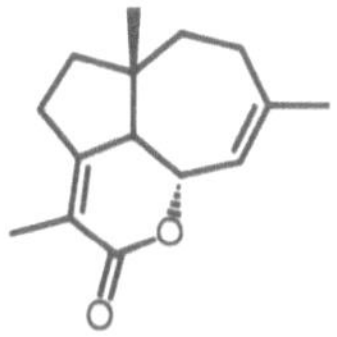

(**327**) Hercinolactone

Chart 21. Daucane (carotane) found in the Hepaticae

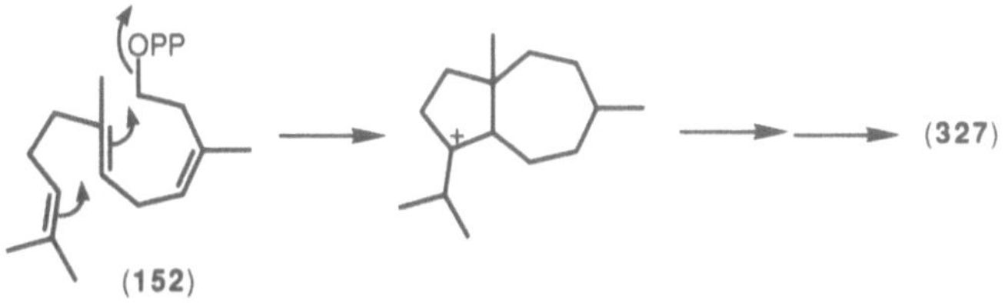

Scheme 27. Possible biogenetic pathways for daucane (carotane)-type sesquiterpenoids

Thus, the original ester was established as drimenyl caffeate (**334a**). The presence of drimenol (**328**) in *B. fauriana* has been detected in the crude extract by GC-MS (*581*). Albicanyl acetate (**330**) has been isolated from *Bazzania japonica*, along with the previously known albicanol (**329**) and albicanyl caffeate (**335**) (*84*). Free albicanol has been found in *Diplophyllum albicans* (Scapaniaceae) (*19*) and its acetate (**330**) in the marine organism, *Cadlina luteomarginata* (*266*). *Diplophyllum serrulatum* contains two new drimane aldehydes, albicanal (**331**) and isoalbicanal (**332**), and albicanic acid (**333**), along with drimenol (**328**) and albicanol (**329**) as well as a monocyclofarnesol (**504**) (*585a*). Possible biogenetic pathways leading to these compounds are shown in Scheme 28.

Makinoa crispata belonging to the Metzgeriales produces drimane-type sesquiterpenoids (*19*). Reinvestigation of the ethyl acetate extract of *M. crispata* resulted in the isolation of three new drimanolides, 7α-chloro-6β-hydroxyconfertifolin (**346**), 6β,7α-dihydroxyconfertifolin (**347a**) and 6β,7β-epoxyconfertifolin (**348**) (*258*). Structure (**346**) was established by analysis of the mass spectrum and by 2D-COSY NMR spectroscopy. The absolute configuration was deduced from the positive Cotton effect at 316 nm of the ketone prepared by Jones oxidation on the basis of the α-axial haloketone rule while the compound obtained on reduction of the halo ketone was identical with (+)-fragrolide isolated from *Cinnamosma fragrans* (*130*). The structure of (**347a**) was established

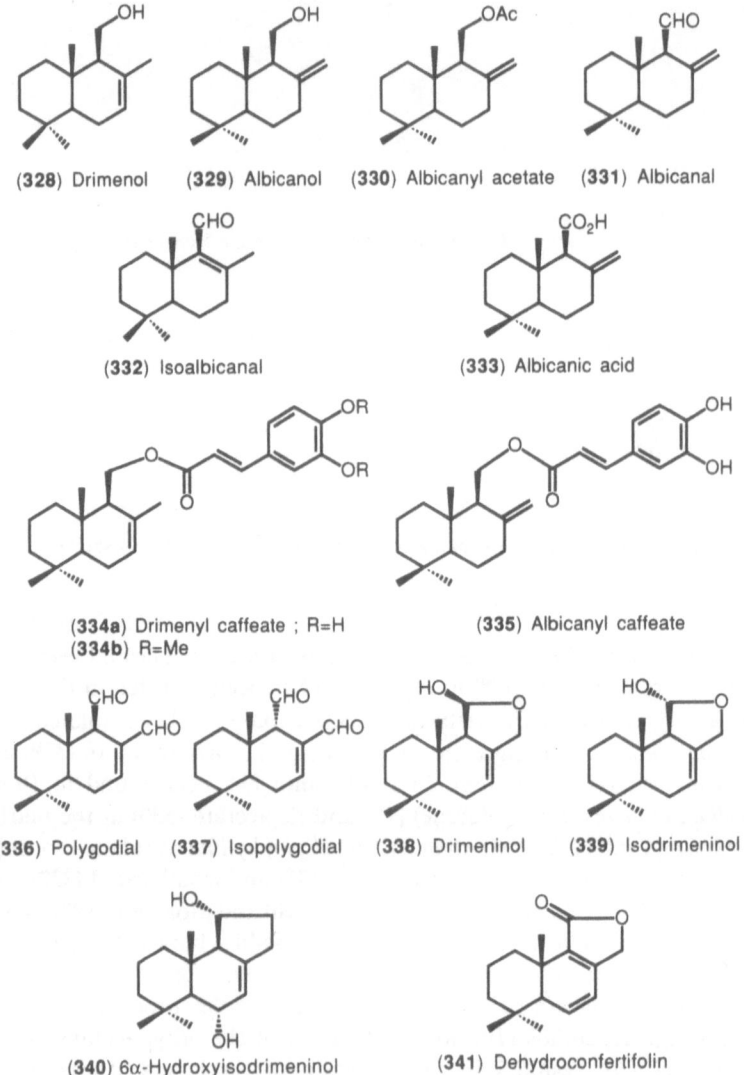

Chart 22a. Drimanes found in the Hepaticae and their derivatives

in the same manner. Its absolute configuration is based on the strong positive Cotton effect at 232 nm of the dibenzoate (**347c**) obtained from (**347a**) by treatment with dimethylaminopyridine and benzoyl chloride in pyridine. Treatment of (**346**) with sodium hydride in dimethylformamide

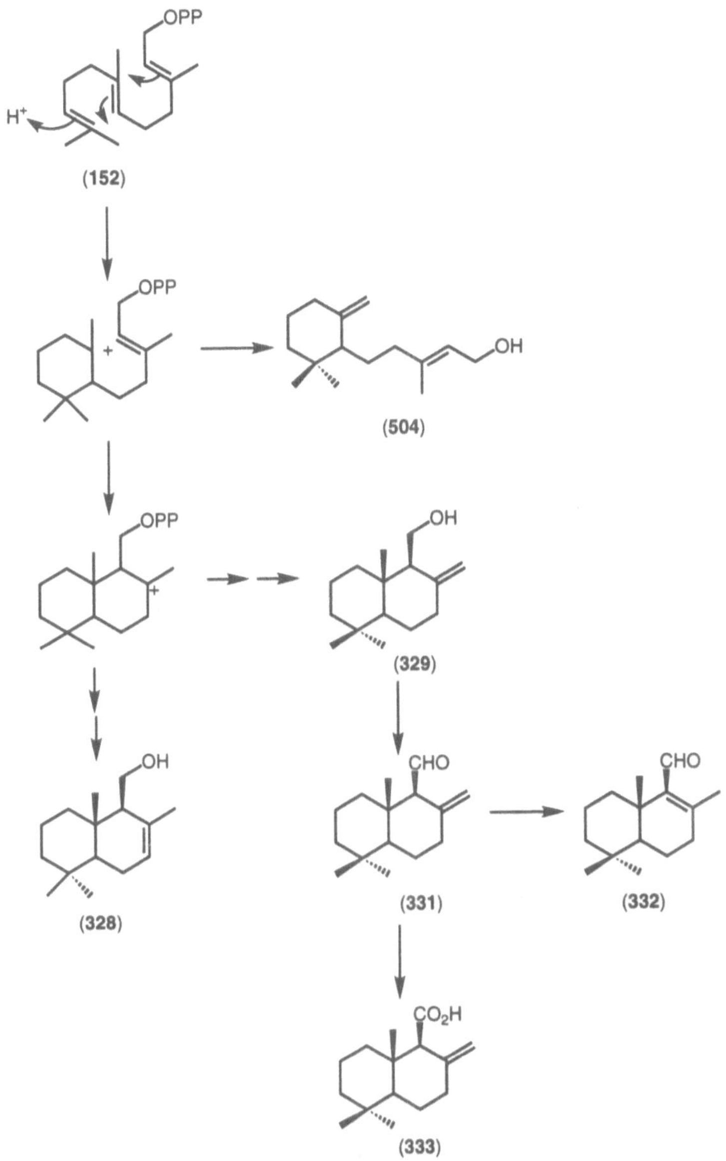

Scheme 28. Possible biogenetic pathways for drimane-type sesquiterpenoids found in
Diplophyllum serrulatum

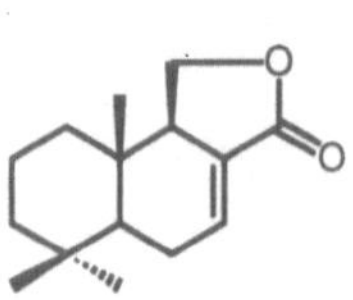

(342) Cinnamolide 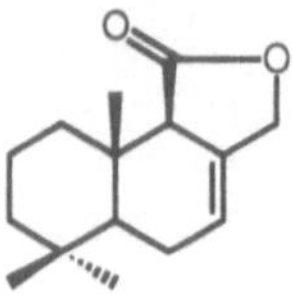   (343) Drimenin 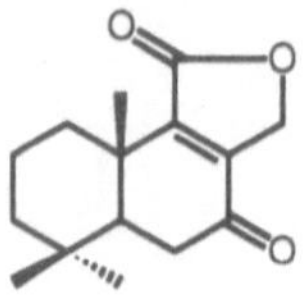   (344) 7-Ketoisodrimenin

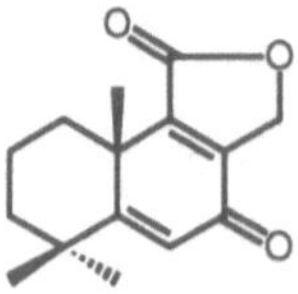

(345) 7-Ketoisodrimenin-5-ene

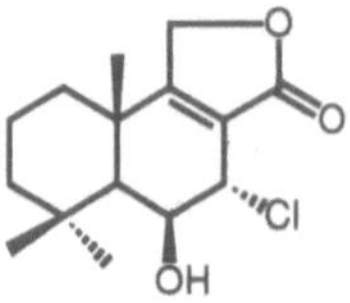

(346) 7α-Chloro-6β-hydroxyconfertifolin

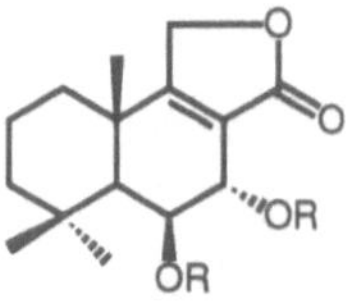

(347a) 6β,7α-Dihydroxyconfertifolin ; R=H
(347b) R=Ac
(347c) R=Bz

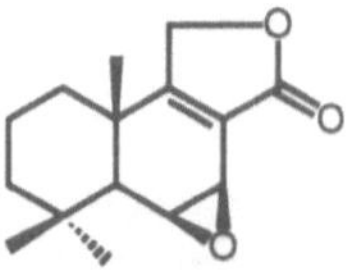

(348) 6β,7β-Epoxyconfertifolin

Chart 22b. Drimanes found in the Hepaticae and their derivatives

at 0–5° yielded an epoxide, presumably 6β,7β-epoxyconfertifolin (348). The configuration at C-7 of (346) is assumed to be S as in the case of (347a) since it is likely that (346) would be formed in the plant by opening of the epoxy ring of (348) due to attack of an anion (Cl$^-$ or OH$^-$) on C-7. Compound (346) could be an artifact, but the isolation of (346) was carried out using chlorine-free solvent and the presence of (346) in the ethyl acetate extract of fresh *M. crispata* was confirmed by GC-MS. This is the first isolation of a chlorine-containing compound from bryophytes.

5.17 Eremophilanes and Eudesmanes

Eremophilane-type sesquiterpenoids are quite rare in Hepaticae. Only three such compounds, eremophilene (**350**), eremofrullanolide (**352**) and its dihydro derivative (**353**) had been found in Jungermanniales prior to 1982 (*19*). Since then eremophilene (**350**) has been isolated from *Frullania serratta* (*51*). Some South American liverworts belonging to Jungermanniales also contain eremophilene (*45*), as do the essential oils from cultured cells of five *Calypogeia* species (*321, 539*) and *Monoclea forsteri* (*513*). *Marsupella emarginata* biosynthesizes longipinane-type sesquiterpenoids (*19*). Further investigation of *M. emarginata* resulted in the isolation of a novel eremophilane-type sesquiterpene, (+)-(4*S**,5*R**,7*S**,8*R**)-eremophila-9,11-dien-8α-ol (**351**) whose stereochemistry was elucidated by extensive NMR spectroscopy (*244*).

Previous to 1982, 46 eudesmane-type sesquiterpenoids had been found in Hepaticae (*19*). Three-fourths of these were eudesmanolides. α-Selinene (**354**) has been newly detected in *Bazzania* (*341*), *Monoclea* (*513*), *Pellia* (*106*), *Scapania* (*622, 638*) and *Plagiochila* (*621, 637*). β-Selinene (**355**) has also been detected in many species not only of Jungermanniales and Metzgeriales but also of Marchantiales as shown in Table IIb. *Isotachis humectata* (*45*) and *Mylia nuda* (*622, 624*), *Scapania ornithopodioides* (*622, 638*) and *S. robusta* (*631*) produce selina-4,11-diene (**356**). δ-Selinene (**358**) has been isolated by low temperature HPLC from *Monoclea gottschei* subsp. *neotropica* (*514*). This hydrocarbon has also been

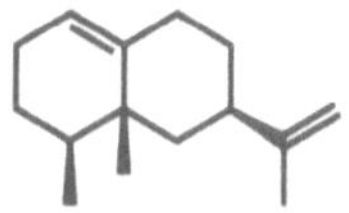

(**350**) Eremophilene

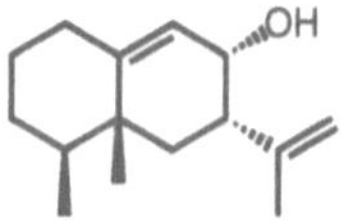

(**351**) (+)-(4*S**,5*R**,7*S**,8*R**)-Eremophila-9,11-dien-8α-ol

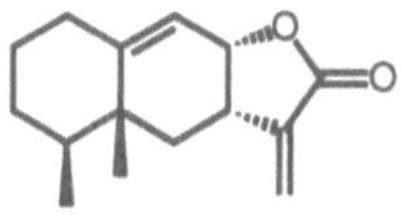

(**352**) Eremofrullanolide

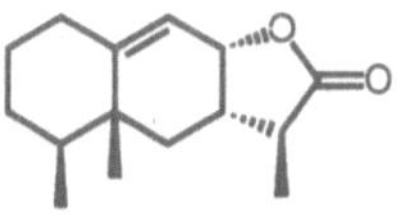

(**353**) Dihydroeremofrullanolide

Chart 23. Eremophilanes found in the Hepaticae

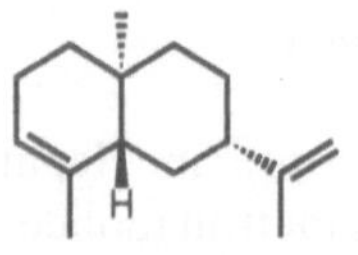

(354) *ent*-α-Selinene 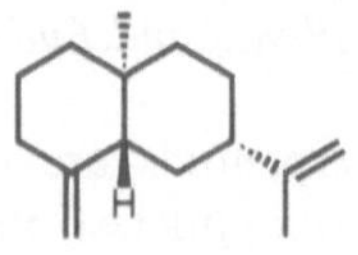   (355) *ent*-β-Selinene 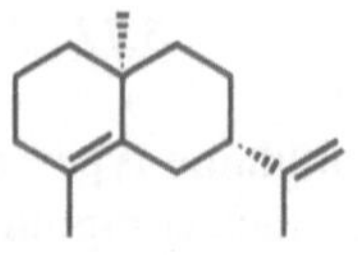   (356) Selina-4,11-diene

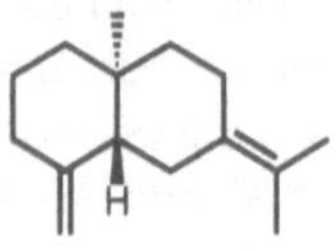

(357) γ-Selinene 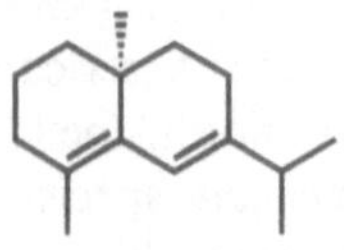   (358) δ-Selinene 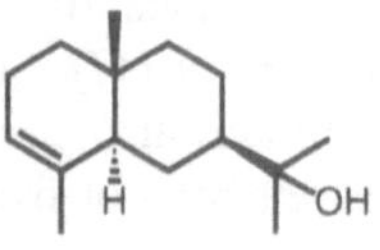   (359) α-Eudesmol

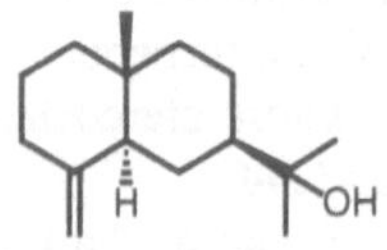

(360) β-Eudesmol 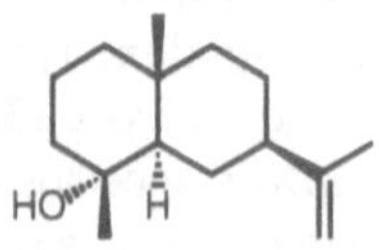   (361) Selina-11-en-4-ol 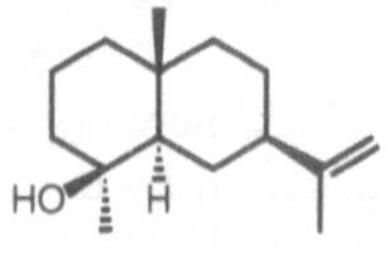   (362) Neointermediol

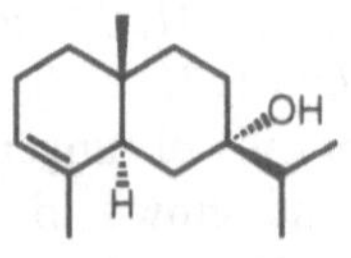

(363) Eudesm-3-en-7α-ol 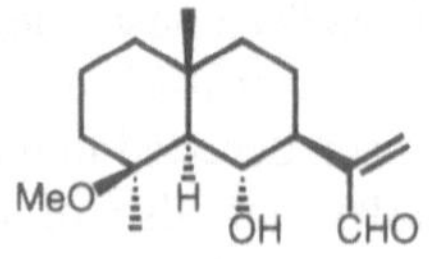   (364) 4β-Methoxyeudesmanal

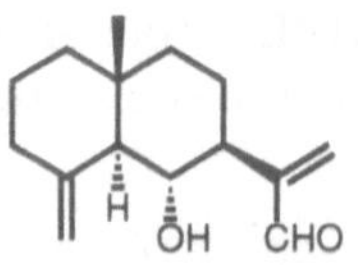

(365) Eudesmanal 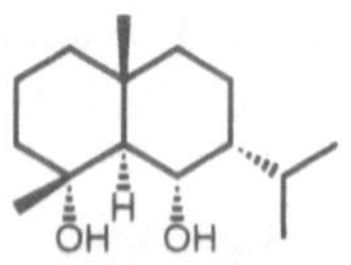   (366) (+)-5α,7β(H)-Eudesm-4α,6α-diol 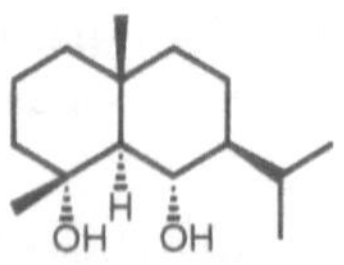   (367) Ajanol

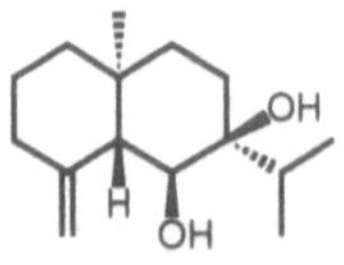

(368) *ent*-Eudesm-4(15)-en-6α,7α-diol 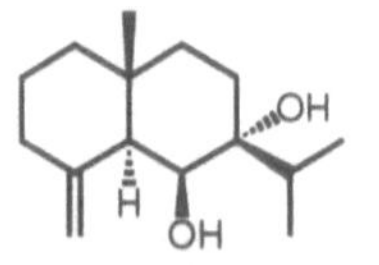   (369) Eudesm-4(15)-en-6β,7α-diol

Chart 24a. Eudesmanes found in the Hepaticae

found in *in vitro* cultured *Symphyogyna brongniartii* (*513*). Treatment of
(−)-α-selinene (**354**) with formic acid in the presence of CH_3SO_3H gave
(−)-δ-selinene (**358**) (*631*).

Fossombronia (*491*) elaborates α-eudesmol (**359**). β-Eudesmol (**360**)
has been found in *Bazzania, Radula, Riccardia, Thysananthus* (*45*) and

Balantiopsis (*65*). As reported earlier (*19*) *ent*-selina-11-en-4-ol (**361**) has been isolated from *Riccardia jackii* (*19*). The same alcohol has since been detected in German *Conocephalum conicum* (*553*). The total synthesis of (±)-(**361**) has been accomplished by KESSELMANS *et al.* (*325*, *326*). *Lepidozia vitrea* is a rich source of eudesmane- as well as bicyclogermacrane-type sesquiterpenoids. The ether extract contains eudesm-3-en-7α-ol (**363**), eudesm-4(15)-en-6β,7α-diol (**369**) and eudesm-3-en-6α-acetoxy-7α-ol (**370**), along with the previously known eudesm-3-en-6β,7α-diol (**371**) (*585a*). *Bazzania tridens* also produces (**363**) and α-cyclogermacrone (**375**) (*623a*). A new methoxyeudesmanal (**364**) has been isolated from the methanol extract of *Frullania tamarisci* subsp. *obscura*; its structure was deduced by spectroscopic comparison with those of the co-occurring eudesmanal (**365**) and eudesmanolide (**390**) (*82*). It is probably an artifact formed during the extraction and purification procedure. The same species produces a new diol (**366**) (*582*), together with costunolide (**441**) and β-cyclocostunolide (**389a**) (*19*); its structure has been settled by spectroscopic and chemical correlation as shown in Scheme 29. The NOE difference spectrum of (**366**) and the positive Cotton effect at 295 nm of (**366b**) established the absolute configuration of (**366**). The C-7 epimer (**367**) of (**366**) has been isolated from a higher plant, *Picea ajamensis* (Piceaceae) (*205*).

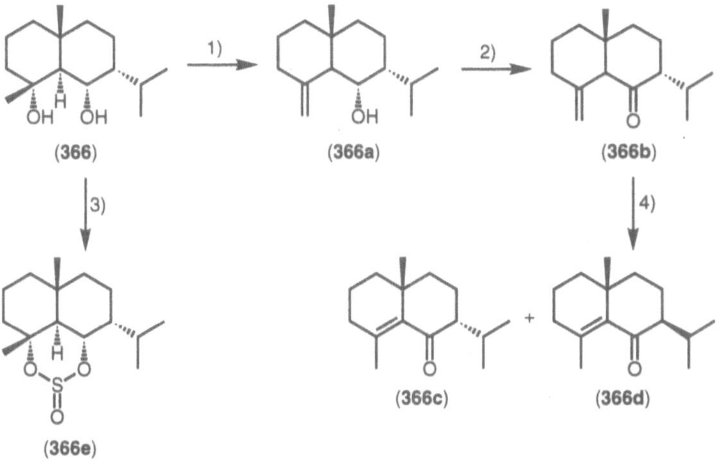

1) *p* -TsOH/C$_6$H$_6$ 2) PCC/CH$_2$Cl$_2$ 3) SOCl$_2$/Py 4) 5% KOH/EtOH

Scheme 29. Reactions of 5α,7β(H)-eudesm-4α,6α-diol

Eudesm-4(15)-en-6β,7β-diol (**368**) and (+)-eudesm-3-en-6β,7α-diol (**371**) have been isolated from *Chiloscyphus pallescens* (*143, 149*) and *Lepidozia reptans* (*146, 147*), respectively; the structures were based on analysis of the ^{1}H- and ^{13}C-NMR spectra. The absolute configurations of both diols have not been determined. *Lophozia ventricosa* elaborates ventricosin A (*286*) which is identical with (−)-*ent*-4(15), 7(11)-eudesmadien-8-one (= β-cyclogermacrone) (**372**) (*287*). *Ent*-4(15), 7(11)-eudesmadien-8-one (**372**) has been isolated previously from higher plants in both enantiomeric forms, the (+)-form from *Atractylodes japonica* (*173*) and the (−)-form from *Asarum caulescens* (*174*) and *Peteravenia schultzii* (*116*). The magnitude and sign of the specific rotation of the enone (**372**) indicate that it belongs to the enantio series.

Further investigation of the chemical constituents of *L. ventricosa* resulted in the isolation of a new eudesmane-type ketone (**373**) whose structure and absolute configuration have been established by ^{1}H-, ^{13}C-NMR, ^{1}H-^{1}H- and ^{13}C-^{1}H- 2D-COSY spectra, by its positive Cotton effect at 298 nm and by formation of *ent*-β-cyclogermacrone (**372**) on acid treatment of (**373**) which is a co-metabolite (*429, 566*). *Ent*-β-cyclogermacrone (**372**) has also been isolated from *Barbilophozia floerkei* (*429, 566*). *Bazzania fauriana* produces a new eudesmenol (**374**), together with α-cyclogermacrone (**375**) which has been prepared from germacrone by cyclization with acidic methanol (*581*). The structure of (**374**) was elucidated by 2D-COSY NMR spectrometry and by oxidation with PPC to the previously known ketone (*446*). The configuration of the hydroxyl group at C-6 is based on the presence of a singlet at δ 3.28 ppm.

A new *ent*-8β-hydroxyeudesm-3,11-diene (**376**) has been isolated from East Malaysian *Bazzania spiralis* along with *ent*-α-selinene (**354**) (*341*). The position and stereochemistry of the axial hydroxyl group in (**376**) were confirmed by its strong resistance to acetylation and by PPC oxidation to a non-conjugated ketone which could be qualified by Al_2O_3 to **375**. The positive Cotton effect of the latter at 295 nm established the absolute configuration. It is interesting to note that *B. spiralis* elaborates *ent*-eudesmane-type sesquiterpenoids (**354, 376**). On the other hand *B. fauriana* produces 6β-hydroxyeudesm-3-ene (**374**) whose absolute configuration is identical with that of many eudesmane-type sesquiterpenoids found in higher plants. *Ent*-cyperone (**377**) has been isolated from German *Marchantia polymorpha* (*71*). This is the first example of the isolation of the enantiomer of (+)-α-cyperone from natural sources. *Lepidozia fauriana* and *L. vitrea* elaborate a unique eudesmane ether, 6β-acetoxyvitranoxide (**378**) (*505a, 623a*). Two known eudesmanes, intermediol (**362**) and furanoeudesma-1,3-diene (**379**) have been detected in the essential oil of *Lophocolea heterophylla* (*527a*).

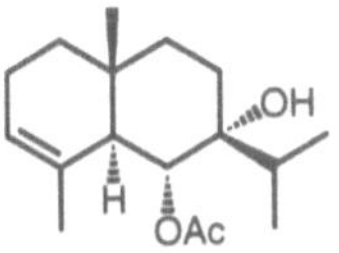

(370) Eudesm-3-en-6α-acetoxy-7α-ol

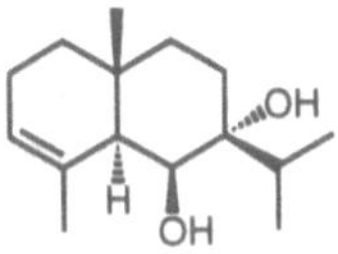

(371) Eudesm-3-en-6β,7α-diol

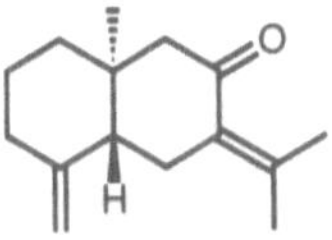

(372) *ent*-4(15),7(11)-Eudemadien-8-one
(= β-Cyclogermacrone)
(= Ventricosin A)

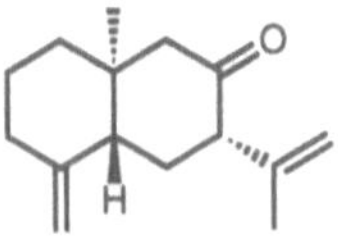

(373) *ent*-4(15),11(12)-Eudemadien-8-one

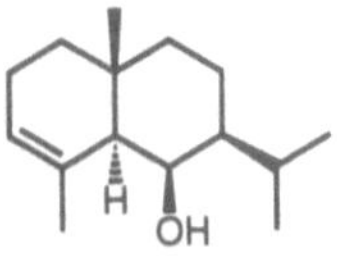

(374) 6β-Hydroxyeudesm-3-ene

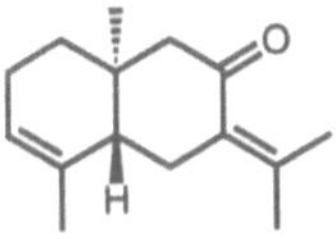

(375) α-Cyclogermacrone

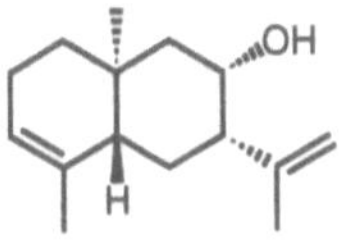

(376) *ent*-8β-Hydroxyeudesm-3,11-diene

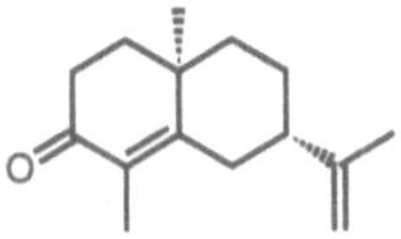

(377) *ent*-α-Cyperone

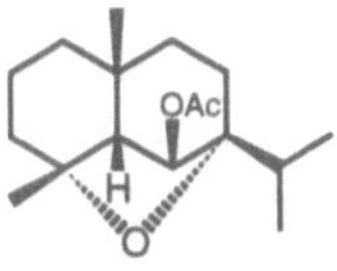

(378) 6β-Acetoxyvitranoxide

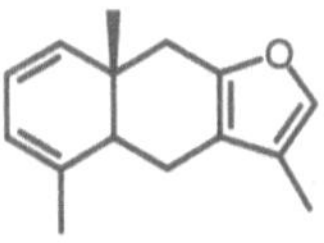

(379) Furanoeudesma-1,3-diene

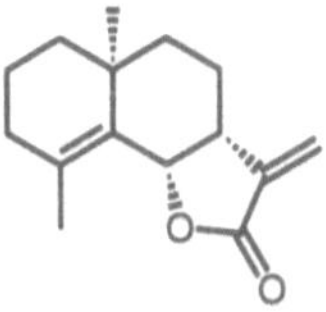

(380a) (+)-Frullanolide

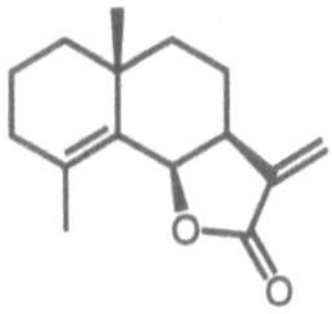

(380b) (-)-Frullanolide

Chart 24b. Eudesmanes found in the Hepaticae

As reported earlier *Frullania* species are rich sources of 12,6-eudesmanolides (*19*). Frullanolide (**380b**) has been newly found in seven *Frullania* species (*47, 60, 63, 513*) Dihydrofrullanolide (**384**) has been also detected in five *Frullania* (*47, 63, 513*) and one *Plagiochila* species (*47*).

While (−)-frullanolide (**380b**) and (±)-frullanolide (**380a + 380b**) have been previously synthesized by two groups (*19*), four additional total syntheses of (±)-frullanolide have been reported (*140, 464, 499, 522*). (−)-Frullanolide (**380b**) has also been isolated from *Grangea maderaspatana* (Compositae) (*475*). The full paper concerning the absolute stereostructures of two eudesmanolides, (+)-β-frullanolide (**385**) and (+)-brothenolide (**386**) has been published (*541*). *Frullania bicornistipula* elaborates dihydro-β-frullanolide (**387**) (*513*). While (+)-arbusculin B (= γ-cyclocostunolide) (**388a**) and its enantiomer (**388b**) have been reported earlier from *Frullania* species (*19*), the former lactone has since been isolated also from *F. serratta* (*51*), *F. hamatiloba* (*588*) and *F. brasiliensis* (*433*). (−)-Arbusculin B has also been isolated from *F. usamiensis*, together with (−)-*ent*-β-cyclocostunolide (**389b**) (*56*). *Conocephalum japonicum* is a rich source not only of germacranolides but also of eudesmanolides. Thus arbusculin A (**391**) and (11*S*)-dihydroarbusculin A (= colartin) (**392**) which have been isolated from some *Artemisia* species (Compositae) (*181*), and (11*S*)-dihydro-β-cyclocostunolide (**393**) has now been found in the dichloromethane extract of *C. japonicum* (*584*).

The presence of β-cyclocostunolide (**389a**) and its dihydro derivative (**393**) has been detected in *Frullania bicornistipula* by GC-MS (*513*). East Malaysian *Wiesnerella denudata* elaborates a new 8α-acetoxy-β-cyclocostunolide (**394**) (*51*) along with the previously known costunolide (**441**), tulipinolide (**443**) and 8α-acetoxyzaluzanin D (**467**) (*19*). Rothin A acetate (**395a**), the C-4(5) double bond isomer of (**394**) has been isolated from an unidentified South American *Frullania* species (*441a*). Compound (**395a**) has been prepared from rothin A (**395b**) by acetylation (*296a*). (−)-α-Santonin (**396**) has been isolated from differentiated cultures of *Fossombronia pusilla*, together with three diterpene dialdehydes, perrottetianals (*491*).

Frullania nepalensis biosynthesizes four new eudesmanolides, nepalensolides A (**397**), B (**398**), C (**399**) and D (**400**), together with *ent*-β-frullanolide (**385**) (*63, 564, 575*). *Frullania serratta* also produces the three nepalensolides (**397–399**), along with β-selinene (**355**) and arbusculin B (**388a**) (*51*). The ^{1}H-^{1}H-, ^{13}C-^{1}H COSY and HMBC spectra of (**397**) suggested the presence of a eudesmane skeleton including a cyclopropane ring and α-methylene-γ-lactone. Since the value of the coupling constants ($J_{7,13} = 3.3$ Hz) of the exomethylene protons was larger than 3 Hz, the lactone ring was deduced to be *trans* if the Samek rule [/4J/(trans-lactone) $\geq$ 3Hz $\geq$ /4J/(cis-lactone)] (*481–484*) applies. On the other hand, NOE spectrometry showed that both H-6 and H-7 were on the same side as the methyl group on C-10 while H-5 was unambiguously assigned as β-axial from the value of the coupling constants of H-6 (*dd*,

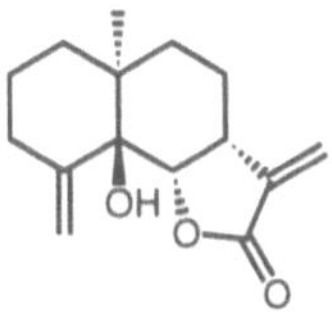

(381) Oxyfrullanolide 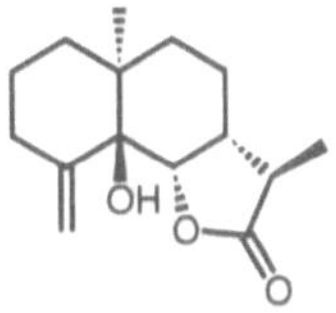   (382) Dihydrooxyfrullanolide 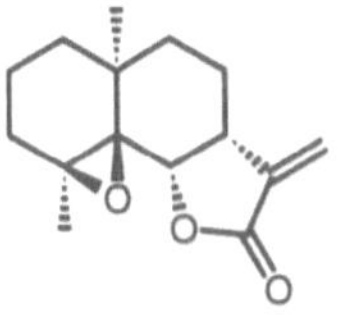   (383)

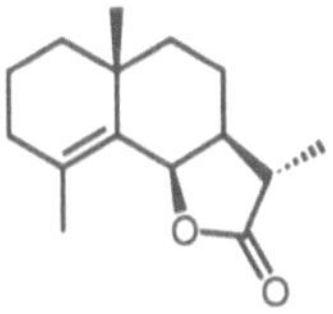

(384) Dihydrofrullanolide 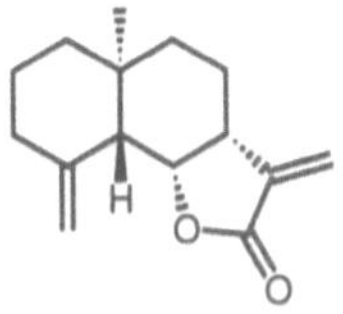   (385) (+)-β-Frullanolide 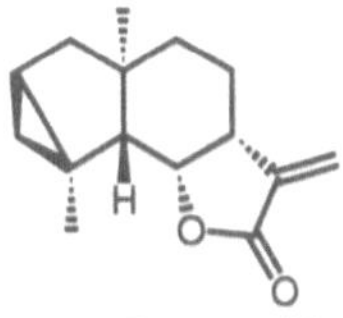   (386) (+)-Brothenolide

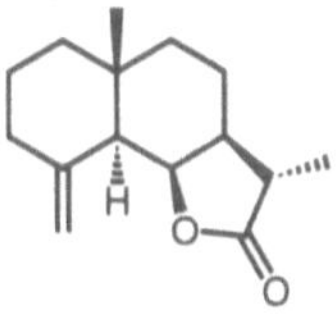

(387) Dihydro-β-frullanolide 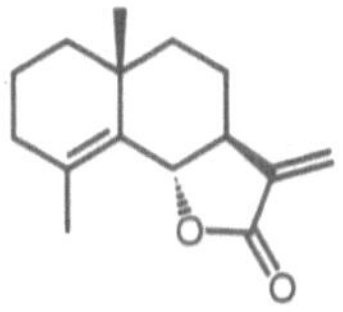   (388a) (+)-Arbusculin B   (388b) (-)-Arbusculin B
                                   (= γ-Cyclocostunolide)

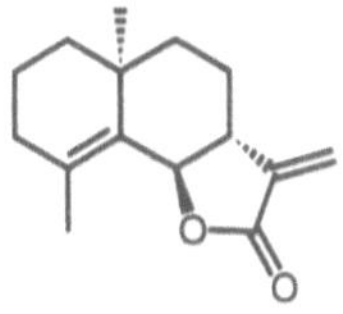

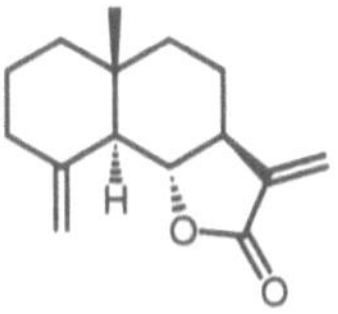

(389a) β-Cyclocostunolide 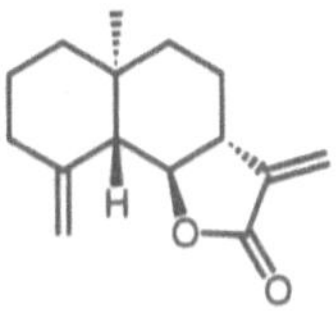   (389b) ent-β-Cyclocostunolide 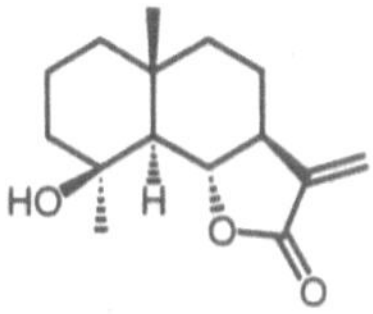   (390) 4-epi-Arbusculin A

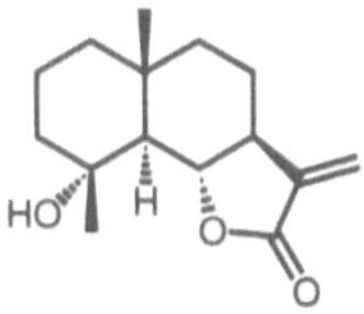

(391) Arbusculin A

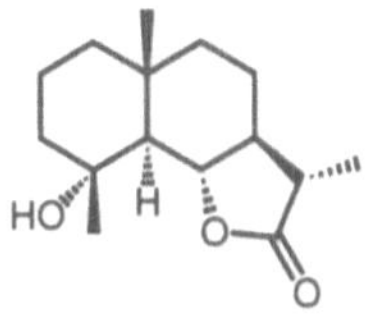

(392) (11S)-Dihydroarbusculin A
      (=Colartin)

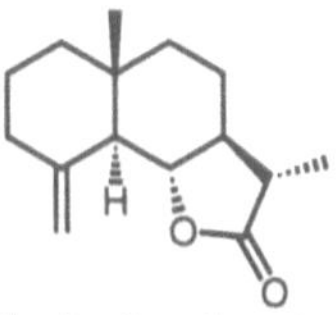

(393) Dihydro-β-cyclocostunolide

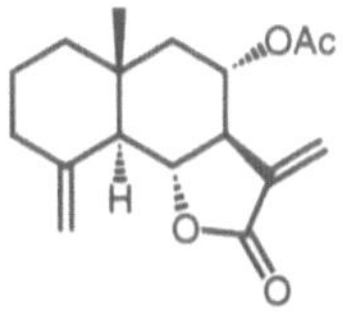

(394) 8α-Acetoxy-β-cyclocostunolide

Chart 24c. Eudesmanes found in the Hepaticae

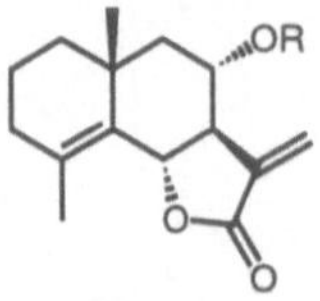

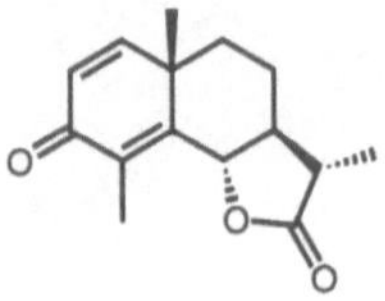

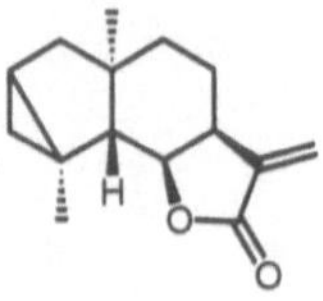

(395a) Rothin A acetate ; R=Ac   (396) α-Santonin   (397) Nepalensolide A
(395b) R=H

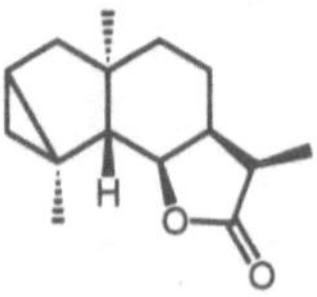

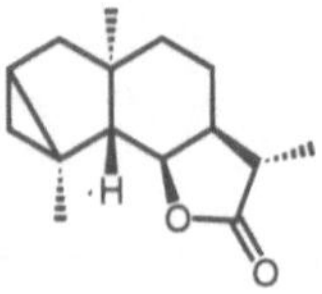

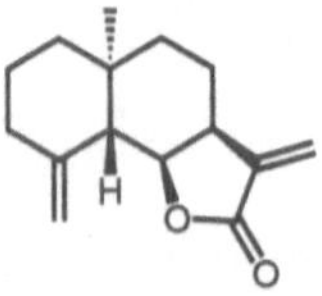

(398) Nepalensolide B   (399) Nepalensolide C   (400) Nepalensolide D

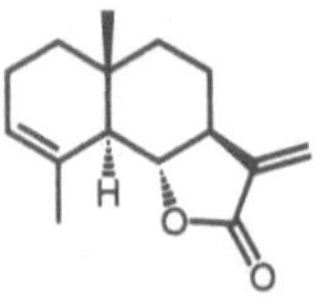

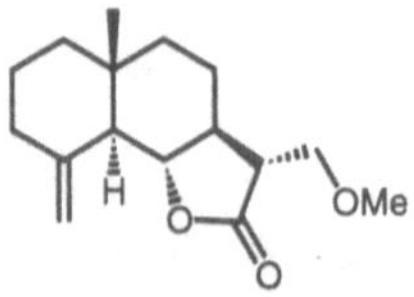

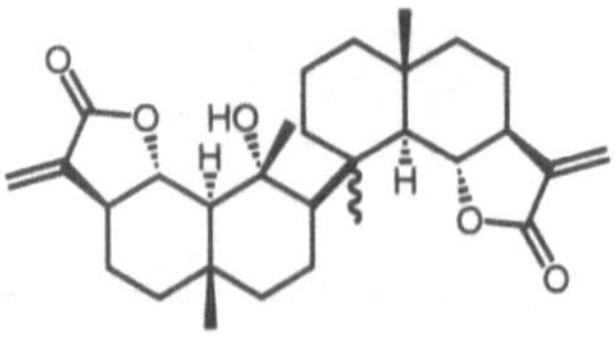

(401) α-Cyclocostunolide   (402) Methoxyfrullanolide   (403a)

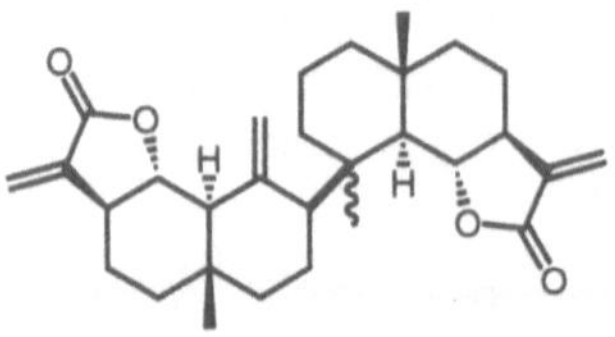

(404a)

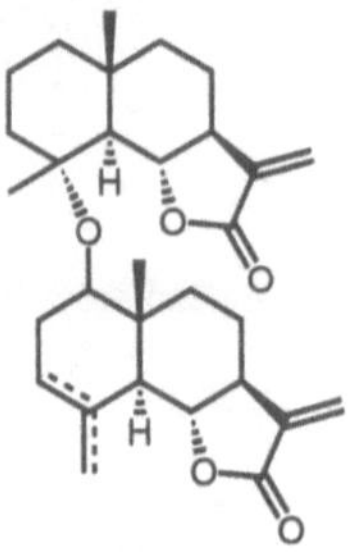

(404b) 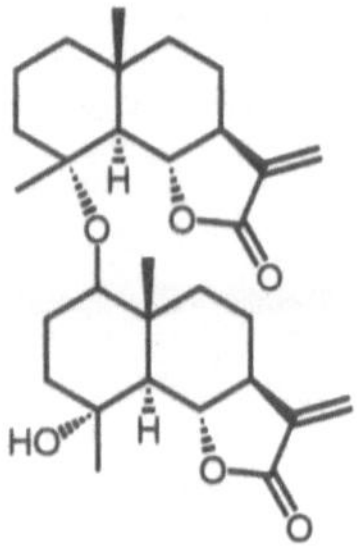   (403b)

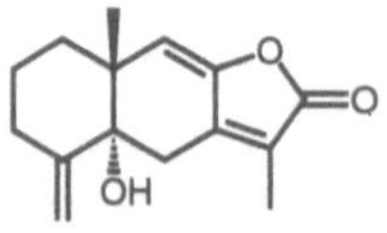

(405) 5α-Hydroxyeudesm-4(15),7(11),8(9)-trien-12,8-olide

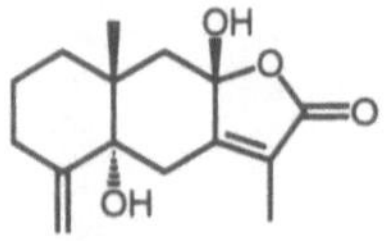

(406) 5α,8β-Dihydroxyeudesm-4(15),7(11)-dien-12,8-olide

Chart 24d. Eudesmanes found in the Hepaticae

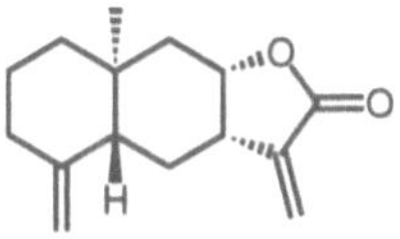

(407) *ent*-Isoalantolactone

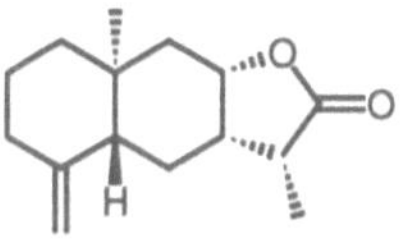

(408) *ent*-Dihydroisoalantolactone

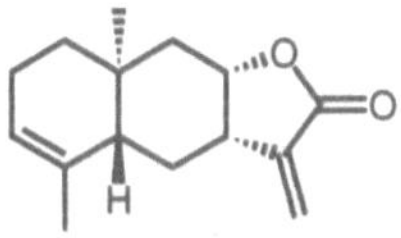

(409) *ent*-Diplophyllolide

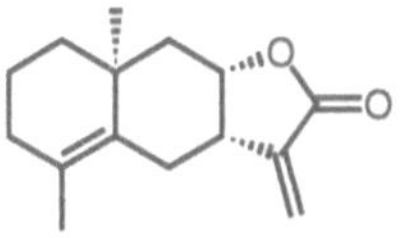

(410) *ent*-Diplophyllin

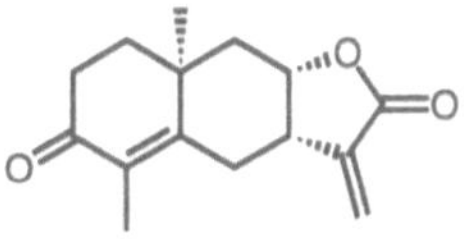

(411) *ent*-3-Oxodiplophyllin

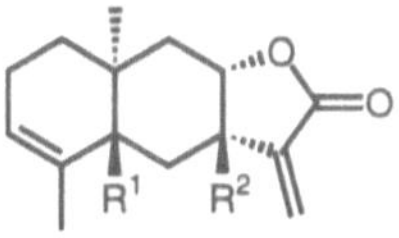

(412) R^1=OH, R^2=H
(413) *ent*-7α-Hydroxydiplophyllolide ; R^1=H, R^2=OH

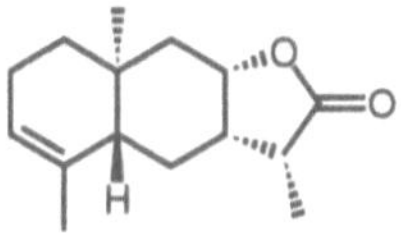

(414) *ent*-Dihydrodiplophyllolide

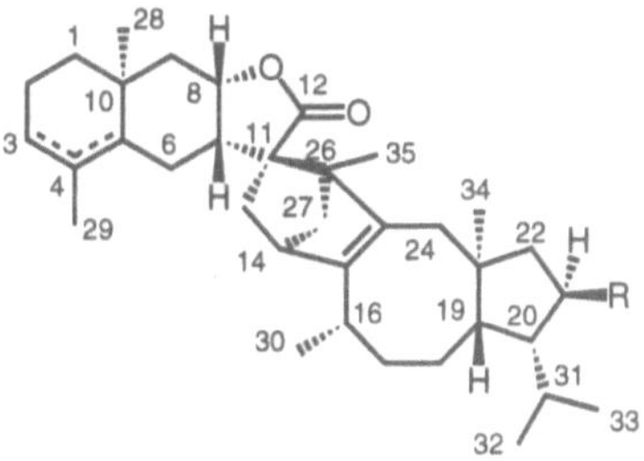

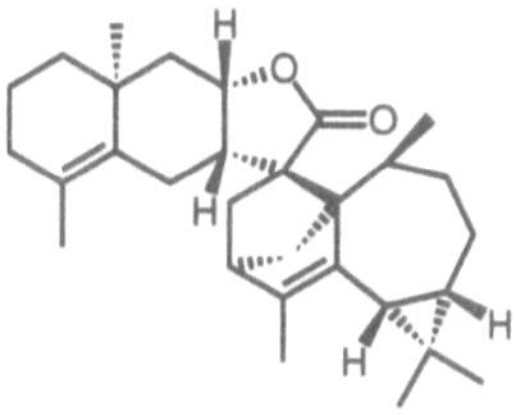

(419) Plagiospirolide E

(415) Plagiospirolide A ; R=H, Δ^3
(416) Plagiospirolide B ; R=H, Δ^4
(417) Plagiospirolide C ; R=OH, Δ^3
(418) Plagiospirolide D ; R=OH, Δ^4

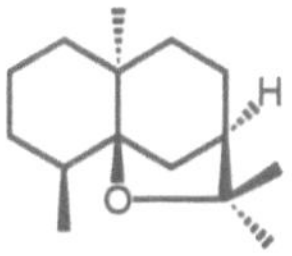

(420) Dihydroagarofuran

Chart 24e. Eudesmanes found in the Hepaticae

$J = 11.6$ and 7.6 Hz) and H-5 (d, $J = 11.6$ Hz). To resolve these contra-dictory conclusions an X-ray crystallographic analysis of (397) was carried out which showed that H-6 and H-7 are on the same side as the methyl group at C-10, which is consistent with the results obtained from the NOE experiments. Thus the Samek rule was not applicable to this case. The absolute configuration ascribed to (397) is based on the positive

Cotton effect at 223 and 245 nm. This is the first instance of a eudesmanolide with a *cis*-fused α-methylene-γ-lactone ring and a large coupling constant between H-13 and H-7.

As nepalensolide A (**397**) did not obey the Samek rule, steric energies and conformations of (**397**) and frullanolide (**380b**), with a *cis*-fused α-methylene-γ-lactone ring, and four types of possible model lactones as well as of some synthetic compounds were calculated by MM2 method and allylic coupling constants between H-7 and the exomethylene protons were estimated to permit evaluation of the Samek rule as shown in Chart 25 (*578*). The results clearly show that great care should be exercised in using the Samek rule, particularly when sesquiterpene

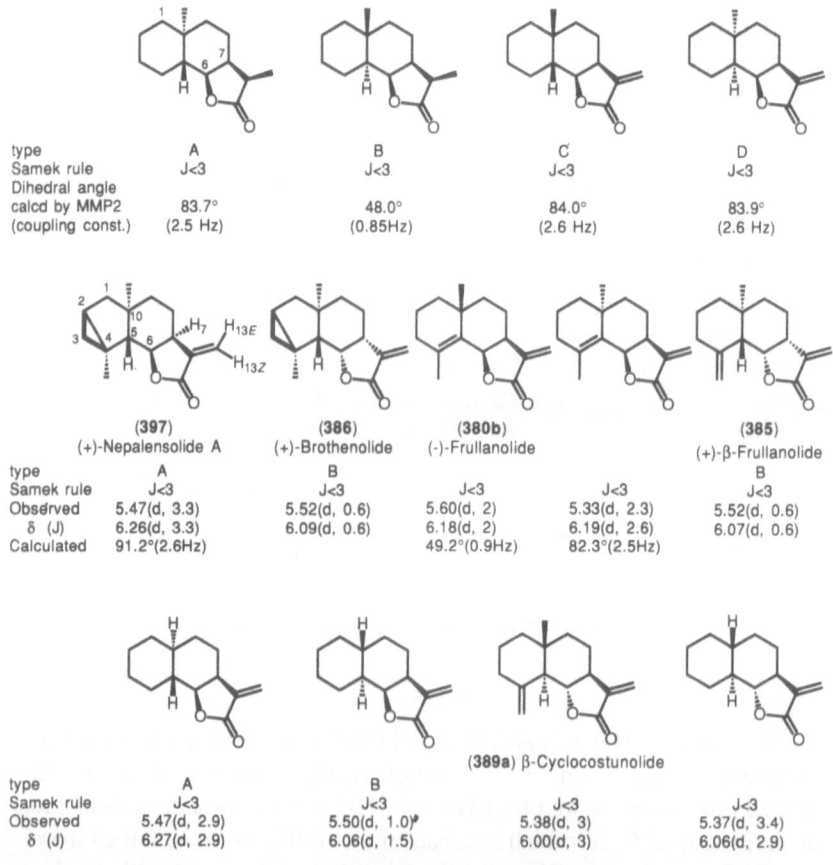

type	A	B	C	D
Samek rule	J<3	J<3	J<3	J<3
Dihedral angle calcd by MMP2	83.7°	48.0°	84.0°	83.9°
(coupling const.)	(2.5 Hz)	(0.85Hz)	(2.6 Hz)	(2.6 Hz)

	(397)	(386)	(380b)		(385)
	(+)-Nepalensolide A	(+)-Brothenolide	(-)-Frullanolide		(+)-β-Frullanolide
type	A	B			B
Samek rule	J<3	J<3	J<3	J<3	J<3
Observed	5.47(d, 3.3)	5.52(d, 0.6)	5.60(d, 2)	5.33(d, 2.3)	5.52(d, 0.6)
δ (J)	6.26(d, 3.3)	6.09(d, 0.6)	6.18(d, 2)	6.19(d, 2.6)	6.07(d, 0.6)
Calculated	91.2°(2.6Hz)		49.2°(0.9Hz)	82.3°(2.5Hz)	

			(389a) β-Cyclocostunolide	
type	A	B		
Samek rule	J<3	J<3	J<3	J<3
Observed	5.47(d, 2.9)	5.50(d, 1.0)*	5.38(d, 3)	5.37(d, 3.4)
δ (J)	6.27(d, 2.9)	6.06(d, 1.5)	6.00(d, 3)	6.06(d, 2.9)

Chart 25. Observed and calculated coupling constants and dihedral angles of various α-methylene-γ-lactones

lactones of type A, C and D are isolated. If two conformations are possible for sesquiterpene lactones with an α-methylene-γ-lactone ring, another method, for example NOE spectrometry, should be used to determine their stereochemistries. Structures of nepalensolides B (398) and C (399) were established by spectroscopic comparison with nepalen-solide A (397).

Frullania tamarisci subsp. *tamarisci* produces methoxyfrullanolide (402) (*249*) and a unique eudesmane-type lactone dimer (403a) whose structure was deduced by ^{1}H- and ^{13}C-NMR spectrometry (*146*). The ^{13}C-NMR spectrum had 30 signals thus establishing, the asymmetrical nature of the dimer. The stereochemistries assigned to the cyclohexane junctions and the lactone rings were based on the coupling constants deduced from the ^{1}H-NMR spectrum and NOE difference spectroscopy. The absolute configuration assigned to each lactone unit was based on the co-occurrence of (−)-frullanolide (380b), (+)-α-cyclocostunolide (401) and (+)-costunolide (441). The proposed mode of formation involves attack of the cationic species formed in the cyclization of costunolide (441) on a molecule of α-cyclocostunolide (401) followed by reaction with water as shown in Scheme 30. The alternative 1,4-linked dimer is less attractive because of steric hindrance. The dimer was isolated from dried *F. tamarisci* subsp. *tamarisci* which had been stored in the laboratory for at least a year which may be significant (*146*), because the dimer has not been isolated from freshly collected material.

Much earlier, two eudesmanolide-type dimers to which structures (403b) and (404b) were assigned, were isolated from Indian *Frullania yunnanensis*, along with (−)-frullanolide (380b) and (−)-dihydrofrul-lanolide (384) (*70*). Since the spectral data of the presumed (403b) are identical with those of dimer (403a) structure (403b) should be abandoned and a presumed structure of the second dimer (404b) should be revised to the dehydration product of 403a, i.e. (404a).

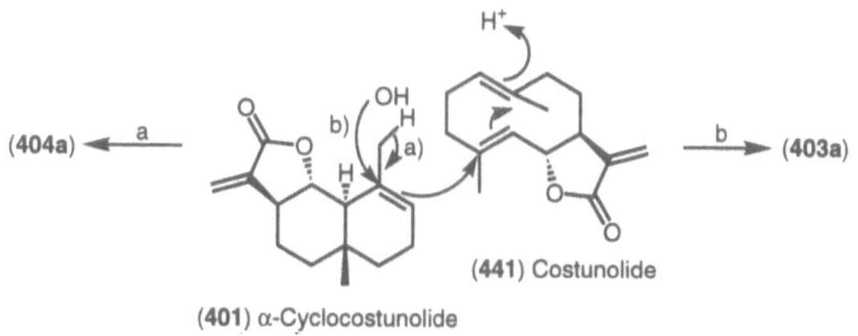

Scheme 30. Formation of dimeric eudesmanolides

Two new eudesmanolides, (405) and (406), have been isolated from *Lophocolea heterophylla* (*586*). The spectral data of the former are similar to those of the latter lactone. Dehydration of both (405) and (406) with $POCl_3$ in pyridine gave the same Δ^4 (15), 5,7 (11) 8-tetradien lactone indicating that both hydroxyl groups at C-5 and C-9 were axial. On the basis of the above results, 2D-COSY NMR data and NOE experiments, the structures of the new eudesmanolides were established as (405) and (406).

European *Lophocolea heterophylla* (*343*) and South American *L. coadunata* (*47*) produce *ent*-isoalantolactone (407). The latter species (*47*) and *L. bidentata* (*527a*) contain dihydroisoalantolactone (408) and diplophyllolide (409), respectively. This is the first known occurrence of *ent*-isoalantolactone in nature. European *Chiloscyphus polyanthos* elaborates the 7,8-*cis*-eudesmanolides (409–411) and a fourth pungent lactone assigned structure (412). However, structure (412) has since been revised to that of *ent*-7α-hydroxydiplophyllolide (413) by using NMR shift reagents and because of the absence of the usual allylic coupling between H-7 and the α-methylene protons (*60*). (±)-3-Oxodiplophyllin (411) has been totally synthesized by CAINE et al. in 9 steps (*126*). Ent-7α-hydroxydiplophyllolide (413) has been isolated from Scottish *Diplophyllum albicans* (*143*) and *Clasmatocolea vermicularis* (*60*). The latter species also elaborates *ent*-diplophyllin (410) (*60*). *Tritomaria quinquedentata* produces *ent*-diplophyllolide (409) and *ent*-dihydrodiplophyllolide (414) (*429, 565, 567a*) which is identical with a dihydro derivative obtained by reduction of (409) with $NaBH_4$ (*81*).

Two unique C-35 terpene lactones containing a sesquiterpene and a diterpene portion, plagiospirolides A (415) and B (416), have been isolated from the Panamanian liverwort *Plagiochila moritziana* together with *ent*-diplophyllolide (409) and *ent*-diplophyllin (410) (*513, 515, 516*). In HPLC, compounds (415) and (416) were represented by one peak each, but in GC-MS, thermal decomposition occurred to give two peaks for each compound. The fragmentation patterns in the mass spectra of one peak from (415) and (416) corresponded to those of diplophyllolide (409) and diplophyllin (410), respectively while the retention times of the second peaks from (415) and (416) were identical and the molecular formula of the second peaks corresponded to that of a diterpene hydrocarbon, $C_{20}H_{32}$ (Scheme 31). The molecular formulas of (415) and (416) thus were $C_{35}H_{52}O_2$, the two oxygens being those of a γ-lactone because of the IR absorption at 1760 cm^{-1} and because of the presence in the ^{13}C NMR spectra of a singlet at δ 182 ppm. In compounds (415) and (416), the C-13 position of (409) and (410) was thought to be attached to the diterpene moiety because the typical signals of the exomethylene

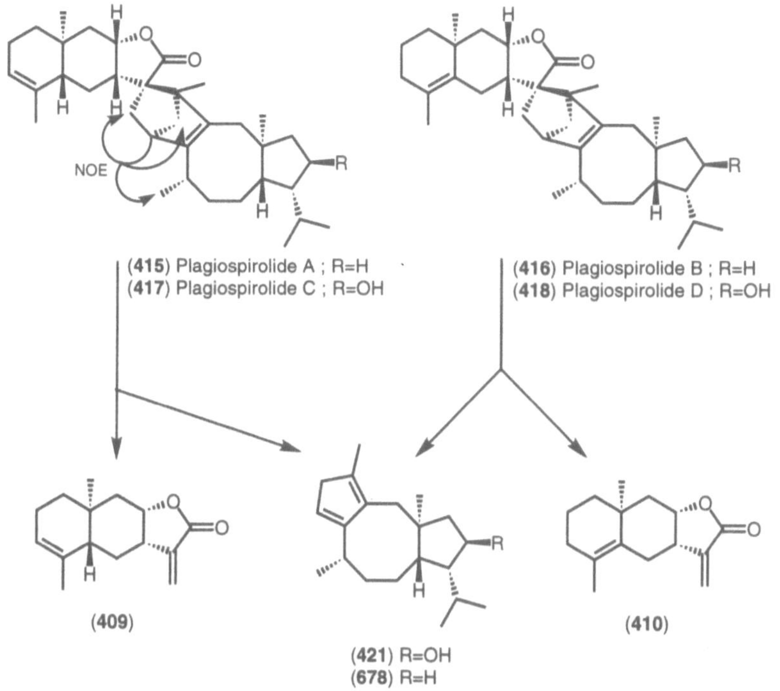

(415) Plagiospirolide A ; R=H
(417) Plagiospirolide C ; R=OH

(416) Plagiospirolide B ; R=H
(418) Plagiospirolide D ; R=OH

(409)

(421) R=OH
(678) R=H

(410)

Scheme 31. Retro-Diels-Alder reaction of plagiospirolides A–D

group were missing from the ^{1}H-NMR spectra while they appeared as a result of the decomposition. The structure of the diterpene part of (415) and (416) was suggested by the ^{1}H-NMR spectral data and by the NOE's shown in the Scheme. That (416) was the double bond isomer of (415), was easily deduced from the NMR spectra. Compounds (415) and (416) might be formed by a Diels-Alder type cycloaddition reaction between two dienophiles (409) and (410) present in *P. moritziana* and the diterpene diene (678).

Further study of the chemical constituents of *P. moritziana* resulted in the isolation of two additional C-35 terpenoids, plagiospirolides C (417) and D (418), and a C-30 terpenoid lactone (sesquiterpene + sesquiterpene), plagiospirolide E (419) (515). On GC analysis, compound (417) and (419) decomposed to give diplophyllolide (409) and diplophyllin (410), respectively, and the diterpene alcohol ($C_{20}H_{32}O$) (421), indicating that the molecular formula of (417) and (418) was $C_{35}H_{52}O_3$. The NMR spectra of (417) and (418), similar to those of (415) and (416), and the molecular formulas showed that (417) and (418) possessed the same

structures with an extra hydroxyl group being located within the
diterpene moiety. The position and the relative stereochemistry of the
hydroxyl group at C-21 were confirmed by an ^{1}H-^{1}H-2D-COSY spectrum and NOE spectrometry. Plagiospirolide E (419) decomposed during
GC analysis to give diplophyllin (410) and a sesquiterpene hydrocarbon,
$C_{15}H_{22}$ ([M]$^+$ 204) the mass spectrum of which was closely related to
that of anastreptene (93), thus suggesting that the sesquiterpene moiety
possessed an aromadendrane skeleton (Scheme 32). Structure (419) was
proposed not only on the basis of the assumption that (419) is formed by a
Diels-Alder type reaction with between an α-methylene-γ-lactone moiety
as dienophile and the aromadendrene derivative (422) as conjugated
diene but also because of the NOE's indicated in the scheme and the ^{1}H-
and ^{13}C-NMR spectra. The absolute configuration of the plagiospirolides A–E (415–419) has not been established; however, it is probable that
the spirolactones should be represented as shown, because the absolute
configuration of (409) and (410) has been established previously. C-35 and
C-20 terpene dimers are very uncommon structures in the plant kingdom
and have so far not been found in other bryophytes. These spirolactones
are not artifacts since they are detected by TLC in the crude extract
immediately after extraction at room temperature (515).

Dihydroagarofuran (420) has been detected in *Fossombronia pusilla*
(491) and *Symphyogyna brasiliensis* by GC-MS (513).

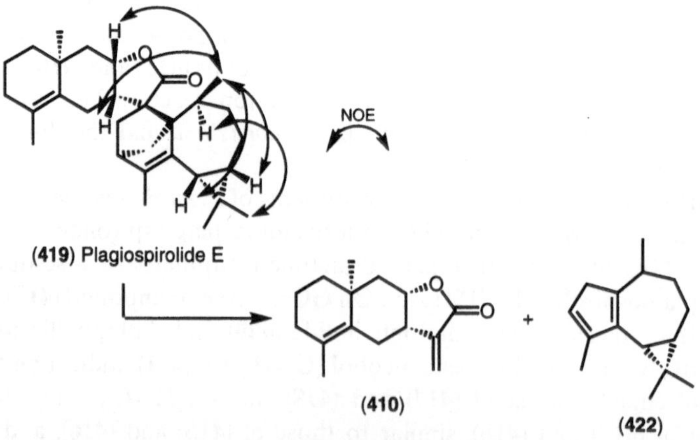

(419) Plagiospirolide E

(410)

(422)

Scheme 32. Degradation of plagiospirolide E

5.18 Farnesanes and Germacranes

Scapania ornithopodioides contains α-farnesene **(423)** *(638)*. *Trans*-β-farnesene **(424)** has been detected in fourteen liverworts collected in South America *(45, 47, 219)*. *Gymnocolea, Lophocolea, Monoclea, Radula* and *Scapania* species also produce *trans*-β-farnesene as shown in Table IIb. The essential oil of *Plagiochila ovalifolia* contains *trans*-farnesol **(425)** *(62)*. *Gymnocolea inflata (432)* and *Lophocolea heterophylla (527a)* biosynthesize (+)-nerolidol **(426)**. The same alcohol has also been isolated from the essential oils of *Plagiochila ovalifolia* and *Wiesnerella denudata (62)*. 4,5-Dehydronerolidol **(427)** has been isolated from a large thalloid liverwort, *Dumortiera hirsuta*, as the major terpenoid component *(585)*. The same compound has been obtained from *Brickellia californica* (Compositae) *(117)*.

Germacrene-B **(428)** has been detected in the essential oil of *Conocephalum japonicum* and *Plagiochila ovalifolia (62)*. Germacrene-D **(429)** has been found in the essential oil of *Conocephalum conicum (62)* and in the crude extracts of *Bazzania praerupta (341)*, *Marchantia foliacea (39)* and South American *Lejeunea* species *(219)*, *Neteroclada confluens*, *Gackstroemia magellanica (45)*, *Lophocolea bidentata* and *L. heterophylla (527a)*. Two new gemacrane-type sesquiterpene alcohols, *ent*-germacra-4(15),5, 10(14)-trien-1β-ol **(431a)** and *ent*-germacra-4(15),5,10(14)-trien-1α-ol **(432)**, have been isolated from *Jackiella javanica (437)*. Acetylation of **(431a)** gave a mono-acetate **(431b)** the ^{13}C-NMR spectrum of which was identical with that of germacra-4(15),5,10(14)-trien-1β-yl acetate **(430b)**

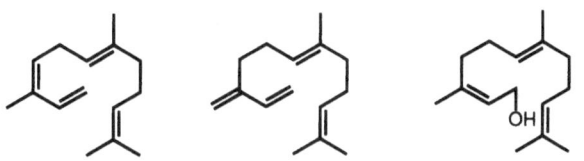

(423) α-Farnesene (424) *trans* -β-Farnesene (425) *trans,cis*-Farnesol

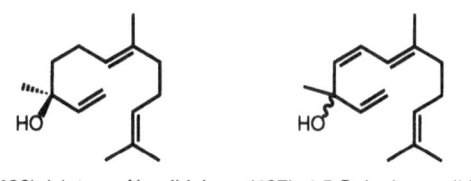

(426) (+)-*trans*-Nerolidol   (427) 4,5-Dehydronerolidol

Chart 26. Farnesanes found in the Hepaticae

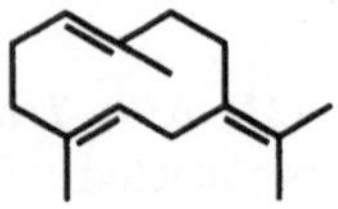

(428) Germacrene-B 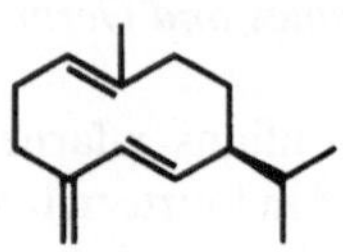   **(429)** Germacrene-D 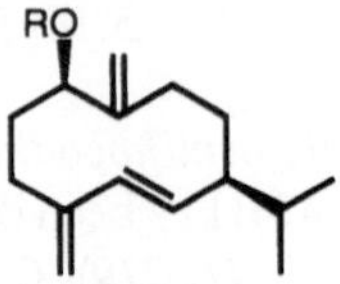   **(430a)** R=H
                                                        (430b) R=Ac

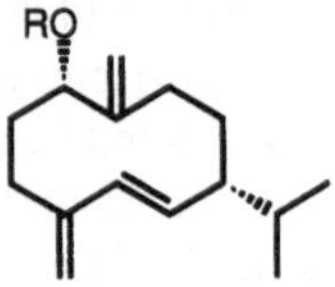

(431a) *ent*-Germacra-4(15),5,10(14)-trien-1β-ol ; R=H
(431b) R=Ac

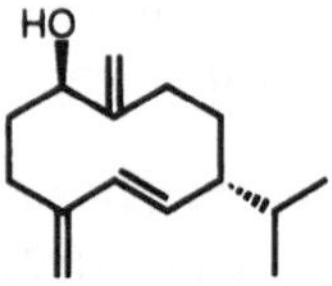

(432) *ent*-Germacra-4(15),5,10(14)-trien-1α-ol

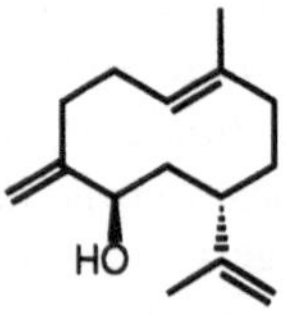

(433) 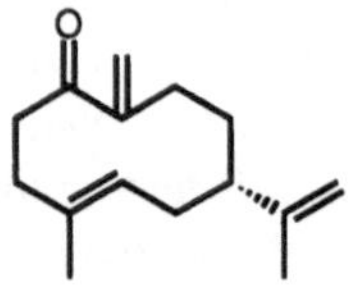   **(434)** 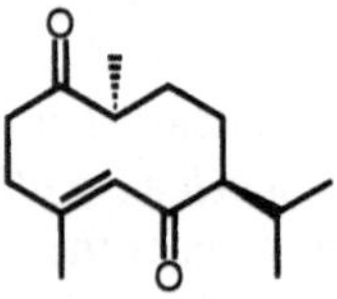   **(435)** 1,6-Diketogermacrene

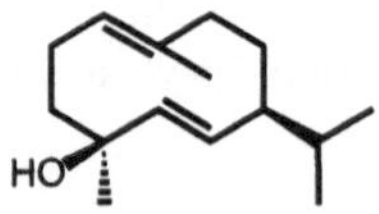

(436) 4β-Hydroxygermacra-1(10),5-diene 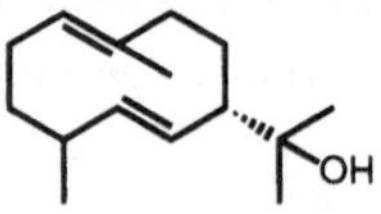   **(437)** *ent*-1(10)*E*, 5*E*-Germacradien-11-ol

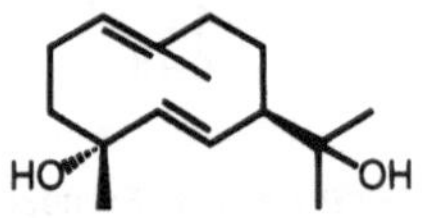

(438) Germacra-1(10),5-dien-4,11-diol 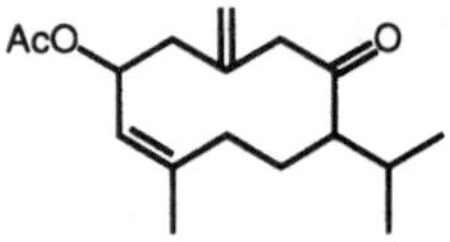   **(439)** 2-Acetoxy-8-keto-germacrene

Chart 27a. Germacranes found in the Hepaticae and their derivatives

isolated from a brown algae (*179*), indicating that **(431a)** was germacra-
4(15),5,10(14)-trien-1β-ol **(430a)** or its enantiomer. All other NMR spectral data supported this structure. The sign of the specific rotation of
(431a) ($[\alpha]_D$ + 146.3) was opposite to that of **(430a)** ($[\alpha]_D$ − 180.3). Thus,
the stereochemistry of **(431a)** at C-1 and C-7 was *R*. All spectral data of

(432) were consistent with those of (431a), except for the chemical shift of the methine proton bearing the hydroxyl group, thus showing that (432) was the C-1 epimer of (431a). Two germacrane-type sesquiterpenoids (433, 434) possessing 7H-β configuration have been isolated from the brown algae, *Dictyopteris* species (*180*). 4β-Hydroxygermacra-1(10),5-diene (436) has been isolated from South American *Marchantia plicata* (*433*), German *Conocephalum conicum* (*553*) and *Porella swartziana* (*569*). Spectral data and the sign of rotation were identical with those of (436) isolated previously from a higher plant, *Pseudobrickellia brasiliensis* (Compositae) (*120*).

Ent-1(10)*E*,5*E*-germacradien-11-ol (437) has been isolated from a large thalloid liverwort, *Dumortiera hirsuta* (*585*). Its enantiomer has been obtained earlier from a higher plant, *Ferula communis* (Umbelliferae) (*607*). *Bryopteris filicina* (*432a*) and *Ptychanthus striatus* (*249a*) contain germacra-1(10),5-dien-4,11-diol (438). Two new ketogermacrenes (435, 439) have been isolated from *Porella swartziana* (*569*) and *Conocephalum conicum* (*585*), respectively. Their gross structures were deduced from extensive spectral data. *Lophocolea heterophylla* produces furanogermacra-1(10),4-diene (= furanodiene) (440) (*527a*) which has been obtained from the higher plant *Curcuma zedoaria* (*206a*).

(+)-Costunolide (441) and dihydrocostunolide (442) were isolated from a few Jungermanniales and Marchantiales species as reported earlier (*19*). The same lactones have since been found in *Clasmatocolea humilis* and *Plagiochila hondurensis* (*48*). *Frullania tamarisci* subsp. *nisquallensis*, *F. californica* (*63*) and *F. serratta* also elaborate costunolide (*51*), while *F. serratta* produces tulipinolide (443) and its (11*R*)-dihydro derivative (444) which are also found in the thalloid liverwort *Wiesnerella denudata* (*19*). (11*S*)-Dihydrotulipinolide (445) has also been isolated from *F. serratta*. These are the first report of (443) and its dihydro derivatives (444, 445) from the Jungermanniales. (11*S*)-Dihydrotulipinolide (445) has been newly isolated from East Malaysian *Wiesnerella denudata* (*51*).

Four new germacranolides, 4α,5β-epoxy-7α,8β,11α-H-germacra-1(10)-en-12,8α-olide (= 11(13)-dihydro-4α,5β-epoxy-8-epi-inunolide) (447), 5-keto-7α-8β,11α-H-germacra -1(10)-en-12,8α-olide (448), 1α-hydroperoxy-4α,5β-epoxygermacra-10(14),11(13)-dien-12,8α-olide (449) and 1β-hydroperoxy-4α,5β-epoxygermacra-10(14),11(13)-dien-12,8α-olide (450), have been isolated from *Porella acutifolia* subsp. *tosana*, together with the previously known 4α,5β-epoxy-8-epi-inunolide (446) (*82, 600*). The last-named epoxide (446) had previously been isolated from the Compositae; its structure rests on X-ray crystallographic analysis (*127*). The ¹H-NMR data of (446), especially the absence or presence of the small coupling constants between i) H-8β and H-9β, and ii) H-5α and H-6α and iii) H-6α

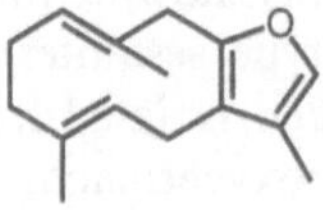

(440) Furanogermacra-1(10),4-diene (=Furanodiene)

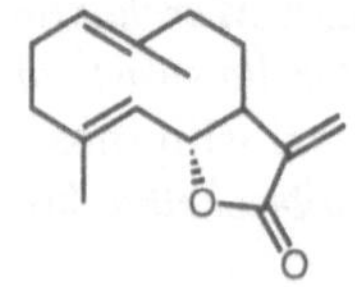

(441) Costunolide 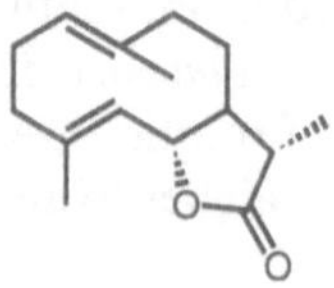   (442) Dihydrocostunolide 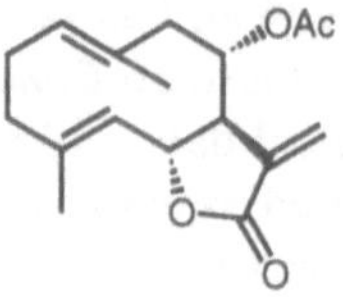   (443) Tulipinolide

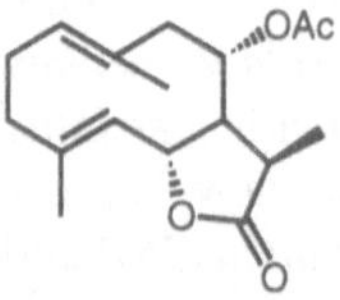

(444) (11R)-Dihydrotulipinolide   (445) (11S)-Dihydrotulipinolide

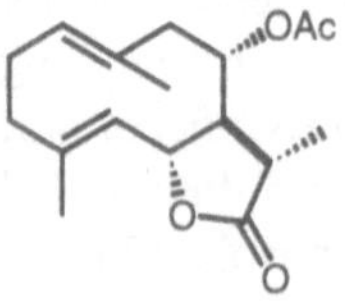

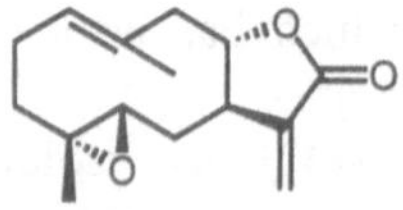

(446) 4α,5β-Epoxy-8-epi-inunolide

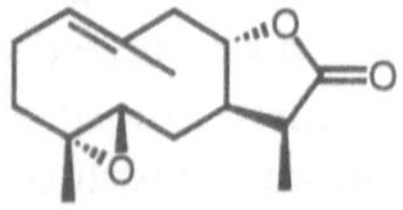

(447) 4α,5β-Epoxy-7α,8β,11α-H-germacra-1(10)-en-12,8α-olide
(=11(13)-Dihydro-4α,5β-epoxy-8-epi-inunolide)

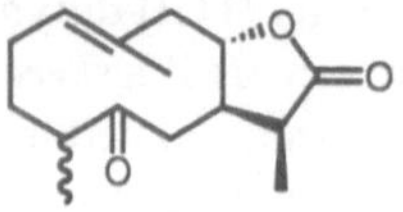

(448) 5-Keto-7α,8β,11α-H-germacra-1(10)-en-12,8α–olide

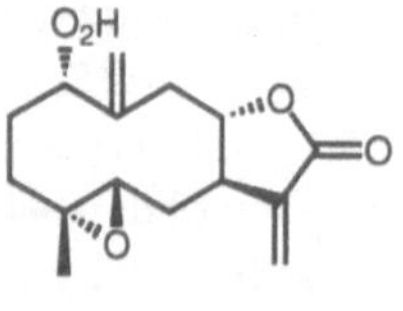

(449) 1α-Hydroperoxy-4α,5β-epoxygermacra-10(14),11(13)-
dien-12,8α-olide

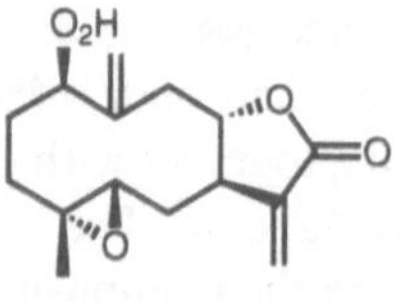

(450) 1β-Hydroperoxy-4α,5β-epoxygermacra-10(14),11(13)-
dien-12,8α-olide

Chart 27b. Germacranes found in the Hepaticae

and H-7α (dihedral angles ca. 90°) were very important for structure determination of the other new germacranolides present in this subspecies. Treatment of (446) with SOCl₂-CHCl₃ gave a guaianolide (446a)

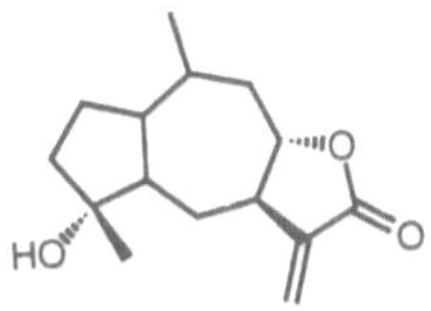

(446a)

which was also isolated from a methanol extract obtained in the absence of light (82). Compound (447) is the C-11(13) dihydro derivative of (446), the presence of an NOE between H-6α and H-13 establishing the orientation of the C-11 methyl group assigned as β. The structure of (448) was also established by ¹H NMR spectrometry, with the stereochemistry of the C-11 methyl group as β-pseudoaxial by the solvent shift method of NARAYANAN (443) (δC₆D₆-δCDCl₃ = 0.52). That compounds (449) and (450) were 1-hydroperoxides of (446) was confirmed by field desorption mass spectrometry, ¹H NMR spectral data, spin decoupling and 2D-COSY NMR data and the color change (colorless to brown) of the solution when 20% potassium iodide solution was added to (449) in ether. A partial synthesis of (449) and (450) which presumably mimics the biogenetic pathway (171) and also furnished the α,β-unsaturated ketone (446b) is shown in Scheme 33.

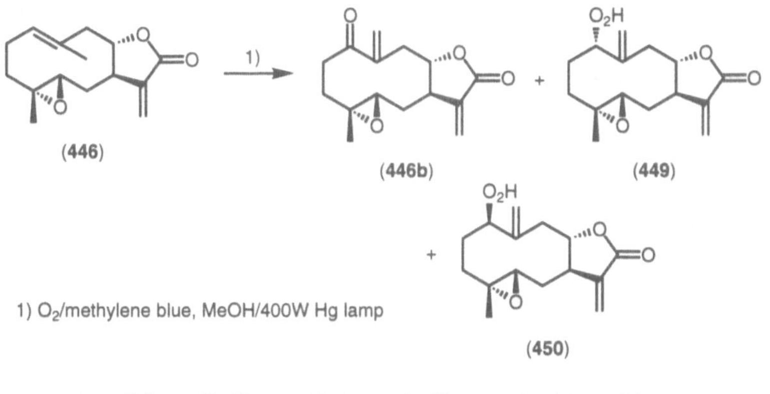

(446)   (446b)   (449)

(450)

1) O₂/methylene blue, MeOH/400W Hg lamp

Scheme 33. Photooxidation of 4α,5β-epoxy-8-epi-inunolide

5.19 Gorgonanes and Guaianes

As reported earlier (*19*) *ent*-maalioxide (**453**) has been isolated from *Jubula japonica* and *Plagiochila sciophila* (= *P. acanthophylla* subsp. *japonica*). Since then the same ether has also been obtained from *Lophozia ventricosa* (*146, 287, 429, 566*) and *Barbilophozia floerkei* (*429*) and *Jubula japonica* (*585b*).

Iso-α-gurjunene (**454**) which has been isolated previously from *Pellia epiphylla* (*19*) has more recently been isolated by low temperature HPLC from *Monoclea gottschei* subsp. *neotropica* and *M. forsteri* along with a new guaiane-type sesquiterpene hydrocarbon, guai-4,11(12)-diene (**455**) (*514*). The structure of (**455**) was elucidated by comparing the ¹H NMR spectrum with that of a co-metabolite iso-α-gurjuene B (**454**). (1S,10R)-Guai-4,6-diene (**456**), the C-10 epimer of (**454**), has also been isolated from *Bryopteris filicina* (*432*). The enantiomers of (**454**) and (**456**) have been prepared from γ-gurjunene (**459**) (*183a*). *Monoclea forsteri* also produces α- (**457**) and β-guaiene (**458**) (*513*). β-Guaiene (**458**) has been found in *Frullania* and *Radula* (*513*). γ-Gurjunene (**459**) has been detected in *Mylia nuda* by GC-MS analysis (*624*). *Porella swartziana* contains two new guaienes, guai-4(15)-en-6-one-1-ol (**460**) and its double bond isomer (**461**) (*249, 574a, 574b*).

Two new guaianolides, isoporelladiolide (**462**) and dehydroisoporella-diolide (**463**) have been isolated from *Porella acutifolia* subsp. *tosana* (*600*), along with the known porelladiolide (**464**) previously obtained from *Porella japonica* (*19*). 8α-Acetoxyzaluzanin D (**467**) isolated from Japanese *Wiesnerella denudata* has also been obtained from East Malaysian *W. denudata*, along with tulipinolide (**443**) and related germacranolides (*51*). A new guaianolide, dihydroestafiatin (**465**) has been isolated from Bolivian *Frullanoides densifolia* (*14, 554, 556*), along with estafiatin (**466**) which was first obtained from *Artemisia mexicana* (Compositae) (*485*). The NMR spectrum of (**465**) was similar to that of (**466**), except for saturation of the exomethylene group. Reduction of (**466**) with NaBH₄ in EtOAc gave (**465**). The configuration of the secondary methyl group at C-11 was established by a combination of NOE spectrometry and Narayanan's ¹H NMR solvent shift method (*443*).

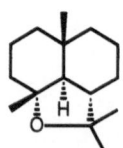

(**453**) *ent*-Maalioxide (=Ventricosin B)

Chart 28. Gorgonane found in the Hepaticae

(454) Iso-α-gurjunene B (455) Guai-4,11(12)-diene (456) (1S,10R)-Guai-4,6-diene
(=10-Epi-iso-α-gurjunene B)

(457) α-Guaiene   (458) β-Guaiene   (459) γ-Gurjunene

(460) (1R*,5S*,7S*,10R*)- Guai-4(15)-en-6-one-1-ol

(461) (1R*,5S*,7S*,10R*)-
Guai-3-en-6-one-1-ol

(462) Isoporelladiolide

(463) Guai-1(10),3,11(13)-trien-14,2β,12,6α-diolide
(=Dehydroisoporelladiolide)

5.20 Himachalanes and Humulanes

α- (474), β- (475) and γ-Himachalenes (476) have been reported previously from several liverworts (19). More recently α-himachalene (474) has been found in four South American *Plagiochila* and β-himachalene (475) from three different South American *Plagiochila* and

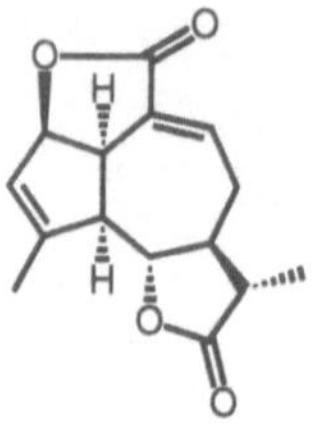

(464) Porelladiolide 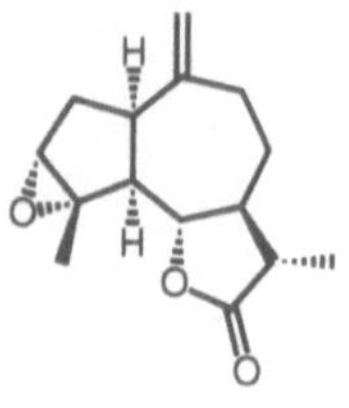   (465) Dihydroestafiatin 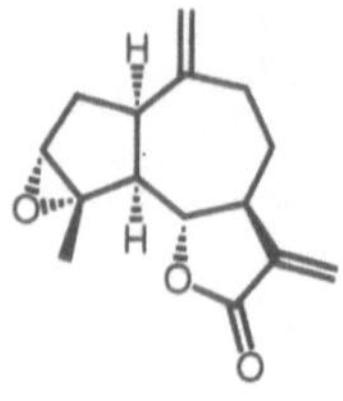   (466) Estafiatin

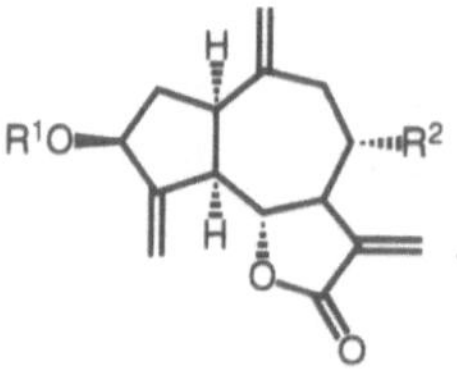

(467) 8α-Acetoxyzaluzanin D ; R¹=Ac, R²=OAc
(468) Zaluzanin D ; R¹=Ac, R²=H
(469) 8α-Acetoxyzaluzanin C ; R¹=H, R²=OAc

Chart 29b. Guaianes found in the Hepaticae

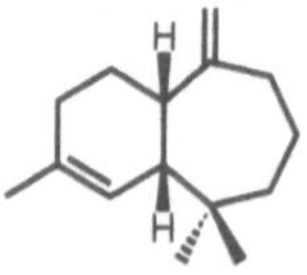

(474) α-Himachalene 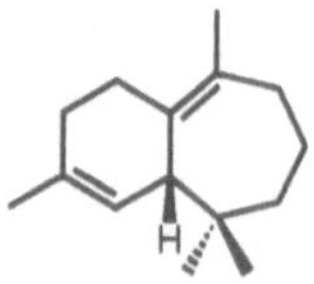   (475) β-Himachalene 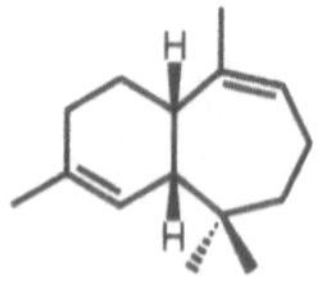   (476) γ-Himachalene

Chart 30. Himachalanes found in the Hepaticae

Triandrophyllum species (*45*). The latter compound has also been detected in *Lophocolea heterophylla* (*527a*) and *Radula boryana* (*513*).

The previous report (*19*) listed five species of Jungermanniales as sources of α-humulene (**477**). Three *Lejeunea* species (*219*), *Scapania robusta* (*631*) and *Lophozia ventricosa* (*429*) also produce α-humulene. (+)-Bicyclohumulenone (**479**) which possesses a strong odor of mushrooms has been isolated from *Plagiochila sciophila* (= *P. acanthophylla* subsp. *japonica*) (*19*). The enantiomer of (**479**) is known from the higher plant *Acritopappus prunifolius* (Compositae) (*119*). Humulenyl acetate (**478**) and 5-hydroxyisobicyclohumulenone (**480**) have been isolated from an unidentified South American *Frullania* species (*249*) and Japanese *Jubula japonica* (*585b*), respectively. (±)-Bicyclohumulenone (**479**) has been synthesized from humulene epoxide (**77**) by a conformationally selective transannular cyclization reaction (*505*). An alternative synthesis of (**479**) has been accomplished by TAKAHASHI et al. (*529*). The first total

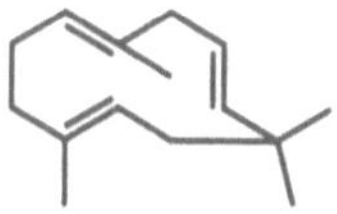

(477) α-Humulene

(478) Humulenyl acetate

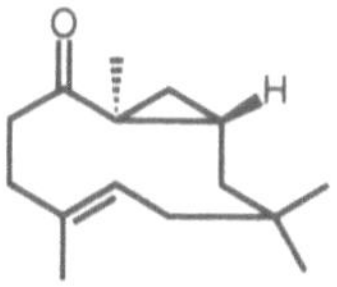

(479) (+)-Bicyclohumulenone

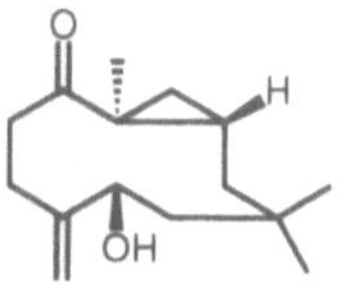

(480) 5-Hydroxyisobicyclohumulenone

Chart 31. Humulanes found in the Hepaticae

synthesis of optically active (+)-bicyclohumulenone (479) has been reported by FUKUYAMA et al. (187) in 12 steps.

5.21 Longifolanes, Longibornanes, Longipinanes and Longicyclanes

Ent-longifolene (481), ent-longiborneol (482), ent-α- (483) and ent-β-longipinene (484) and longipinanol (485) were earlier reported from Scapania undulata together with ent-longicyclene (486) (19). Longifolene (481) and isolongifolene (487) have since been detected in Scapania subalpina and S. uliginosa (289). Mastigophora diclados (621, 635), Plagio-chila moritziana (513) and Scapania robusta (621, 631) also elaborate longifolene (481). Longiborneol (482) has also been obtained from Chiloscyphus pallescens (149). Longiborneol (482) and longicyclene (486) have been totally synthesized as racemates by WELCH et al. (615). Three ent-longipinane-type sesquiterpenoids, marsupellol (488), marsupellone (489) and acetoxymarsupellone (490) which are chemical markers of Marsupellaceae were also previously isolated from Marsupella emargi-nata subsp. tubulosa (19). Further study of the ether extract of German M. aquatica resulted in isolation of a new ent-longipinane derivative whose structure was shown to be (491) by a combination of NMR spectrometry and the transformations shown in Scheme 34 (279). The absolute configuration of (491) has been settled by comparison of its CD spectrum with that of (−)-marsupellone (489) (390). French M. emarginata pro-duces three new longipinane-type sesquiterpenoids, 9,11α,14-triacetoxymarsupellone (492), 9,11β,14-triacetoxy-marsupellone (493)

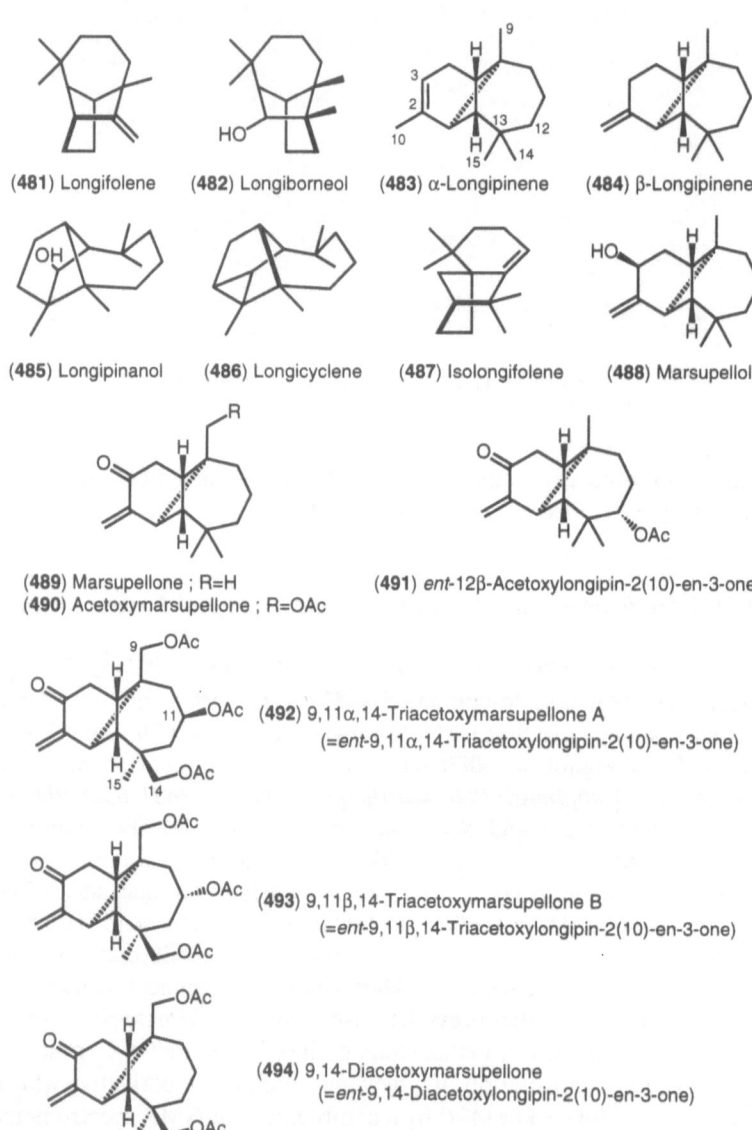

(481) Longifolene   (482) Longiborneol   (483) α-Longipinene   (484) β-Longipinene

(485) Longipinanol   (486) Longicyclene   (487) Isolongifolene   (488) Marsupellol

(489) Marsupellone ; R=H   (491) *ent*-12β-Acetoxylongipin-2(10)-en-3-one
(490) Acetoxymarsupellone ; R=OAc

(492) 9,11α,14-Triacetoxymarsupellone A
        (=*ent*-9,11α,14-Triacetoxylongipin-2(10)-en-3-one)

(493) 9,11β,14-Triacetoxymarsupellone B
        (=*ent*-9,11β,14-Triacetoxylongipin-2(10)-en-3-one)

(494) 9,14-Diacetoxymarsupellone
        (=*ent*-9,14-Diacetoxylongipin-2(10)-en-3-one)

Chart 32. Longifolanes, longibornanes, longipinanes and longicyclanes found in the Hepaticae

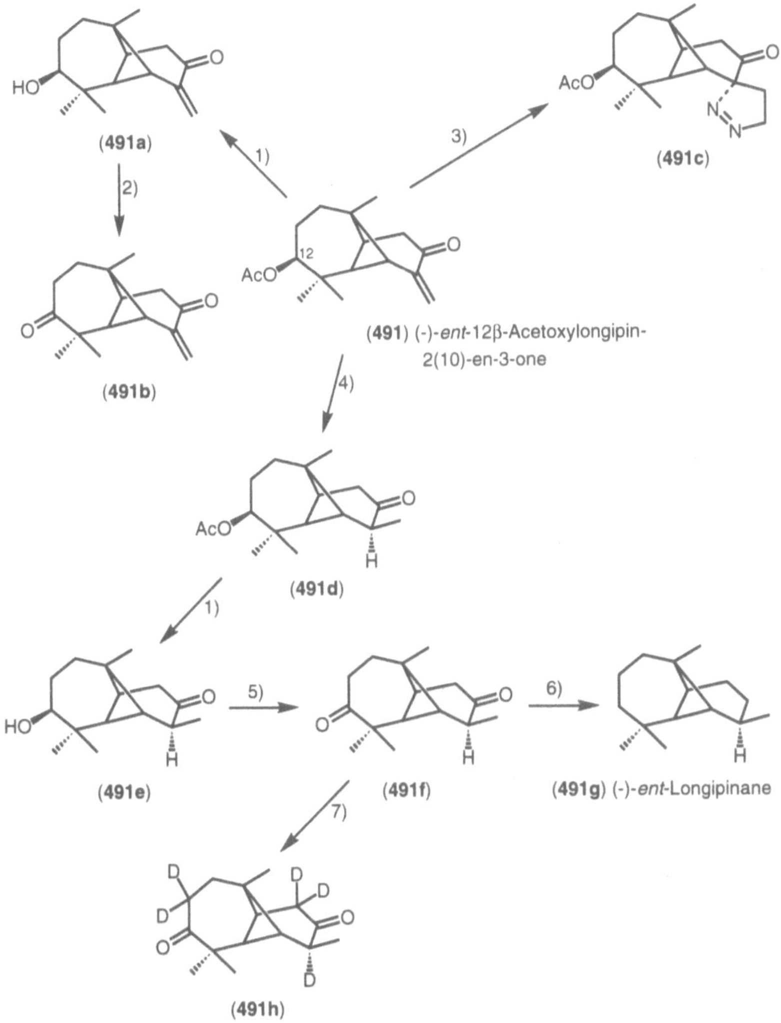

(491a)

(491b)

(491) (-)-*ent*-12β-Acetoxylongipin-2(10)-en-3-one

(491c)

(491d)

(491e)

(491f)

(491g) (-)-*ent*-Longipinane

(491h)

1) KOH/MeOH 2) CrO₃-Py/CH₂Cl₂ 3) 4) H₂/Pd-C 5) Jones oxd./Me₂CO
6) Wolff-Kishner redc. 7) CD₃OD/CD₃ONa

Scheme 34. Reactions of 12β-acetoxylongipin-2(10)-en-3-one

and 9,14-diacetoxymarsupellone **(494)**, together with related longipinenes **(488–490)** *(434, 434a)*. Their stereochemistries were supported by 2D-COSY spectrometry and NOEs. *Plagiochasma rupestre* contains *ent*-marsupellone **(489)** *(243)*. This is the first example of a longipinane-type sesquiterpenoid in the Marchantiales.

5.22 Maalianes and Monocyclofarnesanes

Five *ent*-maaliane-type sesquiterpenes were previously reported in Jungermanniales, particularly in *Plagiochila* species (*19*). β-Maaliene (**495**) has again been detected in *Frullania, Monoclea, Plagiochila, Radula* (*513*) and *Scapania* species (*638*). *Plagiochila dura, P. lecheri* (*46*) and *P. moritziana* (*513, 515*) produce maaliol (**497**). *Ent*-γ-maaliene (**496**) has been isolated from *Riccardia chamedryfolia* (*434*). *Lepidozia vitrea* elaborates *ent*-maalian-5-ol (**498**) (*585a*) previously isolated from *Plagiochila ovalifolia* (*19*). From the ethanol extract of *Mylia taylorii*, a new maaliane-type sesquiterpene alcohol, (+)-*ent*-maali-4(15)-en-1β-ol (**499**) has been isolated together with a few aromadendrane- and secoaromadendrane-type sesquiterpeniods (*386, 531*). Oxidation of (**499**) with Jones reagent gave a cyclohexanone (**499c**). Acetylation of (**499**) gave a mono-acetate (**499a**) whose oxidation with OsO_4 and $NaIO_4$ afforded a keto acetate (**499b**) (Scheme 35). The above chemical evidence and the ^1H-NMR spectrum using a shift reagent [$Eu(fod)_3$] established structure (**499**) for the new alcohol.

The earlier volume (*19*) listed three monocyclofarnesane-type sesquiterpenoids (**500, 502, 503**) from *Ptychanthus striatus* (Lejeuneaceae) which were originally referred to as Ps-1, Ps-2 and Ps-2′ and later named striatene (**500**), striatol (**502**) and β-monocyclonerolidol (**503**), their structures being based on a combination of chemical degradation (Scheme 36) and spectroscopic evidence (*540*). The geometry of 8,9-double bond of (**500**) was determined Z by NOE difference spectrometry. The absolute configuration of (**500**) was established by application of the CD exciton

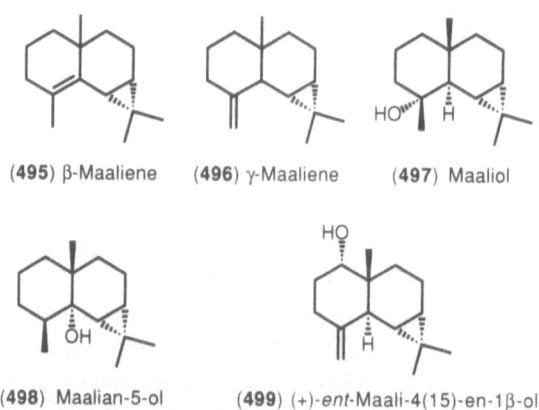

(**495**) β-Maaliene   (**496**) γ-Maaliene   (**497**) Maaliol

(**498**) Maalian-5-ol   (**499**) (+)-*ent*-Maali-4(15)-en-1β-ol

Chart 33. Maalianes found in the Hepaticae

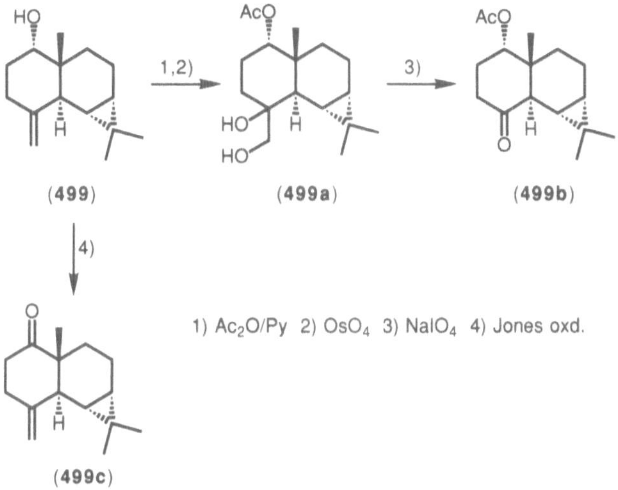

Scheme 35. Reactions of ent-maali-4(15)-en-1β-ol

chirality method to benzoate (500e) (234). The positive Cotton effect at 241 nm (Δε + 3.1) indicated that the exocyclic double bond and the benzoate group constitute a positive chirality. The absolute configuration of the tertiary hydroxyl group at C-9 of (502) was also established by comparing the CD spectrum of the benzoate (2a) of R-linalool (2) with that of the benzoate (502a) of (502) (Scheme 36). Both benzoates exhibited negative Cotton effects at 252 nm (Δε − 0.4 and − 0.5), respectively. The structure of (503) was settled by partial synthesis starting from (±)-ionone (56) (540). Striatene (500), striatol (502) and monocyclo-nerolidol (503) have been detected in twelve, six and one Lejeunea species by GC-MS (219). Monocyclonerolidol (503) and trans-γ-monocyclofar-nesol (504) have also been isolated from Spruceanthus polymorphus belonging to Lejeuneaceae (437) and Diplophyllum serrulatum (585a), respectively, while Porella densifolia subsp. appendiculata and P. densi-folia var. fallax produce striatene (500) and striatol (502) (67). A similar striatane-type sesquiterpene ketone, striatenone (505), has been isolated from North American Porella navicularis and European P. cordaeana (591, 593). The absolute configuration followed from the positive Cotton effect at 319 nm.

A new striatane-type sesquiterpene acid (501a) has been isolated from Malaysian Cheilolejeunea trifaria. The stereostructure was elucidated by 600 MHz NMR and NOESY spectrometry of its methyl ester (501b) (249).

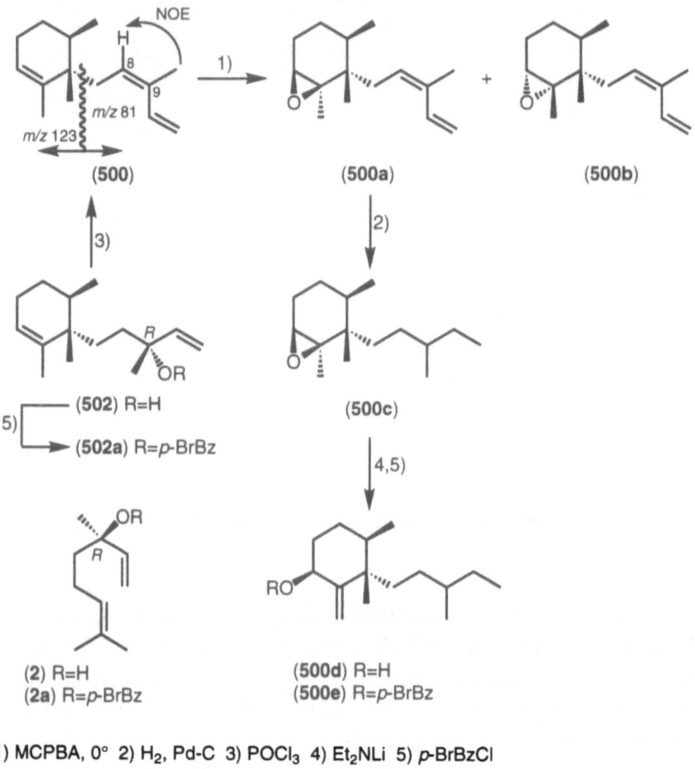
NOE
H
8
9
m/z 81
m/z 123
1)
+
(500)
(500a)
(500b)
3)
2)
R
OR
(502) R=H
5)
(502a) R=p-BrBz
(500c)
4,5)
OR
R
RO
(2) R=H
(2a) R=p-BrBz
(500d) R=H
(500e) R=p-BrBz
1) MCPBA, 0° 2) H2, Pd-C 3) POCl3 4) Et2NLi 5) p-BrBzCl

Scheme 36. Reactions of striatene

Three new monocyclofarnesane-type sesquiterpenoids, ricciocarpins A (**506**), B (**507**) and ricciofuranol (**508**), have been isolated from the dichloromethane extract of an axenic culture of *Ricciocarpos natans* (Ricciaceae) (*639*, *640*). Structures and stereochemistries of these compounds were established by a combination of ^{1}H, ^{13}C-NMR spectral data, ^{1}H-^{1}H- and ^{13}C-^{1}H-2D-COSY NMR and NOE spectrometry. The presence of a β-substituted furan in (**506**) was confirmed by the ^{1}H NMR signals at δ 7.42 (*br d*, J = 1.6 Hz, H-12), 7.39 (*t*, J = 1.6 Hz, H-11) and 6.38 (*br d*, J = 1.6 Hz, H-10) and the mass spectrum. The structure of (**507**) was easily deduced from strong IR absorption bands at 1790 and 1755 cm^{-1} assignable to a γ-lactone and the similarity of the ^{1}H and ^{13}C NMR spectra to that of (**506**), with the signals of the β-substituted γ-butenolide substituted for the β-substituted furan. NOE spectrometry showed that the relative stereochemistry of (**508**) was the same as that of (**506**) and

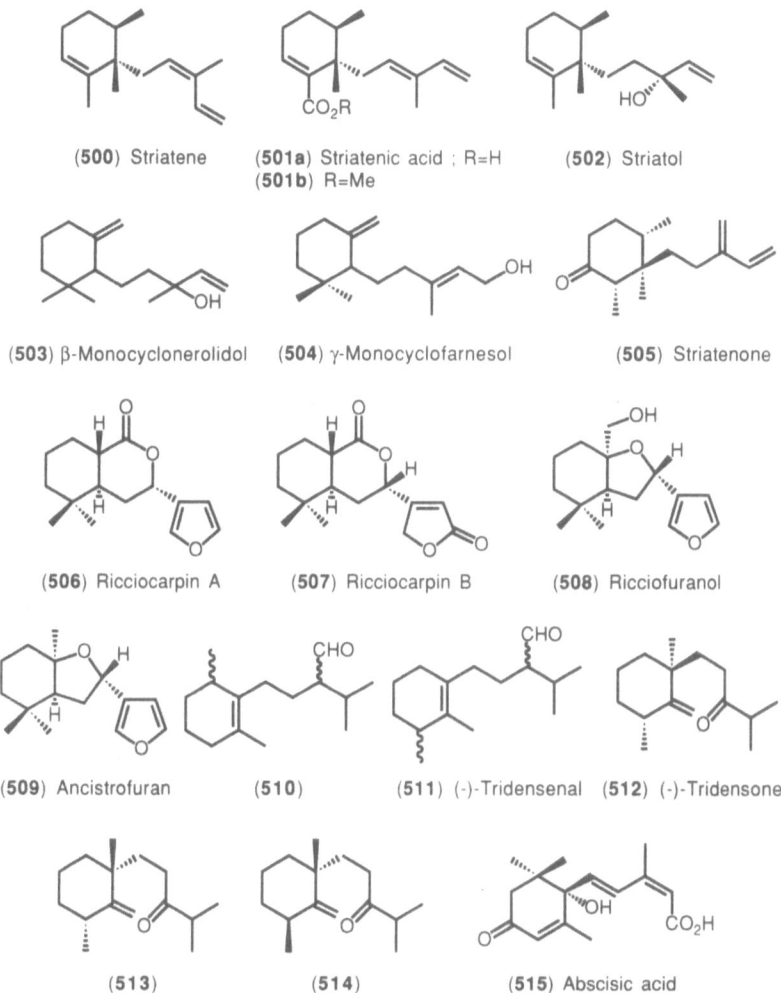

(500) Striatene   (501a) Striatenic acid ; R=H   (502) Striatol
                         (501b) R=Me

(503) β-Monocyclonerolidol   (504) γ-Monocyclofarnesol   (505) Striatenone

(506) Ricciocarpin A   (507) Ricciocarpin B   (508) Ricciofuranol

(509) Ancistrofuran   (510)   (511) (-)-Tridensenal   (512) (-)-Tridensone

(513)   (514)   (515) Abscisic acid

Chart 34. Monocyclofarnesanes found in the Hepaticae

(507). The hydoxyfuran (508) was obtained as a synthetic product prior to its isolation from the natural source (*86*). A similar sesquiterpenoid, ancistrofuran (509), has been isolated from the defense secretion of the West African termite *Ancistrotermes cavithorax* (*85, 86*). The total synthesis of (±)-ricciocarpin A (506) has been achieved by EICHER in 9 steps (*168*).

An aldehyde tridensenal has been isolated from Taiwanese *Bazzania tridens* and the monocyclofarnesane structure (510) proposed for it (*623*).

However, a synthesis of the mixture of diastereoisomers (510) clearly showed that this proposal was in error since the ^{13}C-NMR spectrum of the mixture differed from the spectrum of tridensenal (510) (628). On the other hand, one set of signals in the ^{13}C-NMR spectrum of diastereomer mixture (511) coincided with the ^{13}C signals of tridensenal; hence, the structure of tridensenal was established as (511) except for the stereochemistry at C-1 and C-9 (628). A similar sesquiterpene ketone, (−)-tridensone (512) accompanies (511); the structure (513) proposed for it rests on the ^{1}H- and ^{13}C-NMR spectra including HMBC and NOESY techniques (629). In the mass spectrum the presence of the base peak at M/z 123 resulting from a McLafferty-type cleavage followed by a loss of a hydrogen radical, further supports the proposed structure (513).

Tori et al. (560a) have accomplished the total synthesis of (+)-tridensone (514) and its diastereoisomer (513). Since the naturally occurring ketone is levorotatory it has the absolute configuration depicted in (512). Conocephalum conicum and Marchantia polymorpha contain abscisic acid (515) which is considered to be the dormancy inducing hormone in liverworts (442a).

5.23 Myltaylanes and Cyclomyltaylanes

Mylia species are rich sources not only of aromadendrane- and secoaromadendrane-type sesquiterpenoids but also verrucosane-type diterpenoids. From the ethanol extract of M. taylorii myltaylenol (= myltayl-4(12)-en-15-ol) (516) and cyclomyltaylenol (517) possessing new carbon skeleton have been isolated (533, 535). X-ray crystallographic analysis of a nor-keto benzoate (516b) derived from (516) by osmylation, periodate cleavage and benzoylation demonstrated that the molecule consisted of a norbornane system fused to a cyclohexane ring. The absolute configuration assigned to (516) rests on the sign of the Cotton effect at 291 nm ($\Delta\varepsilon$ + 2.08) of ketone (516a). Structure (517) for cyclomyltaylenol was suggested by acetylation, by comparison with myltaylenol (516) and by ^{1}H-^{1}H- and long range ^{13}C-^{1}H-2D-COSY spectrometry.

(516) Myltaylenol
(=Myltayl-4(12)-en-15-ol)

(516a) R=H
(516b) R=Bz

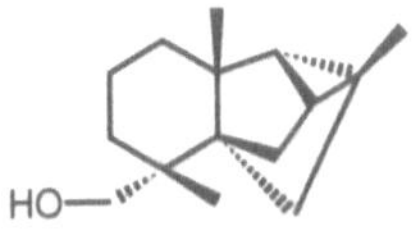

(517) Cyclomyltaylenol

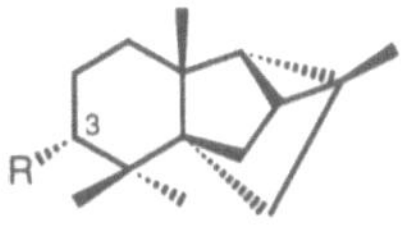

(518a) Cyclomyltaylane-3-ol ; R=OH
(518b) R=OCOC₆H₄Br(p)

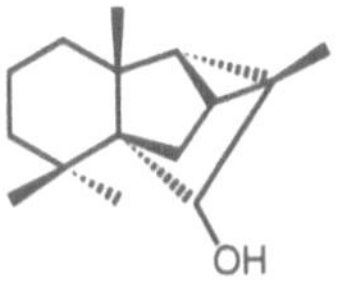

(519) Cyclomyltaylane-5-ol

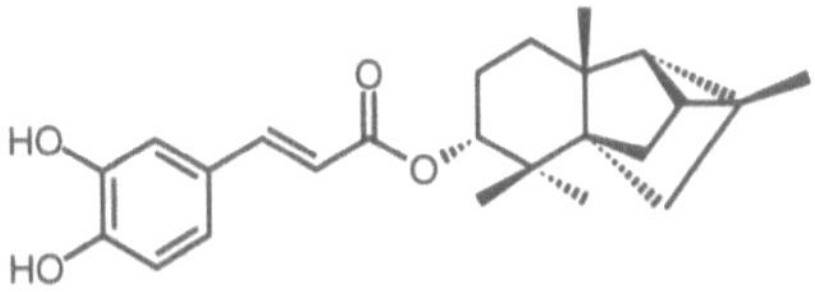

(520) Cyclomyltaylyl-3-caffeate

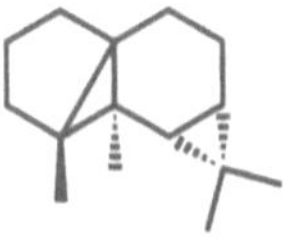

(521a)

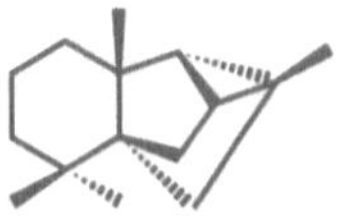

(521b) Cyclomyltaylane
(=Tridensene)

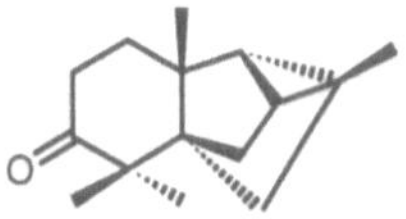

(522)

Chart 35. Myltaylanes and cyclomyltaylanes found in the Hepaticae

The configuration of the -CH$_2$OH group at C-11 has been deduced to be the same as that of myltaylenol using lanthanide shift reagents and NOE spectrometry.

Two new cyclomyltaylane-type sesquiterpenoids, cyclomyltaylan-3-ol (518a) and its caffeate (520) have been isolated from *Bazzania japonica*; relative configurations were established by a combination of various ^{1}H- and ^{13}C-NMR techniques and X-ray crystallographic analysis of a *p*-bromobenzoate (518b) (*84, 599*). The absolute configuration assigned to (518a) is based on the negative Cotton effect at 297 nm of a monoketone (522) prepared from (518a) by oxidation with pyridinium chlorochromate (*84*). This information supports the structure of cyclomyltaylenol (517) proposed by TAKAOKA *et al.* (*535*). Somewhat earlier a compound named tridensene was isolated from *Bazzania tridens* for which structure (521a) was proposed (*623, 628*). However, more recently the structure has been revised to cyclomyltaylane (521b) on the basis on 2D-COSYs and the HMBC technique (*627*). The relative stereochemistry is based on NOE difference spectrometry. *Mannia subpilosa* and *Reboulia hemisphaerica* produce cyclomyltaylane-5-ol (519) (*614*).

It is noteworthy that *Bazzania* (Lepidoziaceae) and *Mylia* (Junger-manniaceae) and *Mannia* and *Reboulia* (Aytoniaceae) produce compounds with the same biogenetically unique cyclomyltaylane-skeleton, although the three families are morphologically quite different.

The myltaylane skeleton may be derived from *trans-cis*-farnesyl pyrophosphate (**211**) through C-3, C-7 cyclization of β-chamigrene (**264**) followed by migration of the C-3 methyl group to the vicinal position (*533*).

5.24 Pacifigorgianes and Patchoulanes

Frullania tamarisci subsp. *tamarisci* produces a sesquiterpene alcohol possessing a strong mossy odor. A small amount of this alcohol named tamariscol MW 222, was isolated from European *F. tamarisci* subsp. *tamarisci* (*38*), and was shown to possess the pacifigorgiane carbon skeleton (**524**) and a positive rotation (+ 19.7°) (*144, 146, 150*). Analysis of the ¹H- and ¹³C-NMR spectroscopic data and the chemical transformations shown in Scheme 37 led to relative configuration (**523**). Formation of (**523c**) and (**523d**) can be rationalized in terms of alternative modes of decomposition of intermediate (**523b**). The enantiomer of (+)-tamariscol, (−)- (**523**), has been isolated from *F. tamarisci* subsp. *obscura* and *F. nepalensis* grown in Asia and *F. tamarisci* subsp. *asagrayana* grown in North America (*63*). The levorotatory isomer, (−)-(**523**) (− 20.5°), was converted to a hydrindanone (**523g**) whose optical rotation was positive and whose CD spectrum exhibited a negative Cotton effect. On the other hand, the same hydrindanone (**525h**) prepared from (−)-carvone (**523h**) had a negative rotation and exhibited a positive Cotton effect (*552, 572, 574*). Thus (−)-tamariscol has the absolute configuration shown in formula (**523**) and the (+)-tamariscol from European *F. tamarisci* is the enantiomer.

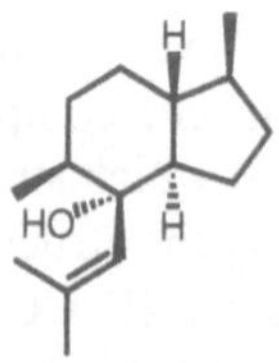

(**523**) Tamariscol 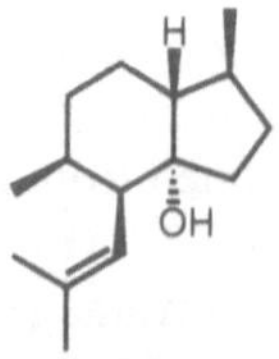   (**524**) Pacifigorgiol

Chart 36. Pacifigorgianes found in the Hepaticae and gorgonian

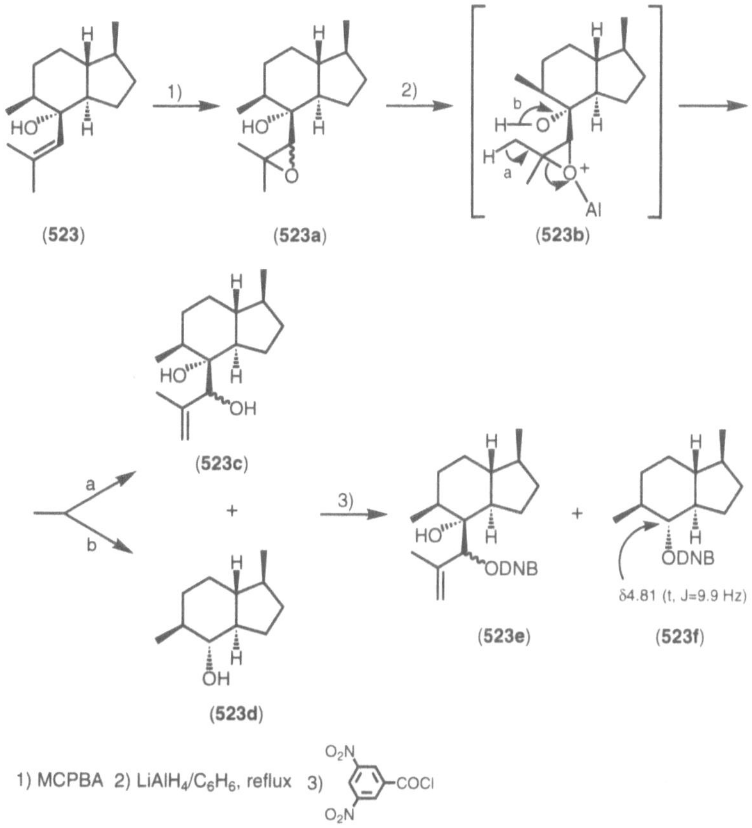

Scheme 37. Reactions of tamariscol

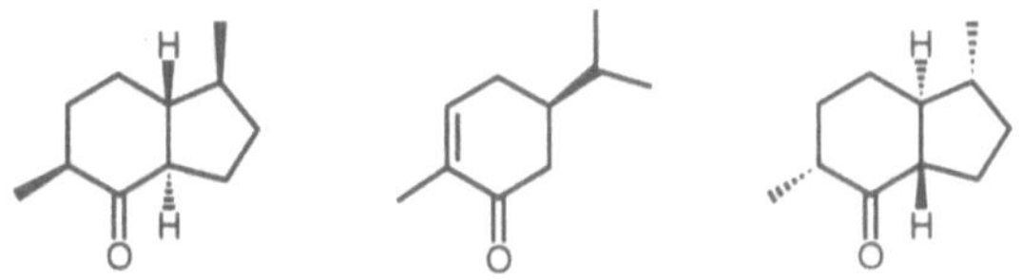

1) MCPBA 2) LiAlH₄/C₆H₆, reflux 3)

The total synthesis of (±)-tamariscol (523) has been accomplished using commercially available p-methoxylacetophenone in 13 steps, (552, 572, 574). CONNOLLY et al. (150) suggest that tamariscol is formed from β-caryophyllene (259) as shown in Scheme 38.

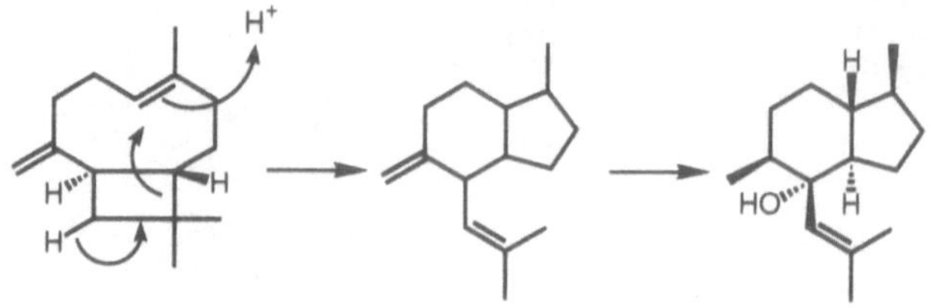

(259) β-Caryophyllene   (526) Pacifigorgiane   (523)

Scheme 38. Possible biogenetic pathway for tamariscol from β-caryophyllene

(527) α-Patchoulene   (528) β-Patchoulene

Chart 37. Patchoulanes found in the Hepaticae

It is interesting that pacifigorgiol (524), an ichthyotoxic substance with the same carbon skeleton has been isolated from a Pacific gorgonian, *Pacifigorgia* cf. *adamsii* (304).

Patchoulane-type sesquiterpenoids in the Hepaticae are very rare. α-Patchoulene (527) has been detected in *Mastigophora diclados* (621, 635), *Plagiochila panamensis* (513) and *Scapania ornithopodioides* (622, 638) by GC-MS. The β-isomer (528) has been detected in *Bazzania fauriana* (622).

5.25 Pinguisanes and Norpinguisanes

Pinguisane-type sesquiterpenoids whose carbon skeleton does not obey the biogenetic isoprene rule have so far not been found in higher plants but are limited to liverworts. Sixteen pinguisane- and four norpinguisane-type sesquiterpenoids were reported previously mainly from Lejeuneaceae, Porellaceae, Trichocoleaceae and Ptilidiaceae in the Jungermanniales (19). *Aneura pinguis* which belongs to the Aneuraceae (= Riccardiaceae) in the Metzgeriales also elaborates pinguisanes (534, 536, 537) (19). Several new pinguisane-type sesquiterpenoids have since been isolated from Lejeuneaceae and Porellaceae, together with previously known pinguisanes and norpinguisanes as shown in Table IIb.

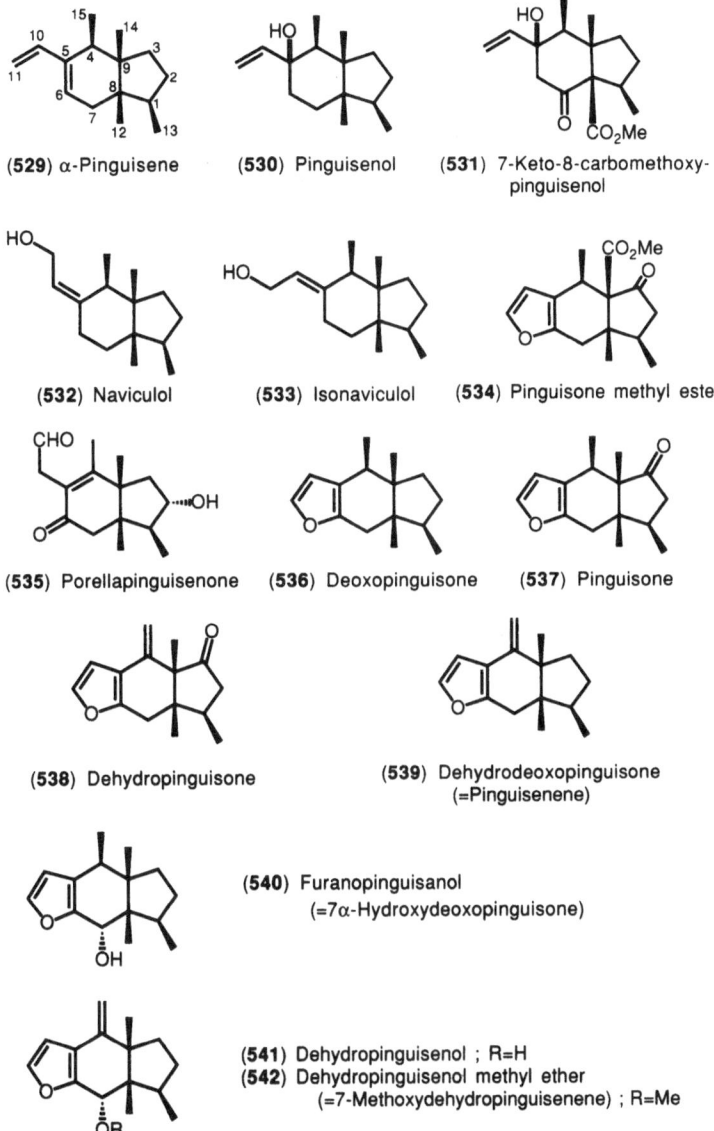

(529) α-Pinguisene

(530) Pinguisenol

(531) 7-Keto-8-carbomethoxy-pinguisenol

(532) Naviculol

(533) Isonaviculol

(534) Pinguisone methyl ester

(535) Porellapinguisenone

(536) Deoxopinguisone

(537) Pinguisone

(538) Dehydropinguisone

(539) Dehydrodeoxopinguisone
(=Pinguisenene)

(540) Furanopinguisanol
(=7α-Hydroxydeoxopinguisone)

(541) Dehydropinguisenol ; R=H
(542) Dehydropinguisenol methyl ether
(=7-Methoxydehydropinguisenene) ; R=Me

Chart 38a. Pinguisanes found in the Hepaticae

Some South American *Plagiochila* species (Plagiochilaceae) elaborate pinguisane-type sesquiterpenoids (*48*). Dehydropinguisone (**538**) has been isolated from New Zealand *Plagiochila retrospectans* and its structure established by X-ray crystallographic analysis of the *p*-bromobenzoate prepared from the derived mono alcohol (*435a*).

Structure (**553**) previously proposed for pinguisanin from *Lejeunea*, *Porella* and *Ptilidium* species (*19*) has been revised to (**554**) by means of 2D-COSYs and NOESY spectra; all carbon signals were assigned by considering substituent effects on the chemical shifts (*40*). In CDCl₃ solution pinguisanin (**554**) is slowly transformed into isopinguisanin (**554a**), a transformation which can be observed in the NMR tube. This appears to be due to acid catalysis as the transformation occurs also on treatment with acid (Scheme 39) while isopinguisanin seems to be stable

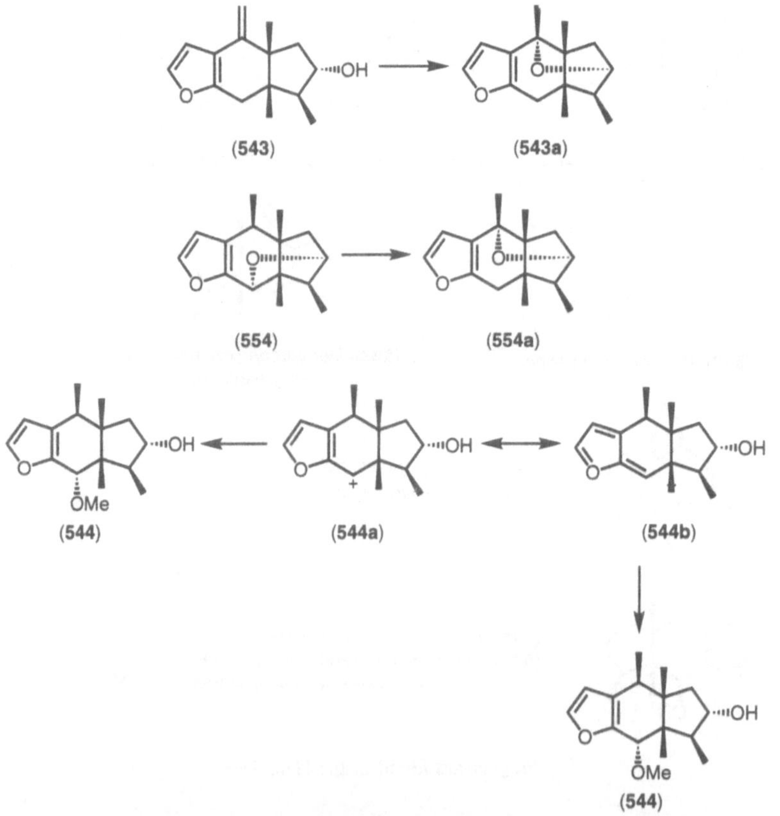

(543)   (543a)

(554)   (554a)

(544)   (544a)   (544b)

(544)

Scheme 39. Isomerization in the pinguisane series

under these conditions. The allylic cation (**544a, 544b**) may be an intermediate in this transformation (*146*). *Trichocoleopsis sacculata* produces deoxopinguisone (**536**) and pinguisone (**537**) (*19*) while *Neotrichocolea bissetii* produces pinguisanin (**554**) (*585a*).

British *Porella platyphylla* elaborates not only pinguisanin (**554**) as major component but also an unstable crystalline pinguisane-type sesquiterpene alcohol (**543**) which is gradually transformed into isopinguisanin (**543a**) in CDCl$_3$ in the NMR tube (Scheme 39) (*146*). Two methoxylated alcohols (**544, 545**) have also been isolated from the methanol extract of *P. platyphylla* and are probably artifacts formed by reaction of (**554**) with methanol either during the extraction or on silica gel column chromatography. Treatment of pinguisanin (**554**) in methanol solution in the presence of silica gel with a trace amount of mineral acid afforded (**554b**) as major product together with (**544, 545**) while (**554c**) was rapidly formed from pinguisanin (**554**) on treatment with DCl/CD$_3$OD in the NMR tube (*146*).

(554b)   (554c)

A pinguisane-type lactone isolated from British *P. platyphylla* had spectroscopic properties which are identical with those of pinguisanolide which was previously assigned structure (**559**) (Chart 38c) (*146*). The structure of pinguisanolide should, however, be revised to (**560**) because of the presence of an epoxide ring and the absence of a trisubstituted double bond. The relative configuration of the lactone ring of pinguisanolide differs from that of ptychanolide (**558**) because irradiation at the frequency of H-10 resulted in NOE's at H-11, H-4 and H-3α. Structure (**561**) previously assigned to isopinguisanolide (*19*) should be revised to (**562**). Pinguisanolide (**560**) has been found in *Neotrichocolea bissetii* (*585a*).

Reinvestigation of the chemical constituents of *Porella elegantula* which is indigenous to New Zealand resulted in isolation of two norpinguisane-type sesquiterpenoids and α-pinguisene (**529**) (*189*). Spectral properties of the first norpinguisane were identical with those of a previously reported norpinguisone methyl ester of presumed structure (**567b**), (see Chart 38d), however, analysis of the long range ^{13}C-^{1}H-2D COSY spectrum showed that formula (**567b**) was in error and that it

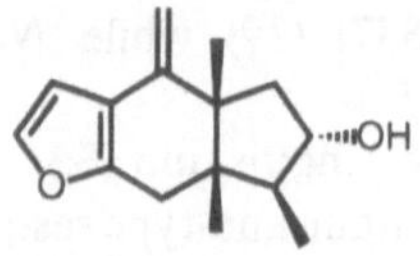

(543) 2-Hydroxypinguisenene

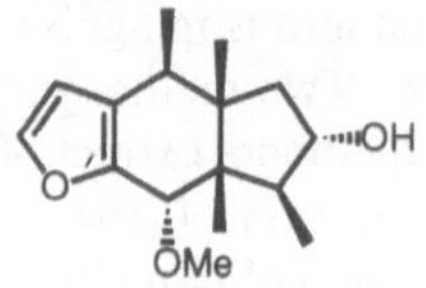

(544) 2-Hydroxy-7-methoxydeoxopinguisone

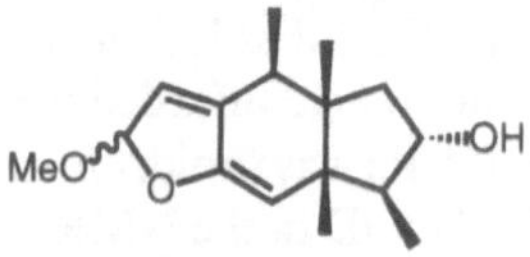

(545) 2-Hydroxy-11ζ-methoxypinguis-5(10),6-diene

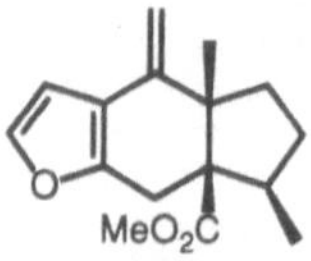

(546) Pinguisenene methyl ester (=Bryopterin A)

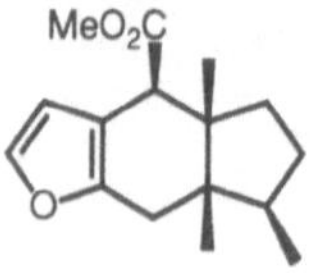

(547) Deoxopinguisone methyl ester

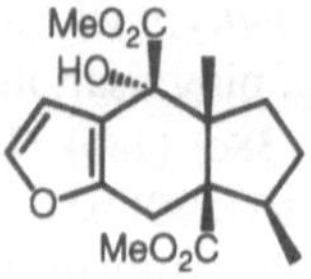

(548) 4-Hydroxydeoxopinguisone-12,15-dimethyl ester (=Bryopterin C)

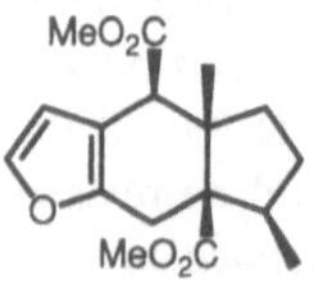

(549) Deoxopinguisone-12,15-dimethyl ester (=Bryopterin B)

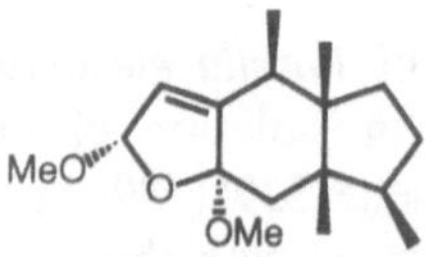

(550) 6α,11α-Dimethoxypinguis-5(10)-ene

Chart 38b. **Pinguisanes found in the Hepaticae**

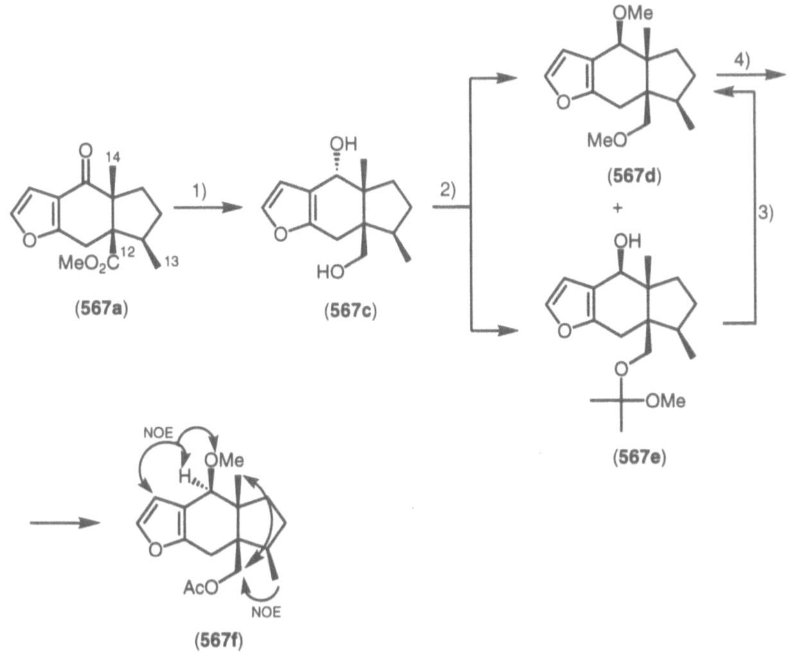

OMe

4)

MeO

(567d)

+

3)

OH

O

OMe

(567e)

O 14

1)

OH

2)

MeO₂C 12 13

HO

(567a)

(567c)

NOE

OMe

H‴

AcO

NOE

(567f)

1) LiAiH₄/Et₂O 2) 2,2-DMP/TsOH 3) 1N HCl/MeOH 4) Ac₂O/Py

Scheme 40. Reactions of norpinguisone methyl ester

should be revised to **(567a)** based on the chemical reactions shown in Scheme 40 and NOE spectrometry. Conversion of **(567c)** to a methyl ether was accompanied by inversion of the configuration on C-4 due to an S_N2 type displacement of the oxonium ion formed between a secondary hydroxyl groups and 2,2-dimethoxypropane in the presence of an acid catalyst by methanol generated *in situ* from the reagent.

The structure of the second norpinguisane, norpinguisanolide, from *Porella elegantula* was shown to be **(569)**, by ^{1}H-, ^{13}C- and 2D-COSY NMR spectrometry and the chemical reactions shown in Scheme 41. The absolute configuration of **(569)** was established by means of the *p*-bromobenzoate **(569b)** whose CD spectrum showed a first negative Cotton effect at 250 nm ($\Delta\varepsilon - 2.5$) arising from interaction between the *p*-bromobenzoyl group at C-4 and the furan chromophore indicating that C-14 was *S* (*189*). The five-membered lactone ring of **(569)** is in a sterically hindered environment due to the two methyl group on C-1 and C-4 so that ring opening of the γ-lactone under the usual basic condition does not occur. The same lactone has been isolated from American *Porella*

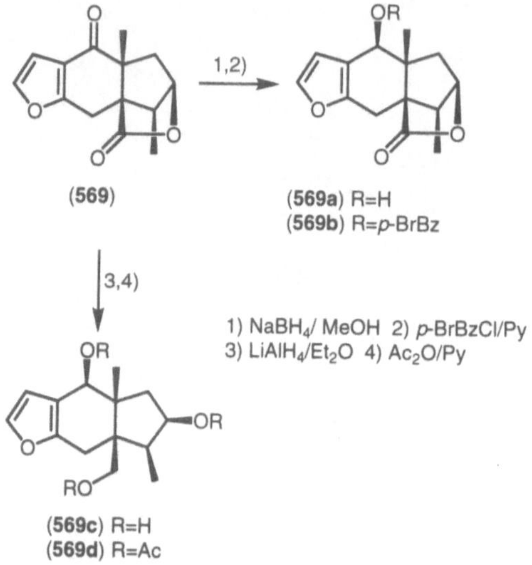

(569)

(569a) R=H
(569b) R=p-BrBz

3,4)

1) NaBH₄/ MeOH 2) p-BrBzCl/Py
3) LiAlH₄/Et₂O 4) Ac₂O/Py

(569c) R=H
(569d) R=Ac

Scheme 41. Reactions of norpinguisanolide

cordaeana (*239a*). Panamanian *Bryopteris filicina* contains bryopterin D which is 2-keto-norpinguisone methyl ester (568) (*432a*).

American *Porella navicularis* produces a new pinguisane-type alcohol, naviculol (532), together with the previously known norpinguisone (566) and norpinguisone methyl ester (567a) (*591*). The structure and relative configuration of (532) were derived by NMR spectrometry while the absolute configuration was settled because of the positive Cotton effect at 290 nm in the CD spectrum of the monoketone formed by ozonolysis of (532).

The stereostructure of ptychanolide (= PS-m-3) (558) (*19*) has been established by degradation to (558a) *via* LiAlH₄ reduction, spectral evidence, periodate cleavage, Bayer-Villiger oxidation of the resulting aldehyde and a second periodate cleavage, and by an X-ray crystallographic analysis (*540*). The absolute configuration ascribed to (558) depends on the positive Cotton effect at 296 nm ($\Delta\varepsilon$ + 2.6) of cyclopen-

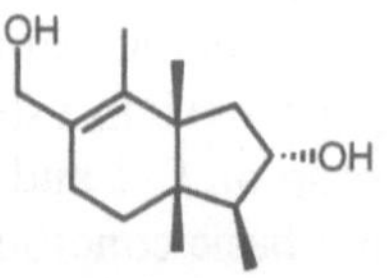

(535a) 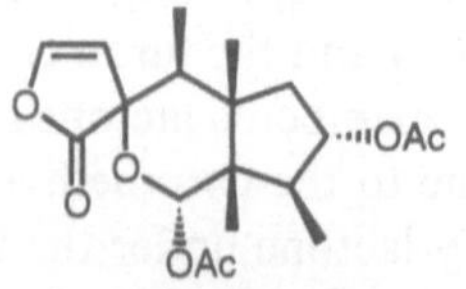   (557a) 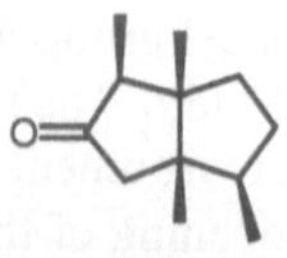   (558a)

tanone (**558a**) and is consistent with the absolute configuration of previously reported pinguisane-type sesquiterpenoids (*19*).

European *Porella cordaeana* contains two highly oxidized sesquiterpene lactones, porellapinguisanolide (**557**) and spiropinguisanin (**565a**), as well as a new ketoaldehyde, porellapinguisenone (**535**), along with pinguisanin (**554**) and norpinguisone methyl ester (**567a**) (*593*). The structure of (**557**) was deduced by extensive NMR spectroscopy. Acetylation of (**557**) gave a rearranged spirolactone diacetate (**557a**) which was formed by opening of the lactone ring followed by relactonization between aldehyde and carboxylic acid in the presence of pyridine.

That (**535**) has the same partial structure as (**557**) was clear from the NMR spectra. The locations of the other functional groups were derived from the NOE difference spectrum and decoupling experiments. Reduction of (**535**) with LiAlH$_4$ afforded an allylic alcohol (**535a**), indicating the presence of a ketone group at C-4 and the aldehyde group at C-10. The spirolactone (**565a**) is structurally similar to ptychanolide (**558**). In the ^{1}H-NMR spectrum of its diacetate (**565b**) the signal pattern was quite similar to that of the co-occurring pinguisanin (**554**), except for the absence of signals corresponding to the furan ring, suggesting that (**565a**) possessed the same C-1 to C-7 partial structure as pinguisanin. The remaining functions consisted of the γ-lactone and two acetoxyl groups, one of which was placed at the hemiacetal carbon (δH 6.29; δC 102.5). Combination of the various paths led to structure and relative configuration of (**565a**) for spiropinguisanin.

Porella acutifolia subsp. *tosana* produces a pinguisane-type keto ester (**531**) whose structure was deduced from the ^{1}H- and ^{13}C-NMR spectral data and spin decoupling experiments, with the position of each functional group being established by ^{13}C- and long range ^{13}C-^{1}H 2D-COSY-NMR spectra and with NOE difference spectrometry confirming the relative stereochemistry. A similar pinguisane-type sesquiterpenoid, pinguisenol (**530**), had previously been isolated from *P. vernicosa* and *P. densifolia* (*19*). Thus, (**531**) is 7-keto-8-carbomethoxypinguisenol.

Six pinguisane-type sesquiterpenoids, deoxopinguisone (**536**), dehydropinguisenol (**541**), pinguisanin (**554**), dehydropinguisanin (**555**), pinguisenal (**556**) and pinguisanolide (**560**) had been isolated previously from *Trocholejeunea sandvicensis* (*19, 82*). Further investigation of the *n*-hexane extract of this species resulted in the isolation of four additional pinguisanes (**540, 542, 550, 551**), together with the known pinguisanes (**541, 554, 555, 558**) (*14, 556*). The structure of furanopinguisanol (= 7α-hydroxy-deoxopinguisone) (**540**) was established by spectroscopic comparison with dehydropinguisenol (**541**), the relative stereochemistry being supported by NOE spectrometry. Catalytic hydrogenation of (**541**) in the

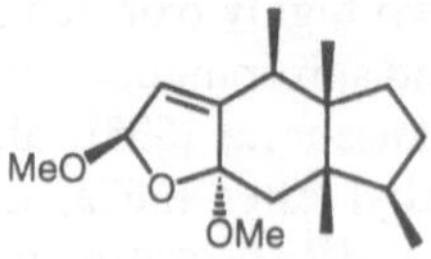

(551) 6α,11β-Dimethoxypinguis-5(10)-ene

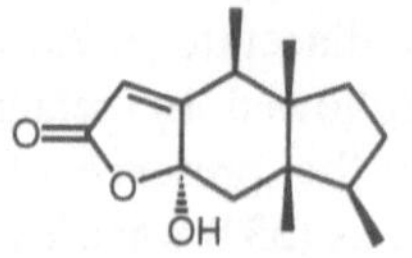

(552) Ptychanolactone

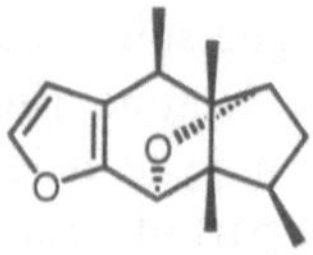

(553)

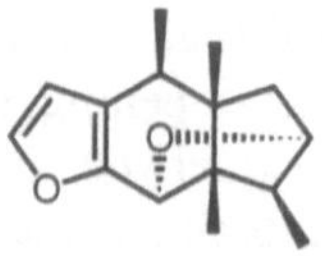

(554) Pinguisanine

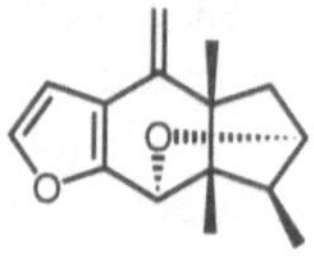

(555) Dehydropinguisanine

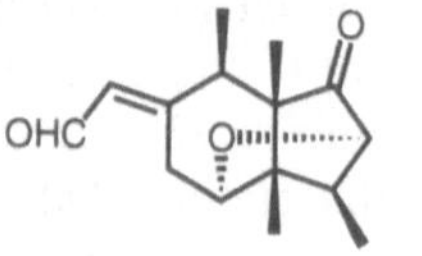

(556) Pinguisenal

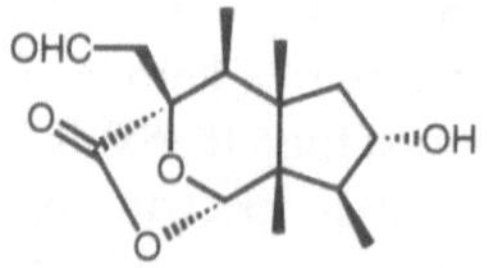

(557) Porellapinguisanolide

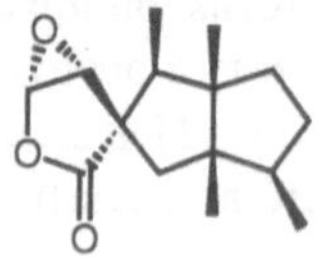

(558) Ptychanolide

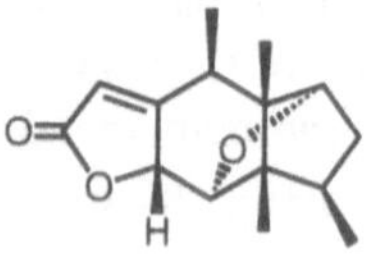

(559)

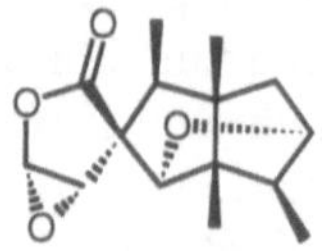

(560) Pinguisanolide

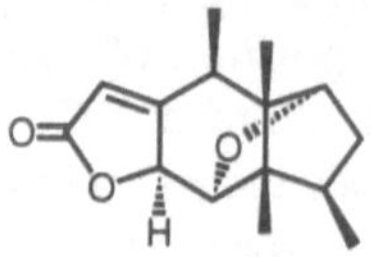

(561)

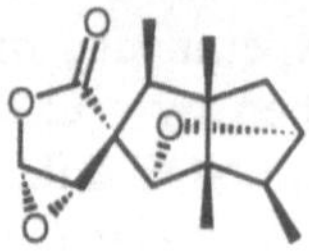

(562) Isopinguisanolide

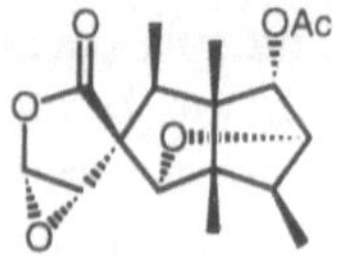

(563) Spirodensifolin A

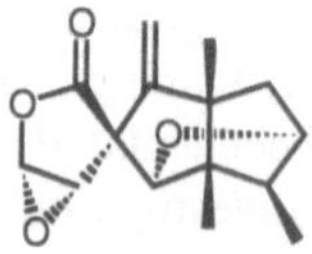

(564) Spirodensifolin B

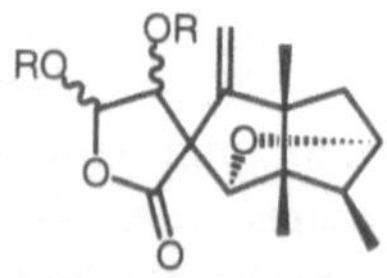

(565a) Spiropinguisanin ; R=H
(565b) R=Ac

Chart 38c. Pinguisanes found in the Hepaticae

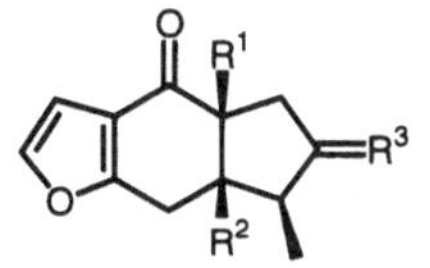

(566) Norpinguisone ; $R^1=R^2=Me$, $R^3=H_2$
(567a) Norpinguisone methyl ester ;
    $R^1=Me$, $R^2=CO_2Me$, $R^3=H_2$
(567b) $R^1=CO_2Me$, $R^2=Me$, $R^3=H_2$
(568) Bryopterin D ; $R^1=Me$, $R^2=CO_2Me$, $R^3=O$

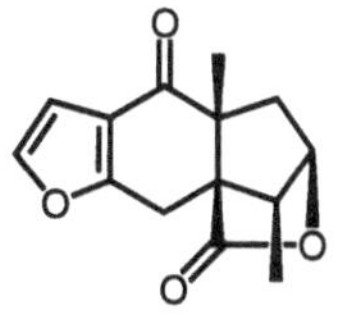

(569) Norpinguisanolide

Chart 38d. Norpinguisanes found in the Hepaticae

presence of 10% Pd-C gave a dihydro derivative identical with (540). Compound (542) was the methyl ether of (540). The structure of (550) was based mainly on ^1H-, ^{13}C-NMR, ^1H-^1H-, ^{13}C-^1H- and long range ^{13}C-^1H 2D-COSY spectrometry, the relative stereochemistry being supported by the NOE technique. The second dimethoxypinguisene (551) was shown to possess the same skeleton as (550) while the presence of an NOE between H-11 and C_6-OMe showed that their *cis* relationship. Compounds (542), (550) and (551) may be artifacts because methanol and chloroform were used as eluents during the chromatographic separation process. Ptychanolactone (552) has been isolated from *Ptychanthus striatus*; the tentative structure assignment was based 2D NMR spectrometry (*249*).

From the ether extract of the Bolivian liverwort *Frullanoides densifolia*, three new oxygenated pinguisanes, isonaviculol (533), spirodensifolin A (563) and spirodensifolin B (564) were isolated, along with naviculol (532) and ptychanolide (558) (*14, 554–556*). The ^1H, ^{13}C NMR and mass spectral data of (533) were closely related to those of naviculol (532), present in the same species and in *P. navicularis*, suggesting that (533) might be a geometrical isomer of (532). This assumption was confirmed by an HMBC experiment with the aid of the ^{13}C-^1H 2D-COSY. The relative stereochemistry was established by the presence of an NOE between H-10 and H-15. The rearranged pinguisane-type skeleton of spirodensifolin A (563) was deduced from a study of the ^1H-, ^{13}C-NMR spectra and the strong IR absorption band at 1785 cm^{-1} indicating the presence of a γ-lactone as well as HMBC and spin decoupling experiments. The stereochemistry followed from the NOE's, indicating that all methyl groups of (**563**) were β-oriented and that the acetoxy group at C-3 was thus α-oriented. As the NOE's did not distinguish between the presence of an α- or β-epoxide, an X-ray crystallographic analysis of (**563**) was carried out which established the stereochemistry as depicted in the formula (*555*).

The spectral data of (**564**) resembled those of (**563**) except for the presence of an exomethylene and the absence of a secondary methyl and a secondary acetoxyl group, indicating that the secondary methyl group and acetoxyl group at C-4 and C-3 in (**563**) were replaced by the exomethylene and the hydrogen atom in (**564**), respectively. The above results together with the co-occurrence of (**563**) in the same liverwort established the structure of spirodensifolin B as (**564**).

It is noteworthy that ptychanolide (**558**), whose structure has been established by X-ray analysis (*540*) has also been found in the liverwort *Ptychanthus striatus*, although the stereochemistry at C-5 is different from the C-5 stereochemistry of (**563**) and (**564**) found in *Frullanoides densifolia*.

Porella vernicosa produces deoxopinguisone (**536**), deoxopinguisone methyl ester (**547**), norpinguisone (**566**) (*19*) and norpinguisone methyl ester (**567a**) (*19, 189*). Cell suspension cultures of *P. vernicosa* also produces the same pinguisane-type sesquiterpenoids (**536, 547, 566, 567a**) among which (**567a**) is the major compound (*458*).

Three new pinguisane methyl esters, bryopterins A (**546**), B (**549**) and C (**548**) have been isolated from Panamanian *Bryopteris filicina*, along with norpinguisone methyl ester (**567a**) (*432a*). The pinguisane structures were proved by analysis of 2D-COSY NMR spectra and by NOE spectrometry of the original compounds and the acetates formed by LiAlH$_4$ reduction and subsequent acetylation.

The biosynthesis of pinguisane- and norpinguisane-type sesquiter-penoids has not been studied so far. TAKEDA et al. (*540*) proposed a possible biogenetic route to deoxopinguisone (**536**), pinguisenene (**539**) and ptychanolide (**558**) from acetal (**570**) as shown in Scheme 42. A possible biogenetic pathway for the formation of rearranged pinguisane-type sesquiterpenoids (**558, 536e**) indicated in Scheme 43 has been proposed by CONNOLLY (*146*). The pinguisane skeleton itself (**579**) might be derived from *trans, cis*-farnesol (**211**) *via* bisabolane (**571, 572**) and acorane (**573, 574**) as shown in Scheme 44 (*14, 556*).

Synthetic interest in pinguisane- and norpinguisane-type sesquiter-penoids arises from the unusual tricyclic skeleton which contains four or three methyl groups located in a *cis* relationship on adjacent carbons within a *cis* hydrindane system. The first total synthesis of pinguisone (**537**) and its C-1 isomer was accomplished in 1981 by BERNASCONI et al. (*103, 104*). The starting material for the synthesis of both furanosesquiter-

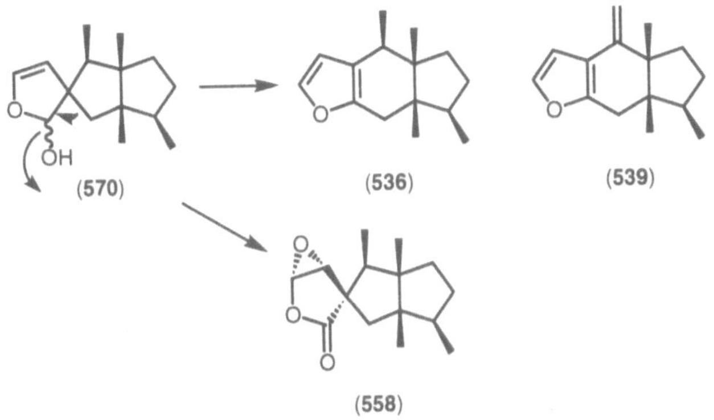

Scheme 42. Possible biogenetic pathways for pinguisanes and spiropinguisanes

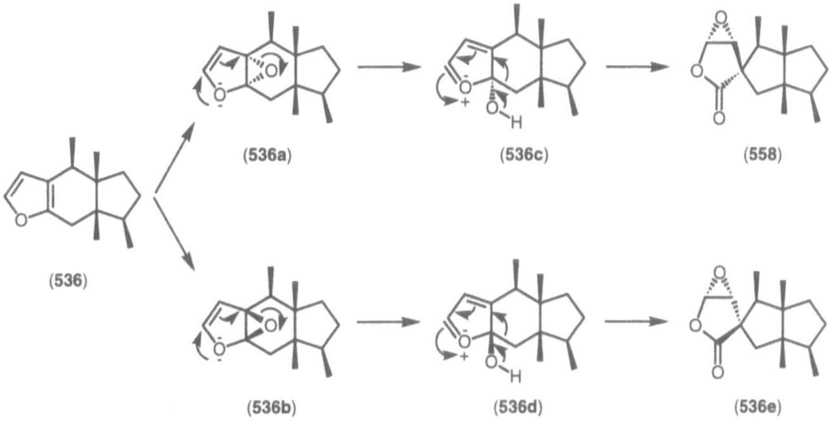

Scheme 43. Possible biogenetic pathways for spiropinguisanes from deoxopinguisone

penoids was S-(+) enantiomer of the well-known 2,3,7,7a-tetrahydro-7a-methylindene-1,5(6H)-dione, derived from 1-methyl-1-(3-ketobutyl)-cyclopenta-2,5-dione by asymmetric aldol cyclization in the presence of a catalytic amount of (S)-(−)-proline (230, 231). An alternative total synthesis of (±)-pinguisone (537) has been achieved by GAMBACORTA et al. (192) and UYEHARA et al. (605, 606). The total synthesis of (±)-deoxopinguisone (536) has been accomplished by UYEHARA et al. (605, 606). Three pinguisanes so far not found in nature, 4-epi-pinguisanol and 3-oxonorpinguisone, have been synthesized by BAKER et al. (87) and MATEOS et al. (375), respectively. The first synthesis of (±)-isoptychanol-ide, a stereoisomer of naturally occurring ptychanolide (558) which differs

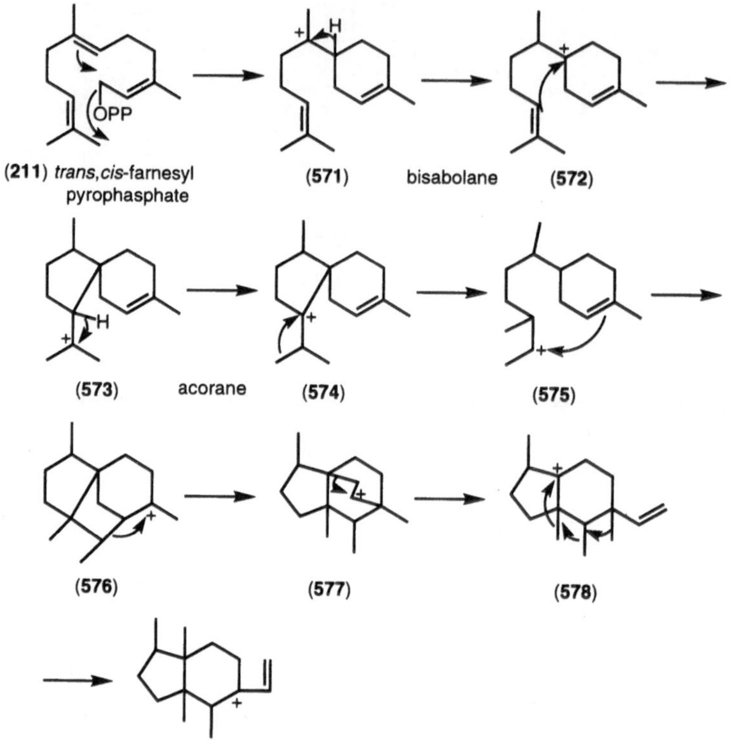

(211) *trans,cis*-farnesyl   **(571)**   bisabolane   **(572)**
        pyrophasphate

(573)   acorane   **(574)**   **(575)**

(576)   **(577)**   **(578)**

(579)

Scheme 44. Possible biogenetic pathways for pinguisanes from *trans,cis*-farnesyl pyrophosphate

in the configuration of the epoxide function, has been accomplished by
SOLAJA *et al.* (*510*).

5.26 Santalanes, Spirovetivanes, Thujopsanes and Valencanes

A new santalane-type sesquiterpene diol **(581)** has been isolated from
Porella caespitans var. *setigera* (*596, 597*). The structure assignment was
based on the similarity of the ¹H- and ¹³C-NMR spectra to those of
tricyclene **(581a)** and α-santalol **(581b)** and chemical transformations.
Acetylation of **(581)** gave a monoacetate whose IR spectrum retained
hydroxyl absorption bands, indicating the presence of a tertiary hydroxyl
group in **(581)**. Oxidation of **(581)** with PCC afforded an α-hydroxyketone

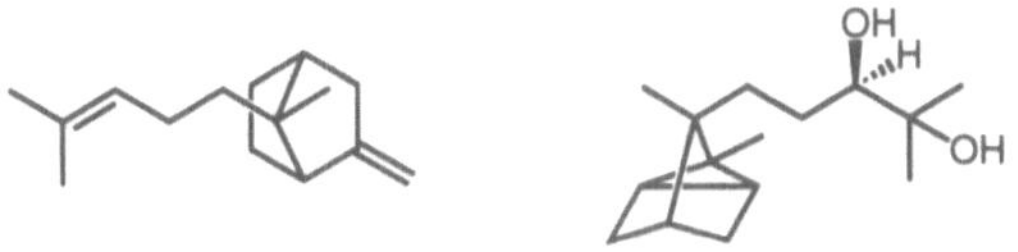

(580) β-Santalene   (581) α-Santalan-12(*R*),13-diol

Chart 39. Santalanes found in the Hepaticae

(581c) and an aldehyde also available from α-santalol (581b) by permanganate oxidation (*224, 346*). The stereochemistry of (581) at C-12 was determined by means of the diol complexation method (*164, 441*). The CD spectrum [342 nm, $\Delta\varepsilon - 0.03$ in CCl_4 employing $Eu(fod)_3$ as complexing agent] of (581) showed the 12R configuration.

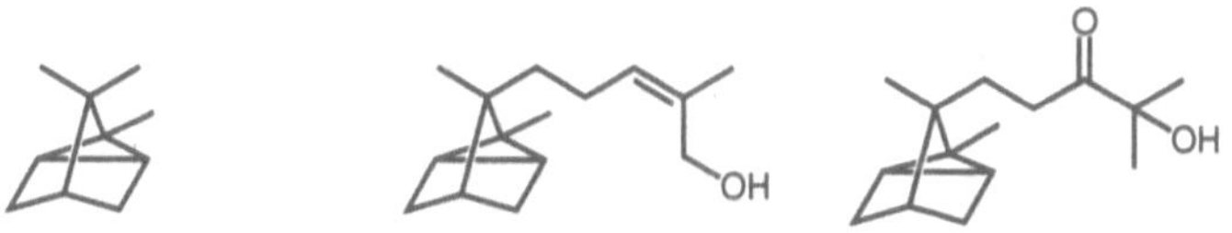

(581a) Tricyclene   (581b) α-Santalol   (581c)

Santalane-type sesquiterpenoids in Hepaticae are otherwise very rare. β-santalene (580) has been detected in the liverworts, *Plagiochila yokogurensis* (*19*) and *Gackstroemia magellanica* (*45*) by GC-MS.

α-Spirovetivene (582) has been isolated from *Scapania robusta* and *S. maxima*, together with the β-isomer (583) (*623, 632*). The NMR signals of (582) and its tetrahydro derivative were identical with those of agarospirene (= spirovetivene) (582) which was obtained from agarospirol (586) (Scheme 45) and agarospirane (587), respectively. Acid treatment of (582) gave (−)-δ-selinene (358) by a sequence of cationic rearrangements. The absolute configuration of (582) was suggested by the negative Cotton effect at 208 nm ($\Delta\varepsilon - 1.1$); however, other possibilities, such as (585), cannot be excluded (*621, 623, 632*). The structure of β-spirovetivene (583) was deduced from the ^{1}H- and ^{13}C-NMR spectra and by comparison

(582) α-Spirovetivene   (583) β-Spirovetivene

Chart 40. Spirovetivanes found in the Hepaticae

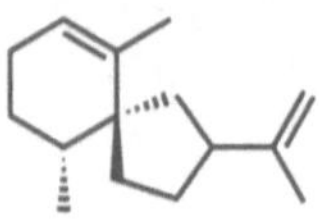

(585)

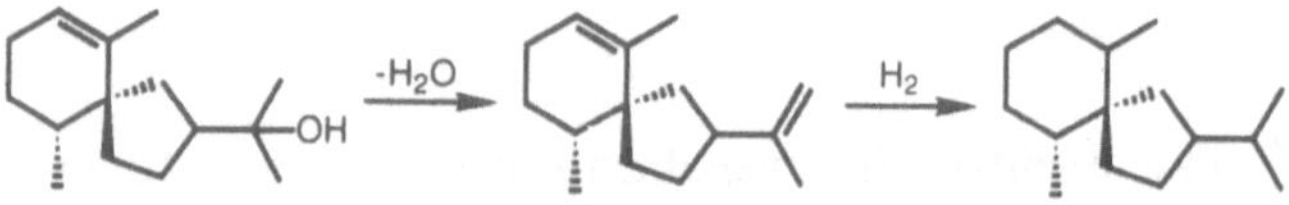

(586) Agarospirol   (582) Agarospirene   (587)
                                (= Spirovetivene)

Scheme 45. Reactions of spirovetivenes and agarospirol

with α-spirovetivene (582). The stereochemistry of (583) remains to be established.

Thujopsane-type sesquiterpenoids in the Hepaticae are also very rare, thujopsene (588) having been previously detected only in *Leucolejeunea xanthocarpa* by GC-MS (*19*). (+)-*Ent*-thujopsene (588) and (−)-*ent*-thujopsenone (589), have since been isolated from Japanese *Marchantia polymorpha* (*74, 382*). The *ent*-configurations of (588) and (589) have been confirmed by the chemical correlation shown in Scheme 46. (−)-Thujopsene (588a) obtained from cedar wood oil was oxidized with Collins reagent to give (+)-thujopsenone (589a), the antipode of *ent*-thujopsenone (589) and (+)-mayurone (591) (*74, 382*). A new thujopsane-type sesquiterpene alcohol (590) has been isolated from *Marchantia polymorpha* together with *ent*-thujopsenone (589) (*71*). Dehydration of (590) in $CHCl_3$ gave (+)-thujopsene (588) (Scheme 46), indicating that (590) was thujopsene with an axial hydroxyl group at C-7. The stereochemistry has been settled by the NOEs shown in the scheme. Thujopsene (588) has also been detected in *Bazzania* (*513, 621, 622*), *Heteroscyphus planus* (*427a*), *Mastigophora* (*621, 635*), *Plagiochila* (*621, 637*) and *Schistochila* species (*621, 630*) by GC-MS.

A valencane-type sesquiterpene alcohol, 7α-hydroxyvalenc-1(10)-ene (592) has been isolated from *Bazzania fauriana* (*581*). Its spectral data were very similar to those of valencene (593) except for the absence of the signals corresponding to the isopropenyl group which suggested, that (592) was valencene with a tertiary hydroxyl group. This assumption and the presence of a 7α-hydroxyl group were confirmed by NMR studies. The absolute configuration of (592) was established by synthesis from

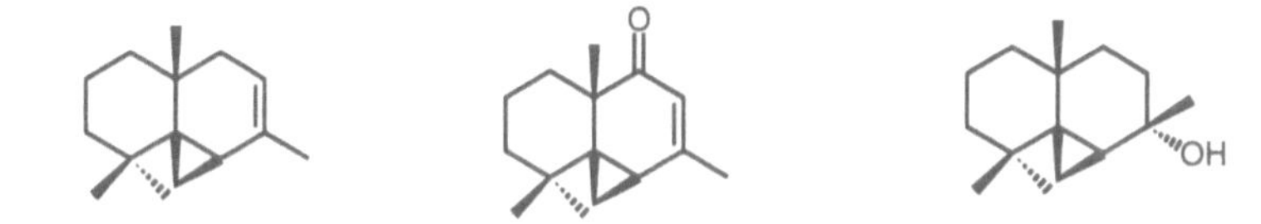

(588) ent-Thujopsene   (589) ent-Thujopsenone   (590) ent-Thujopsan-7β-ol

Chart 41. Thujopsanes found in the Hepaticae

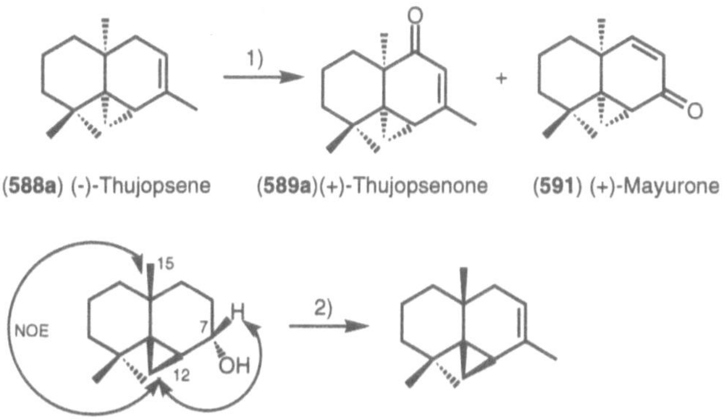

(588a) (-)-Thujopsene   (589a)(+)-Thujopsenone   (591) (+)-Mayurone

(590) ent-Thujopsan-7β-ol   (588)

1) Collins oxd./CH$_2$Cl$_2$ 2) H$^+$

Scheme 46. Formation of (+)-thujopsenone from (−)-thujopsene and dehydration of ent-thujopsan-7β-ol

(+)-nootkatone (*577*). The specific optical rotation of the synthetic product was [α]$_D$ + 76° [lit. + 75°(*581*)]. This is the first example of valencane-type sesquiterpenoid from the Hepaticae.

5.27 Vitranes, Widdranes and Zieranes

The structure and absolute configuration of (+)-vitrenal (**594**) isolated from *Lepidozia vitrea* has been established as (1R,6R,7S,10R)-vitr-4-en-14-al by a combination of chemical degradation (Scheme 47) and X-ray analysis of the di-*p*-bromobenzoate (**594f**) (*19, 389*). The name vitrane has been proposed for the new carbon skeleton (*389*).

The total synthesis of (±)-vitrenal (**594**) has been accomplished in 12 steps by MAGARI et al. (*360*). An alternative total synthesis of (−)-vitrenal

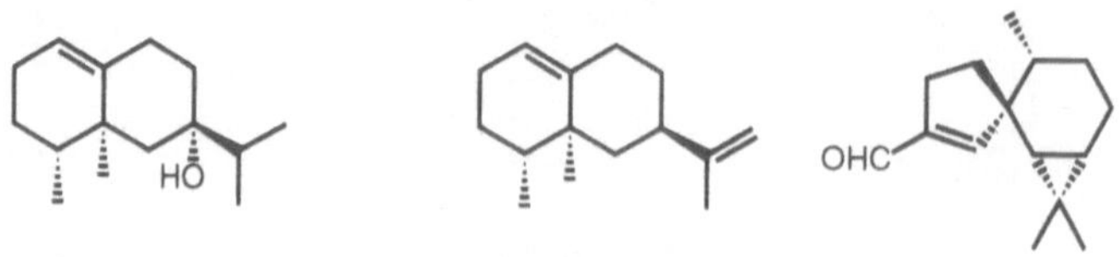

(**592**) 7α-Hydroxyvalenc-1(10)-ene   (**593**) Valencene   (**594**) Vitrenal

Chart 42. Valencanes and vitranes found in the Hepaticae

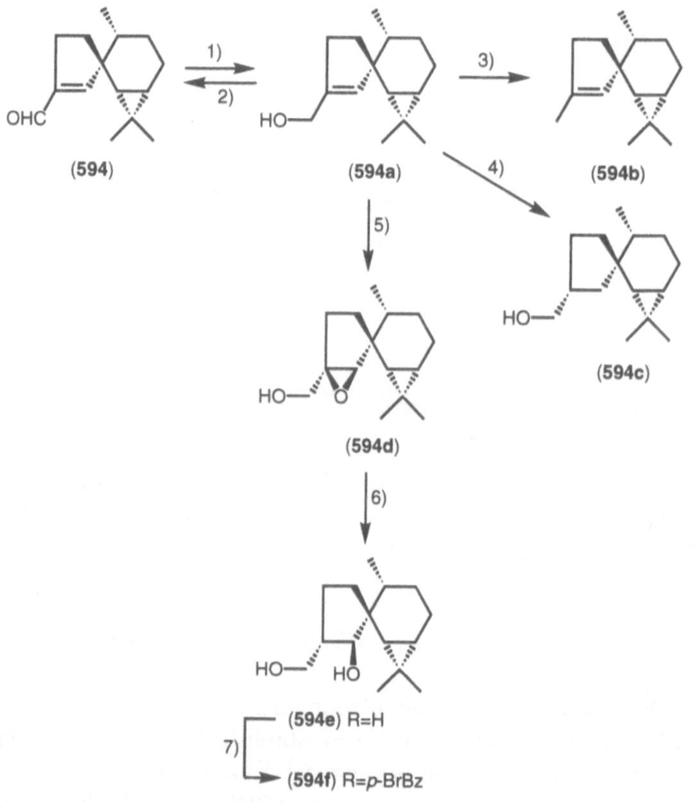

1) LiAlH₄ 2) MnO₂ 3) C₅H₅N-SO₃, LiAlH₄ 4) H₂-PtO₂ 5) MCPBA 6) Li/NH₂(CH₂)₂NH₂
7) *p*-BrBzCl/Py

Scheme 47. Reactions of (+)-vitrenal

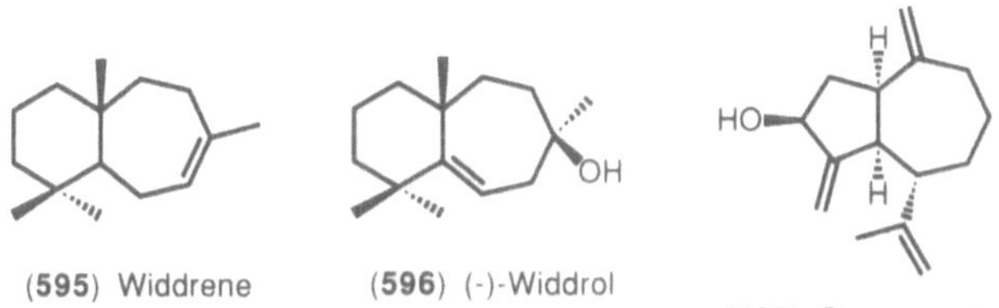

(595) Widdrene   (596) (-)-Widdrol

(597) Saccogynol

Chart 43. Widdranes and zierane found in the Hepaticae

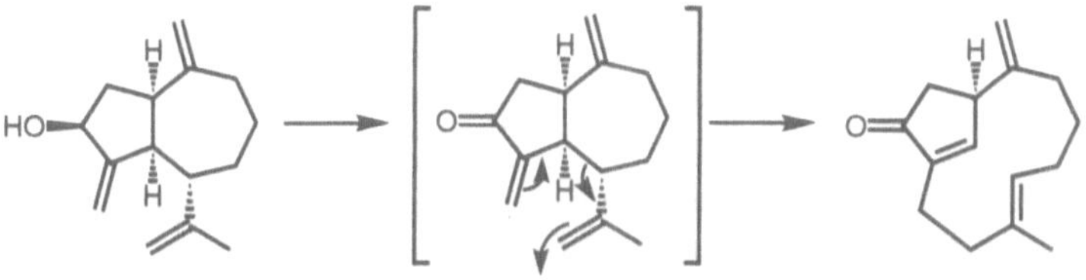

(597)   (597a)   (597b)

Scheme 48. Cope rearrangement of saccogynol

(594), the enatiomer of natural vitrenal, has been performed by KODAMA *et al.* (*335, 336*) using 2-caren-4β-ol.

Widdrane-type sesquiterpenoids are very rare in the Hepaticae. Previously, only widdrene (595) was detected in *Omphalanthus platycoleus* by GC-MS (*19*). (−)-Widdrol (596) has since been isolated from *Marchantia polymorpha* (*74, 382*).

Saccogyna viticulosa produces large amount of an alcohol, named saccogynol (597) and the corresponding hydrocarbon (*146*). The structure of (597) was deduced by detailed analysis of the ^{1}H-NMR spectrum, but the absolute configuration remains to be clarified. Oxidation of (597) did not give the expected ketone (597a) but afforded a product (597b) formed by Cope rearrangement (Scheme 48). The zierane carbon skeleton is rarely found in nature. The only other example is zierone (598) (*209*).

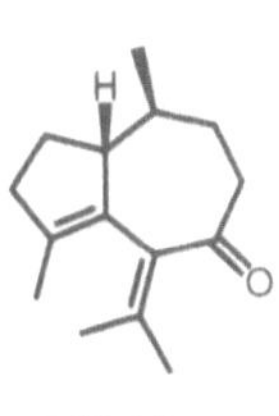

(598) Zierone

5.28 Miscellaneous Sesquiterpenoids

A new rearranged drimane-type sesquiterpene ether, named neo-drimanoxide, has been isolated from *Plagiochila peculiaris*; structure **(599)** was proposed for it by analysis of ^{1}H- and ^{13}C-NMR spectra as well as its mass spectral fragmentation (*623*). However, more recent work using HMBC, NOE and relay COSY NMR spectrometry showed that the structure should be revised to **(600)** which has been named peculiaroxide (*269, 631a*).

Conocephalum conicum produces a new type of sesquiterpene alcohol **(601)** whose structure has been elucidated by extensive NMR spectrometry including 2D-COSYs and the HMBC method (*553*). *Omphalanthus filiformis* produces not only *ent*-chamigrane-type sesquiterpenoids but also the rearranged chamigrane-type sesquiterpene acid **(602a)** named omphalic acid whose structure is based on 600 MHz ^{1}H- and 150 MHz ^{13}C-NMR spectral data and 2D-COSY's of its methyl ester **(602b)** (*441a,*

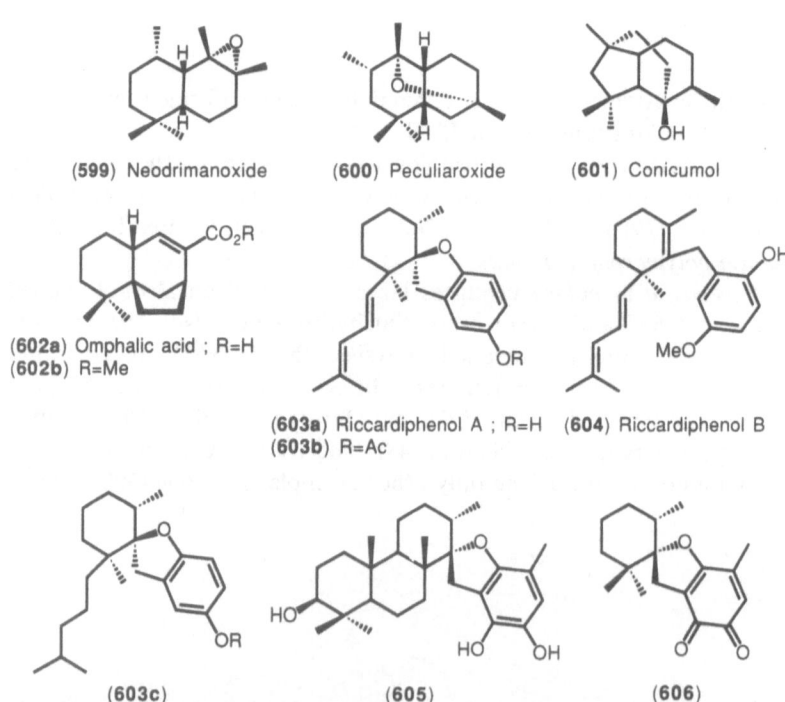

(**599**) Neodrimanoxide   (**600**) Peculiaroxide   (**601**) Conicumol

(**602a**) Omphalic acid ; R=H
(**602b**) R=Me

(**603a**) Riccardiphenol A ; R=H   (**604**) Riccardiphenol B
(**603b**) R=Ac

(**603c**)   (**605**)   (**606**)

Chart 44a. Miscellaneous sesquiterpenoids found in the Hepaticae and their related compounds

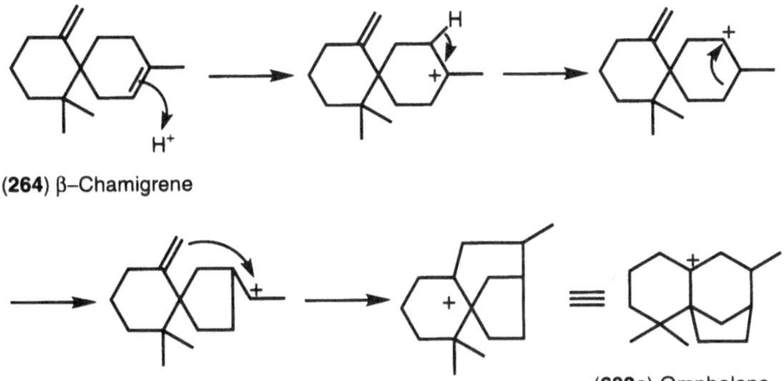

(264) β–Chamigrene

(602c) Omphalane

Scheme 49. Possible biogenetic pathway for omphalane-type sesquiterpenoids from β-chamigrene

571). A possible biogenetic pathway for the formation of omphalane-type sesquiterpenoids is shown in Scheme 49 (*441a*).

The methanol extract of *Riccardia crassa* was chromatographed on silica gel and Sephadex LH-20 to give two new merosesquiterpenes, riccardiphenols A (603a) and B (604) (*583*). Acetylation and catalytic hydrogenation of (603a) gave a monoacetate (603b) and a tetrahydro derivative (603c), respectively, indicating that (603a) was a tricyclic compound with a phenol group and two double bonds. The position and stereochemistry of each functional group was established by a combination of ¹H- and ¹³C-NMR spectroscopic methods, and NOE experiments (*583*). The structure assigned to (604) was based on spectroscopic comparison with (603a). The substitution pattern on the benzene ring was confirmed by the presence of an NOE between the methoxyl group and two aromatic protons. The meroterpene skeleton of (603a) and (604) is unprecedented. An isocopalane-type diterpenoid (605) possessing a similar dihydrobenzofuran group and its analogue (606) have been isolated from the tropical brown alga, *Stypodium zonale* (*179, 180*).

Cheilolejeunea serpentina produces a new sesquiterpene phenol, serpentiphenol (607a) whose relative structure has been determined by use of NMR techniques including 2D-COSY and NOE experiments of the original compound, its dimethyl ether (607b) and its diacetate (607c) (*249*). A monohydroxy analogue (608) has been isolated from the brown alga *Sporochnus bolleanus* (*501a*).

Fractionation of the ether extract of another *Cheilolejeunea trifaria* resulted in the isolation of trifarienols A–E (609–613) containing a new sesquiterpene carbon skeleton (*249*). Structures were deduced by degradation and NMR spectrometry. The relative stereochemistry of (609) was

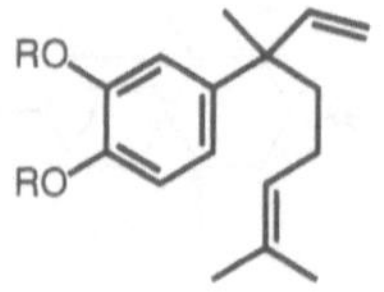

(607a) Serpentiphenol ;R=H   (608) Sporochnol   (609) Trifarienol A
(607b) R=Me
(607c) R=Ac

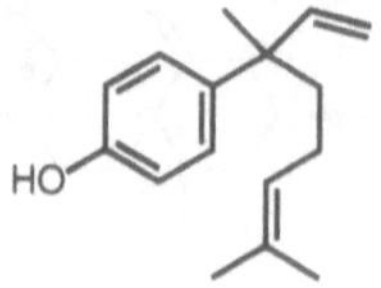

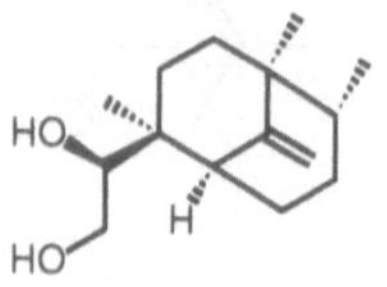

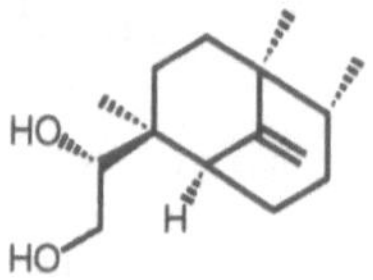

(610) Trifarienol B 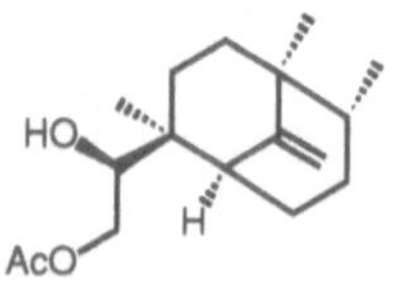   (611) Trifarienol C 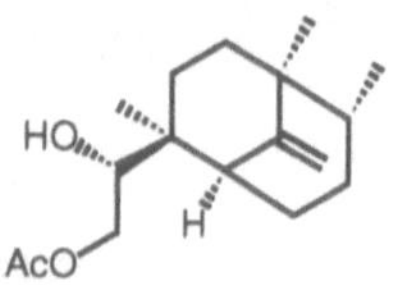   (612) Trifarienol D

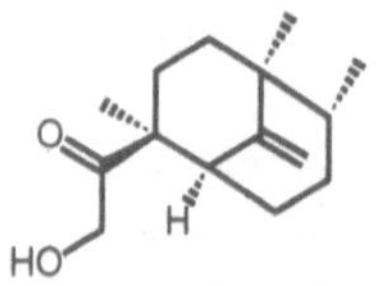

(613) Trifarienol E 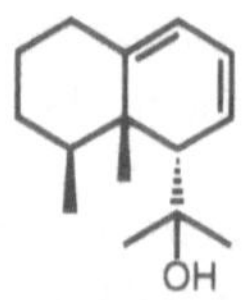   (614) Rebouliadienol 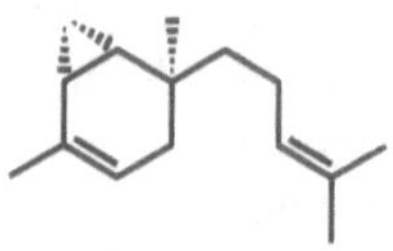   (615) Chenopodene

Chart 44b. Miscellaneous sesquiterpenoids found in the Hepaticae and their related
compounds

established by X-ray crystallographic analysis and the absolute configuration established by means of the CD spectrum [307 nm ($\Delta\varepsilon$ + 53.9)
and 282 nm ($\Delta\varepsilon$ − 55.9)] of (609) using shift reagent [Eu(fod)$_3$] and
that [250 nm ($\Delta\varepsilon$ + 16.1) and 233 nm ($\Delta\varepsilon$ − 5.26)] of the di-*p*-bromo-
benzoate (*135, 233*) as was the absolute configuration of **610** by its CD
spectrum and the CD spectrum of its di-*p*-bromobenzoate. Both of these
exhibited signs opposite to those of (609) and its di-*p*-bromobenzoate.

Two new sesquiterpenoids, rebouliadienol (614) and chenopodene
(615) have been isolated from *Reboulia hemisphaerica* (*249*) and *Marchantia chenopoda* (*553*), respectively. Their structures were elucidated by
600 MHz NMR spectrometry.

6. Diterpenoids

Liverworts are rich sources not only of sesquiterpenoids but also
diterpenoids. The earlier report (*19*) mentioned clerodane-, kaurane-,
labdane-, pimarane-, phytane-, sacculatane- and verrucosane-type diter-

penoids from Hepaticae. In addition to these, abietane-, cembrane-, chettaphanin-, dolabellane-, fusicoccane-, rearranged pimarane-, seco-clerodane-, sphenolobane-, spiroclerodane-, trachylobane- and verticilla-ne-type diterpenoids have been isolated from liverworts in the interval. The absolute configuration of diterpenoids found in liverworts is generally opposite to that found in higher plants, although there are exceptions as in the case of abietane-, clerodane- and labdane-type diterpenoids to be described subsequently.

6.1 Abietanes and Cembranes

Abietatriene (616) has been detected in *Plagiochila peculiaris* by GC-MS (*637*). From the ether extract of *Porella roellii*, dehydroabietic acid (617) has been isolated together with some drimane-type sesquiterpenoids (*249*). This is the first record of the isolation of an abietane-type diterpenoid from bryophytes.

European *Chandonanthus setiformis* (= *Tetralophozia setiformis*) belonging to subgenus *Tetralophozia* produces the cembrane-type diter-penoid setiformenol (618) whose relative structure has been determined by ^{1}H- and ^{13}C-NMR spectrometry as well as by a ^{1}H-^{1}H total correlation spectrum (TOCSY), HMBC and NOESY (*429, 566, 567, 567a*). The stereochemistries of the epoxide and C-12 remain to be clarified. A very similar cembrane-type diterpene epoxide (620) has been found in the marine organism Gorgonian, *Solenopodium stochei* (*113*). Another cembrane-type epoxyketone, chandonanthone (619) has been isolated from Japanese *Chandonanthus hirtellus*; its gross structure is based on IR (1680 cm^{-1}), UV [258 nm, (logε 3.8)] and NMR spectrome-try, the latter including 2D-COSYs (*585*). This is the first record of cembrane-type diterpenoids from bryophytes.

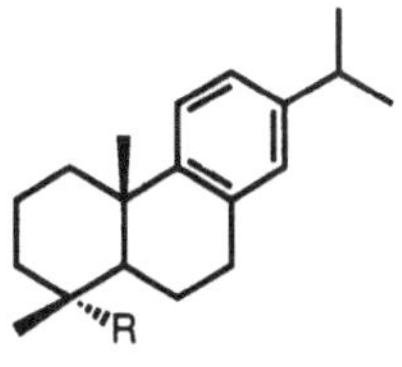

(616) *ar*-Abietatriene ; R=Me
(617) Dehydroabietic acid ; R=CO$_2$H

Chart 45. Abietanes found in the Hepaticae

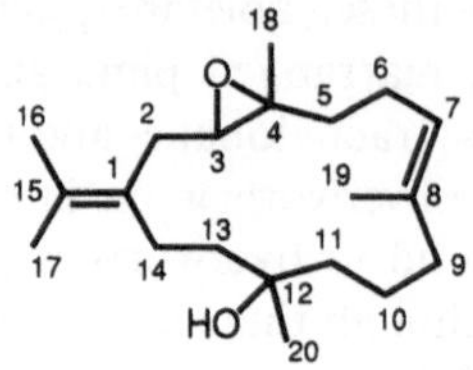

(618) Setiformenol 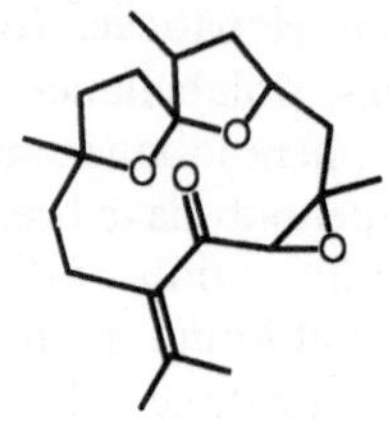   (619) Chandonanthone

Chart 46. Cembranes found in the Hepaticae

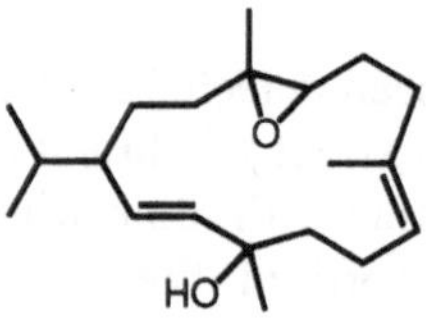

(620)

6.2 Clerodanes, Secoclerodanes and Spiroclerodanes

(−)-Kolavelool (621) has been isolated from *Jungermannia* (*245*), *Macrolejeunea* (*433*), *Nardia subclavata* (*585*), *Pleurozia* (*52, 625*) and *Scapania* species (*377*); it has also been found in the oleoresin of the higher plants *Hardwickia pinnata* (*407*) and *Solidago elongata* (*13*). *Scapania bolanderi* contains a new clerodane ester, (−)-methyl 13-hydroxy-cleroda-3,14-dien-18-carboxylate (624); its structure is based on spectroscopic comparison with kolavelool (621) (*377*) and dimethyl kolavate (622) (*406*).

Jungermannia species are rich sources of diterpenoids. A new clerodane-type diterpenoid, clerod-3,13(16),14-trien-17-oic acid (625a) has been isolated from *J. infusca*; its structure was determined by use of NMR techniques (*592*). As kolavenic acid (623) is easily obtained from *Solidago* species (Compositae), it was also possible to arrive at the structure of (625a) by comparing the NMR spectrum of (625a) with that of (623). Further fractionation of the ethyl acetate extract of *J. infusca* resulted in the isolation of six additional new clerodane diterpenoids (626–631) (*430, 594*). Structures were assigned by extensive 2D-COSY NMR spectroscopy and chemical correlations. The geometry of side chains (C11–C16) of (626–629) was established by NOE experiments and comparing the ^{13}C-NMR data with those of the monoterpene aldehydes neral (6a) and

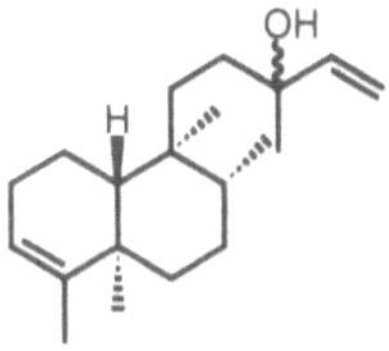

(621) (-)-Kolavelool

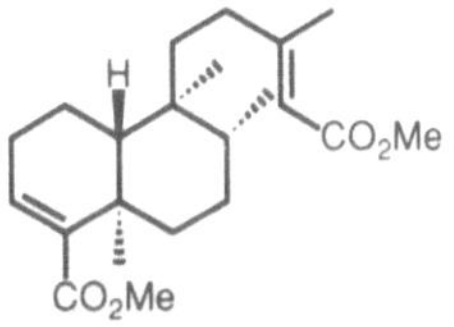

(622) Dimethyl kolavate

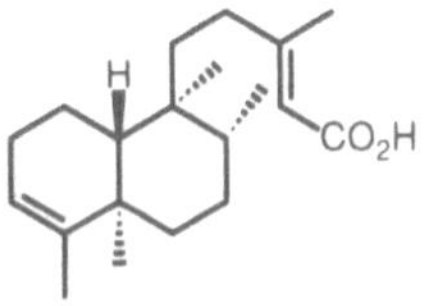

(623) Kolavenic acid

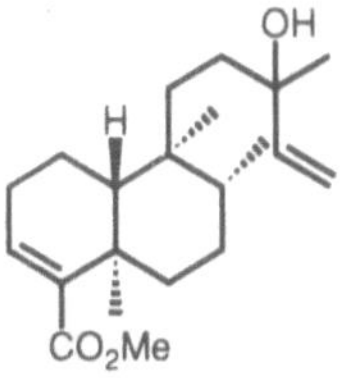

(624) (-)-Methyl 13-hydroxy-clerod-3,14-dien-18-carboxylate

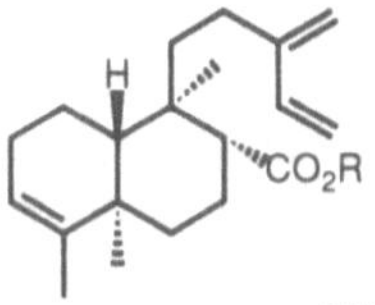

(625a) Clerod-3,13(16),14-trien-17-oic acid (=Infuscaic acid) ; R=H
(625b) R=Me

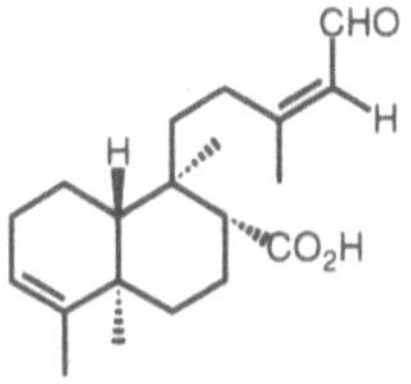

(626) Clerod-3,13Z-dien-15-al-17-oic acid

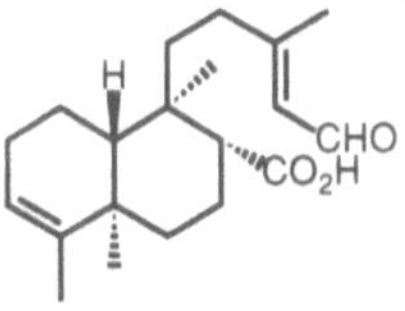

(627) Clerod-3,13E-dien-15-al-17-oic acid

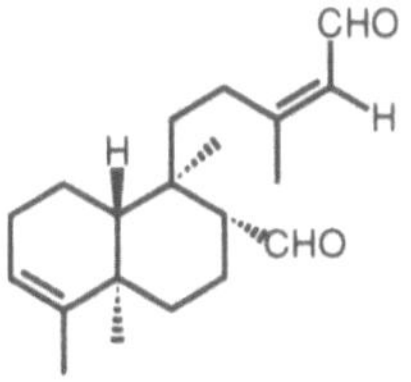

(628) Clerod-3,13Z-dien-15,17-dial

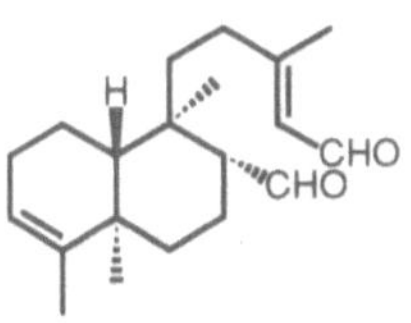

(629) Clerod-3,13E-dien-15,17-dial

Chart 47a. Clerodanes found in the Hepaticae

geranial (**6b**), while the relative stereochemistries of the methyl groups at the ring junctions and the points of attachment of the side chain and the functional group at C-9 were established by NOE difference spectrometry. Compounds (**629**) and (**630**) have also been obtained from a cell culture of *J. infusca* (*457*).

Heteroscyphus bescherellei produces a diterpene acid (**632**) (*83*), whose NMR spectrum is identical with that of (−)-junceic acid (**633**) obtained from *Solidago juncea* (Compositae) (*267*). However, the sign of the specific optical rotation of (**632**) was opposite, hence (**632**) from the liverwort is an enantiomer of the substance from the higher plant.

Jungermannia paroica elaborates not only (−)-kolavelool (**621**) but also the related *ent*-clerodanes (**634–636**) among which (**636**) is new (*245*). Compounds (**634**) and (**635**) were first isolated from *Solidago serotina* (*397*). In the ^{13}C-NMR spectrum, the methyl group on C-16 is more shielded in the (*E*)-isomer (**635**) (δ 17.8) than in the (*Z*)-isomer (**634**) (δ 25.3). Conversely, C-12 is more shielded in the (*Z*)-isomer (δ 26.7) than in the (*E*)-isomer (δ 34.5) (*478*). The structure of (**636**) was readily established by spectroscopic comparison with those of co-metabolites (**621, 634, 635**). The absolute configuration of (**621, 634, 635**) is that of the enantio-series.

From East Malaysian *Schistochila aligera*, a new clerodane (**637a**) was isolated in the form of its methyl ester (**637b**) (*438*). The ^{1}H-NMR spectrum of (**637b**) was similar to that of the methyl ester (**625b**) of (**625a**) from *Jungermannia infusca* (*592*) while 2D-COSY and HMBC experiments and comparison of the ^{13}C-NMR spectrum of (**637b**) with those of (**625b**) and (**638**) (*111*) provided further evidence. The relative stereochemistry assigned to (**637a**) was supported by the NOE difference spectrum. Taiwanese *Schistochila acuminata* produces two novel clerodanes (**639, 640**), together with (**637a**) and its methyl ester (**637b**) (*136, 137*). Relative structures were deduced by analysis of the ^{1}H- and ^{13}C-NMR and NOESY spectra and comparison of the spectral data with those of (**637a**). The presence of the corresponding aldehydes (**641, 642**) and primary alcohols (**643, 644**) of the two major acids (**639**) and (**637a**) in some fractions from *S. acuminata* has been detected by careful analysis of the NMR spectral data of the mixtures (*136*).

Schistochila nobilis grown in New Zealand elaborates two new clerodane-type diterpenoids, schistochilic acids B (**645**) and C (**646**) and a new secoclerodane-type diterpenic acid, schistochilic acid A (**647**), along with verrucosane-type diterpenoids (*561, 563*). Relative configurations assigned to these compounds are based on ^{1}H-^{1}H, ^{13}C-^{1}H, long range ^{13}C-^{1}H-2D-COSY, HMBC, HOHAHA and NOE techniques, except for the C-13 methyl group.

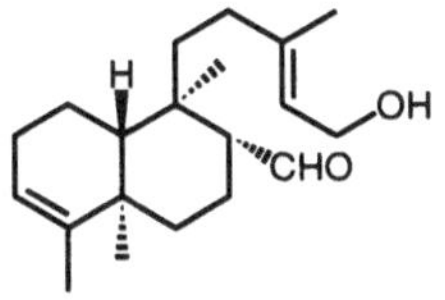

(**630**) Clerod-3,13(14)-dien-15-ol-17-al

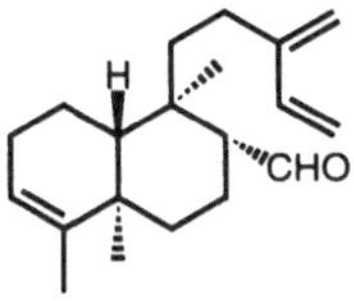

(**631**) Clerod-3,13(16),14-trien-17-al

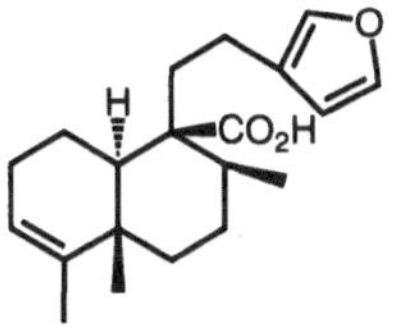

(**632**) *ent*-Junceic acid

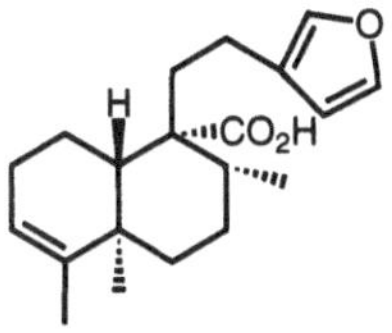

(**633**) Junceic acid

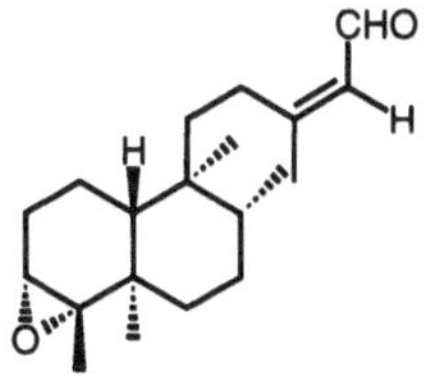

(**634**) *ent*-3β,4β-Epoxyclerod-13*Z*-en-15-al

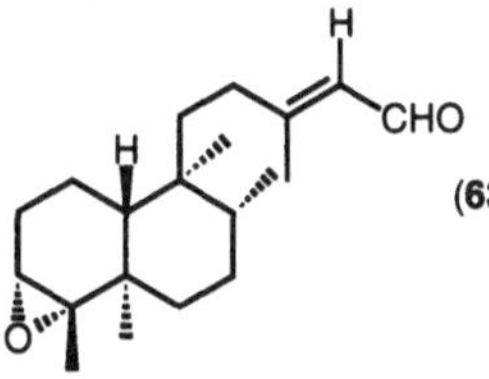

(**635**) *ent*-3β,4β-Epoxyclerod-13*E*-en-15-al

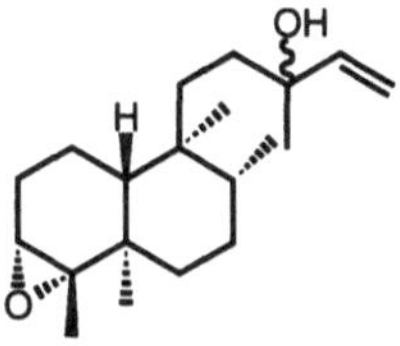

(**636**) *ent*-3β,4β-Epoxyclerod-14-en-13ζ-ol

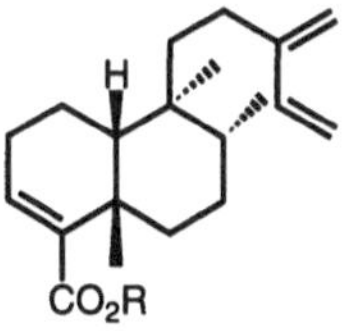

(**637a**) *cis*-Clerod-3,13(16),14-trien-18-oic acid ; R=H
(**637b**) Methyl *cis*-clerod-3,13(16),14-trien-18-oate ; R=Me

Chart 47b. Clerodanes found in the Hepaticae

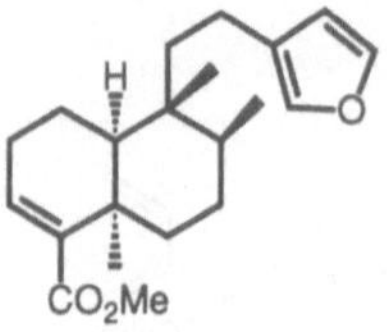

(638)

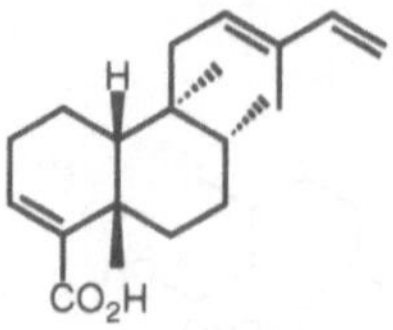

(639) *cis*-Clerod-3,12*E*,14-trien-18-oic acid

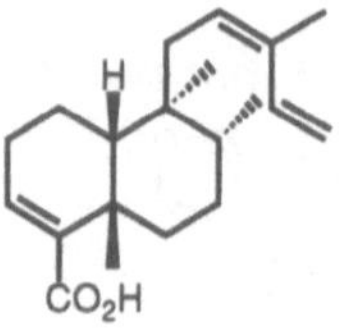

(640) *cis*-Clerod-3,12*Z*,14-trien-18-oic acid

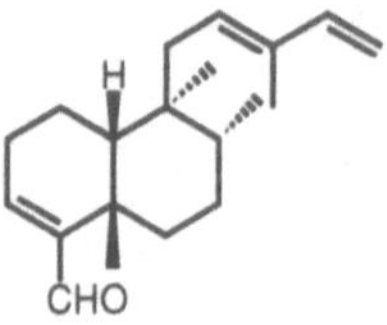

(641) *cis*-Clerod-3,12*E*,14-trien-18-al

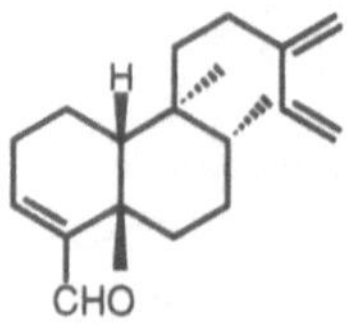

(642) *cis*-Clerod-3,13(16),14-trien-18-al

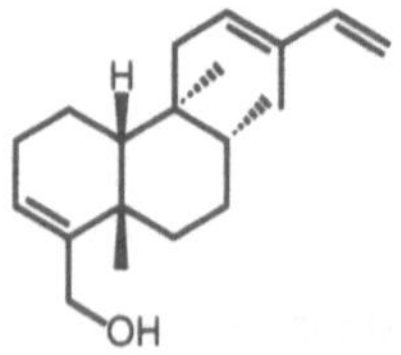

(643) *cis*-Clerod-3,12*E*,14-trien-18-ol

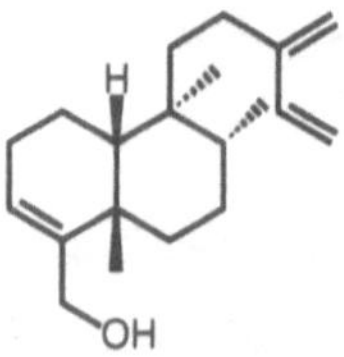

(644) *cis*-Clerod-3,13(16),14-trien-18-ol

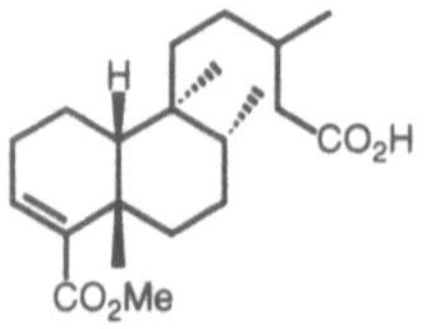

(645) Schistochilic acid B

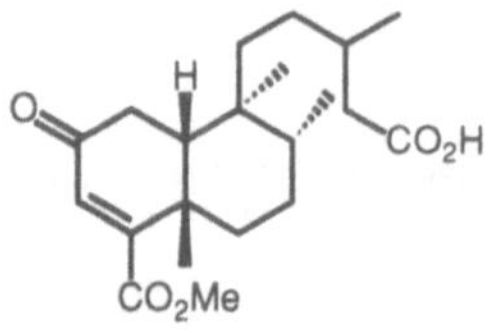

(646) Schistochilic acid C

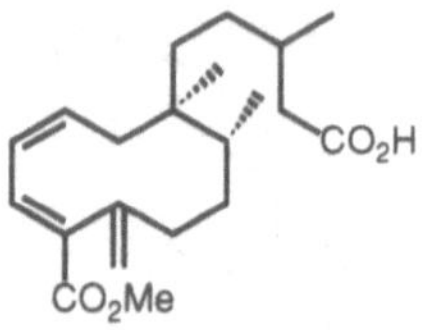

(647) Schistochilic acid A

Chart 47c. Clerodanes and secoclerodane found in the Hepaticae

Gymnocolin (**650**) from *Gymnocolea inflata* was described earlier (*19*). The coupling constants in the ^{1}H-NMR spectrum suggested a *trans*-clerodane structure but X-ray crystallographic analysis showed that (**650**) was a *cis*-clerodane. In the crystalline state gymnocolin has a conformation in which rings A and B are slightly distorted twist boats and ring C is a chair (*272*).

The bitter diterpenes anastreptin and orcadensin from *Anastrepta orcadensis* (*19*) possess structures (651) and (652) (*143*). The relative stereochemistries of (651) and (652) were established by extensive use of NMR spectrometry (*478*). The structure of (652) is similar to that of gymnocolin (650). The genus *Demotarisia* of Jungermanniaceae contains only one species, *D. linguifolia* which produces two new clerodane-type diterpenoids (653, 654) possessing a γ-lactone and a β-substituted furan ring (*431*). The structure of (653) has been deduced from the ^{1}H- and ^{13}C-NMR spectra and 2D-COSY's, the relative stereochemistry being based on the NOE difference spectrum. The structure assigned to (654) rests on spectroscopic comparison with (653) and chemical evidence. Hydrogenation of (654) gave a tetrahydro derivative which was identical with (655) prepared from (653), also by hydrogenation.

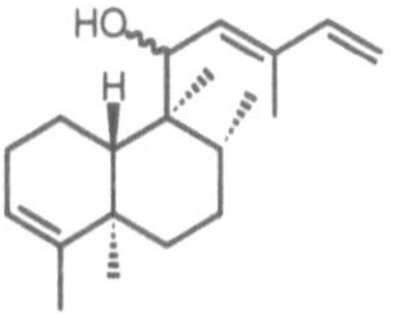

(648) *ent*-Clerod-3,12*E*,14-trien-11ξ-ol 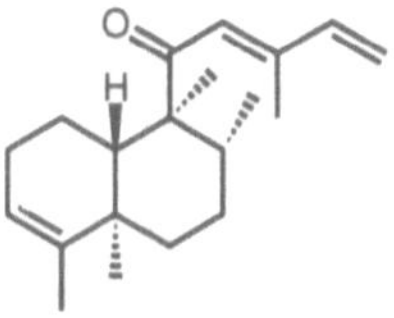   (649)

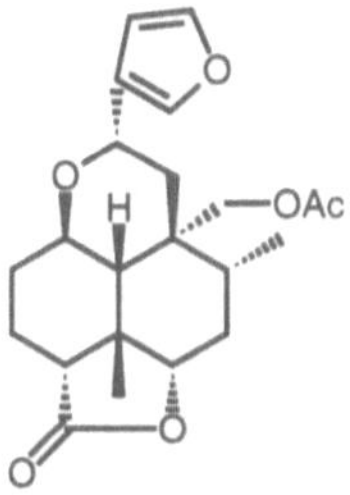

(650) Gymnocolin 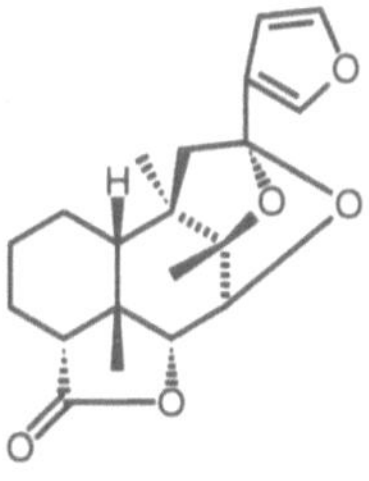   (651) Anastreptin 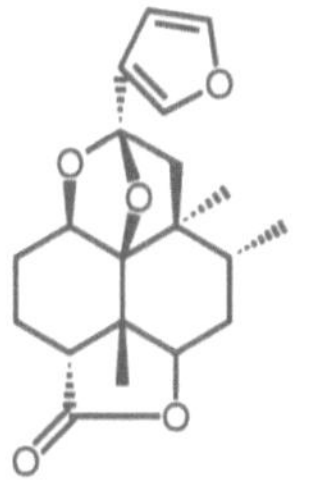   (652) Orcadensin 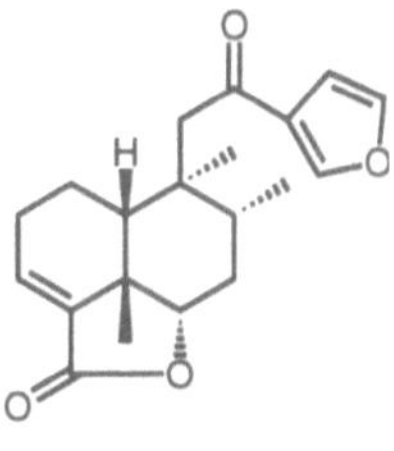   (653) Linguifolide

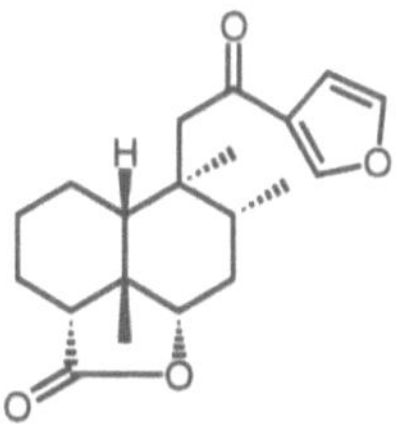

(654) Dihydrolinguifolide 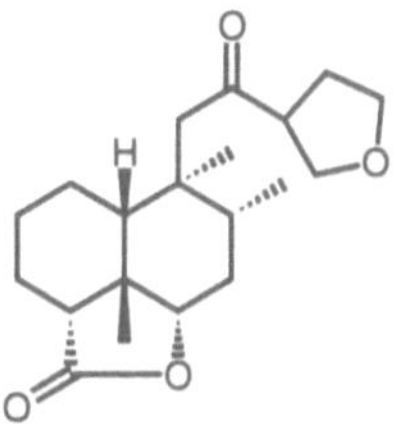   (655) 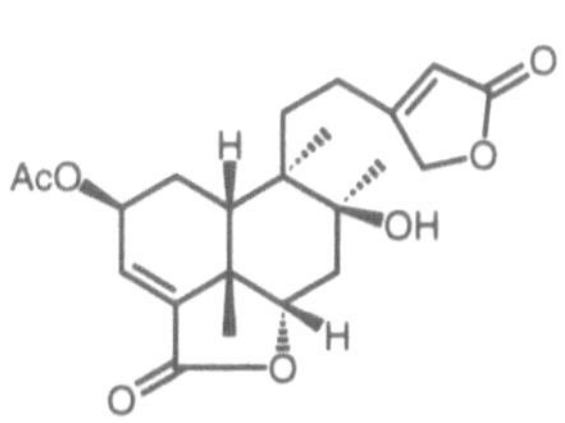   (656) Ventricosenediolide

Chart 47d. Clerodanes found in the Hepaticae

Lophozia ventricosa produces a gorgonane-type sesquiterpenoid (**453**) as well as a clerodane dilactone, ventricosenediolide (**656**), whose structure has been elucidated by a combination of ^{1}H-, ^{13}C-NMR and 2D-COSYs including long range ^{13}C-^{1}H and HMBC techniques (*429, 565, 567a*). The relative stereochemistry has been established by the NOE difference spectrum and X-ray crystallographic analysis.

Jamesoniella autumnalis produces potent bitter furanoditerpenes whose structures had not been established at the time of the previous report (*19*). Since then the dichloromethane and ether extract of *J. autumnalis*, has furnished three novel clerodane diterpene lactones (**657–659**) (*112*). The presence of all functional groups in (**657**) was confirmed by IR, MS and NMR spectrometry. 2D-COSY spectroscopy showed that it was the *cis*-clerodane (**657**). The *cis*-relationship between H-10 and H-19 was confirmed by NOE difference spectrometry which

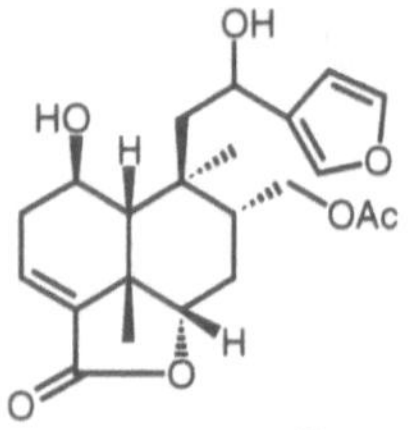

(**657**) 17-Acetoxy-1β,12-dihydroxy-15,16-epoxy-*cis-ent*-clerod-3,13(16),14-trien-6α,18-olide

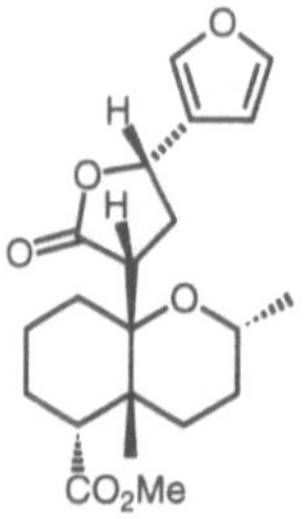

(**658**) Jamesoniellide A 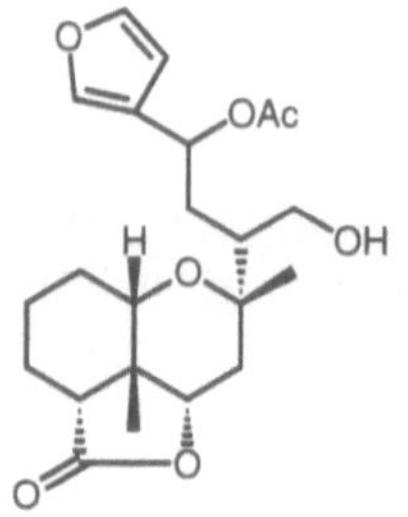   (**659**) Jamesoniellide B 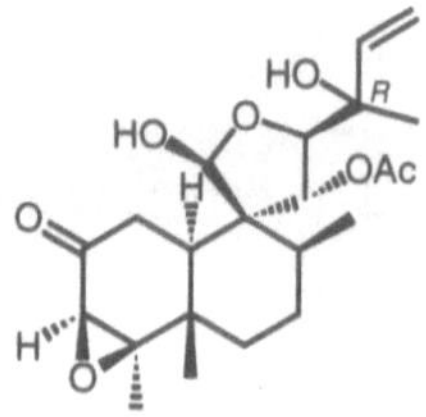   (**660**) Heteroscyphone A

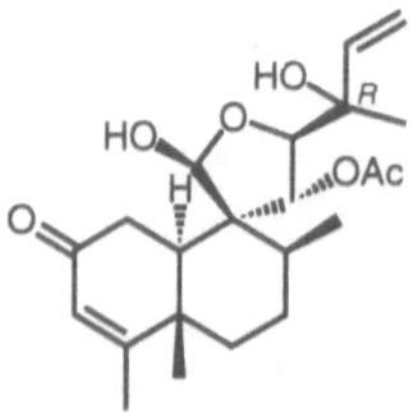

(**661**) Heteroscyphone B 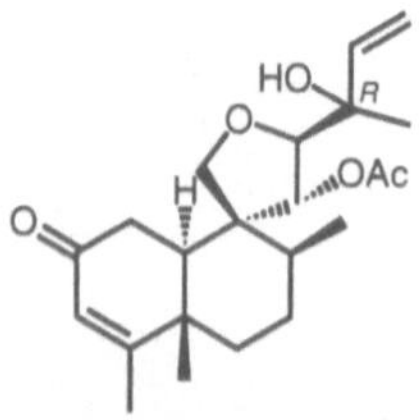   (**662**) Heteroscyphone C 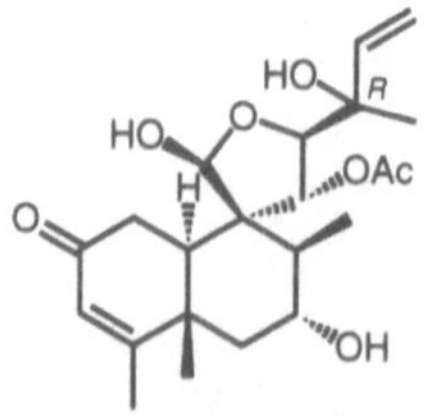   (**663**) Heteroscyphone D

Chart 47e. Clerodanes and spiroclerodanes found in the Hepaticae

also showed that the equatorial acetoxyl at C-8 and the axial methyl at C-9 are on opposite sides of the molecule. Compound (**657**) is structurally related to gymnocolin (**650**).

The secoclerodane skeleton assigned to jamesoniellides A (**658**) and B (**659**) is based on extensive NMR spectrometry including a ^{13}C-^{1}H-2D-COSY. The stereochemistry of (**658**) follows from the presence of an NOE between H-9 and H-12 and between H-9 and H-19. Detailed analysis of the NMR spectral data as well as ^{13}C-^{1}H- and long range ^{13}C-^{1}H-2D-COSY's led to structure (**659**) for jamesoniellide B with stereochemistry being supported by NOE spectrometry. That the two lactones belong to the *ent*-series is suggested by the co-occurrence of *ent*-labdanes in the same liverwort. As *cis*-clerodane lactones with a β-substituted furan ring such as gymnocolin exhibit intense bitterness (*19*), compound (**657**) and the related clerodanes (**658**) and (**659**) may be the bitter principles of *J. autumnalis*.

Chromatography of the ether extract of *Heteroscyphus planus* gave four new highly oxidized spiroclerodane-type diterpenoids, heteroscyphones A–D (**660–663**), as well as an additional new clerodane, heteroscyphol (= cleroda-3,12(*E*),14-trien-11ζ-ol) (**648**) which gave dienone (**649**) by oxidation with PCC (*249, 253*). The relative structure of (**660**) was established by a combination of NMR techniques, chemical transformations (acetylation, conversion of the lactol to a lactone by PPC oxidation) and X-ray crystallographic analysis. The absolute configuration assigned to (**660**) is based on the negative Cotton effect at 298 nm (Δε − 1.81). The structure of (**661**) is based on its preparation from (**660**) by removal of the epoxide (reaction with PhSe$_2$Ph/NaBH$_4$ followed by mild acid treatment). Analysis of the IR, UV and NMR spectra and comparison with those of (**660**) and (**661**) led to the conclusion that the structures of (**662**) and (**663**) were as depicted in Chart 47e. Their absolute configuration is probably the same as that of (**660**).

6.3 Dolabellanes

The earlier report (*19*) mentioned isolation of a diterpenoid bitter principle of unknown structure, barbilycopodin, from *Barbilophozia floerkei* and *B. lycopodioides*. Reinvestigation of the constituents of three European *Barbilophozia* species, *B. attenuata*, *B. floerkei*, and *B. lycopodioides* showed that these species produced dolabellane-type diterpenoids (*143, 273*).

The ether extract of the dried and powdered *B. attenuata*, *B. floerkei* and *B. lycopodioides*, was chromatographed on silica gel to give barbily-

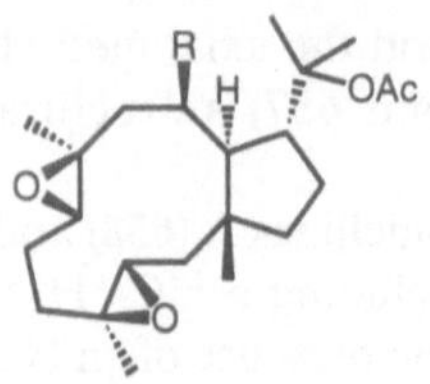

(664) Barbilycopodin (= 10R,18-Diacetoxy-3S,4S,7S,8S-
        diepoxydolabellane) ; R=OAc
(665) 10-Deacetoxybarbilycopodin (= 18-Acetoxy-3S,4S,7S,
        8S-diepoxydolabellane) ; R=H

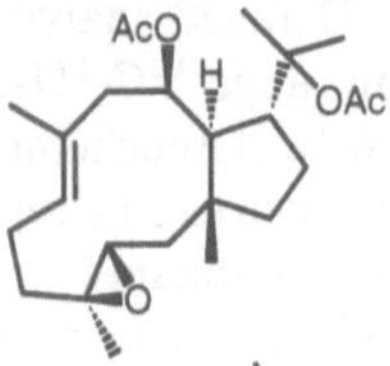

(666) 10R,18-Diacetoxy-3S,4S-epoxydolabell-7E-ene

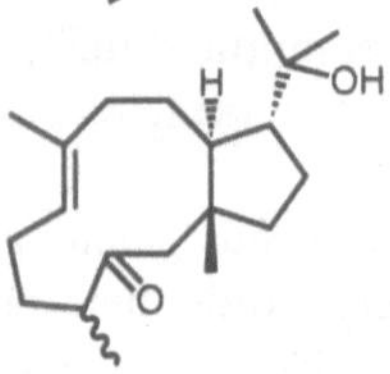

(667) 18-Hydroxydolabell-7E-en-3-one

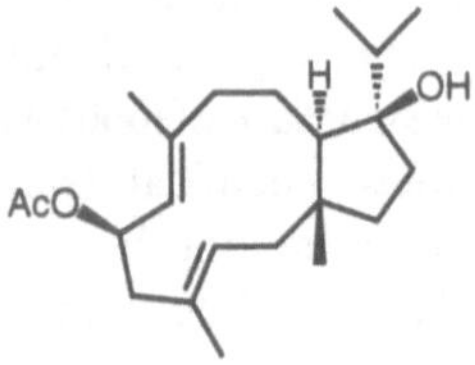

(668) Acetoxyodontoschismenol

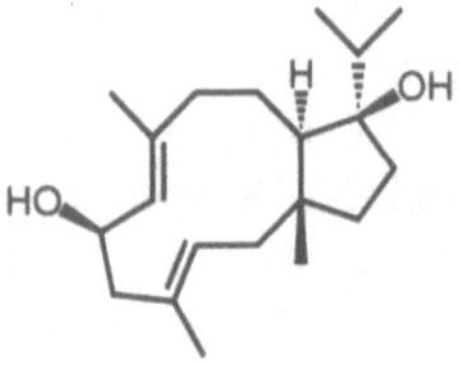

(669) 6,12-Dihydroxydolabella-3E,7E-diene

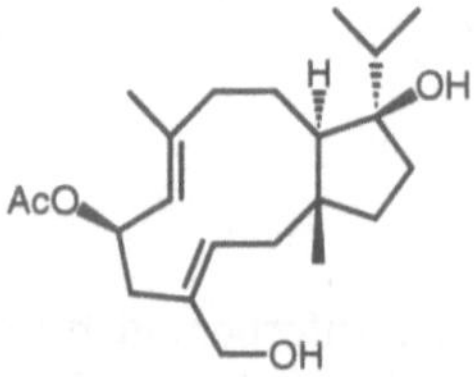

(670) 6-Acetoxy-12,16-dihydroxydolabella-3E,7E-diene

Chart 48a. Dolabellanes found in the Hepaticae

copodin (**664**), while *B. floerkei* produced three additional dolabellane-
type diterpenoids (**665–667**). The structure of (**664**) was deduced by NMR
spectrometry and the chemical transformations shown in Scheme 50.
Alkaline hydrolysis of (**664**) gave (**664a**), whose oxidation with Jones

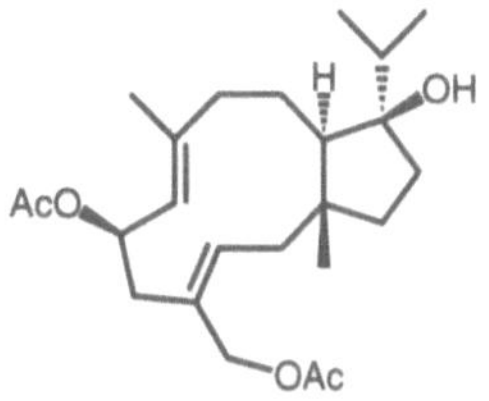

(671) 6,16-Diacetoxy-12-hydroxydolabella-3*E*,7*E*-diene

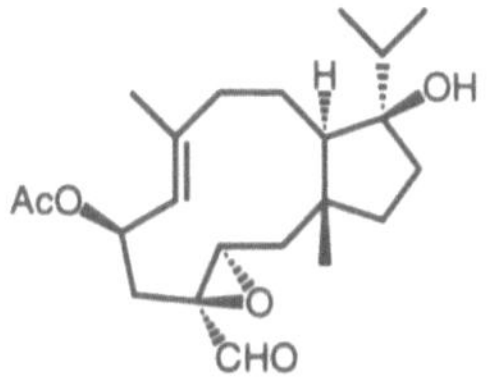

(672) 6-Acetoxy-3,4-epoxy-12-hydroxydolabella-7*E*-en-16-al

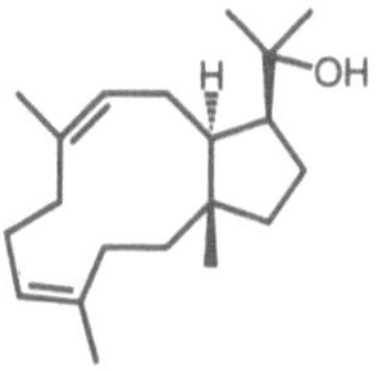

(673) 18-Hydroxy-4,8-dolabelladiene

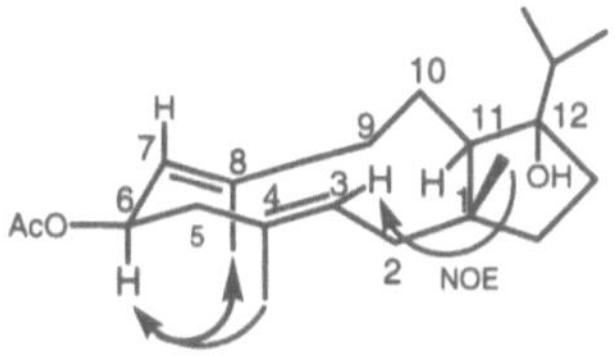

(668)

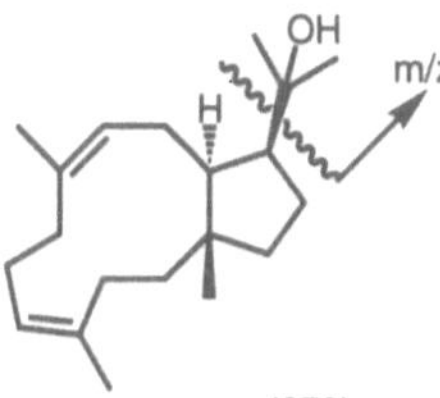

(673)

Chart 48b. Dolabellanes found in the Hepaticae

reagent afforded a ketol (664b). The Eu(fod)$_3$-shifted ^{1}H-NMR spectrum and spin decoupling suggested that (664b) might contain a dolabellane skeleton with one tertiary hydroxyl group, two epoxides and one ketone. This was confirmed by an X-ray analysis of (664a) which also established the relative configuration. The absolute configuration assigned to (664a) is based on the negative Cotton effect of (664b) at 289 nm ($\Delta\varepsilon$ − 0.97). The structure of the monoepoxide (666) was also established by a study of the ^{1}H- and ^{13}C-NMR spectra and by X-ray crystallographic analysis. Treatment of (664) with Zn-Cu in ethanol gave a diacetoxydolabelladiene (664e) and a monoepoxide (666) identical with natural 10*R*,18-diacetoxy-3*S*,4*S*-epoxydolabell-7*E*-ene. Spectroscopic comparison of ^{1}H- and ^{13}C-NMR of (665) and (666) suggested that (665) was 10-deacetoxybarbilycopodin, a suggestion confirmed by an X-ray crystallographic analysis. The

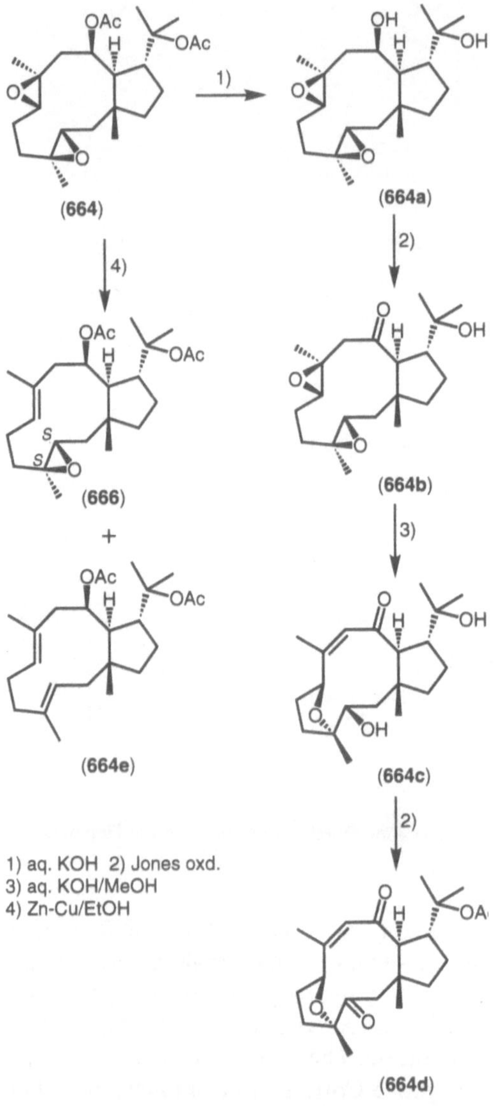

Scheme 50. Reactions of barbilycopodin

absolute configuration of **(665)** is assumed to be the same as that of **(664)**. The structure assigned to the fourth dolabellane, 18-hydroxy-dolabell-7E-en-3-one **(667)**, was arrived at chiefly by a detailed analysis of its spectral properties, although the configuration of the secondary methyl group has not been clarified. The eleven-membered ring has essentially the same conformation in (**664a, 665, 666**), a conformation which has been characterized as a low-energy form of cycloundeca-1,5-diene by molecular-mechanics calculations (*273*). Barbilycopodin (**664**) has also been isolated from *Barbilophozia barbata* (*432*), *B. hatcheri* (*429, 565*) and *Chandonanthus setiformis* (= *Tetralophozia setiformis*) (*286, 647*) which belongs to the Lophoziaceae.

Odontoschisma denudatum of the Lophoziaceae produces five dolabellane-type diterpenoids (+)-acetoxyodontoschismenol (**668**) and the related compounds (**669–672**) (*381, 388*). From the chemical and spectral evidence shown in Scheme 51 and the biogenetic isoprene rule, the carbon skeleton of (**668**) has been deduced to be that of a dolabellane. The relative stereochemistries of C-1 (Me), C-12 (OH) and C-11 were assigned on the basis of a small pyridine-induced solvent shift of C-1 (Me) and a weak lanthanide-induced shift on addition of Eu(dpm)$_3$ and the formation of dolabellatriene (**668a**) shown in Scheme 51. The geometries of the two double bonds and the conformation of the eleven-membered ring were deduced from the ^{13}C-NMR spectrum (δ 16.6 and 17.9 for C-4 and C-8) and NOE spectrometry which showed the absence of an NOE between the vinyl methyls and vinyl protons. The formation of (**668h**) suggested that the tertiary hydroxyl group at C-12 of (**668**) was in the α-configuration and *trans* to the C-1 methyl group. The absolute configuration of

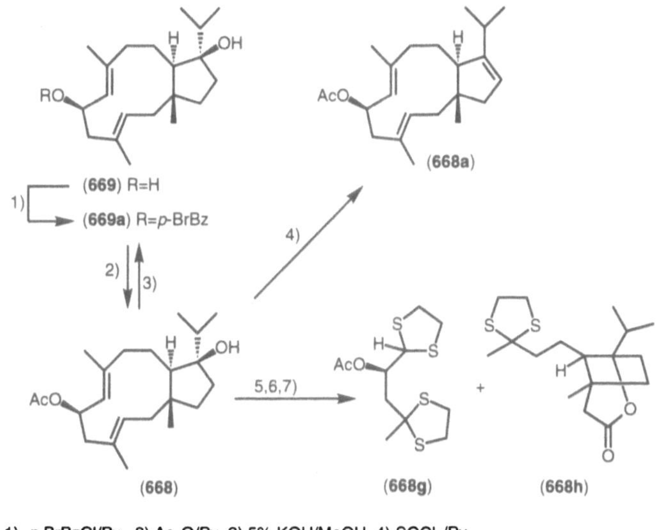

1) *p*-BrBzCl/Py 2) Ac$_2$O/Py 3) 5% KOH/MeOH 4) SOCl$_2$/Py
5) O$_3$ 6) Me$_2$S 7) HS(CH$_2$)$_2$SH

Scheme 51. Reactions of (+)-acetoxyodontoschismenol

(668) was established by combination of an X-ray crystallographic analysis and the exciton chirality method since the CD spectrum of the allylic *p*-bromobenzoate **(669a)** exhibited a first negative Cotton effect ($\Delta\varepsilon_{240\,nm}$ − 8.36). The formation of the δ-lactone **(668h)** must be accompanied by inversion of the configuration at C-12 because the C-12 hydroxyl group is β-oriented and is *cis* to the C-1 methyl group. The solution conformation of **(668)** has been studied by means of Allinger's molecular mechanics (MM2) calculations as well as ¹H-NMR and CD spectrometry *(391)*. The structures of four minor dolabellanoids **(669–672)** have been correlated chemically with **(668)** as shown in Scheme 52. The structure of **(670)** was also confirmed by X-ray crystallographic analysis.

Pleurozia gigantea elaborates dolabellane- **(673)**, fusicoccane-, labdane- and rearranged labdane-type diterpenoids *(52)*. The ¹H- and ¹³C-NMR spectrum as well as HMBC experiments, coupled with the co-occurrence of three fusicoccane-type diterpenoids suggested that **(673)** was a dolabellane-type diterpenoid with a dimethylcarbinyl group [m/z 59 (100%)] and two non-conjugated double bonds. Spin decoupling, ¹H-¹H 2D-COSY and HMBC experiments led to structure **(673)**, with the geometry of the double bonds and the absolute configuration still in question. However, a tentative assignment of absolute stereochemistry

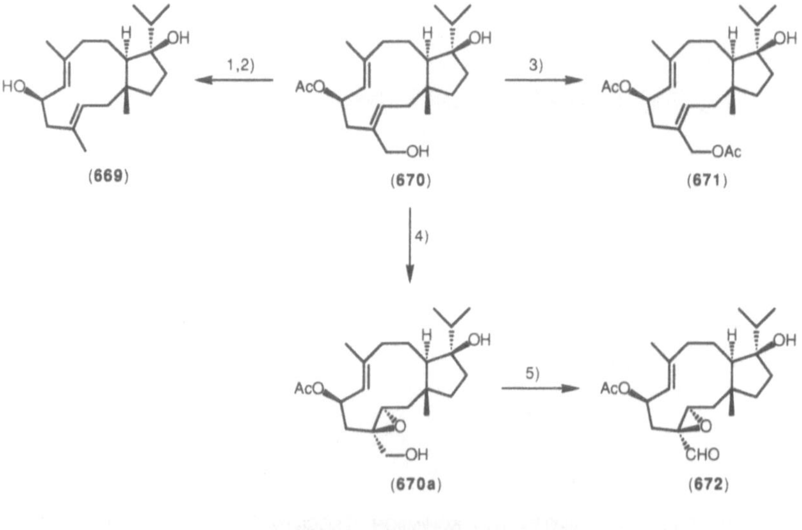

(669)   (670)   (671)

(670a)   (672)

1) Py-SO₃ 2) LiAlH₄ 3) Ac₂O/Py 4) MCPBA/CH₂Cl₂ 5) PDC oxid.

Scheme 52. Reactions of dolabellanoids

can be made by taking into consideration of the co-occurrence of the biogenetically related fusicoccane diterpenoids.

The first dolabellane-type diterpenoid (674) was isolated from the sea hare *Dolabella californica* (296). Dolabellanes (675, 676) similar to (673) have been also isolated from brown algae, but their absolute configurations have not been established (6).

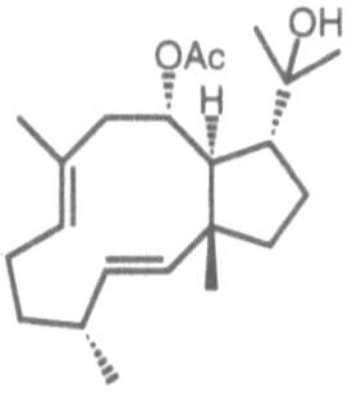

(674) 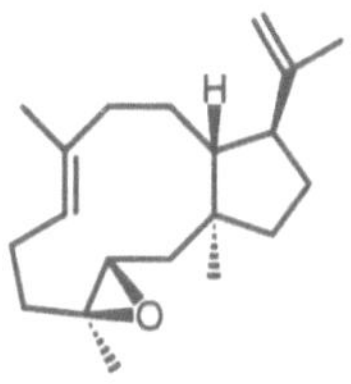   (675) 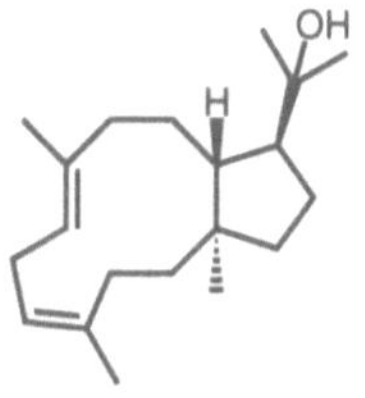   (676)

The biogenesis of the dolabellane-type diterpenoids by cyclization of all-*trans*-geranyl geranyl pyrophosphate (677) can be represented as in Scheme 53 (143, 296). Further cyclization of (677a) in Markownikov

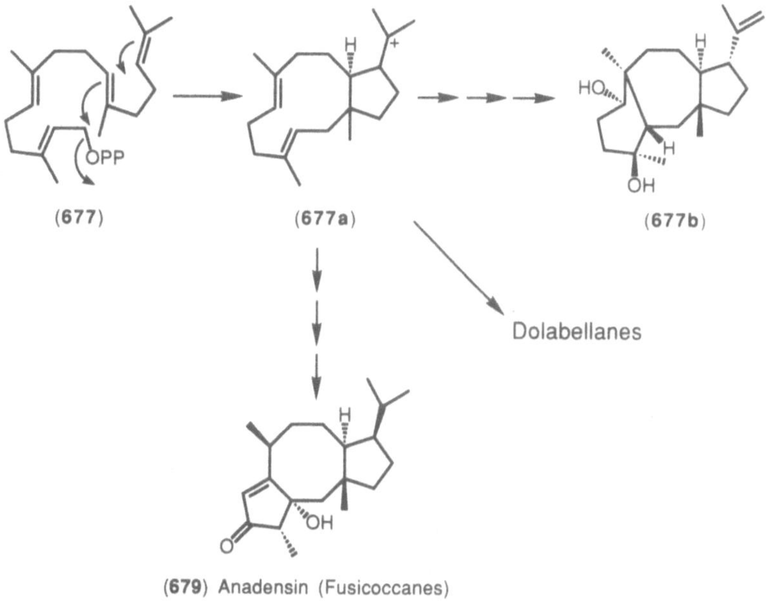

(677)   (677a)   (677b)

Dolabellanes

(679) Anadensin (Fusicoccanes)

Scheme 53. Possible biogenetic pathways for dolabellanes and fusicoccanes from geranyl geranyl pyrophosphate

fashion affords the diterpenoid skeleton (677b) found in marine organisms as shown in Scheme 53 (*143*). A different cyclization mode of
(677a) leads to the fusicoccane-type diterpenoids found in liverworts (see
below).

6.4 Fusicoccanes

Fusicoccane-type diterpenoids are relatively rare in nature although
they have been found in fungi (*88, 89, 92, 487–489*) and higher plants (*2*).
Anadensin (679) which has been isolated from *Anastrepta orcadensis*
(Lophoziaceae) is the first fusicoccane-type diterpenoid found in bryophytes (*143, 274*). The structure of (679) was deduced by analysis of the
^{1}H- and ^{13}C-NMR data including Eu(fod)$_3$-induced shift values. The
final stereochemistry of (679) was established by X-ray crystallographic
analysis. The same compound has been isolated from *Plagiochila ovalifolia* (*435a*).

As mentioned earlier *Plagiochila moritziana* produces a sesquiterpenoid linked to a fusicoccane (*515*). The hydrocarbon, fusicoccadiene (678)
has been found in *Plagiochila geniculata*, *P. moritziana*, *P. panamensis*,
Riccardia andina and *Symphyogyna brasiliensis* (*513*).

Pleurozia gigantea elaborates three fusicoccane diterpenoids, fusico-
gigantones A (680), B (681) and fusicogigantepoxide (682), together with a
biogenetically related dolabellane-type diterpene alcohol (673) discussed
in the previous section (*52*). The presence of a fusicoccane skeleton and
the position of the functional groups in (680) were established by IR and
NMR spectrometry including 2D-COSYs as well as HMBC. Treatment
of (680) with lithium diisopropylamide (LDA) in tetrahydrofuran gave an
α,β-unsaturated ketone identical with anadensin (679) (*274*). Hence, the
absolute configuration of fusicogigantone A was established as (680).
Compound (681) had ^{1}H-, ^{13}C-NMR and mass spectral data similar to
those of (680). The full structure was settled by a combination of 2D-
COSY, HMBC and NOESY techniques, with the absolute stereochemistry presumably the same as that of (680). ^{1}H- and ^{13}C-NMR
spectra as well as 2D-COSY and HMBC experiments of (682) indicated
that it was also a fusicoccane with two epoxy rings in the five membered-
ring, the absolute stereochemistry being presumably the same as (680)
and (681). This was subsequently confirmed by an X-ray crystallographic
analysis of the same material isolated from *Bryopteris filicina* (*432a*). This
was the first report of the co-occurrence of biogenetically related dolabel-
lane- and fusicoccane-type diterpenoids in the plant kingdom. *Plagiochila
corrugata* also elaborates a new fusicoccane diepoxide, fusicorrugatol
(683), together with fusicogigantone (680) (*441a*).

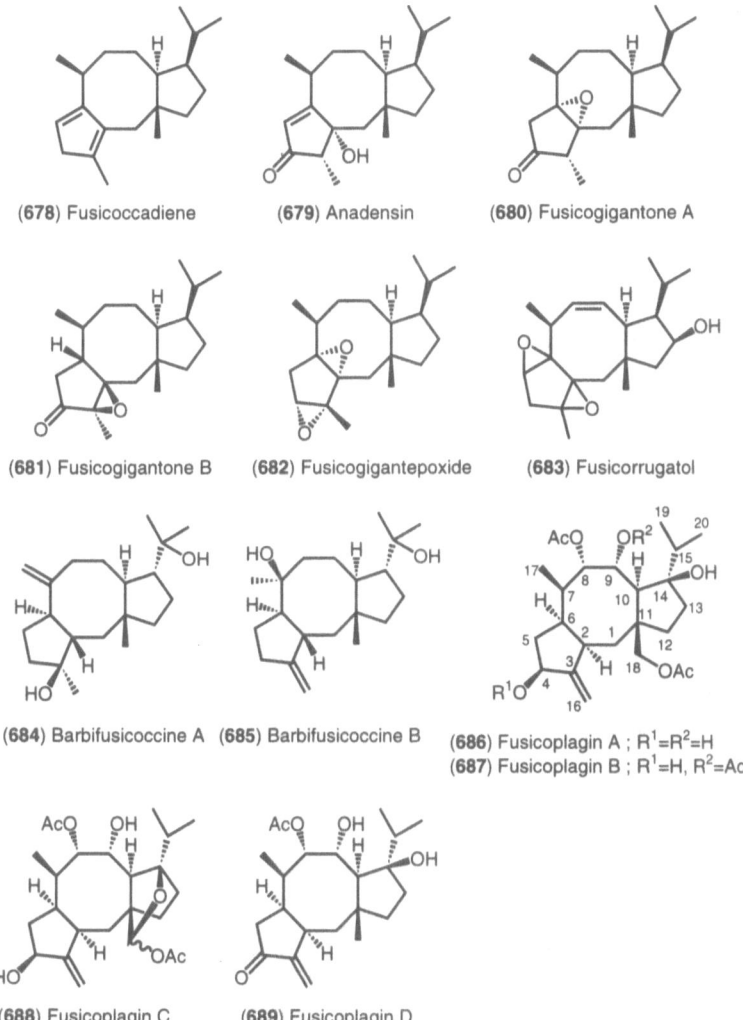

(678) Fusicoccadiene (679) Anadensin (680) Fusicogigantone A

(681) Fusicogigantone B (682) Fusicogigantepoxide (683) Fusicorrugatol

(684) Barbifusicoccine A (685) Barbifusicoccine B

(686) Fusicoplagin A ; R^1=R^2=H
(687) Fusicoplagin B ; R^1=H, R^2=Ac

(688) Fusicoplagin C (689) Fusicoplagin D

Chart 49a. Fusicoccanes found in the Hepaticae

Two new fusicoccanes, barbifusicoccines A (684) and B (685), were isolated from *Barbilophozia floerkei*, along with dolabellanes (664, 665) which might be precursors of the fusicoccanes (429, 565, 567a). Structures of (684) and (685) were elucidated by ^{1}H- and ^{13}C-NMR, 2D-COSY and NOE difference spectra and by considering the nature of the co-occurring dolabellanoids. Treatment of (685) with *p*-bromobenzoyl chloride gave a monobromo- and a dibromobenzoate. The former, with the acyl group on the 8-membered ring, exhibited an NOE between H-18 and the protons on the benzene ring indicating that the hydroxyl and H-18 were *cis*.

Plagiochila sciophila (= *P. acanthophylla* subsp. *japonica*) contains *ent*-sesquiterpenoids as mentioned earlier. Further fractionation of the ethyl acetate soluble portion of a methanol extract of fresh *P. sciophila* resulted in isolation of four new highly oxygenated fusicoccanes, fusico-plagins A (**686**), B (**687**), C (**688**) and D (**689**) (*261*). The stereochemistry of (**686**) was established by X-ray crystallographic analysis of the tetraace-tate prepared by acetylation of the hydroxyl groups on C-4 and C-9. The absolute stereochemistry at C-4 was shown as *S* by the allylic benzoate chirality method (*239*) using the C-4 *p*-bromobenzoate whose CD spectrum showed a positive Cotton effect at 244 nm ($\Delta\varepsilon + 7.51$). That (**687**) was the 9-*O*-acetate of (**686**) was shown by the downfield shift of the H-9 signal and conversion of (**686**) and (**687**) to the same tetraacetate. The structure assigned to (**688**) was based on the presence of an acetal proton at C-18 and by a chemical correlation which involved reduction of the acetylated hemiacetal to a diol followed by conversion to the same tetraacetate furnished by (**686**) and (**687**). The presence of a tertiary methyl group in place of the hydroxymethyl group of (**689**) indicated that C-18 was unsubstituted. The structure of (**689**) was deduced by spectro-scopic comparison with (**686**–**688**).

Plagiochila spinulosa produces two fusicoccanoids, spinuloplagins A (**690**) and B (**691**) together with bibenzyl derivative (**907**) (*478*). The aromatic and terpenoid signals were similar to those of the bibenzyl and

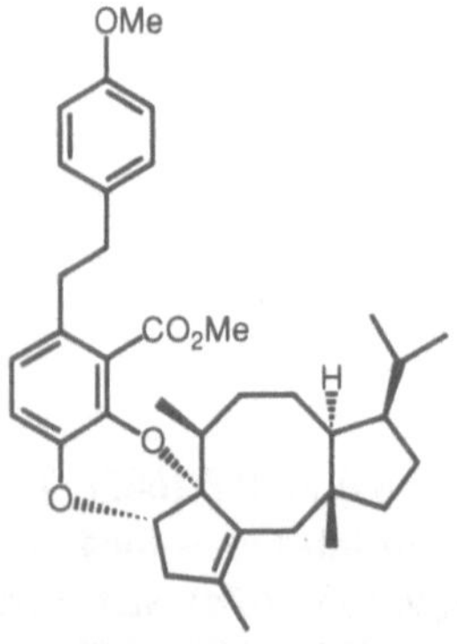

(**690**) Spinuloplagin A

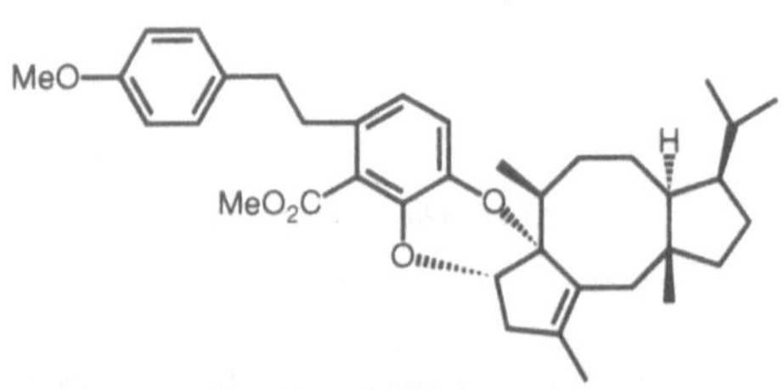

(**691**) Spinuloplagin B

Chart 49b. Fusicoccanes found in the Hepaticae

fusicoccane diterpenoid. The stereostructure of (690) was established by X-ray crystallographic analysis. The second fusicoccane ether (691) was a regioisomer of (690) by spectroscopic comparison with those of (690). The *ortho*-diphenol-olefin conjugate moiety present in these compounds is unusual and new to liverwort constituents.

The proposed biogenesis of fusicoccane-type diterpenoids involves cyclization of dolabellane which in turn has been formed by cyclization of geranyl geranyl pyrophosphate (677), as shown in Scheme 53 (*143*).

6.5 Kauranes

Fourteen *ent*-kaurane-type diterpenes from liverworts distributed over the Jungermanniales in *Anthelia, Jungermannia, Nardia, Porella*, and *Solenostoma* species were described in the earlier report (*19*). Kaurene (692) is the most commonly encountered kaurane-type diterpenoid and has been found in several species (Table IIc).

A new *ent*-kauren-15-one-18-oic acid (696) has been isolated from Indian *Porella densifolia* subsp. *appendiculata*, together with *ent*-18-hydroxykauren-15-one (695) (*19, 67*). Kaurene (692), 15-hydroxykaurene (693) and kauren-15-one (694) have also been detected in the methanol extract of this species. The structure of (696) has been confirmed by its preparation from (695) by Jones oxidation (*67*). The previous report mentioned isolation of *ent*-18-hydroxykauren-15-one (695), (16R)-*ent*-18-hydroxykauran-15-one (697) and *ent*-kauren-18-oic acid (698) from *Porella densifolia* var. *fallax* (*19*). Reinvestigation of the methanol extract of *P. densifolia* var. *fallax* resulted in the isolation of other known *ent*-kaurenes, *ent*-11α-hydroxykauren-15-one (699a) and (16R)-*ent*-11α-hydroxykauran-15-one (700a), together with (698) (*67*) but (697, 698, 699a, 700a) found in *P. densifolia* var. *fallax* were not detected in *P. densifolia* subsp. *appendiculata*. A rare 7-oxygenated kaurane-type diterpenoid (701) has been isolated from *Plagiochila pulcherrima*; its relative stereochemistry was established by spectral data and conversion to an acetonide (*190*) while its absolute configuration is based on the CD spectrum of its dibenzoate (*237*) which showed a positive first Cotton effect at 238 nm ($\Delta\varepsilon$ + 22.8) and a negative second Cotton effect at 222 nm ($\Delta\varepsilon$ − 15.8). The positive sign of the first Cotton effect led to the conclusion that (701) was *ent*-kaur-16-en-7α,15β-diol.

Nardia species are rich sources of kaurane-type diterpenoids. *N. scalaris* and *N. succulenta* produce 15-hydroxykaurene (693), kauren-15-one (694) and (16R)-*ent*-kauran-15-one (702), together with kaurenyl malonates (*vide infra*) (*354a*).

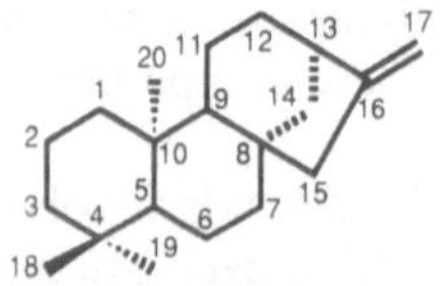

(692) *ent*-Kaurene 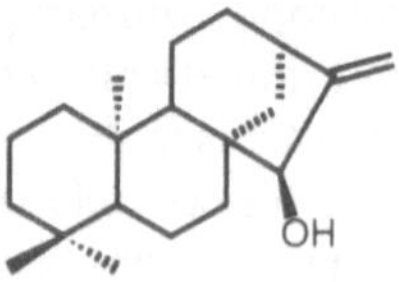   (693) *ent*-15α-Hydroxykaurene 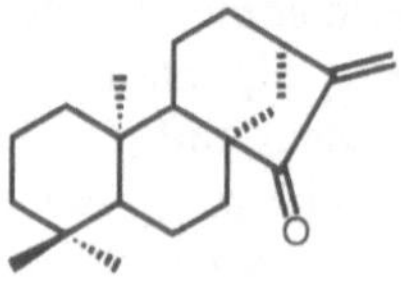   (694) *ent*-Kauren-15-one

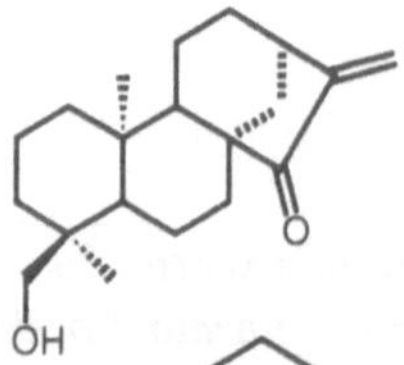

(695) *ent*-18-Hydroxykauren-15-one

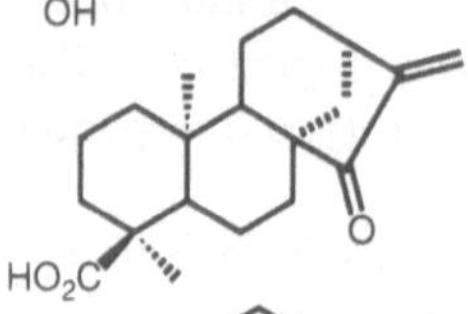

(696) *ent*-Kauren-15-one-18-oic acid

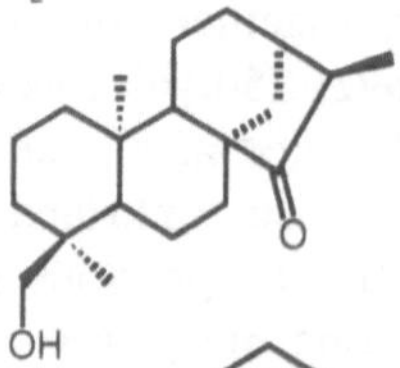

(697) (16*R*)-*ent*-18-Hydroxykaurane-15-one

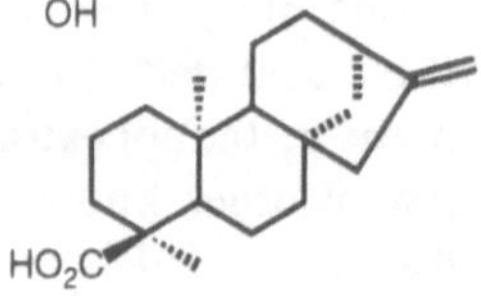

(698) *ent*-Kauren-18-oic acid

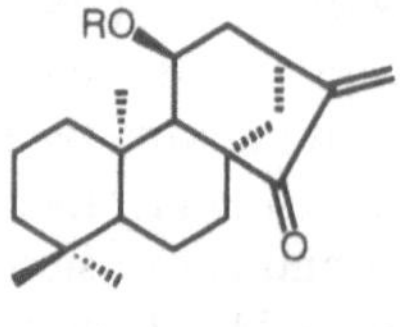

(699a) *ent*-11α-Hydroxykauren-15-one ; R=H
(699b) R=Ac

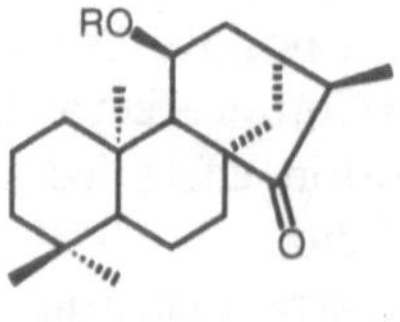

(700a) (16*R*)-*ent*-11α-Hydroxykauran-15-one ; R=H
(700b) R=Ac

Chart 50a. **Kauranes found in the Hepaticae and their derivatives**

Jungermannia infusca gave (16S)-*ent*-11α-hydroxykauran-15-one (**703**) which has been prepared from the (16R)-isomer (**700a**) by isomerization using alkali, along with the previously known *ent*-kaurane-type diterpenoids (**693**, **694**, **699a**, **700a**, **702**) (*594*). Diterpenoids (**699a**, **700a**) have also been isolated from *Jungermannia truncata* (*437*).

Jamesoniella autumnalis produces not only clerodane- and labdane- but also the kaurane-type diterpenoids (**694**, **699a**, **700a**) (*437*) previously isolated from *Jungermannia* species (*19*). *Jungermannia vulcanicola* contains three new *ent*-kaurane-type diterpenoids, *ent*-kauren-3β,15α-diol (**704**), *ent*-15α-hydroxykauren-3β-yl acetate (**705**) and *ent*-3β-hydroxykauran-15-one (**706**) (*437*). The spectral features (2D-COSYs) and comparison of ¹H-NMR spectrum of (**704**) with those of the previously known *ent*-kaurenes from the other *Jungermannia* species showed that (**704**) was a kaurane-type diterpene diol. The position and stereochemistry of the hydroxyl group were confirmed by an HMBC and NOE spectrometry. Acetylation of (**704**) and (**705**) gave the same diacetate. The location of the acetoxy group at C-3 in (**705**) was determined by 2D-COSYs. The presence of a conjugated ketone group in (**706**) was evident from UV maximum at 233 nm. Reduction of (**706**) gave (**704**) through a garryfoline-cuauchichicine rearrangement (see below) while catalytic hydrogenation of (**704**) afforded a dihydro derivative whose CD spectrum exhibited a negative Cotton effect at 308 nm (Δε − 0.59) thus establishing that (**704–706**) were members of the *ent*-kaurane-series.

Two new kaurane-type diterpenoids, *ent*-16β-hydroxykauran-3-one (**708**) and *ent*-kauran-3β, 16β-diol (**709**), have been isolated from *Frullanoides densifolia*, along with *ent*-16β-hydroxykaurane (**707**) (*14*, *554*, *556*). The structure of (**708**) was deduced from the ¹H- and ¹³C-NMR and IR spectra as well as an HMBC experiment. Wolff-Kishner reduction of (**708**) afforded a deoxo product identical with (**707**). Compound (**709**) had a secondary hydroxyl, which was equatorial and β (*dd* at δ 3.19) in place of the ketone function. Reduction of (**708**) with NaBH₄ in methanol gave a diol identical with (**709**) in all respects.

A new *ent*-kaurenol, nardiin (**710**) has been isolated from the chloroform extract of a large amount (4.19 kg) of *Nardia scalaris*. Its structure and stereochemistry were elucidated by analysis of spectral data and by the garryfoline-cuauchichicine type rearrangement accompanying its reduction (Scheme 54) (*101*). Further investigation of the ether extract of *N. scalaris* resulted in isolation of the *ent*-kaurane malonate (**711a**) in the form of its methyl ester (**711b**) (*148*). The relative and absolute configurations were established by analysis of the ¹H- and ¹³C-NMR spectra, and by LiAlH₄ reduction followed by oxidation to a ketone which

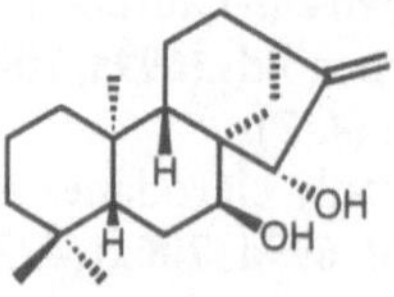

(701) *ent*-Kaur-16-en-7α,15β-diol 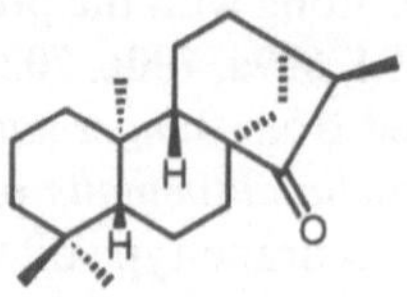   (702) (16*R*)-*ent*-Kauran-15-one

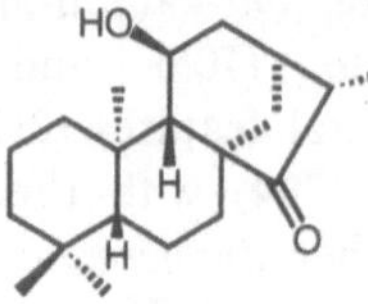

(703) (16*S*)-*ent*-11α-Hydroxykauran-15-one

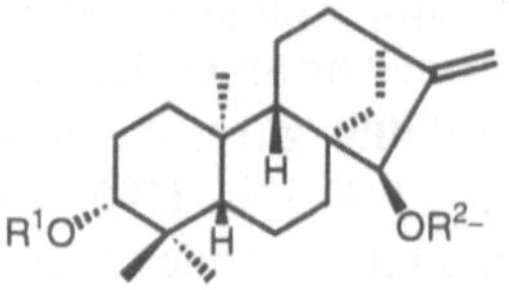

(704) *ent*-Kauren-3β,15α-diol ; R^1=R^2=H
(705) *ent*-15α-Hydroxykauren-3β-yl acetate ; R^1=Ac, R^2=H

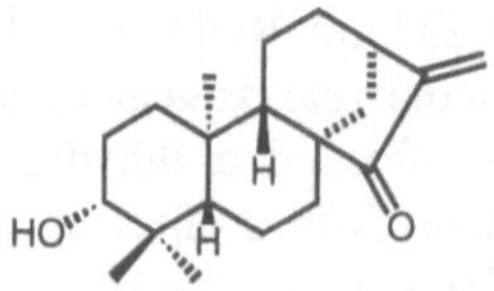

(706) *ent*-3β-Hydroxykauren-15-one 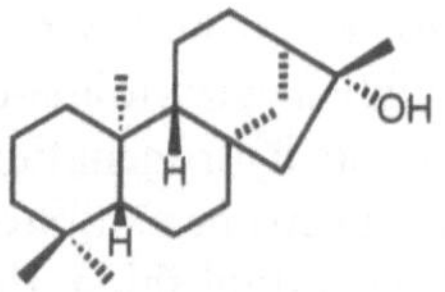   (707) *ent*-16β-Hydroxykaurane

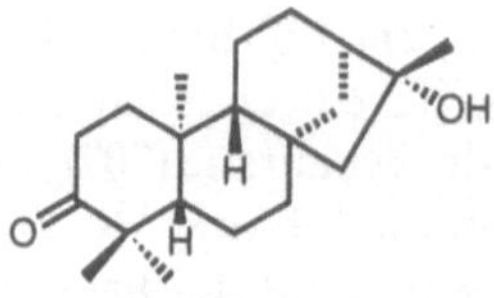

(708) *ent*-16β-Hydroxykauran-3-one 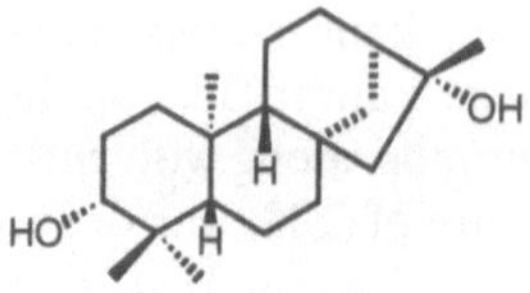   (709) *ent*-Kauran-3β,16β-diol

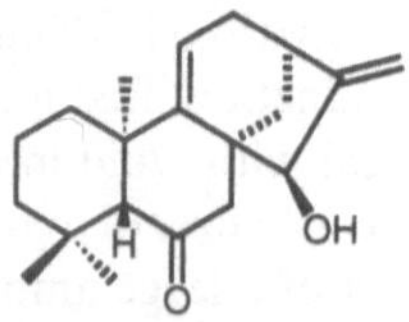

(710) Nardiin

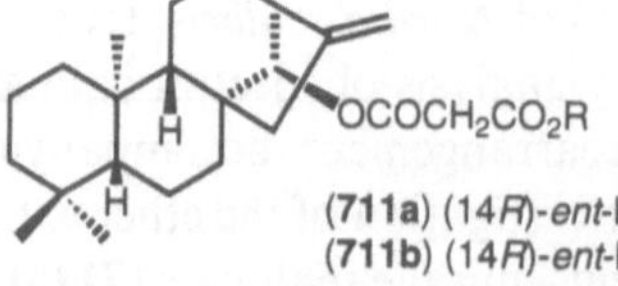

(711a) (14*R*)-*ent*-Kaur-16-en-14-yl hydrogen manolate ; R=H
(711b) (14*R*)-*ent*-Kaur-16-en-14-yl manolate ; R=Me

Chart 50b. Kauranes found in the Hepaticae

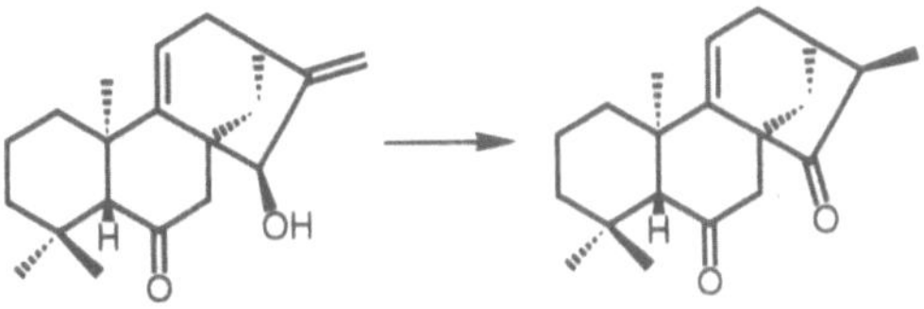

(710)   (710a)

Scheme 54. Reduction of (710)

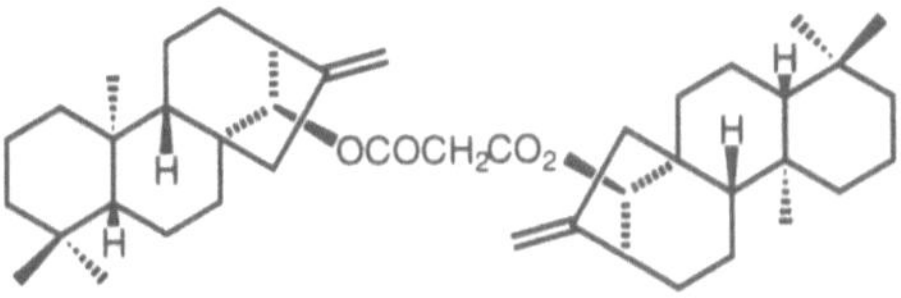

(712) Bis-(14*R*)-*ent*-kaur-16-en-14-yl malonate

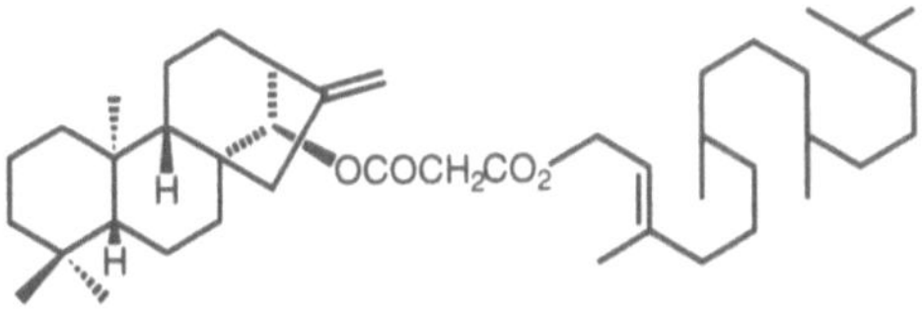

(713) (14*R*)-*ent*-Kaur-16-en-14-yl phytyl malonate

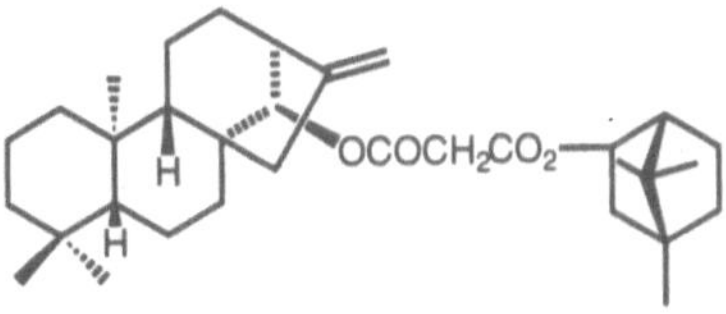

(714) (14*R*)-*ent*-Kaur-16-en-14-yl epi-bornyl malonate

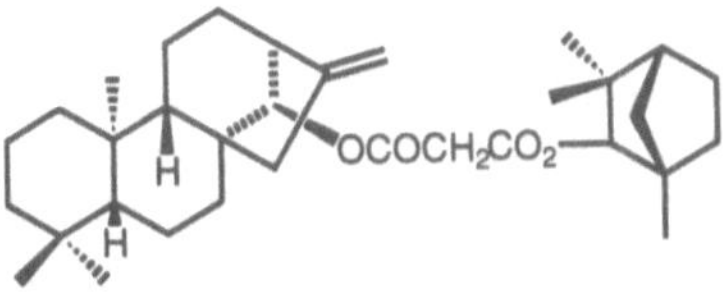

(715) (14*R*)-*ent*-Kaur-16-en-14-yl fenchyl malonate

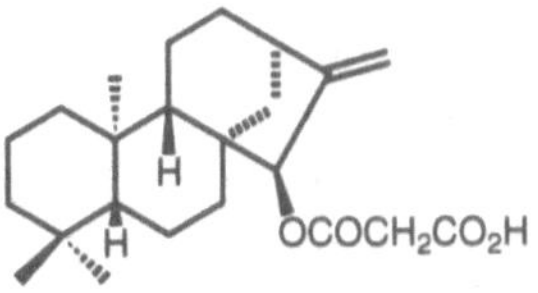

(716) (14*R*)-*ent*-Kaur-16-en-15β-hydrogen malonate

Chart 50c. **Kauranes found in the Hepaticae**

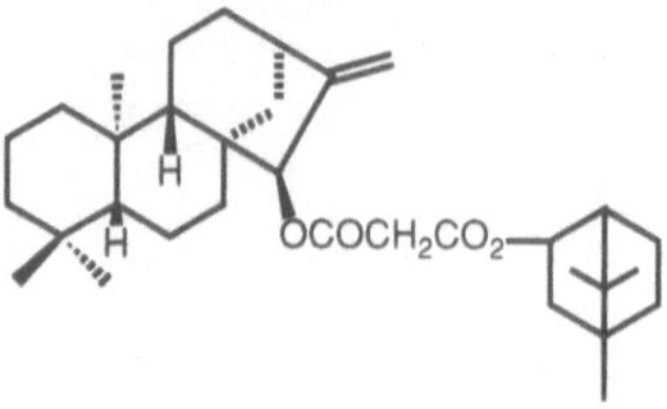

(717) *ent*-Kaur-16-en-15β-yl epi-bornyl malonate

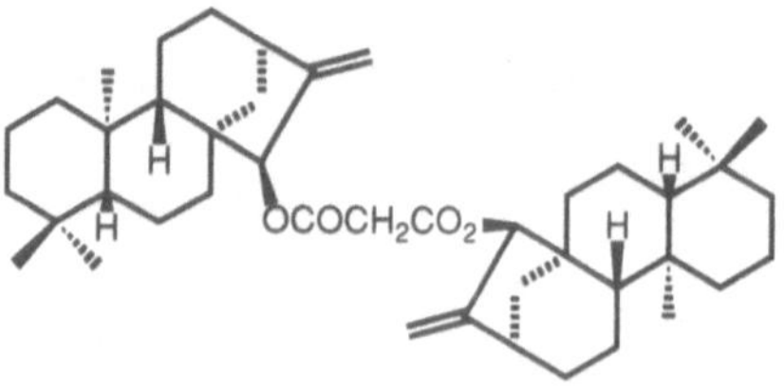

(718) Bis-*ent*-Kaur-16-en-15β-yl malonate

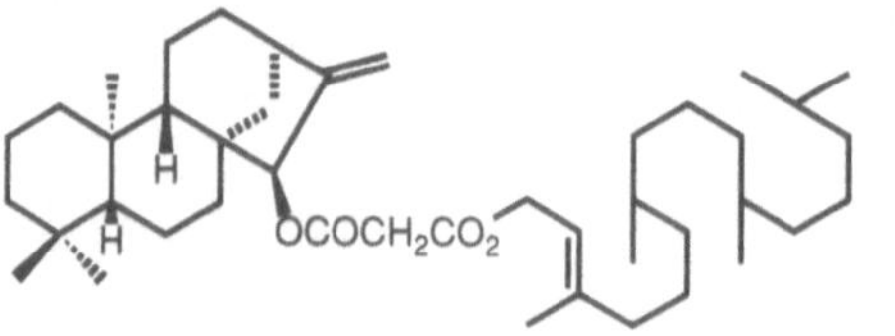

(719) *ent*-Kaur-16-en-15β-yl phytyl malonate

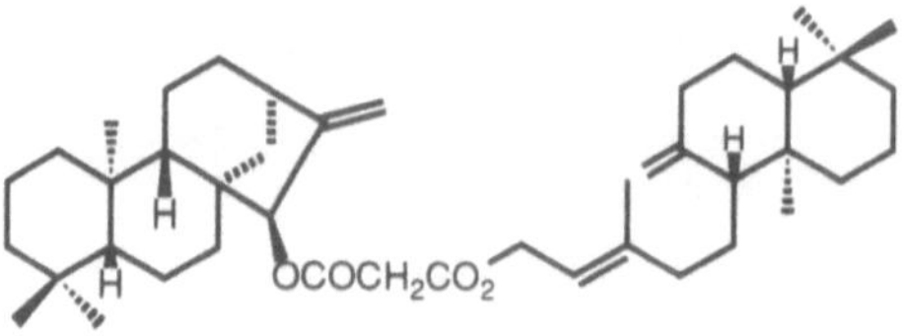

(720) *ent*-Kaur-16-en-15β-yl *ent*-labdan-8(17),13-dien-15-yl malonate

Chart 50d. Kauranes found in the Hepaticae

exhibited a negative Cotton effect at 305 nm ($\Delta\varepsilon - 1.17$). C-14 oxygen-ated kaurane-type diterpenoids are rare in nature (*184, 545*). Two more complex *ent*-kaurene malonates (712) and (713) together with the parent *ent*-14α-kaurenol (726) which was also provided by LiAlH$_4$ reduction of (711a) have been isolated from Japanese *Nardia subclavata*, along with (711a) (*585*), while European *N. scalaris* furnished the malonate ester (714), (715), (716), (717) and (718), along with (711a), (712), (713) and (726) (*354a*). The same kaurene malonates (716, 718) and two new kaurene

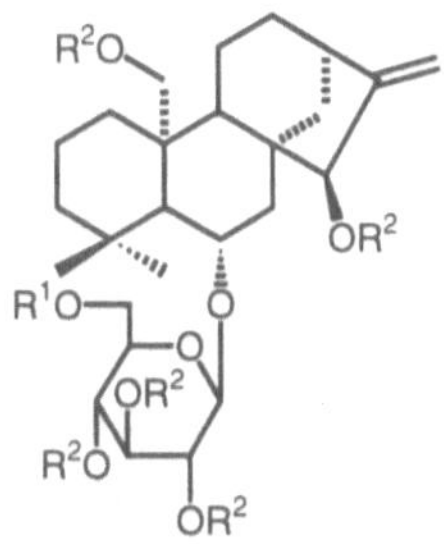

(721a) Infuscaside A (= 6β-(6'-Acetoxy)-β-glucopyranosyl-
ent-15α, 20-dihydroxykaur-16-ene) ; R¹=Ac, R²=H
(721b) R¹=R²=Ac

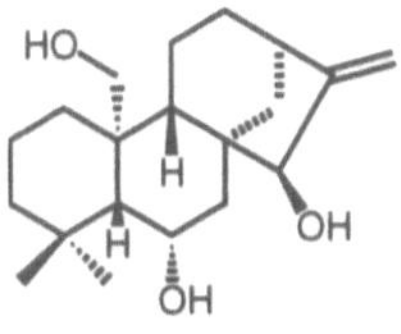

(721c)

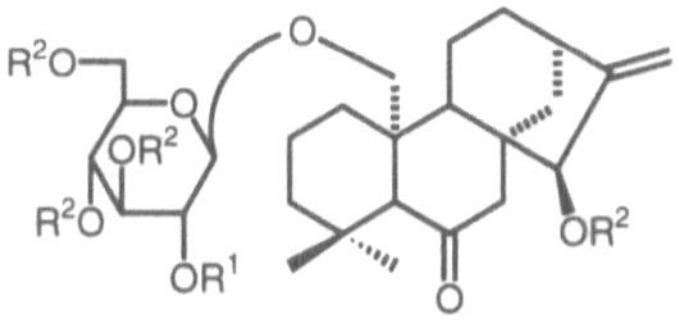

(722a) Infuscaside B
(=20-(2'-Acetoxy)-β-glucopyranosyl-ent-
6-keto-15α-hydroxykaur-16-ene) ;
R¹=Ac, R²=H
(722b) R¹=R²=Ac

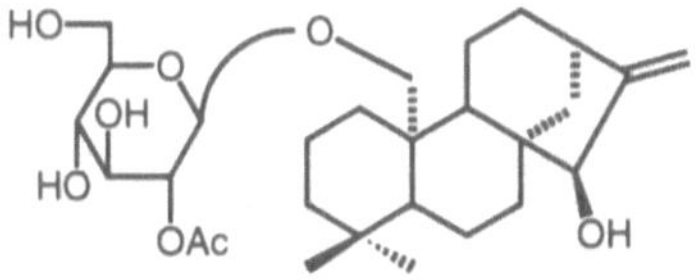

(723) Infuscaside C
(=20-(2'-Acetoxy)-β-glucopyranosyl-ent-
15α-hydroxykaur-16-ene)

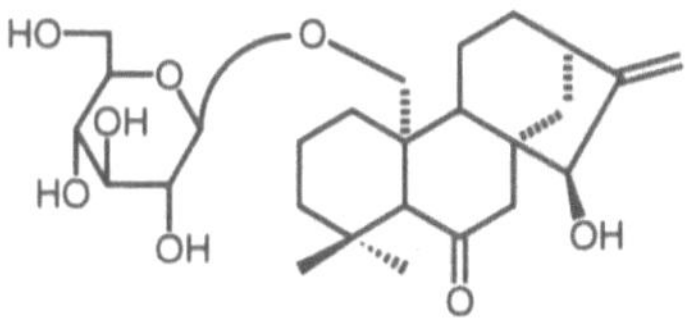

(724) Infuscaside D
(=20-β-Glucopyranosyl-ent-6-keto-
15α-hydroxykaur-16-ene)

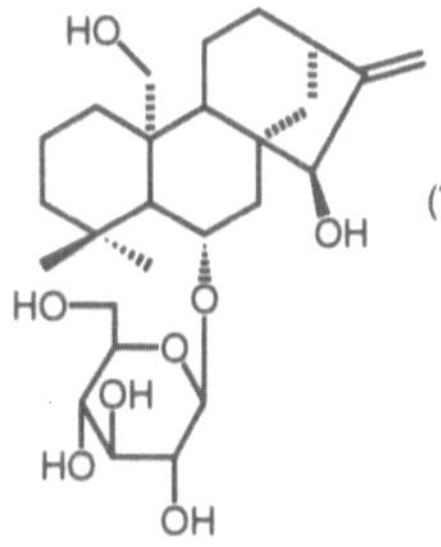

(725) Infuscaside E
(=6β-β-Glucopyranosyl-ent-15α,20-dihydroxykaur-16-ene)

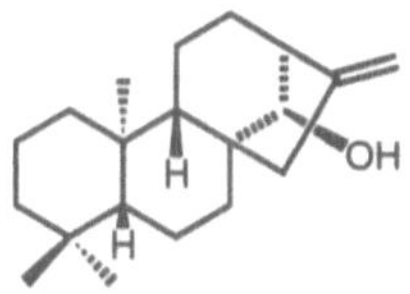

(726) ent-14α-Hydroxykaurene

Chart 50e. Kauranes found in the Hepaticae and their derivatives

malonates, *ent*-kaur-16-en-15β-yl phytyl malonate (**719**) and *ent*-kaur-16-en-15β-yl *ent*-labdan-8(17), 13-dien-15-yl malonate (**720**) have been isolated from European *N. succulenta* (*354a*).

Jungermannia infusca contains very bitter principles. Reinvestigation of the methanol extract by a combination of column chromatography on silica gel, Sephadex LH 20 and HPLC resulted in the isolation of five new kaurane-type glucosides, infuscasides A–E (**721a, 722a, 723–725**) (*436, 594*), along with the known *ent*-15α-hydroxykaurene (**693**) (*19*). Acetylation of (**721a**) gave a hexaacetate (**721b**). Treatment of (**721a**) with cellulase for 3 months gave an aglycone (**721c**), indicating the presence of one acetoxyl group at C-6'. The position of each hydroxyl group and the stereochemistry of (**721c**) were established by the analysis of 2D-COSY and NOESY experiments. That the glucose unit is linked axially was based on the difficulty with which (**721a**) underwent hydrolysis with cellulase while the β-configuration of the glucose at C-6 of (**721a**) was based on the coupling constant ($J = 7.6$ Hz) involving the anomeric proton (H-1').

Acetylation of (**722a**) gave a pentaacetate (**722b**). Location of an acetoxyl group at C-2' in the glucose moiety and of the ketone group at C-6 was established by analysis of the cross peaks in the ^{1}H-^{1}H COSY NMR spectrum of (**722a**). That the glucose unit was linked to C-20 was evident from the chemical shift of H-20, at higher field than the other proton hydroxyl bearing carbon, while the stereochemistry and β-configuration of the glucose unit were established by NOE difference spectrum and the coupling constant ($J = 7.8$ Hz) of the anomeric proton. That infuscaside C (**723**) was the 6-deoxo derivative of (**722a**), became clear on spectroscopic comparison with (**721a**) and (**722a**). The spectral data of infuscaside D (**724**) were almost identical with those of infuscaside B (**722a**) except for the absence of one acetoxyl group. Acetylation of (**724**) also gave pentaacetate (**722b**). The spectral data of infuscaside E (**725**) quite resembled those of infuscaside A (**721a**), except for the absence of an acetoxyl group, suggesting that (**725**) might be deacetoxyinfuscaside A. This presumption has been confirmed by acetylation of (**725**) which gave a hexaacetate identical with hexaacetate (**721b**) from (**721a**). This is the first record of the isolation of terpene glucosides from bryophytes.

6.6 Labdanes and Chettaphanins

As mentioned in the previous report (*19*) six labdane-type diterpenoids had been isolated from the Jungermanniales prior to 1982 three of which belonged to the *ent*-series, while the absolute stereochemistry of the other three labdanes isolated from *Ptychanthus striatus* remained to be clarified. Since then, a various normal and *ent*-labdane-type diterpenoids have been isolated not only from Jungermanniales but also from Marchantiales and the absolute configuration of the previously known labdanes in *P. striatus* has been established.

Taiwanese *Pleurozia acinosa* gave a new labdane-type diterpene, 8-*epi*-sclareol (**727**), together with the clerodane (−)-kolavelool (**621**) (*625*). The structure of (**727**) was established by comparison of the ^{13}C-NMR spectrum with the spectra of sclareol (**728**) and 13-*epi*-sclareol (**729**), except for the absolute stereochemistry. The same diol has also been obtained from East Malaysian *P. gigantea* (*52*) and South American *Plagiochila corrugata* (*441a*).

Porella perrottetiana is a rich source of diterpenoids (*19*). It produces (+)-labda-7,14-dien-13-ol (**730**) and labda-12,14-dien-8α-ol (**731**) first isolated from *Nicotiana tabacum* (*142*) along with the previously known *ent*-labdane-diol (**735**) (*80, 83*). Relative and absolute stereochemistry followed from spectroscopic comparison with (**735**) and the presence of co-metabolite of (**735**). Diterpene (**730**) has also been isolated from *Mylia nuda* (*624*).

Marchantia polymorpha, *M. paleacea* var. *diptera* produces not only *ent*-sesquiterpenoids but also an *ent*-diterpenoid, *ent*-labda-7,13*E*-dien-15-ol (**736**) (*74, 80, 579*). The same compound occurs in *Targionia hypophylla* (*75*) while its enantiomer has been isolated from *Nicotiana setchelli* (*526*). *Mylia nuda* elaborates (+)-manoyl oxide (**737**) (*622*) which occurs in several higher plants. Three previously known labdane-type diterpenoids (**738, 739, 740**) have been isolated from *Jungermannia infusca* (*592*). The first two are gomeraldehyde and *epi*-gomeraldehyde found in higher plants (*207*), the absolute configuration of (+)-isoabienol (**740**) ($[\alpha]_D + 29.1°$) previously isolated from conifers has not been established (*208*). A substance assigned formula (**740**) has been obtained from a *Sideritis* species (*125*); however, the optical rotation ($[\alpha]_D − 0.4°$) differs.

Three labdane-type diterpenoids, haplomitrenolides A (**741**), B (**742**) and C (**743**) have been isolated from the primitive liverwort, *Haplomitrium mnioides* (*77*). The stereochemistry of (**741**) was established by NOE spectrometry. The location of an axial tertiary hydroxyl group at C-9 of (**742**) was deduced from the diamagnetic shifts of C-5 and C-12 (γ-effect) and the paramagnetic shifts of C-9 (α-effect) and C-11 (β-effect) compared with the ^{13}C-NMR spectrum of (**741**). The location of a carbomethoxy group at C-20 of (**743**) was similarly deduced from the diamagnetic shifts of C-3, C-5 and C-19 (γ-effect) and the paramagnetic chemical shift of C-4 (β-effect) compared with (**741**). The absolute configuration of the three

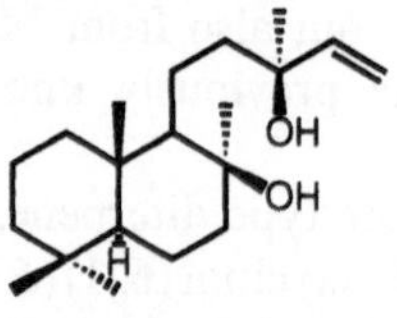

(**727**) 8-epi-Sclareol

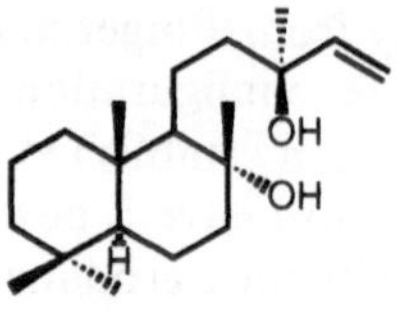

(**728**) Sclareol

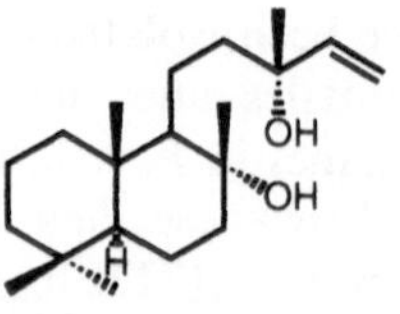

(**729**) 13-epi-Sclareol

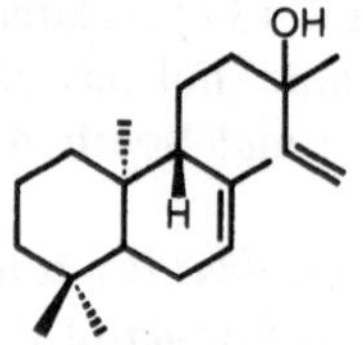

(**730**) Labda-7,14-dien-13-ol

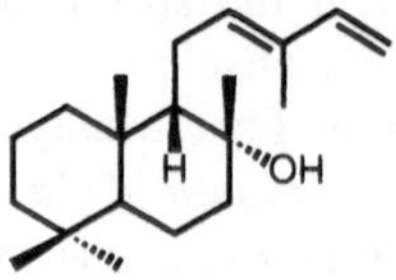

(**731**) Labda-12,14-dien-8α-ol

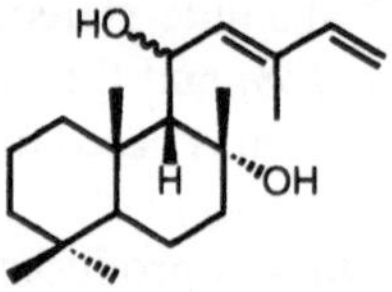

(**732**) Labda-12E,14-dien-8α,11ζ-diol

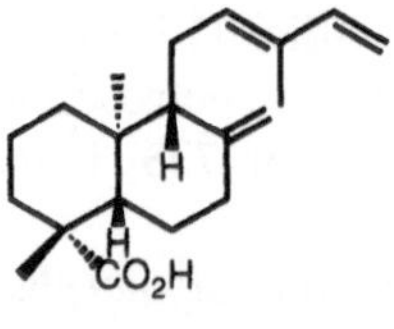

(**733**) (-)-*ent-trans*-Communic acid

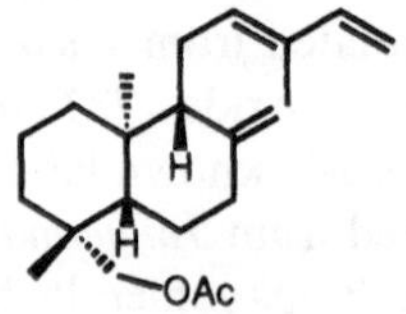

(**734**) *ent-trans*-Cummunol acetate
(=19α-Acetoxy-*ent*-labda-8(17),12E,14-triene)

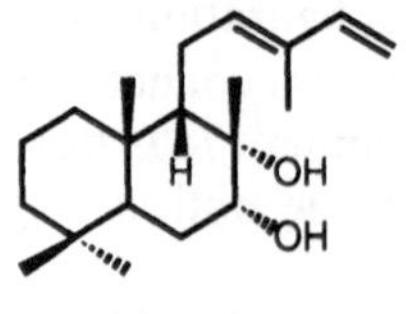

(**735**) Labda-12,14-dien-7,8-diol

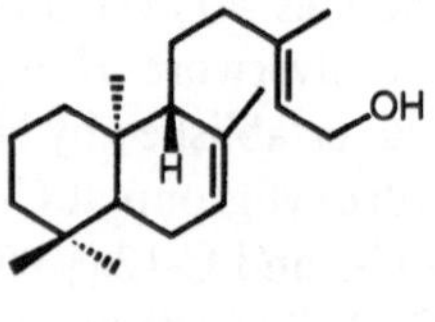

(**736**) Labda-7,13E-dien-15-ol

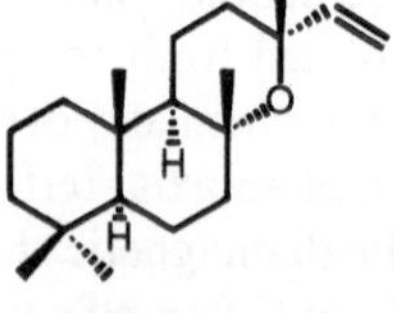

(**737**) Manoyl oxide

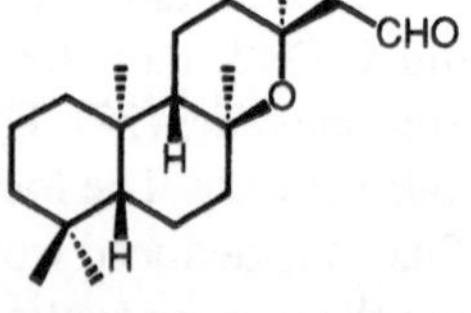

(**738**) Gomeraldehyde

Chart 51a. Labdanes found in the Hepaticae

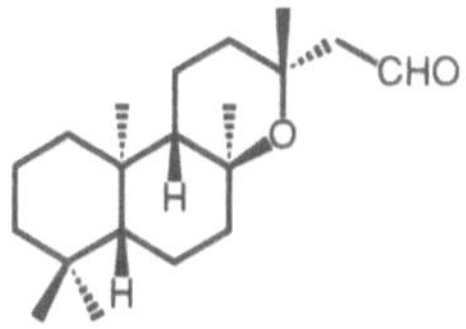

(739) epi-Gomeraldehyde 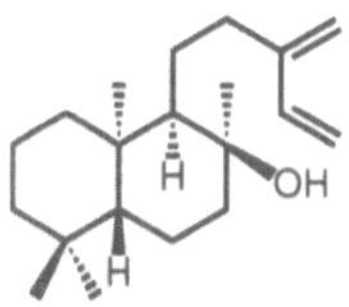   (740) Isoabeinol

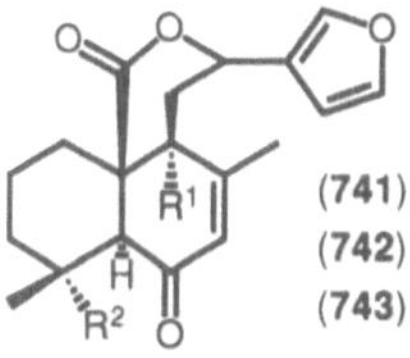

(741) Haplomitrenolide A ; R^1=H, R^2=Me
(742) Haplomitrenolide B ; R^1=OH, R^2=Me
(743) Haplomitrenolide C ; R^1=H, R^2=COOMe

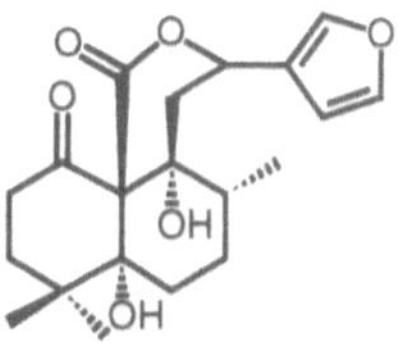

(744)

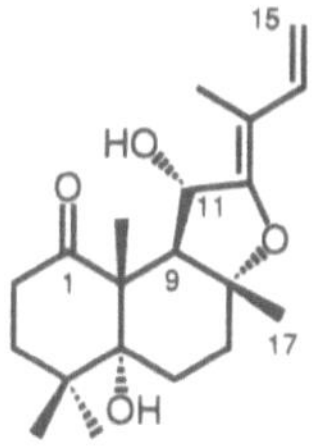

(745) Scapanin A (= Scapanin)
(=(11S,12Z)-8α,12-Epoxy-5α,11α-dihydroxylabda-12,14-dien-1-one)

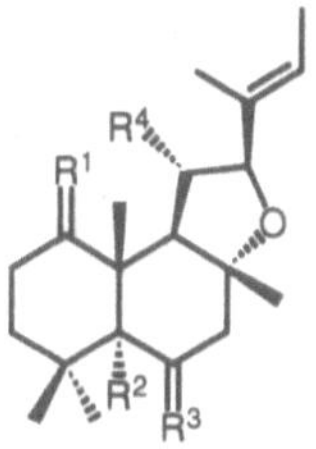

(746) Scapanin B ; R^1=O, R^2=R^4=OH, R^3=H$_2$
(747) R^1=H$_2$, R^2=R^4=H, R^3=H, β-OH
(748) R^1=R^3=H, β-OH, R^2=R^4=H
(749) R^1=R^3=H, β-OH, R^2=H, R^4=OH,
(750) R^1=O, R^2=H, R^3=H, β-OH, R^4=OH
(751) R^1=H, β-OH, R^2=R^4=H, R^3=H$_2$
(752) R^1=O, R^2=H, R^3=H$_2$, R^4=OH
(753) R^1=H, β-OH, R^2=H, R^3=H$_2$, R^4=OH

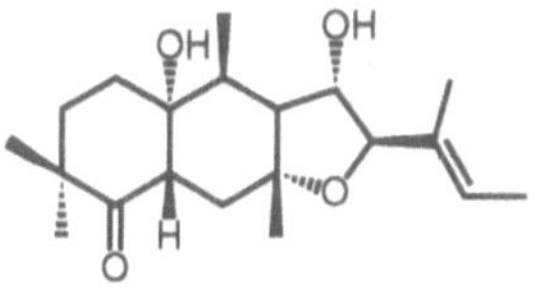

(754) Scapanin G 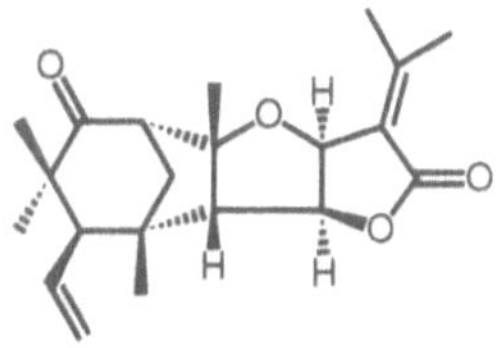   (755) (+)-Pallavicinin

Chart 51b. Labdanes found in the Hepaticae

new labdanes has not been established. A very similar lactone (744), nidorella lactone has been isolated from *Nidorella hottentotica* (Compositae) (*121*).

Scapania undulata is a rich source of both sesquiterpenoids and labdane diterpenoids. Scapanin A (745), originally named scapanin, and its dihydro derivatives, (746) (scapanin B), (747–753), the rearranged labdane (754) and labda-12*E*,14-dien-8α,11ζ-diol (732) have been isolated from *S. undulata* (*143, 144, 277*). Their structures and stereochemistries were established by a combination of chemical transformations some of which are shown in Scheme 55, ^{1}H- and ^{13}C-NMR studies and X-ray analyses of diol (748) and 14,15-dihydroscapanin A (745b). The absolute configurations assigned to (745) and (751) are based on the CD spectra of (745d) and (751a), respectively. Compound (745) is very sensitive to acids and decomposes in CDCl$_3$ solution. Satisfactory spectra were obtained in C$_6$D$_6$ or CD$_3$OD. When dihydroscapanin A (745b) or tetrahydroscapanin A (778) was dissolved in CHCl$_3$ in the presence of dilute aqueous acid, hydration occurred faster than isomerization and crystalline hydrate (745d) or (778a) was obtained. In D$_2$O solution, (745b) or (778) equilibrated rapidly, presumably *via* an oxonium ion intermediate to give a mixture of double bond isomers (745b) and (745h). On addition of D$_2$O, another rapid change occurs to give a 1:1 mixture of (778f) and (778g). Reaction of (745b) or (746) with NaOMe induces a retroaldol reaction followed by an alternative aldol condensation and dehydration to give rearrangement products (745c) or (746e), respectively. A related rearranged labdane, scapanin G (754), has been isolated from Scottish *S. undulata*. German *S. undulata* contains scapanin A (745) and scapanin B (746), but appearance of scapanin B is variable. Extracts from one Scottish site gave compounds (732) and (747–750), while those from a second site, some three miles from the first and across a watershed, furnished scapanin B (746) and compounds (751–753, 754). A new lactone with a rearranged labdane skeleton, (+)-pallavicinin (755), has been isolated from Taiwanese *Pallavicinia subciliata* and its structure established by X-ray crystallographic analysis (*623a*).

As mentioned in the earlier volume (*19*) four labdane-type diterpenoids, ptychantins A (= Ps-3) (756), B (= Ps-4) (757), C (= Ps-5) (758) and D (= Ps-7) (759) have been isolated from *Ptychanthus striatus*; their planar structures were elucidated by spectroscopic evidence. The relative and absolute stereochemistry of (757) has since been established by a combination of chemical reactions, X-ray crystallography, NOE difference spectrometry, the CD spectrum of a dibenzoate (757a) (*80, 250*) and use of the advanced Mosher method (*454*) using the (+)- and (−)-MTPA esters, (757b) and (757c) (*250*). The structure and stereochemistry

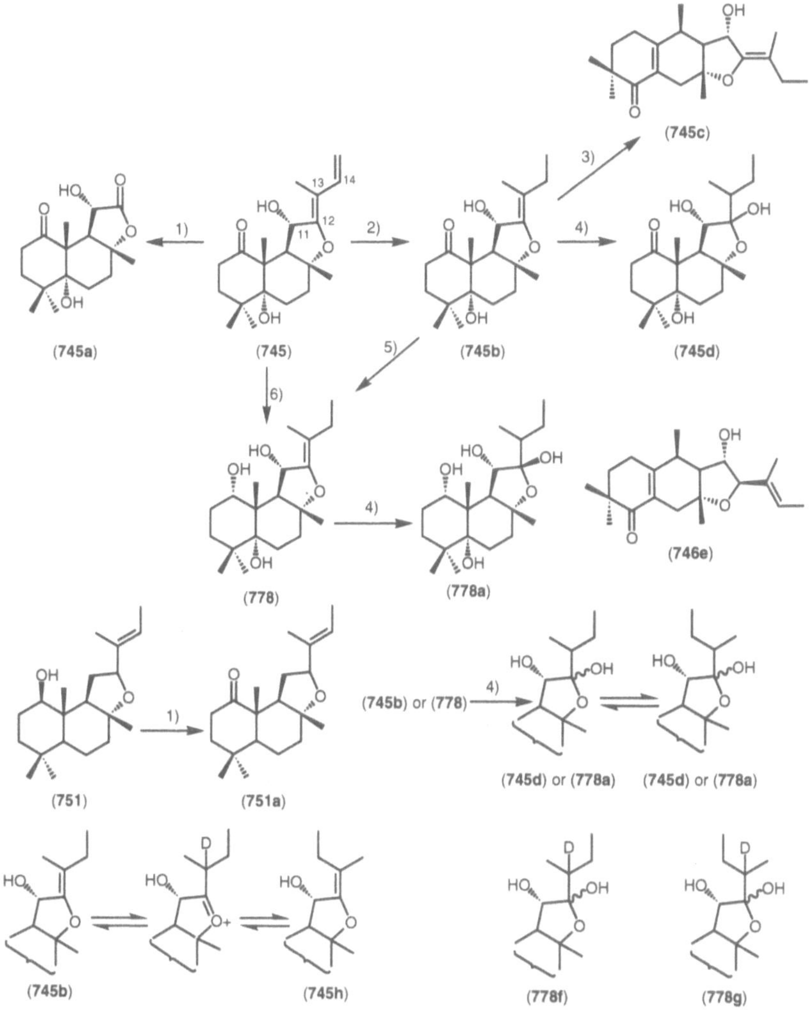

1) Oxidation 2) H₂/Pd-C 3) NaOMe 4) CHCl₃, H₂O, H⁺ 5) LiAlH₄ 6) 2H₂/Pt

Scheme 55. Reactions of scapanins

of (756), (758), and (759) were established by a combination of NOE spectrometry and chemical correlations with (757). Further fractionation of the ether extract of *P. striatus* resulted in the isolation of (760) and (761–763) together with (759) (251). The structure of ptychantin E (760) was deduced by spectroscopic comparison with the previously known

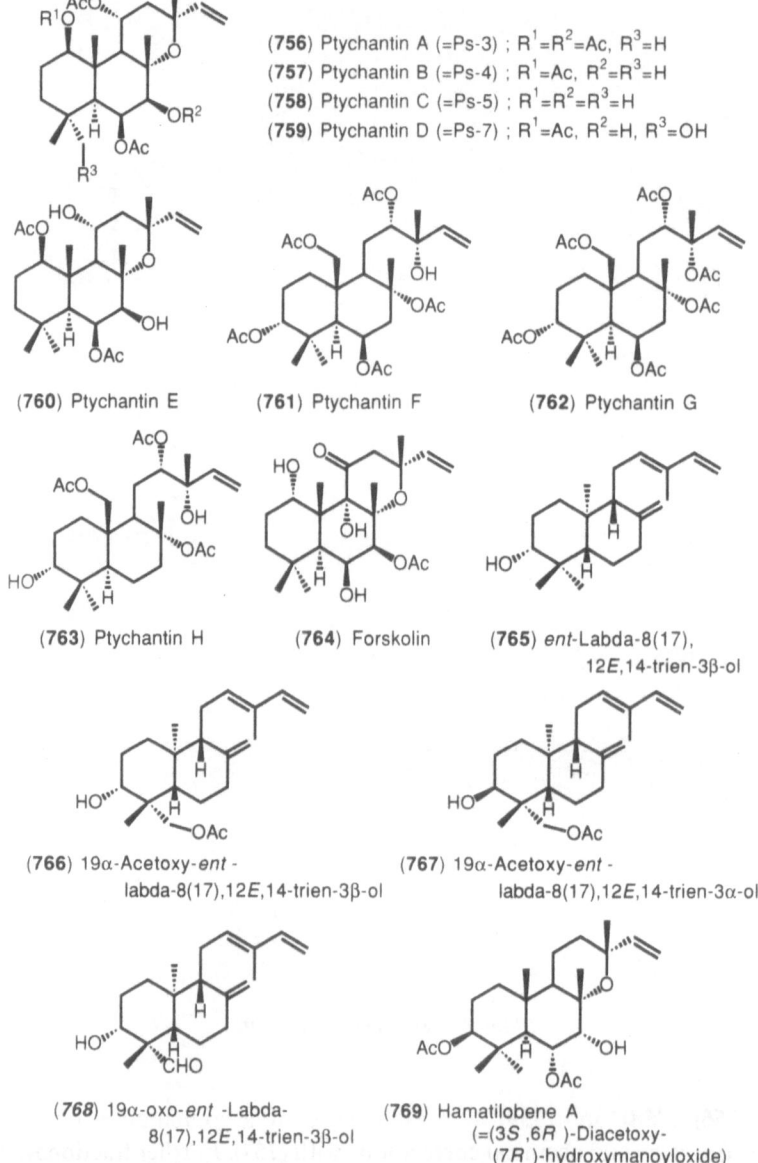

(756) Ptychantin A (=Ps-3) ; R¹=R²=Ac, R³=H
(757) Ptychantin B (=Ps-4) ; R¹=Ac, R²=R³=H
(758) Ptychantin C (=Ps-5) ; R¹=R²=R³=H
(759) Ptychantin D (=Ps-7) ; R¹=Ac, R²=H, R³=OH

(760) Ptychantin E

(761) Ptychantin F

(762) Ptychantin G

(763) Ptychantin H

(764) Forskolin

(765) ent-Labda-8(17),
12E,14-trien-3β-ol

(766) 19α-Acetoxy-ent -
labda-8(17),12E,14-trien-3β-ol

(767) 19α-Acetoxy-ent -
labda-8(17),12E,14-trien-3α-ol

(768) 19α-oxo-ent -Labda-
8(17),12E,14-trien-3β-ol

(769) Hamatilobene A
(=(3S ,6R)-Diacetoxy-
(7R)-hydroxymanoyloxide)

Chart 51c. Labdanes found in the Hepaticae

(757a)　　　　　　　(757b)　　　　　　　(757c)

labdanes (**756–758**) and a chemical correlation with ptychantin B (**757**), both substances giving the same tetraacetate. Structures (**761–763**) were established by a combination of spectral data (^{1}H-, ^{13}C-NMR and NOE different spectrometry) and chemical transformation (*251*). The relative stereochemistry at C-12 and C-13 of (**761**) also rests on NOE's between H-12 and the C-13 methyl, and between H-12 and a β-methyl of an acetonide prepared from (**761**) by hydrolysis (LiAlH$_4$/Et$_2$O) and acetone dimethyl acetal. These diterpenoids isolated from *P. striatus* are structurally very close to forskolin (**764**) which is a potent and direct adenylate cyclase stimulant from Indian *Coleus forskohlii* (Labiatae) (*109*).

3-Oxygenated labdanes of this type are rare in the Hepaticae. *Trichocolea tomentella* collected in Japan produces prenyl 3,4-dimethoxy-benzoates as major components (*19*), but Taiwanese *T. pluma* elaborates a new labdane alcohol, (**765**) (*133, 626*). The same *ent*-labdane has also been isolated from European *Jamesoniella autumnalis* (*112*). The complete assignment of the ^{1}H- and ^{13}C-NMR signals of (**765**) has been achieved (*133*), but the assignment of the methyl groups has been revised by 2D-COSYs (*112*).

Fractionation of the dichloromethane and ether extract of *Jamesoniella autumnalis* resulted in the isolation of six labdanes (**733, 734, 766–768**) (*112*). Compound (**734**) was *ent-trans*-cummunol acetate (= 19α-acetoxy-*ent*-labda-8(17), 12E,14-triene) by comparison with literature data (*93, 448*). The antipode of *ent-trans*-communol acetate has been prepared from *trans*-communol (*328*). The structure of (**766**) was established by ^{1}H-, ^{13}C-NMR and ^{1}H-^{1}H-2D-COSY spectra as well as NOE spectrometry. The corresponding antipodal diol has been found in *Mikania alvimii* (Asteraceae) (*114*). Structure (**767**), the C-3 epimer of (**766**), is based on spectroscopic comparison with (**766**) and NOE spectrometry, while (**733**) was (−)-*ent-trans*-communic acid. The (+)-enantiomer has been isolated from several higher plants (*178, 206*). Structure assignment for (**768**), was based on comparison of the spectroscopic data with those of (**733, 734, 766, 767**). The stereochemistries assigned to the secondary hydroxyl group at C-3 and the formyl group at C-4 are based on the large coupling constant of H-3 ($\delta 3.04\ d$, $J = 10.6$ Hz) which results from a strong hydrogen bond with the aldehyde carbonyl and on the results of NOE spectrometry.

Five highly oxygenated new labdane-type diterpenoids, hamatilobenes A (**769**), B (**770**), C (**771**), D (**772**) and E (**773**), have been isolated from *Frullania hamatiloba* (an erroneous name "*F. hamachiloba*" was cited in the original paper) (*588*). Structures were established by chemical correlations involving PPC oxidation of (**769**) to (**770**), conversion of (**771**) and (**772**) to the same diacetate whose reduction (LiAlH$_4$) gave a triol also prepared by LiAlH$_4$ reduction of (**770**), and NMR spectral data including spin decoupling, 2D-COSYs and NOE experiments. The absolute configurations assigned to (**769–773**) are based on the negative Cotton effect at 263 nm ($\Delta\varepsilon - 11.6$) of the dibenzoate (**770a**) derived from (**770**).

(770a) R=*p*-ClBz

The manoyl oxides (**769–773**) from *F. hamatiloba* and highly oxygenated manoyl oxides from *Ptychanthus striatus* described above have the same absolute configuration as that found in higher plants. On the other hand, (−)-manool, jungermanool, labda-12,14-dien-7,8-diol and labda-7,13(*E*)-dien-15-ol from the other liverworts (*19*) possess configurations enantiomeric to those found in higher plants. Hepaticae produce labdanes of both the normal and enantiomeric series. Two new labdane malonates (**774**) and (**775**) have been isolated from European *Nardia succulenta*, along with the kaurane malonates described earlier (*354a*).

A new chettaphanin-type (= rearranged labdane) diol, pleurodiol (**776**), has been isolated from *Pleurozia gigantea* together with (−)-kolavelool (**621**) (*52*). The presence of the 13-hydroxy-13-methyl-14-ene function in (**776**) was confirmed by epoxidation, opening of the epoxide (LiAlH$_4$) to a 13,14-diol and periodate cleavage to a methyl ketone. The chettaphanin-type skeleton assigned to this substance was based on 2D-COSY and HMBC spectrometry. The stereochemistry has not been

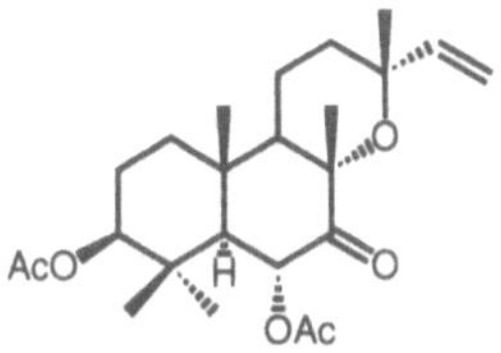

(770) Hamatilobene B
(=(3S,6R)-Diacetoxy-
7-oxomanoyloxide)

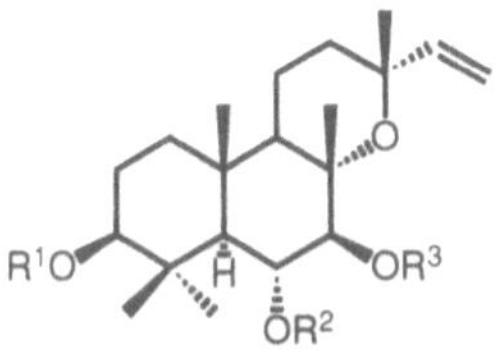

(771) Hamatilobene C
(=(3S,6R)-Diacetoxy-
(7S)-hydroxymanoyloxide) ;
R¹=R²=Ac, R³=H

(772) Hamatilobene D
(=(3S,7S)-Diacetoxy-
(6R)-hydroxymanoyloxide) ;
R¹=R³=Ac, R²=H

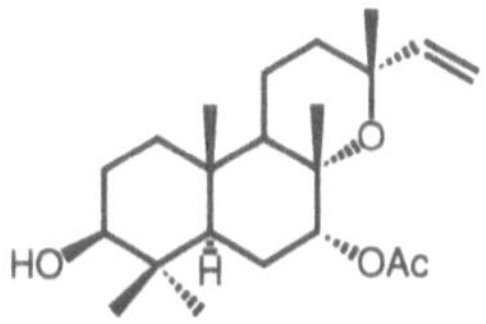

(773) Hamatilobene E
(=(3S)-Hydroxy-(7R)-acetoxymanoyloxide)

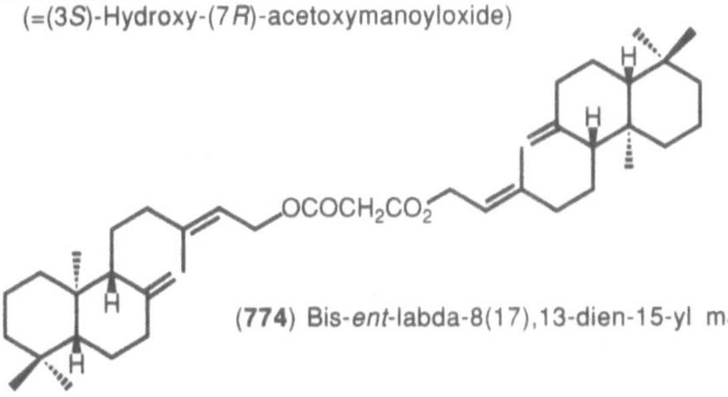

(774) Bis-ent-labda-8(17),13-dien-15-yl malonate

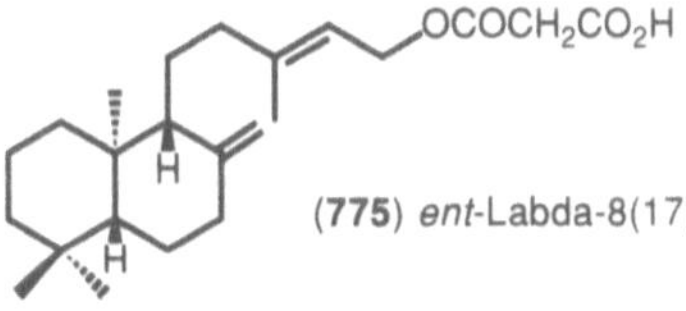

(775) ent-Labda-8(17),13-dien-15-yl hydrogen malonate

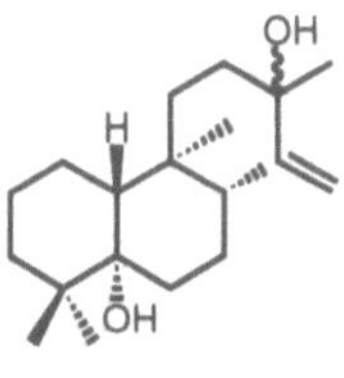

(776) Pleurodiol

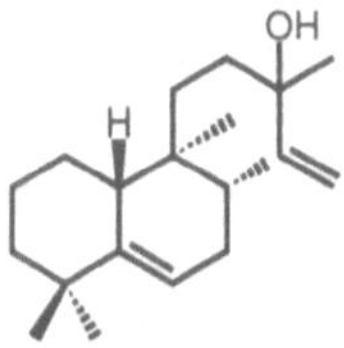

(777) Levierol

Chart 51d. Labdanes and chettaphanins found in the Hepaticae

established; however, considering the co-occurrence of the clerodane-type
diterpenoid (**621**), the tentative stereochemistry shown in the formula has
been proposed. Another chettaphanin-type diterpenoid (**777**) has been
isolated from *Pallavicinia levieri* (*251*). The spectral data of (**777**) were
very similar to those of (**776**). The presence of trisubstituted and terminal
vinyl groups was confirmed by formation of a 4,5-epoxide and a 4,5,14,15-
diepoxide while treatment of (**777**) with $POCl_3$-pyridine gave a 12,13- and
a 13(16)-diene. The planar structure of (**777**) was established by use of 2D-
COSYs and HMBC techniques.

Rearranged labdane-type diterpenoids are rare in nature (*490*). The
above constituents are the first record of chettaphanin-type diterpenoids
from spore-forming plants.

6.7 Phytanes, Pimaranes and Rearranged Pimaranes

As shown in Table IIc, phytol (**780**) has been isolated from or detected
in various liverworts. Phytadienes (**779**) which might be derived from
chlorophyll by pyrolysis during gas chromatography have also been
detected in *Scapania ornithopodioides, Schiffneria hyalina* and *Trichocolea
pluma* (*622*). A phytane-type diterpene lactone, naviculide (**783**) has been
isolated from American *Porella navicularis* (*591*). Structure and stereochemistry, except for the stereochemistry at C-3 and C-13 were deduced
by ^{1}H-^{13}C-NMR and NOE spectrometry and the spectral similarity to
the spectra of linalool (**2**) and nerolidol (**426**). The Z-configuration of the
C-10,11-double bond and the E-configuration of the C-7 double bond
were confirmed by NOE spectrometry and the chemical shift (δ18.4) of
the methyl carbon at C-7.

The ether extract of the primitive liverwort, *Haplomitrium mnioides*
contained a new phytane, haplomitrenone (**784**) (*77*). The structure and
stereochemistry assigned to (**784**) was based on 2D-COSY, NOE experiments and comparison of the ^{13}C-NMR spectrum with that of the
monoterpene hydrocarbon myrcene (**1**). 1,2-Bis-nor-phytone (**785**) has
been isolated from *Nardia scalaris* (*354a*). Phytyl phytenate (**786**) has
been isolated from *Monoclea gottschei* subsp. *neotropica* and its structure
deduced from the IR-, ^{1}H- and ^{13}C-NMR spectra (*513, 514*). Phytol (**780**)
has been found in liverworts, but phytenic acid (**781**) is known only from
some mosses and not from liverworts (*271*).

Ent-pimara-8(14), 15-dien-19-oic acid (**787**) has been isolated from
Mastigophora diclados (*134, 622, 626*). It is interesting that (−)-
sandaracopimaric acid (**794**) with an enantiomeric carbon skeleton was
found in the same species (*623a*). Compound (**787**) and two other

Table IIc *Diterpenoids Found in the Hepaticae*

Structure number	Name of compounds	Formula	m.p. °C	$[\alpha]_D$	Plant source	Order	References	Comments
(616)	*ar*-Abietatriene	$C_{20}H_{30}$			*Plagiochila peculiaris*	J	(637)	
(617)	Dehydroabietic acid	$C_{20}H_{28}O_2$			*Porella roellii*	J	(249)	
(618)	Setiformenol	$C_{20}H_{34}O_2$		$+19.3$	*Chandonanthus setiformis* (= *Tetralophozia setiformis*)	J	(429, 566, 567, 567a)	
(619)	Chandonanthone	$C_{20}H_{30}O_4$	186–9	$+7.9$	*Chandonanthus hirtellus*	J	(585)	
(621)	*ent*-Cleroda-13,14-dien-13ζ-ol [= (−)-Kolavelool]	$C_{20}H_{34}O$		-40.0 -53 -25.7	*Jungermannia paroica* *Macrolejeunea pallescens* *Nardia subclavata* *Pleurozia acinosa* *P. gigantea*	J J J J J	(245) (433) (585) (625) (52)	
(624)	(−)-Methyl 13-hydroxy-cleroda-3,14-dien-18-carboxylate	$C_{21}H_{34}O_3$		-4	*Scapania bolanderi*	J	(377)	
(625a)	Clerod-3,13(16),14-trien-17-oic acid (= Infuscaic acid)	$C_{20}H_{30}O_2$	135–140	-43.6	*Jungermannia infusca*	J	(592)	
(626)	Clerod-3,13Z-dien-15-al-17-oic acid	$C_{20}H_{30}O_3$	103–5	-48	*Jungermannia infusca*	J	(430, 594)	
(627)	Clerod-3,13E-dien-15-al-17-oic acid	$C_{20}H_{30}O_3$		-50	*Jungermannia infusca*	J	(430, 594)	
(628)	Clerod-3,13Z-dien-15,17-dial	$C_{20}H_{30}O_2$		-62.7	*Jungermannia infusca*	J	(430, 594)	

Table IIc (continued)

Structure number	Name of compounds	Formula	m.p. °C	[α]_D	Plant source	Order	References	Comments
(629)	Clerod-3,13E-dien-15,17-dial	$C_{20}H_{30}O_2$		− 72.8	*Jungermannia infusca*	J	(430, 594)	
(630)	Clerod-3,13(14)-dien-15-ol-17-al	$C_{20}H_{32}O_2$			*Jungermannia infusca*	J	(430, 457)	
(631)	Clerod-3,13(16),14-trien-17-al	$C_{20}H_{30}O$		− 63.7	*Jungermannia infusca*	J	(430, 594)	
(632)	*ent*-Junceic acid	$C_{20}H_{28}O_3$		+ 60.8	*Heteroscyphus bescherellei*	J	(83)	
(634)	*ent*-3β,4β-Epoxyclerod-13Z-en-15-al	$C_{20}H_{32}O_2$		− 31.8	*Jungermannia paroica*	J	(245, 478)	
(635)	*ent*-3β,4β-Epoxyclerod-13E-en-15-al	$C_{20}H_{32}O_2$		− 41.0	*Jungermannia paroica*	J	(245, 478)	
(636)	*ent*-3β,4β-Epoxyclerod-14-en-13ξ-ol	$C_{20}H_{34}O_2$		− 29.6	*Jungermannia paroica*	J	(245, 478)	
(637a)	*cis*-Clerod-3,13(16),14-trien-18-oic acid	$C_{20}H_{30}O_2$	153–5	− 16.0	*Schistochila acuminata* *S. aligera*	J J	(136) (438)	
(637b)	Methyl *cis*-clerod-3,13(16),14-trien-18-oate	$C_{21}H_{32}O_2$		− 36.5	*Schistochila acuminata*	J	(136, 137)	as methyl ester
(639)	*cis*-Clerod-3,12E,14-trien-18-oic acid	$C_{20}H_{30}O_2$	86–88	− 21.1	*Schistochila acuminata*	J	(136, 137)	

No.	Compound	Formula	m.p. / rotation		Species		References	CD data
(640)	cis-Clerod-3,12Z,14-trien-18-oic acid	$C_{20}H_{30}O_2$			Schistochila acuminata	J	(136, 137)	
(641)	cis-Clerod-3,12E,14-trien-18-al	$C_{20}H_{30}O$			Schistochila acuminata	J	(136, 137)	
(642)	cis-Clerod-3,13(16),14-trien-18-al	$C_{20}H_{30}O$			Schistochila acuminata	J	(136, 137)	
(643)	cis-Clerod-3,12E,14-trien-18-ol	$C_{20}H_{32}O$			Schistochila acuminata	J	(136, 137)	
(644)	cis-Clerod-3,13(16),14-trien-18-ol	$C_{20}H_{32}O$			Schistochila acuminata	J	(136, 137)	
(645)	Schistochilic acid B	$C_{21}H_{34}O_4$		− 86.2	Schistochila nobilis	J	(561, 563)	
(646)	Schistochilic acid C	$C_{21}H_{32}O_5$		− 62	Schistochila nobilis	J	(561, 563)	
(647)	Schistochilic acid A	$C_{21}H_{32}O_4$		+ 117	Schistochila nobilis	J	(561, 563)	
(648)	Heteroscyphol (= Cleroda-3,12E, 14-trien-11ζ-ol)	$C_{20}H_{32}O$		− 37.5	Heteroscyphus planus	J	(249, 253)	CD 201 nm + 9.5
(650)	Gymnocolin	$C_{22}H_{28}O_6$	196–7		Gymnocolea inflata	J	(143, 273)	CD 202 nm − 3.40 / CD 230 nm + 0.66
(651)	Anastreptin	$C_{20}H_{24}O_5$			Anastrepta orcadensis	J	(143, 478)	
(652)	Orcadensin	$C_{20}H_{24}O_5$			Anastrepta orcadensis	J	(143, 478)	
(653)	Linguifolide	$C_{20}H_{24}O_4$	130–3	− 24	Demotarisia linguifolia	J	(431)	CD 210 nm − 11.3 / CD 248 nm + 7.1
(654)	Dihydrolinguifolide	$C_{20}H_{26}O_4$	158–160	+ 22	Demotarisia linguifolia	J	(431)	CD 215 nm + 0.9 / CD 245 nm + 0.5
(656)	Ventricosenediolide	$C_{22}H_{28}O_7$	172–5	+ 24.6	Lophozia ventricosa	J	(429, 567a, 565)	
(657)	17-Acetoxy-1β,12-dihydroxy-15,16-epoxy-cis-ent-clerod-3,13(16),14-trien-6α,18-olide	$C_{22}H_{28}O_7$		+ 1.7	Jamesoniella autumnalis	J	(112)	

Table IIc (continued)

Structure number	Name of compounds	Formula	m.p. °C	$[\alpha]_D$	Plant source	Order	References	Comments
(658)	Jamesoniellide A	$C_{21}H_{28}O_6$		− 54.1	*Jamesoniella autumnalis*	J	*(112)*	
(659)	Jamesoniellide B	$C_{22}H_{30}H_7$			*Jamesoniella autumnalis*	J	*(112)*	
(660)	Heteroscyphone A	$C_{22}H_{32}O_7$	217.5–220	+ 24.0	*Heteroscyphus planus*	J	*(249, 253)*	CD 298 nm − 1.81
(661)	Heteroscyphone B	$C_{22}H_{32}O_6$	211.5–214	+ 17.1	*Heteroscyphus planus*	J	*(249, 253)*	CD 329 nm − 1.41
(662)	Heteroscyphone C	$C_{22}H_{32}O_5$	72–4	− 21.4	*Heteroscyphus planus*	J	*(249, 253)*	CD 329 nm − 1.73
(663)	Heteroscyphone D	$C_{22}H_{32}O_7$	138–140.5	− 13.8	*Heteroscyphus planus*	J	*(249, 253)*	CD 333 nm − 1.29
(664)	Barbilycopodin (= 10R,18-Diacetoxy-3S,4S,7S,8S-diepoxydolabellane)	$C_{24}H_{38}O_6$			*Barbilophozia attenuata*	J	*(273)*	
					B. floerkei	J	*(273)*	
				− 32.7	*B. floerkei*	J	*(429, 565, 567a)*	
				− 37.3	*B. hatcheri*	J	*(429, 565, 567a)*	
			197–8	− 37.8	*B. lycopodioides*	J	*(143, 273)*	
					Chandronanthus setiformis	J	*(651)*	
(665)	10-Deacetoxy-barbilycopodin (= 18-Acetoxy-3S,4S,7S,8S-diepoxydolabellane)	$C_{22}H_{36}O_4$	89–91	− 16.8	*Barbilophozia barbata*	J	*(432)*	
					B. floerkei	J	*(273)*	
				− 52.9	*B. floerkei*	J	*(429, 565, 567a)*	
(666)	10R,18-Diacetoxy-3S,4S-epoxydolabella-7E-ene	$C_{24}H_{38}O_5$	161–2	− 75	*Barbilophozia floerkei*	J	*(273)*	
(667)	18-Hydroxydolabell-7E-en-3-one	$C_{20}H_{34}O_2$		+ 52.9	*Barbilophozia floerkei*	J	*(273)*	
(668)	Acetoxyodonto-schismenol	$C_{22}H_{36}O_3$	75–6	+ 66.2	*Odontoschisma denudatum*	J	*(381, 388, 391)*	

No.	Name	Formula	mp	[α]	Species		Ref.	CD
(669)	6,12-Dihydroxy-dolabella-3E,7E-diene	$C_{20}H_{34}O_2$	139–140	+ 56.2	Odontoschisma denudatum	J	(381)	
(670)	6-Acetoxy-12,16-dihydroxydolabella-3E,7E-diene	$C_{22}H_{36}O_4$	103.5–104.5	+ 36.0	Odontoschisma denudatum	J	(381)	
(671)	6,16-Diacetoxy-12-hydroxydolabella-3E,7E-diene	$C_{24}H_{38}O_5$	76–7	+ 2.9	Odontoschisma denudatum	J	(381)	
(672)	6-Acetoxy-3,4-epoxy-12-hydroxy-dolabella-7E-en-16-al	$C_{22}H_{34}O_5$		− 23.3	Odontoschisma denudatum	J	(381)	
(673)	18-Hydroxy-4,8-dolabelladiene	$C_{20}H_{34}O$			Pleurozia gigantea	J	(52)	
(678)	Fusicoccadiene	$C_{20}H_{32}$			Plagiochila geniculata	J	(513)	
					P. moritziana	J	(513, 515)	
					P. panamensis	J	(513)	
					Riccardia andina	Me	(513)	
					Symphyogyna brasiliensis	J	(513)	
(679)	Anadensin	$C_{20}H_{32}O_2$	171–2		Anastrepta orcadensis	J	(143, 274)	CD 194 nm + 1.63; CD 206 nm − 5.22; CD 239 nm + 17.6; CD 322 nm − 8.5
(680)	Fusicogigantone A	$C_{20}H_{32}O_2$		+ 28	Plagiochila ovalifolia	J	(435a)	CD 266 nm + 0.17; CD 275 nm − 0.38; CD 330 nm + 0.34
					Plagiochila corrugata	J	(441a)	
(681)	Fusicogigantone B	$C_{20}H_{32}O_2$		+ 5.9	Pleurozia gigantea	J	(52)	
					Pleurozia gigantea	J	(52)	
(682)	Fusicogigantepoxide	$C_{20}H_{32}O_2$		+ 47	Bryopteris filicina	J	(432a)	
					Pleurozia gigantea	J	(52)	

Table IIc (continued)

Structure number	Name of compounds	Formula	m.p. °C	$[\alpha]_D$	Plant source	Order	References	Comments
(683)	Fusicorrugatol	$C_{20}H_{30}O_3$			*Plagiochila corrugata*	J	*(441a)*	
(684)	Barbifusicoccine A	$C_{20}H_{34}O_2$		-2.0	*Barbilophozia floerkei*	J	*(429, 565, 567a)*	
(685)	Barbifusicoccine B	$C_{20}H_{34}O_2$		$+45.0$	*Barbilophozia floerkei*	J	*(429, 565, 567a)*	
(686)	Fusicoplagin A	$C_{24}H_{38}O_7$			*Plagiochila sciophila* (= *P. acanthophylla* subsp. *japonica*)	J	*(261)*	
(687)	Fusicoplagin B	$C_{26}H_{40}O_8$			*Plagiochila sciophila* (= *P. acanthophylla* subsp. *japonica*)	J	*(261)*	
(688)	Fusicoplagin C	$C_{24}H_{36}O_7$			*Plagiochila sciophila* (= *P. acanthophylla* subsp. *japonica*)		*(261)*	
(689)	Fusicoplagin D	$C_{22}H_{34}O_5$			*Plagiochila sciophila* (= *P. acanthophylla* subsp. *japonica*)	J	*(261)*	
(690)	Spinuloplagin A	$C_{37}H_{48}O_5$			*Plagiochila spinulosa*	J	*(478)*	
(691)	Spinuloplagin B	$C_{37}H_{48}O_5$			*Plagiochila spinulosa*	J	*(478)*	
(692)	*ent*-Kaurene	$C_{20}H_{32}$			*Bazzania spiralis*	J	*(341)*	
					Isotachis humectata	J	*(45)*	
					Jensenia erythropus	J	*(47)*	
					Jungermannia subulata	J	*(453)*	
					Lepidolejeunea ornata	J	*(219)*	
					Plagiochila dura	J	*(46)*	
					P. pulcherrima	J	*(190)*	

					Species		Ref.
(693)	ent-15α-Hydroxy-kaurene	$C_{20}H_{32}O$	91–7	−57.6	Porella densifolia subsp. appendiculata	J	(67)
					P. densifolia var. fallax	J	(67)
					Jungermannia infusca	J	(594)
					Nardia scalaris	J	(354a)
					N. succulenta	J	(354a)
					Porella densifolia subsp. appendiculata	J	(67)
(694)	ent-Kauren-15-one	$C_{20}H_{30}O$	102–3.5	−116.7	Jamesoniella autumnalis	J	(437)
					Jungermannia infusca	J	(594)
					Nardia scalaris	J	(354a)
					N. succulenta	J	(354a)
(695)	ent-18-Hydroxy-kauren-15-one	$C_{20}H_{30}O_2$			Porella densifolia subsp. appendiculata	J	(67)
(696)	ent-Kauren-15-one-18-oic acid	$C_{20}H_{28}O_3$	227–8	−110	Porella densifolia subsp. appendiculata	J	(67)
(698)	ent-Kauren-18-oic acid	$C_{20}H_{30}O_2$			Porella densifolia var. fallax	J	(67)
(699a)	ent-11α-Hydroxy-kauren-15-one	$C_{20}H_{30}O_2$			Jamesoniella autumnalis	J	(437)
					Jungermannia infusca	J	(594)
					J. truncata	J	(437)
					Porella densifolia var. fallax	J	(67)
(700a)	(16R)-ent-11α-Hydroxy-kauran-15-one	$C_{20}H_{32}O_2$			Jamesoniella autumnalis	J	(437)
					Jungermannia infusca	J	(594)
					Porella densifolia var. fallax	J	(67)
(701)	ent-Kaur-16-en-7α,15β-diol	$C_{20}H_{32}O_2$	206–8	−35.8	Plagiochila pulcherrima	J	(190)

Table IIc (continued)

Structure number	Name of compounds	Formula	m.p. °C	[α]$_D$	Plant source	Order	References	Comments
(702)	(16R)-ent-Kauran-15-one	$C_{20}H_{32}O$	146–8	– 71.2	*Jungermannia infusca* *Nardia scalaris* *N. succulenta*	J J J	(594) (354a) (354a)	
(703)	(16S)-ent-11α-Hydroxykauran-15-one	$C_{20}H_{32}O_2$			*Jungermannia infusca*	J	(594)	
(704)	ent-Kauren-3β,15α-diol	$C_{20}H_{32}O_2$	154–6	– 58	*Jungermannia vulcanicola*	J	(437)	
(705)	ent-15α-Hydroxykauren-3β-yl acetate	$C_{22}H_{34}O_3$	183–5	– 69	*Jungermannia vulcanicola*	J	(437)	
(706)	ent-3β-Hydroxy-kauren-15-one	$C_{20}H_{30}O_2$	154.5–156.5	– 153	*Jungermannia vulcanicola*	J	(437)	CD 350 nm – 0.39
(707)	ent-16β-Hydroxy-kaurane	$C_{20}H_{34}O$		– 21.2	*Anthelia julacea* *Frullanoides densifolia*	J J	(435) (14, 554, 556)	
(708)	ent-16β-Hydroxy-kauran-3-one	$C_{20}H_{32}O_2$	162–3	– 71.1	*Frullanoides densifolia*	J	(14, 554, 556)	CD 290 nm – 1.07
(709)	ent-Kauran-3β,16β-diol	$C_{20}H_{34}O_2$	202–3	– 36.8	*Frullanoides densifolia*	J	(14, 554, 556)	
(710)	Nardiin	$C_{20}H_{28}O_2$	177–8	– 29	*Nardia scalaris*	J	(101)	CD 215 nm + 14.33 CD 360 nm – 2.76
(711a)	(14R)-ent-Kaur-16-en-14-yl-hydrogen malonate	$C_{23}H_{34}O_4$		– 48.3	*Nardia scalaris* *N. subclavata*	J J	(148, 354a) (585)	

(712)	Bis-(14R)-ent-kaur-16-en-14-yl malonate	$C_{43}H_{64}O_4$			Nardia scalaris N. subclavata	J J	(354) (585)
(713)	(14R)-ent-kaur-16-en-14-yl phytyl malonate	$C_{43}H_{72}O_4$		−42.2	Nardia scalaris N. subclavata	J J	(354a) (585)
(714)	(14R)-ent-Kaur-16-en-14-yl epi-bornyl malonate	$C_{33}H_{50}O_4$		−53	Nardia scalaris	J	(354a)
(715)	(14R)-ent-Kaur-16-en-14-yl fenchyl malonate	$C_{33}H_{50}O_4$		−22.1	Nardia scalaris	J	(354a)
(716)	(14R)-ent-Kaur-16-en-15β-hydrogen malonate	$C_{23}H_{34}O_4$	147.5–9	−79.6	Nardia scalaris N. succulenta	J J	(354a) (354a)
(717)	ent-Kaur-16-en-15β-yl epi-bornyl malonate	$C_{33}H_{50}O_4$		−87.7	Nordia scalaris	J	(354a)
(718)	Bis-ent-Kaur-16-en-15β-yl malonate	$C_{43}H_{64}O_4$		−81.5	Nardia scalaris N. succulenta	J J	(354a) (354a)
(719)	ent-Kaur-16-en-15β-yl phytyl malonate	$C_{43}H_{72}O_4$		−46.7	Nardia succulenta	J	(354a)
(720)	ent-Kaur-16-en-15β-yl ent-labdan-8(17),13-dien-15-yl malonate	$C_{43}H_{64}O_4$		−62.0	Nardia succulenta	J	(354a)
(721a)	Infuscaside A (= 6β-(6'-Acetoxy)-β-glucopyranosyl-ent-15α,20-dihydroxy-kaur-16-ene)	$C_{28}H_{44}O_9$	199–122	−7.7	Jungermannia infusca	J	(436, 594)

Table IIc (*continued*)

Structure number	Name of compounds	Formula	m.p. °C	$[\alpha]_D$	Plant source	Order	References	Comments
(722a)	Infuscaside B (= 20-(2'-Acetoxy)-β-glucopyranosyl-ent-6-keto-15α-hydroxykaur-16-ene)	$C_{28}H_{42}O_9$	115–7	– 4.2	*Jungermannia infusca*	J	*(436, 594)*	
(723)	Infuscaside C (= 20-(2'-Acetoxy)-β-glucopyranosyl-ent-15α-hydroxykaur-16-ene)	$C_{28}H_{44}O_8$		– 25	*Jungermannia infusca*	J	*(436, 594)*	
(724)	Infuscaside D (= 20-β-Glucopyranosyl-ent-6-keto-15α-hydroxy-kaur-16-ene)	$C_{26}H_{40}O_8$		– 37	*Jungermannia infusca*	J	*(436, 594)*	
(725)	Infuscaside E (= 6β-β-Glucopyranosyl-ent-15α,20-dihydroxy-kaur-16-ene)	$C_{26}H_{42}O_8$		– 15	*Jungermannia infusca*	J	*(436, 594)*	

No.	Name	Formula	mp	[α]	Species		Ref.
(726)	ent-14α-Hydroxy-kaurene	$C_{20}H_{32}O$		−47.1	*Nardia scalaris*	J	(354a)
					N. subclavata	J	(585)
(727)	8-epi-Sclareol	$C_{20}H_{36}O_2$			*Plagiochila corrugata*	J	(441a, 569)
					Pleurozia acinosa	J	(622, 625)
					P. gigantea	J	(52)
(730)	Labda-7,14-dien-13-ol	$C_{20}H_{34}O$		+20.3	*Mylia nuda*	J	(622, 624)
(731)	Labda-12,14-dien-8α-ol	$C_{20}H_{34}O$			*Porella perrottetiana*	J	(80, 83)
					Porella perrottetiana	J	(80, 83)
(732)	Labda-12E,14-dien-8α,11ζ-diol	$C_{20}H_{34}O_2$			*Scapania undulata*	J	(143, 144, 277)
(733)	(−)-ent-trans-Communic acid	$C_{20}H_{30}O_2$		−22.5	*Jamesoniella autumnalis*	J	(112)
(734)	ent-trans-Communol acetate (= 19α-Acetoxy-ent-labda-8(17),12E,14-triene)	$C_{22}H_{34}O_2$		−9.5	*Jamesoniella autumnalis*	J	(112)
(735)	Labda-12,14-dien-7,8-diol	$C_{20}H_{34}O_2$			*Porella perrottetiana*	J	(80, 83)
(736)	Labda-7,13E-dien-15-ol	$C_{20}H_{34}O$		−12.1	*Marchantia paleacea* var. *diptera*	M	(74, 80, 579)
					M. polymorpha	M	(74, 80, 579)
					Targionia hypophylla	M	(75)
(737)	Manoyl oxide	$C_{20}H_{34}O$		+14.4	*Mylia nuda*	J	(622, 624)
(738)	Gomeraldehyde	$C_{20}H_{34}O_2$			*Jungermannia infusca*	J	(592)
(739)	epi-Gomeraldehyde	$C_{20}H_{34}O_2$			*Jungermannia infusca*	J	(592)
(740)	Isoabienol	$C_{20}H_{34}O$			*Jungermannia infusca*	J	(592)
(741)	Haplomitrenolide A	$C_{20}H_{24}O_4$	178–180	+76.7	*Haplomitrium mnioides*	C	(77) CD 244 nm + 10.9

Table IIc (continued)

Structure number	Name of compounds	Formula	m.p. °C	$[\alpha]_D$	Plant source	Order	References	Comments
(742)	Haplomitrenolide B	$C_{20}H_{24}O_5$	194–6	− 37.5	Haplomitrium mnioides	C	(77)	
(743)	Haplomitrenolide C	$C_{21}H_{24}O_6$	195–6	+ 121.2	Haplomitrium mnioides	C	(77)	
(745)	Scapanin A (= 11S,12Z)-8α,12-Epoxy-5α,11α-dihydroxylabda-12,14-dien-1-one)	$C_{20}H_{30}O_4$			Scapania undulata	J	(143, 144, 277)	
(746)	Scapanin B	$C_{20}H_{32}O_4$			Scapania undulata	J	(143, 144, 277)	
(747)	Scapanin B-type labdane	$C_{20}H_{34}O_2$			Scapania undulata	J	(143, 144, 277)	
(748)	Scapanin B-type labdane	$C_{20}H_{32}O_3$			Scapania undulata	J	(143, 144, 277)	
(749)	Scapanin B-type labdane	$C_{20}H_{34}O_4$			Scapania undulata	J	(143, 144, 277)	
(750)	Scapanin B-type labdane	$C_{20}H_{32}O_4$			Scapania undulata	J	(143, 144, 277)	
(751)	Scapanin B-type labdane	$C_{20}H_{34}O_2$			Scapania undulata	J	(143, 144, 277)	
(752)	Scapanin B-type labdane	$C_{20}H_{32}O_3$			Scapania undulata	J	(143, 144, 277)	
(753)	Scapanin B-type labdane	$C_{20}H_{34}O_3$			Scapania undulata	J	(143, 144, 277)	
(754)	Scapanin G	$C_{20}H_{32}O_4$			Scapania undulata	J	(143, 144, 277)	
(755)	Pallavicinin	$C_{20}H_{28}O_4$			Pallavicinia subciliata	Me	(623a)	

No.	Compound	Formula	mp	$[\alpha]$	Species		Ref.
(756)	Ptychantin A (= Ps-3)	$C_{28}H_{42}O_9$		−47.1	*Ptychanthus striatus*	J	(19, 251)
(757)	Ptychantin B (= Ps-4)	$C_{26}H_{40}O_8$			*Ptychanthus striatus*	J	(19, 80, 250, 454)
(758)	Ptychantin C (= Ps-5)	$C_{24}H_{38}O_7$		−54.0	*Ptychanthus striatus*	J	(19, 251)
(759)	Ptychantin D (= Ps-7)	$C_{26}H_{40}O_9$		−68.9	*Ptychanthus striatus*	J	(19, 249, 250, 251)
(760)	Ptychantin E	$C_{24}H_{38}O_7$		−17.7	*Ptychanthus striatus*	J	(249, 250, 251)
(761)	Ptychantin F	$C_{30}H_{46}O_{11}$		−11.0	*Ptychanthus striatus*	J	(251)
(762)	Ptychantin G	$C_{32}H_{48}O_{12}$		−6.8	*Ptychanthus striatus*	J	(251)
(763)	Ptychantin H	$C_{28}H_{44}O_{10}$		+4.8	*Ptychanthus striatus*	J	(251)
(765)	*ent*-Labda-8(17),12E,14-trien-3β-ol	$C_{20}H_{32}O$		−11.9	*Jamesoniella autumnalis*	J	(112)
				−21	*Trichocolea pluma*	J	(133, 626)
(766)	19α-Acetoxy-*ent*-labda-8(17),12E,14-trien-3β-ol	$C_{22}H_{34}O_3$		−12.5	*Jamesoniella autumnalis*	J	(112)
(767)	19α-Acetoxy-*ent*-labda-8(17),12E,14-trien-3α-ol	$C_{22}H_{34}O_3$		+6.0	*Jamesoniella autumnalis*	J	(112)
(768)	19α-Oxo-*ent*-labda-8(17),12E,14-trien-3β-ol	$C_{20}H_{30}O_2$		+18.1	*Jamesoniella autumnalis*	J	(112)
(769)	Hamatilobene A (= 3S,6R-Diacetoxy-7R-hydroxy-manoyloxide)	$C_{24}H_{38}O_6$	136–8	+38.9	*Frullania hamatiloba*	J	(588)

Table IIc (continued)

Structure number	Name of compounds	Formula	m.p. °C	$[\alpha]_D$	Plant source	Order	References	Comments
(770)	Hamatilobene B (= 3S,6R-Diacetoxy-7-oxo-manoyloxide)	$C_{24}H_{36}O_6$	154–6	+ 37.8	*Frullania hamatiloba*	J	*(588)*	
(771)	Hamatilobene C (= 3S,6R-Diacetoxy-7S-hydroxy-manoyloxide)	$C_{24}H_{38}O_6$	94–5	+ 19.1	*Frullania hamatiloba*	J	*(588)*	
(772)	Hamatilobene D (= 3S,7S-Diacetoxy-6R-hydroxy-manoyloxide)	$C_{24}H_{38}O_6$	144–6	+ 46.4	*Frullania hamatiloba*	J	*(588)*	
(773)	Hamatilobene E (= 3S-Hydroxy-7R-acetoxymanoyl-oxide)	$C_{22}H_{36}O_4$	90–92	± 0	*Frullania hamatiloba*	J	*(588)*	
(774)	Bis-*ent*-labda-8(17), 13-dien-15-yl malonate	$C_{43}H_{68}O_4$		− 24.4	*Nardia succulenta*	J	*(354a)*	
(775)	*ent*-Labda-8(17)-13-dien-15-yl hydrogen malonate	$C_{23}H_{36}O_4$		− 20.3	*Nardia succulenta*	J	*(354a)*	
(776)	Pleurodiol	$C_{20}H_{36}O_2$		+ 3	*Pleurozia gigantea*	J	*(52)*	
(777)	Levierol	$C_{20}H_{34}O$	46–9	+ 71.5	*Pallavicinia levieri*	Me	*(251)*	

(779)	Phytadienes	$C_{20}H_{38}$		
			Makinoa crispata	Me	(622)

			Makinoa crispata	Me	(622)
			Scapania ornithopodioides	J	(622)
			Schiffneria hyalina	J	(622)
			Trichocolea pluma	J	(622)
			Blepharolejeunea incongrua	J	(219)
(780)	Phytol	$C_{20}H_{40}O$	Bryopteris diffusa	J	(219)
			Isotachis haematodes	J	(48)
			Jungermannia subulata	J	(453)
			Lophocolea bidentata	J	(527a)
			Makinoa crispata	Me	(622)
			Marchantia paleacea var. diptera	M	(74)
			M. palmata	M	(72)
			M. polymorpha	M	(61, 74)
			Plagiochila engelii	J	(46)
			Porella densifolia subsp. appendiculata	J	(67)
			P. elegantula	J	(39)
			P. perrottetiana	J	(83)
			Radula perrottetii	J	(585a)
			Riccardia jackii	Me	(380)
			Riccardia sp.	Me	(48)
			Ricciocarpos natans	Me	(639, 640)
			Scapania ornithopodioides	J	(622)
			Schiffneria hyalina	J	(622)
			Schistochila laminigera	J	(45)
			Stictolejeunea squamata	J	(219)
			Thysananthus amazonicus	J	(219)
			Trichocolea tomentella	J	(622)
			Wiesnerella denudata	M	(51)

Table IIc (continued)

Structure number	Name of compounds	Formula	m.p. °C	$[\alpha]_D$	Plant source	Order	References	Comments
(783)	Naviculide	$C_{20}H_{30}O_3$			*Porella navicularis*	J	(591)	
(784)	Haplomitrenone	$C_{20}H_{30}O_2$		+ 7.8	*Haplomitrium mnioides*	C	(77)	
(785)	(1,2)-Bis-nor-phytone	$C_{18}H_{36}O$		− 8.0	*Nardia scalaris*	J	(354a)	
(786)	Phytyl phytenate	$C_{40}H_{76}O_2$			*Monoclea gottschei* subsp. *neotropica*	Mo	(513, 514)	
(787)	*ent*-Pimara-8(14),15-dien-19-oic acid	$C_{20}H_{30}O_2$	166-4	− 102.6	*Mastigophora diclados*	J	(134, 622, 626)	
(788)	*ent*-Pimara-8(14),15-dien-19-ol	$C_{20}H_{32}O$			*Jungermannia thermarum*	J	(19)	
(789)	Termarol	$C_{20}H_{34}O$			*Jungermannia thermarum*	J	(19)	
(790)	Aligerol	$C_{20}H_{34}O$		+ 39.7	*Schistochila aligera*	J	(438)	
(794)	(−)-Sandara-copimaric acid	$C_{20}H_{30}O_2$			*Mastigophora diclados*	J	(623a)	
(795)	Sacculatal	$C_{20}H_{30}O_2$	65-6	− 31.4	*Pallavicinia levieri*	Me	(249, 251)	
					Pellia endiviifolia	Me	(457)	
					P. neesiana	Me	(457)	
					Riccardia lobata var. *yakushimensis*	Me	(42)	
(796)	Isosacculatal	$C_{20}H_{30}O$			*Riccardia lobata* var. *yakushimensis*	Me	(42)	
(797)	1β-Hydroxy-isosacculatal	$C_{20}H_{30}O_3$		− 71.1	*Pellia endiviifolia*	Me	(249, 260)	CD 292 nm − 0.91
(798)	Sacculatanolide	$C_{20}H_{30}O_2$		− 38.8	*Pellia endiviifolia*	Me	(249, 260)	
(799a)	1β-Hydroxy-sacculatanolide	$C_{20}H_{30}O_3$		− 20.3	*Pellia endiviifolia*	Me	(249, 260)	

(800)	11α-Hydroxy-sacculatanolide	$C_{20}H_{30}O_3$		−65.4	*Pellia endiviifolia*	Me	(249, 260)
(801)	1β,11β-Dihydroxy-sacculatanolide	$C_{20}H_{30}O_4$		−27.7	*Pellia endiviifolia*	Me	(249, 260)
(803)	12-Deoxo-1β,11α-dihydroxy-sacculatanolide	$C_{20}H_{32}O_3$			*Pellia endiviifolia*	Me	(249)
(804a)	1β,11α-Dihydroxy-sacculatenolide	$C_{20}H_{28}O_5$			*Pellia endiviifolia*	Me	(249, 260)
(805)	Perrottetianal A	$C_{20}H_{30}O_2$			*Fossombronia pusilla*	Me	(491)
					Plagiochila hondurensis	J	(48, 513)
					Porella acutifolia subsp. *tosana*	J	(600)
					P. caespitans var. *setigera*	J	(597)
					P. cordaeana	J	(593)
					P. elegantula	J	(39, 189)
					P. navicularis	J	(591)
					P. perrottetiana	J	(83)
					Symphyogyna brongniartii	J	(513)
(806)	Perrottetianal B	$C_{20}H_{30}O_3$			*Fossombronia pusilla*	Me	(94, 491)
(807)	8-Hydroxy-9-hydro-perrottetianal	$C_{20}H_{32}O_3$			*Fossombronia pusilla*	Me	(94, 491)
(808)	Sacculaplagin	$C_{24}H_{38}O_7$	184–5	+25.9	*Plagiochila sciophila* (= *P. acanthophylla* subsp. *japonica*)	J	(256)
(809)	Sacculaporellin	$C_{20}H_{32}O_2$			*Porella perrottetiana*	J	(83)
(810)	3α,4α-Epoxy-5α-acetoxy-18-hydroxysphenoloba-13E,16E-diene	$C_{22}H_{34}O_4$	46–8	+22.3	*Anastrophyllum minutum*	J	(106, 108)

Table IIc (continued)

Structure number	Name of compounds	Formula	m.p. °C	$[\alpha]_D$	Plant source	Order	References	Comments
(811)	3α,4α-Epoxy-5α-18-dihydroxysphenoloba-13E,16E-diene	$C_{20}H_{32}O_3$		28.3	*Anastrophyllum minutum*	J	*(106, 108)*	
(812)	3α,4α-Epoxy-5α-acetoxy-18-hydroxy-sphenoloba-13Z,16E-diene	$C_{22}H_{34}O_4$		− 11.5	*Anastrophyllum minutum*	J	*(106, 108)*	
(813)	3α,4α-Epoxy-5α-acetoxysphenoloba-13E,16E,18-triene	$C_{22}H_{32}O_2$		+ 13.0	*Anastrophyllum minutum*	J	*(106, 108)*	
(814)	3α,4α-Epoxy-5α-acetoxysphenoloba-13Z,16E,18-triene	$C_{22}H_{32}O_3$		+ 3.2	*Anastrophyllum minutum*	J	*(106, 108)*	
(815)	3α,4α-Epoxy-5α-hydroxy-sphenoloba-13Z,16E,18-triene	$C_{20}H_{30}O_2$	77–9	− 51.1	*Anastrophyllum minutum*	J	*(106, 108)*	
(818)	3α,18-Dihydroxy-trachyloban-19-oic acid	$C_{20}H_{30}O_4$			*Jungermannia exsertifolia* subsp. *cordifolia*	J	(240)	
(819)	Methyl 3α,18-diacetoxy-trachyloban-19-oate	$C_{25}H_{36}O_6$		− 37.1	*Jungermannia exsertifolia* subsp. *cordifolia*	J	(240)	
(821)	2β,9α,13β-Trihydro-xyverrucosane	$C_{20}H_{34}O_3$	243–244.5	− 72.6	*Mylia verrucosa*	J	(534)	

No.	Name	Formula	mp	[α]	Species		Ref.
(822)	9α-Acetoxy-2β,13β-dihydroxyverrucosane	$C_{22}H_{36}O_4$	234.5–235.5	−46.2	Mylia verrucosa	J	(534)
(823)	2β,13β-Dihydroxy-9-oxo-verrucosane	$C_{20}H_{32}O_3$	222.5–223.0	−107.2	Mylia verrucosa	J	(534)
(824)	2β-Hydroxy-verrucosane	$C_{20}H_{34}O$	76.5–78	−58	Scapania bolanderi	J	(377)
(825)	9α-Acetoxy-2β-hydroxy-verrucosane	$C_{22}H_{36}O_3$		−83	Scapania bolanderi	J	(377)
(826)	2β-Acetoxy-11α-hydroxy-verrucosane	$C_{22}H_{36}O_3$	203–4	−103	Scapania bolanderi	J	(377)
(827)	2β,9α-Dihydroxy-verrucosane	$C_{20}H_{34}O_2$	153–4	−72	Mylia anomala	J	(102, 246)
					Scapania bolanderi	J	(377)
(828)	2β-Hydroxy-9-oxo-verrucosane	$C_{20}H_{32}O_2$	111–2	−103	Scapania bolanderi	J	(377)
(829)	Neoverrucosan-5β-ol	$C_{20}H_{34}O$	174–5	−10	Schistochila acuminata	J	(136, 626)
					Scapania bolanderi	J	(377)
(830)	2β,9α-Dihydroxy-verrucos-13-ene	$C_{20}H_{32}O_2$	169–170	−26	Scapania bolanderi	J	(377)
(831)	2β,8β-Dihydroxy-verrucosane	$C_{20}H_{34}O_2$	158.5–160	−108.4	Gyrothyra underwoodiana	J	(350)
(832)	(15S,16S)-2β,16-Epoxy-verrucosan-16-ol	$C_{20}H_{32}O_2$	143–5	−19.69	Mylia taylorii	J	(246)
(833a)	13-epi-Neoverrucosan-5β-ol	$C_{20}H_{34}O$	157–9	+54.9	Heteroscyphus planus	J	(249, 253)
			151–3.5	+45.5	Plagiochila stephensoniana	J	(188)
					Schistochila nobilis	J	(54)
(834)	13-epi-Homo-verrucosan-5β-ol	$C_{20}H_{34}O$	123–4	+47.1	Schistochila nobilis	J	(54)

Table IIc (continued)

Structure number	Name of compounds	Formula	m.p. °C	$[\alpha]_D$	Plant source	Order	References	Comments
(835)	Homoverrucosan-5β-ol	$C_{20}H_{34}O$			Schistochila acuminata	J	(136, 626)	
(836)	13-epi-Neoverrucosan-5β,20-diol	$C_{20}H_{34}O_2$	220–221.5	+ 56.3	Heteroscyphus planus	J	(253)	
(837)	ent-Verticillol	$C_{20}H_{34}O$			Jackiella javanica	J	(247, 437)	
(838)	ent-13-epi-Verticillol	$C_{20}H_{34}O$			Jackiella javanica	J	(247, 437)	
(839)	ent-Verticillanediol	$C_{20}H_{34}O_2$			Jackiella javanica	J	(247, 437)	
(840)	ent-13-epi-Verticillanediol	$C_{20}H_{34}O_2$			Jackiella javanica	J	(247, 437)	
(841)	ent-Isoverticillenol	$C_{20}H_{32}O$			Jackiella javanica	J	(247, 437)	
(842)	Verticillene	$C_{20}H_{32}$			Jackiella javanica	J	(432)	
(843)	Isoverticillene	$C_{20}H_{32}$			Jackiella javanica	J	(432)	

C: Calobryales; J: Jungermanniales; M: Marchantiales; Me: Metzgeriales; Mo: Monocleales

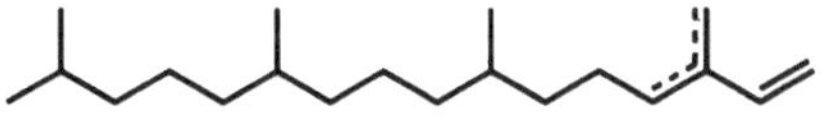

(**779**) Phytadienes

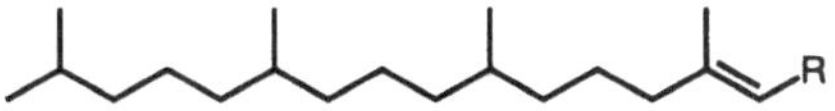

(**780**) Phytol ; R=CH$_2$OH
(**781**) Phytenic acid ; R=CO$_2$H

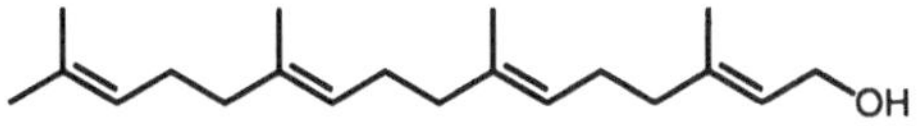

(**782**) Geranyl geraniol

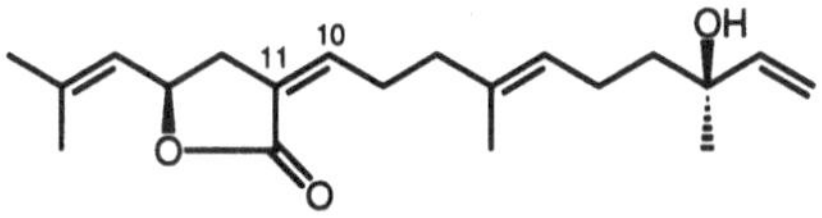

(**783**) Naviculide

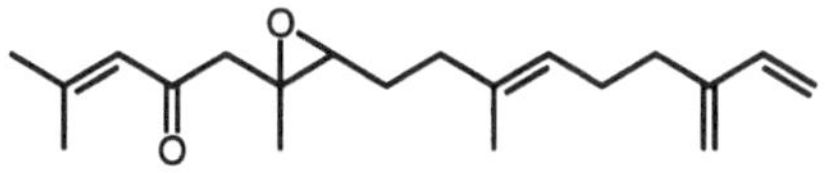

(**784**) Haplomitrenone

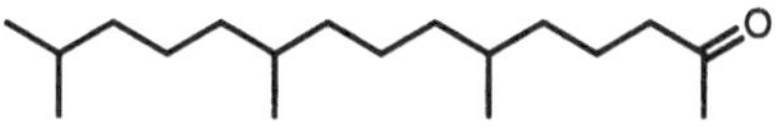

(**785**) (1,2)-Bis-norphytone

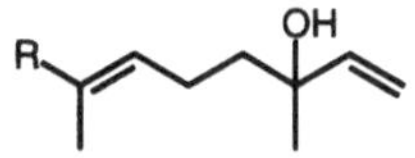

(**2**) R=Me
(**426**) R=(CH$_2$)$_2$CH=C(Me)$_2$

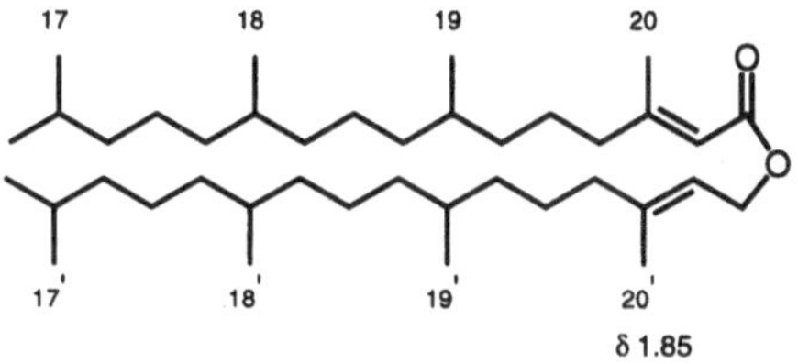

δ 1.85

(**786**) Phytyl phytenate

Chart 52. Phytanes found in the Hepaticae and the Musci

pimaranes, (**788**) and termarol (**789**), had already been found in *Junger-mannia thermarum* (*19*).

East Malaysian *Schistochila aligera* produces a new substance (**790**) with what appears to be a dolabrane carbon skeleton (*438*). The skeleton was deduced by various NMR techniques. Dehydration of (**790**) with POCl$_3$ in pyridine gave (**791**), ([α]$_D$ + 86°), whose spectral data closely

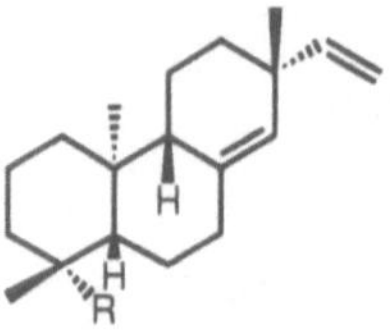

(787) *ent*-Pimara-8(14),15-dien-19-oic acid ; R=CO$_2$H
(788) *ent*-Pimara-8(14),15-dien-19-ol ; R=CH$_2$OH

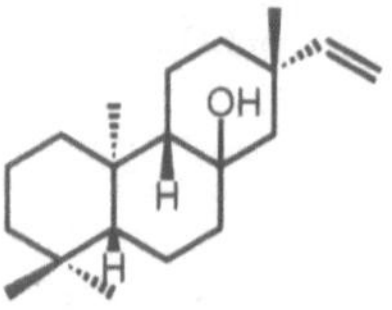

(789) Termarol 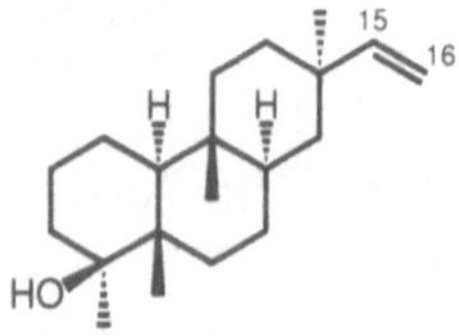   (790) Aligerol 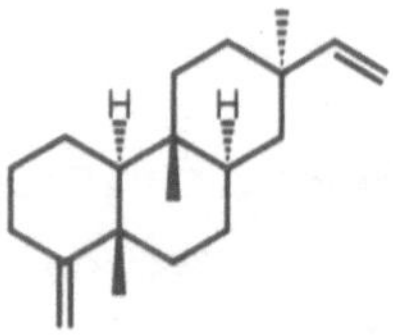   (791) Isodolabradiene

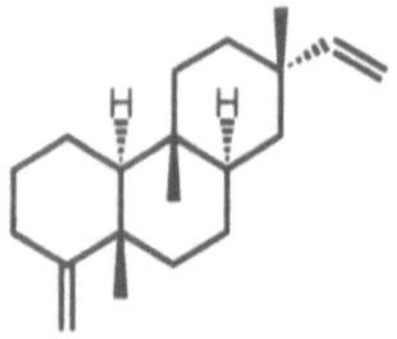

(792) Dolabradiene 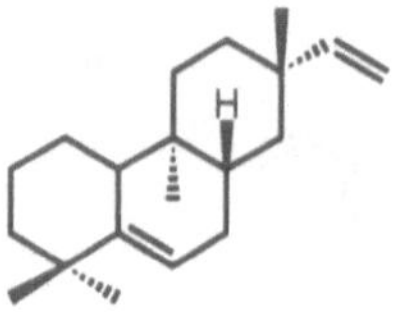   (793) *ent*-Rosa-5,15-diene 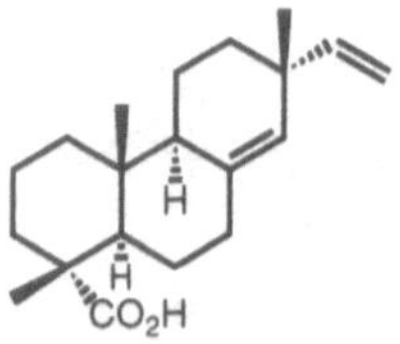   (794) (-)-Sandaracopimaric acid

Chart 53. Pimaranes and rearranged pimarane found in the Hepaticae and their related compounds

resembled dolabradiene (792) ([α]$_D$ − 70°) (*327*), except for the ^{13}C-NMR data. Although the ^{13}C-NMR signals of C-15 and C-16 in (791) appeared at δ151.3 and 108.4, those of (792) were observed at δ147.2 and 114.4 (*124*). On the other hand, the ^{13}C-NMR spectrum of *ent*-rosa-5,15-diene (793) (*194*) whose relative stereochemistry at C-8, C-9 and C-13 is opposite of that of (791), exhibits almost the same chemical shifts of C-15 (δ151.3) and C-16 (δ108.4) as do (790) and (791). Thus, (791) may be the C-13 epimer of (792). The relative stereochemistry of the tertiary hydroxyl group at C-4 of (790) has been established as equatorial, since dehydration of (790) gave (791) as the major product and there was no NOE between H-18 and H-19.

6.8 Sacculatanes

The earlier volume (*19*) listed eleven new sacculatane-type diterpenoids from liverworts. Sacculatal (795) was originally isolated from *Trichocoleopsis sacculata* (Jungermanniales) and *Pellia endiviifolia* (Metzgeriales) has also been isolated from *Riccardia lobata* var. *yakushimensis* (*42*) and *Pallavicinia levieri* (*251*) both of which belong to Metzgeriales. The former species also contains isosacculatal (**796**) as a minor component. Cell suspension cultures of *Pellia endiviifolia* and *P. neesiana* produce sacculatal (**795**). It is interesting that *in vitro* cultures of *P. endiviifolia* do not produce isosacculatal (**796**) which has been isolated from plant material (*457*).

Naturally occurring (−)-sacculatal (**795**) has been synthesized from (4aS,5S,8aS)-(−)-5β,8aβ-dimethyl-5α-(4-methyl-3-pentenyl)-3,4,4a,5, 6,7,8,8a-octahydro-1(1H)-naphthalenone (*227, 228*).

Pellia endiviifolia is a rich source of sacculatane-type diterpenoids. Further fractionation of the *n*-hexane extract of *P. endiviifolia* resulted in the isolation of six sacculatanes (**797, 799a, 800, 801a, 803, 804a**), together with sacculatal (**795**) and its C-9 epimer (**796**) (*260*). The spectral data of (**797**) were similar to those of isosacculatal (**796**), except for the presence of a secondary hydroxyl group. The position and stereochemistry of the hydroxyl group were deduced by the ^{1}H- and ^{13}C-NMR spectra with those of isosacculatal (**796**) and the NOE difference spectrum which showed the presence of NOE's between H-1 and H-5, and H-1 and H-11 of the formyl group. The absolute configuration ascribed to (**797**) is based on the negative Cotton effect at 292 nm (Δε − 0.98). The structure of (**799a**) was deduced by comparing the spectral data with those of the previously known sacculatanolide (**798**). The β-configuration of the secondary hydroxyl group at C-1 was established by the formation of a monoacetate (**799b**) and the presence of NOE's between H-1 and H-9, and H-1 and H-5, respectively. Except for the presence of a hemiacetal group, compound (**800**) was spectroscopically similar to (**798**) and (**799a**). Acetylation of (**800**) gave a monoacetate (**801b**) whose CD spectrum exhibited a positive Cotton effect at 239 nm (Δε + 1.53). The stereochemistry of the hydroxyl group at C-11 of (**800**) was assigned as α because of the presence of an NOE between H-11 and H13-Me. The absolute configuration of (**800**) has also been established by preparation of the known diol (**802**) from (**800**) with LiAlH$_4$ and by its positive Cotton effect at 243 nm (Δε + 2.75).

Spectral data of (**801a**) resembled those of (**800**), except for the presence of a proton (δ 5.72, *d*, J = 6.3 Hz) on a carbon-bearing hydroxyl group. Acetylation of (**801a**) gave a monoacetate (**801c**). Analysis of ^{1}H-^{1}H, ^{13}C-^{1}H, long range ^{13}C-^{1}H 2D-COSY spectra and NOE spectrometry led to the structure shown. NOE's were observed between H-1 and H-9, and H-1 and H-5, respectively. The CD spectrum of (**801a**) showed a positive Cotton effect at 237 nm (Δε + 1.54), indicating that the absolute configuration was the same as that of (**800**). Compound (**804a**) was

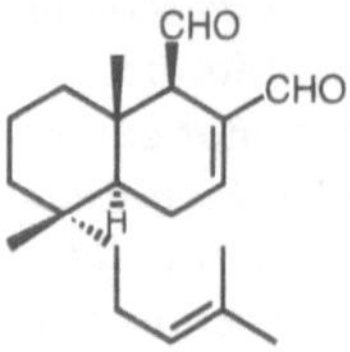

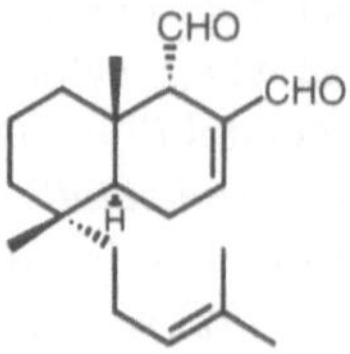

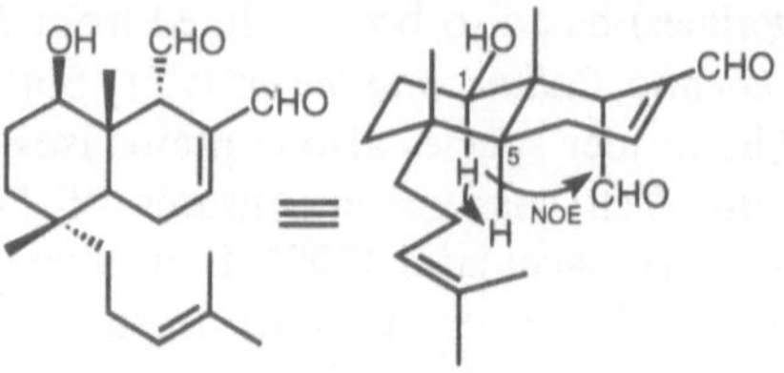

(795) Sacculatal   (796) Isosacculatal   (797) 1β-Hydroxyisosacculatal

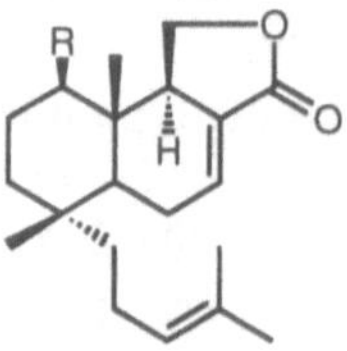

(798) Sacculatanolide ; R=H
(799a) 1β-Hydroxysacculatanolide ; R=OH
(799b) R=OAc

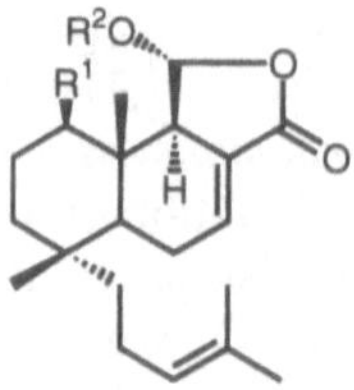

(800) 11α-Hydroxysacculatanolide ;
      R¹=R²=H
(801a) 1β,11β-Dihydroxysacculatanolide;
       R¹=OH, R²=H
(801b) R¹=H, R²=Ac
(801c) R¹=OH, R²=Ac

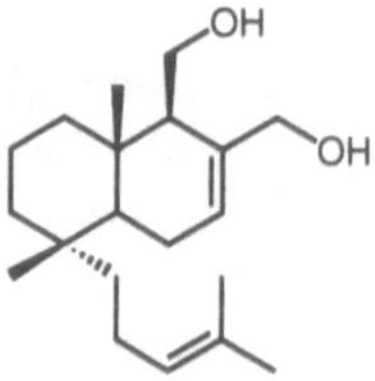

(802)

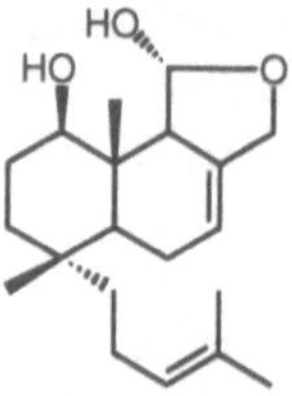

(803) 12-Deoxo-1β,11α-dihydroxysacculatanolide

Chart 54a. Sacculatanes found in the Hepaticae and their derivatives

isolated as the acetate (804b) which had spectral properties very similar to those of (801c) except for the presence of an exomethylene group conjugated with a ketone. The location of the conjugated exomethylene group at C-18 was confirmed by 2D-COSYs spectra. The relative and absolute stereochemistry of (804b) was deduced from the NOE difference spectrum and the positive Cotton effect at 238 nm (Δε + 1.33).

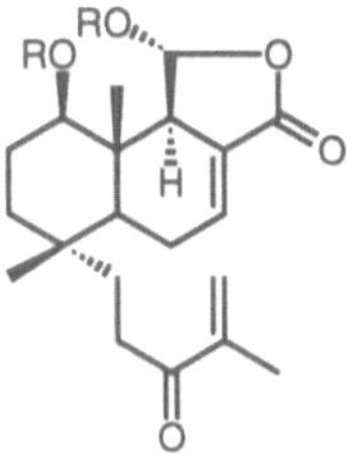

(804a) 1β,11α-Dihydroxysacculatenolide ; R=H
(804b) R=Ac

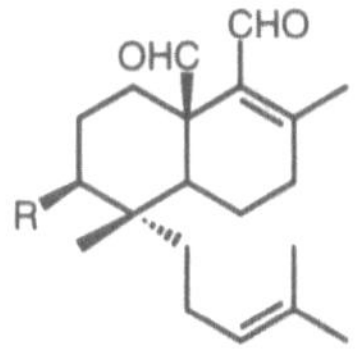

(805) Perrottetianal A ; R=H
(806) Perrottetianal B ; R=OH

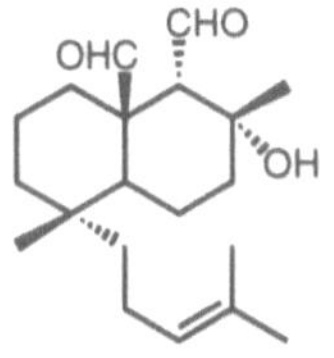

(807) 8-Hydroxy-9-hydro-
perrottetianal

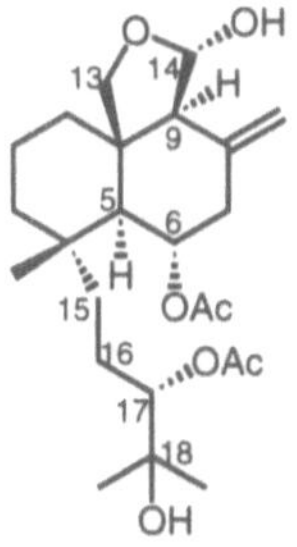

(808) Sacculaplagin

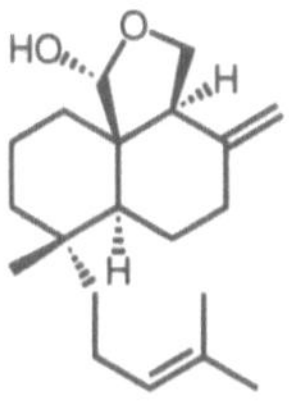

(809) Sacculaporellin

Chart 54b. Sacculatanes found in the Hepaticae and their derivative

As mentioned in the previous review (*19*) perrottetianal A (**805**) occurs not only in Metzgeriales and in Jungermanniales, particularly in the Porellaceae. Since then the same diterpene dialdehyde has also been isolated from *Fossombronia pusilla* (*491*), *Plagiochila hondurensis* (*48, 513*), *Porella acutifolia* subsp. *tosana* (*600*), *P. caespitans* var. *setigera* (*597*), *P. cordaeana* (*593*), *P. elegantula* (*39, 189*), *P. navicularis* (*591*) and from *in vitro* cultured *Symphyogyna brongniartii* (*513*). Surface and liquid cultures of *Fossombronia pusilla* furnished a new sacculatane-type diterpene, 8-hydroxy-9-hydroperrottetianal (**807**) together with perrottetin A (**805**) and B (**806**) (*94, 491*). The amounts of (**805**) and (**806**) from the cultures and field collected plants are almost the same. The structure of (**807**) was deduced by comparing the NMR spectral data with those of

(805) and (806). The relative stereochemistry of (807) is based on NOE spectrometry.

The total synthesis of (+)-perrottetianal (805) has been accomplished by HAGIWARA et al. (226, 229) starting from an optically active Wieland-Miescher ketone analogue (544) in 17 steps.

A new highly oxygenated sacculatane-type diterpene hemiacetal, sacculaplagin (808), has been isolated from P. sciophila (= P. acantho-phylla subsp. japonica) (256). The sacculatane structure for (808) was deduced by 2D-COSY, NOESY and 2D-INADEQUATE techniques on the triacetate prepared by acetylation of (808). The stereochemistry at C-11 was α, as the H-11 signal was a singlet thus showing that the H-9, 11 dihedral angle was close to 90°. The absolute configuration assigned to (808) was based on the negative Cotton effect at 231 nm (Δε − 16.3) of the lactone obtained from (808) by chromic acid oxidation. Finally the absolute configuration at C-17 was shown to be S from the following evidence. The triacetate was converted to the 14-methoxy-17,18-diol, the CD spectrum of which in CCl_4 in the presence of $Eu(fod)_3$ exhibited a positive Cotton effect at 322 nm (Δε + 7.13). It is interesting from the biogenetic point of view that both fusicoccane- and sacculatane-type diterpenoids have been isolated from Plagiochila species.

Porella perrottetiana elaborates perrottetianal (805) as well as a new sacculatane-type diterpene hemiacetal, sacculaporellin (809) (83). The relative and absolute configuration (809) has been settled by a combination of ^{1}H-, ^{13}C-, and ^{1}H-^{1}H COSY spectral data as well as NOE spectrometry and by considering that (805) is a co-metabolite.

6.9 Sphenolobanes

Six diterpene epoxides (810–815) with a novel carbon skeleton for which the name sphenolobane has been proposed have been isolated from European Anastrophyllum minutum. Structures were established by means of NMR spectrometry including 2D-COSYs, the relative configurations being deduced from NOE difference spectra (106, 108). Compound (811) is identical with a substance produced by $LiAlH_4$ reduction of (810). Compound (813) was prepared from (810) by dehydration with $POCl_3$.

The sphenolobanes can be considered to be isoprenologues of the daucane (carotanes)-type sesquiterpenoid mentioned earlier and presumably arise by folding of geranyl geranyl pyrophosphate (677) as shown in Scheme 56. It is noteworthy that a similar diterpene alcohol (816) has been isolated from an Okinawan sponge, Halichondria panicea (pallas) (440). It is also of interest that hercynolactone (327), the only daucane

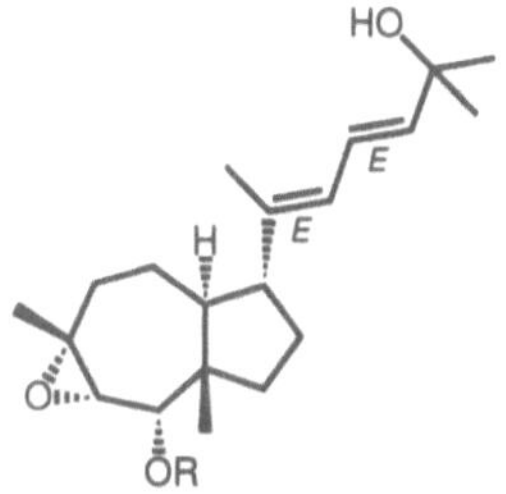

(810) 3α,4α-Epoxy-5α-acetoxy-18-hydroxysphenoloba-
13E,16E-diene ; R=Ac
(811) 3α,4α-Epoxy-5α,18-dihydroxysphenoloba-
13E,16E-diene ; R=H

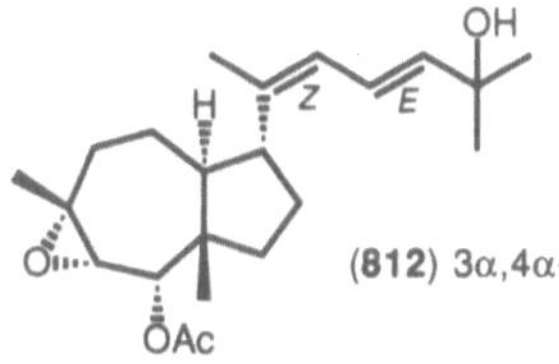

(812) 3α,4α-Epoxy-5α-acetoxy-18-hydroxysphenoloba-13Z,16E-diene

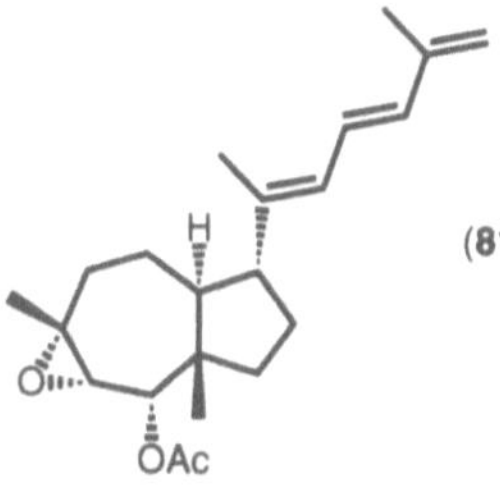

(813) 3α,4α-Epoxy-5α-acetoxysphenoloba-13E,16E,18-triene

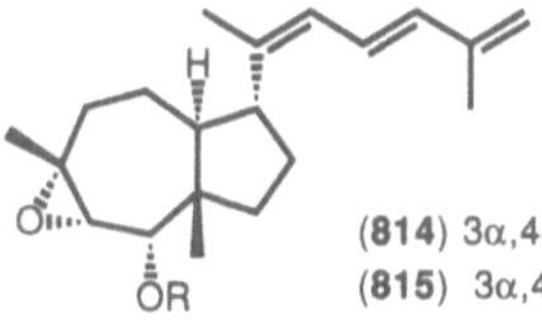

(814) 3α,4α-Epoxy-5α-acetoxysphenoloba-13Z,16E,18-triene ; R=Ac
(815) 3α,4α-Epoxy-5α-hydroxysphenoloba-13Z,16E,18-triene ; R=H

Chart 55. Sphenolobanes found in the Hepaticae

reported up to now from the Hepaticae, occurs in the closely related
species *Barbilophozia lycopodioides* and *B. hatcheri* (275).

6.10 Trachylobanes and Verrucosanes

A novel trachylobane diterpenoid, 3α,18-dihydroxytrachyloban-19-
oic acid (818) has been isolated in the form of its diacetate methyl ester
(819) from *Jungermannia exsertifolia* subsp. *cordifolia* (240). The NMR
spectrum of (819) indicated the presence of a tetrasubstituted cyclopropane, two tertiary methyls, a hydroxymethyl, a secondary acetoxyl and a

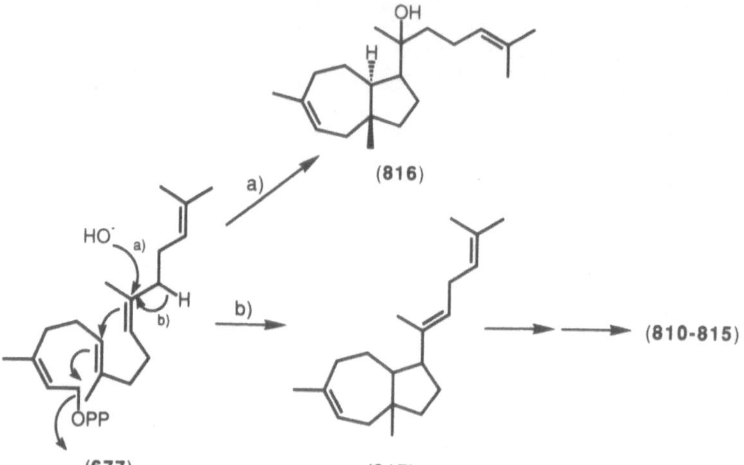

Scheme 56. Possible biogenetic pathways for sphenolobanes from geranyl geranyl pyrophosphate

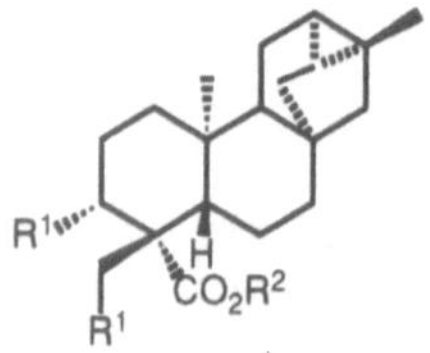

(818) 3α,18-Dihydroxytrachyloban-19-oic acid ; R¹=OH, R²=H
(819) Methyl 3α,18-diacetoxytrachyloban-19-oate ; R¹=OAc, R²=Me
(820) Methyl trachyloban-19-oate ; R¹=H, R²=Me

Chart 56. Trachylobanes found in the Hepaticae

carbomethoxyl group, indicating that (819) is a member of the trachylobane class. Comparison of the chemical shifts with those for methyl trachyloban-19-oate (820) showed that the oxidized methyl groups are those of C-18 and C-19, and that the secondary acetoxyl group is at C-3. That the latter is equatorial followed from the coupling constants of H-3. The remaining relative stereochemistry of (819) was based on the NOE difference spectrum. This is the first instance of a trachylobane-type diterpenoid from the Hepaticae although similar trachylobanes have been found in higher plants (210).

Mylia species belonging to the Jungermanniales are rich sources of verrucosane- and neoverrucosane-type diterpenoids as mentioned earlier (*19*). In addition to the previously known verrucosanes, three new verrucosane diols (**821–823**) have been isolated from *M. verrucosa* (*534*). Structures were assigned by a combination of spectral data and chemical interconversions and degradation of (**823**) to (**823c**) shown in Scheme 57.

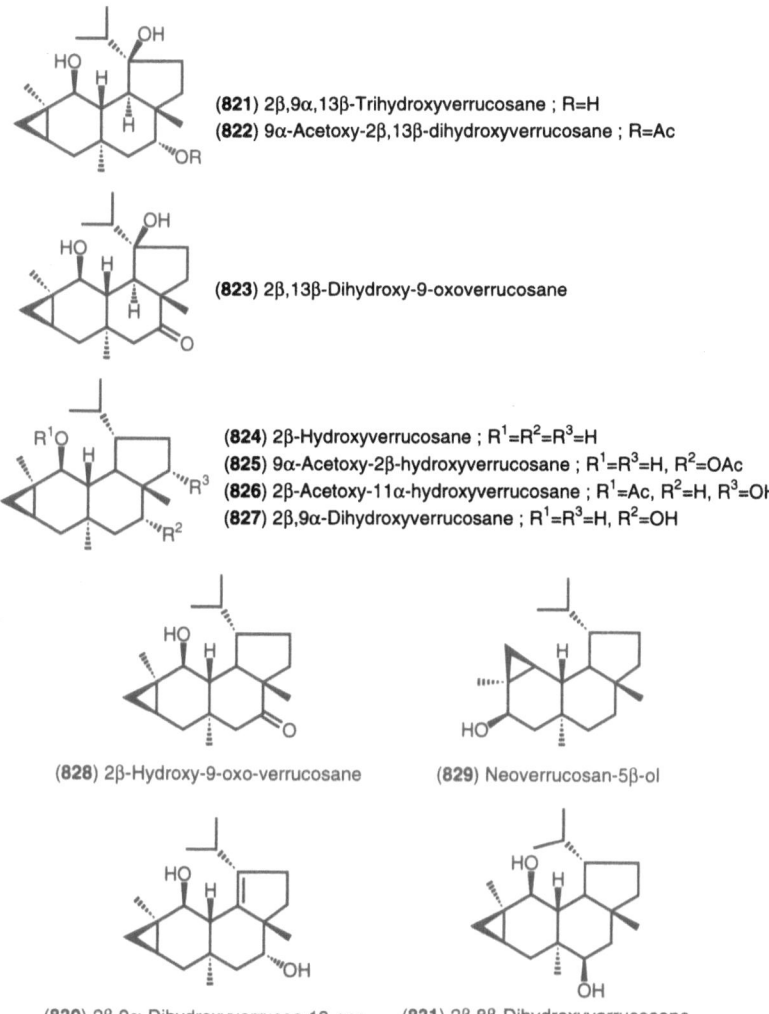

(**821**) 2β,9α,13β-Trihydroxyverrucosane ; R=H
(**822**) 9α-Acetoxy-2β,13β-dihydroxyverrucosane ; R=Ac

(**823**) 2β,13β-Dihydroxy-9-oxoverrucosane

(**824**) 2β-Hydroxyverrucosane ; R¹=R²=R³=H
(**825**) 9α-Acetoxy-2β-hydroxyverrucosane ; R¹=R³=H, R²=OAc
(**826**) 2β-Acetoxy-11α-hydroxyverrucosane ; R¹=Ac, R²=H, R³=OH
(**827**) 2β,9α-Dihydroxyverrucosane ; R¹=R³=H, R²=OH

(**828**) 2β-Hydroxy-9-oxo-verrucosane

(**829**) Neoverrucosan-5β-ol

(**830**) 2β,9α-Dihydroxyverrucos-13-ene

(**831**) 2β,8β-Dihydroxyverrucosane

Chart 57a. Verrucosanes found in the Hepaticae

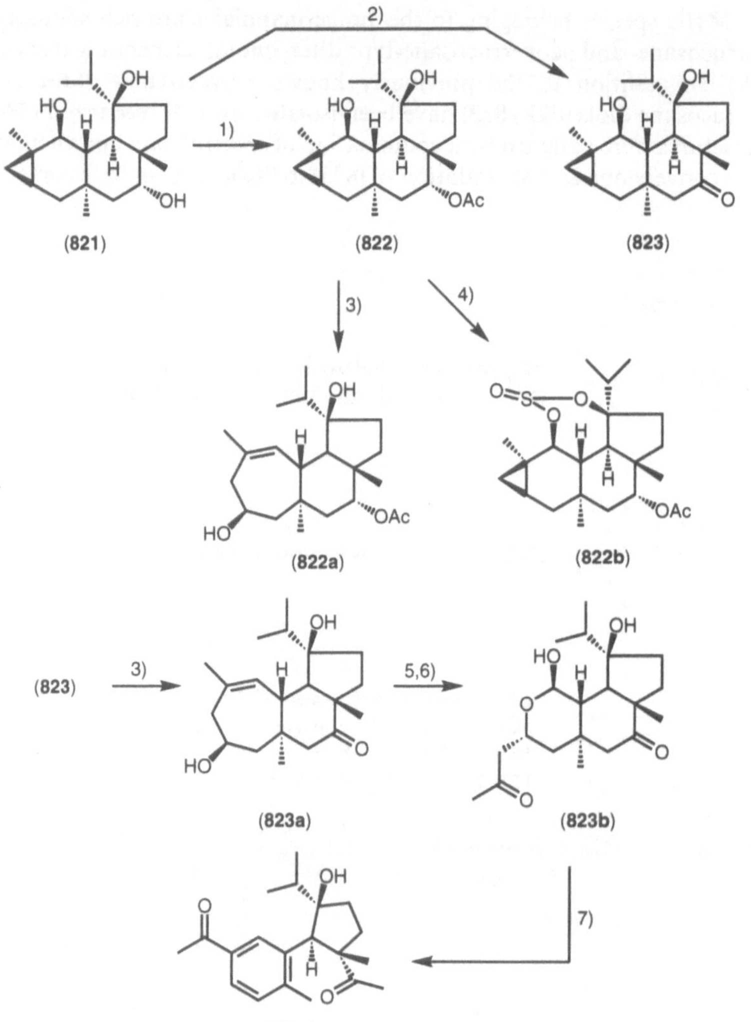

1) Ac$_2$O/Py 2) Jones oxd./Me$_2$CO 3) 0.5N H$_2$SO$_4$/Me$_2$CO
4) SOCl$_2$/Py 5) OsO$_4$/Py 6) NaIO$_4$ 7) 5% KOH-MeOH

Scheme 57. Reactions of 2β,9α,13β-trihydroxyverrucosane

The configuration of the C-13 hydroxyl group in (821–823) has been
deduced from solvent shifts in the NMR spectra and the formation of a
sulfite from (822). Among the three tertiary methyls the signal of the
methyl on C-10 experienced much larger deshielding solvent shifts than

the other effects [Δ − 0.37, − 0.34 and − 0.14 in (**821**), (**822**) and (**823**): $\Delta = \delta(CDCl_3) − \delta(C_5D_5N)$].

Scapania bolanderi furnished a new verrucosane (**830**) in addition to the five previously known verrucosanes (**824–828**) and neoverrucosane (**829**) (*377*). The structure assigned to (**830**) depends on spectroscopic comparison (^{1}H-NMR and mass spectra) with (**821**) (*19*). The structure and relative stereochemistry of a new verrucosane (**831**) from the ether extract of fresh *Gyrothyra underwoodiana* was established by a 2D-COSY experiment and X-ray diffraction (*350*). The absolute configuration was deduced by using the dibenzoate chirality method on its bis(*p*-bromobenzoate) derivative (**831a**) of (**831**) as shown below. The CD spectrum of (**831a**) has a positive Cotton effect at 251 nm ($\Delta\varepsilon$ + 22.4) and a negative Cotton effect at 234 nm ($\Delta\varepsilon$ − 2.7). The positive sign of first Cotton effect showed that the exciton chirality between transition moment of two *p*-bromobenzoate chromophores was positive.

(831a)

BENESOVA *et al.* (*102*) isolated a diterpene diol from *M. anomala* whose structure remained to be clarified. It was later shown to be identical with 2β,9α-dihydroxyverrucosane (**827**) (*246*). Further fractionation of the ether extract of *Mylia taylorii* resulted in the isolation of a new hemiacetal (**832**) together with the previously known sesquiterpenoids, myliol (**98**) and taylorione (**111**). The structure of the new verrucosane with a hemiacetal linkage between C-2 and C-16 was established by analysis of the ^{1}H- and ^{13}C-NMR spectra (*246*). Wolff-Kishner reduction of (**832**) gave (**824**) (*19*), thus confirming the gross structure of (**832**) as well as defining the absolute stereochemistry, except for the configuration of C-15 and C-16. The absolute configuration at these two centers, could, however, be established by the NOE difference spectrum of (**832**) and lactone (**832a**) prepared from (**832**) with PDC. NOE's were observed between H-16 and the secondary methyl at C-15 in (**832**) and between H-15 and H-2 in (**832a**), as shown below (Scheme 58).

Plagiochila stephensoniana furnished a new verrucosane (**833a**). The absolute stereochemistry was established by a combination of spectral

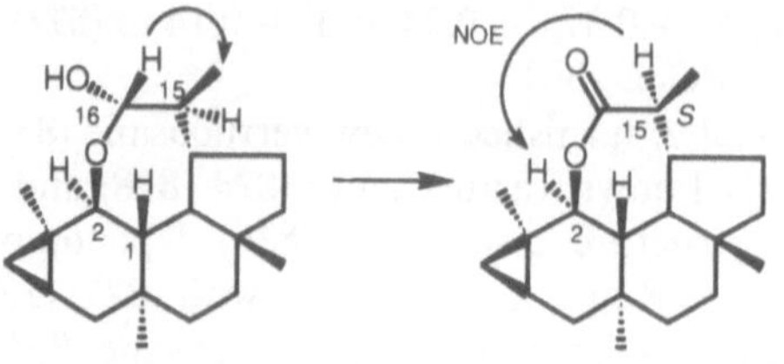

Scheme 58. NOE difference spectroscopy of 2β,16-epoxyverrucosane-16-ol and the derived lactone

data and X-ray crystallographic analysis of its *p*-bromobenzoate (**833b**) (*188*). Compound (**833a**) has a β-oriented isopropyl group on C-13 whereas all previously known verrucosane and neoverrucosane-type diterpenoids contain an α-isopropyl group at this position.

The New Zealand liverwort *Schistochila nobilis* elaborates a new verrucosane (**834**) together with 13-*epi*-neoverrucosan-5β-ol (**833a**) (*54*). The spectral data of (**834**) were very similar to those of (+)-homoverrucosan-5β-ol (**835**) obtained from neoverrucosan-5β-ol (**829**) by treatment with acid (*19*) except for the melting point and the value of the specific rotation, indicating that (**834**) might be the stereoisomer of (**835**). This was confirmed by conversion of (**833a**) with dilute sulfuric acid to (**834**). *Schistochila acuminata* (previously referred to as *S. rigidula*) contains neoverrucosan-5β-ol (**829**) and homoverrucosan-5β-ol (**835**) (*626*). *Heteroscyphus planus* elaborates not only clerodanes but also a new verrucosane (**836**), along with the known 13-*epi*-neoverrucosan-5β-ol (**833a**) (*253*). The structure assigned to the new verrucosane was based on the resemblance of the spectral data to those of (**833a**) and by chemical correlation involving reduction of the primary toluenesulfonate of (**836**) to **833a** with NaI/Zn. It is noteworthy that neoverrucosan-5β-ol (**829**), 13-*epi*-neoverrucosan-5β-ol (**833a**) and homoverrucosan-5β-ol (**835**) have been found in an Okinawan sponge, *Halichondria panicea* (*440*). The verrucosanes may be formed by a further modification of the dolabellane skeleton (*143*).

6.11 Verticillanes

Investigation of *Jackiella javanica* collected in Yaku Island (Japan) has led to isolation of a series of diterpenoids, *ent*-verticillol (**837**), *ent*-13-

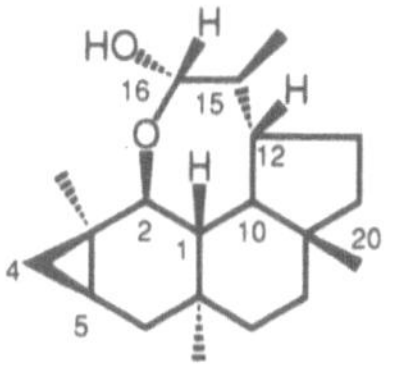

(832) (15S,16S)-2β,16-Epoxyverrucosane-16-ol

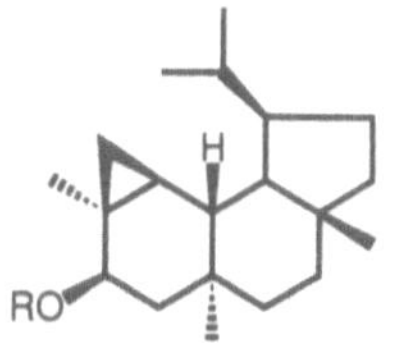

(833a) 13-epi-Neoverrucosan-5β-ol ; R=H
(833b) R=p-BrBz

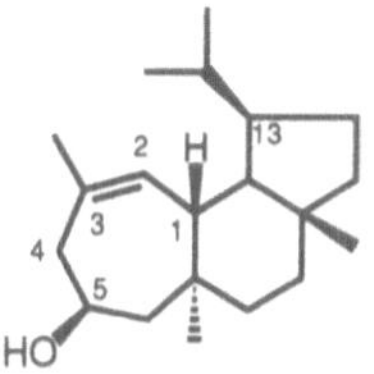

(834) 13-epi-Homoverrucosan-5β-ol

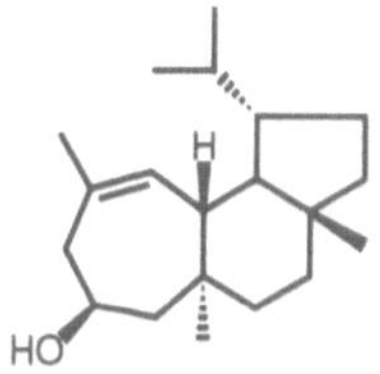

(835) Homoverrucosan-5β-ol

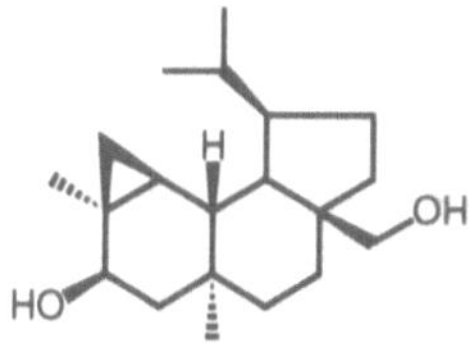

(836) 13-epi-Neoverrucosan-5β,20-diol

Chart 57b. Verrucosanes found in the Hepaticae and their derivatives

epi-verticillol (**838**), *ent*-verticillanediol (**839**), *ent*-13-*epi*-verticillanediol (**840**) and *ent*-isoverticillenol (**841**) belonging to the rare verticillane class (*247, 437*). Structures were elucidated by a combination of spectral data and an X-ray crystallographic analysis of (**839**). Monoalcohol (**837**) is the enantiomer of (+)-verticillol isolated from the conifer *Sciadopitys verticillata* (Taxodiaceae) (*311*); hence, the absolute configurations of (**837–841**) are as shown. Dehydration of (**837**) with SOCl₂ gave two hydrocarbons, verticillene (**842**) and isoverticillene (**843**) which have been found in the ether extract of *J. javanica* (*432*). The *Jackiella* Shiffn. appears to be a very isolated genus, because no verticillane diterpenoids have been found in other liverworts so far.

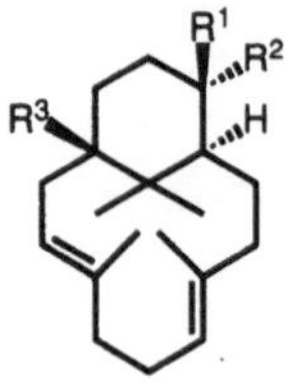

(837) *ent*-Verticillol ; R¹=Me, R²=OH, R³=H
(838) *ent*-13-epi-Verticillol ; R¹=OH, R²=Me, R³=H
(839) *ent*-Verticillanediol ; R¹=Me, R²=R³=OH
(840) *ent*-13-epi-Verticillanediol ; R¹=R³=OH, R²=Me

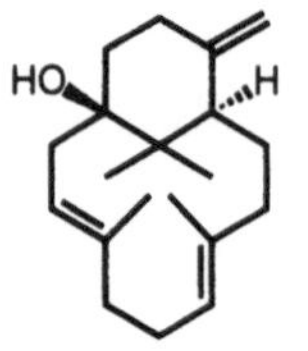

(841) *ent*-Isoverticillenol 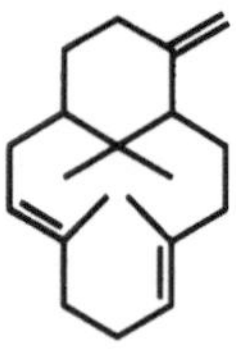   (842) Verticillene 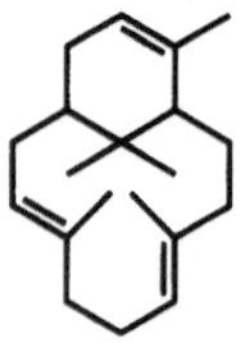   (843) Isoverticillene

Chart 58. Verticillanes found in the Hepaticae

7. Steroids and Triterpenoids

Almost all liverworts produce campesterol (= 24α-methylcholes-terol) (**844a**), stigmasterol (= 24α-ethylcholesta-5,22-dien-3β-ol) (**846a**) and sitosterol (= 24α-ethylcholest-5-en-3β-ol) (**848a**). Cholesterol (**849**), brassicasterol (**850**) and campesteryl behenate (**845**) have previously been found in Hepaticae (*19*). Species which have been shown to contain sterols since the earlier review are listed in Table IId.

Lipids of *Pallavicinia lyellii* were extracted with acetone and the extract saponified with 5% KOH-MeOH. The sterol fractions were purified by HPLC. The ¹H NMR spectra of each component showed the presence of a mixture of campesterol (**844a**) (16%), 24β-methylcholesterol (= dihydrobrassicasterol) (**844b**) (36%), sitosterol (**848a**) (33%) and 24ζ-ethylcholest-5,22-dien-3β-ol (**846a**, **846b**) (15%) (*3*). The sterol composi-tion of *Marchantia polymorpha* has also been investigated (*463*); it contains 24α-(**844a**) and 24β-methylcholesterol (**844b**) (in a 1:1 ratio) and 24α-(**848a**) and 24β-ethylcholesterol (**848b**) (1:1). Cʜɪᴜ *et al.* (*138*) sug-gested that 24β-methylcholesterol (**844b**) might be formed from 24β-methyl-Δ²⁵(27)-intermediates because both cyclolaudenol (**1234**) and 31-nor-cyclolaudenol (**1235**) have been detected in bryophytes (*271*). 24-Methylene sterols are also suggested as a precursors for 24α-meth-ylcholesterol (**844a**) and 24α-(**848a**) and 24β-ethylcholesterol (**848b**) (*138*) because 24-methylenecycloartanol (**1236**), cycloeucalenol (**1237**) and ob-tusifoliol (**1238**) have been found in bryophytes (*271*).

Table IId. *Steroids and Triterpenoids Found in the Hepaticae*

Structure number	Name of compounds	Formula	m.p. °C	$[\alpha]_D$	Plant source	Order	References	Comments
(844a)	Campesterol	$C_{28}H_{48}O$			*Acrolejeunea mariana*	J	(219)	
					A. pusilla	J	(219)	
					A. torulosa	J	(219)	
					Adelanthus lindenbergianus	J	(45)	
					Anastrepta orcadensis	J	(284)	
					Archilejeunea olivacea	J	(219)	
					A. parviflora	J	(45)	
					Balantiopsis erinacea	J	(45)	
					Barbilophozia hatcheri	J	(284)	
					Bazzania sp.	J	(138)	
					B. spiralis	J	(341)	
					Blepharolejeunea densifolia	J	(219)	
					B. incongrua	J	(219)	
					Bryopteris diffusa	J	(219)	
					Clasmatocolea humilis	J	(45)	
					Conocephalum conicum	M	(138)	
					Dumortiera hirsuta	M	(47)	
					Frullania brasiliensis	J	(47)	
					F. davurica	J	(420)	
					F. falciloba	J	(64)	
					F. jackii	J	(420)	
					Gackstroemia magellanica	J	(45)	
					Isotachis haematodes	J	(47)	
					I. humectata	J	(45)	
					Jungermannia exsertifolia subsp. *cordifolia*	J	(284)	

Table IId (continued)

Structure number	Name of compounds	Formula	m.p. °C	[α]_D	Plant source	Order	References	Comments
					Lejeunea albescens	J	(219)	
					L. discreta	J	(219)	
					L. glaucescens	J	(219)	
					Lejeunea sp.	J	(47)	
					Lepidozia reptans	J	(147, 284)	
					Leucolejeunea aff. decurrens	J	(219)	
					Lopholejeunea howei	J	(219)	
					Lophozia ventricosa	J	(284)	
					Makinoa crispata	Me	(622)	
					Mannia fragrans	M	(276)	
					Marchantia berteroana	M	(39)	
					M. diptera	M	(138)	
					M. foliacea	M	(39)	
					M. palmata	M	(72)	
					M. plicata	M	(47)	
					M. polymorpha	M	(61, 71, 72, 463)	
					Marchesinia brachiata	J	(219)	
					Mastigolejeunea humilis	J	(219)	
					Mastigophora diclados	J	(138)	
					M. undulata	J	(219)	
					Mylia taylorii	J	(532)	
					M. verrucosa	J	(532)	
					Neteroclada confluens	J	(45)	
					Pallavicinia lyellii	Me	(3, 138)	
					Plagiochila dura	J	(46)	

P. engelii	J	(46)
P. falcata	J	(48)
P. friabilis	J	(48)
P. fuegiensis	J	(346)
P. gayana	J	(46)
P. neesiana	J	(46)
P. parvidens	J	(46)
Pleurozia gigantea	J	(52)
Porella densifolia subsp. appendiculata	J	(67)
P. elegantula	J	(39)
P. platyphylla	J	(284)
P. squamurifera	J	(47)
P. swartziana	J	(47)
Radula kojana	J	(50)
Schiffneria hyalina	J	(622)
Schiffneriolejeunea nymannii	J	(219)
S. omphalanthoides	J	(219)
Stictolejeunea squamata	J	(219)
Symbiezidium transversale var. hookeriana	J	(219)
Triandrophyllum subtrifidum var. trifidum	J	(45)
T. subtrifidum	J	(219)
Thysananthus convolutus	J	(219)
T. fruticosus	J	(219)
T. mollis	J	(219)
Tylimanthus urvilleanus	J	(45)
Bazzania sp.	M	(138)
Conocephalum conicum	M	(138)
Marchantia diptera	M	(138)

(844b) Dihydrobrassicasterol $C_{28}H_{48}O$

Table IId (continued)

Structure number	Name of compounds	Formula	m.p. °C	$[\alpha]_D$	Plant source	Order	References	Comments
					M. polymorpha	M	(138, 463)	
					Mastigophora diclados	J	(138)	
					Pallavicinia lyellii	Me	(3, 138)	
(846a)	Stigmasterol (24α-ethyl)	$C_{29}H_{48}O$			Acrolejeunea pusilla	J	(219)	
					A. torulosa	J	(219)	
					Adelanthus lindenbergianus	J	(45)	
					Anastrepta orcadensis	J	(284)	
					Aneura latissima	Me	(47)	
					Archilejeunea mariana	J	(219)	
					A. olivacea	J	(219)	
					A. parviflora	J	(45)	
					Balantiopsis erinacea	J	(45)	
					Barbilophozia hatcheri	J	(284)	
					Bazzania sp.	J	(138)	
					B. spiralis	J	(341)	
					Blepharolejeunea densifolia	J	(219)	
					B. incongrua	J	(219)	
					Bryopteris diffusa	J	(219)	
					B. filicina	J	(432)	
					B. trinitensis	J	(219)	
					Clasmatocolea humilis	J	(45)	
					C. vermicularis	J	(47)	
					Conocephalum conicum	M	(138)	
					Dumortiera hirsuta	M	(47)	
					Frullania brasiliensis	J	(47)	

Species		
F. davurica	J	(420)
F. falciloba	J	(64)
F. jackii	J	(420)
Gackstroemia magellanica	J	(45)
Isotachis haematodes	J	(47)
I. humectata	J	(45)
Jungermannia exsertifolia subsp. *cordifolia*	J	(284)
J. subulata	J	(453)
Lejeunea albescens	J	(219)
L. discreta	J	(219)
L. glaucescens	J	(219)
Lejeunea sp.	J	(47)
Lepidolejeunea ornata	J	(219)
Lepidozia reptans	J	(147, 284)
Leucolejeunea aff. *decurrens*	J	(219)
Lophocolea coadunata	J	(47)
Lopholejeunea howei	J	(219)
L. subfusca	J	(219)
Lophozia ventricosa	J	(284)
Makinoa crispata	Me	(622)
Mannia fragrans	M	(276)
Marchantia berteroana	M	(39)
M. diptera	M	(138)
M. foliacea	M	(39)
M. palmata	M	(72)
M. piicata	M	(47)
M. polymorpha	M	(61, 71)
Marchesinia brachiata	J	(219)
Mastigolejeunea humilis	J	(219)
M. undulata	J	(219)

Table IId (continued)

Structure number	Name of compounds	Formula	m.p. °C	[α]_D	Plant source	Order	References	Comments
					Mastigophora diclados	J	(138)	
					Metzgeria albinea	Me	(47)	
					Monoclea forsteri	Mo	(513)	
					M. gottschei	Mo	(514)	
					subsp. neotropica	Mo	(514)	
					Mylia taylorii	J	(532)	
					M. verrucosa	J	(532)	
					Neteroclada confluens	J	(45)	
					Pallavicinia lyellii	Me	(3)	
					Plagiochila asplenioides	J	(432)	
					P. cipaconensis	J	(48)	
					P. cucullata	J	(48)	
					P. dichotoma	J	(48)	
					P. dilatata	J	(48)	
					P. dura	J	(48)	
					P. engelii	J	(48)	
					P. falcata	J	(48)	
					P. friabilis	J	(48)	
					P. fuegiensis	J	(46)	
					P. gayana	J	(46)	
					P. guayrapurinensis	J	(48)	
					P. neesiana	J	(46)	
					P. parvidens	J	(46)	
					P. pulcherrima	J	(190)	
					P. rosariensis	J	(48)	

	P. tenerrima	J	(48)
	Pleurozia gigantea	J	(52)
	Porella densifolia subsp. appendiculata	J	(67)
	P. elegantula	J	(39)
	P. platyphylla	J	(284)
	P. squamurifera	J	(47)
	P. swartziana	J	(47)
	Radula kojana	J	(50)
	Reboulia hemisphaerica	M	(411)
	Riccardia prehensilis	Me	(45)
	Roivainenia jacquinotii	J	(45)
	Schiffneria hyalina	J	(622)
	Schiffneriolejeunea nymannii	J	(219)
	S. omphalanthoides	J	(219)
	Stictolejeunea balfourii var. bekkei	J	(219)
	S. squamata	J	(219)
	Symbiezidium barbiflorum	J	(219)
	S. transversale var. hookeriana	J	(219)
	Syzygiella anomala	J	(47)
	Triandrophyllum subtrifidum var. trifidum	J	(45)
	Tritomaria quinquedentata	J	(429, 567a)
	Thysananthus amazonicus	J	(219)
	T. convolutus	J	(219)
	T. fruticosus	J	(219)
	T. mollis	J	(219)
	Tylimanthus urvilleanus	J	(45)
	Isotachis japonica	J	(76)
	Jungermannia infusca	J	(436)
(847)	Stigmasteryl glucoside $C_{35}H_{58}O_6$		

Table IId (continued)

Structure number	Name of compounds	Formula	m.p. °C	[α]_D	Plant source	Order	References	Comments
(848a)	Sitosterol (24α-ethyl)	C$_{29}$H$_{50}$O			Acrolejeunea pusilla	J	(219)	
					A. torulosa	J	(219)	
					Anastrepta orcadensis	J	(284)	
					Aneura latissima	J	(47)	
					Archilejeunea mariana	J	(219)	
					A. olivacea	J	(219)	
					Barbilophozia hatcheri	J	(284)	
					Bazzania sp.	J	(138)	
					B. spiralis	J	(341)	
					Blepharolejeunea densifolia	J	(219)	
					B. incongrua	J	(219)	
					Bryopteris diffusa	J	(219)	
					Clasmatocolea humilis	J	(45)	
					Conocephalum conicum	M	(138, 463)	
					Dumortiera hirsuta	M	(47)	
					Frullania brasiliensis	J	(47)	
					F. davurica	J	(420)	
					F. falciloba	J	(64)	
					F. jackii	J	(420)	
					Gackstroemia magellanica	J	(45)	
					Jungermannia exsertifolia subsp. cordifolia	J	(284)	
					Lejeunea discrenata	J	(219)	
					L. glaucescens	J	(219)	
					Lejeunea sp.	J	(47)	
					Lepidozia reptans	J	(147, 284)	

Leucolejeunea aff. decurrens	J	(219)
Lophozia ventricosa	J	(284)
Makinoa crispata	Me	(622)
Mannia fragrans	M	(276)
Marchantia berteroana	M	(39)
M. diptera	M	(138)
M. foliacea	M	(391)
M. palmata	M	(72)
M. polymorpha	M	(61, 71, 463)
Marchesinia brachiata	J	(219)
Mastigolejeunea humilis	J	(219)
Mastigophora diclados	J	(138)
M. undulata	J	(219)
Metzgeria albinea	Me	(47)
Mylia taylorii	J	(532)
M. verrucosa	J	(532)
Neteroclada confluens	J	(45)
Pallavicinia lyellii	Me	(3)
Plagiochila chacabucensis	J	(46)
P. dura	J	(46)
P. engelii	J	(46)
P. parvidens	J	(46)
Pleurozia gigantea	J	(52)
Porella densifolia subsp. appendiculata	J	(67)
P. elegantula	J	(39)
P. platyphylla	J	(284)
Riccardia prehensilis	Me	(45)

Table IId (*continued*)

Structure number	Name of compounds	Formula	m.p. °C	[α]$_D$	Plant source	Order	References	Comments
					Radula kojana	J	(50)	
					Schiffneria hyalina	J	(622)	
(848b)	Clionasterol (24β-ethyl)	C$_{29}$H$_{50}$O			*Marchantia polymorpha*	M	(138, 463)	
					Mastigophora diclados	J	(138)	
(849)	Cholesterol	C$_{27}$H$_{46}$O			*Mylia taylorii*	J	(532)	
					M. verrucosa	J	(532)	
(850)	Brassicasterol	C$_{28}$H$_{46}$O			*Mylia taylorii*	J	(532)	
					M. verrucosa	J	(532)	
(851)	Squalene	C$_{30}$H$_{50}$			*Leptoscyphus liebmanianus*	J	(47)	
					Makinoa crispata	Me	(622)	
					Marchantia plicata	M	(47)	
					Pallavicinia lyellii	Me	(47)	
					Plagiochila beskeana	J	(48)	
					P. cipaconensis	J	(48)	
					P. corniculata	J	(48)	
					P. cristatissima	J	(48)	
					P. cucullata	J	(48)	
					P. dichotoma	J	(48)	
					P. dilatata	J	(48)	
					P. excisa	J	(48)	
					P. falcata	J	(48)	
					P. friabilis	J	(48)	
					P. guayrapurinensis	J	(48)	
					P. kroneana	J	(48)	
					P. oresitropha	J	(48)	
					P. oxyphylla	J	(48)	

No.	Compound	Formula	mp	[α]	Species		Ref.
					P. parvitexta	J	(48)
					P. tambillensis	J	(48)
					P. tenerrima	J	(48)
					P. trabeculata	J	(589)
					Porella cordaeana	J	(593)
					P. swartziana	J	(47)
					Syzygiella anomala	J	(47)
					Triandrophyllum subtrifidua var. *trifidum*	J	(45)
					Thysananthus amazonicus	J	(219)
					T. convolutus	J	(219)
					T. mollis	J	(219)
					Wiesnerella denudata	M	(51)
(852)	(+)-21α-Methoxyserrat-14-en-3-one	$C_{31}H_{50}O_2$	212–5		*Nardia scalaris*	J	(101)
(853)	Friedelin	$C_{30}H_{50}O$			*Bazzania japonica*	J	(84)
(854)	Cycloartenol	$C_{30}H_{50}O$	114.5–116	+46.3	*Lophozia ventricosa*	J	(284)
					Mastigophora diclados	J	(138)
					Mylia taylorii	J	(532)
(855)	Cycloart-23-en-3β,25-diol	$C_{30}H_{50}O_2$			*Plagiochila peculiaris*	J	(626)
(856)	Diploptene (= Hop-22(29)-ene)	$C_{30}H_{50}$			*Adelanthus lindenbergianus*	J	(45)
					Plagiochila bispinosa	J	(46)
					P. elata	J	(46)
					P. lecheri	J	(46)
(857)	Diplopterol (= Hop-22-ol)	$C_{30}H_{52}O$	256–9		*Conocephalum conicum*	M	(553)

Table IId (continued)

Structure number	Name of compounds	Formula	m.p. °C	$[\alpha]_D$	Plant source	Order	References	Comments
					C. japonicum	M	(584)	
					Diplasiolejeunea patelligera	J	(219)	
(858)	α-Zeorin (= Hop-6α,22-diol)	$C_{30}H_{52}O_2$	224-6	+ 56	Conocephalum japonicum	M	(584)	
					Plagiochasma rupestre	M	(243)	
			222-4	+ 50	Reboulia hemisphaerica	M	(411, 412)	
(859)	Hop-22,23-diol	$C_{30}H_{52}O_2$			Conocephalum japonicum	M	(585)	
(860)	Hop-12β,22-diol	$C_{30}H_{52}O_2$			Mannia subpilosa	M	(613)	
(861)	Hop-6α,11α,22-triol	$C_{30}H_{52}O_3$			Reboulia hemisphaerica	M	(249)	
(862)	Methyl 3α-hydroxy- 18-oleanen-28-oate	$C_{31}H_{50}O_3$			Frullania sp.	J	(441a)	

J: Jungermanniales; M: Marchantiales; Me: Metzgeriales; Mo: Monocleales

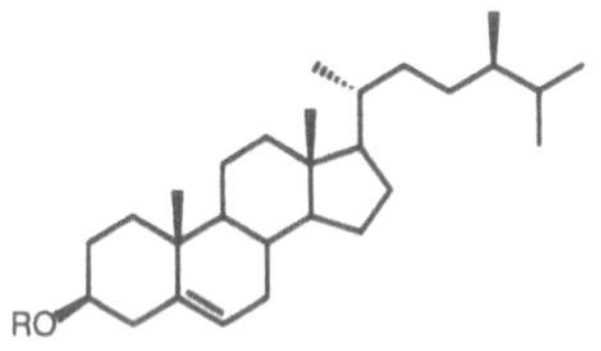

(844a) Campesterol (24α Methyl) ; R=H
(844b) Dihydrobrassicasterol (24β Methyl) ; R=H
(845) Campesteryl behenate ; R=C$_{21}$H$_{43}$CO

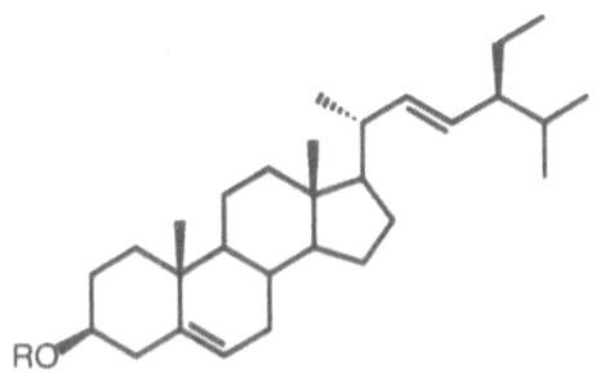

(846a) Stigmasterol (24α-Ethyl) ; R=H
(846b) Stigmasterol (24β-Ethyl) ; R=H
(847) Stigmasteryl glucoside ; R=Glc.

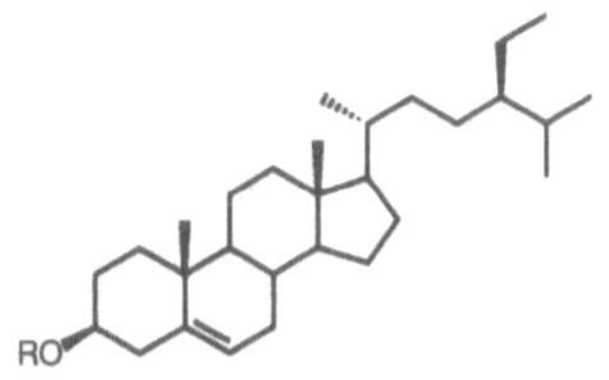

(848a) Sitosterol (24α-Ethyl) ; R=H
(848b) Clionasterol (24β-Ethyl) ; R=H

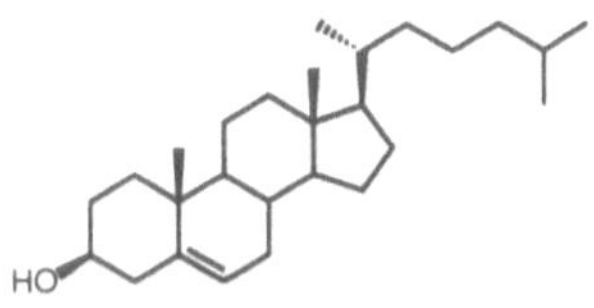

(849) Cholesterol

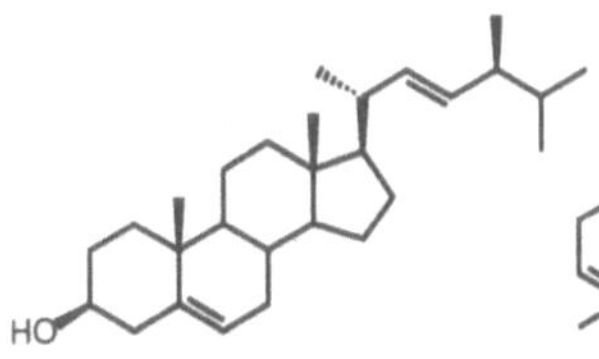

(850) Brassicasterol

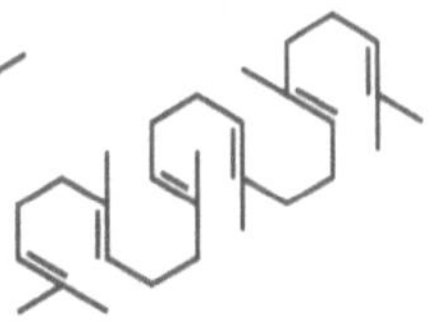

(851) Squalene

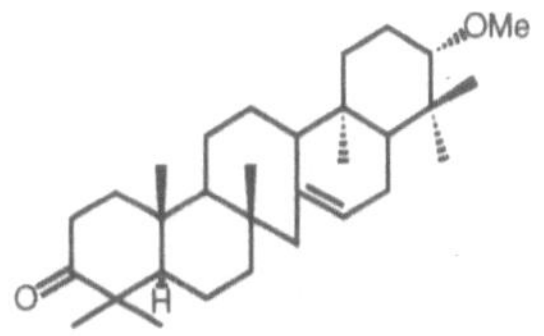

(852) (+)-21α-Methoxyserrat-14-en-3-one

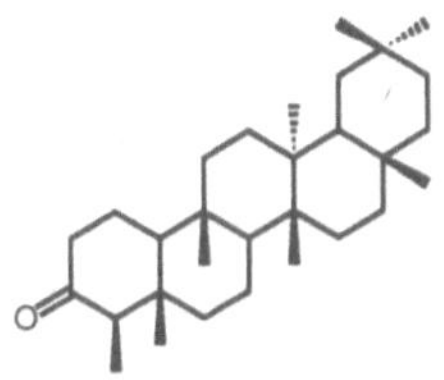

(853) Friedelin

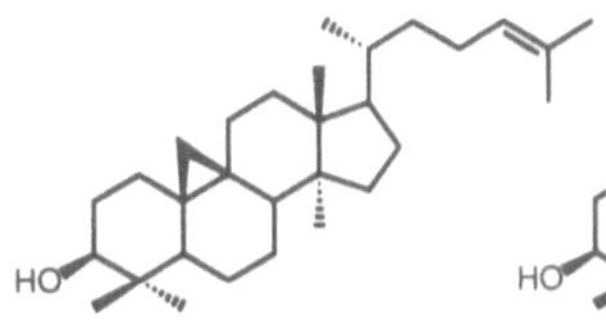

(854) Cycloartenol

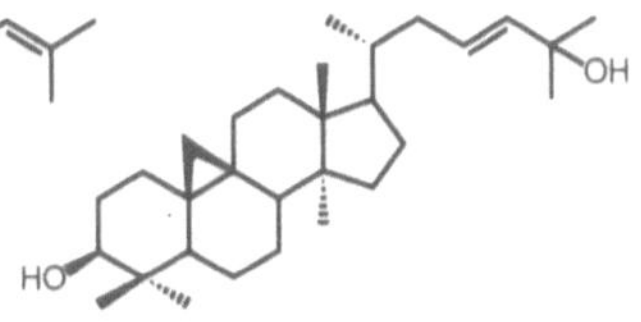

(855) Cycloart-23-en-3β,25-diol

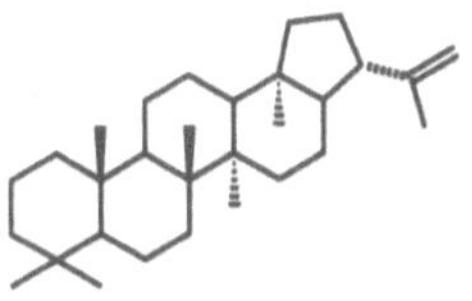

(856) Diploptene (=Hop-22(29)-ene)

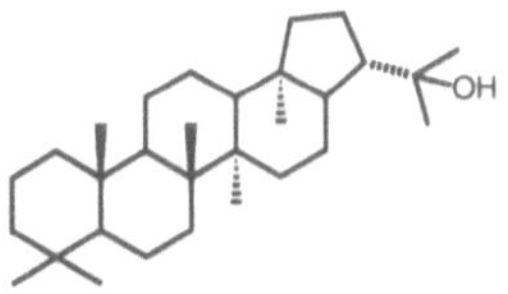

(857) Diplopterol (=Hop-22-ol)

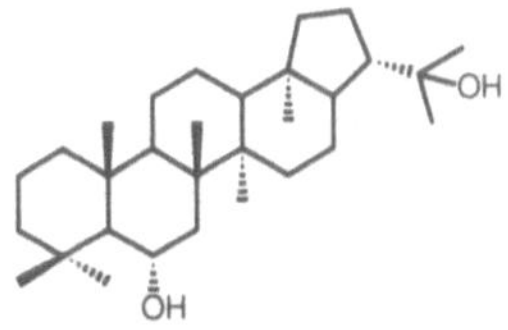

(858) α-Zeorin

Chart 59a. Steroids and triterpenoids found in the Hepaticae

Mylia taylorii contained steryl esters in 14% yield of the ethanol extract while from *M. verrucosa*, a mixture of steryl esters was isolated in 7.9% yield (*532*). Each mixture was subjected to methanolysis to obtain fatty acid methyl esters and sterols. The composition of the fatty acid methyl esters was determined by GC and GC-MS. The sterols were obtained by LiAlH$_4$ reduction of each steryl ester and analyzed by GC and GC-MS.

Campesterol (**844a**), stigmasterol (**846a**), cholesterol (**847**), sitosterol (**848a**) and brassicasterol (**850**), were detected in *M. taylorii*, while *M. verrucosa* contained the same sterols save cholesterol. *Calypogeia* species contain sterols esterified with oleic, linoleic and linolenic acids (*315*). The level of C-20 polyunsaturated fatty acids was significantly small. The liverworts studied so far are very different from the mosses with respect to the proportion of C20 polyunsaturated fatty acids in steryl esters. The function of steryl esters is still obscure.

CHIU et al. (*138*) investigated the sterol composition of four liverworts, a *Bazzania* species, *Conocephalum conicum*, *Marchantia diptera* and *Mastigophora diclados*. The major sterols were 24-methylcholesterol (**844a, 844b**), 24-ethylcholest-5,22-dien-3-ol (**846a, 846b**) and 24-ethylcholest-5-en-3-ol (**848a, 848b**). The 24α- and 24β-epimers of several sterols have been distinguished by high resolution ^{1}H NMR spectral data and capillary GC. The 24-methylcholesterols in all liverworts are a mixture of campesterol (the 24α-epimer) (**844a**) and 22-dihydrobrassicasterol (the 24β-epimer) (**844b**) with campesterol making up 40–80% of the 24-methylcholesterol fraction. Sitosterol (**848a**), the 24α-epimer, was the only 24-ethylcholesterol in *Conocephalum conicum*, but clionasterol (**848b**), the 24β-epimer, has been found in *Marchantia diptera* (40%) and in *Mastigophora diclados* (10%). On the other hand, stigmasterol (**846a**), 24α-epimer, was the only epimer of 24-ethyl-cholest-5,22-dien-3β-ol. More than 30 other species of the liverworts have been studied; however, clionasterol (**848b**) has not been previously reported in liverworts (*138*). Stigmasteryl-3β-glucoside (**847**) has been isolated from *Isotachis japonica* (*76*) and *Jungermannia infusca* (*436*). This is the first time that a steryl glucoside has been reported from bryophytes.

Reports of triterpenoids in the Hepaticae are comparatively rare. The last review (*19*) contained only references to one ursane-, three friedelane- and three hopane-type triterpenoids from Hepaticae (*19*); since then chemical constituents of about 700 species of Hepaticae have been investigated by GC-MS. Almost all species contain squalene (**851**) (*37*). BENES et al. (*101*) reported the presence of (+)-21α-methoxyserrat-14-en-3-one (**852**) in *Nardia scalaris*. The very low content of (**852**) (approx. 10 ppm) suggests that it might originate from admixture of humus to the liverwort collection. Triterpenoids of this type were originally isolated from clubmosses (Lycopodiaceae), ferns (Polypodiaceae) and quite often from conifers.

From *Bazzania japonica*, friedelin (**853**) has been obtained in the pure state (*84*). Cycloartenol (**854**) made up more than 60% of the methyl sterols of *Mastigophora diclados* (*138*). Some *Bazzania* species (*138*), *Lophozia ventricosa* (*284*) and *Mylia taylorii* (*532*), produce cycloartenol (**854**). Wu (*626*) reported the presence of cycloart-23-en-3β,25-diol (**855**) in *Plagiochila peculiaris*. Diploptene [= hop-22(29)-ene] (**856**) has been found in four South American liverworts, *Adelanthus lindenbergianus*, *Plagiochila bispinosa*, *P. elata* and *P. lecheri* (*46*). *Conocephalum conicum* (*553*), *C. japonicum* (*584*) and *Diplasiolejeunea patelligera* (*219*) furnished diplopterol (= hop-22-ol) (**857**). α-Zeorin (= 6α,22-dihydroxyhopane) (**858**) which has been obtained from various lichens (*152*), has also been isolated from *Conocephalum japonicum* (*584*), *Plagiochasma rupestre* (*243*) and European *Reboulia hemisphaerica* (*411, 412*). *C. japonicum* also yielded hop-22,23-diol (**859**) (*585*). Hop-12β,22-diol (**860**) has been isolated from *Mannia subpilosa* (*613*). Japanese *R. hemisphaerica* produces hop-6α,11α,22-triol (**861a**) whose structure has been determined by a combination of 600 MHz NMR spectrometry and chemical reaction [preparation of diacetate (**861b**), dibenzoate (**861c**) and 8,11-diketone (**861d**)] (*249*). A new oleanane-type triterpene, methyl 3α-hydroxy-18-oleanen-28-oate (**862**) has been isolated from an unidentified *Frullania*

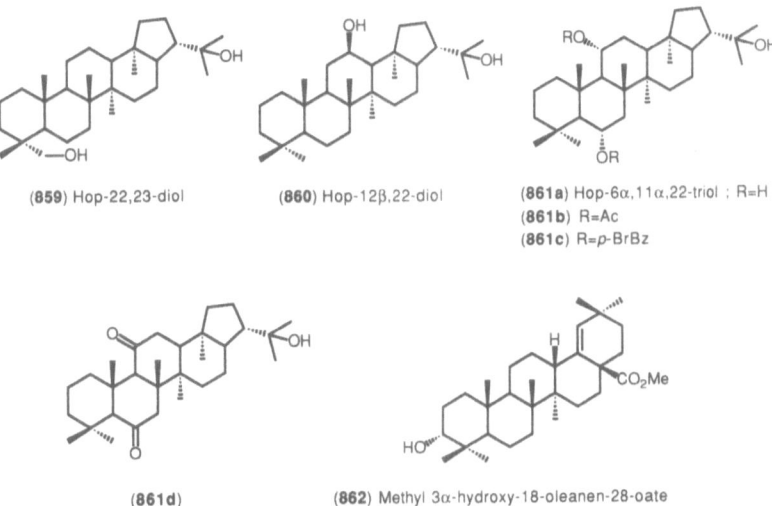

(**859**) Hop-22,23-diol   (**860**) Hop-12β,22-diol   (**861a**) Hop-6α,11α,22-triol ; R=H
(**861b**) R=Ac
(**861c**) R=*p*-BrBz

(**861d**)   (**862**) Methyl 3α-hydroxy-18-oleanen-28-oate

Chart 59b. Triterpenoids found in the Hepaticae and their derivatives

species collected in Venezuela and its stereochemistry determined by 600 MHz NMR spectrometry (*441a*).

8. Aromatic Compounds

8.1 Benzoic Acid and Cinnamic Acid Derivatives

The earlier review (*19*) listed three substituted benzoic acid and thirteen benzoates from liverworts. Since then 2-methoxybenzyl benzoate (**863**) has been isolated from *Balantiopsis rosea*, together with benzyl benzoate (**864**) and β-phenylethyl benzoate (**865**) (*65*). The occurrence of (**863**) in nature is very rare; it has also been found in *Uvaria purpurea* (Annonaceae) (*337*). South American *Isotachis haematodes, Balantiopsis erinacea* and *B. cancellata* also elaborate benzyl benzoate (**864**) (*45*). *B. erinacea* contains β-phenylethyl benzoate (**865**) (*45*).

The three prenyl benzoates, trichocolein (**867**), tomentellin (**868**) and isotomentellin (**869**) which have been isolated previously from Japanese and European *Trichocolea tomentella* (*19*) have more recently been obtained from East Malaysian *T. pluma*, along with methyl 3,4-dimethoxybenzoate (**866**) (*51*). Benzoate (**868**) has also been isolated from Taiwanese *T. pluma* (*133*) while Taiwanese *T. tomentella* contained the benzoates (**867, 868**) (*622*). New Zealand *T. mollissima* produces tomentellol (**870**) (*585a*). Tomentellin (**868**) has been synthesized from geranyl acetate (**8**) (*462*).

Eight cinnamic acid derivatives from liverworts were listed earlier (*19*). Four of these were cinnamates from *Isotachis japonica*. South American *I. humectata* and *I. haematodes* also furnished benzyl *trans*-cinnamate (**873**) (*45, 47*) while phenyl cinnamate (**872**) was detected in *I. humectata* by GC-MS (*45*). South American *Balantiopsis erinacea* gave β-phenylethyl cinnamate (**875**) (*45*). New Zealand *B. rosea* produced benzyl *trans*-cinnamate (**873**), benzyl *cis*-cinnamate (**874**), β-phenylethyl dihydrocinnamate (**876**) and β-phenylethyl cinnamate (**875**) (*65*) which were isolated earlier from Japanese *Isotachis japonica* (*19*). Methyl cinnamate (**871**) has been isolated from the essential oil of the male thallus of *Conocephalum conicum* (*580*). Thus benzoic and cinnamic acid derivatives appear to be restricted to a few of the liverworts examined so far.

8.2 Bibenzyls

The earlier review listed fifteen bibenzyls lacking a prenyl group and twelve bibenzyls with a prenyl group as having been isolated from

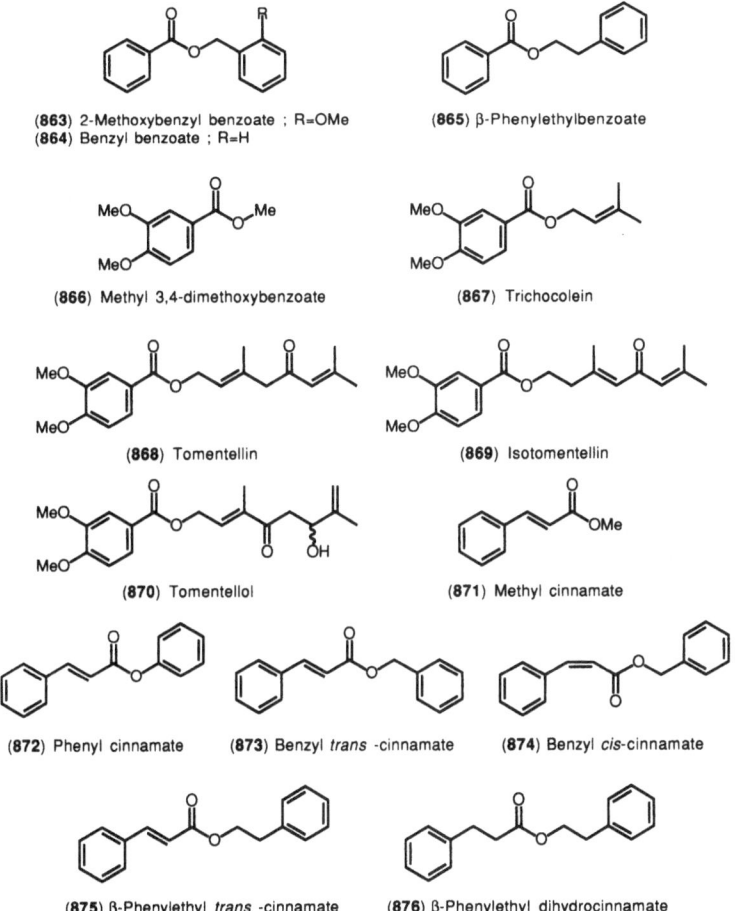

(**863**) 2-Methoxybenzyl benzoate ; R=OMe
(**864**) Benzyl benzoate ; R=H

(**865**) β-Phenylethylbenzoate

(**866**) Methyl 3,4-dimethoxybenzoate

(**867**) Trichocolein

(**868**) Tomentellin

(**869**) Isotomentellin

(**870**) Tomentellol

(**871**) Methyl cinnamate

(**872**) Phenyl cinnamate   (**873**) Benzyl *trans* -cinnamate   (**874**) Benzyl *cis*-cinnamate

(**875**) β-Phenylethyl *trans* -cinnamate   (**876**) β-Phenylethyl dihydrocinnamate

Chart 60. Benzoic and cinnamic acid derivatives found in the Hepaticae

liverworts (*19*). A number of different bibenzyls have since been isolated not only from Jungermanniales but also Marchantiales as shown in Table IIe.

Lunularic acid (**877**), a dormancy inducing bibenzyl derivative, is widely distributed in liverworts. This carboxylic acid was also isolated from an axenic cultured *Ricciocarpos natans* (*351*) and suspension cultures of *Marchantia polymorpha*, *Jungermannia subulata* and *Lophocolea heterophylla* grown in Murashige-Skoog's-2 medium (*1, 450*). The callus of *M. polymorpha* and *Calypogeia tosana* also produces (**877**). The

Table IIe. *Aromatic Compounds Found in the Hepaticae*

Structure number	Name of compounds	Formula	m.p.°C	$[\alpha]_D$	Plant source	Order	References	Comments
(863)	2-Methoxybenzyl benzoate	$C_{15}H_{14}O_3$			*Balantiopsis rosea*	J	(65)	
(864)	Benzyl benzoate	$C_{14}H_{12}O_2$			*Balantiopsis cancellata*	J	(45)	
					B. erinacea	J	(45)	
					B. rosea	J	(65)	
					Isotachis haematodes	J	(47)	
(865)	β-Phenylethyl benzoate	$C_{15}H_{14}O_2$			*Balantiopsis erinacea*	J	(45)	
					B. rosea	J	(65)	
(866)	Methyl 3,4-dimethoxy-benzoate	$C_{10}H_{12}O_4$			*Trichocolea pluma*	J	(51)	
(867)	Trichocolein	$C_{14}H_{18}O_4$			*Trichocolea pluma*	J	(51, 133, 626)	
(868)	Tomentellin	$C_{19}H_{24}O_5$			*Trichocolea pluma*	J	(51)	
					T. tomentella	J	(622)	
(869)	Isotomentellin	$C_{19}H_{24}O_5$			*Trichocolea pluma*	J	(51)	
					T. tomentella	J	(622)	
(870)	Tomentellol	$C_{19}H_{24}O_6$			*Trichocolea mollissima*	J	(585a)	
(871)	Methyl cinnamate	$C_{10}H_{10}O_2$			*Conocephalum conicum*	M	(580)	
					Trichocolea pluma	J	(51)	
(872)	Phenyl cinnamate	$C_{15}H_{12}O_2$			*Isotachis humectata*	J	(45)	
(873)	Benzyl trans-cinnamate	$C_{16}H_{14}O_2$			*Balantiopsis rosea*	J	(65)	
					Isotachis haematodes	J	(47)	
					I. humectata	J	(45)	
(874)	Benzyl cis-cinnamate	$C_{16}H_{14}O_2$			*Balantiopsis rosea*	J	(65)	

No.	Compound	Formula	Rotation	Species	Type	References
(875)	β-Phenylethyl trans-cinnamate	$C_{17}H_{16}O_2$		Balantiopsis erinacea	J	(45)
				B. rosea	J	(65)
(876)	β-Phenylethyl dihydro-cinnamate	$C_{17}H_{18}O_2$		Isotachis humectata	J	(45)
				Balantiopsis rosea	J	(65)
(877)	Lunularic acid	$C_{15}H_{14}O_4$		Calypogeia tosana	J	(1, 450)
				Jungermannia subulata	J	(1, 450)
				Lophocolea heterophylla	J	(1, 450)
				Marchantia polymorpha	M	(1, 450)
				Ricciocarpos natans	M	(351)
				Dumortiera hirsuta	M	(585)
				Marchantia berteroana	M	(39)
				M. chenopoda	M	(553)
				M. paleacea var. diptera	M	(74)
				M. polymorpha	M	(71, 74)
				Ricciocarpos natans	M	(639, 640)
(878)	Lunularin	$C_{14}H_{14}O_2$		Marchantia polymorpha	M	(449, 450, 453)
(879a)	Prelunularic acid	$C_{15}H_{16}O_5$	− 41.6	Ricciocarpos natans	M	(351)
(881)	5,4'-Dihydroxy-bibenzyl-2-O-β-D-glucopyranoside	$C_{20}H_{24}O_8$				
(882)	5,3',4'-Trihydroxy-bibenzyl-2-O-β-D-glucopyranoside	$C_{20}H_{24}O_9$	− 39	Ricciocarpos natans	M	(351)
(883)	2-Carboxy-4,3'-dihydroxybibenzyl-3-O-β-D-glucopyranoside	$C_{21}H_{24}O_{10}$	− 20.4	Ricciocarpos natans	M	(351)
(884)	4-p-(Methoxyphenyl-ethyl)-cyclohex-2-en-1-one	$C_{15}H_{18}O_2$		Plagiochila longispina	J	(506)

Table IIe (continued)

Structure number	Name of compounds	Formula	m.p.°C	$[\alpha]_D$	Plant source	Order	References	Comments
(885)	3-Hydoxybibenzyl	$C_{14}H_{14}O$			Radula frondescens	J	(47)	
(886)	3-Methoxybibenzyl	$C_{15}H_{16}O$			Radula buccinifera	J	(66)	
					R. frondescens	J	(47)	
					R. javanica (= R. variabilis)	J	(49)	
(887)	3-Hydroxy-4-methoxybibenzyl	$C_{15}H_{16}O_2$			Radula frondescens	J	(47)	
(888)	3-Methoxy-4-hydroxybibenzyl	$C_{15}H_{16}O_2$			Radula frondescens	J	(47)	
(889)	3,5-Dimethoxy-bibenzyl	$C_{16}H_{18}O_2$			Radula javanica (= R. variabilis)	J	(49, 66)	
(890)	3-Hydroxy-4,5-methylenedioxy-bibenzyl	$C_{15}H_{14}O_3$			Radula javanica (= R. variabilis)	J	(49, 66)	
(891)	3-Methoxy-4'-hydroxybibenzyl	$C_{15}H_{16}O_2$			Frullania davurica	J	(56, 420)	
					F. jackii	J	(420)	
					Plagiochila stephensoniana	J	(39, 385a)	
					P. subdura	J	(46)	
					Radula frondescens	J	(47)	
(892)	3,4'-Dimethoxy-bibenzyl	$C_{16}H_{18}O_2$			Frullania davurica	J	(56)	
					F. falciloba	J	(64)	
					F. jackii	J	(420)	
					Plagiochila stephensoniana	J	(39)	

No.	Name	Formula	mp	Species		Ref.
(893)	3-Hydroxy-4'-methoxybibenzyl	$C_{15}H_{16}O_2$		*P. subdura*	J	*(46)*
				Radula buccinifera	J	*(66)*
				R. complanata	J	*(66)*
				R. frondescens	J	*(47)*
				Frullania davurica	J	*(585)*
(894)	3,4'-Dimethoxy-4-hydroxybibenzyl	$C_{16}H_{18}O_3$		*Frullania davurica*	J	*(420, 585)*
				F. jackii	J	*(420)*
(895)	3-Hydroxy-4,3'-dimethoxybibenzyl	$C_{16}H_{18}O_3$	81–2	*Frullania falciloba*	J	*(64)*
(896)	Pellepiphyllin (= 2-Hydroxy-3,4'-dimethoxybibenzyl)	$C_{16}H_{18}O_3$		*Frullania davurica*	J	*(420)*
				F. jackii	J	*(420)*
(897)	3,4,4'-Trimethoxy-2-hydroxybibenzyl	$C_{17}H_{20}O_4$		*Plagiochila fuegiensis*	J	*(46)*
(898)	3,4-Methylenedioxy-3'-methoxybibenzyl	$C_{16}H_{16}O_3$	49–50	*Frullania ericoides*	J	*(56)*
				F. falciloba	J	*(64)*
				Trocholejeunea sandvicensis	J	*(82)*
(899)	3,3'-Dimethoxy-4,5-methylenedioxy-4'-hydroxybibenzyl	$C_{17}H_{18}O_5$	62–3	*Frullania bonincola*	J	*(56)*
(900)	3,3'-Dimethoxy-4,5-methylenedioxy-bibenzyl	$C_{17}H_{18}O_4$	135–7	*Frullania ericoides*	J	*(56)*
(901)	3-Methoxy-4,5-methylenedioxy-4'-hydroxybibenzyl	$C_{16}H_{16}O_4$		*Plagiochila chacabucensis*	J	*(46)*
(902)	3,4;3',4'-Dimethylene-dioxybibenzyl	$C_{16}H_{14}O_4$	132–3	*Frullania parvistipula*	J	*(56)*

Table IIe (*continued*)

Structure number	Name of compounds	Formula	m.p.°C	$[\alpha]_D$	Plant source	Order	References	Comments
(903)	3,3'-Dihydroxy-4,5,4',5'-dimethylenedioxy-bibenzyl	$C_{16}H_{14}O_6$			*Frullania parvistipula*	J	(56)	
(904)	3,4,5,3',4'-Penta-methoxybibenzyl	$C_{19}H_{24}O_5$			*Frullania serratta*	J	(51)	
(905)	Brittonin B	$C_{19}H_{22}O_6$			*Frullania serratta*	J	(51)	
(907)	1-Carbomethoxy-2,3-dihydroxy-4'-methoxybibenzyl	$C_{17}H_{18}O_5$			*Plagiochila spinulosa*	J	(478)	
(908)	3,4-Dihydroxy-5,3'-dimethoxybibenzyl	$C_{16}H_{18}O_4$			*Stictolejeunea balfourii* var. *bekkei*	J	(219)	
(913)	3-Hydroxy-4-(3-methyl-2-butenyl)bibenzyl	$C_{19}H_{22}O$			*Radula oyamensis*	J	(66)	
(914)	3-Hydroxy-5-methoxy-4-(3-methyl-2-butenyl)bibenzyl	$C_{20}H_{24}O_2$			*Radula oyamensis*	J	(66)	
(915)	3,5-Dihydroxy-4-(3-methyl-2-butenyl)bibenzyl	$C_{19}H_{22}O_2$			*Radula buccinifera* R. *complanata*	J J	(66) (66)	
(917)	3,5-Dihydroxy-4-(2,3-epoxy-3-methylbutyl)bibenzyl	$C_{19}H_{22}O_3$			*Radula complanata*	J	(66)	

No.	Name	Formula	mp	Source		Ref.
(918)	2-Geranyl-3,5-dihydroxybibenzyl	$C_{24}H_{30}O_2$	57–8	Radula complanata	J	(43, 66)
				R. frondescens	J	(47)
				R. javanica (= R. variabilis)	J	(49)
				R. kojana	J	(50)
				R. oyamensis	J	(66)
				R. tokiensis	J	(66)
				R. voluta	J	(47)
(920)	2-Geranyl-3,5,4'-tri-hydroxybibenzyl	$C_{24}H_{30}O_3$		Radula frondescens	J	(47)
(921a)	3,5-Dihydroxy-4-geranylbibenzyl	$C_{24}H_{30}O_2$		Radula sp.	J	(50)
(922)	2-(3,7-Dimethyl-2,7-octadienyl)-3,5-dihydroxy-bibenzyl	$C_{24}H_{30}O_2$		Radula oyamensis	J	(66)
(924)	3,4'-Dihydroxy-4-(3-methyl-2-butenyl)bibenzyl	$C_{19}H_{22}O_2$		Radula oyamensis	J	(66)
(925)	3-Hydroxy-4-(3-methyl-2-butenyl)-4'-methoxy-bibenzyl	$C_{20}H_{24}O_2$		Radula oyamensis	J	(66)
(926)	3-Methoxy-4-(3-methyl-2-butenyl)-4'-hydroxy-bibenzyl	$C_{20}H_{24}O_2$		Radula kojana	J	(50)
				R. oyamensis	J	(66)
(927)	3,5-Dihydroxy-2-(3-methyl-2-butenyl)bibenzyl	$C_{19}H_{22}O_2$	91.5–92.5	Lepidozia vitrea	J	(585a)
				Radula complanata	J	(43)
				R. frondescens	J	(47)
				R. kojana	J	(50)

Table IIe (continued)

Structure number	Name of compounds	Formula	m.p.°C	$[\alpha]_D$	Plant source	Order	References	Comments
(928)	3-Methoxy-5-hydroxy-2-(3-methyl-2-butenyl)bibenzyl	$C_{20}H_{24}O_2$	63–4		R. perrottetii R. voluta Radula kojana	J J J	(43) (47) (50)	
(929)	3-Hydroxy-5-methoxy-2-(3-methyl-2-butenyl)bibenzyl	$C_{20}H_{24}O_2$	40–1		Radula kojana	J	(50)	
(931)	3,5-Dihydroxy-2-(2,3-epoxy-3-methylbutyl)bibenzyl	$C_{19}H_{22}O_3$	90–1		Radula kojana	J	(50)	
(932)	3-Hydroxy-5-methoxy-2-(3-hydroxy-3-methyl-butyl)bibenzyl	$C_{20}H_{26}O_3$	97–8		Radula kojana	J	(50)	
(933)	2,2-Dimethyl-7-hydroxy-5-(2-phenylethyl)chromene	$C_{19}H_{20}O_2$			Radula kojana R. perrottetii	J J	(50) (43)	
(934)	2,2-Dimethyl-7-methoxy-5-(2-phenylethyl)chromene	$C_{20}H_{22}O_2$			Radula kojana	J	(43)	

No.	Name	Formula	mp	$[\alpha]$	Species		Ref.	CD data
(935)	2,2-Dimethyl-7,8-dihydroxy-5-(2-phenylethyl)chromene	$C_{19}H_{20}O_3$			*Radula perrottetii*	J	(43)	
(937)	2,2-Dimethyl-5-hydroxy-7-(2-phenyl-ethyl)chromene	$C_{19}H_{20}O_2$			*Radula kojana*	J	(50)	
(938)	2,2-Dimethyl-5-methoxy-7-(2-phenylethyl)chromene	$C_{20}H_{22}O_2$			*Radula kojana*	J	(50)	
(939)	2,2-Dimethyl-5-hydroxy-6-carboxy-7-(2-phenylethyl)chromene	$C_{20}H_{20}O_4$	122–3		*Radula kojana*	J	(50)	
(940)	2(S)-2-Methyl-2-(4-methyl-3-pentenyl)-7-hydroxy-5-(2-phenylethyl)chromene	$C_{24}H_{28}O_2$		+ 21.4	*Radula kojana*	J	(50)	CD 262 nm + 0.6 CD 295 nm + 0.3
(942)	2(S)-2-Methyl-2-(4-methyl-3-pentenyl)-6-carboxy-7-hydroxy-5-(2-phenylethyl)chromene	$_{25}H_{28}O_4$		+ 26.5	*Radula kojana*	J	(50)	CD 206 nm − 2.0 CD 250.5 nm + 1.3 CD 280 nm + 1.1
(943a)	6-Hydroxy-4-(2-phenylethyl)benzofuran	$C_{16}H_{14}O_2$	217–8		*Radula kojana*	J	(50)	

Table IIe (continued)

Structure number	Name of compounds	Formula	m.p.°C	$[\alpha]_D$	Plant source	Order	References	Comments
(945)	2(R)-Isopropenyl-6-hydroxy-4-(2-phenyl-ethyl)dihydro-benzofuran	$C_{19}H_{20}O_2$	109–110.5	−13.5	Radula perrottetii	J	(43)	
(946a)	Perrottetin D	$C_{19}H_{20}O_3$	82–3	−9.0	Radula perrottetii	J	(43)	CD 207 nm +0.29
					R. voluta	J	(47)	CD 235 nm −0.02
								CD 265 nm +0.06
(947)	3,5-Dihydroxy-6-carbomethoxy-2-(3-methyl-2-butenyl)bibenzyl	$C_{21}H_{24}O_4$	95–8		Radula perrottetii	J	(43)	
(952a)	Perrottetin A	$C_{19}H_{22}O_3$	99–100		Radula perrottetii	J	(43, 66)	
(953a)	Perrottetin B	$C_{19}H_{22}O_4$			Radula perrottetii	J	(43, 66)	
(954a)	Perrottetin C	$C_{19}H_{24}O_4$			Radula perrottetii	J	(43, 66)	
(957)	Tylimanthin A	$C_{20}H_{22}O_2$			Tylimanthus urvilleanus	J	(45)	
(958)	Tylimanthin B	$C_{19}H_{20}O_2$			Tylimanthus urvilleanus	J	(45)	
(959)	Tylimanthin C	$C_{21}H_{24}O_3$			Tylimanthus urvilleanus	J	(45)	
(960)	Radulanin A	$C_{19}H_{20}O_2$			Radula buccinifera	J	(66)	
					R. complanata	J	(19, 43)	
					R. javanica (= R. variabilis)	J	(49)	
					R. voluta		(47)	
(962)	Radulanin C	$C_{19}H_{20}O_3$			Radula buccinifera	J	(66)	

							CD
(963a)	Radulanin H	$C_{20}H_{20}O_4$	122–3		Radula complanata R. javanica (= R. variabilis)	J J	(66) (49)
(964)	4'-Hydroxy-radulanin H	$C_{20}H_{20}O_5$			Radula complanata	J	(542)
(965a)	Radulanin L	$C_{19}H_{20}O_3$	125–6		Radula complanata	J	(43)
(966)	Radulanin I (= 3-Methyl-6-hydroxy-8-(2-phenylethyl)-3,5-cyclopropano-chroman)	$C_{19}H_{20}O_2$		−12.8	Radula javanica (= R. variabilis)	J	(49)	CD 214 nm −0.28 CD 237 nm +0.07
(967)	Radulanin J (= 3-Methyl-6-methoxy-8-(2-phenylethyl)-3,5-cyclopropano-chroman)	$C_{20}H_{22}O_2$		+47.1	Radula javanica (= R. variabilis)	J	(49)	CD 215 nm +0.47 CD 230 nm −0.16
(968a)	Radulanin K (= 3-Methyl-6-hydroxy-7-carboxy-8-(2-phenyl-ethyl)-3,5-cyclo-propanochroman)	$C_{20}H_{20}O_4$	147–9	+7.6	Radula javanica (= R. variabilis)	J	(49)	CD 223 nm +0.57 CD 267 nm −0.21
(969)	Perrottetinene	$C_{24}H_{28}O_2$		−121.3	Radula perrottetii	J	(585a)
(985a)	Riccardin A	$C_{29}H_{26}O_4$			Riccardia multifida	Me	(19, 78, 79)

Table IIe (continued)

Structure number	Name of compounds	Formula	m.p. °C	$[\alpha]_D$	Plant source	Order	References	Comments
(986)	Riccardin F	$C_{29}H_{26}O_4$			*Blasia pusilla*	Me	(249)	
					Marchantia tosana	M	(74)	
(987a)	Riccardin B	$C_{28}H_{24}O_4$			*Preissia quadorata*	M	(249a)	
					Riccardia multifida	Me	(19, 78, 79)	
(988)	Riccardin C	$C_{28}H_{24}O_4$			*Blasia pusilla*	Me	(249)	
					Marchantia palmata	M	(72)	
					M. polymorpha	M	(61, 72)	
					Monoclea forsteri	Mo	(590)	
					Reboulia hemisphaerica	M	(55)	
					Ricciocarpos natans	M	(351)	
(989a)	Riccardin D	$C_{28}H_{24}O_4$			*Monoclea forsteri*	Mo	(590)	
(990)	Riccardin E	$C_{29}H_{26}O_4$			*Monoclea forsteri*	Mo	(590)	
(991)	Riccardin G	$C_{29}H_{26}O_4$			*Marchantia chenopoda*	M	(553)	
(992a)	Isoriccardin C	$C_{28}H_{24}O_4$			*Marchantia palmata*	M	(72)	
					M. polymorpha	M	(72)	
(993)	Marchantin A	$C_{28}H_{24}O_5$			*Marchantia paleacea* var. *diptera*	M	(74)	
					M. plicata	M	(433)	
					M. polymorpha	M	(22, 71, 72, 74, 78, 579)	
					M. tosana	M	(74)	
					Wiesnerella demudata	M	(51)	
(994)	Marchantin B	$C_{28}H_{24}O_6$			*Marchantia palmata*	M	(72)	
					M. polymorpha	M	(22, 71, 74, 78, 579)	
					M. tosana	M	(74)	

No.	Compound	Formula	Species		Reference
(995)	Marchantin C	$C_{28}H_{24}O_4$	Wiesnerella denudata	M	(51)
			Dumortiera hirsuta	M	(585)
			Marchantia chenopoda	M	(553)
			M. paleacea var. diptera	M	(74)
			M. palmata	M	(72)
			M. polymorpha	M	(22, 61, 71, 72, 74, 78, 579)
			M. tosana	M	(74)
			Monoclea forsteri	Mo	(513)
			M. gottschei subsp. neotropica	Mo	(514)
			Plagiochila sciophila (= P. acanthophylla subsp. japonica)	J	(259)
			Reboulia hemisphaerica	M	(614)
			Schistochila glaucescens	J	(562)
(996)	Marchantin C dimethyl ether	$C_{30}H_{28}O_4$	Reboulia hemisphaerica	M	(249)
(997)	Marchantin C monomethyl ether (= Marchantin O)	$C_{29}H_{26}O_4$	Reboulia hemisphaerica	M	(249, 623a)
(998)	Marchantin P	$C_{29}H_{26}O_4$	Marchantia chenopoda	M	(553)
(999)	Marchantin D	$C_{28}H_{24}O_6$	Marchantia paleacea var. diptera	M	(74)
			M. polymorpha	M	(21, 71, 72, 74, 78, 553, 579)
(1000a)	Marchantin E	$C_{29}H_{26}O_6$	Marchantia paleacea var. diptera	M	(74)
			M. polymorpha	M	(21, 71, 72, 74, 78, 579)

Table IIe (continued)

Structure number	Name of compounds	Formula	m.p. °C	[α]D	Plant source	Order	References	Comments
(1001)	Marchantin F	$C_{28}H_{24}O_7$			Marchantia paleacea var. diptera	M	(74)	
					M. polymorpha	M	(21, 71, 74, 78, 579)	
(1002)	Marchantin G	$C_{28}H_{22}O_6$			Marchantia paleacea var. diptera	M	(74)	
					M. palmata	M	(72)	
					M. polymorpha	M	(21, 71, 72, 74, 78, 579)	
(1003)	Marchantin H	$C_{28}H_{24}O_5$			Marchantia diptera	M	(620)	
					M. polymorpha	M	(61)	
					Plagiochasma intermedium	M	(550, 576)	
					Plagiochila sciophila (= P. acanthophylla subsp. japonica)	J	(259)	
(1004)	Marchantin I	$C_{29}H_{26}O_4$			Riccardia multifida	Me	(550, 576)	
(1005a)	Marchantin J	$C_{30}H_{28}O_6$			Marchantia polymorpha	M	(71)	
(1006a)	Marchantin K	$C_{29}H_{26}O_7$			Marchantia polymorpha	M	(71)	
(1007a)	Marchantin L	$C_{28}H_{24}O_6$			Marchantia polymorpha	M	(71)	
(1008)	Marchantinquinone	$C_{28}H_{22}O_5$	112		Mannia subpilosa	M	(612)	
(1009)	Marchantin M	$C_{29}H_{26}O_5$			Mannia subpilosa	M	(613, 623a)	
(1010)	Marchantin N	$C_{29}H_{24}O_6$			Mannia subpilosa	M	(613, 623a)	
(1011a)	Isomarchantin C	$C_{28}H_{24}O_4$	216–8		Bryopteris fricina	J	(432)	
					Marchantia palmata	M	(72)	
					M. polymorpha	M	(72)	
					Mylia nuda	J	(622, 624)	

No.	Name	Formula	mp	Species		Ref.
(1012)	Pakyonol	$C_{29}H_{26}O_4$	185–6	Mannia fragrans	M	(280)
(1013a)	Neomarchantin A	$C_{28}H_{24}O_4$		Monoclea forsteri	Mo	(513)
				M. gottschei subsp. neotropica	Mo	(614)
				Preissia quadorata	M	(249a)
				Schistochila glaucescens		(562)
(1014a)	Neomarchantin B	$C_{28}H_{24}O_5$		Schistochila glaucescens	J	(562)
(1015a)	Plagiochin A	$C_{29}H_{26}O_6$		Plagiochila sciophila (= P. acanthophylla subsp. japonica)	J	(259)
(1016a)	Plagiochin B	$C_{29}H_{26}O_5$		Plagiochila sciophila (= P. acanthophylla subsp. japonica)	J	(259)
(1017a)	Plagiochin C	$C_{29}H_{26}O_5$		Plagiochila sciophila (= P. acanthophylla subsp. japonica)	J	(259)
(1018a)	Plagiochin D	$C_{29}H_{26}O_4$		Plagiochila sciophila (= P. acanthophylla subsp. japonica)	J	(259)
(1019)	Isoplagiochin A	$C_{28}H_{22}O_4$		Plagiochila fruticosa	J	(249, 251a, 262b, 262c)
(1020)	Isoplagiochin B	$C_{28}H_{22}O_5$		Plagiochila fruticosa	J	(249, 251a, 262b, 262c)
(1021)	Isoplagiochin C	$C_{28}H_{22}O_4$		Plagiochila fruticosa	J	(262b)
(1022a)	Isoplagiochin D	$C_{28}H_{24}O_4$		Plagiochila fruticosa	J	(262b)
(1023a)	Perrottetin E	$C_{28}H_{26}O_4$		Jungermannia comata	J	(437)
				Marchantia polymorpha	M	(72)
				Monoclea forsteri	Mo	(513, 590)
				M. gottschei subsp. neotropica	Mo	(514)
				Nardia subclavata	J	(585)

Table IIe (continued)

Structure number	Name of compounds	Formula	m.p. °C	$[\alpha]_D$	Plant source	Order	References	Comments
					Pellia endiviifolia	Me	(254)	
					P. epiphylla	Me	(159a)	
					Plagiochila sciophila (= *P. acanthophylla* subsp. *japonica*)	J	(259)	
(1024a)	Perrottetin F	$C_{28}H_{26}O_5$			*Radula kojana*	J	(50)	
					R. perrottetii	J	(43, 598)	
					Lumularia cruciata	M	(249, 255)	
(1025)	Perrottetin G	$C_{29}H_{28}O_5$			*Radula kojana*	J	(50)	
					R. perrottetii	J	(43, 598)	
(1026)	Perrottetin E-11'-methyl ether	$C_{29}H_{28}O_4$			*Pellia endiviifolia*	Me	(254)	
(1027)	14-Hydroxy-perrottetin E-11'-methyl ether	$C_{29}H_{28}O_5$			*Pellia endiviifolia*	Me	(254)	
(1028)	Perrottetin H	$C_{24}H_{26}O_6$			*Jubula japonica*	J	(585b)	
(1029)	14-Hydroxy-perrottetin E	$C_{28}H_{26}O_5$			*Pellia epiphylla*	Me	(159a)	
(1030)	14'-Hydroxy-perrottetin E	$C_{28}H_{26}O_5$			*Pellia epiphylla*	Me	(159a)	
(1031)	14,14'-Dihydroxy-perrottetin E	$C_{28}H_{26}O_6$			*Pellia epiphylla*	Me	(159a)	
(1032)	7',8'-Dehydro-perrottetin F	$C_{29}H_{24}O_5$			*Lumularia cruciata*	M	(255)	

	Name	Formula	Species		Ref.
(1033)	Isoperrottetin A	$C_{28}H_{26}O_4$	*Radula perrottetii*	J	(585a)
(1034)	Paleatin A	$C_{30}H_{30}O_7$	*Marchantia paleacea var. diptera*	M	(249, 255)
(1035)	Paleatin B	$C_{29}H_{28}O_6$	*Marchantia paleacea var. diptera*	M	(249, 255)
(1036)	Cruciatin	$C_{56}H_{48}O_{10}$	*Lunularia cruciata*	M	(249)
(1037a)	Pusilatin A	$C_{56}H_{46}O_8$	*Blasia pusilla*	Me	(262a–c)
(1038a)	Pusilatin B	$C_{56}H_{46}O_8$	*Blasia pusilla*	Me	(262a–c)
(1039a)	Pusilatin C	$C_{56}H_{46}O_8$	*Blasia pusilla*	Me	(262a–c)
(1040a)	Pusilatin D	$C_{56}H_{46}O_8$	*Blasia pusilla*	Me	(262a–c)
(1043)	3-Undecylphenol	$C_{17}H_{28}O$	*Schistochila appendiculata*	J	(53)
(1044)	6-Undecylsalicylic acid	$C_{18}H_{28}O_3$	*Schistochila appendiculata*	J	(53)
(1045)	Potassium 6-undecyl salicylate	$C_{18}H_{27}O_3K$	*Schistochila appendiculata*	J	(53)
(1046)	3-Tridecylphenol	$C_{19}H_{32}O$	*Schistochila appendiculata*	J	(53)
(1047)	6-Tridecylsalicylic acid	$C_{20}H_{32}O_3$	*Schistochila appendiculata*	J	(53)
(1048)	Potassium 6-tridecyl salicylate	$C_{20}H_{31}O_3K$	*Schistochila appendiculata*	J	(53)
(1049)	3-Pentadecylphenol	$C_{21}H_{36}O$	*Schistochila appendiculata*	J	(53)
(1050)	6-Pentadecylsalicylic acid	$C_{22}H_{36}O_3$	*Schistochila appendiculata*	J	(53)
(1051)	Potassium 6-pentadecyl salicylate	$C_{22}H_{35}O_3K$	*Schistochila appendiculata*	J	(53)
(1052)	6-Undecylcatechol	$C_{17}H_{28}O_2$	*Schistochila appendiculata*	J	(53)
(1053)	3-Heptadecenyl-phenol	$C_{23}H_{38}O$	*Schistochila appendiculata*	J	(53)

Table IIe (*continued*)

Structure number	Name of compounds	Formula	m.p. °C	$[\alpha]_D$	Plant source	Order	References	Comments
(1054)	5-Heptadeca-8Z,11Z,14Z-trienyl-resorcinol monomethylether	$C_{24}H_{36}O_2$			*Omphalanthus filiformis*	J	(*441a, 571*)	
(1056)	Naphthalene	$C_{10}H_8$			*Plagiochila subdura*	J	(*46*)	GC-MS
					Triandrophyllum subtrifidum var. trifidum	J	(*45*)	
(1057)	2,4,7-Trimethoxy-naphthalene	$C_{13}H_{14}O_3$			*Wettsteinia inversa*	J	(*623a*)	
(1058)	Wettstein A	$C_{13}H_{12}O_4$			*Wettsteinia schusterana*	J	(*37*)	
(1059)	Wettstein B	$C_{13}H_{12}O_4$			*Wettsteinia schusterana*	J	(*37*)	
(1060)	Wettstein C	$C_{13}H_{14}O_3$			*Wettsteinia schusterana*	J	(*37*)	
(1061)	8-Hydroxy-6,7-dimethoxy-3-methyl-isocoumarin	$C_{12}H_{12}O_5$			*Wettsteinia inversa*	J	(*623a*)	
(1062)	Inversin	$C_{12}H_{10}O_5$			*Wettsteinia inversa*	J	(*623a*)	
(1063)	Dihydroinversin	$C_{12}H_{12}O_5$			*Wettsteinia inversa*	J	(*623a*)	
(1064)	Wettsteinolide	$C_{13}H_{14}O_6$			*Wettsteinia inversa*	J	(*37*)	
					W. schusterana	J	(*37*)	
(1065a)	Scapaniapyrone A	$C_{17}H_{10}O_8$			*Wettsteinia schusterana*	J	(*421*)	
(1066)	6,7-Dihydroxy-4-(3,4-dihydroxyphenyl)naphthalene-2-carboxylic acid	$C_{17}H_{12}O_6$	325–7		*Scapania undulata*	J		
					Pellia epiphylla	Me	(*474*)	

No.	Name	Formula	$[\alpha]$ / mp	Species	Me/J	Ref.
(1067)	(1R,2S)-2,3-Dicarboxy-6,7-dihydroxy-1-(3',4'-dihydroxy)-phenyl-1,2-dihydronaphthalene	$C_{18}H_{14}O_8$	−130.77	Pellia epiphylla	Me	(159a)
(1068)	2,3-Dicarboxy-6,7-dihydroxy-1-(3',4'-dihydroxy)-phenyl-1,2-dihydro-naphthalene-10-methyl ester	$C_{19}H_{16}O_8$	−112.10	Pellia epiphylla	Me	(159a)
(1069)	2,3-Dicarboxy-6,7-dihydroxy-1-(3',4'-dihydroxy)-phenyl-1,2-dihydro-naphthalene-9,5''-shikimic acid ester	$C_{25}H_{22}O_{12}$	−122.19	Pellia epiphylla	Me	(159a)
(1070)	(−)-Licarin A	$C_{20}H_{22}O_4$	−43	Jackiella javanica	J	(437)
(1071)	2-Hydroxy-3,4,7-trimethoxy-9,10-dihydrophenanthrene	$C_{17}H_{18}O_4$		Plagiochila spinulosa	J	(144)
(1072)	3,4,7-Trimethoxy-9,10-dihydrophenanthrene	$C_{17}H_{18}O_3$		Plagiochila spinulosa	J	(144)
(1073)	3,4-Dimethoxy-5-hydroxy-9,10-dihydro-phenanthrene	$C_{16}H_{16}O_3$	125.5–126.5	Riccardia jackii	Me	(380)

Table IIe (continued)

Structure number	Name of compounds	Formula	m.p. °C	$[\alpha]_D$	Plant source	Order	References	Comments
(1074)	2,5-Dimethoxy-3-hydroxy-phenanthrene	$C_{16}H_{14}O_3$			Marchantia tosana	M	(74)	
(1075)	2-Hydroxy-3,6-dimethoxy-phenanthrene	$C_{16}H_{14}O_3$			Marchantia paleacea var. diptera	M	(74)	
(1076)	2-Hydroxy-3,7-dimethoxy-phenanthrene	$C_{16}H_{14}O_3$	159–160		Marchantia polymorpha	M	(72)	
(1078)	3-(4'-Methoxybenzyl)-5,6-dimethoxy-phthalide	$C_{18}H_{18}O_5$	78–80		Frullania falciloba	J	(64)	
(1079)	3-(3',4'-Dimethoxy-benzyl)-7-hydroxy-5-methoxyphthalide	$C_{18}H_{18}O_6$	131–2		Balantiopsis rosea	J	(65)	
(1080)	Radulanolide	$C_{20}H_{18}O_4$	103–4		Radula complanata	J	(66)	
(1081)	3-(3'-Methoxy-4',5'-methylenedioxy-benzyl)-5,7-dimethoxyphthalide	$C_{19}H_{18}O_7$			Trocholejeunea sandvicensis	J	(82)	
(1082)	3-(3',4',5'-Trimethoxy-benzyl)-5,7-dimethoxy-phthalide	$C_{20}H_{22}O_7$			Trocholejeunea sandvicensis	J	(82)	

No.	Compound	Formula	mp	Species		Ref.
(1083)	m-Hydroxy-benzaldehyde	$C_7H_6O_2$		*Marchantia polymorpha*	M	(71)
(1084)	p-Hydroxy-benzaldehyde	$C_7H_6O_2$		*Marchantia paleacea* var. *diptera*	M	(74)
				M. polymorpha	M	(61, 71)
				Wiesnerella denudata	M	(51)
(1085)	Shikimic acid	$C_7H_{10}O_5$		*Conocephalum conicum*	M	(214)
(1086)	Ellagic acid	$C_{14}H_6O_8$		*Lophocolea bidentata*	J	(214)
				Plagiochila asplenioides	J	(214)
(1087)	Phenylacetoaldehyde	C_8H_8O		*Lophocolea heterophylla*	J	(527a)
(1088)	Orellinic acid methyl ester	$C_9H_{10}O_4$		*Blasia pusilla*	Me	(249)
(1089)	Methyl 2-methyl-3,4-methylenedioxy-6-methoxybenzoate	$C_{11}H_{12}O_5$		*Plagiochila spinulosa*	J	(144)
(1090)	4-Vinylguaiacol	$C_9H_{10}O_2$		*Lophocolea bidentata*	J	(527a)
(1091)	2,4,5-Trimethoxy-styrene	$C_{11}H_{14}O_3$		Unidentified Jamaican liverwort	J	(144)
(1092)	3,4-Dimethoxystyrene	$C_{10}H_{12}O_2$		*Asterella* sp.	M	(249)
(1093)	2,4,5-Trimethoxyallyl-benzene	$C_{12}H_{16}O_3$		*Conocephalum conicum*	M	(585)
				Unidentified Jamaican liverwort	J	(144)
(1094)	4-Hydroxy-3,5-dimethoxy-allylbenzene or 3-Hydroxy-4,5-dimethoxy allylbenzene	$C_{11}H_{14}O_3$		*Marchesinia brachiata*	J	(219)
(1096)	1-(3,4-Dihydroxy-5-methoxybenzyl)-3-methylbut-2-ene	$C_{12}H_{16}O_3$	94–5	*Plagiochila rutilans*	J	(278)

Table IIe (continued)

Structure number	Name of compounds	Formula	m.p.°C	$[\alpha]_D$	Plant source	Order	References	Comments
(1097)	6-(3-Methyl-2-butenyl) indole	$C_{13}H_{15}N$			Riccardia chamedryfolia	Me	(435)	
					R. multifida	Me	(78)	
(1098)	7-(3-Methyl-2-butenyl) indole	$C_{13}H_{15}N$			Riccardia chamedryfolia	Me	(435)	
(1099)	Skatole	C_9H_9N			Asterella sp.	M	(249)	
(1100)	Indole acetic acid	$C_{10}H_9O_2N$			Conocephalum conicum	M	(442a)	
					Marchantia polymorpha	M	(442a)	
(1102)	2,5-Dihydroxy-4-methyl-6-methoxy-acetophenone	$C_{10}H_{12}O_4$			Trocholejeunea sandvicensis	J	(22)	
(1103)	2,5-Dihydroxy-4-formyl-6-methoxy-acetophenone	$C_{10}H_{10}O_5$			Trocholejeunea sandvicensis	J	(22)	
(1104)	Trocholejeunin	$C_{19}H_{18}O_8$			Trocholejeunea sandvicensis	J	(22)	
(1105)	δ-Tocopherol	$C_{27}H_{46}O_2$			Radula perrottetii	J	(585a)	
(1106a)	Isotachioside (= 2-Methoxy-4-hydroxyphenyl-1β-O-glucoside)	$C_{13}H_{18}O_8$			Isotachis japonica	J	(76)	

(1107)	Salidroside	$C_{14}H_{20}O_7$	Ricciocarpos natans	M	(502)
(1108)	β-(3,4-Dihydroxy-phenyl)-ethyl-O-β-D-glucoside	$C_{14}H_{20}O_8$	Ricciocarpos natans	M	(502)
(1109)	1-O-β-D-(6'-Caffeoyl)-gluco-pyranosyl glycerol	$C_{18}H_{24}O_{11}$	Frullania muscicola	J	(345b)
(1110)	1-O-β-D-(4'-Caffeoyl)-gluco-pyranosyl glycerol	$C_{18}H_{24}O_{11}$	Frullania muscicola	J	(345b)
(1111)	1-O-β-D-(3'-Caffeoyl)-gluco-pyranosyl glycerol	$C_{18}H_{24}O_{11}$	Frullania muscicola	J	(345b)

J: Jungermanniales; M: Marchantiales; Me: Metzgeriales; Mo: Monocleales

content of (877) in the *M. polymorpha* cultures changed significantly during growth, ranging from 1 to 7 μg/mg dry weight, and increased dramatically to 160% of the control when the medium was deficient in phosphate. In nitrogen or glucose deficient medium, the amount of (877) decreased to 54% or 27% of the control.

Lunularin (878), the decarboxy derivative of (877), has been isolated from or detected in *Dumortiera hirsuta* (585), New Zealand *M. berteroana* (39), South American *Marchantia chenopoda* (553), *M. paleacea* var. *diptera* (74), German and Japanese *M. polymorpha* (71, 74) and *Ricciocarpos natans* (640).

A new labile aromatic compound (879a) named prelunularic acid because it possesses a "prearomatic" structure has been isolated from the ethanol extract of suspension cultured cells of *Marchantia polymorpha*

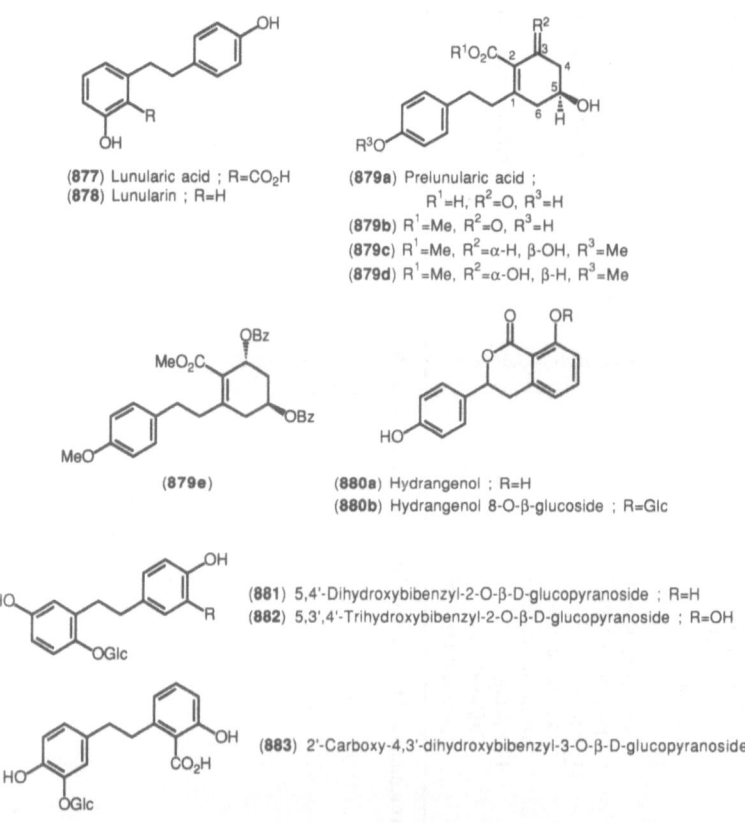

(877) Lunularic acid ; R=CO₂H
(878) Lunularin ; R=H

(879a) Prelunularic acid ;
      R¹=H, R²=O, R³=H
(879b) R¹=Me, R²=O, R³=H
(879c) R¹=Me, R²=α-H, β-OH, R³=Me
(879d) R¹=Me, R²=α-OH, β-H, R³=Me

(879e)

(880a) Hydrangenol ; R=H
(880b) Hydrangenol 8-O-β-glucoside ; R=Glc

(881) 5,4'-Dihydroxybibenzyl-2-O-β-D-glucopyranoside ; R=H
(882) 5,3',4'-Trihydroxybibenzyl-2-O-β-D-glucopyranoside ; R=OH

(883) 2'-Carboxy-4,3'-dihydroxybibenzyl-3-O-β-D-glucopyranoside

Chart 61a. Bibenzyls found in the Hepaticae and their derivatives

(*449, 450, 453*). The UV spectrum of (**879a**) at pH 12 indicated time-dependent changes with a clear set of isosbestic points at 247, 284 and 307 nm, while the final spectrum was identical with that of lunularic acid (**877**) [(λ_{max} 238 nm (log ε 4.25) and 293 (3.71)] after 160 min at the same pH, suggesting that (**879a**) was directly converted to lunularic acid (**877**) under basic conditions. The orientation of each functional group in the B-ring was confirmed by analysis of the NMR spectrum. Reduction of prelunularic acid (**879a**) with NaBH$_4$, followed by methylation with diazomethane afforded two epimeric diol methyl esters, (**879c**) and (**879d**) (4:1), the ^{1}H-NMR spectra of which showed that the two hydroxyl groups at C-3 and C-5 were equatorial-equatorial and axial-equatorial, respectively. Esterification of (**879d**) with *p*-dimethylaminobenzoyl chloride in tetrahydrofuran and triethylamine under reflux afforded a bis-(*p*-dimethylamino) benzoate (**879e**) whose CD spectrum exhibited typical exciton-split Cotton effects at 321 nm (Δε + 46.3) and 295 nm (Δε − 14.8), thus establishing that the absolute configurations at C-3 and C-5 were *R* and *S*, respectively. Hydrangenol (**880a**), a dihydrostilbene derivative isolated from higher plant *Hydrangea* species has been thought to be the direct precursor of lunularic acid (**877**) (*19*), but prelunularic acid (**879a**) is a more plausible precursor of lunularic acid than hydrangenol because (**880a**) has never been detected in liverworts. Prelunularic acid is also present in intact *M. polymorpha*. Compound (**879a**) is the first example of an intermediate possessing a "prearomatic" structure in the phenylpropanoid-polymalonate biosynthetic pathway shown in Scheme 59.

The methanol extract of an axenic cultured *Ricciocarpos natans* was partitioned between EtOAc and water followed by fractionation of the organic layer to give three new bibenzyl glycosides: 5,4'-dihydroxy-bibenzyl-2-O-β-D-glucopyranoside (**881**), 5,3',4'-trihydroxybibenzyl-2-O-β-D-glucopyranoside (**882**) and 2'-carboxy-4,3'-dihydroxybibenzyl-3-O-β-D-glucopyranoside (**883**), together with lunularic acid (**877**) and a cyclic bis(bibenzyl), riccardin C (**988**) (*351*). The sugar moiety of (**881**) and (**882**) was β-D-glucose. The substitution pattern of the two benzene rings and the location of the glucose in the two glycosides (**881**, **882**) was confirmed by a combination of NOE and long range ^{13}C-^{1}H 2D-COSY spectrometry. Structure (**883**) has been proposed by analysis of NMR and NOE data and comparison of the ^{13}C NMR spectra with those of lunularic acid (**877**). This is the first report of bibenzyl glycosides in bryophytes although their aglycones are widely distributed in liverworts.

As lunularic acid (**877**) possesses interesting biological activity (see Chapter V), many synthetic methods have been reported (*19*); however, in

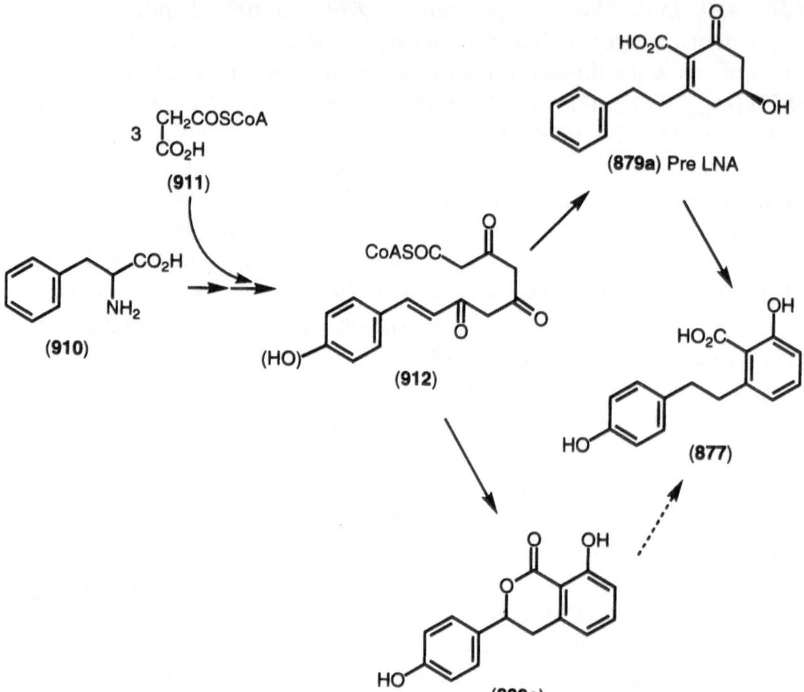

Scheme 59. Possible biogenetic pathways for prelunularic and lunularic acids

all of them the yield of (**877**) is very poor. On the other hand, hydrolysis of hydrangenol-8-O-β-glucoside (**880b**) which is easily obtained from the leaves of *Hydrangea macrophylla* var. *otaksa* (Saxifragaceae) with dilute acid affords a good yield of hydrangenol (**880a**) whose reduction with $NaBH_4$-$PdCl_2$ produces (**877**) quantitatively (Scheme 60) (*257*). In this manner hundred gram quantities of (**877**) may easily be prepared. Three alternative efficient and convenient total syntheses of (**877**) have been reported by EICHER *et al.* (*168, 169*). The starting material is 6-methyl-salicylic acid ethyl ester (*263*) prepared from crotonaldehyde and ethyl acetoacetate and 10–100g of lunularic acid can be prepared in this way.

An Ecuadorian liverwort, *Plagiochila longispina*, produces an unusual aromatic compound whose structure was established as 4-(*p*-methoxyphenylethyl)cyclohex-2-en-1-one (**884**) by a combination of UV (237 nm), IR (1660, 1640 cm^{-1}), MS (*m/z* 121, 109, 95), ^{1}H- and ^{13}C-NMR spectrometry (*506*). The total synthesis has been accomplished by SIEGEL *et al.* in nine steps (*506*). The C-4 position represents a stereogenic center;

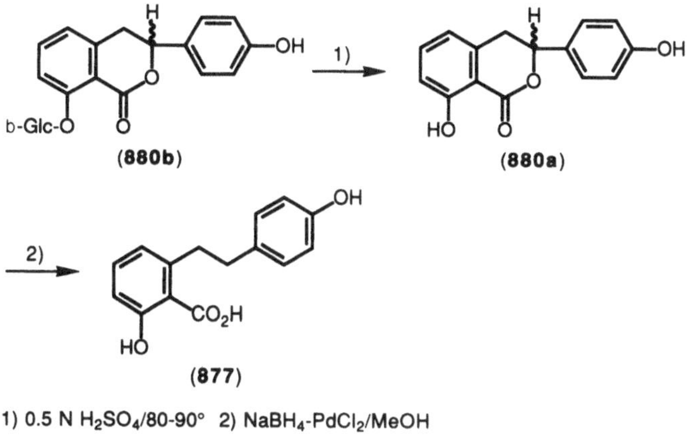

Scheme 60. Formation of lunularic acid from hydrangenol-8-O-β-glucoside

1) 0.5 N H_2SO_4/80-90° 2) $NaBH_4$-$PdCl_2$/MeOH

however, no optical activity was observed indicating that (**884**) must be a racemate.

Radula species in particular are rich sources of bibenzyl derivatives (*19, 66*). In subsequent work 3-hydroxybibenzyl (**885**), 3-hydroxy-4-methoxybibenzyl (**887**) and 3-methoxy-4-hydroxybibenzyl (**888**) have been detected in *R. frondescens* (*47*), together with 3-methoxybibenzyl (**886**) which has also been found in *R. buccinifera* (*66*) and *R. javanica* (= *R. variabilis*) (*49*). 3,5-Dimethoxybibenzyl (**889**) and 3-hydroxy-4,5-methylenedioxybibenzyl (**890**) have been detected in *Radula javanica* (*49, 66*). 3-Methoxy-4'-hydroxybibenzyl (**891**) is widely distributed in Jungermanniales (*19*) and has more recently been found in two *Plagiochila* (*39, 46, 358a*) and one *Radula* species (*47*). Bibenzyl (**891**) and 3,4'-dimethoxybibenzyl (**892**) had already been detected in *Frullania davurica* by GC-MS (*19, 56*). Further investigation of the ether extract of *F. davurica* resulted in actual isolation of these two compounds (*56*) as well as 3-hydroxy-4'-methoxybibenzyl (**893**) and 3,4'-dimethoxy-4-hydroxybibenzyl (**894**) (*420, 585*). The structures of (**891**) and its isomer (**893**) were established by total synthesis (*39*). Dimethoxybibenzyl (**892**) has also been found in *Frullania* (*56*), *Gongylanthus*, *Leucolejeunea*, and *Riccardia* species (*19*) as well as in two *Plagiochila* species (*39, 46*), three *Radula* species (*47, 66*) and *Frullania falciloba* (*64*). The 4-methyl ether (**897**) of pellepiphyllin (**896**) (*19*) has been detected in South American *Plagiochila fuegiensis* (*46*). A similar bibenzyl, 3-hydroxy-4,3'-dimethoxybibenzyl (**895**) has been isolated from *Frullania falciloba* (*64*) together with 3,4-methylenedioxy-3'-methoxybibenzyl (**898**) which had previously been detected in four *Frullania* species (*19*) and

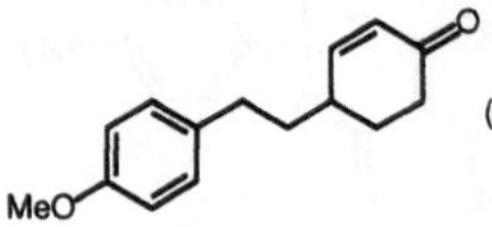

(884) 4-*p*-(Methoxyphenylethyl)cyclohex-2-en-1-one

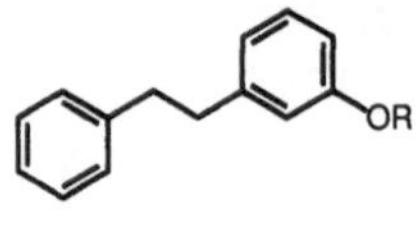

(885) 3-Hydroxybibenzyl ; R=H
(886) 3-Methoxybibenzyl ; R=Me

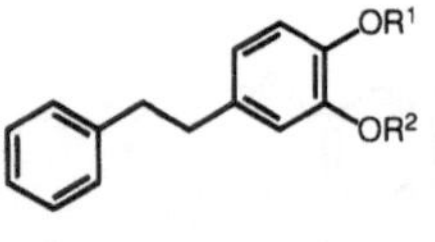

(887) 3-Hydroxy-4-methoxybibenzyl ;
     R^1=Me, R^2=H
(888) 3-Methoxy-4-hydroxybibenzyl ;
     R^1=H, R^2=Me

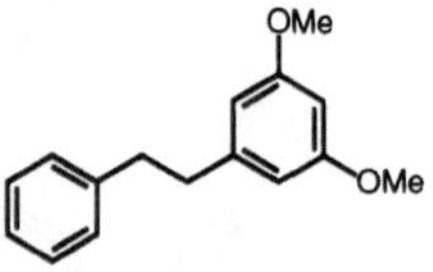

(889) 3,5-Dimethoxybibenzyl

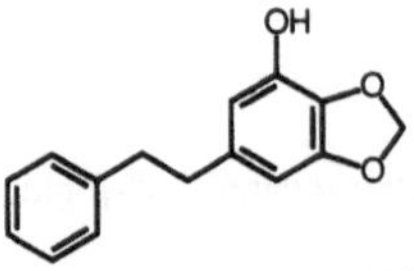

(890) 3-Hydroxy-4,5-methylenedioxybibenzyl

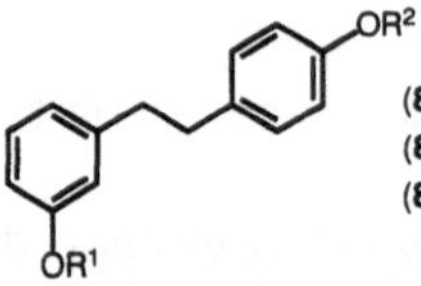

(891) 3-Methoxy-4'-hydroxybibenzyl ; R^1=Me, R^2=H
(892) 3,4'-Dimethoxybibenzyl ; R^1=R^2=Me
(893) 3-Hydroxy-4'-methoxybibenzyl ; R^1=H, R^2=Me

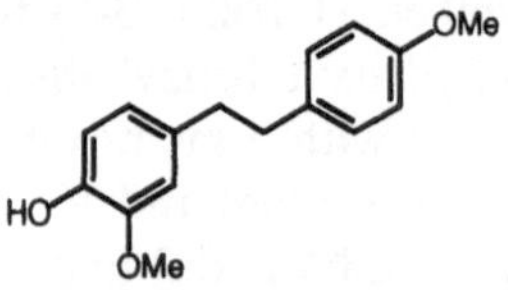

(894) 3,4'-Dimethoxy-4-hydroxybibenzyl

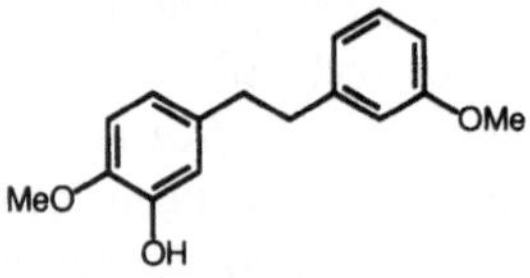

(895) 3-Hydroxy-4,3'-dimethoxybibenzyl

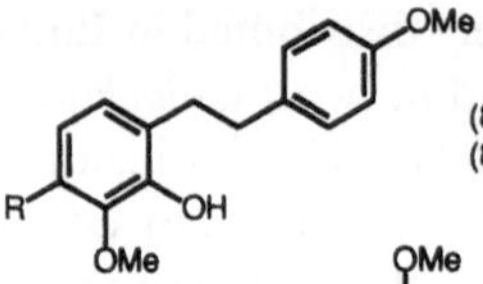

(896) Pellepiphyllin (=2-Hydroxy-3,4'-dimethoxybibenzyl) ; R=H
(897) 3,4,4'-Trimethoxy-2-hydroxybibenzyl ; R=OMe

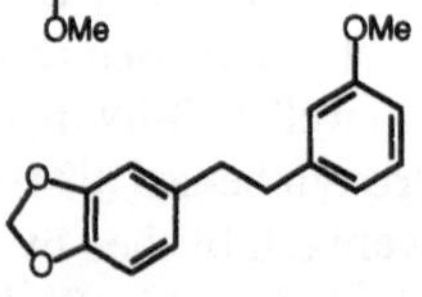

(898) 3,4-Methylenedioxy-3'-methoxybibenzyl

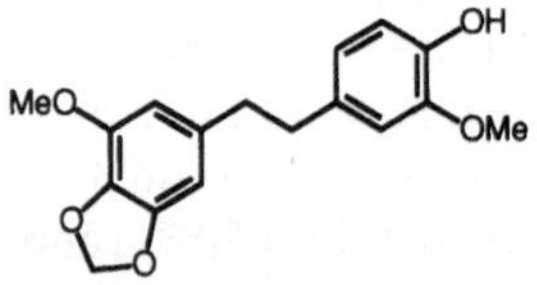

(899) 3,3'-Dimethoxy-4,5-methylenedioxy-4'-hydroxybibenzyl

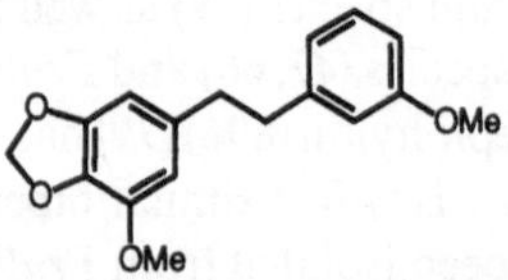

(900) 3,3'-Dimethoxy-4,5-methylenedioxybibenzyl

Chart 61b. Bibenzyls found in the Hepaticae

Trocholejeunea sandvicensis (82). Structures of (**895**) and (**898**) were assigned from the ^{1}H-NMR and NOE measurements. Bibenzyl (**898**) has been synthesized starting from *m*-methoxybenzaldehyde and 3,4-dimethoxybenzyltriphenylphosphonium bromide *(64)*. Lunularin (**878**), (**898**) and 3,3'-dimethoxy-4,5-methylenedioxy-4'-hydroxybibenzyl (**899**) had already been detected in *Frullania bonincola (19)*. Since then bibenzyl (**899**) has been isolated from the same species; its structure was established by the synthesis *(56)*. *F. ericoides* produces a new bibenzyl, 3,3'-dimethoxy-4,5-methylenedioxybibenzyl (**900**), together with (**898**) *(56)*. The structure of (**900**) has been established by total synthesis *(56)*.

Two new bibenzyls possessing two methylenedioxy groups have been isolated from *Frullania parvistipula* and their structures established as 3,4;3',4'-dimethylenedioxybibenzyl (**902**) and 3,3'-dihydroxy-4,5;4',5'-dimethylenedioxybibenzyl (**903**) on the basis of spectral data and by their total synthesis *(56)*. A new bibenzyl, 3,4,5,3',4'-pentamethoxybibenzyl (**904**) has been isolated from *Frullania serratta*, together with brittonin B (**905**) *(19)*; the structure was verified by comparison with an authentic sample *(51)*. Three new bibenzyls, 1-carbomethoxy-2,3-dihydroxy-4'-methoxybibenzyl (**907**), 3-methoxy-4,5-methylenedioxy-4'-hydroxybibenzyl (**901**) and 3,4-dihydroxy-5,3'-dimethoxybibenzyl (**908**) have been obtained from *Plagiochila spinulosa (478)*, *P. chacabucensis (46)* and *Stictolejeunea balfourii* var. *bekkei (219)*, respectively. A similar bibenzyl, notholanic acid (**906**) has been isolated from the frond exudate of two gymnogrammoid ferns *(427)*. The brominated bibenzyl (**909**) has also been isolated from a red alga *(211)*.

Radula species are also rich sources of bibenzyls with prenyl groups. Twelve such bibenzyls from three *Radula* species were listed in the previous review *(19)*. In addition to 3-hydroxy-5-methoxy-4(3-methyl-2-butenyl)bibenzyl (**914**) and 2-geranyl-3,5-dihydroxybibenzyl (**918**) [corrected from the previously assigned structure 3,5-dihydroxy-4-geranylbibenzyl (**921a**) (*vide infra*)], five new bibenzyls, 3-hydroxy-4-(3-methyl-2-butenyl)bibenzyl (**913**), 3,4'-dihydroxy-4-(3-methyl-2-butenyl)bibenzyl (**924**), 3-hydroxy-4-(3-methyl-2-butenyl)-4'-methoxybibenzyl (**925**), 3-methoxy-4-(3-methyl-2-butenyl)-4'-hydroxybibenzyl (**926**), 2-(3,7-dimethyl-2,7-octadienyl)-3,5-dihydroxybibenzyl (**922**) [corrected from the previous assignment 3,5-dihydroxy-4(3,7-dimethyl-2,7-octadienyl)bibenzyl (**923**)] have been detected in *Radula oyamensis (66)*. Structures of the new bibenzyls were based on the ^{1}H-NMR and mass spectra and spectroscopic comparison with previously known bibenzyls as well as total synthesis. *Radula buccinifera* also produces the previously known prenylbibenzyl (**915**), radulanin A (**960**) and radulanin C (**962**) while radulanin A (**960**) has also been detected in *R. voluta (47)*. Fourteen

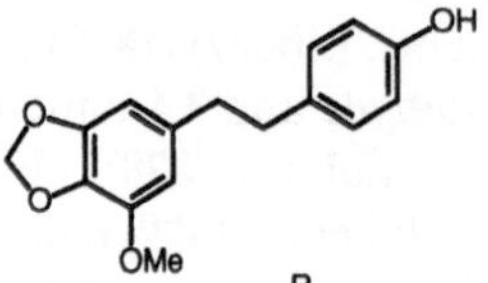

(901) 3-Methoxy-4,5-methylenedioxy-4'-hydroxybibenzyl

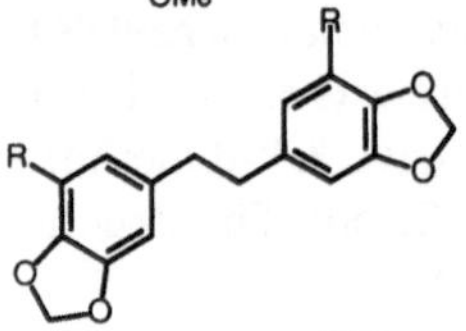

(902) 3,4;3',4'-Dimethylenedioxybibenzyl ; R=H
(903) 3,3'-Dihydroxy-4,5;4',5'-dimethylenedioxybibenzyl ; R=OH

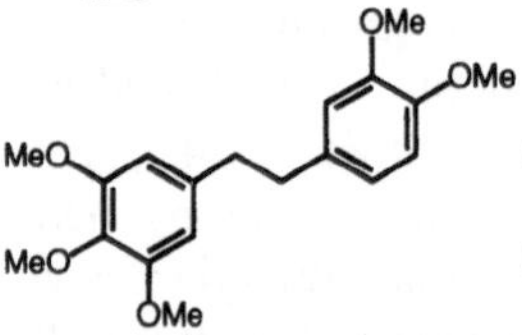

(904) 3,4,5,3',4'-Pentamethoxybibenzyl

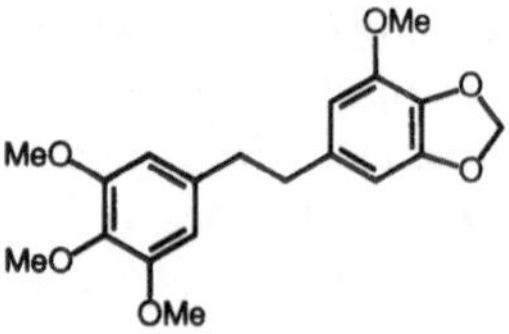

(905) Brittonin B

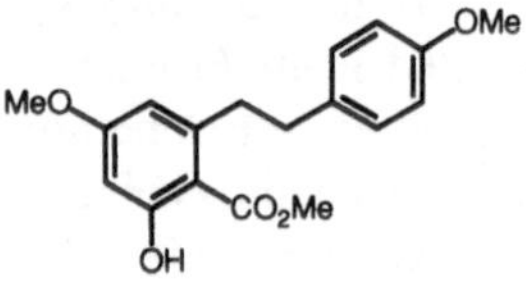

(906) Notholanic acid

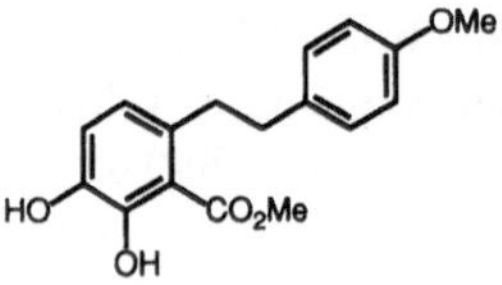

(907) 1-Carbomethoxy-2,3-dihydroxy-4'-methoxybibenzyl

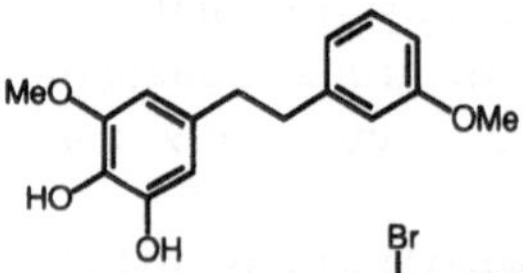

(908) 3,4,-Dihydroxy-5,3'-dimethoxybibenzyl

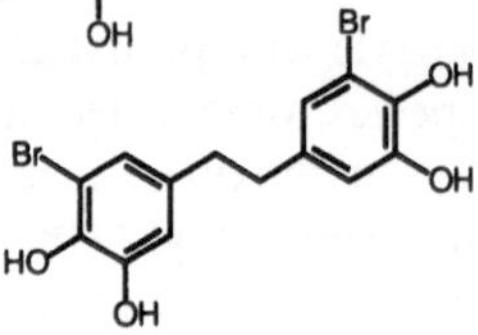

(909) 3,3'-Dibromo-4,4',5,5'-tetrahydroxybibenzyl

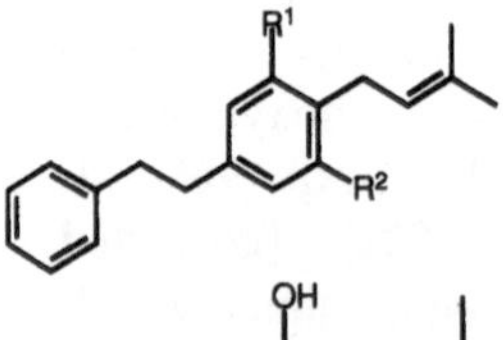

(913) 3-Hydroxy-4-(3-methyl-2-butenyl)bibenzyl ; R^1=OH, R^2=H
(914) 3-Hydroxy-5-methoxy-4-(3-methyl-2-butenyl)bibenzyl ; R^1=OH, R^2=OMe
(915) 3,5-Dihydroxy-4-(3-methyl-2-butenyl)bibenzyl ; R^1=R^2=OH
(916) 3,5-Dimethoxy-4-(3-methyl-2-butenyl)bibenzyl ; R^1=R^2=OMe

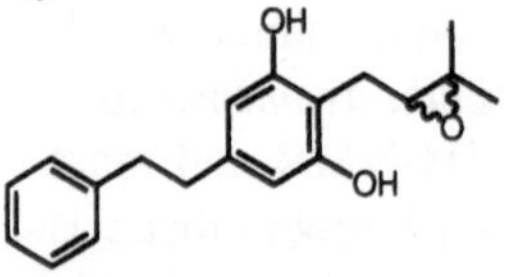

(917) 3,5-Dihydroxy-4-(2,3-epoxy-3-methylbutyl)bibenzyl

Chart 61c. Bibenzyls found in the Hepaticae and dibromobibenzyl (909) in a red alga

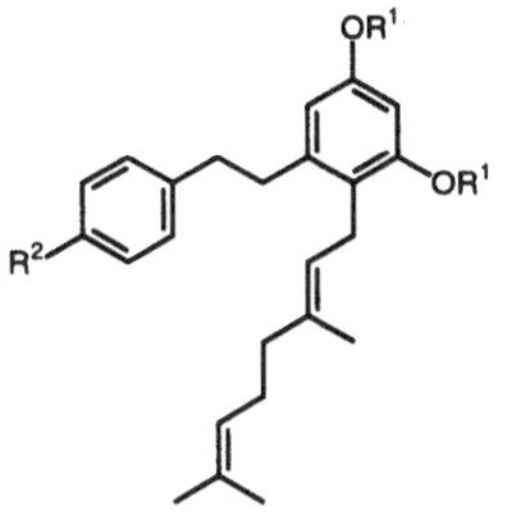

(918) 2-Geranyl-3,5-dihydroxybibenzyl ; $R^1=R^2=H$
(919) 2-Geranyl-3,5-dimethoxybibenzyl ; $R^1=Me, R^2=H$
(920) 2-Geranyl-3,5,4'-trihydroxybibenzyl ; $R^1=H, R^2=OH$

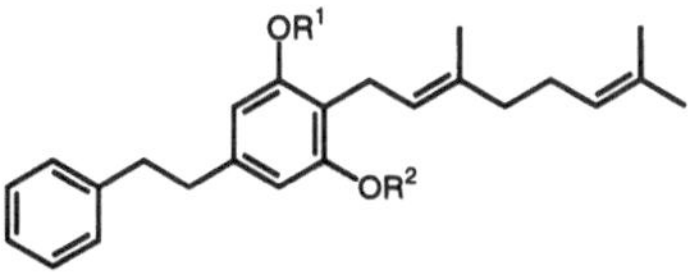

(921a) 3,5-Dihydroxy-4-geranylbibenzyl (=3,5-Dihydroxy-4-
(3,7-dimethyl-2,6-octadienyl) bibenzyl) ; $R^1=R^2=H$
(921b) $R^1=H, R^2=Me$
(921c) $R^1=R^2=Me$

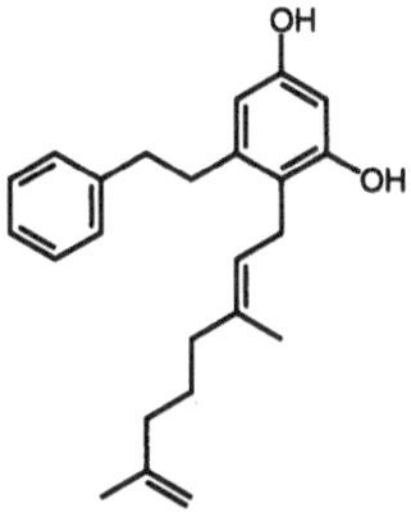

(922) 2-(3,7-Dimethyl-2,7-octadienyl)-3,5-dihydroxybibenzyl

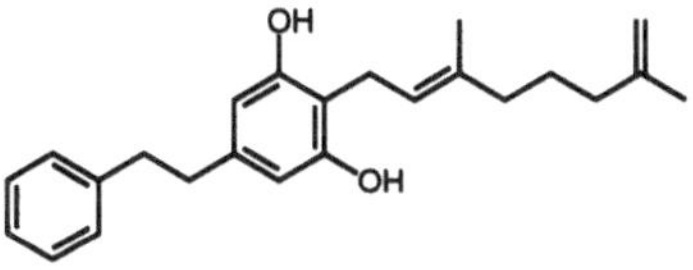

(923) 3,5-Dihydroxy-4-(3,7-dimethyl-2,7-octadienyl) bibenzyl

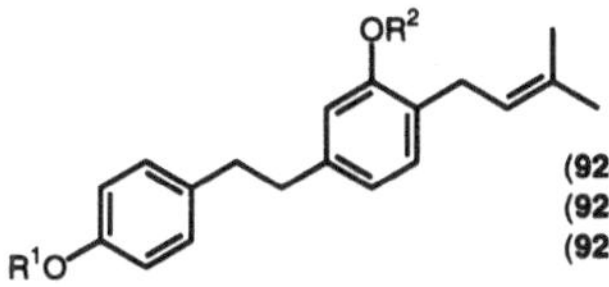

(924) 3,4'-Dihydroxy-4-(3-methyl-2-butenyl)bibenzyl ; $R^1=R^2=H$
(925) 3-Hydroxy-4-(3-methyl-2-butenyl)-4'-methoxybibenzyl ; $R^1=Me, R^2=H$
(926) 3-Methoxy-4-(3-methyl-2-butenyl)-4'-hydroxybibenzyl ; $R^1=H, R^2=Me$

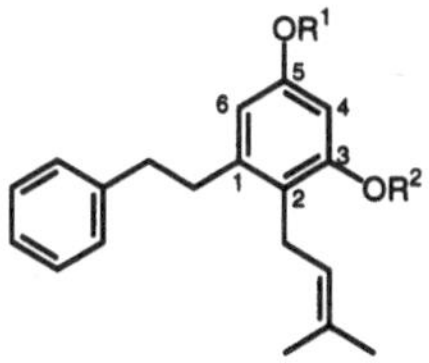

(927) 3, 5-Dihydroxy-2-(3-methyl-2-butenyl)bibenzyl ; $R^1=R^2=H$
(928) 3-Methoxy-5-hydroxy-2-(3-methyl-2-butenyl)bibenzyl ; $R^1=H, R^2=Me$
(929) 3-Hydroxy-5-methoxy-2-(3-methyl-2-butenyl)bibenzyl ; $R^1=Me, R^2=H$
(930) 3, 5-Dimethoxy-2-(3-methyl-2-butenyl)bibenzyl ; $R^1=R^2=Me$

Chart 61d. Bibenzyls found in the Hepaticae and their derivatives

new bibenzyl derivatives (**918**, **927–929**, **931–934**, **937**, **938**, **940–942**, **943a**) have been isolated from *Radula kojana*, together with two bis(bibenzyls), perrottetin E (**1023a**) and perrottetin F (**1024a**); structures were established by various NMR spectroscopic techniques together with chemical transformations (methylation, dehydration and epoxidation) and synthesis (*50*). The 2-prenylbibenzyl (**927**) has been isolated from *Lepidozia vitrea* (**585a**) and detected in *R. frondescens* and *R. voluta* (*47*). For synthesis of the bibenzyls (**913–915**, **918**, **921a**, **927–929**), pinosylvin (3,5-dihydroxy stilbene) (**980a**) and pinosylvin monomethyl ether (3-hydroxy-5-methoxystilbene) (**980b**) which occur *inter al.* in *Alnus sieboldiana* (Betulaceae) (*16*) were hydrogenated followed by prenylation to give the prenylated derivatives (**914**), (**918**), (**921a–c**), **927–930** (*50*). Two alternative syntheses of (**929**), its methyl ether (**930**) and the methyl ether (**919**) of (**918**) have been reported (*50*).

The prenylated bibenzyls found in *Radula* species are very similar to those isolated from *Cannabis sativa* (*158*). Crombie et al. (*159*) reported that the structure of the presumed (**921a**) from *Radula variabilis* (= *R. javanica*) later also isolated from *R. complanata*, *oyamensis* and *tokiensis* (*66*) should be revised to (**918**) on the basis of the ¹H NMR spectra of authentic (**921a**) (*p*-cannabigerol) and (**918**) (*o*-cannabigerol) prepared by synthesis (*159*) while (**921a**) was identical with a prenylbibenzyl from *Helichrysum umbraculigerum* (Compositae) as originally proposed (*115*). This was indeed the case. Another synthesis of (**921a**) has been reported by Zinsmeister et al. (*647*). 2-Geranyl-3,5-dihydroxybibenzyl (**918**) has been detected in *R. frondescens* and *R. voluta* (*47*). 2-Geranyl-3,5,4′-trihydroxybibenzyl (**920**) has been detected in *R. frondescens* (*47*) while 3,5-dihydroxy-4-geranylbibenzyl (= *p*-cannabigerol) (**921a**) has been found in some South American *Radula* species (*50*). The methyl ether (**938**) of (**937**) was prepared from 3-hydroxy-5-methoxyprenylbibenzyl (**914**). The structure assigned to (**937**) was supported by NOE difference spectrum of its methyl ether and comparison of the spectral data with (**937**) and (**938**).

The absolute configuration of C-2′ in the new bibenzyls, (**940**) and (**941**) from *R. perrottetii* was shown to be *S* because of positive Cotton effects at 262 nm ($\Delta\varepsilon$ + 0.6) and 250.5 nm ($\Delta\varepsilon$ + 1.3), respectively (*546*). Crombie et al. (*159*) have synthesized (±)-*o*-cannabichromene (**940**) and its *p*-isomer by base catalyzed condensation of 3,5-dihydroxybibenzyl (**980d**) with citral. The spectral data of (**940**) from the liverwort were identical with those of synthetic *o*-cannabichromene. The synthesis of methyl ether (**941**) from 5-methyl ether of (**918**) by dehydrogenation with DDQ has been reported (*50*). A similar chromene derivative (**944**) has been isolated from the ferns *Sceptridium ternatum* var. *ternatum* and *S.*

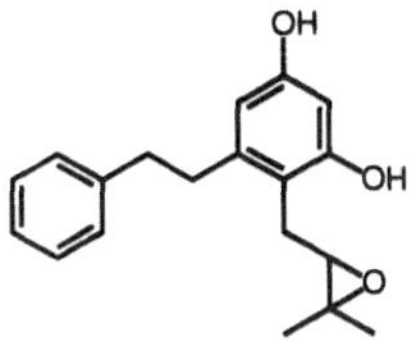

(931) 3,5-Dihydroxy-2-(2,3-epoxy-3-methylbutyl)bibenzyl

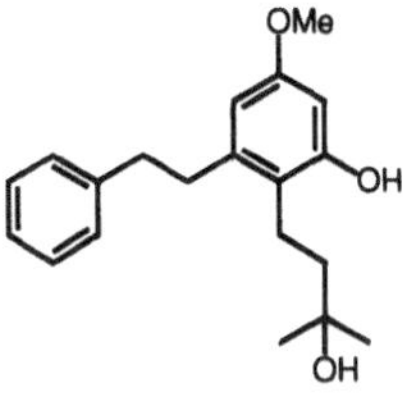

(932) 3-Hydroxy-5-methoxy-2-(3-hydroxy-3-methylbutyl)bibenzyl

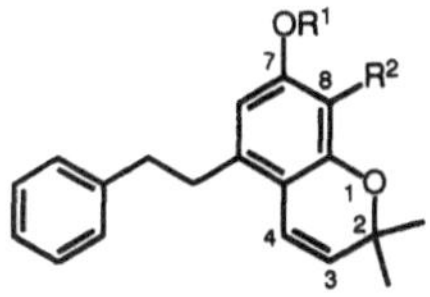

(933) 2,2-Dimethyl-7-hydroxy-5-(2-phenylethyl)chromene ; $R^1=R^2=H$
(934) 2,2-Dimethyl-7-methoxy-5-(2-phenylethyl)chromene ; $R^1=Me$, $R^2=H$
(935) 2,2-Dimethyl-7,8-dihydroxy-5-(2-phenylethyl)chromene ; $R^1=H$, $R^2=OH$
(936) 2,2-Dimethyl-7,8-dimethoxy-5-(2-phenylethyl)chromene ; $R^1=Me$, $R^2=OMe$

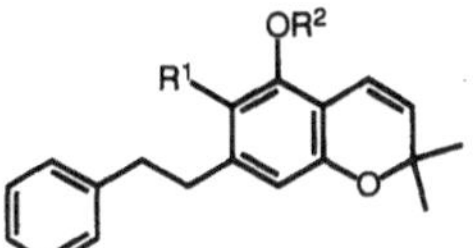

(937) 2,2-Dimethyl-5-hydroxy-7-(2-phenylethyl)chromene ; $R^1=R^2=H$
(938) 2,2-Dimethyl-5-methoxy-7-(2-phenylethyl)chromene ; $R^1=H$, $R^2=Me$
(939) 2,2-Dimethyl-5-hydroxy-6-carboxy-7-(2-phenylethyl)chromene ;
$\quad R^1=CO_2H$, $R^2=H$

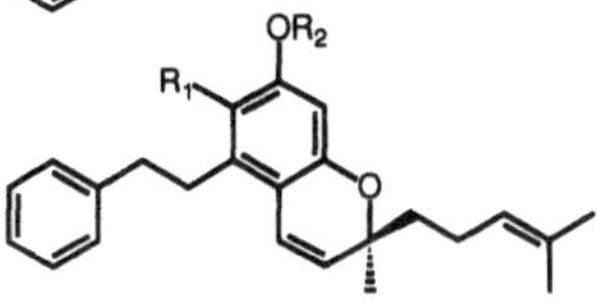

(940) 2(S)-2-Methyl-2-(4-methyl-3-pentenyl)-7-hydroxy-
5-(2-phenylethyl)chromene ; $R^1=R^2=H$
(941) 2(S)-2-Methyl-2-(4-methyl-3-pentenyl)-7-methoxy-
5-(2-phenylethyl)chromene ; $R^1=H$, $R^2=Me$
(942) 2(S)-2-Methyl-2-(4-methyl-3-pentenyl)-6-carboxy-
7-hydroxy-5-(2-phenylethyl)chromene ; $R^1=CO_2H$, $R^2=H$

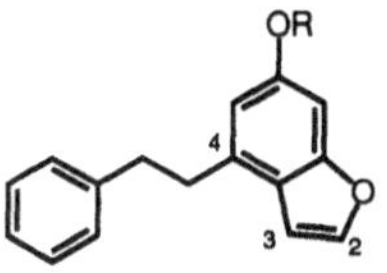

(943a) 6-Hydroxy-4-(2-phenylethyl)benzofuran ; R=H
(943b) R=Me

Chart 61e. Bibenzyls found in the Hepaticae and their derivatives

japonicum (427, 546). The structure of a new benzofuran **(943a)** was elucidated by a combination of spin decoupling experiments and NOE spectrometry of its methyl ether **(943b)** *(50)*.

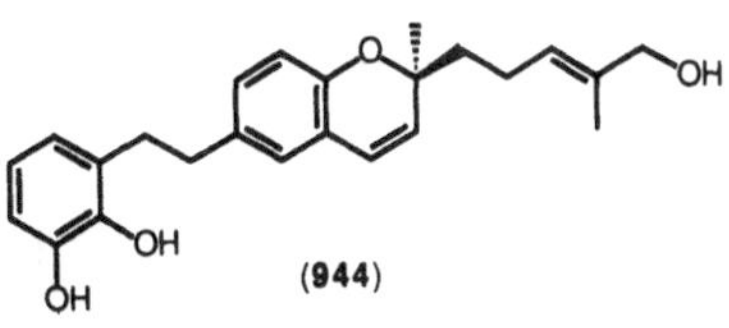

(944)

Two new prenylated bibenzyls 2(R)-isopropenyl-6-hydroxy-4(2-phenylethyl) dihydrobenzofuran (945) and 2,2-dimethyl-7,8-dihydroxy-5-(2-phenylethyl)chromene (935) have been isolated from *Radula perrottetii*, along with the previously known bibenzyls (927, 933, 946a, 947) (43) and the bis(bibenzyls), perrottetins E (1023a), F (1024a) and G (1025). Conclusive evidence for structures (945) and (935) was obtained by syntheses of (945), its methyl ether (946b) and their isomers (958) and (957), and the dimethyl ether (936) of (935), (43). On the basis of model reaction, acetates (914a) and (927a) were cyclized with Ca(OCl)$_2$ to yield (946b) and (957), with the former being identical with the methyl ether of the natural product (945) (43). The bibenzyls (946b) and (957) were similarly synthesized from 3-hydroxy-5-methoxystilbene, while dihydro-pinosylvin gave naturally occurring dihydrobenzofuran (945) and isomer (958). The absolute configuration of (945) at C-2 was R because of the positive Cotton effect of (946b) (298, 299). Biosynthesis of (945) might proceed from the coexisting prenyl bibenzyl (927). A synthesis of (936) proceeded from 3-hydroxy-4,5-dimethylmethoxybibenzyl which was prenylated by 2,2-dimethylallylbromide. The product was dehydrogenated with DDQ to afford a chromene derivative (43), identical with (936) prepared from (935).

(914a)   (927a)

The earlier report (19) noted that five prenyl bibenzyls, 3,5-dihydroxy-4-(3-methyl-2-butenyl)bibenzyl (915), perrottetin A–D (949–951, 946d) and a prenyldihydrochalcone (955) had been isolated from *R. perrottetii*, their structures based on 60 MHz NMR spectrometry. However, these assignment had to be revised to (927), (952a), (953a), (954a), (946a) and (947) as the result of high resolution NMR spectrometry and synthesis of the methyl ethers of the natural bibenzyls (43). The presumed (915) was identical with synthetic (927) while the structure of perrottetin A was revised to (952a) because of the NOE difference spectrum of the derived trimethyl ether (952b) and the total synthesis (43). An alternative synthesis of (952a) has been reported by ZINSMEISTER et al. (647). Structures (950) and (951) originally assigned to perrottetin B and C were revised to (953a)

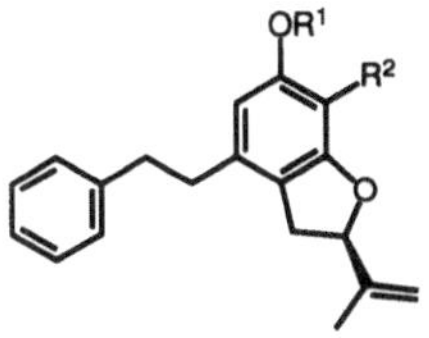

(945) 2(*R*)-Isopropenyl-6-hydroxy-4-(2-phenylethyl)dihydrobenzofuran ; $R^1=R^2=H$
(946a) Perrottetin D ; $R^1=H$, $R^2=OH$
(946b) $R^1=Me$, $R^2=H$
(946c) $R^1=Me$, $R^2=OMe$

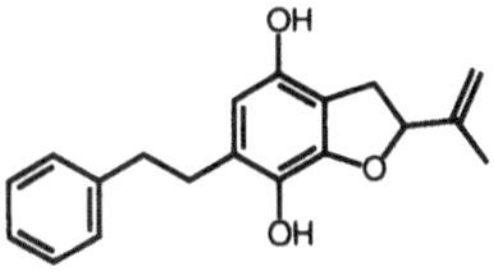

(946d)

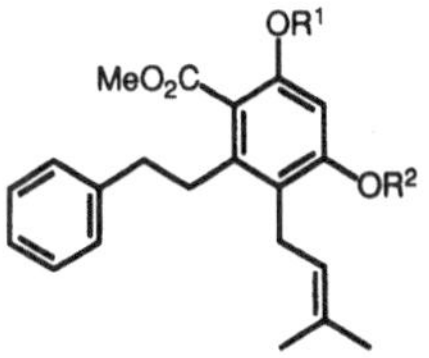

(947) 3, 5-Dihydroxy-6-carbomethoxy-2-(3-methyl-2-butenyl)bibenzyl ; $R^1=R^2=H$
(948) 3, 5-Dimethoxy-6-carbomethoxy-2-(3-methyl-2-butenyl)bibenzyl ; $R^1=R^2=Me$

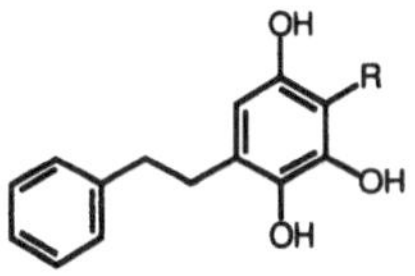

(949) R= 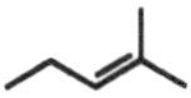
(950) R= 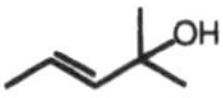
(951) R=

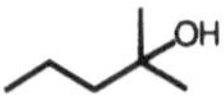

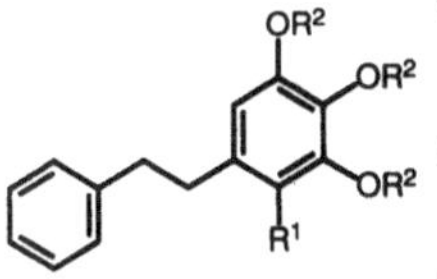

(952a) Perrottetin A ; $R^1=$ 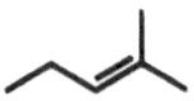 $R^2=H$
(952b) $R^1=$ 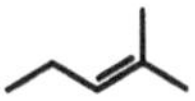 $R^2=Me$
(953a) Perrottetin B ; $R^1=$ 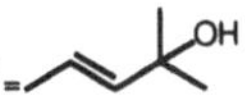 $R^2=H$

(953b) $R^1=$ 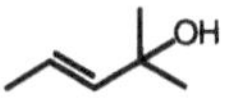 $R^2=Me$
(954a) Perrottetin C ; $R^1=$ 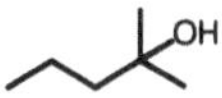 $R^2=H$
(954b) $R^1=$ 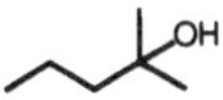 $R^2=Me$

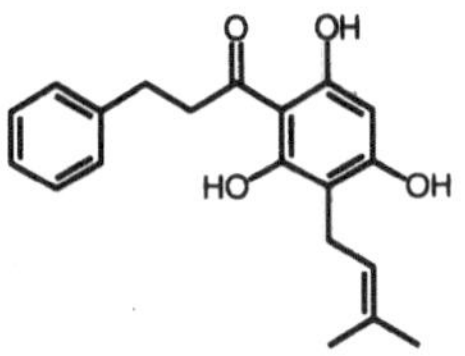

(955)

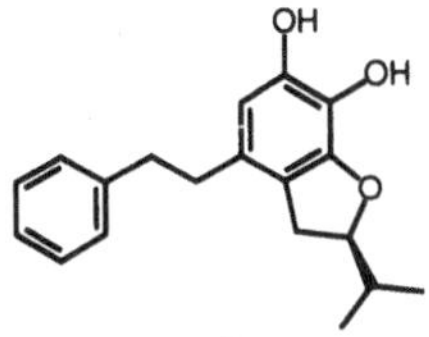

(956)

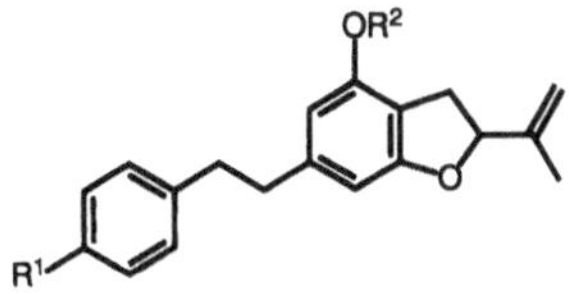

(957) Tylimanthin A ; $R^1=H$, $R^2=Me$
(958) Tylimanthin B ; $R^1=R^2=H$
(959) Tylimanthin C ; $R^1=OMe$, $R^2=Me$

Chart 61f. Bibenzyls found in the Hepaticae and their derivatives

and (**954a**), because hydrogenation of the trimethyl ether (**953b**) to (**954b**), followed by dehydration gave (**952b**).

On the basis of its ^{1}H NMR spectrum, structure (**946d**) was originally proposed for perrottetin D from *R. perrottetii* (*66*) and *R. voluta* (*47*). Its dimethyl ether (**946c**) exhibited NOE's between H-α and H-5, and H-5 and OMe-4. This, however, did not exclude the alternative structure (**946a**). The ^{1}H- and ^{13}C-NMR spectra of (**946a**) and (**946c**) were almost identical with those of (**945**) and its dimethyl ether (**946b**), indicating that the structure of perrottetin D might actually be (**946a**). This was confirmed by total synthesis (*43, 647*). The absolute configuration at C-2 was established as *R* as a result of the positive Cotton effects exhibited by (**946a**) and (**946c**) and the dihydro derivative (**956**) (*298, 299*). *Tylimanthus urvilleanus* produces similar dihydrobenzofuran derivatives, tylimanthins A (**957**), B (**958**) and C (**959**) (*45*).

A compound originally assigned structure (**955**) from *R. perrottetii* (*66*) was later revised to (**947**) as a result of NOE spectrometry and further transformations. The positions of one hydroxyl group and a prenyl group were confirmed by formation of a 2,2-dimethylchroman from (**947**) on treatment with acid. Compound (**947**) exhibited an NOE between H-α and H-1′ but not between any other proton and the OMe group while methyl ether (**948**) exhibited NOEs between H-α and H-1′ as well as H-4 and OMe-3. These data, the occurrence of the C-4 signal at high field [δ 102.3 in (**947**) and 93.6 in (**948**)] and the similarity of the NMR spectra of (**947**) and (**948**) to those of (**927**) and (**929**), forced revision of the structure to (**947**).

The earlier review (*19*) also mentioned the isolation of what was thought to be (**915**) and its monomethyl ether (**914**) from European *Radula complanata*. Further fractionation of the crude extract of *R. complanata* yielded 3,5-dihydroxy-2-(3-methyl-2-butenyl)bibenzyl (**927**) and two new bibenzyls, 4′-hydroxyradulanin H (**964**) (*542*) and radulanin L (**965a**) (*43*). The Rts of (**927**) and the presumed (**915**) were the same while the MS spectra were almost identical. However, while the chemical shifts of H-2 and H-6, C-2 and C-6, and C-3 and C-5 in (**915**) are the same, but the same protons and carbons in (**927**) show different chemical shifts (*159*).

(**927b**)

In order to obtain the natural bibenzyl (**915**), an attempt was made to demethylate (**916**) and (**914**) using BBr$_3$ or EtSNa in HMPA; however, under these conditions, chroman (**927b**) and deprenylated material were obtained in place of (**915**) (*43*). Synthesis of (**915**) has, however, been reported by EICHER (*169*) and ZINSMEISTER *et al.* (*647*).

Radulanin A (**960**) has been isolated from *R. javanica* and *R. complanata* (*19*). The latter species also produces radulanin H (**963a**) and 3,5-dihydroxy-4-(2,3-epoxy-3-methylbutyl)bibenzyl (**917**) (*66*). Confirmation for structure (**960**) was provided by the spectral data and a 2D-COSY as well as the NOE difference spectrum of its methyl ether, radulanin B (**961**) (*19*). The structure assigned to (**963a**) was supported by the NOE difference spectrum of its methyl ether (**963b**), the presence of a hydrogen bonded carboxyl group (1650 cm^{-1}) and the NMR spectra of the related compounds (**960, 961, 963b**). The structure of 4′-hydroxyradulanin H (**964**) was suggested by comparison of the spectral data with those of the co-metabolite, radulanin H (**963a**) (*542*). The structure of radulanin L (**965a**) from *R. complanata* has been established by an extensive NMR study (*49*). The presence of a 1,2-disubstituted benzene ring and the locations of the functional groups in the two benzene rings was evident from the coupling patterns in the ^{1}H NMR spectra and the NOE difference spectra of (**965a**) and its methyl ether (**965b**).

As mentioned earlier *Radula javanica* (= *R. variabilis*) produces several bibenzyls possessing a dihydrooxepin skeleton (*19*). Further fractionation of the methanol extract of *R. javanica* has yielded three unusual new cyclopropanochroman derivatives, radulanin I (**966**), J (**967**) and K (**968a**), together with the previously known radulanins A (**960**) and H (**963a**) (*19*) and 2-geranyl-3,5-dihydroxybibenzyl (**918**) while the additional presence of 3-methoxybibenzyl (**886**) and 3,5-dimethoxybibenzyl (**889**) was detected by GC and GC-MS (*49*). The presence of a 1,1,2-trisubstituted cyclopropane ring and two protons on the carbon bearing the ether oxygen of (**966**) and (**967**) was deduced from the ^{1}H- and ^{13}C-NMR spectra. Methylation of (**966**) gave a monomethyl ether which was identical with (**967**). The substitution pattern and the position of each functional group on the benzene ring of (**966**) was established by the NOE difference spectrum of (**967**). Comparison of the spectral data of (**966**) and radulanin K (**968a**) showed that the compounds differed only with respect to presence or absence of the carboxyl, leading to two possible structures (**968a**) or (**968b**) for radulanin K. Structure (**968a**) was favored as the closely related compounds (**966**) and (**967**) occur in the same species. Radulanin J (*967*) and radulanin K (**968a**) are optically active with a positive optical rotations and negative Cotton effects [(230 nm in (**967**) and 263 nm in (**968a**)]. On the other hand, radulanin I (**966**) and its methyl

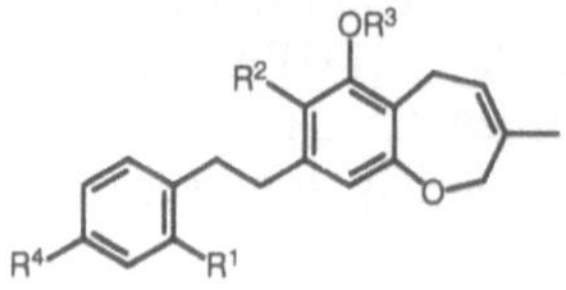

(960) Radulanin A ; $R^1=R^2=R^3=R^4=H$
(961) Radulanin B ; $R^1=R^2=R^4=H$, $R^3=Me$
(962) Radulanin C ; $R^1=R^2=R^3=H$, $R^4=OH$
(963a) Radulanin H ; $R^1=R^3=R^4=H$, $R^2=CO_2H$
(963b) $R^1=R^2=CO_2Me$, $R^3=Me$, $R^4=H$
(964) 4'-Hydroxyradulanin H ; $R^1=R^3=H$, $R^2=CO_2H$, $R^4=OH$
(965a) Radulanin L ; $R^1=OH$, $R^2=R^3=R^4=H$
(965b) $R^1=OMe$, $R^2=R^4=H$, $R^3=Me$

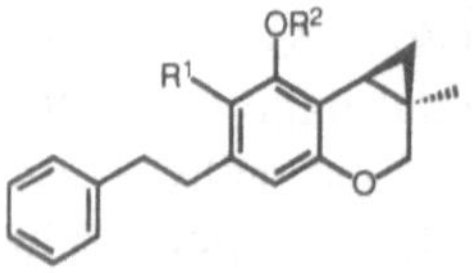

(966) Radulanin I (=3-Methyl-6-hydroxy-8-(2-phenylethyl)-
3, 5-cyclopropanochroman) ; $R^1=R^2=H$
(967) Radulanin J (=3-Methyl-6-methoxy-8-(2-phenylethyl)-
3, 5-cyclopropanochroman) ; $R^1=H$, $R^2=Me$
(968a) Radulanin K (=3-Methyl-6-hydroxy-7-carboxy-8-(2-phenylethyl)-
3, 5-cyclopropanochroman) ; $R^1=CO_2H$, $R^2=H$

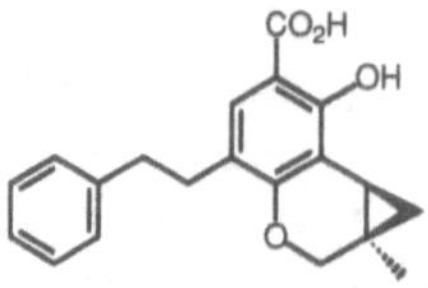

(968b)

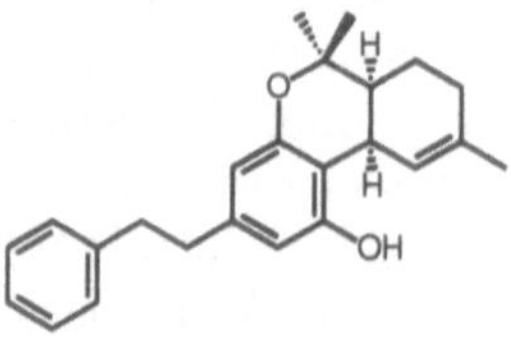

(969) Perrottetinene

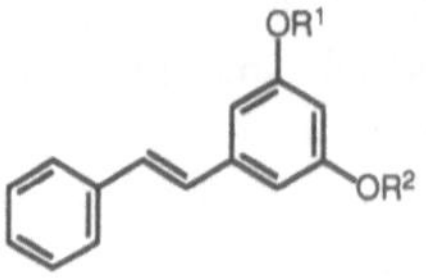

(980a) Pinosylvin ; $R^1=R^2=H$
(980b) Pinosylvin methylether ; $R^1=Me$, $R^2=H$

Chart 61g. Bibenzyls found in the Hepaticae and their derivatives

ether (967) exhibited negative optical rotations and positive Cotton effects (237 nm in (966) and 230 nm in 967). The absolute value of the optical rotations and the amplitudes of the CD curves of (966) and (967) were smaller than those of radulanin J (967) which suggested that (966) as isolated might be a mixture of the *R* and *S* forms. However, the ¹H-NMR spectrum of (966) in the presence of a chiral shift reagent exhibited the signals of only one enantiomer, hence the difference in optical rotation and the amplitudes of the CD curves had to be ascribed to a small impurity in (966). The presence of both *R* and *S* cyclopropanochroman derivatives in *R. javanica* is noteworthy and also represents the first isolation of a cyclopropanochroman from natural sources. *Radula perrottetii* biosynthesizes a unique bibenzyl, perrottetinene (969): Its structure and relative configuration are based on 600 MHz NMR spectrometry (585a).

8.3 Bis-Bibenzyls

The Hepaticae are rich sources of bis-bibenzyl derivatives. The earlier review mentioned two such compounds, riccardin A (**985a**) and marchantin A (**993**) from *Riccardia multifida* (Metzgeriales) and *Marchantia polymorpha* (Marchantiales), respectively as well as marchantin B, an unidentified bis-bibenzyl from *M. polymorpha* (*19*). Since then riccardin B (**987a**) has been isolated from *R. multifida* together with riccardin A (**985a**) (*79*). European *Preissia quadorata* produces riccardin B (**987a**) (*249a*). Riccardin A (**985a**) contains an 18-membered ring whose benzene units are connected by an ether oxygen between benzene rings A and B and by a biphenyl bond between the benzene rings C and D. In the ^{1}H-NMR spectrum a strongly shielded one proton doublet at δ 5.33 (1H) has been assigned to an inner proton (H-3) on benzene ring A which lies over the plane of benzene ring B.

The IR and ^{1}H-NMR spectra of riccardin B (**987a**), its dimethoxy derivative (**987b**) and its diacetate (**987c**) quite resembled those of (**985a**), its dimethyl ether (**985b**) and its diacetate (**985c**), indicating that riccardin B might be a desmethoxy derivative (**987a**) of (**985a**) with an additional ether linkage between C-12 and C-11′ in place of the biphenyl bond. The structures assigned to riccardin A and B have been confirmed by total synthesis. That of riccardin A (**985a**) has been achieved by GOTTSEGEN *et al.* (*216*) by mean of Wittig reaction. Riccardin B (**987a**) has been synthesized in three different laboratories. The macrocyclic bis-bibenzyl framework was constructed by an intermolecular Wadsworth-Emmons olefination of phosphonate which was prepared from diethyl[4-[2-methoxy-5-(1,3-dioxan-2-yl)-phenoxy]benzyl] phosphate and methyl-3-methoxy(3-formylphenoxy) benzoate (*334, 504*). A second efficient and short-step synthesis of riccardin B (**987a**) by mean of nickel-catalyzed intramolecular cyclization of acyclic precursors has been accomplished by IYODA *et al.* (*302*). GOTTSEGEN *et al.* (*216*) and NÓGRADI *et al.* (*447*) synthesized riccardin B (**987a**) *via* Ullmann coupling, Wittig reaction and an intermolecular Wurtz reaction using sodium and tetraphenylethene. Riccardin C (**988**), demethylated riccardin A (**985a**), first isolated from *Reboulia hemisphaerica* (*55*) has also been obtained from South African (*61*) and Indian *Marchantia polymorpha* (*72*) and Indian *M. palmata* (*72*) and New Zealand *Monoclea forsteri* (*590*), *Ricciocarpos natans* (*351*) and Japanese *Blasia pusilla* (*249*). Methylation of (**988**) gave a trimethyl ether identical with the dimethyl ether (**985b**) of riccardin A (**985a**). The structure of riccardin C (**988**) has been confirmed by total synthesis (*215, 216*).

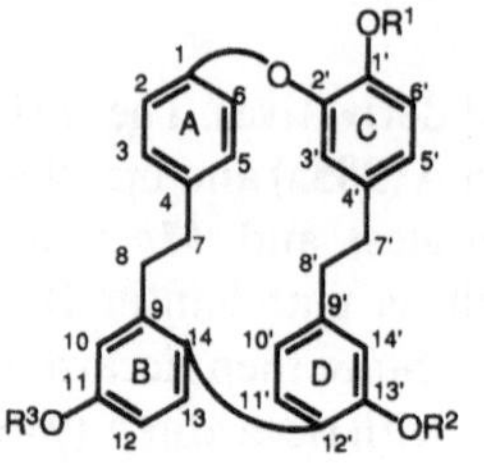

(985a) Riccardin A ; R¹=R²=H, R³=Me
(985b) R¹=R²=R³=Me
(985c) R¹=R²=Ac, R³=Me
(986) Riccardin F ; R¹=Me, R²=R³=H

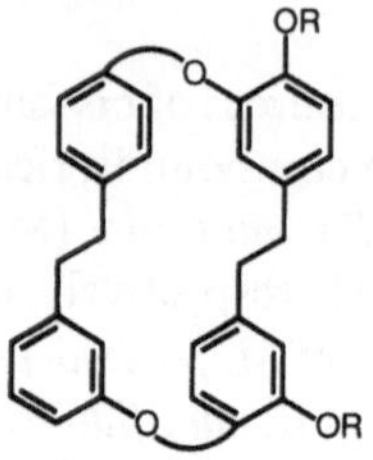

(987a) Riccardin B ; R=H
(987b) R=Me
(987c) R=Ac

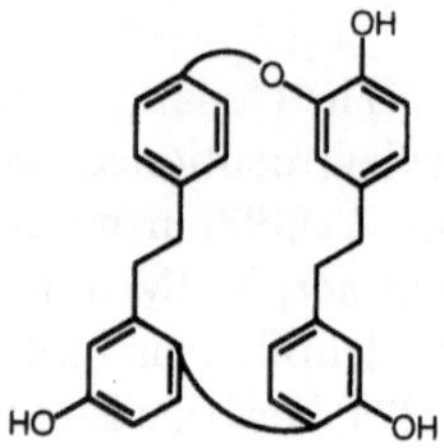

(988) Riccardin C

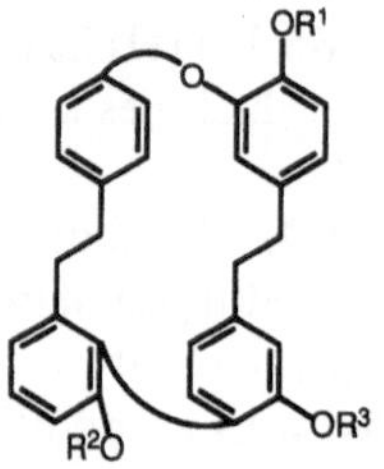

(989a) Riccardin D ; R¹=R²=R³=H
(989b) R¹=R²=R³=Me
(990) Riccardin E ; R¹=R³=H R²=Me
(991) Riccardin G ; R¹=Me R²=R³=H

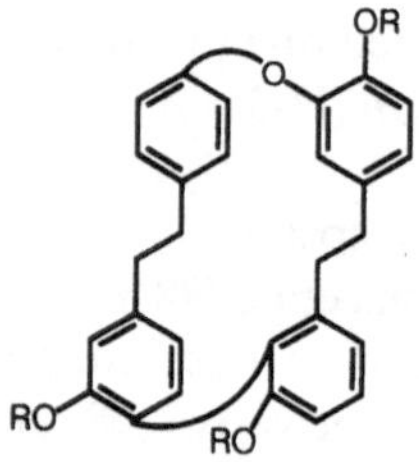

(992a) Isoriccardin C ; R=H
(992b) R=Me

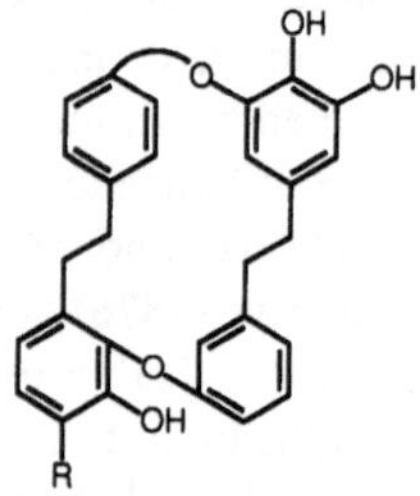

(993) Marchantin A ; R=H
(994) Marchantin B ; R=OH

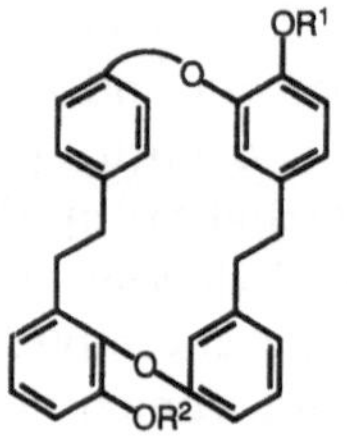

(995) Marchantin C ; R¹=R²=H
(996) Marchantin C dimethyl ether ; R¹=R²=Me
(997) Marchantin C monomethyl ether
    (= Marchantin O) ; R¹=H, R²=Me
(998) Marchantin P ; R¹=Me, R²=H

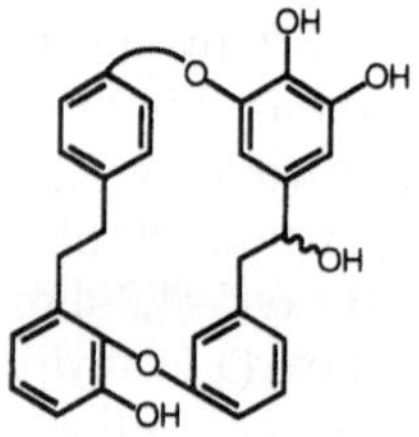

(999) Marchantin D

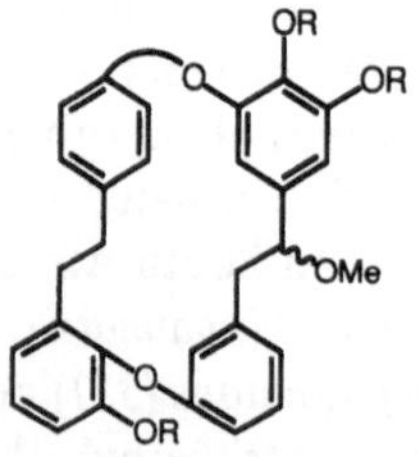

(1000a) Marchantin E ; R=H
(1000b) R=Me

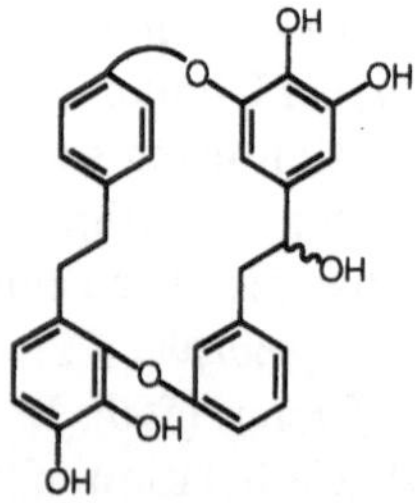

(1001) Marchantin F

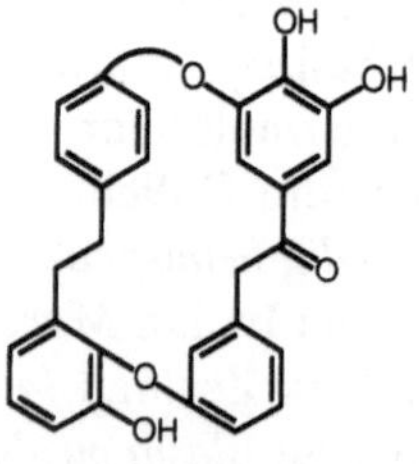

(1002) Marchantin G

Chart 62a. Bis-bibenzyls found in the Hepaticae and their derivatives

Biosynthesis of the riccardins may occur *via* lunularic acid (**877**) or lunularin (**878**) which is widely distributed in leafy and thalloid liverworts.

Two new bis-bibenzyl derivatives, riccardins D (**989a**) and E (**990**) have been isolated from the New Zealand thalloid liverwort *Monoclea forsteri*, along with riccardin C (**988**) and perrottetin E (**1023a**) (*590*). Structure elucidation of (**989a**) and (**990**) was carried out by extensive spin decoupling and NOE spectrometry on the naturally occurring compounds and their tri- and dimethyl ethers (**989b**) and by spectroscopic comparison with riccardin A (**985a**). Birch reduction of (**989b**) gave the non-cyclic bis-bibenzyl derivative (**989c**), which incorporated a *p*-methoxybenzyl and a *p*-hydroxyl benzyl group (Scheme 61). The location of these three methoxyls and the single hydroxyl group was established by spin decoupling experiments. Riccardin F (**986**) previously named riccardin D, has been isolated from *Marchantia tosana* (*74*) and *Blasia pusilla* (*249*) and its structure established by comparing its spectral data with those of riccardins A (**985a**) and C (**988**). *Marchantia chenopoda* contains riccardin G (**991**), the isomer of (**990**) (*553*).

The common thalloid liverwort, *Marchantia polymorpha* belonging to Marchantiaceae (Marchantiales) grows on wet soil. It was known that extracts of this species had inhibitory activity against Gram positive bacteria, produced allergenic contact dermatitis and were diuretic (see Chapter V). The extract was chromatographed on silica gel and then Sephadex LH-20 ($CHCl_3$-MeOH 1:1) to give seven macrocyclic bis-bibenzyls, marchantins A (**993**), B (**994**), C (**995**), D (**999**), E (**1000a**), F (**1001**) and G (**1002**) among which marchantin A was the major component (*22, 74, 78, 579*). Marchantin A (**993**) and the related compounds (**994–1002, 1005a, 1006a, 1007a**) have been isolated from French (*74*),

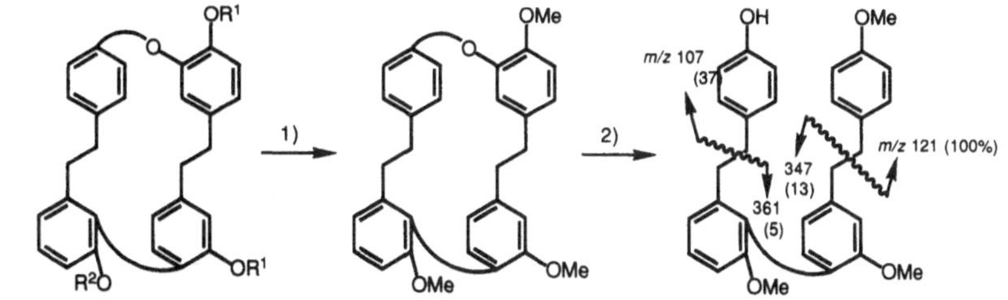

(**989a**) Riccardin D ; $R^1 = R^2 = H$
(**990**) Riccardin E ; $R^1 = H$, $R^2 = Me$
(**989b**)
(**989c**)

1) MeI/Me$_2$CO/K$_2$CO$_3$ 2) Na/liq. NH$_3$

Scheme 61. Reactions of riccardins D and E

German (*71*), Indian (*72*), Malaysian (*553*) and South African *M. poly-morpha* (*61*), Ecuadorian *M. chenopoda* (*553*), *M. plicata* (*433*), Japanese *M. paleacea* var. *diptera* and *M. tosana* (*74, 78*) as shown in Table IIe. Marchantins A, B and C were found not only in the thallus (female, male and sterile), but also in the capsule with peduncle of *M. polymorpha* and *M. paleacea* var. *diptera* (*78*), while the marchantins of *M. tosana* were found only in its sterile thallus (*78*). Marchantins A and B were also isolated from Malaysian *Wiesnerella denudata* belonging to the Conoce-phalaceae (Marchantiales) (*51*) while marchantin C (**995**) has been isolated from *Dumortiera hirsuta* (Marchantiaceae) (*585*) and *Reboulia hemisphaerica* (Aytoniaceae) (*614*). The latter species also produces marchantin C di- (**996**) (*614*) and monomethyl ether (= marchantin O) (**997**) (*249*). Marchantin P (**998**), the isomer of (**997**), has been isolated from *Marchantia chenopoda* (*553*). It is interesting that marchantin C (**995**) is distributed not only in the Marchantiales, including *Monoclea forsteri* (*513*) and *M. gottschei* subsp. *neotropica* (*514*), but also in the Jungerman-niales, such as *Plagiochila sciophila* (= *P. acanthophylla* subsp. *japonica*) (*259*) and *Schistochila glaucescens* (*562*).

The yield of marchantin A (**993**) depends on the species. For example, pure marchantin A (**993**) has been isolated in 100–120 g quantity from Japanese *M. paleaceae* var. *diptera* (from 2 kg of dried material) whereas the total extract of German *M. polymorpha* contains *ca* 20% of marchan-tin A (**993**) (*71*). Its structure was established by a combination of UV, IR, ^{1}H- and ^{13}C-NMR spectrometry, the chemical transformations shown in Scheme 62, as well as an X-ray crystallographic analysis of the trimethyl ether (**993a**) (*22, 579*). The ^{1}H-NMR spectrum of (**993**) contains signals of four benzylic methylenes, signals of three OH protons which disappear on addition of D_2O and thirteen protons on benzene rings. That marchantin A (**993**) possessed three phenolic hydroxyl groups followed from methylation with methyl iodide which gave a trimethyl ether (**993a**) and acetylation which gave a triacetate (**993b**). Treatment of (**993**) with methylene iodide in dimethylsulfoxide in the presence of cupric oxide gave a methylene dioxide (**993c**), indicating two vicinal phenolic hydrox-yls. The remaining two oxygen atoms were ether oxygens since the IR spectrum of (**993a**) contained neither carbonyl nor hydroxyl absorption bands. Hence (**993**) was a cyclic bis-bibenzyl in which the two bibenzyls units were linked by two ether oxygens. Hydrogenation of (**993a**) in the presence of platinum oxide gave a hydrogenolysis product (**993d**), which afforded a monoacetate (**993e**) and a tetramethyl ether (**993f**), respective-ly. Methylation of (**993d**) with CD_3I gave a trideuterated methyl ether (**993g**). In the mass spectra of (**993d–993g**), base peaks were observed at *m/z* 167, 167, 181 and 184, together with an intense fragment ion at *m/z* 91,

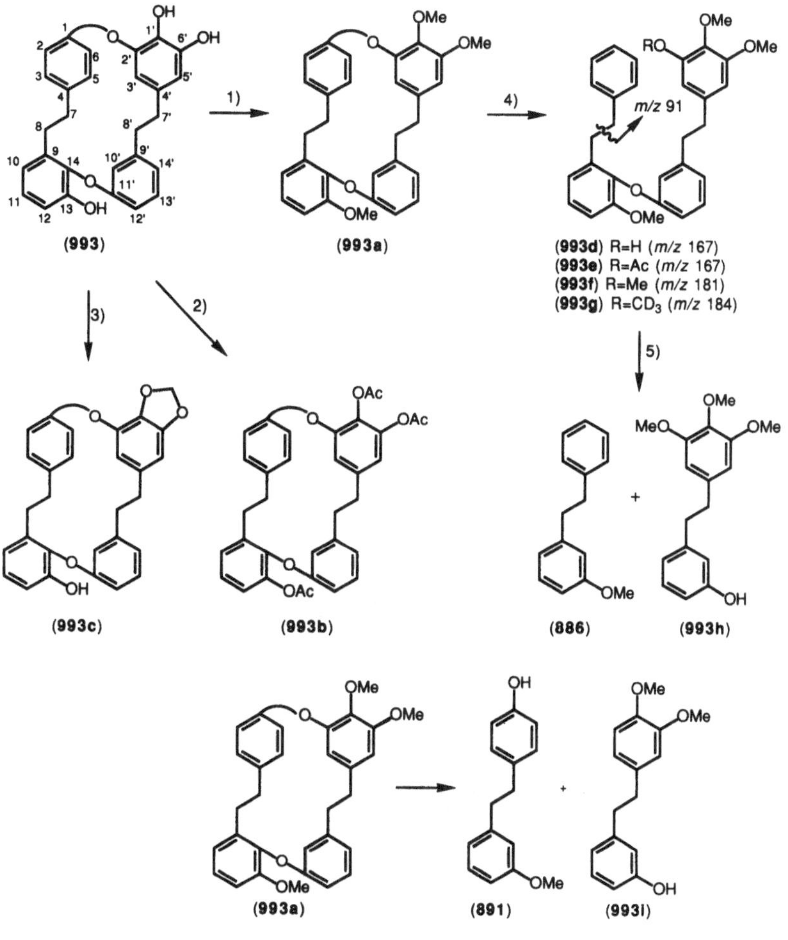

1) MeI/OH 2) Ac$_2$O/Py 3) CH$_2$I$_2$/CuO/DMSO 4) H$_2$/PtO$_2$ 5) Na/liq. NH$_3$

Scheme 62. Reactions of marchantin A

the base peak of **(993e)** being due to fragment ion which had lost acetate. This implied that cleavage of one of the ether linkages had occurred at C-4. Treatment of **(993f)** with sodium in liquid ammonia furnished a monomethoxybibenzyl **(886)** and a monohydroxytrimethoxybibenzyl **(993h)**. In the ^{1}H-NMR spectrum of **(886)**, A$_2$B$_2$ signals were absent, indicating that the methoxyl group was placed at C-13 or C-14. In the ^{1}H-NMR spectrum of **(993h)**, signals of *meta*-coupled protons were observed at δ6.20 and 6.40, suggesting that C-1', C-2' and C-6' were substituted.

The two bibenzyls were synthesized by Wittig reactions of benzyltriphen-ylphosphonium bromide with m-methoxybenzaldehyde and of 3,4,5-trimethoxybenzyltriphenylphosphonium bromide with m-hydroxybenz-aldehyde, respectively (*36a, 579*). Hence, one ether oxygen linked C-2′ and the A-ring and another oxygen linked between C-11′ and B-ring.

Birch reduction of (**993a**) gave two bibenzyl derivatives (**891**) and (**993i**) whose structures were also established by synthesis (*36a, 579*). This showed that one of the ether oxygens linked C-1 and C-2′ and another oxygen linked C-11′ and the B-ring. That the second ether oxygen was also attached to C-14 was established by NMR spectrometry including the ^{13}C-^{1}H and long range ^{13}C-^{1}H-2D-COSY and NOE experiments.

(993i)

These results coupled with the Birch reduction of (**993a**) and (**993f**) established the substitution patterns of the four benzene rings leading to (**993**) as the full structure of marchantin A. Marchantin A (**993**) and its derivatives are viscous gums, however, the trimethyl ether (**993a**) furnished crystals suitable for X-ray analysis. In the ^{1}H-NMR spectra of (**993**) and (**993a**), H-3′ appears at unusually high field ($\delta 5.13$). This is understandable as the result of the paramagnetic effect of two benzene rings A and D between which H-3′ is sandwiched. The total synthesis of marchantin A (**993**) in 12 steps has been accomplished by KODAMA *et al.* (*333, 334*). Structures of the other marchantins B-G (**994, 995, 999, 1000a, 1001, 1002**) have been established in a similar manner and chemical correlation. A total synthesis of marchantin B (**994**) was carried out by HA *et al.* (*225*).

^{1}H- and ^{13}C-NMR spectra of the marchantins and their derivatives have been completely assigned using NOE difference, ^{13}C-^{1}H-2D-COSY and long range proton selective decoupling (LSPD) spectra; this permitted derivation of the structures of marchantin H (**1003**) from *Plagiochasma intermedium* (Marchantiales) and marchantin I (**1004**) from *Riccardia multifida* (Metzgeriales) (*550, 576*). The former bis (bibenzyl) had been isolated from South African *Marchantia polymorpha* (*61*) and *M. diptera* (*623*) and *Plagiochila sciophila* (= *P. acanthophylla* subsp.

japonica) (*259*). Marchantins H (**1003**) and I (**1004**) were synthesized by HA *et al.* (*225*) and DIENES *et al.* (*163*), respectively.

A collection of *Marchantia polymorpha*, from Germany furnished the new marchantins J (**1005a**), K (**1006a**) and L (**1007a**), together with marchantins A–E (**993, 994, 995, 999, 1000a**) and G (**1002**) (*71*). Methylation and acetylation of (**1005a**) gave a trimethyl ether (**1005b**) and a triacetate (**1005c**), indicating the presence of three phenolic hydroxyl groups. The presence of an ethoxyl group at the benzylic position were established by the ^{1}H- and ^{13}C-NMR spectra. The ^{13}C-NMR spectrum of (**1005a**) was almost identical with that of marchantin E trimethyl ether (**1000b**), except for replacement of one of the methoxyls by an ethoxyl group indicating that (**1005a**) was monoethoxymarchantin A. This was confirmed by NMR spectrometry of (**1005b**) including NOE difference spectra. This constitutes the first record of an ethoxylated compound from bryophytes.

Methylation and acetylation of (**1006a**) gave a pentamethyl ether (**1006b**) and a tetraacetate (**1006c**), respectively. The ^{1}H-NMR signals of (**1006b**) were similar to those of marchantin F (**1001**), except for the presence of the methoxyl groups, suggesting that (**1006b**) possessed the same skeleton as marchantin F (**1001**). The NOE difference spectrum and spin decoupling experiments clarified the position not only of the methoxyl groups but also the substitution pattern in the four benzene rings of (**1006b**). Marchantin L (**1007a**) was methylated and acetylated to afford a trimethyl ether (**1007b**) and a tetraacetate (**1007c**), respectively, indicating that (**1007a**) possessed three phenolic hydroxyl groups and a hydroxyl group on an aliphatic carbon atom. The ^{1}H- and ^{13}C-NMR spectra of (**1007b**) resembled those of marchantin A trimethyl ether (**993a**), except for the presence of one secondary hydroxyl group on a benzylic carbon atom, indicating that (**1007b**) was marchantin A trimethyl ether with a hydroxyl group on C-7, 8, 7' or 8'. The location of the hydroxyl group at C-7 was established by extensive spin decoupling as well as NOE difference spectrometry which permitted complete assignment of the proton and carbon signals of (**1007b**).

Indian *Marchantia polymorpha* elaborates not only marchantins A (**993**), C–E (**994, 999, 1000a**), G (**1002**), riccardin C (**988**) and perrottetin E (**1023a**), but also two new macrocyclic bis-bibenzyls, isoriccardin C (**992a**) and isomarchantin C (**1011a**) (*72*). Isoriccardin C (**992a**), isomarchantin C (**1011a**), riccardin C (**988**), marchantins C (**995**) and G (**1002**) have also been isolated from Indian *M. palmata* (*72*), while isomarchantin C (**1011a**) has been isolated from *Mylia nuda* (Jungermanniales) (*622*) and *Bryopteris filicina* (*432*). Gross structures of (**992a**) and (**1011a**) have been proposed by analysis of the ^{1}H- and ^{13}C-NMR spectra with substitution

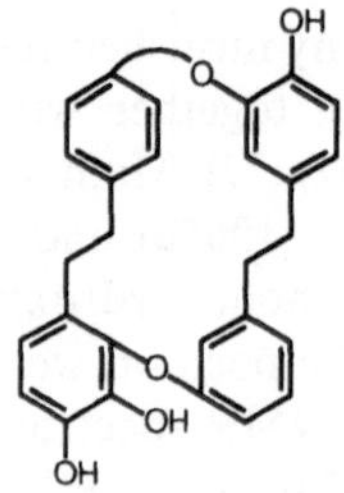

(1003) Marchantin H

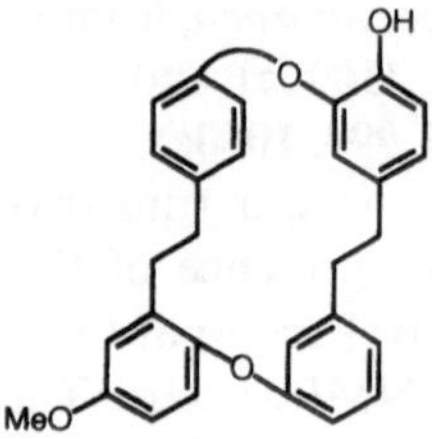

(1004) Marchantin I

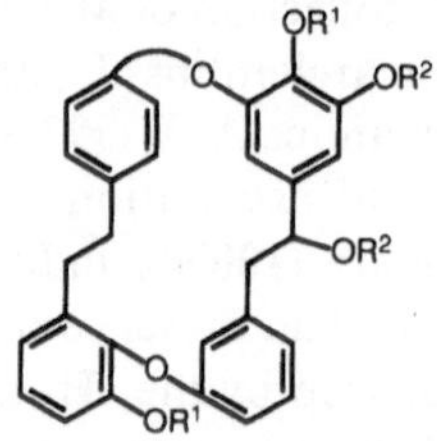

(1005a) Marchantin J ; R^1=H, R^2=Et
(1005b) R^1=Me, R^2=Et
(1005c) R^1=Ac, R^2=Et

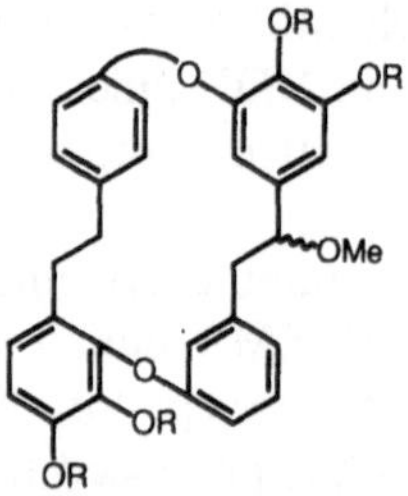

(1006a) Marchantin K ; R=H
(1006b) R=Me
(1006c) R=Ac

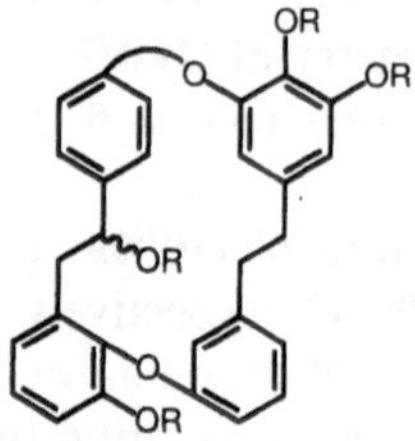

(1007a) Marchantin L ; R=H
(1007b) R=Me
(1007c) R=Ac

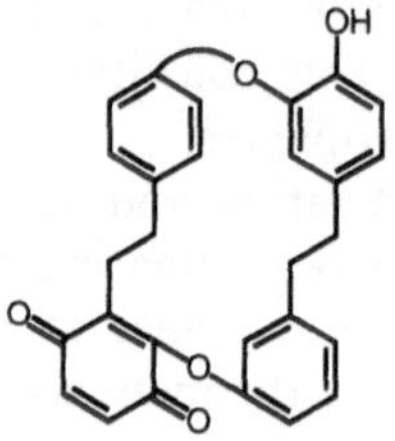

(1008) Marchantinquinone

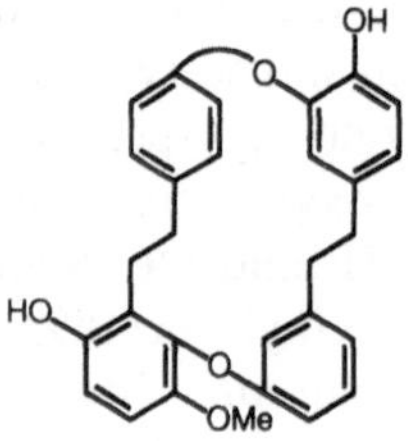

(1009) Marchantin M

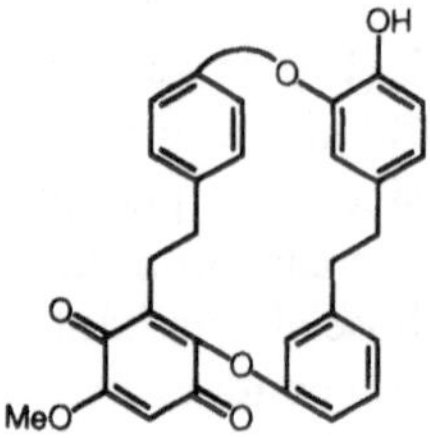

(1010) Marchantin N

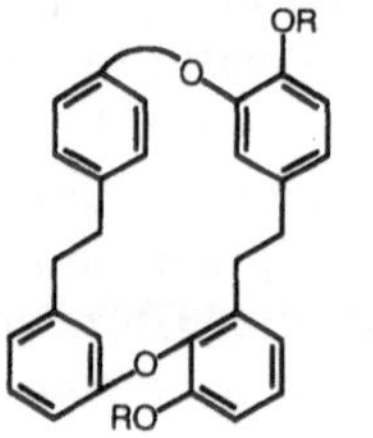

(1011a) Isomarchantin C ; R=H
(1011b) R=Me

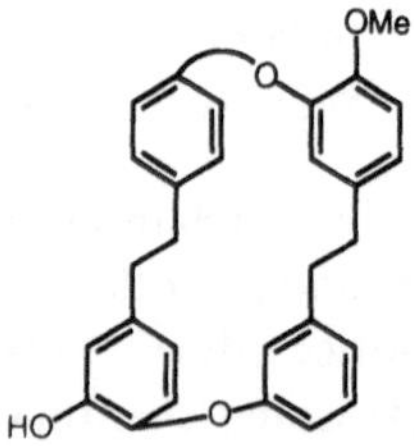

(1012) Pakyonol

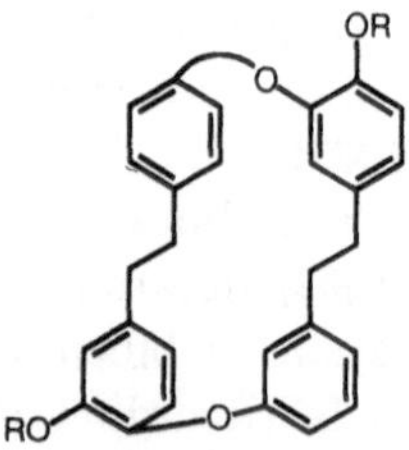

(1013a) Neomarchantin A ; R=H
(1013b) R=Me

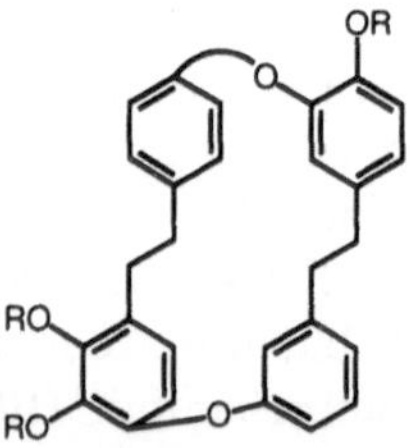

(1014a) Neomarchantin B ; R=H
(1014b) R=Me

Chart 62b. Bis-bibenzyls found in the Hepaticae and their derivatives

pattern on each benzene ring and the position of each functional group being established by spin decoupling and NOE experiments on (**992a**) and (**1011a**), and their methyl ethers (**992b**, **1011b**).

Taiwanese *Mannia subpilosa* produces marchantinquinone (**1008**) (*612*) as well as marchantin M (**1009**) and marchantin N (**1010**) (*613, 623a*). Analysis of the ^{1}H- and ^{13}C-NMR spectra of (**1008**) as well as NOE and spin decoupling experiments showed that (**1008**) was a marchantin-type cyclic bis-bibenzyl with one phenolic hydroxyl group at C-1' and 1,2-disubstituted quinone in the B-ring. Presence of quinone was also evidenced by the UV (373 nm) and IR adsorption bands (1652 and 1672 cm^{-1}) and by a comparison of the ^{13}C-NMR spectrum of (**1008**) with that of 2-methyl-1,4-quinone and 2-methoxy-5-methyl-1,4-quinone. Further support for the structure was provided by the presence in the mass spectrum of intense fragment ions at *m/z* 211 (60%) and 107 (35%) corresponding to double benzylic cleavage and the stable *p*-hydroxytropylium ion. Within Aytoniaceae (= Grimaldiaceae) which belong to the Marchantiales, *Mannia*, *Plagiochasma* and *Reboulia* species produce common macrocyclic bis-bibenzyls.

Mannia fragrans collected in North Korea elaborates not only cuparane-type sesquiterpenoids but also a cyclic bis-bibenzyl, pakyonol (**1012**), whose structure was elucidated by analysis of NMR spectral data (*280*).

Two new cyclic bis-bibenzyls, neomarchantins A (**1013a**) and B (**1014a**) have been isolated from New Zealand *Schistochila glaucescens* (*562*). Structures were elucidated utilizing the ^{1}H- and ^{13}C-NMR and NOE difference spectra of di- (**1013b**) and trimethyl ether (**1014b**). Neomarchantin A (**1013a**) is identical with demethylated pakyonol (**1012**) (*280*). Neomarchantin A has also been isolated from *Monoclea forsteri* (*513*), and *M. gottschei* subsp. *neotropica* (*514*) and *Preissia quadorata* (*249a*).

Plagiochila sciophila (= *P. acanthophylla* subsp. *japonica*) biosynthesizes not only the previously known marchantins C (**995**), H (**1003**) and perrottetin E (**1023a**), the latter a member of the perrottetin class, which contains single *o,p*-ether linkage between two bis-bibenzyl groups, but also the new plagiochins, plagiochins A–D (**1015a**, **1016a**, **1017a**, **1018a**) which possess an additional *ortho* biphenyl linkage between the two benzyl groups (*259*). The structure and stereochemistry of (**1015a**) were established by chemical reactions (methylation and acetylation), ^{1}H-NMR spectrometry and NOE studies on the pentamethyl ether (**1015b**) as well as by an X-ray crystallographic analysis of the latter (**1015b**). This showed that ring A is perpendicular to ring C and parallel with ring D. The proton at C-3' is strongly shielded by both rings A and D, causing a

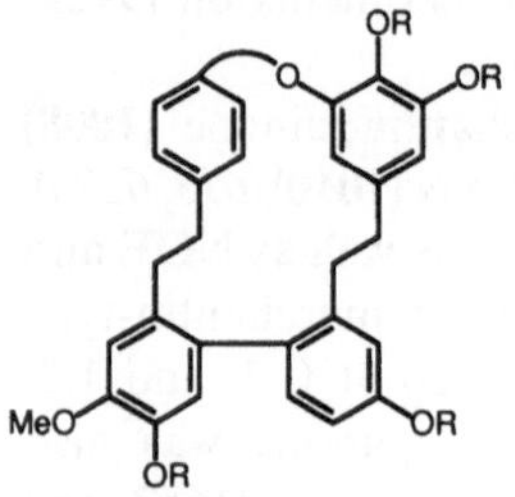

(**1015a**) Plagiochin A ; R=H
(**1015b**) R=Me
(**1015c**) R=Ac

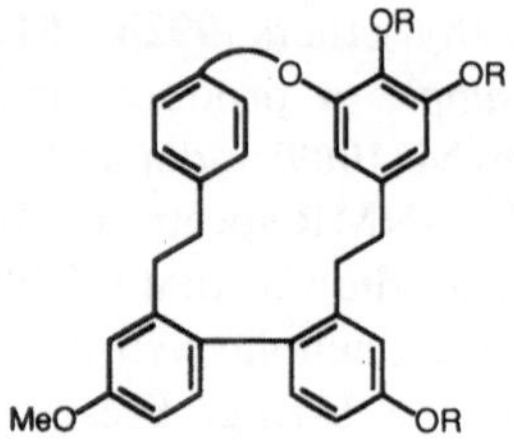

(**1016a**) Plagiochin B ; R=H
(**1016b**) R=Ac

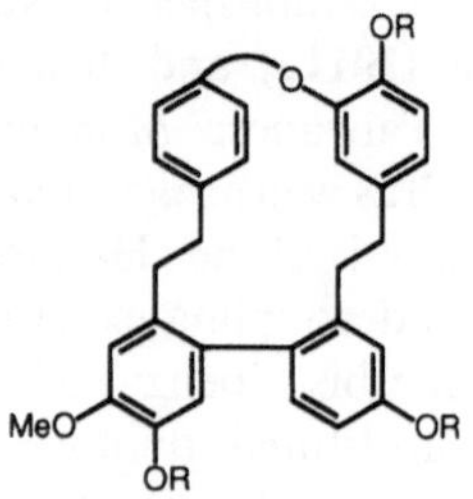

(**1017a**) Plagiochin C ; R=H
(**1017b**) R=Ac

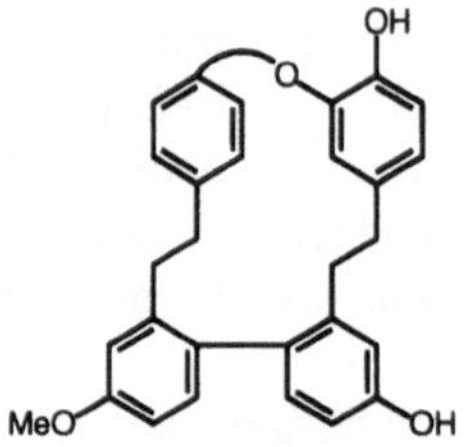

(**1018a**) Plagiochin D ; R=H
(**1018b**) R=Ac

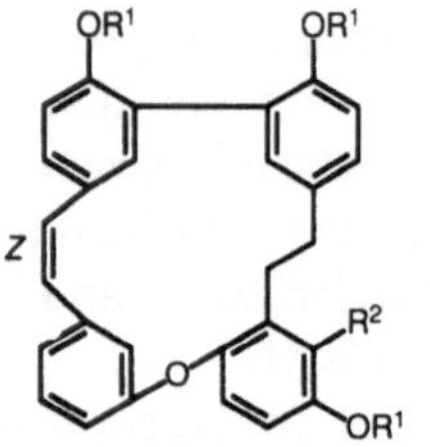

(**1019**) Isoplagiochin A ; $R^1=R^2=H$
(**1019a**) $R^1=Ac, R^2=H$
(**1019b**) $R^1=Me, R^2=H$
(**1020**) Isoplagiochin B ; $R^1=H, R^2=OH$

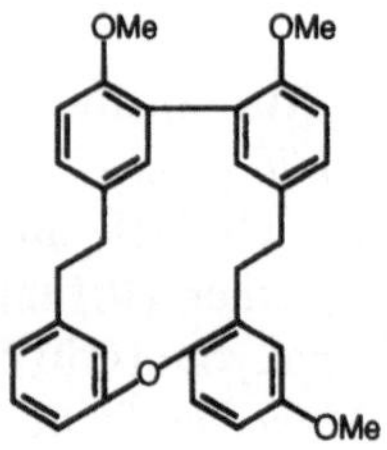

(**1019c**)

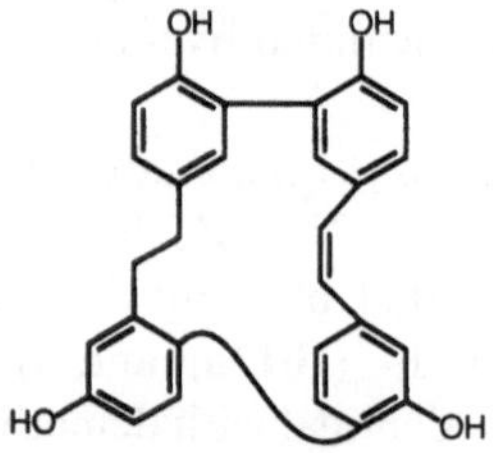

(**1021**) Isoplagiochin C

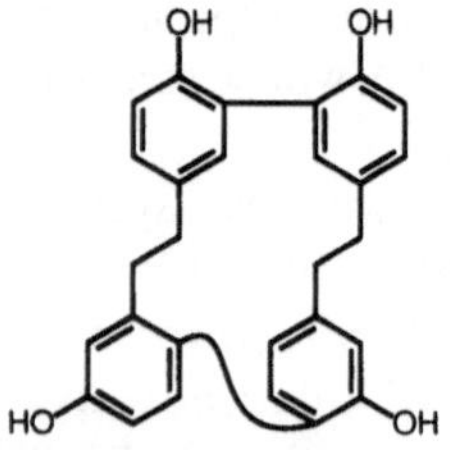

(**1022a**) Isoplagiochin D ; R=H
(**1022b**) R=Ac

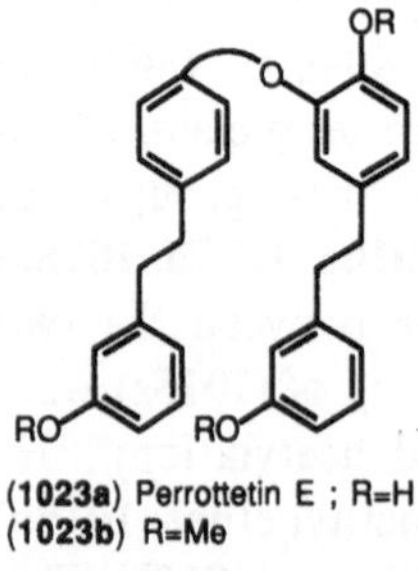

(**1023a**) Perrottetin E ; R=H
(**1023b**) R=Me

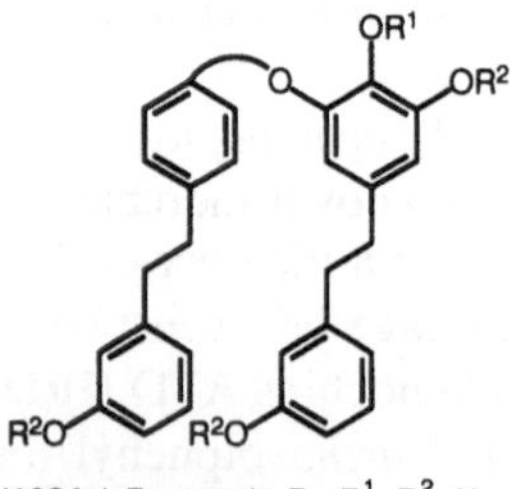

(**1024a**) Perrottetin F ; $R^1=R^2=H$
(**1024b**) $R^1=R^2=Me$
(**1024c**) $R^1=R^2=Ac$
(**1025**) Perrottetin G ; $R^1=Me, R^2=H$

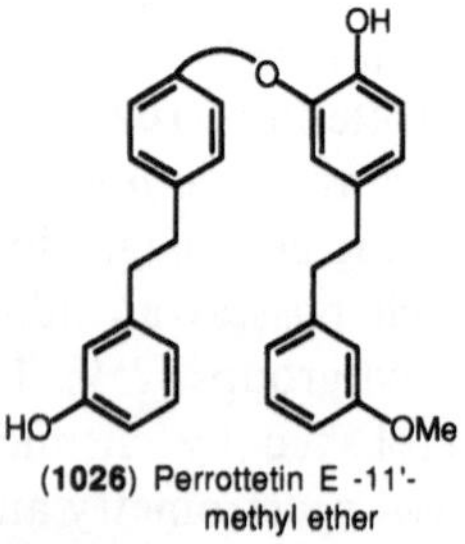

(**1026**) Perrottetin E -11'-
          methyl ether

Chart 62c. Bis-bibenzyls found in the Hepaticae and their derivatives

high field shift to δ4.83 in (**1015a**). This phenomenon has been also encountered in the series of marchantins and riccardins described earlier (*550, 576*). Structures of the remaining cyclic bis-bibenzyls (**1016a, 1017a, 1018a**) were also deduced from the double resonance and NOE difference spectra of acetylated derivatives (**1016b, 1017b, 1018b**). Plagiochins C (**1017a**) and D (**1018a**) have been synthesized by KESERÜ *et al.* (*323*).

Plagiochila fruticosa produces not only 2,3-secoaromadendranes (*19*) but also two new cyclic bis-bibenzyls, isoplagiochin A (**1019**) and isoplagiochin B (**1020**) (*249*). The structure of (**1019**) was established by a combination of 2D-COSY techniques and NOE spectrometry on triacetate (**1019a**) and trimethyl ether (**1019b**) and the preparation of dihydro-derivative (**1019c**) and X-ray crystallographic analysis of (**1019a**). The structure of (**1020**) was based on spectroscopic comparison with (**1019**). Further fractionation of the methanol extract of *P. fruticosa* yielded two additional bis-bibenzyls, isoplagiochin C (**1021**) and isoplagiochin D (**1022a**) whose structures are based on spectroscopic comparison with (**1019**) and (**1019c**) (*262b*).

Radula species are rich sources not only of bibenzyls but also bis-bibenzyls. From *R. perrottetii* (*598*) and *R. kojana* (*50*), the new perrottetin E (**1023a**) and perrottetin F (**1024a**) have been isolated. The former species elaborates an additional new bis-bibenzyl, perrottetin G (**1025**) (*598*). Perrottetin F (**1024a**) has also been isolated from *Lunularia cruciata* (*255*). Structure (**1023a**) was deduced by extensive NMR studies of (**1023a**) and its trimethyl ether (**1023b**) and was confirmed by total synthesis of (**1023a**) (*598, 647*). The presence of one additional hydroxyl group at C-2′ in perrottetin F (**1024a**) was deduced by NMR studies of its tetramethyl ether (**1024b**) and confirmed by total synthesis (*403*). Methylation of perrottetin G (**1025**) gave a tetramethyl ether which was identical with (**1024b**) prepared from (**1024a**), indicating that (**1025**) was a monomethyl ether of (**1024a**). The position of the methoxyl group of (**1025**) on 1′ was based on the absence of an NOE on irradiation of the methoxyl group. Conclusive evidence for the structure of (**1025**) was also obtained by its total synthesis (*403*).

Perrottetin E (**1023a**) has also been isolated from *Jungermannia comata* (*437*), *Marchantia polymorpha* (*72*), *Monoclea forsteri* (*513, 590*), *M. gottschei* subsp. *neotropica* (*514*), *Nardia subclavata* (*585*), *Pellia endiviifolia* (*254*), and *Plagiochila sciophila* (= *P. acanthophylla* subsp. *japonica*) (*259*). Two new bibenzyl ethers, perrottetin E-11′-methyl ether (**1026**) and 14-hydroxyperrottetin E-11′-methyl ether (**1027**) have been isolated from *Pellia endiviifolia*, along with perrottetin E (**1023a**). Structures were established by spectroscopic methods including NOE experiments, the usual alkylations, acylations and hydrogenolysis reactions and

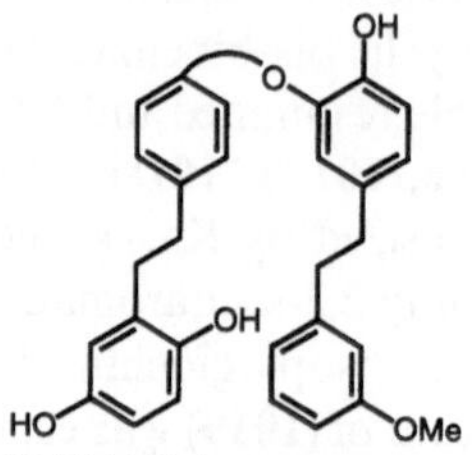

(1027) 14-Hydroxyperrottetin E -
11'-methyl ether

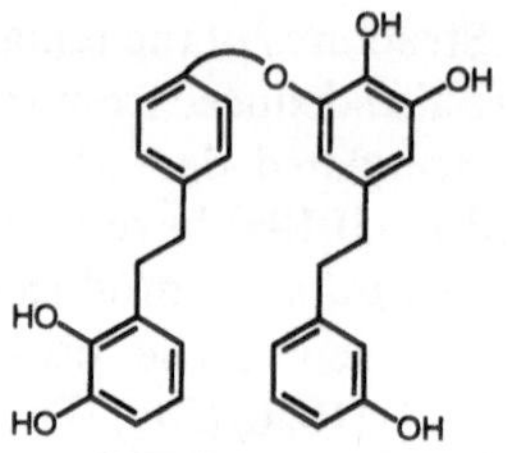

(1028) Perrottetin H

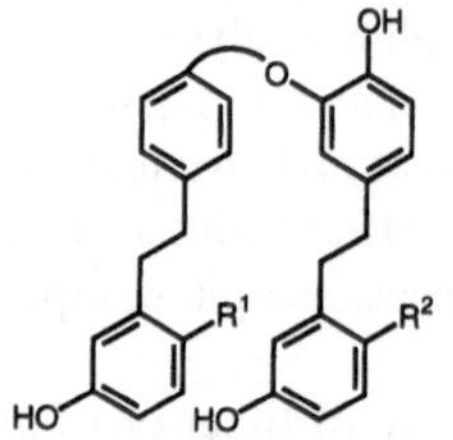

(1029) 14-Hydroxyperrottetin E ; R^1=OH R^2=H
(1030) 14'-Hydroxyperrottetin E ; R^1=H R^2=OH
(1031) 14,14'-Dihydroxyperrottetin E ; R^1=R^2=OH

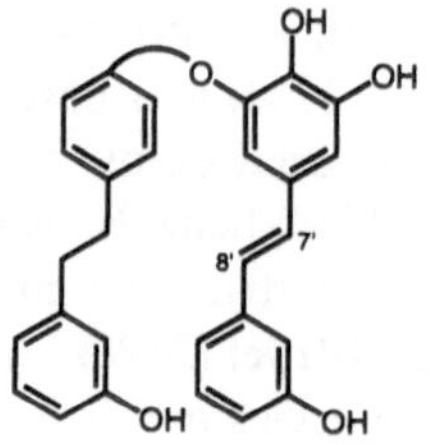

(1032) 7',8'-Dehydroperrottetin F

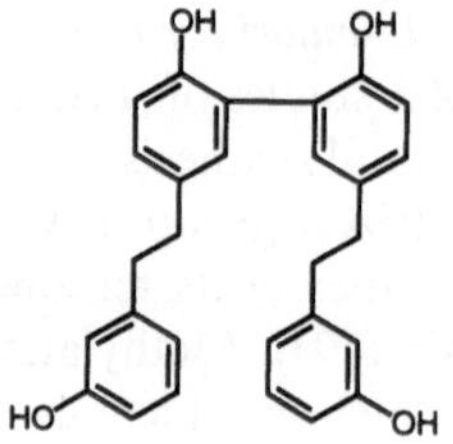

(1033) Isoperrotetin A

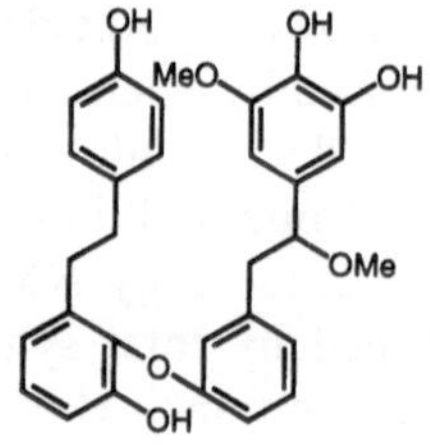

(1034) Paleatin A

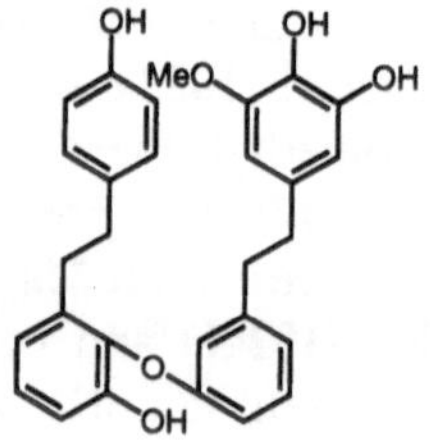

(1035) Paleatin B

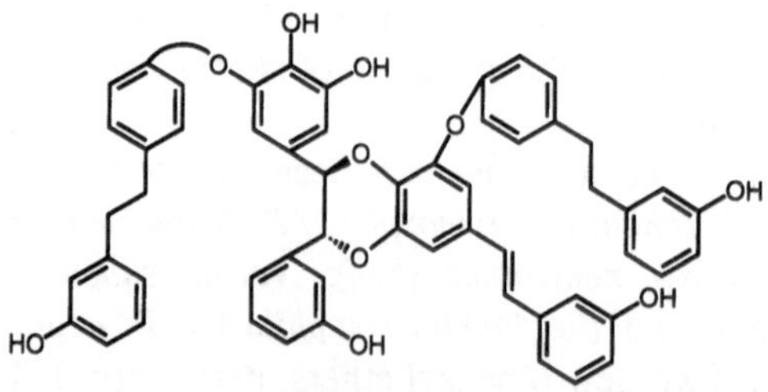

(1036) Cruciatin

Chart 62d. Bis-bibenzyls found in the Hepaticae

by total syntheses (*254*). *Jubula japonica* elaborates a new bis-bibenzyl, perrottetin H (**1028**) (= 14-hydroxyperrottetin F) (*585b*) while *Pellia epiphylla* contains three new bis-bibenzyls, 14-hydroxyperrottetin E (**1029**), 14′-hydroxyperrottetin E (**1030**) and 14,14′-dihydroxyperrottetin E (**1031**), along with perrottetin E (**1023a**), whose structures were elucidated by spectroscopic comparison with 14-hydroxyperrottetin E-11′-methyl ether (**1027**) and NOE spectrometry (*159a*). A new bis-bibenzyl, 7′,8′-dehydroperrottetin F (**1032**) has been isolated from *Lunularia cruciata*, along with perrottetin F (**1024a**) (*255*). The structure assigned to (**1032**) is based on spectral data and correlation with perrottetin F (**1024a**) by hydrogenation. The position of the double bond at C7′–C8′ was confirmed by NOE spectrometry of the tetramethyl ether. *Radula perrottetii* contains isoperrottetin A (**1033**) (*585a*), a new bis-bibenzyl, biogenetically correlated with isoplagiochin D (**1022a**) (*262b*).

Further fractionation of the methanol extract of *Marchantia paleacea* var. *diptera* resulted in the isolation of the new paleatin A (**1034**) and paleatin B (**1035**) (*255*). The structure of (**1034**) was established by ^{13}C-^{1}H and long range ^{13}C-^{1}H-2D-COSY techniques, NOE spectrometry on hexamethyl ether (**1034b**) and tetraacetate (**1034a**), by comparing spectral data of (**1034b**) with those of the pentamethyl ether (**1000c**) prepared from marchantin E (**1000a**) as well as by the chemical transformations and correlation shown in Scheme 63. Structure (**1035**) is based on spectroscopic comparison with (**1034**) and on the correlation also shown in Scheme 63. These phenolic compounds are of interest because they are the linear analogues of the macrocyclic bis-bibenzyl ethers which have been found in Jungermanniales, Marchantiales, Metzgeriales and Monocleales and are possible biogenetic precursors of the marchantins, plagiochins and riccardins.

In addition to (**1024a**) and (**1032**), Japanese *L. cruciata* also yielded a highly unusual bis-bibenzyl dimer, cruciatin (**1036**) (*249*). The structure of (**1036**) was established by analysis of ^{1}H- and ^{13}C-NMR spectra as well as by the chemical degradation outlined in Scheme 64. Methylation and acetylation of (**1036**) gave a hexaacetate (**1036a**) and hexamethyl ether (**1036b**), respectively, indicating that (**1036**) possessed six phenolic hydroxyl groups. Birch reduction of the hexamethyl ether (**1036b**) afforded two bibenzyls (**891**) and (**1042h**) and bis-bibenzyl (**1024f**) all these of which were synthesized from *p*-benzyloxybenzaldehyde (**1084a**) and 3,4-dimethoxybenzaldehyde (**1042f**) by a Wittig reaction and from naturally occurring perrottetin F (**1024a**) by dimethylation, respectively. Cruciatin (**1036**) is a dimer of perrottetin F (**1024a**), two phenolic hydroxyl groups of which are linked to two benzylic methylenes at C-7′ and C-8′ of another perrottetin E molecule.

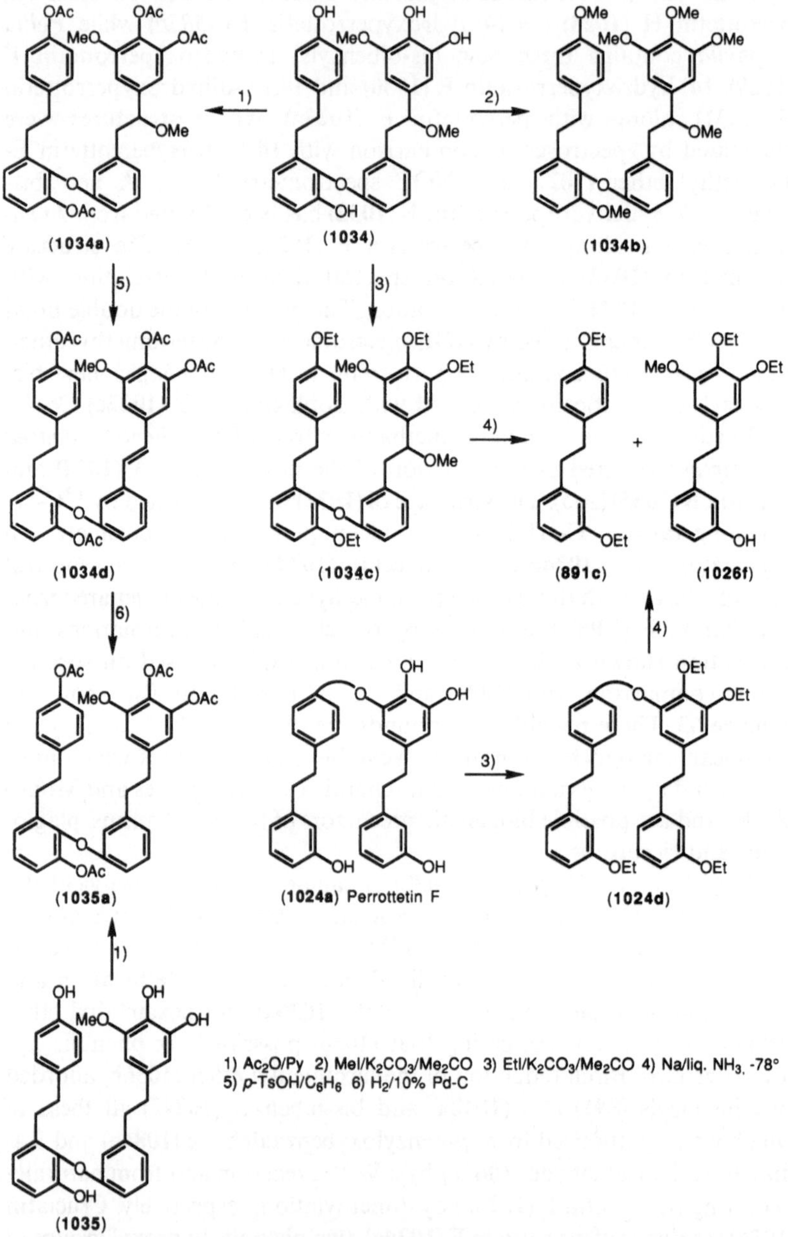

Scheme 63. Reactions of paleatins A and B and correlation with perrottetin E

1) Ac₂O/Py 2) MeI/K₂CO₃/Me₂CO 3) EtI/K₂CO₃/Me₂CO 4) Na/liq. NH₃, -78°
5) p-TsOH/C₆H₆ 6) H₂/10% Pd-C

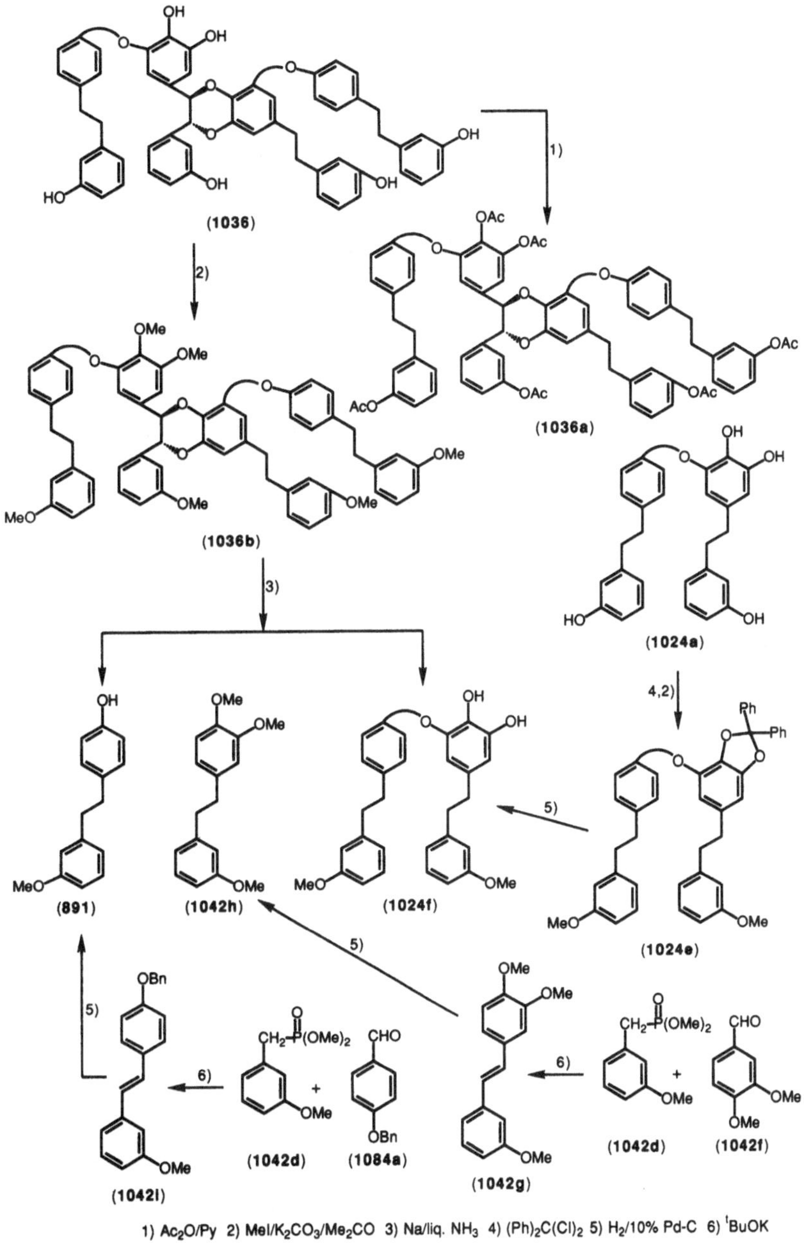
(1036)
(1036a)
(1036b)
(1024a)
(891)
(1042h)
(1024f)
(1024e)
(1042i)
(1042d)
(1084a)
(1042g)
(1042d)
(1042f)
1) Ac_2O/Py 2) MeI/K_2CO_3/Me_2CO 3) Na/liq. NH_3 4) $(Ph)_2C(Cl)_2$ 5) H_2/10% Pd-C 6) ^tBuOK

Scheme 64. Reactions of cruciatin and correlation with perrottetin E

As mentioned earlier, *Blasia pusilla* contained phytosterols, lunularic acid **(877)** and lunularin **(878)** (*19*). Further investigation of the methanol extract of *B. pusilla* resulted in the isolation of four new cyclic bis-bibenzyl dimers, pusilatins A–D **(1037a, 1038a, 1039a, 1040a)**. The structure of **(1037a)** has been established by a combination of 600 MHz NMR spectrometry of the parent compound and its hexaacetate **(1037b)** and X-ray crystallographic analysis of the latter derivative. Structures of the remaining compounds are based on ^{1}H and ^{13}C NMR spectrometry of their permethylated or peracetylated derivatives and spectroscopic comparison with those of **(1037a)** (*262a–c*).

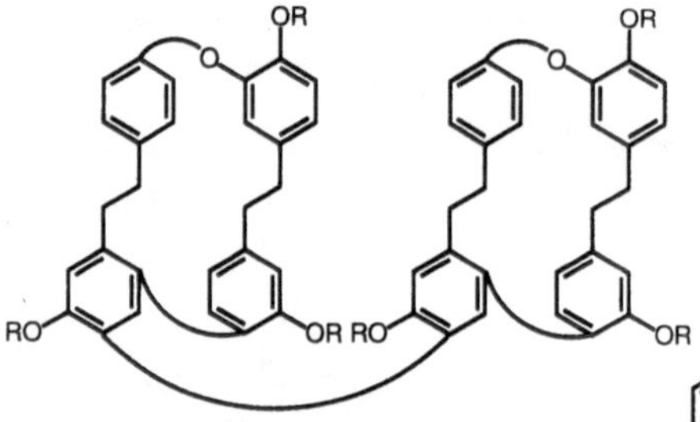

(**1037a**) Pusilatin A ; R=H
(**1037b**) R=Ac

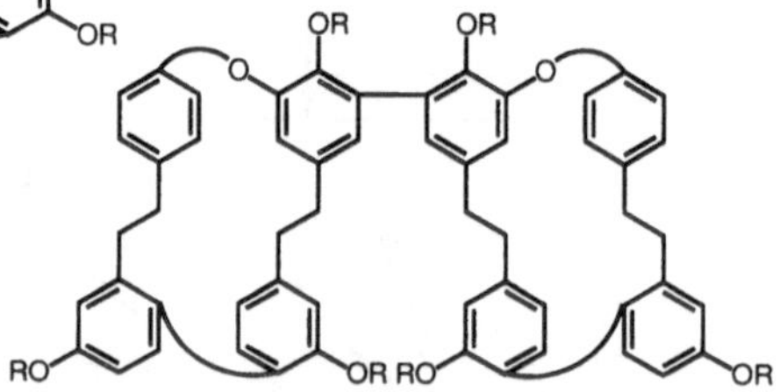

(**1038a**) Pusilatin B ; R=H
(**1038b**) R=Ac
(**1038c**) R=Me

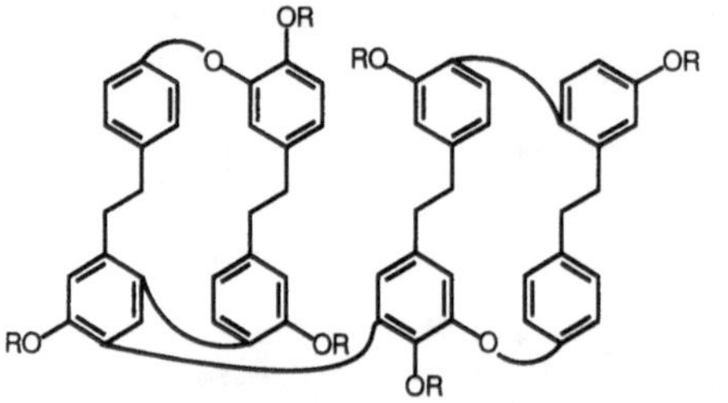

(**1039a**) Pusilatin C ; R=H
(**1039b**) R=Ac
(**1039c**) R=Me

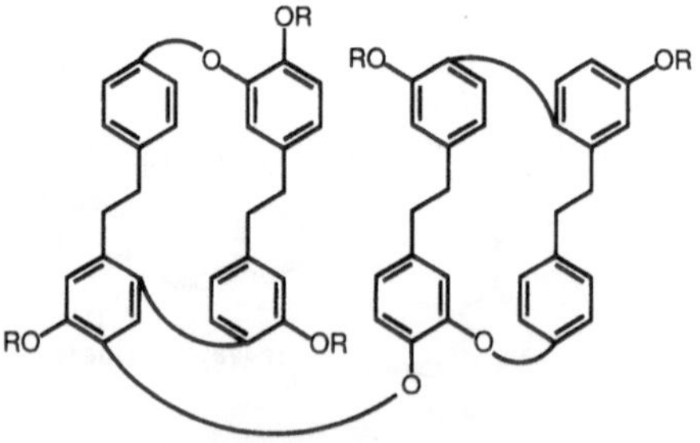

(**1040a**) Pusilatin D ; R=H
(**1040b**) R=Ac

Chart 62e. Bis-bibenzyls found in the Hepaticae and their derivatives

The minimum strain energies of all possible cyclic bis-bibenzyl skeletons derivable from a hypothetical acyclic precursor, perrottetin E (**1023a**) by oxidative cyclization have been calculated using the DTMM and MM2 programs and compared with the natural marchantin, isomarchantin, neomarchantin, riccardin and plagiochin series (*324*). This indicated that riccardin B (**987a**) is of very low energy (total energy, 1.2 Kcal mol^{-1}; free enthalpy change (ΔG^*) $-$ 26.3 Kcal mol^{-1}) compared with the marchantin and plagiochin series which have energies of 10.2 and 35.1 Kcal mol^{-1} and ΔG^*, $-$ 17.3 and 5.2 Kcal mol^{-1}, respectively.

8.4 Long Chain Alkyl Phenols

Schistochila appendiculata contains the long chain alkylphenols (**1043**–**1051**) (*53*). The presence of phenols (**1046, 1049**), salicylic acids (**1047, 1050**) and potassium salicylates (**1048, 1051**) as minor components of a mixture with (**1043**), (**1044**) and (**1045**), respectively, was deduced from GC-MS of methylated products of (**1043**–**1045**). GC-MS analysis further indicated that *S. appendiculata* also contained 6-undecylcatechol (**1052**) and 3-heptadecenylphenol (**1053**).

3-Undecylphenol (**1043**), 6-undecylsalicylic acid (**1044**) and the potassium salt of the long chain alkylsalicylic acid have not been found previously in nature although a crystalline salicylic acid derivative named anagigantic acid has been isolated from *Anacardium gigantheum* (Anacardiaceae) and its structure tentatively proposed as 3-undecylsalicylic acid (**1044**) (*501*). 3-Tridecylphenol (**1046**) and 6-tridecylsalicylic acid (**1047**) have been found in the brown algae, *Caulocystis* species (*322*). The fruit of *Ginkgo biloba* produces 6-pentadecenyl- and 6-pentadecylsalicylic acid (**1050**) (*162, 177*) and many Anacardiaceae species elaborate various types of *n*-C$_{15}$ and *n*-C$_{17}$ alkyl phenols (*162, 177*). However, isolation of (**1043**–**1045**) is the first record of long chain alkylphenols in the bryophytes. These long chain alkyl phenols therefore appear to be significant chemical markers of *Schistochila appendiculata*. As most species belonging to Jungermanniales produce terpenoids as major markers, *S. appendiculata* is one of the chemically most distinct liverworts.

Omphalanthus filiformis produces chamigrane-type sesquiterpenoids and an alkylresorcinol, 5-heptadeca-8(Z),11(Z),14(Z)-trienylresorcinol (**1054**) (*441a, 571*) which has also been isolated from the higher plant *Philodendron scandens* subsp. *oxycardium* (Araceae) (*470*). The brown alga, *Cystophora torulosa* elaborates a very similar alkylresorcinol (**1055**) (*179*).

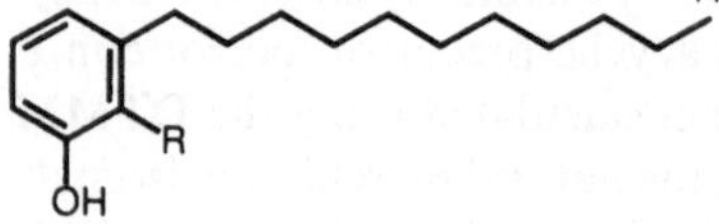

(1043) 3-Undecylphenol ; R=H
(1044) 6-Undecylsalicylic acid ; R=CO$_2$H
(1045) Potassium 6-undecylsalicylate ; R=CO$_2$K

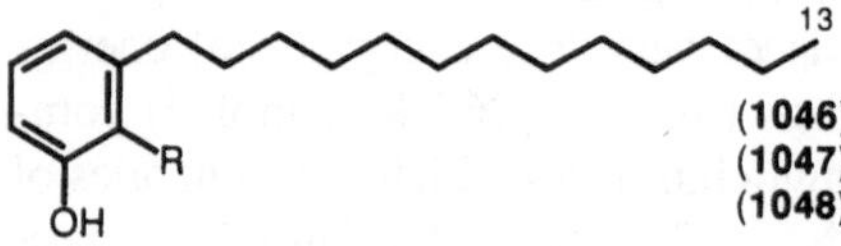

(1046) 3-Tridecylphenol ; R=H
(1047) 6-Tridecyl salicylic acid ; R=CO$_2$H
(1048) Potassium 6-tridecyl salicylate ; R=CO$_2$K

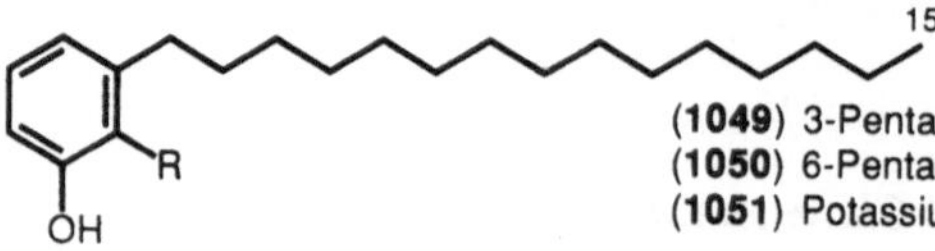

(1049) 3-Pentadecylphenol ; R=H
(1050) 6-Pentadecylsalicylic acid ; R=CO$_2$H
(1051) Potassium 6-pentadecyl salicylate ; R=CO$_2$K

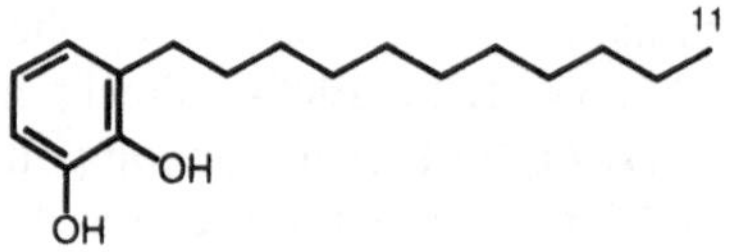

(1052) 6-Undecyl catechol

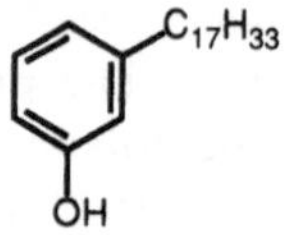

(1053) 3-Heptadecenylphenol

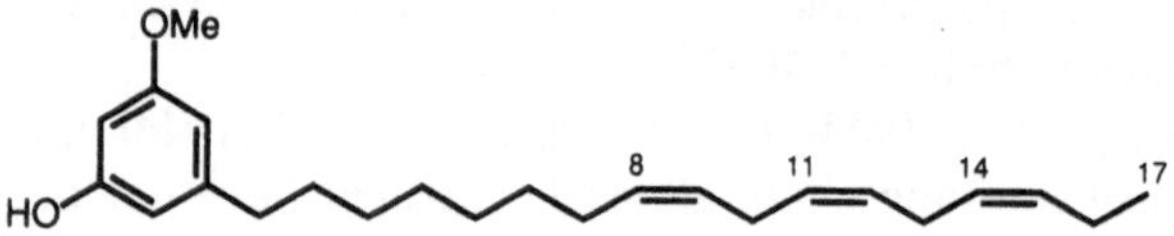

(1054) 5-Heptadeca-8(Z),11(Z),14(Z)-trienylresorcinol monomethyl ether

Chart 63. Long chain alkyl phenols found in the Hepaticae

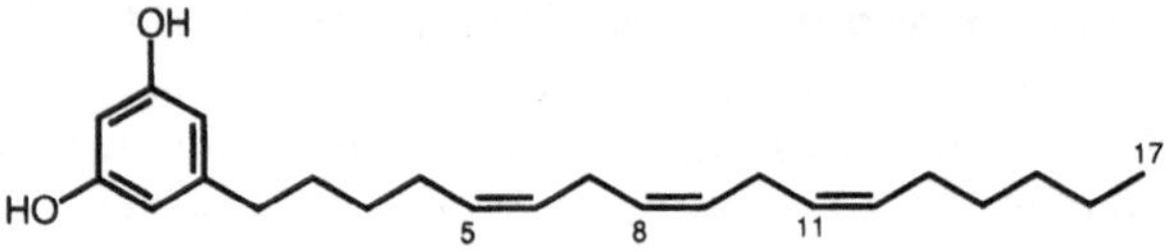

(1055)

8.5 Naphthalenes and Isocoumarins

The presence of naphthalene (1056) has been detected in South American *Triandrophyllum subtrifidum* and *Plagiochila subdura* by GC-MS (46). Taiwanese *Wettsteinia inversa* produces 2,4,7-trimethoxy-naphthalene (1057), along with three new isocoumarins, 8-hydroxy-6,7-dimethoxy-3-methylisocoumarin (1061), inversin (= 6-methoxy-7,8-methylenedioxy-3-methylisocoumarin) (1062) and dihydroinversin (1063)

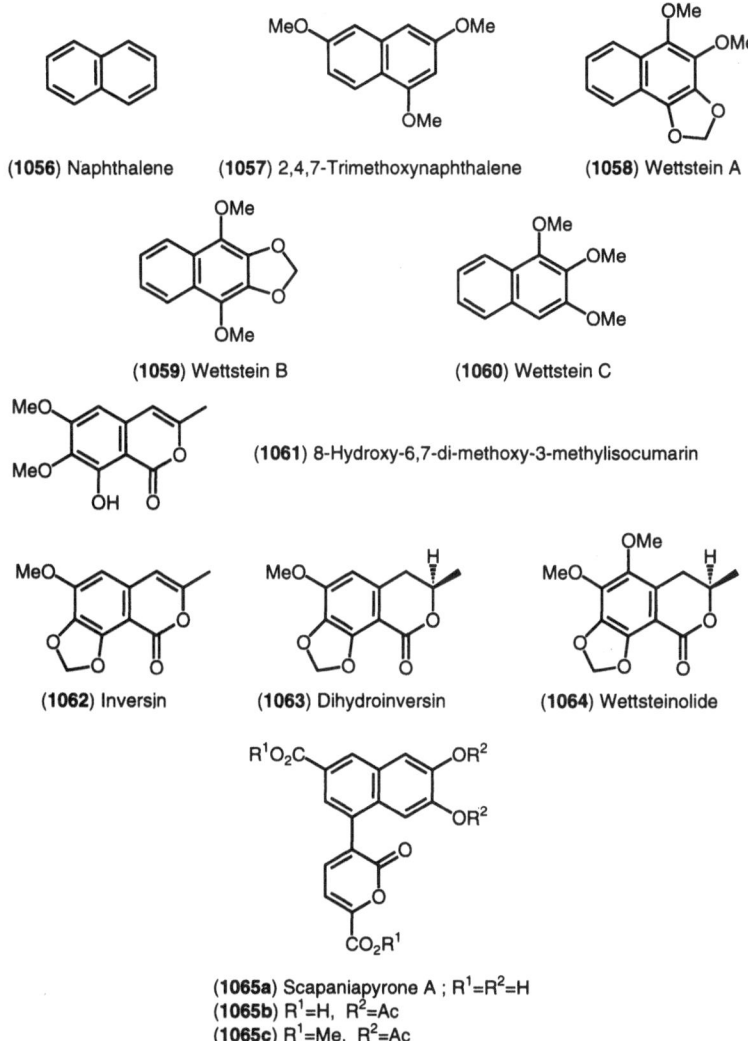

(**1056**) Naphthalene   (**1057**) 2,4,7-Trimethoxynaphthalene   (**1058**) Wettstein A

(**1059**) Wettstein B   (**1060**) Wettstein C

(**1061**) 8-Hydroxy-6,7-di-methoxy-3-methylisocumarin

(**1062**) Inversin   (**1063**) Dihydroinversin   (**1064**) Wettsteinolide

(**1065a**) Scapaniapyrone A ; R^1=R^2=H
(**1065b**) R^1=H, R^2=Ac
(**1065c**) R^1=Me, R^2=Ac

Chart 64a. Naphthalenes and isocoumarins found in the Hepaticae and their derivatives

while New Zealand *W. schusterana* biosynthesizes three new naphthalene derivatives, wettsteins A–C (**1058**–**1060**) and a new isocoumarin, wettsteinolide   [= 3(*R*)-methyl-5,6-dimethoxy-7,8-methylenedioxydihydroisocoumarin] (**1064**), together with dihydroinversin (**1063**) (*37*). These structures were established by various techniques of 600 MHz NMR spectrometry (2D-COSY, HMBC, HMQC and NOE). The (*R*)-configurations at C-3 of (**1063**) and (**1064**) are based on the negative Cotton effect [317 nm and 269.5 nm in (**1063**) and 321 and 267 nm in (**1064**) and the positive Cotton effect [248 nm in (**1063**) and 245 nm in (**1064**)] (*256a*). *Scapania undulata* produces not only various sesquiterpenoids but also a new naphthalene derivative, scapaniapyrone A (**1065a**) (*421*) which contains two phenolic hydroxyl and two carboxylic groups because it forms a diacetate (**1065b**) and dimethyl ester (**1065c**). The structure was deduced by analysis of ^{13}C-^{1}H and long range ^{13}C-^{1}H-2D-COSY spectra.

An aqueous extract of air-dried gametophytic tissue of *Pellia epiphylla* furnished the unusual naphthalene derivative (**1066**) (*474*). Its structure was suggested by analysis of the spectroscopic data and by considering the possible biosynthesis of such a phenyl-substituted naphthalene system from two caffeic acid molecules, a common phenolic unit in

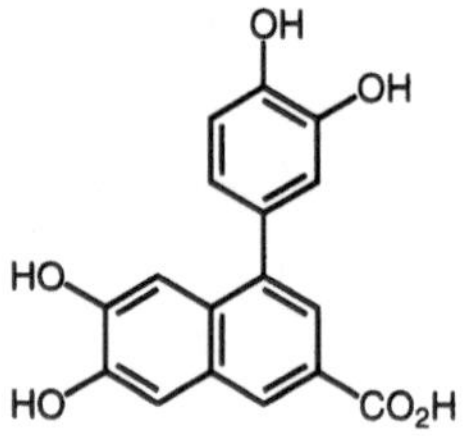

(**1066**) 6,7-Dihydroxy-4-(3,4-dihydroxyphenyl)naphthalene-2-carboxylic acid

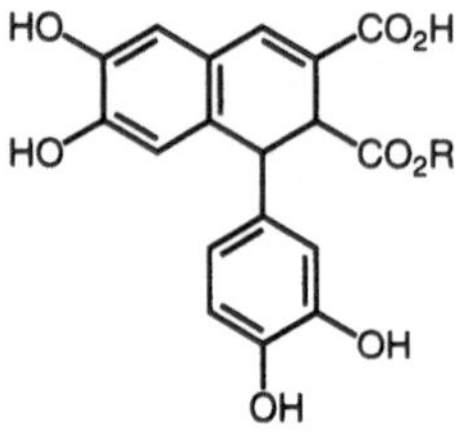

(**1067**) (1*R*,2*S*)-2,3-Dicarboxy-6,7-dihydroxy-1-(3',4'-dihydroxy)-phenyl-1,2-dihydronaphthalene ; R=H
(**1068**) 2,3-Dicarboxy-6,7-dihydroxy-1-(3',4'-dihydroxy)-phenyl-1,2-dihydronaphthalene-10-methyl ester ; R=Me

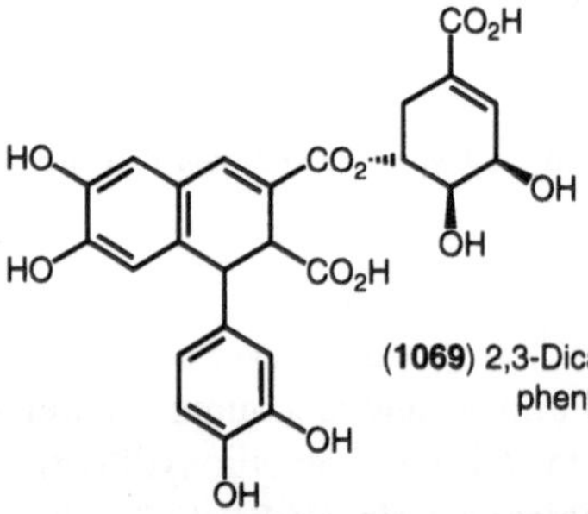

(**1069**) 2,3-Dicarboxy-6,7-dihydroxy-1-(3',4'-dihydroxy)-phenyl-1,2-dihydronaphathalene-9,5''-shikimic acid ester

Chart 64b. Naphthalenes found in the Hepaticae

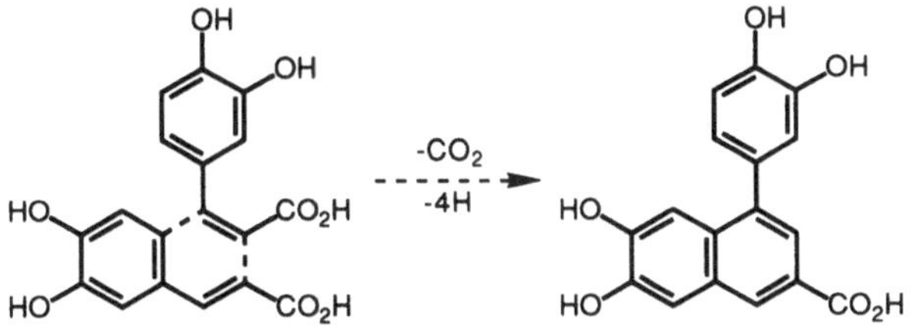

Scheme 65. Possible biogenesis of napthalene derivative

liverworts, as shown in Scheme 65 (*474*). The proposed structure was confirmed by total synthesis of (**1066**) (*168, 474, 647*). Further investigation of the methanol extract of gametophytes of the same species gave three new phenyldihydronaphthalene derivatives, (**1067**), (**1068**) and (**1069**) (*159a*). The structures of (**1067**) and (**1068**) are based on a combination of negative FAB and CI mass spectra and NMR spectroscopic comparison with the phenyldihydronaphthalene moiety of rabdosiin from *Rabdosia japonica* (*3a*). The structure of (**1069**) was also established by the negative CI mass spectrum and comparison of the ^{1}H and ^{13}C NMR spectra with (**1067**) and (**1068**) as well as shikimic acid (**1085**) and 5-O-caffeoyl shikimate.

8.6 Neolignans, Phenanthrenes and Phthalides

A neolignan, (−)-licarin A (**1070**) has been isolated from *Jackiella javanica* (*437*). In the original paper, the structure of (**1070**) was reproduced erroneously, one methoxyl group being missing. The same substance has been isolated from the higher plant *Urbanodendron verrucosum* (*4*); (+)-licarin A, the enantiomer of (**1070**), was found in *Magnolia kachirachirai* (*172*). This is the first record of a neolignan from bryophytes.

Plagiochila spinulosa contains two phenanthrene derivatives, 2-hydroxy-3,4-7-trimethoxy-9,10-dihydrophenanthrene (**1071**) and 3,4,7-trimethoxy-9,10-dihydrophenanthrene (**1072**) (*144*) whose structures have been assigned by study of their NMR spectra. *Riccardia jackii* elaborates a dihydrophenanthrene whose structure has been established as 3,4-dimethoxy-5-hydroxy-9,10-dihydrophenanthrene (**1073**) by X-ray crystallographic analysis (*380*). 2,5-Dimethoxy-3-hydroxyphenanthrene (**1074**) has been isolated from *Marchantia tosana* and 2-hydroxy-3,6-dimethoxyphenanthrene (**1075**) from *M. paleacea* var. *diptera* (*74*). A similar trisubstituted phenanthrene derivative (**1076**) has been isolated

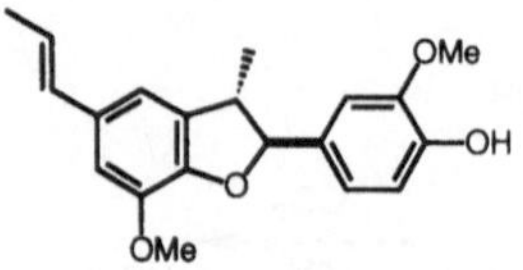

(1070) (-)-Licarin A

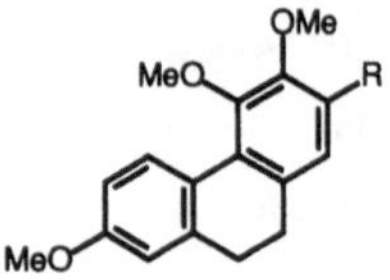

(1071) 2-Hydroxy-3,4,7-trimethoxy-9,10-dihydrophenanthrene ; R=OH
(1072) 3,4,7-Trimethoxy-9,10-dihydrophenanthrene ; R=H

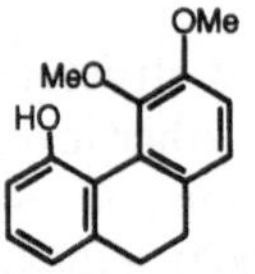

(1073) 3,4-Dimethoxy-5-hydroxy-9,10-dihydrophenanthrene

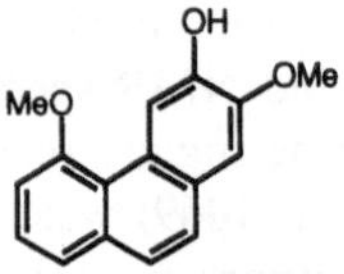

(1074) 2,5-Dimethoxy-3-hydroxyphenanthrene

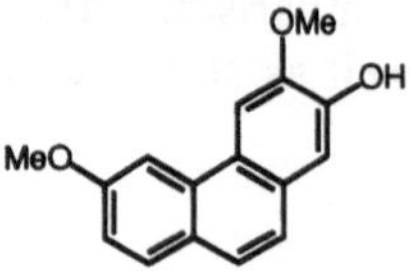

(1075) 2-Hydroxy-3,6-dimethoxyphenanthrene

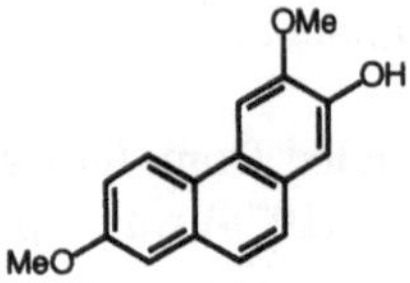

(1076) 2-Hydroxy-3,7-dimethoxyphenanthrene

Chart 65. Neolignan and phenanthrenes found in the Hepaticae

from Indian *M. polymorpha*, together with various macrocyclic bis-bibenzyls as described earlier (*72*). The position of each functional group in the phenanthrenes from *Marchantia* species was deduced by studying NOE of the original compounds and their methyl ethers. A similar 9,10-dihydrophenanthrene derivative (**1077**) has been found in the Senegalese red algae, *Polysyphonia ferulacea* (*5*).

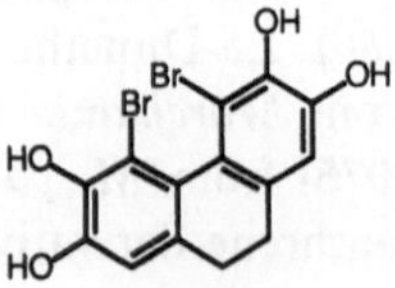

(1077)

A new phthalide, 3-(4'-methoxybenzyl)-5,6-dimethoxyphthalide (**1078**), has been isolated from *Frullania falciloba*, together with the two bibenzyls mentioned earlier (*64*).

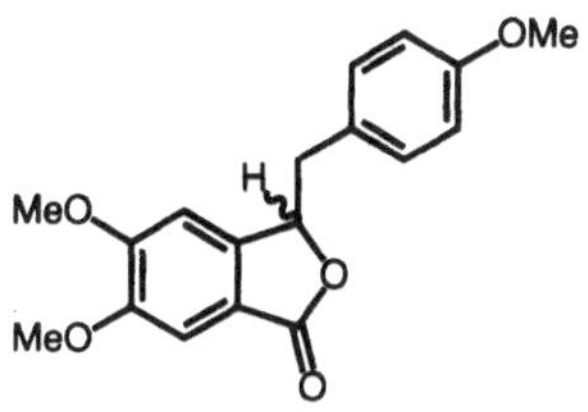

(**1078**) 3-(4'-Methoxybenzyl)-5,6-dimethoxyphthalide

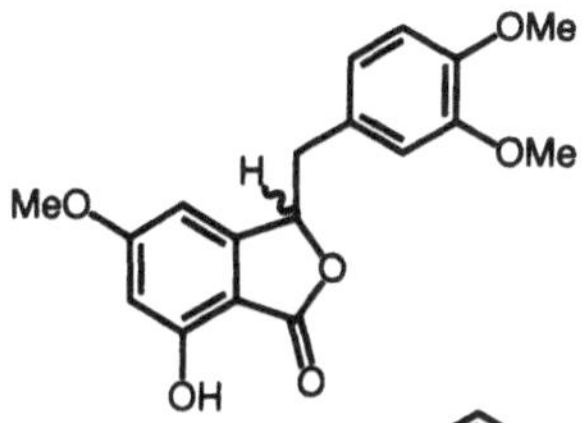

(**1079**) 3-(3',4'-Dimethoxybenzyl)-7-hydroxy-5-methoxy-phthalide

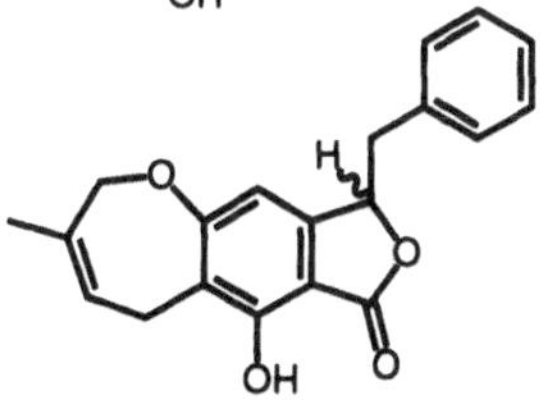

(**1080**) Radulanolide

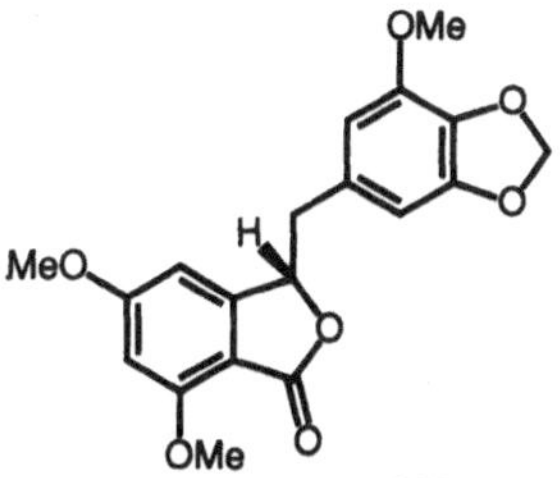

(**1081**) 3-(3'-Methoxy-4',5'-methylenedioxybenzyl)-5,7-dimethoxyphthalide

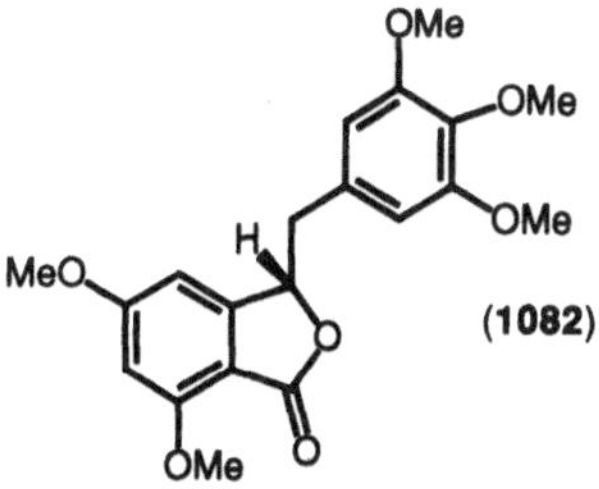

(**1082**) 3-(3',4',5'-Trimethoxybenzyl)-5,7-dimethoxyphthalide

Chart 66. Phthalides found in the Hepaticae

The arrangement of the functional groups on the two benzene rings was established by NOE spectrometry. *Balantiopsis rosea* produces a new phthalide named balantiolide, 3-(3′,4′-dimethoxybenzyl)-7-hydroxy-5-methoxyphthalide (**1079**) (*65*). The presence of the partial structure, a 3-substituted 7-hydroxy-5-methoxy-phthalide, was deduced from the intense UV (252 nm) and IR absorption band (1735 cm^{-1}) assignable to a chelated phthalide. The arrangement of the three methoxyl groups on the two benzene rings was deduced by analysis of the ^{1}H-NMR spectrum, spin decoupling and the NOE difference spectrum. *Radula complanata* contained a new phthalide named radulanolide (**1080**) together with various bibenzyl derivatives (*66*). The presence of a 7-hydroxyphthalide was deduced from the intense IR absorption band at 1735 cm^{-1} while the complete structure rests on analysis of ^{1}H-NMR and mass spectral data and spin decoupling experiments. Two new phthalides (**1081**) and (**1082**) have been isolated from *Trocholejeunea sandvicensis*; their structures are based on an analysis of the ^{1}H- and ^{13}C-NMR spectra (*82*).

8.7 Miscellaneous Aromatic Compounds

Shikimic acid (**1085**) has been obtained from *Conocephalum conicum* (*214*). *Plagiochila asplenioides* and *Lophocolea bidentata* produce ellagic acid (**1086**) (*214*). The latter species contains 4-vinylguaiacol (**1090**) (*527a*). *Blasia pusilla* produces orellinic acid methyl ester (**1088**) (*249*). Phenylace-toaldehyde (**1087**) has been detected in the essential oil of *Lophocolea heterophylla* (*527a*). *p*-Hydroxybenzaldehyde (**1084**) has been isolated from *Marchantia paleacea* var. *diptera* (*74*), *M. polymorpha* (*61, 71*) and *Wiesnerella denudata* (*51*). *M. polymorpha* also produces *m*-hydroxybenz-aldehyde (**1083**) (*71*). *Plagiochila spinulosa* elaborates the dihydrophen-anthrenes (**1071**) and (**1072**) as well as methyl 2-methyl-3,4-methylenedioxy-6-methoxybenzoate (**1089**) (*144*). From an unidentified Jamaican liverwort, CONNOLLY (*144*) isolated two aromatic compounds, (**1091**) and (**1093**). 3,4-Dimethoxystyrene (**1092**) was isolated from Malay-sian *Asterella* species and callus of *Conocephalum conicum* (*585*). The former liverwort elaborates skatole (**1099**) (*249*). From *Marchesinia brachiata*, an allylbenzene has been isolated whose structure is either (**1094**) or (**1095**) (*219*). 1-(3,4-Dihydroxy-5-methoxybenzyl)-3-methylbut-2-ene (**1096**) has been isolated from *Plagiochila rutilans*; the structure was established by synthesis (*278*).

The prenylindole derivatives, 6-(3-methyl-2-butenyl)indole (**1097**) and 7-(3-methyl-2-butenyl) indole (**1098**) have been isolated from European *Riccardia chamedryfolia* (*19, 435*). The same indole derivative has been isolated from Japanese *R. multifida* (*78*). *Conocephalum conicum* and

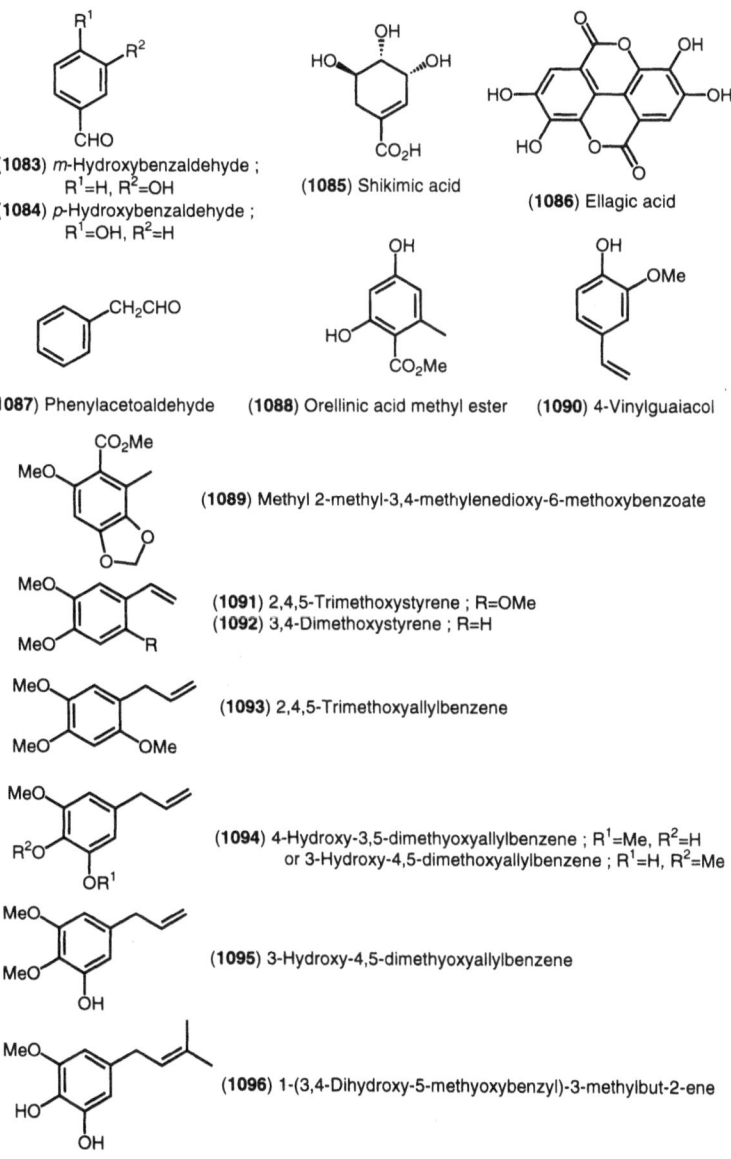

(1083) *m*-Hydroxybenzaldehyde ;
R¹=H, R²=OH
(1084) *p*-Hydroxybenzaldehyde ;
R¹=OH, R²=H

(1085) Shikimic acid

(1086) Ellagic acid

(1087) Phenylacetoaldehyde

(1088) Orellinic acid methyl ester

(1090) 4-Vinylguaiacol

(1089) Methyl 2-methyl-3,4-methylenedioxy-6-methoxybenzoate

(1091) 2,4,5-Trimethoxystyrene ; R=OMe
(1092) 3,4-Dimethoxystyrene ; R=H

(1093) 2,4,5-Trimethoxyallylbenzene

(1094) 4-Hydroxy-3,5-dimethyoxyallylbenzene ; R¹=Me, R²=H
or 3-Hydroxy-4,5-dimethoxyallylbenzene ; R¹=H, R²=Me

(1095) 3-Hydroxy-4,5-dimethyoxyallylbenzene

(1096) 1-(3,4-Dihydroxy-5-methyoxybenzyl)-3-methylbut-2-ene

Chart 67a. Miscellaneous aromatic compounds found in the Hepaticae

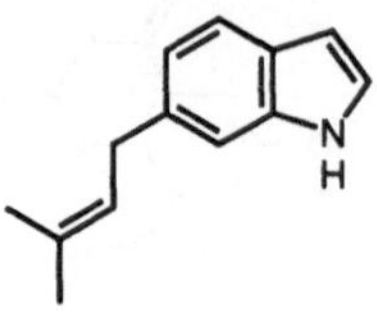

(**1097**) 6-(3-Methyl-2-butenyl) indole

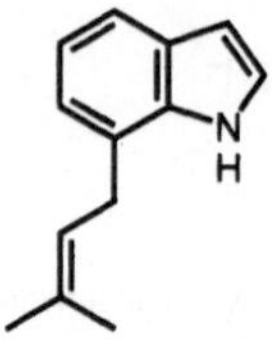

(**1098**) 7-(3-Methyl-2-butenyl) indole

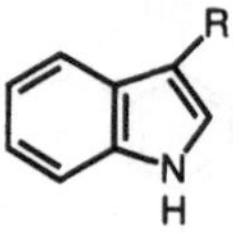

(**1099**) Skatole ; R=Me
(**1100**) Indole acetic acid ; R=CH₂CO₂H
(**1101**) Indole 3-acetonitrile ; R=CH₂CN

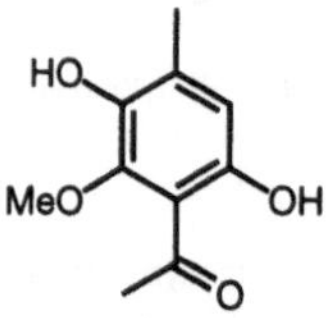

(**1102**) 2,5-Dihydroxy-4-methyl-6-methoxyacetophenone

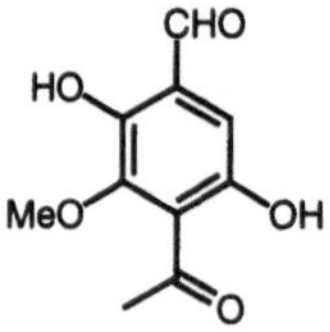

(**1103**) 2,5-Dihydroxy-4-formyl-6-methoxyacetophenone

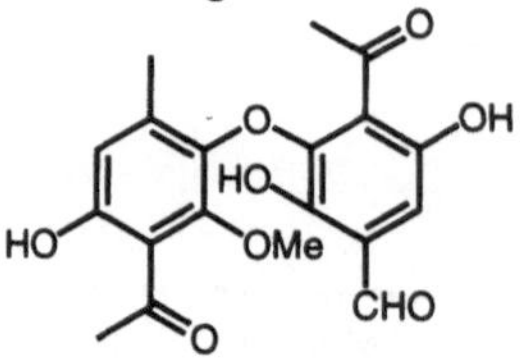

(**1104**) Trocholejeunin

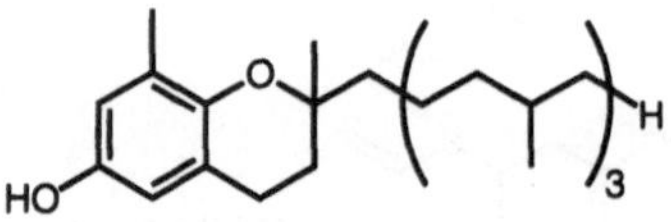

(**1105**) δ-Tocopherol

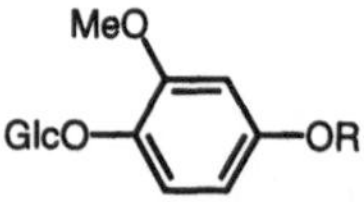

(**1106a**) Isotachioside (=2-Methoxy-4-hydroxyphenyl-1β-glucoside) ;
R=H
(**1106b**) R=Me

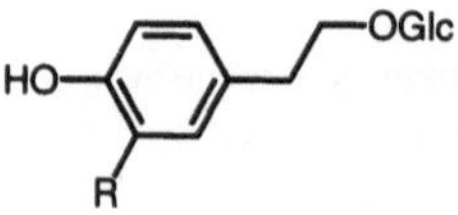

(**1107**) Salidroside ; R=H
(**1108**) β-(3,4-Dihydroxyphenyl)-ethyl-O-β-D-glucoside ; R=OH

Chart 67b. Miscellaneous aromatic compounds found in the Hepaticae

Marchantia polymorpha produce indoleacetic acid (**1100**) (*442a*). *Trocho-lejeunea sandvicensis* elaborates not only various pinguisane-type sesqui-terpenoids but also highly oxidized acetophenone derivatives (**1102–1104**) which are shown in Chart 67b. These structures were based on analysis of spectroscopic data (*22*). *Radula perrottetii* contains δ-tocopherol (**1105**) (*585a*).

Isotachis japonica contained a phenyl glucoside (**1106a**) whose struc-ture is based on the ^{1}H-, ^{13}C-NMR and NOE difference spectra and mass spectral fragmentation [m/z 140 (base)] of (**1106a**) and its methyl ether (**1106b**) (*76*). The same glucoside (**1106a**) has been isolated from the higher plant *Berchemia racemosa* (*292*). Two phenylethanoid glucosides salidro-side (**1107**) (*408*) and β-(3,4-dihydroxyphenyl)-ethyl-O-β-D-glucoside (**1108**) (*502*) have been isolated from an axenic culture of *Ricciocarpos natans*, together with three new bibenzyl glucosides (**881, 882, 883**) (*351*). Three new glycerol glucosides (**1109–1111**) have been isolated from aqueous alcoholic extract of *Frullania muscicola* and their structures elucidated by a combination of chemical degradation (alkaline hydroly-sis) and FAB-MS as well as NMR spectrometry (*345b*). The absolute configuration at C-2 for (**1109–1111**) has not been determined. Isolation of (**1106a**), (**1107**) and (**1108**) constitutes the first record of such glucosides from the bryophytes, although a number of flavonoid glucosides have been found in the bryophytes (see the following section).

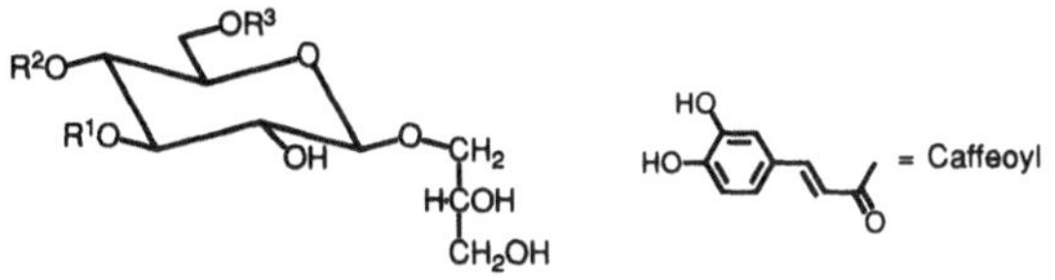

(**1109**) 1-O-β-D-(6'-Caffeoyl)-glucopyranosyl glycerol ; R^1=R^2=H R^3=Caffeoyl
(**1110**) 1-O-β-D-(4'-Caffeoyl)-glucopyranosyl glycerol ; R^1=R^3=H R^2=Caffeoyl
(**1111**) 1-O-β-D-(3'-Caffeoyl)-glucopyranosyl glycerol ; R^1=Caffeoyl R^2=R^3=H

Chart 67c. Miscellaneous aromatic compounds found in the Hepaticae

8.8 Flavonoids

Flavonoids are widely distributed in Hepaticae. Previous reviews (*19, 271*) listed the presence of flavonoids in 58 liverwort species from 32 genera (*19*) and 55 species from 31 genera (*271*). A total of 103 flavonoids from liverworts were listed in the 1982 review (*19*). The review of MARKHAM in "The Flavonoids Advances in Research Since 1980" with tables provide a good listing of flavonoids up to 1988 (*362*). In this review, flavonoids isolated from or detected in 106 liverwort species are included and their distribution in 6 orders of the Hepaticae is also discussed. Other review articles dealing with flavonoids of bryophytes have appeared (*363, 416, 648*). The distribution of C-glycosylflavones in 22 species of the liverworts has been investigated (*413*). All compounds were isolated from an 80% aq. methanol extract. No flavones could be detected in *Herbertus*, *Mastigophora* and *Ptilidium* species. The distribution of flavonoids in *Trichocolea* and *Porella* species is not uniform. Some *Trichocolea* species produce flavonoids while others do not. *Porella* species elaborate flavones which differ qualitatively and quantitatively depending on the species. For example, the flavonoids pattern of *P. platyphylla* collected in 16 different localities from various parts of Europe is surprisingly different (*413*).

444 liverwort and four hornworts have been analyzed for occurrence of flavonoids by two-dimensional thin layer chromatography (2D-TLC) (*416*) and some basic information on isolation and identification techniques of flavonoids by 2D-TLC has been reviewed (*417*). ^{1}H- and ^{13}C-NMR spectrometry may be of use in structure determination of flavone C-6 or C-8 glycosides because the presence of a C-8-linked hexose results in the existence of rotamers, in which the hexose and B-ring interact sterically to give isomers distinguishable by NMR techniques (*367*). The number of species screened for flavonoids and their polar more derivatives mainly by 2D-TLC has been reported by MARKHAM (*363*). In members of Marchantiidae which have been investigated, flavonoids are generally present. The main flavonoid types present in Marchantiidae are flavone O-glucuronides followed by flavone C-glycosides, while flavonols are very rare. Table IIf shows the distribution of flavonoids in the Hepaticae. Apigenin (**1112**) and luteolin (**1143**) derivatives are common.

Flavone C- and O-glycosides, flavonols, dihydroflavones, dihydrochalcones, and aurones have all been found in one or more liverworts, but isoflavones, chalcones, biflavones, anthocyanins, and proanthocyanidins have not been detected. The order Bryales of Musci contains flavone C- and O-glycosides, biflavones, aurones, isoflavones, and deoxyanthocyanins; however, proanthocyanidins and 3-hydroanthocyanins has not yet been detected. Flavonoids in bryophytes are distributed over 8 orders (*363*). Those orders lacking flavonoids are the Anthocerotales (Anthocerotae), Andreaeidae, Polytrichales, Sphagnales and Tetraphidales (Musci). Less than half of the Metzgeriales and Jungermanniales (Hepaticae) investigated biosynthesize flavonoids, while on the other hand, the distribution of flavonoids in Marchantiales and Sphaerocarpales is estimated to be 70–100%. It is obvious from the data in Table IIf that the

Table III. *Flavonoids Found in the Hepaticae*

Structure number	Name of compounds	Formula	m.p.°C	$[\alpha]_D$	Plant source	Order	References	Comments
(1112)	Apigenin	$C_{15}H_{10}O_5$			*Blasia pusilla*	Me	(419)	
					Frullania brachyclada	J	(645)	
					F. ecklonii	J	(645)	
					F. riojaneirensis	J	(645)	
(1113)	Apigenin-7,4′-dimethyl ether	$C_{17}H_{14}O_5$	177–9		*Frullania davurica*	J	(420)	
					F. jackii	J	(420)	
					F. vethii	J	(56)	
					Monoclea gottschei subsp. neotropica	Mo	(218, 345a)	
					Plagiochasma rupestre	M	(243)	
					Reboulia hemisphaerica	M	(411)	
(1115)	Apigenin-4′-O-glucoside				*Blasia pusilla*	Me	(419)	
					Monoclea gottschei subsp. neotropica	Mo	(218, 345a)	
(1116)	Apigenin-7,4′-di-O-galacturonide				*Blasia pusilla*	Me	(419)	
(1117)	Apigenin-7-O-glucoside-4′-O-galacturonide				*Blasia pusilla*	Me	(419)	
(1118)	Apigenin-7,4′-di-O-glucuronide				*Blasia pusilla*	Me	(419)	
(1119)	Apigenin-7-O-rhamnosyl-glucuronide-4′-O-glucuronide or Apigenin-7-O-glucuronide-4′-O-rhamnosylglucuronide				*Monoclea gottschei subsp. neotropica*	Mo	(218, 345a)	

Table IIf *(continued)*

Structure number	Name of compounds	Formula	m.p.°C	$[\alpha]_D$	Plant source	Order	References	Comments
(1120)	Apigenin-7-O-neo-hesperidoside				*Cavicularia densa*	Me	*(419)*	
(1121)	Apigenin-7-O-glucoside				*Blasia pusilla*	Me	*(419)*	
					Frullania arecae	J	*(645)*	
					F. brachyclada	J	*(645)*	
					F. confertiloba	J	*(645)*	
					F. ecklonii	J	*(645)*	
					F. riojaneirensis	J	*(645)*	
					F. sphaerocephala	J	*(645)*	
					F. wallichiana	J	*(645)*	
(1122)	Acacetin-7-O-rhamno-arabinosyl galacturonide				*Monoclea gottschei* subsp. *neotropica*	Mo	*(218, 345a)*	
(1123)	Acacetin-7-O-rhamno-xylosyl galacturonide				*Monoclea gottschei* subsp. *neotropica*	Mo	*(218, 345a)*	
(1124)	Acacetin-7-O-rhamno-glucosyl glucuronide				*Monoclea gottschei* subsp. *neotropica*	Mo	*(218, 345a)*	
(1125)	Acacetin-7-O-diglucuronide				*Monoclea gottschei* subsp. *neotropica*	Mo	*(218, 345a)*	
(1126)	Acacetin-7-O-di-galacturonide				*Monoclea gottschei* subsp. *neotropica*	Mo	*(218, 345a)*	
(1127)	Apigenin-6,8-di-C-glucoside (= Vicenin-2)				*Eupleurozia simplicissima*	J	*(422)*	
					Frullania brachyclada	J	*(645)*	
					Mylia anomala	J	*(424, 419)*	
					M. nuda	J	*(424, 419)*	
					M. taylorii	J	*(424, 419)*	
					M. verrucosa	J	*(424, 419)*	

No.	Name	Formula	Species		Ref.
(1128)	6-Hydroxyapigenin (= Scutellarein)	$C_{15}H_{10}O_6$	Frullania davurica	J	(420)
(1129)	6-Hydroxyapigenin (= Scutellarein)-7-O-diglucoside		Frullania davurica	J	(420)
			F. muscicola	J	(345b)
(1130)	Scutellarein-7-O-glucoside-6″-malonate		Frullania muscicola	J	(345b)
(1131)	Scutellarein-7-O-xyloside		Frullania davurica	J	(420)
(1132)	Scutellarein-6-O-xyloside-7-O-glucoside		Frullania davurica	J	(420)
(1133)	Scutellarein-7-O-gluco-xylglucoside		Frullania davurica	J	(420)
(1134)	Scutellarein-7-O-glucoside-6″-hydroxy-3-methyl-glutarate		F. muscicola	J	(345b)
(1135)	Scutellarein-6-O-glucoside-7-O-rhamnoglucoside		Frullania jackii	J	(420)
(1136)	Apigenin-6-C-rhamnoside-rhamnoside-7-O-glucoside (= Isofurcatain-7-O-glucoside)		Metzgeria furcata var. uvula	Me	(369, 549)
(1138)	Apigenin-6,7,4′-trimethyl ether	$C_{18}H_{16}O_6$	Monoclea gottschei subsp. neotropica	Mo	(218, 345a)
(1139)	Pectolinarigenin-7-O-glucuronide		Monoclea gottschei subsp. neotropica	Mo	(218, 345a)
(1140)	Pectolinarigenin-7-O-galacturonide		Monoclea gottschei subsp. neotropica	Mo	(218, 345a)

Table IIf (*continued*)

Structure number	Name of compounds	Formula	m.p.°C	$[\alpha]_D$	Plant source	Order	References	Comments
(1141)	4'-Methoxyscutellarein-7-O-glucronyl-rhamnoside				*Monoclea gottschei* subsp. *neotropica*	Mo	(218, 345a)	
(1142)	5-Hydroxy-7,8,4'-trimethoxyflavone		218–222		*Reboulia hemisphaerica*	M	(411)	
(1143)	Luteolin	$C_{15}H_{10}O_6$			*Blasia pusilla*	Me	(419)	
					Brachiolejeunea phyllorhiza	J	(217)	
					Eupleurozia paradoxa	J	(422)	
					Frullania africana	J	(645)	
					F. arecae	J	(645)	
					F. brachyclada	J	(645)	
					F. confertiloba	J	(645)	
					F. ecklonii	J	(645)	
					F. gibbosa	J	(645)	
					F. gradsteinii	J	(645)	
					F. obscura	J	(645)	
					F. pluricarinata	J	(645)	
					F. riojaneirensis	J	(645)	
					F. sphaerocephala	J	(645)	
					F. tolimana	J	(645)	
					F. wallichiana	J	(645)	
					Marchantia paleacea var. *diptera*	M	(457)	suspension culture
					M. polymorpha	M	(71)	

No.	Compound	Formula	Species		Ref.
(1144)	Luteolin-3',4'-dimethyl ether		*Frullania davurica*	J	*(420)*
			F. jackii	J	*(420)*
			Monoclea gottschei subsp. *neotropica*	Mo	*(218, 345a)*
(1145)	Luteolin-7,3'-dimethyl ether (= Velutin)	$C_{17}H_{14}O_6$	*Monoclea gottschei* subsp. *neotropica*	Mo	*(218, 345a)*
(1146)	Luteolin-7,3',4'-trimethyl ether		*Frullania davurica*	J	*(420)*
			F. jackii	J	*(420)*
(1147)	Luteolin-7-O-glucoside		*Blasia pusilla*	Me	*(419)*
			Brachiolejeunea laxifolia	J	*(217)*
			B. phyllorhiza	J	*(217)*
			Frullania africana	J	*(645)*
			F. arecae	J	*(645)*
			F. brachyclada	J	*(645)*
			F. confertiloba	J	*(645)*
			F. ecklonii	J	*(645)*
			F. gibbosa	J	*(645)*
			F. gradsteinii	J	*(645)*
			F. laxiflora	J	*(645)*
			F. muscicola	j	*(345b)*
			F. obscura	J	*(645)*
			F. ovistipula	J	*(645)*
			F. pluricarinata	J	*(645)*
			F. riojaneirensis	J	*(645)*
			F. sphaerocephala	J	*(645)*
			F. tolimana	J	*(645)*
			F. wallichiana	J	*(645)*

Table IIf *(continued)*

Structure number	Name of compounds	Formula	m.p.°C	$[\alpha]_D$	Plant source	Order	References	Comments
(1148)	Luteolin-7-O-glucoside-6″-malonate				*F. arecae*	J	*(645)*	
					F. confertiloba	J	*(645)*	
					F. ecklonii	J	*(645)*	
					F. sphaerocephala	J	*(645)*	
					F. wallichiana	J	*(645)*	
(1149)	Luteolin-7-O-glucuronide				*Monoclea gottschei* subsp. *neotropica*	Mo	*(218, 345a)*	
(1150)	Luteolin-7-O-gentiobiose				*Frullania brachyclada*	J	*(645)*	
					F. confertiloba	J	*(645)*	
					F. ecklonii	J	*(645)*	
(1151)	Luteolin-3′-O-glucoside				*Blasia pusilla*	Me	*(419)*	
(1152)	Luteolin-4′-O-glucoside				*Blasia pusilla*	Me	*(419)*	
(1153)	Luteolin-4′-O-glucuronide				*Eupleurozia paradoxa*	J	*(422)*	
(1154)	Luteolin-7-O-glucoside-6″-hydroxy-3-methyl-glutarate				*Frullania muscicola*	J	*(345b)*	
(1155)	Luteolin-7,4′-di-O-glucuronide				*Blasia pusilla*	Me	*(419)*	
(1156)	Luteolin-7,3′-di-O-glucoside				*Blasia pusilla*	Me	*(419)*	
(1157)	Luteolin-4′-O-rhamnosyl-β-D-galacturonide				*Monoclea gottschei* subsp. *neotropica*	Mo	*(218, 345a)*	
(1158)	Crysoeriol-7-O-glucoside				*Blasia pusilla*	Me	*(419)*	
					Cavicularia densa	Me	*(419)*	

No.	Compound	Species		Ref.
(1159)	Crysoeriol-7-O-glucuronide	Cavicularia densa	Me	(419)
		Monoclea gottschei subsp. neotropica	Mo	(218, 345a)
(1160)	Crysoeriol-7-O-neohesperidoside	Cavicularia densa	Me	(419)
(1161)	Crysoeriol-7,4'-di-O-glucoside	Cavicularia densa	Me	(419)
(1162)	3',4'-Dimethoxyluteolin-7-O-rhamnoarabinosyl galacturonide	Monoclea gottschei subsp. neotropica	Mo	(218, 345a)
(1163)	3',4'-Dimethoxyluteolin-7-O-rhamnoxylosyl galacturonide	Monoclea gottschei subsp. neotropica	Mo	(218, 345a)
(1164)	Luteolin-6,8-di-C-glucoside (= Lucenin 2)	Eupleurozia paradoxa	J	(422)
		E. simplicissima	J	(422)
		Frullania africana	J	(645)
		F. arecae	J	(645)
		F. brachyclada	J	(645)
		F. confertiloba	J	(645)
		F. ecklonii	J	(645)
		F. obscura	J	(645)
		F. riojaneirensis	J	(645)
		F. sphaerocephala	J	(645)
		F. wallichiana	J	(645)
		Mylia anomala	J	(419, 424)
		M. nuda	J	(419, 424)
		M. taylorii	J	(419, 424)
		M. verrucosa	J	(419, 424)
		Pleurozia acinosa	J	(422)
		P. articulata	J	(422)
		P. caledonica	J	(422)
		P. conchifolia	J	(422)

Table IIf (continued)

Structure number	Name of compounds	Formula	m.p.°C	$[\alpha]_D$	Plant source	Order	References	Comments
(1165)	Luteolin-6,8-di-C-arabinoside				Plagiochasma rupestre	M	(491a)	
(1166)	6-Hydroxyluteolin	$C_{15}H_{10}O_7$			Frullania jackii	J	(420)	
					F. tolimana	J	(645)	
(1167)	Luteolin-6,7-dimethyl-ether (= Cirsilol)	$C_{17}H_{14}O_7$			Monoclea gottschei subsp. neotropica	Mo	(218, 345a)	
(1168)	Luteolin-6,3',4'-trimethyl ether (= Eupatilin)				Monoclea gottschei subsp. neotropica	Mo	(218, 345a)	
(1169)	6-Hydroxyluteolin-7-O-sophoroside				Frullania davurica	J	(420)	
					F. jackii	J	(420)	
(1170)	6-Hydroxyluteolin-7-O-glucoside				Frullania ecklonii	J	(645)	
					F. gibbosa	J	(645)	
					F. gradsteinii	J	(645)	
					F. laxiflora	J	(645)	
					F. muscicola	J	(345b)	
					F. pluricarinata	J	(645)	
					F. tolimana	J	(645)	
(1171)	Nodifloretin-7-monomethyl ether (= Pedalitin)	$C_{16}H_{12}O_7$			Frullania jackii	J	(420)	
					Monoclea gottschei subsp. neotropica	Mo	(218, 345a)	
(1172)	6-Hydroxyluteolin-7,3'-dimethyl ether	$C_{17}H_{14}O_7$			Monoclea gottschei subsp. neotropica	Mo	(218, 345a)	
(1173)	6-Hydroxyluteolin-6-O-xyloside-7-O-glucoside				Frullania davurica	J	(420)	
					F. jackii	J	(420)	

(1174)	6-Hydroxyluteolin-6-O-glucoside-7-O-xylo-glucoside	*Frullania davurica*	J	(420)
		F. jackii	J	(420)
(1175)	Nodifloretin-7-O-rhamnoside-4'-O-glucuronide	*Monoclea gottschei* subsp. *neotropica*	Mo	(218, 345a)
(1176)	Nodifloretin-7-O-xyloglucoside	*Frullania davurica*	J	(420)
(1177)	Nodifloretin-6-O-xyloside-7-O-glucoside	*Frullania davurica*	J	(420)
(1178)	Nodifloretin-6-O-glucoside-7-O-xyloside	*Frullania davurica*	J	(420)
(1179)	Nodifloretin-7-O-glucoside-6''-hydroxy-3-methyl-glutarate	*Frullania polysticta*	J	(345b)
(1180)	6-Hydroxyluteolin-6-O-di-glucoside	*Frullania jackii*	J	(420)
(1181)	6-Hydroxyluteolin-7-O-rhamnoside	*Frullania jackii*	J	(420)
(1182)	6-Hydroxyluteolin-7-O-glucoside-6''-hydroxy-3-methylglutarate	*Frullania muscicola*	J	(345b)
(1183)	6-Hydroxyluteolin-7-O-glucoside-4''-hydroxy-3-methylglutarate	*Frullania muscicola*	J	(345b)
(1184)	6-Hydroxyluteolin-7-O-glucoside-3''-hydroxy-3-methylglutarate	*Frullania muscicola*	J	(345b)
(1185)	6-Hydroxyluteolin-3'-methyl ether (= Nodifloretin- or Batatifolin-7'-O-sophoroside)	*Frullania davurica*	J	(420)
		F. jackii	J	(420)

Table IIf (continued)

Structure number	Name of compounds	Formula	m.p.°C	$[\alpha]_D$	Plant source	Order	References	Comments
(1186)	Isoorientin-7-O-glucoside (= Lutonarin)				*Lejeunea cavifolia*	J	(217)	
					Stictolejeunea squamata	J	(217)	
(1187)	Luteolin-6,8-di-C-glucoside-3'-glucoside (= Lucenin-2 3'-glucoside)				*Pleurozia acinosa*	J	(422)	
					P. articulata	J	(422)	
					P. caledonica	J	(420)	
(1188)	Luteolin-6-C-glucoside-8-C-arabinoside (= Carlinoside)				*Pleurozia conchifolia*	J	(422)	
(1189)	Luteolin-6-C-arabinoside-8-C-glucoside (= Isocarlinoside)				*Pleurozia conchifolia*	J	(422)	
(1190)	Onopordin-7,4'-di-O-poly-saccharide				*Monoclea forsteri*	Mo	(218, 345a)	
(1191)	Kaempferol-3-methyl ether	$C_{16}H_{12}O_6$			*Frullanoides densifolia*	J	(347)	
(1193)	Kaempferol-6-C-glucoside				*Metzgeria* sp.	Me	(363)	
(1194)	Kaempferol-6-C-glucoside-3-O-glucoside				*Metzgeria* sp.	Me	(363)	
(1199)	Tricetin-6,8-di-C-glucoside				*Eupleurozia simplicissima*	J	(422)	
					Frullania polysticta	J	(345b)	
					Lejeunea cavifolia	J	(217)	
					Plagiochila jamesonii	J	(491a)	
					Pleurozia acinosa	J	(422)	

No.	Compound	Species		Ref.
(1200)	Tricetin-6-C-glucoside (= Isoaffinetin)	P. articulata	J	(422)
		P. caledonica	J	(422)
		P. conchifolia	J	(422)
		Frullania polysticta	J	(345b)
(1201)	Tricetin-7-O-glucoside-3'-O-glucoside-6'''-hydroxy-3-methyl-glutarate	Frullania polysticta	J	(345b)
(1202)	Tricetin-6,8-di-C-arabinoside	Plagiochila jamesonii	J	(491a)
(1203)	Tricetin-6-C-arabinoside-8-C-glucoside	Metzgeria furcata	Me	(367)
(1204)	Selgin-6,8-di-C-glucoside (= Stellarin-2)	Eupleurozia simplicissima	J	(422)
		Mylia anomala	J	(419)
		M. taylorii	J	(419)
		M. verrucosa	J	(419)
		Pleurozia acinosa	J	(422)
		P. articulata	J	(422)
		P. caledonica	J	(422)
		P. conchifolia	J	(422)
(1205)	Tricetin-6-C-glucoside-8-C-arabinoside	Metzgeria furcata var. uvula	Me	(369, 549)
		Plagiochila jamesonii	J	(491a)
(1206)	Tricin-6,8-di-C-glucoside	Eupleurozia simplicissima	J	(422)
		Frullania danurica	J	(420)
		F. jackii	J	(420)
		Mylia verrucosa	J	(419)

J: Jungermanniales; M: Marchantiales; Me: Metzgeriales; Mo: Monocleales

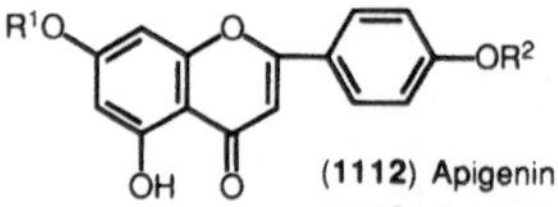

(1112) Apigenin ; $R^1=R^2=H$
(1113) Apigenin-7,4'-dimethyl ether ; $R^1=R^2=Me$
(1114) Acacetin ; $R^1=H$, $R^2=Me$
(1115) Apigenin-4'-O-glucoside ; $R^1=H$, $R^2=Glc$
(1116) Apigenin-7,4'-di-O-galacturonide ; $R^1=R^2=Gal A$
(1117) Apigenin-7-O-glucoside-4'-O-galacturonide ; $R^1=Glc$, $R^2=Gal A$
(1118) Apigenin-7,4'-di-O-glucuronide ; $R^1=R^2=GA$
(1119) Apigenin-7-O-rhamnosylglucuronide-4'-O-glucuronide or
        Apigenin-7-O-glucuronide-4'-O-rhamnosylglucuronide ;
        $R^1=Rha-GA$, $R^2=GA$ or $R^1=GA$, $R^2=Rha-GA$
(1120) Apigenin-7-O-neohesperidoside ; $R^1=Neoh$, $R^2=H$
(1121) Apigenin-7-O-glucoside ; $R^1=Glc$, $R^2=H$
(1122) Acacetin-7-O-rhamnoarabinosyl galacturonide ;
        $R^1=Rha-Ara-Gal A$, $R^2=Me$
(1123) Acacetin-7-O-rhamnoxylosyl galacturonide ;
        $R^1=Rha-Xyl-Gal A$, $R^2=Me$
(1124) Acacetin-7-O-rhamnoglucosyl glucuronide ;
        $R^1=Rha-Glc-GA$, $R^2=Me$
(1125) Acacetin-7-O-diglucuronide ; $R^1=GA-GA$, $R^2=Me$
(1126) Acacetin-7-O-di-galacturonide ; $R^1=Gal A-Gal A$, $R^2=Me$

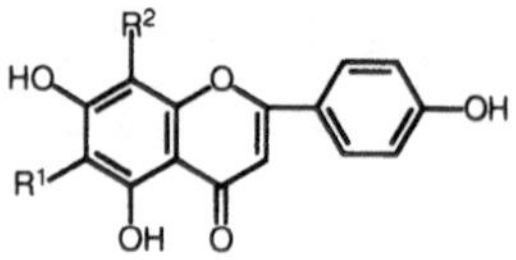

(1127) Apigenin-6,8-di-C-glucoside (=Vicenin-2) ; $R^1=R^2=Glc$
(1128) 6-Hydroxyapigenin (=Scutellarein) ; $R^1=OH$, $R^2=H$

Chart 68a. Flavones found in the Hepaticae

predominant flavonoid type in all orders is represented by flavone derivatives. The Metzgeriales are characterized by flavone-C-glycosides, the Jungermanniales by flavone-C- and flavone-O-glycosides, and the Marchantiales, Sphaerocarpales and Monocleales by flavone-O-glucuronides. The Takakiales and Bryales (Musci) produce a wider range of flavone glycosides than the other orders, in particular 25% of the Bryales contain biflavonoids (*362, 363*) (Chapter III).

The Monocleales and Sphaerocarpales can be distinguished from the Marchantiales by the complete lack of flavone C-glycosides. The Marchantiales produce flavone C-glycosides (11%); these are of the biosynthetically simple apigenin-6,8-di-C-glucoside (**1127**) and luteolin 6,8-di-C-glucoside (**1164**) type and distinct from the flavonoid C-glycosides found in the Jungermanniales and the Metzgeriales in which oxidation of the B-ring in flavones is characteristic and in which many different types of sugars (glucose, rhamnose, arabinose and xylose) have been detected.

The flavonoids profile of the Jungermanniales is similar to that of the Metzgeriales.

To proceed to more specific cases, the distribution of flavonoids in 14 species of the subgenus *Chonanthelia* in the genus *Frullania* has been investigated leading to isolation and identification of 18 flavonoids three of which are aglycones and 15 of which are glycosides (*645*). The free aglycones are apigenin (**1112**), luteolin (**1143**) and 6-hydroxyluteolin (**1166**). Luteolin-7-O-glucoside (**1147**) was the only glycoside found in all species, thus luteolin and its 7-O-glucoside are chemical markers for species within subgenus *Chonanthelia*. Both compounds had already been isolated previously from *F. dilatata* (*425*), whereas this is the first report of the co-occurrence of apigenin (**1112**) and 6-hydroxyluteolin (**1166**) as free aglycones in any *Frullania* species. As a result of comparative flavonoid analysis, two basic species groups can be distinguished by their flavonoid distribution. This result is in good accordance with their morphological classification into two sections, *Cladocarpicae* and *Chonanthelia*. *F. ecklonii* exhibits a flavonoid pattern intermediate between that of sect. *Cladocarpicae* and that of *Chonanthelia*. *Frullania jackii* and *F. davurica* are definitely distinguished by different chromosome numbers. The chromosome number of the former is $n = 17$ and that of the latter is $n = 9$ for the female and $n = 8$ for the male plant. The pattern of lipophilic compounds proved to be very similar in both taxa; more specially, the bibenzyl constituents of *F. jackii* are qualitatively and quantitatively identical with those of *F. davurica*. Twenty different flavonoids were isolated from both taxa; luteolin-7,3',4'-trimethyl ether (**1146**) and luteolin-3',4'-dimethyl ether (**1144**) being common to both. The occurrence of all three flavone aglycones in both taxa is taxonomically highly significant and suggests that they are closely related to each other. The major flavone O-glycoside of both species is the very rare 6-hydroxyluteolin-7-O-sophoroside (**1169**) which has been found for the first time in any bryophytes (*420*). Most O-glycosides of *F. davurica* are based on 6-hydroxyapigenin (= scutellarein) (**1128**) and 6-hydroxy-luteolin-3'-methyl ether (**1185**), whereas 6-hydroxyluteolin (**1166**) is the aglycone of most *F. jackii*-O-glycosides. Rhamnosides were detected only among the glycosides of *F. jackii*, xylosides with one exception only in the glycosides of *F. davurica* (*420*). Thus the flavonoid patterns of *F. jackii* and *F. davurica* suggest that they are closely related but nevertheless different taxa.

Brachiolejeunea phyllorhiza and *B. laxifolia* (Lejeuneaceae) produce luteolin-7-O-glucoside (**1147**). The former also contains luteolin (**1143**) and its 7-O-acylglucoside as well as 6-hydroxyluteolin-7-O-acylglucoside (*217*). Isoorientin-7-O-glucoside (lutonarin) (**1186**) was isolated from

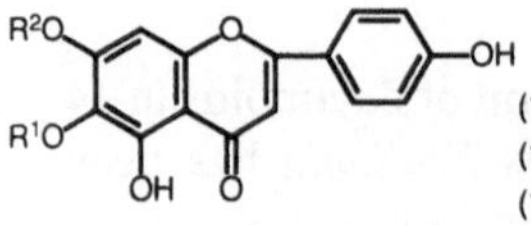

(1129) 6-Hydroxyapigenin (=Scutellarein)-7-O-diglucoside ; R^1=H, R^2=Glc-Glc
(1130) Scutellarein-7-O-glucoside-6"-malonate ; R^1=H, R^2=6-Malonyl Glc
(1131) Scutellarein-7-O-xyloside ; R^1=H, R^2=Xyl
(1132) Scutellarein-6-O-xyloside-7-O-glucoside ; R^1=Xyl, R^2=Glc
(1133) Scutellarein-7-O-glucoxylglucoside ; R^1=H, R^2=Glc-Xyl-Glc
(1134) Scutellarein-7-O-glucoside-6"-hydroxy-3-methylglutarate ;
  R^1=H, R^2=6-Hydroxy-3-methylglutaryl Glc
(1135) Scutellarein-6-O-glucoside-7-O-rhamnoglucoside ; R^1=Glc, R^2=Rha-Glc

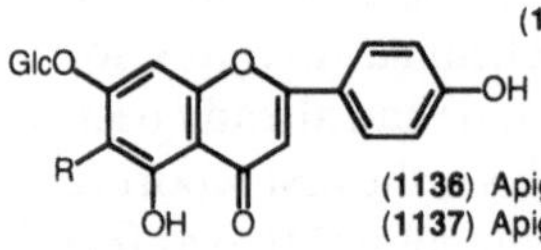

(1136) Apigenin-6-C-rhamnoside-7-O-glucoside (=Isofurcatain-7-O-glucoside) ; R=Rha
(1137) Apigenin-6-C-glucoside-7-O-glucoside (=Saponarin) ; R=Glc

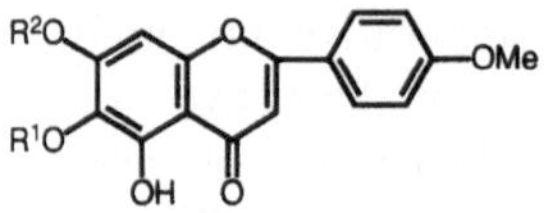

(1138) Apigenin-6,7,4'-trimethyl ether ; R^1=R^2=Me
(1139) Pectolinarigenin-7-O-glucuronide ; R^1=Me, R^2=GA
(1140) Pectolinarigenin-7-O-galacturonide ; R^1=Me, R^2=Gal A
(1141) 4'-Methoxyscutellarein-7-O-glucronylrhamnoside ;
  R^1=H, R^2=GA-Rha

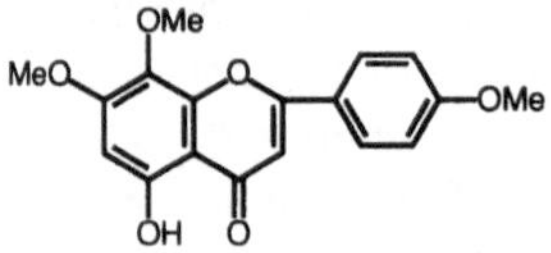

(1142) 5-Hydroxy-7,8,4'-trimethoxyflavone

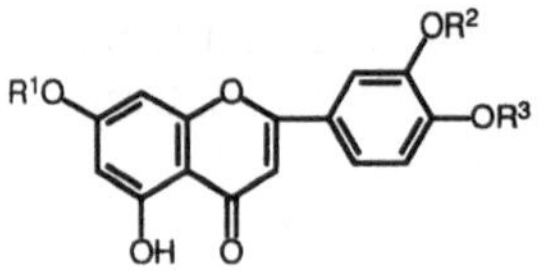

(1143) Luteolin ; R^1=R^2=R^3=H
(1144) Luteolin-3',4'-dimethyl ether ; R^1=H, R^2=R^3=Me
(1145) Luteolin-7,3'-dimethyl ether (=Velutin) ; R^1=R^2=Me, R^3=H
(1146) Luteolin-7,3',4'-trimethyl ether ; R^1=R^2=R^3=Me
(1147) Luteolin-7-O-glucoside ; R^1=Glc, R^2=R^3=H
(1148) Luteolin-7-O-glucoside-6"-malonate ; R^1=6-Malonyl Glc, R^2=R^3=H
(1149) Luteolin-7-O-glucuronide ; R^1=GA, R^2=R^3=H
(1150) Luteolin-7-O-gentiobiose ; R^1=Ge, R^2=R^3=H
(1151) Luteolin-3'-O-glucoside ; R^1=R^3=H, R^2=Glc
(1152) Luteolin-4'-O-glucoside ; R^1=R^2=H, R^3=Glc
(1153) Luteolin-4'-O-glucuronide ; R^1=R^2=H, R^3=GA
(1154) Luteolin-7-O-glucoside-6"-hydroxy-3-methylglutarate ;
  R^1=6-Hydroxy-3-methylglutaryl Glc, R^2=R^3=H
(1155) Luteolin-7,4'-di-O-glucuronide ; R^1=R^3=GA, R^2=H
(1156) Luteolin-7,3'-di-O-glucoside ; R^1=R^2=Glc, R^3=H
(1157) Luteolin-4'-O-rhamnosyl-β-D-galacturonide ;
  R^1=R^2=H, R^3=Rha-Gal A
(1158) Crysoeriol-7-O-glucoside ; R^1=Glc, R^2=Me, R^3=H
(1159) Crysoeriol-7-O-glucuronide ; R^1=GA, R^2=Me, R^3=H
(1160) Crysoeriol-7-O-neohesperidoside ; R^1=Neoh, R^2=Me, R^3=H
(1161) Crysoeriol-7,4'-di-O-glucoside ; R^1=R^3=Glc, R^2=Me
(1162) 3',4'-Dimethoxyluteolin-7-O-rhamnoarabinosyl galacturonide ;
  R^1=Rha-Ara-Gal A, R^2=R^3=Me
(1163) 3',4'-Dimethoxyluteolin-7-O-rhamnoxylosyl galacturonide ;
  R^1=Rha-Xyl-Gal A, R^2=R^3=Me

Chart 68b. Flavones found in the Hepaticae

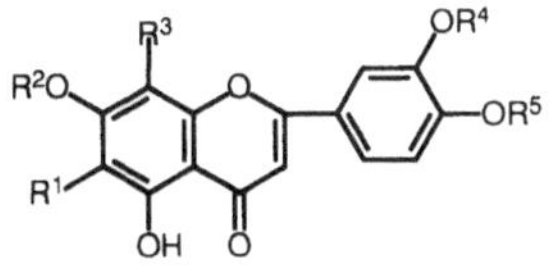

(1164) Luteolin-6,8-di-C-glucoside (=Lucenin-2) ; $R^1=R^3=$Glc, $R^2=R^4=R^5=$H

(1165) Luteolin-6,8-di-C-arabinosie ; $R^1=R^3=$Ara, $R^2=R^4=R^5=$H

(1166) 6-Hydroxyluteolin ; $R^1=$OH, $R^2=R^3=R^4=R^5=$H

(1167) Luteolin-6,7-dimethyl ether (=Cirsilol) ; $R^1=$OMe, $R^2=$Me, $R^3=R^4=R^5=$H

(1168) Luteolin-6,3',4'-trimethyl ether (=Eupatilin) ; $R^1=$OMe, $R^2=R^3=$H, $R^4=R^5=$Me

(1169) 6-Hydroxyluteolin-7-O-sophoroside ; $R^1=$OH, $R^2=$Soph, $R^3=R^4=R^5=$H

(1170) 6-Hydroxyluteolin-7-O-glucoside ; $R^1=$OH, $R^2=$Glc, $R^3=R^4=R^5=$H

(1171) Nodifloretin-7-monomethyl ether (=Pedalitin) ; $R^1=$OH, $R^2=$Me, $R^3=R^4=R^5=$H

(1172) 6-Hydroxyluteolin-7,3'-dimethyl ether ; $R^1=$OH, $R^2=R^4=$Me, $R^3=R^5=$H

(1173) 6-Hydroxyluteolin-6-O-xyloside-7-O-glucoside ; $R^1=$O-Xyl, $R^2=$Glc, $R^3=R^4=R^5=$H

(1174) 6-Hydroxyluteolin-6-O-glucoside-7-O-xyloglucoside ; $R^1=$O-Glc, $R^2=$Xyl-Glc, $R^3=R^4=R^5=$H

(1175) Nodifloretin-7-O-rhamnoside-4'-O-glucuronide ;
        $R^1=$OH, $R^2=$Rha, $R^3=$H, $R^4=$Me, $R^5=$GA

(1176) Nodifloretin-7-O-xyloglucoside ; $R^1=$OH, $R^2=$Xyl-Glc, $R^3=R^5=$H, $R^4=$Me

(1177) Nodifloretin-6-O-xyloside-7-O-glucoside ; $R^1=$O-Xyl, $R^2=$Glc, $R^3=R^5=$H, $R^4=$Me

(1178) Nodifloretin-6-O-glucoside-7-O-xyloside ; $R^1=$O-Glc, $R^2=$Xyl, $R^3=R^5=$H, $R^4=$Me

(1179) Nodifloretin-7-O-glucoside-6''-hydroxy-3-methylglutarate ;
        $R^1=$OH, $R^2=$6-Hydroxy-3-methylglutaryl Glc, $R^3=R^5=$H, $R^4=$Me

(1180) 6-Hydroxyluteolin-6-O-diglucoside ; $R^1=$O-Glc-Glc, $R^2=R^3=R^4=R^5=$H

(1181) 6-Hydroxyluteolin-7-O-rhamnoside ; $R^1=$OH, $R^2=$Rha, $R^3=R^4=R^5=$H

(1182) 6-Hydroxyluteolin-7-O-glucoside-6''-hydroxy-3-methylglutarate ;
        $R^1=$OH, $R^2=$6-Hydroxy-3-methylglutarylGlc, $R^3=R^4=R^5=$H

(1183) 6-Hydroxyluteolin-7-O-glucoside-4''-hydroxy-3-methylglutarate ;
        $R^1=$OH, $R^2=$4-Hydroxy-3-methylglutarylGlc, $R^3=R^4=R^5=$H

(1184) 6-Hydroxyluteolin-7-O-glucoside-3''-hydroxy-3-methylglutarate ;
        $R^1=$OH, $R^2=$3''-Hydroxy-3-methylglutaryl-Glc, $R^3=R^4=R^5=$H

(1185) 6-Hydroxyluteolin-3'-methyl ether (=Nodifloretin or Batatifolin)-7-O-sophoroside ;
        $R^1=$OH, $R^2=$Soph, $R^3=R^5=$H, $R^4=$Me

(1186) Isoorientin-7-O-glucoside (=Lutonarin) ; $R^1=R^2=$Glc, $R^3=R^4=R^5=$H

(1187) Luteolin-6,8-di-C-glucoside-3'-glucoside (=Lucenin-2 3'-glucoside) ;
        $R^1=R^3=R^4=$Glc, $R^2=R^5=$H

(1188) Luteolin-6-C-glucoside-8-C-arabinoside (=Carlinoside) ; $R^1=$Glc, $R^2=R^4=R^5=$H, $R^3=$Ara

(1189) Luteolin-6-C-arabinoside-8-C-glucoside (=Isocarlinoside) ;
        $R^1=$Ara, $R^2=R^4=R^5=$H, $R^3=$Glc

(1190) Onopordin-7,4'-di-O-polysaccharide ; $R^1=R^4=$H, $R^2=R^5=$Polysaccharide, $R^3=$OMe

Chart 68c. Flavones found in the Hepaticae

Stictolejeunea squamata and *Lejeunea cavifolia.* The latter also yielded tricetin-6,8-di-C-β-D-glucopyranoside (**1199**) (*217*). Twenty-two species of Lejeuneaceae, subfamily Ptychanthoideae were studied for flavonoids and cinnamic acid derivatives. Almost all species produce both types of compounds (*347*). *Frullanoides densifolia* (Lejeuneaceae) produces kaempferol 3-methyl ether (**1191**) which rarely occurs in higher plants and had not previously been isolated from bryophytes (*347*). A *Metzgeria* species also produces kaempferol-6-C-glucoside (**1193**) and its 3-O-glucoside (**1194**) which are very rare in plant kingdom (*363*).

Mylia anomala, *M. nuda*, *M. taylorii* and *M. verrucosa* contain the flavone 6,8-di-C-β-glucopyranosides vicenin-2 (**1127**) and lucenin-2 (**1164**) which are widely distributed in liverworts (*424*). The former seems to be one of the flavone C-glycosides frequently found in Jungermanniales and Metzgeriales.

There are four dendroid liverworts, *Jensenia connivens* (subfamily Pallavicinioideae), *Symphyogyna podophylla* (subfamily Symphyogynoideae), *Hymenophyton flavellatum* and *H. leptopodum* (Hymenophytaceae), all belonging to the order Metzgeriales, in New Zealand. The four species differ remarkably different in flavonoid patterns, the two *Hymenophyton* species being differentiated from *Jensenia* and *Symphyogyna* by high levels of flavonoids. *Hymenophyton leptopodum* may be a more advanced species than *H. flavellatum* since the latter species elaborates kaempferol (**1192**) glycosides with the common flavonoids, apigenin (**1112**) and acacetin (**1114**) glycosides being found in both *H. flavellatum* and *H. leptopodum* (*129*). *Plagiochila jamesonii* produces a new tricetin glycoside, tricetin-6,8-di-C-α-L-arabinoside (**1202**), along with the known tricetin-6,8-di-C-β-D-glucopyranoside (**1199**) and tricetin-6-C-β-D-glucosyl-8-C-α-L-arabinoside (**1205**) (*441a*).

An unexpected doubling of signals occurs in the ^{13}C-NMR spectra of tricetin-6,8-di-C-glucopyranoside (**1199**) and tricetin-6-C-arabinopyranoside-8-C-glucoside (**1203**) isolated from or detected in *Metzgeria furcata* containing 8-C-hexosyl substituents at room temperature, but not in the spectra of those containing 8-C-arabinosyl substituents (*367*). This suggests that interaction occurs between C-linked (β)-monohexose at C-8 and the B-ring. As the phenomenon has not been observed in 8-C-pentopyranosides, the primary hydroxyl group of the hexose would appear to be the function interacting with the B-ring. This results in restricted rotation of the B-ring and/or the hexose, giving rise to a mixture of two NMR-distinguishable isomers. Rotational isomers should be interconvertible at high temperatures and indeed the doubling disappeared when spectra of lucenin-2 (**1164**) and selgin-6,8-di-C-glucoside (= stellarin-2) (**1204**) were measured at 90°. The absence of signal doubling in 8-C-hexosylisoflavonoids is due to the greater distance of the B-ring from the glycosyl moiety at C-8 which eliminates the steric interaction.

Further fractionation of an 80% aqueous methanolic extract of *Metzgeria furcata* var. *uvula* resulted in the isolation of a new flavone glycoside, apigenin-6-C-α-L-rhamnoside (= isofurcatain)-7-O-β-D-glucopyranoside (**1136**) and tricetin-6-C-β-D-glucopyranoside-8-C-α-L-arabinopyranoside (**1205**) (*369, 549*). The 7-O-linked glucose of the former compound is considered to be β-linked to apigenin-6-C-rhamnoside since H-1 of the glucose unit appears as a doublet ($J = 9.8$ Hz) centered at δ5.26 in the ^{1}H-NMR spectrum. Furthermore, the sugar carbon signals associated with the O-linked glucose resemble those observed in the ^{13}C-NMR spectrum of apigenin-6-C-β-D-glucopyranoside-7-O-β-D-glucopyranoside (**1137**).

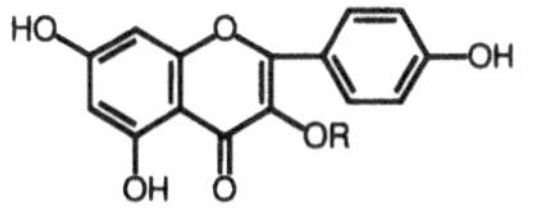

(**1191**) Kaempferol-3-methyl ether ; R=Me
(**1192**) Kaempferol ; R=H

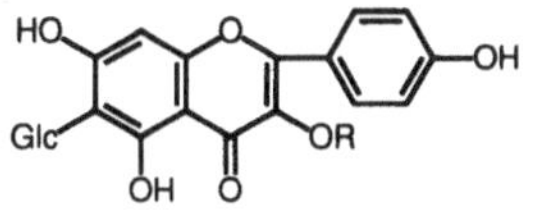

(**1193**) Kaempferol-6-C-glucoside ; R=H
(**1194**) Kaempferol-6-C-glucoside-3-O-glucoside ; R=Glc

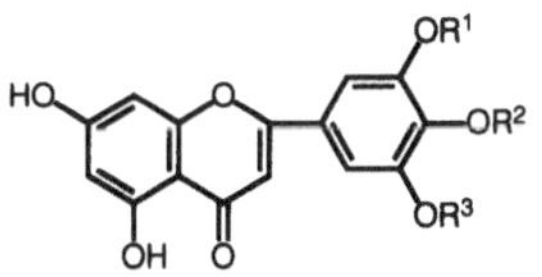

(**1195**) Tricetin ; $R^1=R^2=R^3=H$
(**1196**) Apometzgerin ; $R^1=R^2=Me$, $R^3=H$
(**1197**) Tricin ; $R^1=R^3=Me$, $R^2=H$
(**1198**) Selgin ; $R^1=Me$, $R^2=R^3=H$

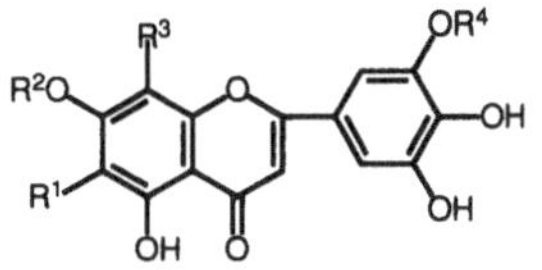

(**1199**) Tricetin-6,8-di-C-glucoside ; $R^1=R^3=Glc$, $R^2=R^4=H$
(**1200**) Tricetin-6-C-glucoside (=Isoaffinetin) ; $R^1=Glc$, $R^2=R^3=R^4=H$
(**1201**) Tricetin-7-O-glucoside-3'-O-glucoside-6''-hydroxy-3-methylglutarate ; $R^1=R^3=H$, $R^2=Glc$, $R^4=6$-Hydroxy-3-methylglutaryl Glc
(**1202**) Tricetin-6,8-di-C-arabinoside ; $R^1=R^3=Ara$, $R^2=R^4=H$
(**1203**) Tricetin-6-C-arabinoside-8-C-glucoside ; $R^1=Ara$, $R^2=R^4=H$, $R^3=Glc$
(**1204**) Selgin-6,8-di-C-glucoside (=Stellarin-2) ; $R^1=R^3=Glc$, $R^2=H$, $R^4=Me$
(**1205**) Tricetin-6-C-glucoside-8-C-arabinoside ; $R^1=Glc$, $R^2=R^4=H$, $R^3=Ara$

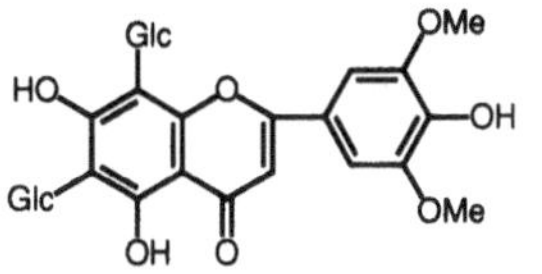

(**1206**) Tricin-6,8-di-C-glucoside

Chart 68d. Flavones and flavonols found in the Hepaticae

Flavonoids of two *Eupleurozia* and eight *Pleurozia* species (Pleurozi-aceae) were analyzed by 2D-TLC (*422*). *Pleurozia acinosa, P. articulata* and *P. caledonica* produce a new flavone glycoside, lucenin-2 3'-O-glucoside (**1187**), together with lucenin-2 (**1164**), stellarin-2 (**1204**) and tricetin-6,8-di-C-glucoside (**1199**). The flavonoid pattern of *P. gigantea, P. giganteoides* and *P. heterophylla* is similar to that of the above three species, except for the absence of lucenin-2 3'-glucoside (**1187**). *P. conchifolia* contains (**1164, 1199, 1204**) as well as carlinoside (**1188**) and isocarlinoside (**1189**). Luteolin-4'-O-glucuronide (**1153**) and luteolin (**1143**) were isolated from *Eupleurozia paradoxa* along with lucenin-2 (**1164**). *E. simplicissima* produces (**1164, 1199, 1204**) as well as vicenin-2 (**1127**) and tricin-6,8-di-C-glucoside (**1206**).

Cell suspension cultures of *Marchantia paleacea* var. *diptera* produces luteolin (**1143**) (*457*). Apigenin-7,4'-dimethylether (**1113**) were isolated from *Reboulia hemisphaerica* (*411*) and *Plagiochasma rupestre* (Aytoni-aceae) (*243*). The latter species contains luteolin-6,8-di-C-arabinoside (**1165**) (*491a*). *R. hemisphaerica* also produces 5-hydroxy-7,8,4'-tri-methoxyflavone (**1142**) (*411*).

Total syntheses of tricetin (**1195**), apometzgerin (**1196**), tricin (**1197**) and selgin (**1198**) which occur as flavonoid components in bryophytes have been accomplished by EICHER *et al.* (*168, 355, 647*).

9. Lipids

9.1 n-Alkanes and Related Compounds

The earlier review (*19*) listed the distribution of *n*-alkanes in 40 species of liverworts. The relative amounts of odd and even numbered *n*-alkanes in some species are close to equal in contrast to the situation encountered in higher plants in which odd carbon numbered *n*-alkanes (normally $C_{27}-C_{33}$) are predominant.

Lophocolea heterophylla contains *n*-alkanes (C15–C25), *n*-nonacosan-10-one (**1207**) and cerides which were transesterified with MeOH to give methyl esters of fatty acids (C12–C26) and alcohols (C18–C26). Palmitic and stearic acids and docosanol are predominant. The main ceride in the wax is docosanyl hexadecanoate. More polar esters which were trans-esterified with MeOH yielded primary aliphatic alcohols (C13–C16), nonacosan-10-ol (**1208**) which is the major component, and normal fatty acids (C12–C24) (*343*).

The essential oils of *Conocephalum conicum, Plagiochila ovalifolia* and *Wiesnerella denudata* (*62*) and *Lophocolea heterophylla* and *L. bidentata*

(*527a*) contain 1-octen-3-ol (**1213**) and 1-octen-3-yl acetate (**1214**) as well as monoterpenoids. Optically pure (*S*)-(+)-1-octen-3-ol (**1213**) and (*R*)-(+)-1-octen-3-yl acetate (**1214**) have been isolated from *Conocephalum japonicum* (*585*). The latter compound was also isolated from *Frullania davurica* (*585*). *C. japonicum* and *Plagiochila sciophila* (= *P. acanthophylla* subsp. *japonica*) contain octan-3-yl acetate (**1215**) and 3-octanone (**1216**), respectively (*62*). *n*-Nonanal (**1217**) and 2-methylbutanol (**1218**) have been detected in *C. japonicum* (*62*) and *Porella cordaeana* (*593*), respectively. The former aldehyde has been detected in *Lophocolea bidentata* (*527a*). The essential oil of *Lophocolea heterophylla* contains *n*-hexanol (**1209**), *n*-hexanal (**1210**), *n*-heptanal (**1211**), 1-octanol (**1212**), *n*-nananal (**1217**) and *n*-decanal (**1219**) (*527a*). Octanol, octenol, octanal and octanone have also been detected in some mosses (see Chapter III).

9.2 Fatty Acids

Most liverworts contain various free fatty acids (*19, 271*). A new acetylenic acid, 9-octadecen-6-ynoic acid (**1220a**) has been isolated from the thalli of *Riccia fluitans* (Marchantiales), together with the previously known 9,12-octadecadien-6-ynoic (**1221**) and 9,12,15-octadecatrien-6-ynoic acids (**1222**) (*340*). The triglycerides were transesterified with 5% H_2SO_4 in MeOH and the resulting methyl esters were purified by preparative GC to give three methyl esters of each fatty acid (**1220a, 1221, 1222**). Hydrogenation of the methyl ester (**1220b**) of (**1220a**) gave stearic acid methyl ester. The IR spectrum of (**1220b**) had a band at 2310 cm^{-1}, assignable to a nonterminal acetylenic group. The stereochemistry at the double bond and the position of the functional groups of (**1220b**) were based on the ^{1}H-NMR spectrum and spin decoupling experiments. This is the first record of acetylenic fatty acids in liverworts.

Investigation of twelve other *Riccia* species resulted in the detection of acetylenic fatty acids in all species (*339*). Acetylenic fatty acids were components not only of the triacylglycerols but also, to a remarkable degree, in the glycolipids, specially in the monogalactosyl diglycerides. In species belonging to subgenus *Ricciella*, 9-octadecen-6-ynoic acid (**1220a**) was the predominant component. On the other hand, the major fatty acid of the species of subgenus *Euriccia* was 9,12,15-octadecatrien-6-ynoic acid (**1222**).

The family Ricciaceae consists of two genera, *Ricciocarpos* and *Riccia*. *Ricciocarpos* is monotypic; the sole member is *R. natans* which is free of acetylenic acids. The fatty acid composition of *R. natans* is similar to that of Marchantiaceae with linolenic acid (= 18:3ω3) as the major component together with large amounts of arachidonic acid (ARA) (= 5Z,8Z,11Z,14Z-eicosatetraenoic acid) (20:4ω6) (**1223**) and eicosapentaenoic acid (EPA) (= 5Z,8Z,11Z,14Z,17Z-eieosapentaenoic acid) (20:5ω3) (**1224**) (*339*). *Marchantia*, *Corsinia* and *Oximitra* species are free of acetylenic fatty acids (*339*). *Monoclea forsteri* belonging to Monocleaceae contains 9-octadecen-6-ynoic acid (**1220a**), along with unidentified acetylenic acids.

Further investigation of the methanol extract of the largest thalloid liverwort, *Monoclea forsteri* collected in New Zealand, resulted in the isolation of two new fatty acids, 10-keto-8E-octadecen-6-ynoic acid (**1225**) and 10-hydroxy-octadec-6-yn-8E-enoic acid (**1226**) whose structures were elucidated by chemical and spectral evidence (*590*). The position of the ene-yne-one system and the location of the ketone group in (**1225**) followed from IR, UV (267 nm), ^{1}H- and ^{13}C-NMR spectrometry as well as the mass fragmentation of hexahydro derivative (**1225a**) shown in Scheme 66. In addition to the acetylenic and intense carbonyl bands (2200 and 1680 cm^{-1}), the IR spectrum exhibited a more intense band characteristic of a *trans*-ethylenic double bond (1590 cm^{-1}). The arithmetic difference in frequencies between two absorption bands is 90 cm^{-1}, indicating that the enone system has the *s-cis* conformation. Oxidation of (**1226**) with PCC gave (**1225**) thus establishing its structure. Highly unsaturated fatty acids with an ene-yne system are widespread in the Musci. Acetylenic fatty acids are also frequently found in mosses, but are limited to a few families or genera, such as the Ditrichaceae, Dicranaceae, Bryum and Fontinalis (*307, 338*) (see Chapter III).

Shinmen *et al.* (*503*) have reported that *Marchantia polymorpha* is a rich source of both ARA and EPA. A culture of *M. polymorpha* contained high amounts of ARA and EPA (92 and 48 mg l^{-1}, respectively) under photomixotrophic conditions. The cell culture of *M. polymorpha* shows rapid growth in Murashige-Skoog's medium (MSK)-2 or MSK-12. ARA

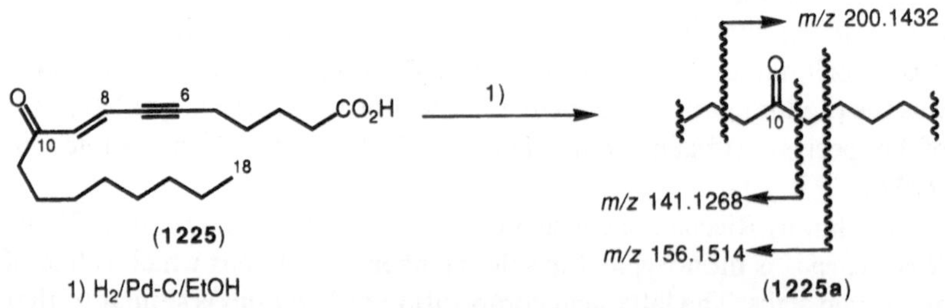

Scheme 66. Hydrogenation of monocleic acid and mass fragments of hexahydro derivative

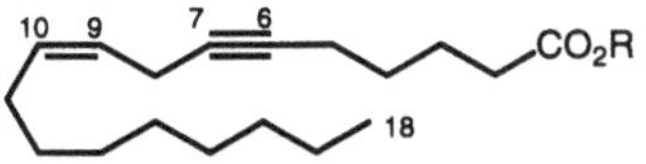

(**1220a**) 9-Octadecen-6-ynoic acid ; R=H   (**1221**) 9,12-Octadecadien-6-ynoic acid
(**1220b**) R=Me

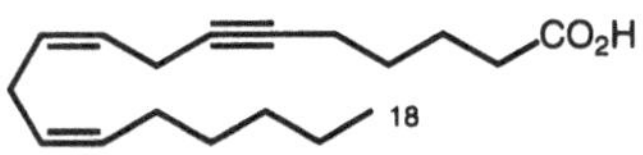

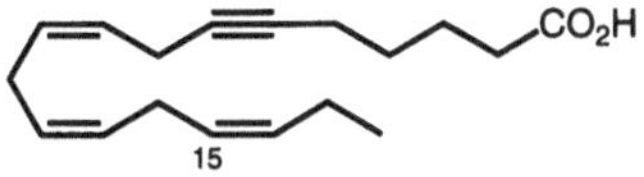

(**1222**) 9,12,15-Octadecatrien-6-ynoic acid

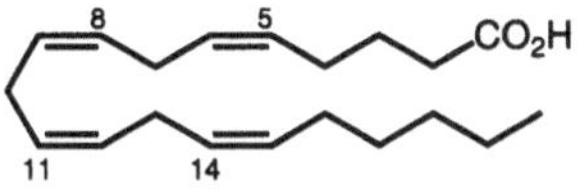

(**1223**) Arachidonic acid (=5Z,8Z,11Z,14Z-Eicosatetraenoic acid)

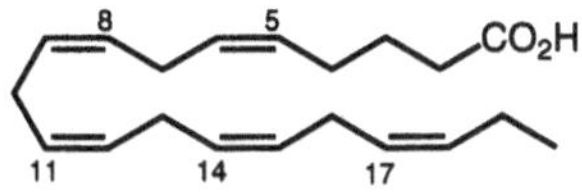

(**1224**) Eicosapentaenoic acid (=5Z,8Z,11Z,14Z,17Z-Eicosapentaenoic acid)

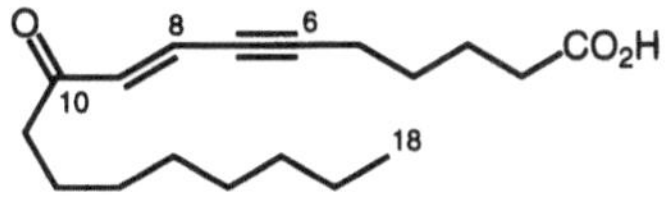

(**1225**) Monocleic acid (=10-Keto-8E-octadecen-6-ynoic acid)

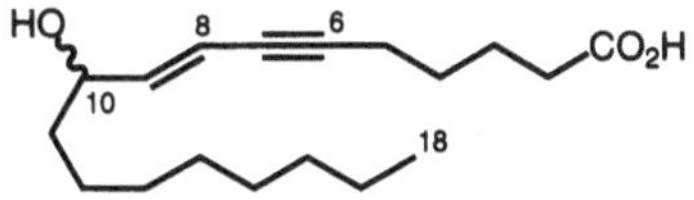

(**1226**) Monocleolic acid (=10-Hydroxy-8E-octadecen-6-ynoic acid)

Chart 69. Fatty acids found in the Hepaticae

and EPA were found mainly as components of diacylglycerides. In addition to ARA and EPA, 16:0 (palmitic), 16:1 (palmitoleic), 16:2, 16:3, 18:0 (stearic), 18:1 (oleic), 18:2 (linoleic), 18:3α (α-linolenic), 18:3γ (γ,γ-linolenic), 18:4ω3, 20:3ω6 and 20:4ω3 acids were found in the triglyceride, fatty acid, diglyceride and steryl ester portions of *M. polymorpha* culture, 16:0, 16:3 and 18:3 being predominant.

Steryl ester fraction of *Mylia taylorii* and *M. verrucosa* contains oleic, linolenic and linoleic acids (*532*). The level of C-20 polyunsaturated fatty acids is significantly low. In general the liverworts studied so far are very different from the mosses with respect to the proportion of steryl esters.

10. Sulfur-Containing Compounds

A strong seaweed-like odor is produced on drying of liverworts. Dimethyl sulfide (**1227**) which is one of the important flavors of seaweeds has been detected in European *Porella cordaeana* by GC-MS and FPD detector used for GC (*593*). *Pellia endiviifolia* and *Plagiochila sciophila* (= *P. acanthophylla* subsp. *japonica*) also produce dimethyl sulfide (*62*).

Isotachis japonica produces simple benzoates and cinnamates (*19*). Reinvestigation of the chemical constituents of *I. japonica* resulted in the isolation of two new sulfur-containing esters, isotachin A (**1228**) and isotachin B (**1229**) whose structures were elucidated by a combination of ^{1}H- and ^{13}C-NMR and high resolution mass spectrometry (*76*). A new *trans*-β-methylthioacrylate, isotachin C (**1230**), has been isolated from *Balantiopsis rosea*, along with previously known benzoates and cinnamates as well as isotachins A (**1228**) and B (**1229**) and 2-methoxybenzyl benzoate (**863**) (*65*). The position of the methoxy group on the benzene ring was deduced from the NMR spectrum and observation of an NOE only between H-3 and the methoxyl group. Sulfur-containing compounds have not been found previously in bryophytes.

CH$_3$SCH$_3$
(**1227**) Dimethylsulfide

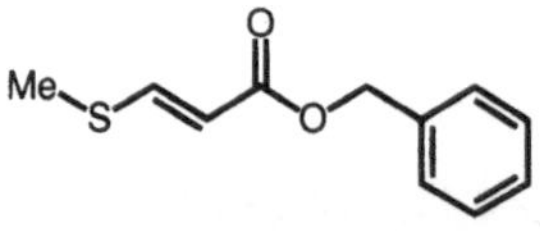

(**1228**) Isotachin A
(=Benzyl *trans*-β-methyl thioacrylate)

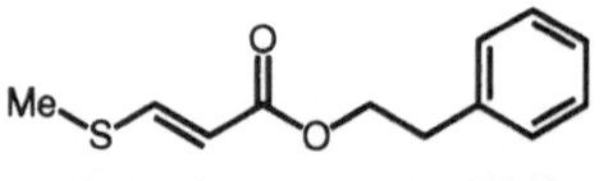

(**1229**) Isotachin B
(=β-Phenylethyl *trans*-β-methyl thioacrylate)

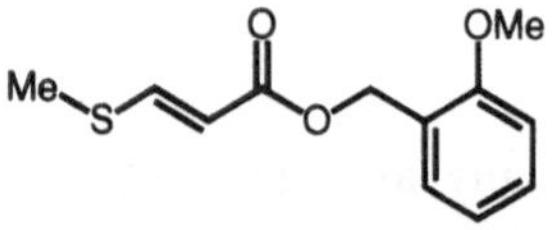

(**1230**) Isotachin C
(=2-Methoxybenzyl *trans*-β-methyl thioacrylate)

Chart 70. Sulfur-containing compounds found in the Hepaticae

11. Carbohydrates

Earlier reports (*19, 271*) mentioned the detection of many carbohydrates in bryophytes. Five water soluble carbohydrates, (−)-L-bornesitol,

Table IIg. *Chemical Constituents (Lipids and Miscellaneous) Found in the Hepaticae*

Structure number	Name of compounds	Formula	m.p.°C	$[\alpha]_D$	Plant source	Order	References	Comments
(1207)	Nonacosan-10-one	$C_{29}H_{58}O$			*Lophocolea heterophylla*	J	(343)	
(1208)	Nonacosan-10-ol	$C_{29}H_{60}O$			*Lophocolea heterophylla*	J	(343)	
(1209)	*n*-Hexanol	$C_6H_{14}O$			*Lophocolea heterophylla*	J	(527a)	
(1210)	*n*-Hexanal	$C_6H_{12}O$			*Lophocolea heterophylla*	J	(527a)	
(1211)	*n*-Heptanal	$C_7H_{14}O$			*Lophocolea heterophylla*	J	(527a)	
(1212)	*n*-Octanol	$C_8H_{18}O$			*Lophocolea heterophylla*	J	(527a)	
(1213)	(S)-(+)-1-Octen-3-ol	$C_8H_{16}O$			*Conocephalum conicum*	M	(62, 580)	
					Lophocolea bidentata	J	(527a)	
					L. heterophylla	J	(527a)	
					Plagiochila ovalifolia	J	(62)	
					Wiesnerella denudata	M	(62)	
(1214)	(R)-(+)-1-octen-3-yl acetate	$C_{10}H_{18}O_2$		+ 4.8	*Conocephalum conicum*	M	(62)	
					C. japonicum	M	(585)	
					Lophocolea bidentata	J	(527a)	
					L. heterophylla	J	(527a)	
					Plagiochila ovalifolia	J	(62)	
					Wiesnerella denudata	M	(62)	
(1215)	Octan-3-yl acetate	$C_{10}H_{20}O_2$			*Conocephalum japonicum*	M	(62)	
					Plagiochila sciophila (= *P. acanthophylla* subsp. *japonica*)	J	(62)	
(1216)	3-Octanone	$C_8H_{16}O$			*Conocephalum japonicum*	M	(62)	
					Plagiochila sciophila (= *P. acanthophylla* subsp. *japonica*)	J	(62)	

Table IIg (continued)

Structure number	Name of compounds	Formula	m.p.°C	$[\alpha]_D$	Plant source	Order	References	Comments
(1217)	Nonanal	$C_9H_{18}O$			Conocephalum japonicum	M	(62)	
					Lophocolea bidentata	J	(527a)	
					L. heterophylla	J	(527a)	
(1218)	2-Methyl butanol	$C_5H_{12}O$			Porella cordaeana	J	(593)	
(1219)	n-Decanal	$C_{10}H_{20}O$			Lophocolea heterophylla	J	(527a)	
(1220a)	9-Octadecen-6-ynoic acid	$C_{18}H_{30}O_2$			Monoclea forsteri	Mo	(393)	
(1221)	9,12-Octadecadien-6-ynoic acid	$C_{18}H_{28}O_2$			Riccia fluitans	M	(340)	
					Riccia fluitans	M	(340)	
(1222)	9,12,15-Octadecatrien-6-ynoic acid	$C_{18}H_{26}O_2$			Riccia fluitans	M	(340)	
(1223)	Arachidonic acid (= 5Z,8Z,11Z,14Z-Eicosatetraenoic acid)	$C_{20}H_{32}O_2$			Barbilophozia barbata	J	(500)	
					Marchantia polymorpha	M	(503)	
					Ptilidium ciliare	J	(500)	
					Riccia duplex	M	(339)	
					Ricciocarpos natans	M	(339)	
					Tritomaria quinquedentata	J	(500)	
(1224)	Eicosapentaenoic acid (= 5Z,8Z,11Z,14Z,17Z-Eicosapentaenoic acid)	$C_{20}H_{30}O_2$			Barbilophozia barbata	J	(500)	
					Marchantia polymorpha	M	(503)	
					Ptilidium ciliare	J	(500)	
					Riccia duplex	M	(339)	
					Ricciocarpos natans	M	(339)	
					Tritomaria quinquedentata	J	(500)	

No.	Compound	Formula	Species		Ref.
(1225)	Monocleic acid (= 10-Keto-8E-octadecen-6-ynoic acid)	$C_{18}H_{28}O_3$	Monoclea forsteri	Mo	(590)
(1226)	Monocleolic acid (= 10-Hydroxy-8E-octadecen-6-ynoic acid)	$C_{18}H_{30}O_3$	Monoclea forsteri	Mo	(590)
	Arabinose		Mylia verrucosa	J	(387)
			Calypogeia neesiana subsp. subalpina	J	(387)
	(−)-L-Bornesitol		Mylia taylorii	J	(87)
			M. verrucosa	J	(387)
	Fructose		Calypogeia neesiana subsp. subalpina	J	(387)
			Jungermannia infusca	J	(387)
			J. torticalyx	J	(387)
			Marsupella emarginata var. patens	J	(387)
			Mylia taylorii	J	(387)
			M. verrucosa	J	(387)
			Scapania parvidens	J	(387)
	Fucose		Calypogeia neesiana subsp. subalpina	J	(387)
	D-Galactose		Calypogeia neesiana subsp. subalpina	J	(387)
			Jungermannia torticalyx	J	(387)
			Marsupella emarginata var. patens	J	(387)
			Mylia taylorii	J	(387)

Table IIg (continued)

Structure number	Name of compounds	Formula	m.p.°C	$[\alpha]_D$	Plant source	Order	References	Comments
	(+)-D-Glucose				Calypogeia neesiana subsp. subalpina	J	(387)	
					Jungermannia infusca	J	(387)	
					J. torticalyx	J	(387)	
					Marsupella emarginata var. patens	J	(387)	
					Mylia taylorii	J	(387)	
					M. verrucosa	J	(387)	
					Scapania parvidens	J	(387)	
	Mannitol				Mylia taylorii	J	(387)	
					M. verrucosa	J	(387)	
	myo-Inositol				Jungermannia infusca	J	(387)	
					Mylia taylorii	J	(387)	
					M. verrucosa	J	(387)	
	Rhamnose				Calypogeia neesiana subsp. subalpina	J	(387)	
	(+)-D-Saccharofructose				Jungermannia infusca	J	(387)	
					J. torticalyx	J	(387)	
					Marsupella emarginata var. patens	J	(387)	
					Mylia taylorii	J	(387)	
					Scapania parvidens	J	(387)	
					S. stephanii	J	(387)	

Compound	Species		
(+)-Trehalose	*Calypogeia neesiana* subsp. *subalpina*	J	(387)
	Jungermannia infusca	J	(387)
	J. torticalyx	J	(387)
	Marsupella emarginata var. *patens*	J	(387)
	Mylia taylorii	J	(387)
	Scapania parvidens	J	(387)
	Marsupella emarginata var. *patens*	J	(387)
Xylose	*Scapania stephanii*	J	(387)

J: Jungermanniales; M: Marchantiales; Mo: Monocleales

Table IIh. *Sulfur-Containing Compounds Found in the Hepaticae*

Structure number	Name of compounds	Formula	m.p.°C	[α]$_D$	Plant source	Order	References	Comments
(1227)	Dimethyl sulfide	C$_2$H$_6$S			*Pellia endiviifolia*	Me	(62)	GC-MS
					Plagiochila sciophila	J	(62)	GC-MS
					(= *P. acanthophylla* subsp. *japonica*)			
					Porella cordaeana	J	(593)	GC-MS
(1228)	Isotachin A	C$_{11}$H$_{12}$O$_2$S			*Balantiopsis rosea*	J	(65)	
	(= Benzyl *trans*-β-methyl thioacrylate)				*Isotachis japonica*	J	(76)	
(1229)	Isotachin B	C$_{12}$H$_{14}$O$_2$S			*Balantiopsis rosea*	J	(65)	
	(= β-Phenylethyl *trans*-β-methyl thioacrylate)				*Isotachis japonica*	J	(76)	
(1230)	Isotachin C	C$_{12}$H$_{14}$O$_3$S			*Balantiopsis rosea*	J	(65)	
	(= 2-Methoxybenzyl *trans*-β-methyl thioacrylate)							

J: Jungermanniales; Me: Metzgeriales

(−)-D-mannitol, (+)-glucose, (+)-sucrose and (+)-trehalose have been isolated from *Mylia taylorii*, in the form of their peracetyl derivatives (*387*). The absolute configuration of these sugars is the same as that found in higher plants, although many sesqui- and diterpenoids isolated from liverworts are enantiomers of those found in higher plants. The soluble carbohydrates from seven other liverwort species are listed in Table IIg.

III. Chemical Constituents of Musci (Mosses)

1. Mono- and Diterpenoids

Lower terpenoids are rare in mosses. That entomophilous species of the Splachnaceae possess a characteristic odor has been known for a long time. Flies attracted by the odor of *Splachnum rubrum* serve as spore carriers to fresh dung. The volatile components of ten species of *Splachnum* have been analyzed by GC and GC-MS (*344, 469*). Phellandrene (α- or β-form) (**12a** or **12b**), Δ³-carene (**25**) and α-pinene (**27**), have been detected in female and male gametophytes of *S. rubrum*, together with octane derivatives. 3-Carene has also been detected in gametophytes of *S. luteum* and seta of *S. rebrum*. This is the first report of monoterpenoids in mosses.

Ent-16β-hydroxykaurane (**707**) has been isolated from *Saelania glaucescens* (*445*) as the first diterpenoid from a moss. Compound (**707**) is thought to contribute to the bluish tint of *S. glaucescens*. It is also the major component in the waxy coating of the liverworts *Anthelia juratzkana* and *A. julacea* (Hepaticae) (*19*). *Hypnum plumaeforme* produces momilactone A (**1231**) and momilactone B (**1232** (*448a*) which have been isolated from the seed husk of *Oryza sativa* (*321a*). Chamaecydin (**1233**) which has been found in the seed of the conifer tree, *Chamaecyparis obtusa* (*268a*) has been isolated from *Thuidium kanedae* (*448a*).

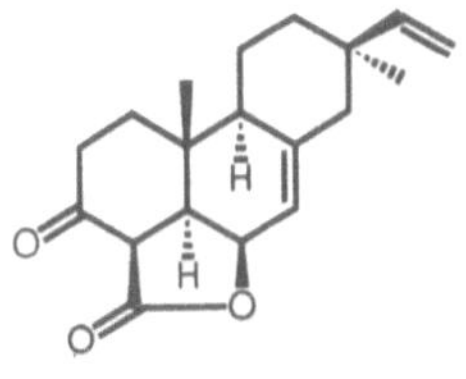

(**1231**) Momilactone A 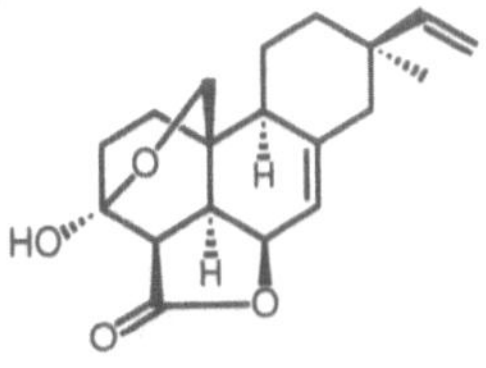   (**1232**) Momilactone B 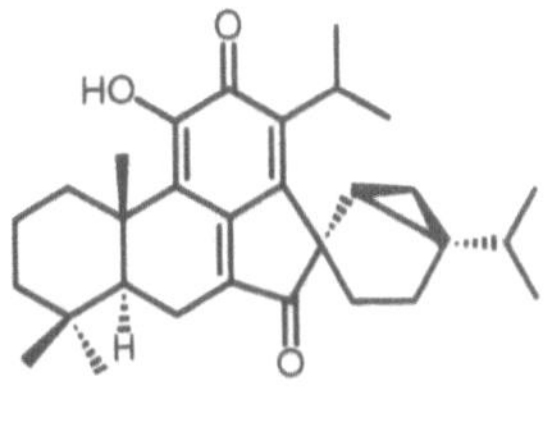   (**1233**) Chamaecydin

Chart 71. Diterpenoids found in the Musci

2. Steroids, Triterpenoids and Carotenoids

Mosses contain mainly campesterol (24α-epimer) (**844a**), stigmasterol (24α-epimer) (**846a**) and sitosterol (24α-epimer) (**848a**), together with cholesterol (**849**), brassicasterol (= 24-methyl-5,22-cholestadienol) (**850**) and 24-methyl-5,7,22-cholestatrienol (**1239**) as the minor component as shown in Tables IIIa (*138, 271*). The sterol composition of *Plagiomnium succulentum* and *Sphagnum palustre* has been investigated by high resolution ¹H-NMR spectroscopy and capillary GC (*138, 436*). The data indicate that the 24-methylcholesterol of the two species is a mixture of the 24α- (**844a**) and 24β-epimers (**844b**), whilst the 24-ethyl-5-cholesterol of *S. palustre* is a mixture of the 24α- (**848a**) and 24β-epimers (**848b**) and that of *P. succulentum* is only the 24α-epimer. Both species elaborate only the 24α-epimer (**846a**) of 24-ethyl-5,22-cholestadienol. Similar phenomena were observed in the sterol composition of liverworts (see Chapter II, Table IId). Cholesterol (**849**) was also found in the sterol fractions of both species.

Triterpenoids are widely distributed in mosses. Table IIIa lists ursane-, fernane-, friedelane-, hopane-, lupane-, taraxane-, cycloeucalane-, cycloartane-, cyclolaudane-, 24-methylenecycloartane-, norcyclolaudane- and obtusifolane-type triterpenoids. *Sphagnum* species (peat moss) produce α-amyrin (**1240**), taraxerol (**1244**) and taraxerone (**1245**) (*300*). *Thuidium kanedae* elaborates two fernane-type triterpenoids, fern-7-ene (**1246**) and fern-9(11)-ene (**1247**) (*70*). The latter triterpene has also been obtained from *Scleropodium touretii* (*132*), together with cyclolaudenol (**1234**), and cycloeucalenol (**1237**) which has also been found in *Brachythecium rivulare* and *Campylopus introflexus* (*132*). 3α-Friedelinol (**1248**) has been found in *Leucobryum glaucum* (*288*). The most common triterpenoid in mosses in diploptene [= hop-22(29)-ene] (**856**) which has been detected in *Abietinella* (*270, 372*), *Brachythecium, Campylopus* (*132*), *Climacium* (*372*), *Ctenidium* (*132*), *Hypnum, Neckera* (*372*), *Pseudoscleropodium* (*373*), *Rhytidiadelphus* (*288, 372*), *Thamnium* (*371*) and *Thuidium* species (*70, 132*). Lupeol (**1249**) has been found only in *Thuidium kanedae* (*70*). *Hypnum, Pseudoscleropodium* and *Rhytidiadelphus triquetrus* var. *typicus* elaborate ursolic acid (**1241**) which is widespread in higher plants (*372, 373*). *Dicranum elongatum* produces cycloartenol (**854**) (*170, 319*). Cyclolaudenol (**1234**) and 31-norcyclolaudenol (**1235**) are relatively common; the former has been found in 10 species and the latter detected in 7 species including *Sphagnum teres* (*271*). *Dicranum elongatum* (*170*) and *Thuidium tamariscifolium* (*371*) elaborate 24-methylenecycloartanol (**1236**). *Racomitrium lanuginosum* biosynthesizes obtusifoliol (**1238**) (*132, 288*). *R. japonicum* produces β-amyrin (**1242**) and 11,12-dehydroursolic

Table IIIa. Mono-, Di- and Triterpenoids, Steroids and Carotenoids Found in the Musci

Structure number	Name of compounds	Formula	m.p.°C	$[\alpha]_D$	Plant source	Subclass	References	Comments
(12a or 12b)	Phellandrene	$C_{10}H_{16}$			Splachnum luteum	B	(344, 469)	GC-MS
					S. rubrum	B	(344, 469)	GC-MS
(25)	Δ^3-Carene	$C_{10}H_{16}$			Splachnum luteum	B	(344, 469)	GC-MS
					S. rubrum	B	(344, 469)	GC-MS
(27)	α-Pinene	$C_{10}H_{16}$			Splachnum rubrum	B	(344, 469)	GC-MS
(707)	ent-16β-Hydroxy-kaurane	$C_{20}H_{34}O$	214-5	-45	Saelania glaucescens	B	(445)	
(780)	Phytol	$C_{20}H_{40}O$			Dicranum elongatum	B	(315)	Hydrolized compd.
					Hygrohypnum luridum	B	(315)	
					Mnium cuspidatum	B	(315)	
					Paraleucobryum longifolium	B	(315)	
					Sphagnum fuscum	S	(315)	
(781)	Phytenic acid	$C_{20}H_{38}O_2$			Dicranum elongatum	B	(315)	Hydrolized compd.
					Hygrohypnum luridum	B	(315)	
					Hylocomium splendens var. alaskanum	B	(315)	
					Mnium cuspidatum	B	(315)	
					M. medium	B	(315)	
					Paraleucobryum longifolium	B	(315)	
					Pleurozium schreberi	B	(315)	
					Polytrichum commune	P	(315)	
					Sphagnum fuscum	S	(315)	
(782)	Geranyl geraniol	$C_{20}H_{34}O$			Dicranum elongatum	B	(315)	Hydrolized compd.

Table IIIa (continued)

Structure number	Name of compounds	Formula	m.p.°C	[α]_D	Plant source	Subclass	References	Comments
(844a)	Campesterol (24α-epimer)	$C_{28}H_{48}O$			Atrichum undulatum	P	(463)	
					Brotherella recurvans	B	(463)	
					Dicranum flagilifolium	B	(468)	
					D. polycetum	B	(463)	
					D. scoparium	B	(463)	
					Plagiomnium succulentum	B	(463)	
					Pleurozium schreberi	B	(463)	
					Polytrichum commune	P	(463)	
					Sphagnum palustre	S	(138)	
(844b)	Dihydrobrassicasterol	$C_{28}H_{48}O$			Plagiomnium succulentum	B	(463)	
					Sphagnum palustre	S	(138)	
(846a)	Stigmasterol (24α-epimer)	$C_{29}H_{48}O$			Atrichum undulatum	P	(463)	
					Brotherella recurvans	B	(463)	
					Dicranum flagilifolium	B	(468)	
					D. polycetum	B	(463)	
					D. scoparium	B	(463)	
					Plagiomnium succulentum	B	(463)	
					Pleurozium schreberi	B	(463)	
					Polytrichum commune	P	(463)	
					Sphagnum palustre	S	(138)	
(848a)	Sitosterol (24α-epimer)	$C_{29}H_{50}O$			Atrichum undulatum	P	(463)	
					Brotherella recurvans	B	(463)	
					Dicranum flagilifolium	B	(468)	
					D. polycetum	B	(463)	
					D. scoparium	B	(463)	

No.	Compound	Formula	Species		Ref.
			Plagiomnium succulentum	B	(436)
			Pleurozium schreberi	P	(463)
			Polytrichum commune	P	(463)
			Sphagnum palustre	S	(138)
(848b)	Clionasterol	$C_{29}H_{50}O$	*Plagiomnium succulentum*	B	(463)
			Sphagnum palustre	S	(138)
(849)	Cholesterol	$C_{27}H_{46}O$	*Plagiomnium succulentum*	B	(138)
			Sphagnum palustre	S	(138)
(850)	Brassicasterol (= 24-methyl-5,22-cholestadienol)	$C_{28}H_{46}O$	*Plagiomnium succulentum*	B	(138)
			Sphagnum palustre	S	(138)
(851)	Squalene	$C_{30}H_{50}$	*Racomitrium japonicum*	B	(448a)
(853)	Friedelin	$C_{30}H_{50}O$	*Racomitrium japonicum*	B	(448a)
(854)	Cycloartenol	$C_{30}H_{50}O$	*Dicranum elongatum*	B	(170, 319)
(856)	Diploptene (= Hop-22(29)-ene)	$C_{30}H_{50}$	*Abietinella abietina*	B	(270)
			Brachythecium rivulare	B	(132)
			Campylopus introflexus	B	(132)
			Climacium dendroides	B	(372)
			Ctenidium molluscum	B	(132)
			Hypnum cupressiforme subsp. *imponens*	B	(372)
			Neckera crispa	B	(372)
			Pseudoscleropodium purum	B	(373)
			Racomitrium japonicum	B	(448a)
			Rhytidiadelphus squarrosus subsp. *squarrosus*	B	(372)
			R. triquetrus	B	(288)
			R. triquetrus var. *typicus*	B	(372)
			Thamnium alopecurum subsp. *eualopecurum*	B	(371)
			Thuidium kanedae	B	(70)

Table IIIa (continued)

Structure number	Name of compounds	Formula	m.p.°C	$[\alpha]_D$	Plant source	Subclass	References	Comments
(1231)	Momilactone A	$C_{19}H_{24}O_3$			*Hypnum plumaeforme*	B	(448a)	
(1232)	Momilactone B	$C_{19}H_{24}O_4$			*Hypnum plumaeforme*	B	(448a)	
(1233)	Chamaecydin	$C_{30}H_{40}O_3$			*Thuidium kanedae*	B	(448a)	
(1234)	Cyclolaudenol	$C_{31}H_{52}O$			*Brachythecium rivulare*	B	(132)	
					Campylopus introflexus	B	(132)	
					Neckera crispa	B	(372)	
					Polytrichum formosum subsp. *euformosum* var. *typicum*	P	(371)	
					Pseudoscleropodium purum	B	(373)	
					Racomitrium lanuginosum	B	(132, 288)	
					Rhytidiadelphus squarrosus subsp. *eusquarrosus*	B	(371)	
					R. triquetrus var. *typicus*	B	(132, 371)	
					Scleropodium touretii	B	(132)	
					Sphagnum teres	S	(372)	
(1235)	31-Norcyclolaudenol	$C_{30}H_{50}O$			*Andrea rupestris*	A	(288)	
					Brachythecium rivulare	B	(132)	
					Hypnum cupressiforme subsp *imponens*	B	(371)	
					Racomitrium lanuginosum	B	(132, 288)	
					Rhytidiadelphus squarrosus subsp. *eusquarrosus*	B	(371)	
					Sphagnum teres	S	(371)	

No.	Compound	Formula	Species		Ref.
(1236)	24-Methylene-cycloartanol	$C_{30}H_{50}O$	Dicranum elongatum	B	(170, 319)
(1237)	Cycloeucalenol	$C_{30}H_{50}O$	Thuidium tamariscifolium	B	(371a)
			Brachythecium rivulare	B	(132)
			Campylopus introflexus	B	(132)
			Scleropodium touretii	B	(132)
(1238)	Obtusifoliol	$C_{30}H_{50}O$	Racomitrium lanuginosum	B	(132, 288)
			Thuidium tamariscifolium	B	(132, 371a)
(1239)	24-Methyl-5,7,22-cholestatrienol	$C_{28}H_{44}O$	Campylopus introflexus	B	(132)
(1240)	α-Amyrin	$C_{30}H_{50}O$	Ctenidium molluscum	B	(132)
			Racomitrium japonicum	B	(448a)
			Sphagnum sp.	S	(300)
(1241)	Ursolic acid	$C_{30}H_{48}O_3$	Hypnum cupressiforme subsp. imponens	B	(372)
			Rhytidiadelphus triquetrus var. typicus	B	(373)
			Pseudoscleropodium purum	B	(372)
			Sphagnum teres	S	(372)
			Thuidium kanedae	B	(70, 132)
(1242)	β-Amyrin	$C_{30}H_{50}O$	Racomitrium japonicum	B	(448a)
(1243)	11,12-Dehydroursolic acid	$C_{30}H_{46}O_3$	Racomitrium japonicum	B	(448a)
(1244)	Taraxerol	$C_{30}H_{50}O$	Racomitrium japonicum	B	(448a)
			Sphagnum sp.	S	(300)
(1245)	Taraxerone	$C_{30}H_{48}O$	Sphagnum sp.	S	(300)
(1246)	Fern-7-ene	$C_{30}H_{50}$	Thuidium kanedae	B	(70)
(1247)	Fern-9(11)-ene	$C_{30}H_{50}$	Scleropodium touretii	B	(132)
			Thuidium kanedae	B	(70)
(1248)	3α-Friedelinol	$C_{30}H_{52}O$	Leucobryum glaucum	B	(288)
(1249)	Lupeol	$C_{30}H_{50}O$	Racomitrium japonicum	B	(448a)
			Thuidium kanedae	B	(70)
	Antheraxanthin		Polytrichum commune	P	(271)

Table IIIa (continued)

Structure number	Name of compounds	Formula	m.p.°C	$[\alpha]_D$	Plant source	Subclass	References	Comments
	Auroxanthin							
	α-Carotene				Fontinalis antipyretica	B	(271)	
					Aulacomnium palustre	B	(271)	
					Bryum ventricisum	B	(271)	
					Chamberlainia acuminata	B	(271)	
					Dawsonia superba	B	(271)	
					Ditrichum lineare	B	(271)	
					Entodon seductrix	B	(271)	
					Fontinalis anipyretica	B	(271)	
					Hygrohypnum smithii	B	(271)	
					Mnium cuspidatum	B	(271)	
					Philonotis fontana	B	(271)	
					Polytrichum commune	P	(271)	
					Sphagnum sp.	S	(271)	
					Thuidiopsis furfurosa	B	(271)	
					Thuidium recognitum	B	(271)	
	β-Carotene				Aulacomnium palustre	B	(271)	
					Chamberlainia acuminata	B	(271)	
					Dawsonia superba	B	(271)	
					Ditrichum lineare	B	(271)	
					Entodon seductrix	B	(271)	
					Fontinalis antipyretica	B	(271)	
					Hygrohypnum smithii	B	(271)	
					Mnium cuspidatum	B	(271)	
					Philonotis fontana	B	(271)	
					Pohlia longicollis	B	(271)	

Compound	Species		Ref.
Cryptoxanthin	Polytrichum commune	P	(271)
	Sphagnum sp.	B	(271)
Epoxylutein	Thuidiopsis furfurosa	S	(271)
Lutein	Thuidium recognitum	B	(271)
	Bryum ventricisum	B	(271)
	Entodon seductrix	B	(271)
	Bryum ventricisum	B	(271)
	Fontinalis antipyretica	B	(271)
	Aulacomnium palustre	B	(271)
	Bryum ventricisum	B	(271)
	Chamberlainia acuminata	B	(271)
	Dawsonia superba	B	(271)
	Ditrichum lineare	B	(271)
	Entodon seductrix	B	(271)
	Fontinalis antipyretica	B	(271)
	Hygrohypnum smithii	B	(271)
	Mnium cuspidatum	B	(271)
	Philonotis fontana	B	(271)
	Pohlia longicollis	B	(271)
	Polytrichum commune	P	(271)
	Sphagnum sp.	S	(271)
Neo-β-carotene U	Thuidiopsis furfurosa	B	(271)
	Thuidium recognitum	B	(271)
Neoxanthin	Fontinalis antipyretica	B	(271)
	Aulacomnium palustre	B	(271)
	Bryum ventricisum	B	(271)
	Dawsonia superba	B	(271)
	Ditrichum lineare	B	(271)
	Entodon seductrix	B	(271)
	Fontinalis antipyretica	B	(271)
	Hygrohypnum smithii	B	(271)

Table IIIa *(continued)*

Structure number	Name of compounds	Formula	m.p.°C	$[\alpha]_D$	Plant source	Subclass	References	Comments
					Philonotis fontana	B	(271)	
					Pohlia longicollis	B	(271)	
					Sphagnum sp.	S	(271)	
					Thuidiopsis furfurosa	B	(271)	
	Neoxanthin neo A				*Polytrichum commune*	P	(271)	
	Violaxanthin				*Aulacomnium palustre*	B	(271)	
					Bryum ventricisum	B	(271)	
					Chamberlainia acuminata	B	(271)	
					Dawsonia superba	B	(271)	
					Ditrichum lineare	B	(271)	
					Entodon seductrix	B	(271)	
					Hygrohypnum smithii	B	(271)	
					Mnium cuspidatum	B	(271)	
					Philonotis fontana	B	(271)	
					Pohlia longicollis	B	(271)	
					Polytrichum commune	P	(271)	
					Sphagnum sp.	B	(271)	
					Thuidiopsis furfurosa	B	(271)	
					Thuidium recognitum	B	(271)	

Zeaxanthin

Aulacomnium palustre	B	(271)
Bryum ventricosum	B	(271)
Chamberlainia acuminata	B	(271)
Dawsonia superba	B	(271)
Ditrichum lineare	B	(271)
Entodon seductrix	B	(271)
Fontinalis antipyretica	B	(271)
Hygrohypnum smithii	B	(271)
Philonotis fontana	B	(271)
Pohlia longicollis	B	(271)
Polytrichum commune	P	(271)
Sphagnum sp.	B	(271)
Thuidiopsis furfurosa	B	(271)
Thuidium recognitum	B	(271)

A: Andreaeidae; B: Bryidae; P: Polytrichidae; S: Sphagnidae

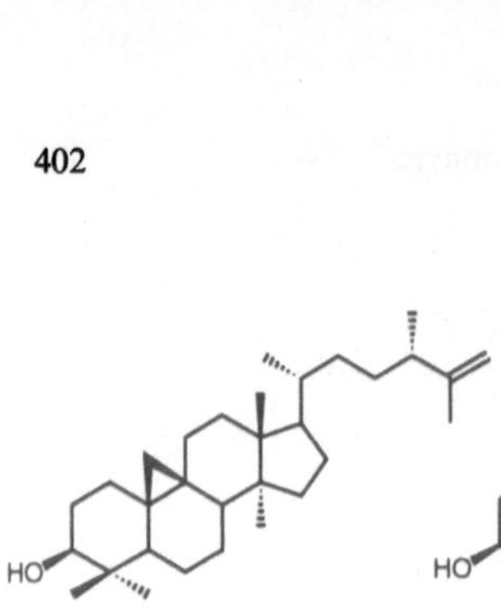

(**1234**) Cyclolaudenol

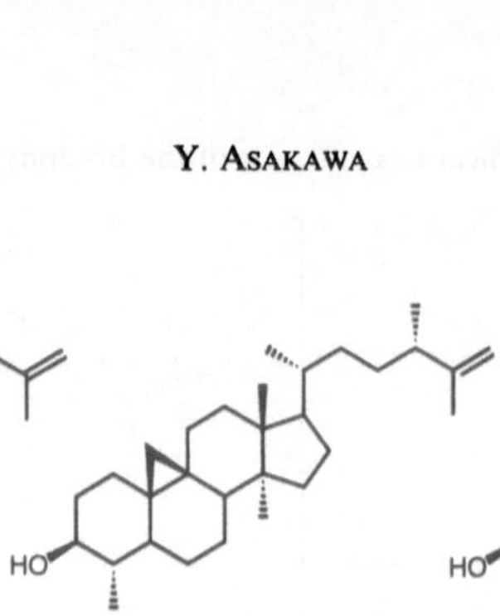

(**1235**) 31-Norcyclolaudenol

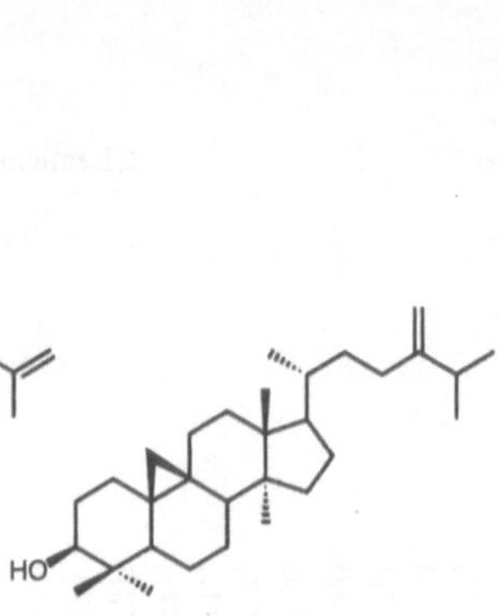

(**1236**) 24-Methylenecycloartanol

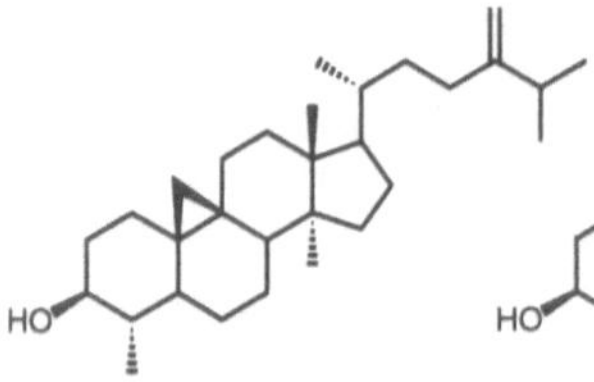

(**1237**) Cycloeucalenol

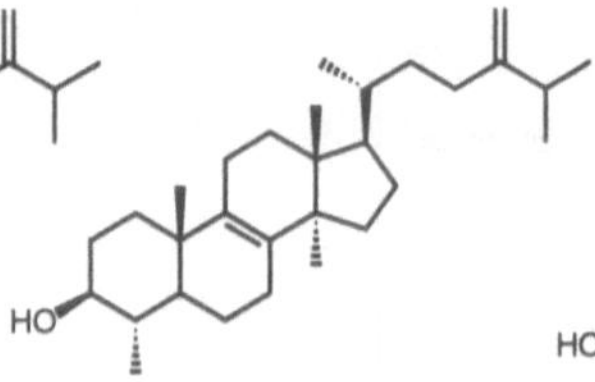

(**1238**) Obtusifoliol

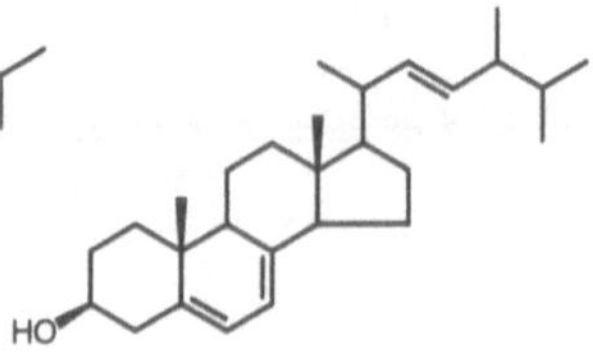

(**1239**) 24-Methyl-5,7,22-cholestatrienol

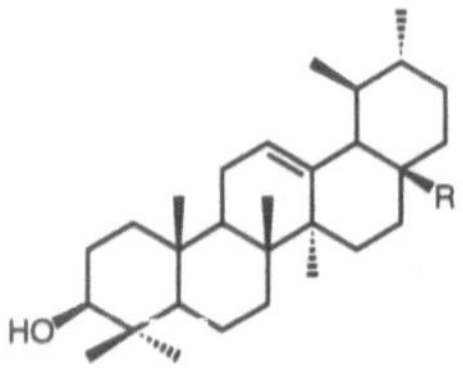

(**1240**) α-Amyrin ; R=Me
(**1241**) Ursolic acid ; R=CO$_2$H

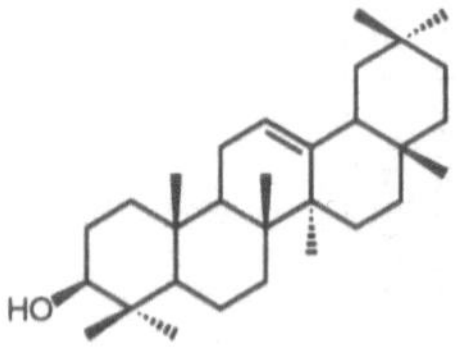

(**1242**) β-Amyrin

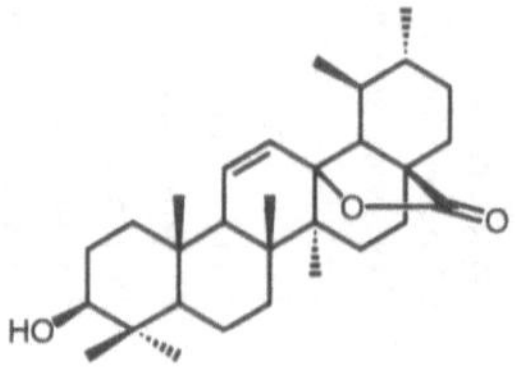

(**1243**) 11,12-Dehydroursolic acid lactone

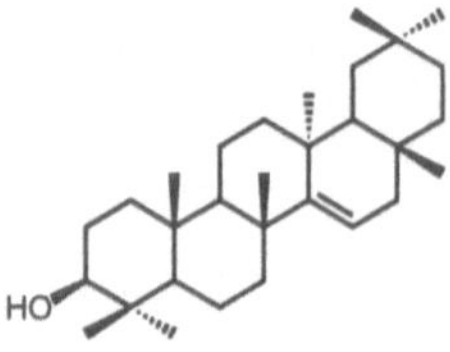

(**1244**) Taraxerol

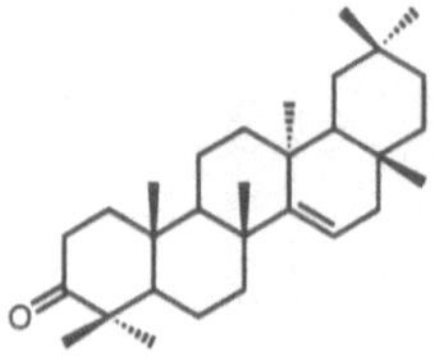

(**1245**) Taraxerone

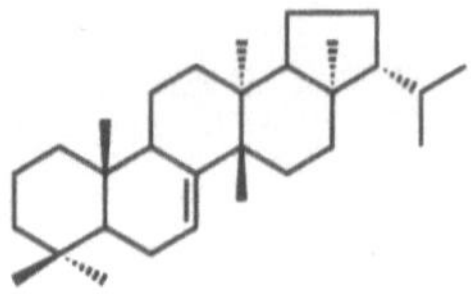

(**1246**) Fern-7-ene

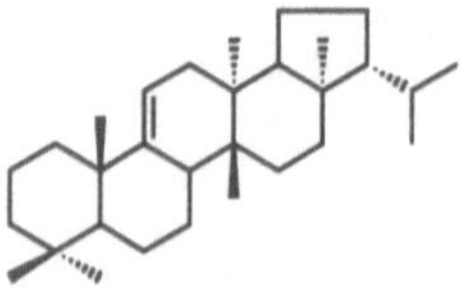

(**1247**) Fern-9(11)-ene

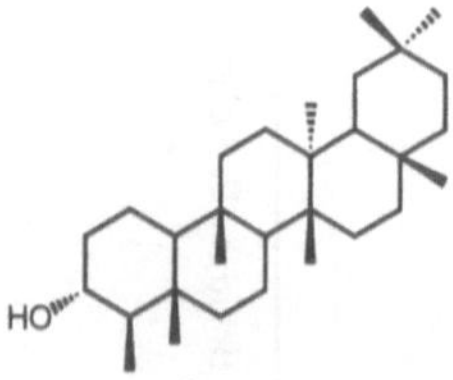

(**1248**) 3α-Friedelinol

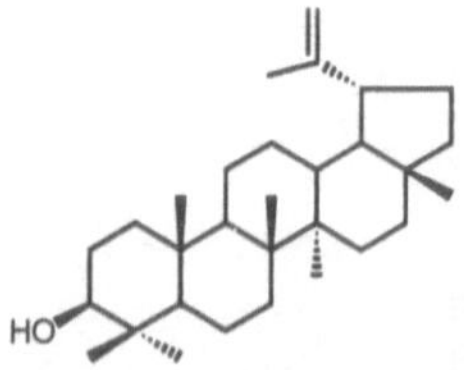

(**1249**) Lupeol

Chart 72. Steroids and triterpenoids found in the Musci

acid lactone (**1243**), together with squalene (**851**), friedelin (**853**), hop-22(29)-ene (**856**), α-amyrin (**1240**), taraxerol (**1244**) and lupeol (**1249**) (*448a*). All triterpenoids listed above have so far not found in liverworts, with the exception of cycloartenol (**854**), diplotene (**856**) and 3α-friedelinol (**1248**).

The distribution of carotenoids in several mosses has been investigated (*167*). Mosses contain antheraxanthin, auroxanthin, α-carotene, β-carotene, β,β-carotene-4-one (= echinenone), cryptoxanthin, epoxylutein, lutein, neo-β-carotene, neoxanthin, neoxanthin neo A, violaxanthin and zeaxanthin as shown in Table IIId (*271*). Antheraxanthin, α- and β-carotenes, lutein, neo-β-carotene U, neoxanthin, violaxanthin and zeaxanthin have also been found in Hepaticae (*19*).

3. Aromatic Compounds

3.1 Benzoic and Cinnamic Acid Derivatives

A reliable and precise procedure for isolation, separation and determination of monophenolic substances in *Sphagnum* species by HPLC has been described by WILSCHKE and RUDOLPH (*618*). The benzoic acid derivatives, *p*-hydroxybenzaldehyde (**1084**), *p*-hydroxybenzoic acid (**1251**), vanillic acid (**1253**) and vanillin (**1254**) and the cinnamic acid derivatives, *p*-coumaric acid (**1256**), caffeic acid (**1257**), ferulic acid (**1258**), *o*-coumaric acid (**1259**), *m*-coumaric acid (**1260**), sphagnic acid (= sphagnum acid) (**1261**) and 2,5-dihydro-5-hydroxy-4-(4'-hydroxyphenyl)-2-furanone (**1262**) have been detected as shown in Table IIIb. *Racomitrium japonicum* contains *p*-hydroxybenzaldehyde (**1084**) (*448a*). Caffeoyl diglycoside in *Pohlia wahlenbergii* has been detected by MÅRTENSSON and NILSSON (*374*).

Splachnum and *Aplodon* species produce benzoic acid (**1250**), phenylacetic acid (**1263**) and phenylacetylene (**1264**) (*344*). Benzyl alcohol (**1228b**), phenylethyl alcohol (**1229b**), phenol (**1265**), phenylacetoaldehyde (**1087**), benzophenone (**1266**), indole (**1267**) and pyrrolidine (**1268**) have been detected in a few *Splachnum* species by GC and GC-MS (*344*).

DAVIDSON et al. (*160*) investigated the phenolic components of two mosses, *Brachythecium rutabulum* and *Mnium hornum*. The alcoholic extract of shoots of both species contained gallic acid (**1252**) and protocatechuic acid (**1042**). *p*-Hydroxybenzoic acid (**1251**) and vanillic acid (**1253**) were identified in immature capsules and shoot extracts of *Mnium hornum* after alkaline hydrolysis. *Trans*-cinnamic acid (**1255**) has been detected in the shoots of *B. rutabulum* and in the extract *M. hornum*

Table IIIb. Aromatic Compounds Found in the Musci

Structure number	Name of compounds	Formula	m.p.°C	$[\alpha]_D$	Plant source	Subclass	References	Comments
(1042)	Protocatechuic acid	$C_7H_6O_4$			*Brachythecium rutabulum*	B	*(160)*	
(1084)	p-Hydroxy-benzaldehyde	$C_7H_6O_2$			*Mnium hornum*	B	*(160)*	
					Racomitrium japonicum	B	*(448a)*	
(1087)	Phenylacetoaldehyde	C_8H_8O			*Sphagnum* sp.	S	*(618)*	
(1228b)	Benzyl alcohol	C_7H_8O			*Splachnum rubrum*	B	*(344)*	
					Splachnum sphaericum	B	*(344)*	
(1229b)	Phenylethyl alcohol	$C_8H_{10}O$			*Splachnum rubrum*	B	*(344)*	
(1250)	Benzoic acid	$C_7H_6O_2$			*Aplodon wormskioldii*	B	*(344)*	
					Splachnum rubrum	B	*(344)*	
					S. sphaericum	B	*(344)*	
					S. vasculosum	B	*(344)*	
(1251)	p-Hydroxybenzoic acid	$C_7H_6O_3$			*Mnium hornum*	B	*(160)*	
(1252)	Gallic acid	$C_7H_6O_5$			*Sphagnum* sp.	S	*(618)*	
(1253)	Vanillic acid	$C_8H_8O_4$			*Brachythecium rutabulum*	B	*(160)*	
					Mnium hornum	B	*(160)*	
(1254)	Vanillin	$C_8H_8O_3$			*Mnium hornum*	B	*(160)*	
(1255)	trans-Cinnamic acid	$C_9H_8O_2$			*Sphagnum* sp.	S	*(618)*	
					Brachythecium rutabulum	B	*(160)*	
(1256)	p-Coumaric acid	$C_9H_8O_3$			*Mnium hornum*	B	*(618)*	
					Sphagnum sp.	S	*(618)*	

(1257)	Caffeic acid	$C_9H_8O_4$	Brachythecium rutabulum	B	(160)
			Mnium hornum	B	(160)
			Sphagnum sp.	S	(618)
(1258)	Ferulic acid	$C_{10}H_{10}O_4$	Sphagnum sp.	S	(618)
(1259)	o-Coumaric acid	$C_9H_8O_3$	Sphagnum sp.	S	(618)
(1260)	m-Coumaric acid	$C_9H_8O_3$	Mnium hornum	B	(160)
(1261)	Sphagnic acid (= Sphagnum acid)	$C_{11}H_{10}O_5$	Sphagnum aongstroemii	S	(477)
			S. balticum	S	(477)
			S. centrale	S	(477)
			S. compactum	S	(477)
			S. contortum	S	(477)
			S. cuspidatum	S	(477)
			S. fallax	S	(477)
			S. fimbriatum	S	(477)
			S. fuscum	S	(477)
			S. imbricatum	S	(477)
			S. jensenii	S	(477)
			S. lindbergii	S	(477)
			S. magellanicum	S	(477, 602–604, 618)
			S. majus	S	(477)
			S. molle	S	(477)
			S. nemoreum	S	(477)
			S. obtusum	S	(477)
			S. palustre	S	(477)
			S. papillosum	S	(477)
			S. quinquefarium	S	(477)
			S. riparium	S	(477)
			S. rubellum	S	(477)
			S. russowii	S	(477)

Table IIIb (continued)

Structure number	Name of compounds	Formula	m.p.°C	$[\alpha]_D$	Plant source	Subclass	References	Comments
					S. squarrosum	S	(477)	
					S. subfulvum	S	(477)	
					S. subsecundum	S	(477)	
					S. temellum	S	(477)	
					S. teres	S	(477)	
					S. warnstorfii	S	(477)	
(1262)	2,5-Dihydro-5-hydroxy-4-(4'-hydroxyphenyl)-2-furanone	$C_{10}H_8O_4$			Sphagnum magellanicum	S	(618)	
(1263)	Phenylacetic acid	$C_8H_8O_2$			Aplodon wormskioldii	B	(344)	
					Splachnum rubrum	B	(344)	
					S. sphaericum	B	(344)	
					S. vasculosum	B	(344)	
(1264)	Phenylacetylene	C_8H_6			Aplodon worskioldii	B	(344)	
					Splachnum sphaericum	B	(344)	
					S. vasculosum	B	(344)	
(1265)	Phenol	C_6H_6O			Splachnum sphaericum	B	(344)	
(1266)	Benzophenone	$C_{13}H_{10}O$			Splachnum luteum	B	(344)	
(1267)	Indole	C_8H_7N			Splachnum rubrum	B	(344)	
(1268)	Pyrrolidine	C_4H_9N			Splachnum rubrum	B	(344)	

B: Bryidae; S: Sphagnidae

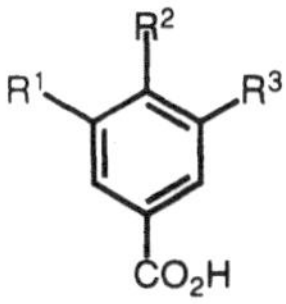

(1250) Benzoic acid ; $R^1=R^2=R^3=H$
(1251) p-Hydroxybenzoic acid ; $R^1=R^3=H$, $R^2=OH$
(1252) Gallic acid ; $R^1=R^2=R^3=OH$
(1253) Vanillic acid ; $R^1=H$, $R^2=OH$, $R^3=OMe$

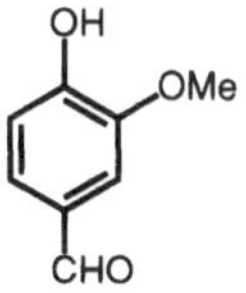

(1254) Vanillin

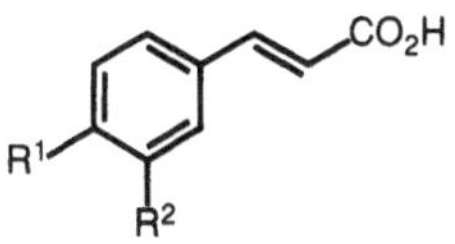

(1255) trans-Cinnamic acid ; $R^1=R^2=H$
(1256) p-Coumaric acid ; $R^1=OH$, $R^2=H$
(1257) Caffeic acid ; $R^1=R^2=OH$
(1258) Ferulic acid ; $R^1=OH$, $R^2=OMe$

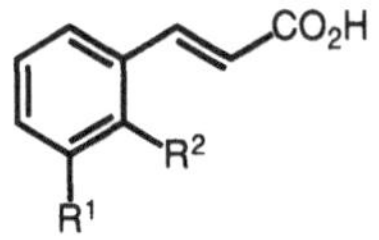

(1259) o-Coumaric acid ; $R^1=H$, $R^2=OH$
(1260) m-Coumaric acid ; $R^1=OH$, $R^2=H$

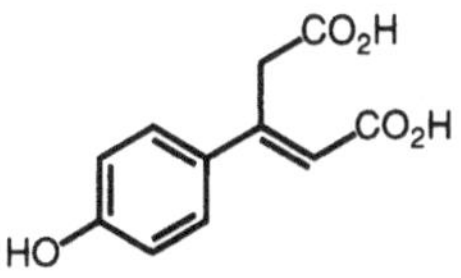

(1261) Sphagnic (=Sphagnum) acid

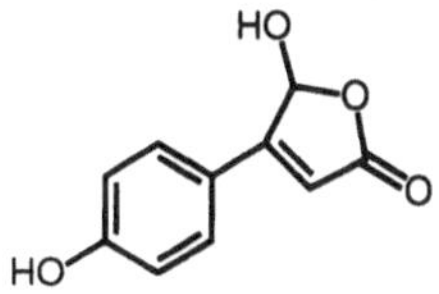

(1262) 2,5-Dihydro-5-hydroxy-4-(4'-hydroxyphenyl)-2-furanone

Chart 73. Benzoic and cinnamic acid derivatives found in the Musci

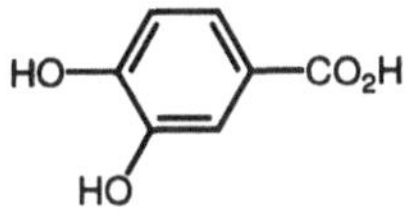

(1042)

capsule after hydrolysis. *M. hornum* shoots and immature capsule and immature capsules of *B. rutabulum* also produce caffeic acid (1257). *p*-Coumaric (1256) and *m*-coumaric acids (1260) have also been found in *B. rutabulum* and *M. hornum*, respectively.

The Millon-positive component of the cell walls of *Sphagnum magellanicum* has been characterized as *p*-hydroxy-β-(carboxymethyl)-cinnamic acid (= sphagnic acid or sphagnum acid) (1261) (*476, 602–604*).

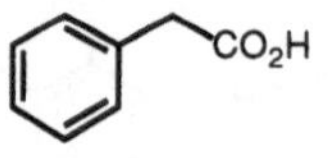

(**1263**) Phenylacetic acid 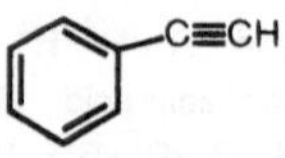   (**1264**) Phenylacetylene 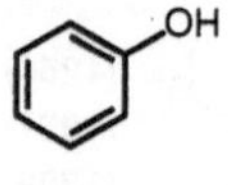   (**1265**) Phenol

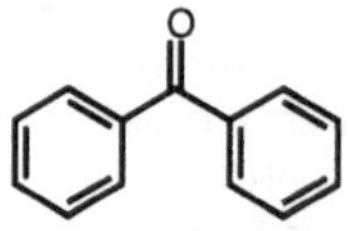

(**1266**) Benzophenone 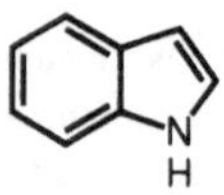   (**1267**) Indole    (**1268**) Pyrrolidine

Chart 74. Miscellaneous aromatic compounds and nitrogen-containing compounds found in the Musci

Degradation of sphagnic acid with 2N NaOH at 350° for 5 min gives *p*-hydroxybenzoic acid (**1251**) (*602*). RUDOLPH and SAMLAND (*477*) studied the distribution of sphagnic acid in 51 mosses, 9 Hepaticae and *Anthoceros punctatus*; the acid was detected in all 30 *Sphagnum* species (Sphagnales) but not in the others. Synthesis of sphagnic acid which is biosynthesized *via* the shikimate pathway was inhibited by *N*-(phosphonomethyl)glycine and accumulated shikimate. The high content of sphagnic acid and other phenolic compounds in the cell walls of *Sphagnum* species may be partly responsible for the resistance of *Sphagnum* to decomposition.

WILSCHKE *et al.* (*617*) have investigated the biosynthesis of sphagnic acid (**1261**) using L(U-^{14}C)-phenylalanine, L-(2,3,4,5,6-^{3}H)-phenylalanine and (U-^{14}C)-acetate as tracers. The labeled hydrogen was located in the aromatic ring, while the *p*-hydroxybenzoic acid isolated after degradation of sphagnic acid labeled with sodium (U-^{14}C) acetate was not radioactive. These results indicate that the *p*-hydroxycinnamic acid and the carboxymethyl side chain are formed from shikimate and malonate, respectively.

3.2 Flavonoids

3.2.1 Flavones

The review by MARKHAM in "Flavonoid Advances in Research Since 1980" lists flavonoids up to 1988 and also discusses their distribution in the Musci (*362*). Most of the published work on flavonoids of mosses deals with the Bryales (subclass Bryidae). Species belonging to this order contain flavone C- and O-glycosides, biflavones, aurones, isoflavones and deoxyanthocyanins as shown in Table IIIc.

Hypnobryales produce not only dihydroflavonol, but also its dimer, bi-dihydroflavonol. Proanthocyanidins and 3-hydroanthocyanins have so far not been detected in Musci.

HPLC using a photodiode detector is a useful technique for simple and rapid separation system of complex flavonoid mixtures (*399*). In this manner a well separated fingerprint chromatogram of a crude extract of *Bryum capillare* has been obtained. The complete analysis including extraction required only two hours and permitted identification of 16 pure flavonoids including biflavonoids (*vide infra*) (*508*). *Bryum capillare* produces flavone glucoside malonyl esters such as luteolin-7-O-glucoside-6″-malonate **(1148)**, diosmetin-7-O-glucoside-6″-malonate **(1275)**, and 6-hydroxyluteolin-7-O-glucoside-6″-malonate **(1279)** (*508, 519*) as well as isoflavone glucoside malonyl esters, both types being new in the plant kingdom.

Rhizomnium magnifolium and *R. pseudopunctatum* produce the same flavonoid glucuronides, the first to have been isolated from mosses, namely apometzgerin-7-O-glucuronide (= tricetin-3′,4′-dimethyl ether-7-O-monoglucuronide) **(1295)**, selgin-7,5′-di-O-glucuronide **(1296)** and tricetin-7,3′-di-O-glucuronide **(1297)** which are new as natural products, together with luteolin-7-O-glucuronide **(1149)**, crysoeriol-7-O-glucuronide **(1159)**, luteolin-7,3′-di-O-glucuronide **(1276)** (*423*). *Rhizomnium punctatum* contained no flavonoids but other unidentified phenolic compounds. On the other hand, *Bryum pseudotriquetrum* yielded more than twenty different phenolic compounds, fifteen of which were identified (*520*). Among these apigenin-7-O-neohesperidoside-6″-malonyl ester **(1269)**, luteolin-7-O-neohesperidoside-6″-malonyl ester **(1278)**, scutellarein-7-O-glucoside-6″-malonyl ester **(1130)** and kaempferol-3-O-galactoside-4′-O-glucoside **(1288)** were new. This is the first record of flavonols in mosses. *Plagiomnium elatum* produces a new flavone glucoside, luteolin-6-C-β-D-glucopyranoside-8-C-α-L-rhamnoside (= elatin) **(1280)**, together with luteolin-6-β-D-glucoside (= isoorientin) **(1281)** and luteolin-8-C-β-D-glucoside (= orientin) **(1282)**, and a biflavone, 5′-dihydroxyamentoflavone **(1308)** (see later) (*10*). While *P. cuspidatum* elaborates lutonarin **(1186)**, isoscoparin-7-O-glucoside **(1283)** and saponarin **(1137)**, together with four biflavonoids (*10*). Isoorientin-O-glycosides are widely distributed in the plant kingdom, particularly in the genus *Gentiana* (Gentianaceae).

Nineteen flavonoids, including apigenin **(1112)**, luteolin **(1143)** and their glucosides, isoscutellarein-7-O-glucoside **(1285)** and hypolaetin-7-O-glucoside **(1287)** have been detected in antarctic *Bryum argenteum*

Table IIIc. *Aromatic Compounds (Flavones, Isoflavones, Biflavones, Aurones, Anthocyanins and Benzonaphthoxanthenones) Found in the Musci*

Structure number	Name of compounds	Formula	m.p.°C	$[\alpha]_D$	Plant source	Subclass	References	Comments
(1112)	Apigenin	$C_{15}H_{10}O_5$			*Bryum argenteum*	B	*(365)*	
					Plagiomnium cuspidatum	B	*(10, 95)*	
(1120)	Apigenin-7-O-neo-hesperidoside				*P. elatum*	B	*(197)*	
					Bryum pseudo-triquetrum	B	*(520)*	
					B. schleicheri	B	*(521)*	
					Plagiomnium elatum	B	*(10)*	
(1121)	Apigenin-7-O-glucoside				*Bryum argenteum*	B	*(365)*	
					Bryum capillare	B	*(519)*	
(1127)	Apigenin-6,8-di-C-glucoside (= Vicenin 2)				*Bryum pallescens*	B	*(521)*	
					B. pseudo-triquetrum	B	*(520)*	
					B. schleicheri	B	*(521)*	
(1130)	Scutellarein-7-O-glucoside-6''-malonate Saponarin				*Bryum pseudotriquetrum*	B	*(520)*	
(1137)					*Plagiomnium cuspidatum*	B	*(10)*	
(1143)	Luteolin	$C_{15}H_{10}O_6$			*Bryum argenteum*	B	*(365)*	
					B. capillare	B	*(199, 508, 519)*	
					B. schleicheri	B	*(521)*	
					Dicranum robustum	B	*(364)*	
					Plagiomnium cuspidatum	B	*(10)*	
					P. elatum	B	*(197)*	

No.	Compound	Formula	Species		Ref.
(1147)	Luteolin-7-O-glucoside		*Bryum argenteum*	B	(365)
			B. capillare	B	(508, 519)
			B. schleicheri	B	(521)
(1148)	Luteolin-7-O-glucoside-6''-malonate		*Bryum argenteum*	B	(365)
			B. capillare	B	(508, 519)
			B. schleicheri	B	(521)
(1149)	Luteolin-7-O-glucuronide		*Rhizomnium magnifolium*	B	(423)
			R. pseudopunctatum	B	(423)
(1159)	Crysoeriol-7-O-glucuronide		*Rhizomnium magnifolium*	B	(423)
			R. pseudopunctatum	B	(423)
(1164)	Luteolin-6,8-di-C-glucoside (= Lucenin 2)		*Bryum pallescens*	B	(521)
			B. pseudotriquetrum	B	(520)
			B. schleicheri	B	(521)
			Plagiomnium elatum	B	(10)
(1170)	6-Hydroxyluteolin-7-O-glucoside		*Bryum capillare*	B	(508, 519)
(1186)	Lutonarin		*B. schleicheri*	B	(521)
			Plagiomnium cuspidatum	B	(10)
(1191)	Kaempferol-3-methyl ether	$C_{16}H_{12}O_6$	*Bryum pseudotriquetrum*	B	(520)
(1204)	Selgin-6,8-di-C-glucoside (= Stellarin 2)		*Bryum pallescens*	B	(521)
			B. pseudotriquetrum	B	(520)
(1269)	Apigenin-7-O-neohesperidoside-6''-malonate		*Bryum pseudotriquetrum*	B	(520)

Table IIIc (*continued*)

Structure number	Name of compounds	Formula	m.p.°C	$[\alpha]_D$	Plant source	Subclass	References	Comments
(1270)	Apigenin-7-O-glucoside-6″-malonate				*Bryum argenteum*	B	(365)	
					B. capillare	B	(519)	
(1271)	Shaftoside				*Bryum schleicheri*	B	(521)	
(1272)	Isoshaftoside				*Bryum schleicheri*	B	(521)	
(1273)	Vitexin-2″-rhamnoside				*Bryum schleicheri*	B	(521)	
(1274)	Diosmetin-7-O-glucoside				*Bryum capillare*	B	(508, 519)	
(1275)	Diosmetin-7-O-glucoside-6″-malonate				*Bryum capillare*	B	(508, 519)	
(1276)	Luteolin-7,3′-di-O-glucuronide				*Rhizomnium magnifolium*	B	(423)	
					R. pseudopunctatum	B	(423)	
(1277)	Luteolin-7-O-neohesperidoside				*Bryum pseudotriquetrum*	B	(520)	
					B. schleicheri	B	(521)	
(1278)	Luteolin-7-O-neohesperidoside-6″-malonate				*Bryum pseudotriquetrum*	B	(520)	
					B. schleicheri	B	(521)	
(1279)	6-Hydroxyluteolin-7-O-glucoside-6″-malonate				*Bryum capillare*	B	(508, 519)	
					B. schleicheri	B	(521)	
(1280)	Luteolin-6-C-glucoside-8-C-rhamnoside (= Elatin)				*Plagiomnium elatum*	B	(10)	

(1281)	Luteolin-6-C-glucoside (= Isoorientin)	*Plagiomnium elatum*	B	(10)
(1282)	Luteolin-8-C-glucoside (= Orientin)	*Plagiomnium elatum*	B	(10)
(1283)	Isoscoparin-7-O-glucoside	*Plagiomnium cuspidatum*	B	(10, 365)
(1285)	Isoscutellarein-7-O-glucoside	*Bryum argenteum*	B	(365)
(1287)	Hypolaetin-7-O-glucoside	*Bryum argenteum*	B	(365)
(1288)	Kaempferol-3-O-galactoside-4'-O-glucoside	*Bryum pseudotriquetrum*	B	(520)
(1289)	Kaempferol-3-O-glucoside	*Bryum pallescens*	B	(521)
		B. pseudo-triquetrum	B	(520)
(1290)	Kaempferol-3-O-galactoside	*Bryum pseudo-triquetrum*	B	(520)
(1291)	Kaempferol-3-O-neohesperidoside	*Bryum pallescens*	B	(521)
(1292)	Kaempferol-3,4'-di-O-glucoside	*Bryum pallescens*	B	(521)
		B. pseudo-triquetrum	B	(520)
(1293)	Kaempferol-3-O-glucoside-6''-malonate	*Bryum pseudo-triquetrum*	B	(520)
(1294)	Kaempferol-3-O-rhamnosyl-glucoside	*Bryum pseudo-triquetrum*	B	(520)
(1295)	Apometzgerin-7-O-glucuronide (= Tricetin-3',4'-dimethyl ether-7-O-mono-glucuronide)	*Rhizomnium magnifolium*	B	(423)
		R. pseudo-punctatum	B	(423)

Table IIIc (continued)

Structure number	Name of compounds	Formula	m.p.°C	$[\alpha]_D$	Plant source	Subclass	References	Comments
(1296)	Selgin-7,5'-di-O-glucuronide				*Rhizomnium magnifolium*	B	(423)	
					R. pseudo-punctatum	B	(423)	
(1297)	Tricetin-7,3'-di-O-glucuronide				*Rhizomnium magnifolium*	B	(423)	
					R. pseudo-punctatum	B	(423)	
(1298)	3,5,7,4'-Tetrahydroxy-3'-(3"-formyl-6"-hydroxyphenyl) flavanone				*Hypnum cupressiforme*	B	(509)	
(1300)	Orobol				*Bryum capillare*	B	(12, 199, 508, 519)	
					B. schleicheri	B	(521)	
(1301)	Orobol-7-O-glucoside				*Bryum capillare*	B	(12, 508, 519)	
					B. schleicheri	B	(521)	
(1302)	Orobol-7-O-di-glucoside				*Bryum schleicheri*	B	(521)	
(1303)	Orobol-7-O-glucoside-6"-malonate				*Bryum capillare*	B	(12, 508, 519)	
					B. schleicheri	B	(521)	
(1304)	Pratensein	$C_{16}H_{12}O_6$			*Bryum capillare*	B	(12, 508, 519)	
(1305)	Pratensein-7-O-glucoside				*Bryum capillare*	B	(12, 508, 519)	

No.	Compound	Formula	Species		Ref.
(1306)	Pratensein-7-O-glucoside-6''-malonate		Bryum capillare	B	(12, 519)
(1308)	5'-Hydroxyamento-flavone	$C_{30}H_{18}O_{11}$	Plagiomnium elatum	B	(10, 197)
			Rhytidiadelphus squarrosus	B	(498)
(1309)	5',3'''-Dihydroxy-amento-flavone (= 5',8''-Biluteolin)	$C_{30}H_{18}O_{12}$	Bryum schleicheri	B	(521)
			Campylopus clavatus	B	(198a)
			C. holomitrium	B	(198a)
			Dicranoloma robustum	B	(364)
			D. scoparium	B	(357, 459)
			Philonotis fontana	B	(198)
			Racomitrium lanuginosum	B	(197)
			Rhytidiadelphus squarrosus	B	(498)
(1311)	5'-Hydroxyrobusta-flavone	$C_{30}H_{18}O_{11}$	Rhytidiadelphus squarrosus	B	(498)
(1312)	5',3'''-Dihydroxy-robustaflavone (= 5',6''-Biluteolin)	$C_{30}H_{18}O_{12}$	Antitrichia curtipipendula	B	(197)
			Bryum capillare	B	(508)
			Campylopus clavatus	B	(198a)
			C. holomitrium	B	(198a)
			Dicranoloma robustum	B	(364)
			Hylocomium splendens	B	(95)
			Philonotis fontana	B	(198)

Table IIIc (continued)

Structure number	Name of compounds	Formula	m.p.°C	$[\alpha]_D$	Plant source	Subclass	References	Comments
(1313)	Bryoflavone	$C_{30}H_{18}O_{12}$			*Plagiomnium cuspidatum*	B	*(10, 11, 95)*	
(1314)	Heterobryoflavone	$C_{30}H_{18}O_{12}$			*Racomitrium lanuginosum*	B	*(197)*	
(1315)	Dicranolomin (= 2',6''-Biluteolin)	$C_{30}H_{18}O_{12}$			*Bryum capillare*	B	*(199, 508)*	
					Bryum capillare	B	*(199, 508)*	
					B. schleicheri	B	*(521)*	
					Dicranoloma robustum	B	*(364)*	
(1316)	2,3-Dihydro-dicranolomin (= 2,3-Dihydro-2',6''-biluteolin)	$C_{30}H_{20}O_{12}$			*Philonotis fontana*	B	*(198)*	
					Dicranoloma robustum	B	*(364)*	
(1317)	2'',3''-Dihydro-5',6''-biluteolin	$C_{30}H_{20}O_{12}$			*Dicranoloma robustum*	B	*(364)*	
(1320)	2,3-Dihydro-5-hydroxy-amentoflavone	$C_{30}H_{20}O_{11}$			*Plagiomnium cuspidatum*	B	*(10, 11)*	
					Rhytidiadelphus squarrosus	B	*(498)*	
(1321)	2,3-Dihydro-5',3'''-dihydroxy-amentoflavone	$C_{30}H_{20}O_{12}$			*Philonotis fontana*	B	*(198)*	
					Plagiomnium cuspidatum	B	*(10, 11)*	
(1322)	2,3-Dihydro-5'-hydroxy-robustaflavone	$C_{30}H_{20}O_{11}$			*Plagiomnium cuspidatum*	B	*(10, 11)*	

No.	Name	Formula		Species		Ref.
(1323)	Philonotisflavone (= 2',8''-Biluteolin)	$C_{30}H_{18}O_{12}$		Dicranoloma robustum	B	(364)
(1324)	2,3-Dihydrophilonotis-flavone	$C_{30}H_{20}O_{12}$		Philonotis fontana	B	(198)
				Dicranoloma robustum	B	(364)
(1325)	Hypnogenol A	$C_{30}H_{22}O_{12}$		Philonotis fontana	B	(198)
			−9.04	Hypnum cupressiforme	B	(509)
(1326)	Hypnogenol B	$C_{30}H_{22}O_{11}$		Hypnum cupressiforme	B	(509)
(1327)	3,5,7,4',3'',5'',7'',3''',4'''-Nonahydroxy-3',6''-biflavone	$C_{30}H_{22}O_{13}$		Hypnum cupressiforme	B	(509)
(1328)	3,5,7,4',3'',5'',7''-Heptahydroxy-3'-O-4'''-biflavone	$C_{30}H_{22}O_{12}$		Hypnum cupressiforme	B	(509)
(1329)	1''',2'',3''',4'''-Tetrahydro-3,3'',5,5'',7,7''-hexahydroxy-4'''-keto-3',4'-O-2'''-biflavone	$C_{30}H_{22}O_{12}$		Hypnum cupressiforme	B	(509)
(1330)	3,3'''-Binaringenin	$C_{30}H_{22}O_{10}$		Homalothecium lutescens	B	(497a)
(1331)	2,3-Dihydro-3,3'''-biapigenin	$C_{30}H_{20}O_{10}$		Homalothecium lutescens	B	(497a)
(1332)	Campylopusaurone	$C_{30}H_{20}O_{12}$		Campylopus clavatus	B	(198a)
(1335)	Bracteatin	$C_{15}H_{10}O_{6}$		C. holomitrium	B	(198a)
				Funaria hygrometrica	B	(614a)

Table IIIc *(continued)*

Structure number	Name of compounds	Formula	m.p.°C	$[\alpha]_D$	Plant source	Subclass	References	Comments
(1336)	Luteolinidin-5-O-glucoside				*Bryum cryophilum*	B	*(99)*	
					B. rutilans	B	*(98)*	
					B. weigelii	B	*(98)*	
					Funaria hygrometrica	B	*(614a)*	
(1337)	Luteolinidin-5-O-diglucoside				*Bryum cryophilum*	B	*(99)*	
					B. rutilans	B	*(98)*	
					B. weigelii	B	*(98)*	
(1338)	Sphagnorubin A (= 2-(3′,4′-Di-hydroxyphenyl)-8,11-dihydroxy-9H-phenanthro[2,1-b]-pyran-9-one	$C_{23}H_{14}O_6$			*Sphagnum magellanicum*	S	*(476, 610)*	
					S. nemoreum	S	*(476)*	
					S. plum	S	*(476)*	
					S. rubellum	S	*(401, 402)*	
(1339)	Sphagnorubin B (= 2-(3′,4′-Dihydroxy-phenyl)-8,9-dihydroxy-11-methoxyphenanthro-[2,1-b]pyrylium-chloride)	$C_{24}H_{17}O_6Cl$	> 350		*Sphagnum magellanicum*	S	*(476, 610)*	
					S. nemoreum	S	*(476)*	
					S. plum	S	*(476)*	
					S. rubellum	S	*(401, 402)*	
(1340)	Sphagnorubin C (= 8,9-Dihydroxy-2-(4′-hydroxy-3′-methoxy-phenyl)-11-methoxy-phenanthro[2,1-b]-pyryliumchloride)	$C_{25}H_{19}O_6Cl$	> 350		*Sphagnum magellanicum*	S	*(476, 610)*	
					S. nemoreum	S	*(476)*	
					S. plum	S	*(476)*	
					S. rubellum	S	*(401, 402)*	

(1342)	Ohioensin A	$C_{23}H_{16}O_5$	274–5	+ 37	Polytrichum ohioense	P	(645a, 645b)	CD 211 nm + 2.06 CD 228 nm − 22.76 CD 255 nm − 5.76 CD 285 nm + 13.60 CD 298 nm + 15.19 CD 360 nm − 1.64
(1343)	Ohioensin B	$C_{24}H_{18}O_5$	246–47		Polytrichum ohioense	P	(645a, 645b)	CD 209 nm + 1.87 CD 225 nm − 20.35 CD 248 nm − 1.30 CD 284 nm + 5.38 CD 310 nm + 0.90 CD 345 nm − 1.40
(1344)	Ohioensin C	$C_{23}H_{16}O_5$	230–1	− 18	Polytrichum ohioense	P	(645a, 645b)	CD 208 nm + 11.60 CD 230 nm − 26.37 CD 289 nm + 10.30 CD 310 nm + 3.50 CD 360 nm − 0.82
(1345)	Ohioensin D	$C_{24}H_{18}O_6$	244–5	− 59	Polytrichum ohioense	P	(645a, 645b)	CD 203 nm + 0.97 CD 217 nm − 12.72 CD 235 nm − 3.19 CD 285 nm + 6.33 CD 295 nm + 5.34 CD 345 nm − 2.80
(1346)	Ohioensin E	$C_{25}H_{20}O_6$	226–8	− 42	Polytrichum ohioense	P	(645a, 645b)	CD 209 nm + 8.82 CD 227 nm − 27.10 CD 283 nm + 12.92 CD 340 nm − 3.78

B: Bryidae; P: Polytrichidae; S: Sphagnidae

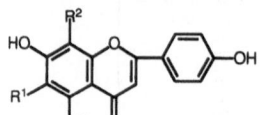

(1269) Apigenin-7-O-neohesperidoside-6"-malonate ; R=6-Malonyl neoh
(1270) Apigenin-7-O-glucoside-6"-malonate ; R=6-Malonyl Glc

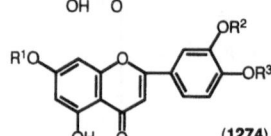

(1271) Shaftoside ; R^1=Glc, R^2=Ara
(1272) Isoshaftoside ; R^1=Ara, R^2=Glc
(1273) Vitexin-2"-rhamnoside ; R^1=H, R^2=Glc-2"-Rha

(1274) Diosmetin-7-O-glucoside ; R^1=Glc, R^2=H, R^3=Me
(1275) Diosmetin-7-O-glucoside-6"-malonate ; R^1=6-Malonyl Glc, R^2=H, R^3=Me
(1276) Luteolin-7,3'-di-O-glucuronide ; R^1=R^2=GA, R^3=H
(1277) Luteolin-7-O-neohesperidoside ; R^1=Neoh, R^2=R^3=H
(1278) Luteolin-7-O-neohesperidoside-6"-malonate ; R^1=6-Malonyl neoh, R^2=R^3=H

(1279) 6-Hydroxyluteolin-7-O-glucoside-6"-malonate ; R^1=OH, R^2=6-malonyl Glc, R^3=H
(1280) Luteolin-6-C-glucoside-8-C-rhamnoside (=Elatin) ; R^1=Glc, R^2=H, R^3=Rha
(1281) Luteolin-6-C-glucoside (=Isoorientin) ; R^1=Glc, R^2=R^3=H
(1282) Luteolin-8-C-glucoside (=Orientin) ; R^1=R^2=H, R^3=Glc

(1283) Isoscoparin-7-O-glucoside

(1284) Isoscutellarein ; R=H
(1285) Isoscutellarein-7-O-glucoside ; R=Glc

(1286) Hypolaetin ; R=H
(1287) Hypolaetin-7-O-glucoside ; R=Glc

Chart 75a. Flavones found in the Musci

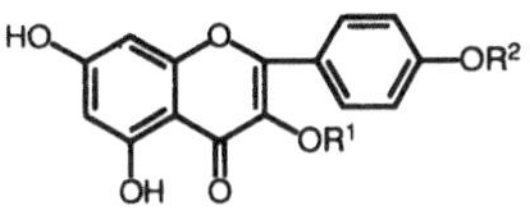

(1288) Kaempferol-3-O-galactoside-4'-glucoside ; R^1=Gal, R^2=Glc
(1289) Kaempferol-3-O-glucoside ; R^1=Glc, R^2=H
(1290) Kaempferol-3-O-galactoside ; R^1=Gal, R^2=H
(1291) Kaempferol-3-O-neohesperidoside ; R^1=Neoh, R^2=H
(1292) Kaempferol-3,4'-di-O-glucoside ; R^1=R^2=Glc
(1293) Kaempferol-3-O-glucoside-6''-malonate ; R^1=6-Malonyl Glc, R^2=H
(1294) Kaempferol-3-O-rhamnosylglucoside ; R^1=Rha-Glc, R^2=H

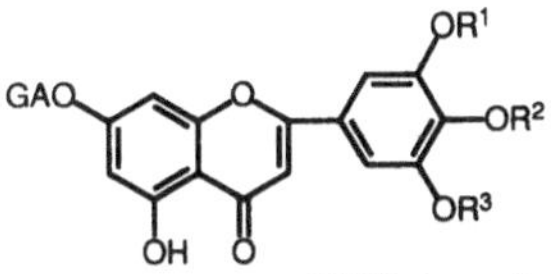

(1295) Apometzgerin-7-O-glucuronide (=Tricetin-3',4'-dimethyl ether-7-O-monoglucuronide ;
R^1=R^2=Me, R^3=H
(1296) Selgin-7,5'-di-O-glucuronide ; R^1=Me, R^2=H, R^3=GA
(1297) Tricetin-7,3'-di-O-glucuronide ; R^1=GA, R^2=R^3=H

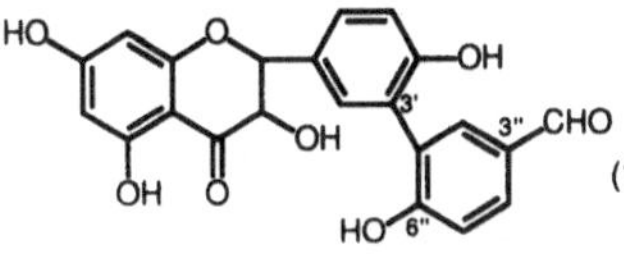

(1298) 3,5,7,4'-Tetrahydroxy-3'-(3''-formyl-6''-hydroxyphenyl)flavanone

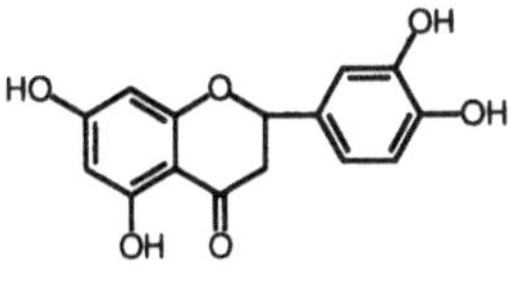

(1299) Eriodictyol

Chart 75b. Flavones, flavonols, flavanones and flavanols found in the Musci

(365). Isoscutellarein (1284) and hypolaetin (1286) and their 7-O-gluco-sides had not been found previously in bryophyte. Twenty flavonoids including biflavones were detected in *B. schleicheri* while kaempferol-3-O-glucoside (1289) and kaempferol-3-O-neohesperidoside (1291) were detected in *Bryum pallescens (521)*.

The flavonoid patterns of the gametophytes and sporophytes of *B. capillare* are different. The pattern of the adult gametophyte undergoes several changes in the course of development. During maturation of the spore, new flavonoids appear and apigenin-7-O-glucoside (1121) and its 6''-malonate (1270) disappear (519).

It is suggested that *Tetraphis pellucida* (Tetraphidales) produces dihydroflavonols (362). Actually, a new dihydroflavonol, 3,5,7,4'-tetrahydroxy-3'-(3''-formyl-6''-hydroxyphenyl)flavone (1298) was isolated from *Hypnum cupressiforme* (Hypnobryales), together with five new bi-dihydroflavonols (*509*).

3.2.2 Isoflavones

While the most widespread flavonoids in Musci and Hepaticae are flavones which occur mainly as O- and C-glycosides as discussed in the previous section isoflavones in bryophytes are very rare. ANHUT *et al.* (*12*) were the first to report isolation of isoflavonoids from a moss, *Bryum capillare*. These are 5,7,3',4'-tetrahydroxyisoflavone (= orobol)-7-O-glucoside (**1301**) and 7-O-glucoside-6''-malonate (**1303**) and 5,7,3'-trihydroxy-4'-methoxyisoflavone (= pratensein-7-O-glucoside) (**1305**) and 7-O-glucoside-6''-malonate (**1306**) together with the aglycones orobol (**1300**) and pratensein (**1304**) (*12, 519*). *B. schleicheri* also contains a new isoflavone glucoside (**1302**) and the same isoflavones (**1300, 1301, 1303**) as those found in *B. capillare* (*521*). The isoflavonoids were detected in 80% MeOH extracts of fresh and air-dried plant material. When the fresh plant was extracted with ice cold acetone, the aglycones were absent which suggests that the aglycones in *Bryum* species arise by chemical and enzymatic hydrolysis during the extraction. *Bryum capillare* is the first non-vascular plant in which isoflavonoids were encountered. More than 90% of known isoflavones has been found in subfamily Leguminosae and the remainder in other angiosperm and gymnosperm families.

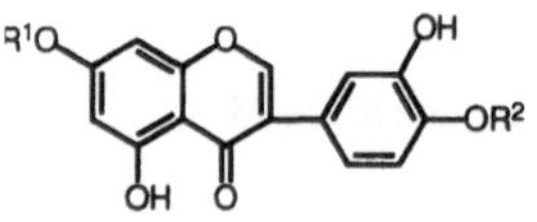

(**1300**) Orobol ; $R^1=R^2=H$
(**1301**) Orobol-7-O-glucoside ; $R^1=Glc$, $R^2=H$
(**1302**) Orobol-7-O-diglucoside ; $R^1=Glc-Glc$, $R^2=H$
(**1303**) Orobol-7-O-glucoside-6''-malonate ; $R^1=6$-Malonyl Glc, $R^2=H$
(**1304**) Pratensein ; $R^1=H$, $R^2=Me$
(**1305**) Pratensein-7-O-glucoside ; $R^1=Glc$, $R^2=Me$
(**1306**) Pratensein-7-O-glucoside-6''-malonate ; $R^1=6$-Malonyl Glc, $R^2=Me$

Chart 76. Isoflavonoids found in the Musci

3.2.3 Biflavones

For a long time, 5',3'''-dihydroxyamentoflavone (= 5',8''-biluteolin) (**1309**) from the moss *Dicranum scoparium* (*357, 459*) was the only known biflavonoid from bryophytes. Since then this compound and other biflavonoids have been isolated from the other mosses (Table IIIc). The most common moss biflavonoids are dimers of luteolin and their 2,3- or 2'',3''-dihydro derivatives.

It has been suggested that biflavonoids are part of the cell wall phenolics of the mosses. In spite of extensive studies on flavonoids of

Hepaticae and in Anthocerotae, biflavonoids have not so far been found in these two classes, whereas in Musci, on the other hand, biflavonoids occur frequently and have been isolated from seven distinct families, Bartramiaceae, Bryaceae, Dicranaceae, Grimmiaceae, Hylocomiaceae, Leucodontaceae and Mniaceae (196). The polar moss biflavonoids are found in the hydrophilic cell walls, whereas the more lipophilic gymnosperm biflavonoids are accumulated in the cutinized layer. Biflavonoids are extracted from mosses by percolation with aq. MeOH or acetone after defatting with $CHCl_3$ or CH_2Cl_2.

Most of the naturally occurring biflavonoids possess C-C interflavonoid linkages and their structure determination is difficult, particularly whether C-6 or C-8 is involved in the C-C linkage. ^{13}C-NMR spectroscopy has proved to be an excellent method for structure elucidation of biflavonoids (368). For example, each signal in the ^{13}C-NMR spectrum of carbon atom of 5′,8″-biluteolin (1309) isolated from *Dicranum scoparium* could be assigned thus leading to determination of the structure (459).

Hylocomium splendens (95) and *Antitrichia curtipipendula* (197) produce a biflavone, 5′,3‴-dihydroxyrobustaflavone (1312), and *Racomitrium lanuginosum* elaborates 5′,8″-biluteolin (1309) together with biflavone (1312) (197). Presence of a 3′,6″ linkage was suggested by the similarity of the UV spectrum to that of luteolin (1143) and the mass spectrum of underivatized compound and its octamethyl ether. The site of the interflavonyl linkage was established by ^{13}C-NMR spectrometry, the quaternary carbon at C-6″ appearing at δ108.9, close to that (δ109.3) of robustaflavone (1310).

The gametophytic tissue of *Bryum capillare* yielded two new biflavonoids, bryoflavone (1313) and heterobryoflavone (1314) which might be formed by oxidative coupling of a flavone and isoflavone moiety. Structures were characterized by a combination of ^{1}H-, ^{13}C-NMR, FD-MS and UV spectral data and comparison of the ^{1}H-, ^{13}C-NMR and UV spectra with those of luteolin (1143) and orobol (1300) which were found in the same species (199). Investigation of the flavonoids of *Dicranum robustum* has led to the finding of luteolin (1143), the rare 5′,8″- (1309) and 5′,6″-biluteolins (1312) as well as dicranolomin (= 2′,6″-biluteolin) (1315), 2,3-dihydrodicranolomin (= 2,3-dihydro-2′,6″-biluteolin) (1316) and 2″,3″-dihydro-5′,6″-biluteolin (1317) (364). These biflavonoids were obtained by a combination of droplet counter-current chromatography (DCCC), preparative TLC and HPLC. Compound (1316) is the first record of a dihydrobiluteolin as a natural product and compounds (1315) and (1316) are representatives of a unique biluteolin series with a 2′,6″-interflavone linkage.

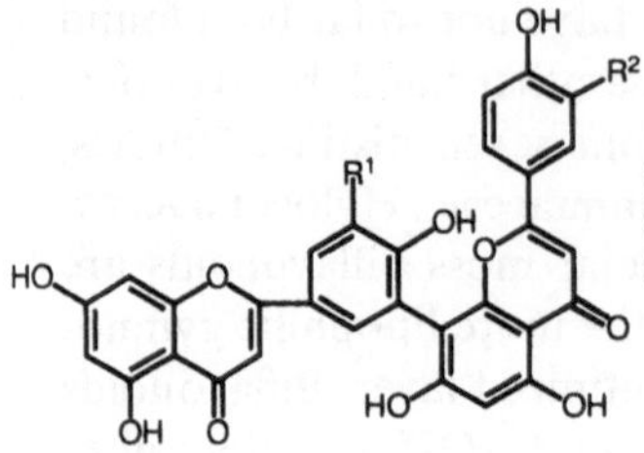

(1307) Amentoflavone ; $R^1=R^2=H$
(1308) 5'-Hydroxyamentoflavone ; $R^1=OH$, $R^2=H$
(1309) 5',3'''-Dihydroxyamentoflavone (=5',8"-Biluteolin) ; $R^1=R^2=OH$

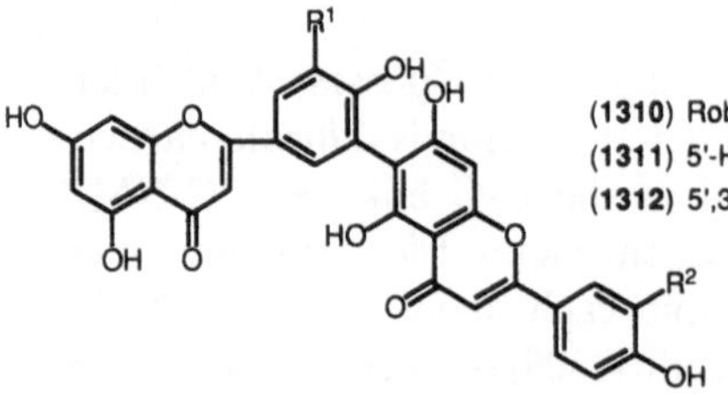

(1310) Robustaflavone ; $R^1=R^2=H$
(1311) 5'-Hydroxyrobustaflavone ; $R^1=OH$, $R^2=H$
(1312) 5',3'''-Dihydroxyrobustaflavone (=5',6"-Biluteolin) ; $R^1=R^2=OH$

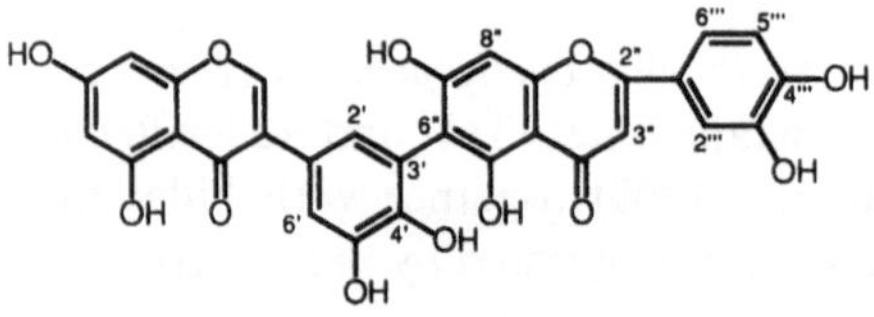

(1313) Bryoflavone

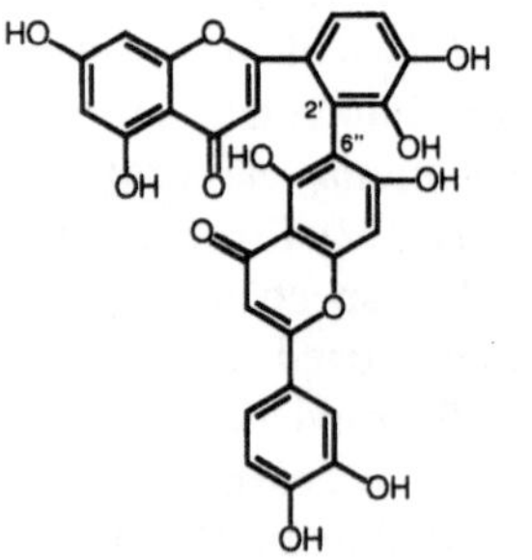

(1315) Dicranolomin (=2',6"-Biluteolin)

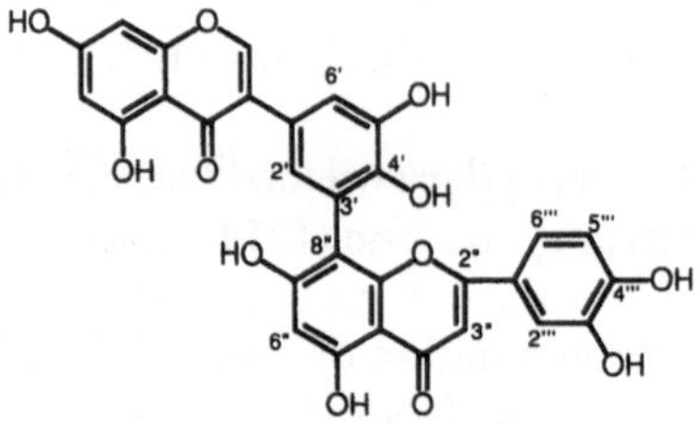

(1314) Heterobryoflavone

Chart 77a. Biflavonoids found in the Musci and their related compound

Plagiomnium elatum produces a new biflavone, 5'-hydroxyamentofla-vone (1308) (*10, 197*). The FD-MS of (1308) showed a molecular ion at m/z 554 indicating the presence of a heptahydroxy-biflavone whose structure was suggested by the co-occurrence of apigenin (1112) and luteolin (1143) and the UV spectrum of an equimolar mixture of (1112) and (1143) which was superimposable on that of (1308). Comparison of the ^{1}H-NMR spectrum with those of amentoflavone (1307) and 5',8"-biluteolin (1309) led to the complete structure. The 4′,5′-dimethyl ether (1318) of (1308), named 5′-methoxybilovetin has been isolated previously from *Ginkgo biloba* (*308*). Thus, 5′-hydroxyamentoflavone (1308) is the first example of a biflavone in tracheophytes as well as bryophytes.

Three new dihydrobiflavonoids, 2,3-dihydro-5′-hydroxyamentoflavone (1320), 2,3-dihydro-5′,3‴-dihydroxyamentoflavone (1321) and 2,3-dihydro-5′-hydroxyrobustaflavone (1322), have been isolated from *Plagiomnium cuspidatum* (*11*), together with 5′,3‴-dihydroxyrobustaflavone (= 5′,6″-biluteolin) (1312) as well as apigenin, luteolin and crysoeriol glucosides (*10, 95*). The ¹H NMR spectrum of (1320) exhibits a typical three proton ABX system at δ2.78, 3.18 and 5.46 as well as two *meta* coupled aromatic doublet at δ5.91 and 5.91 which are assignable to the protons at C-2 and C-3, and C-6 and C-8 of the dihydroflavone, respectively. The signals of H-3″,6″,2‴,3‴,5‴ and 6‴ are found at almost the same position and are of the same multiplicity as in the spectra of amentoflavone (1307) and 2,3-dihydroamentoflavone (1319) (*368*). Therefore, the second flavonoid moiety is apigenin (1112), linked *via* C-8. Two additional *meta*-coupled doublets at δ6.87 and 7.04 are assignable to the C-2′ and C-6′ protons of the flavone moiety, eriodictyol (1299). The structure of (1321) was established by comparing the ¹H- and ¹³C-NMR spectra with those of (1320) and 5′,8″-biluteolin (5′,3‴-dihydroxyamentoflavone) (1309) (*357, 459*).

Gametophytes of *Philonotis fontana* produce two novel biflavonoids, philonotisflavone (= 2′,8″-biluteolin) (1323) and 2,3-dihydrophilonotisflavone (1324), together with previously known four biflavonoids, 5′,3‴-dihydroxyamentoflavone (1309), 5′,3‴-dihydroxyrobustaflavone (1312), dicranolomin (1315), and 2,3-dihydro-5′,3‴-dihydroxyamentoflavone (1321) (*198*). Structures of the new biflavonoids were deduced by comparing the ¹H-NMR spectrum with that of dicranolomin (1315) and by analysis of the ¹³C-NMR spectrum. *Dicranoloma robustum* also contains philonotisflavone (1323) (*364*) and its 2,3-dihydro derivative (1324) (*196*).

Rhytidiadelphus squarrosus (Hylocomiaceae) also produces a new biflavonoid, 5′-hydroxyrobustaflavone (1311), together with three known biflavonoids, 5′-hydroxyamentoflavone (1308), 5′,3‴-dihydroxyamentoflavone (1309) and 2,3-dihydro-5′-hydroxyamentoflavone (1320) (*498*). The structure assigned to (1311) was based on a combination of FAB-MS, ¹H-, ¹³C-NMR spectrometry and UV data and comparison of the spectral data with those of the previously known biflavone, 5′,3‴-dihydroxyrobustaflavone (1312).

Five new bi-dihydroflavonols, hypnogenol A (1325) and hypnogenol B (1326), and their derivatives (1327–1329), possessing an aromadendrin

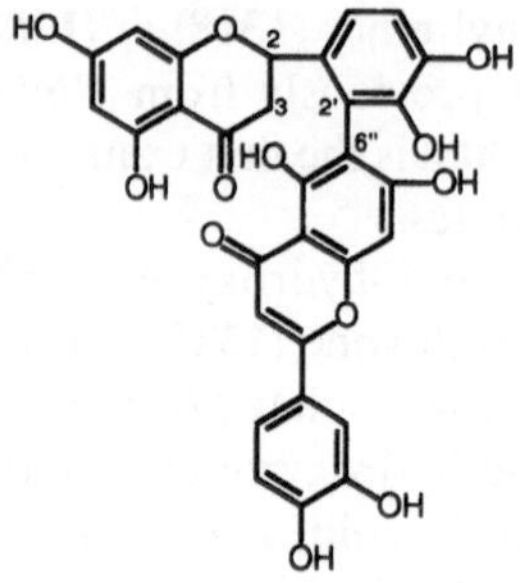

(1316) 2,3-Dihydrodicranolomin
(=2,3-Dihydro-2',6"-biluteolin)

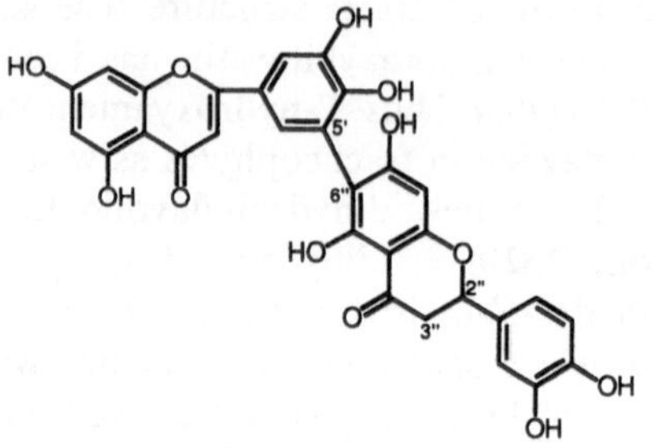

(1317) 2",3"-Dihydro-5',6"-biluteolin

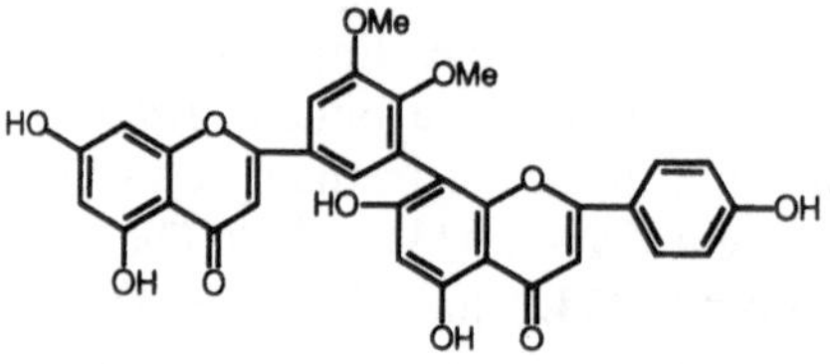

(1318) 5'-Methoxybilovetin
(=5'-Methoxyamentoflavone-4'-methyl ether)

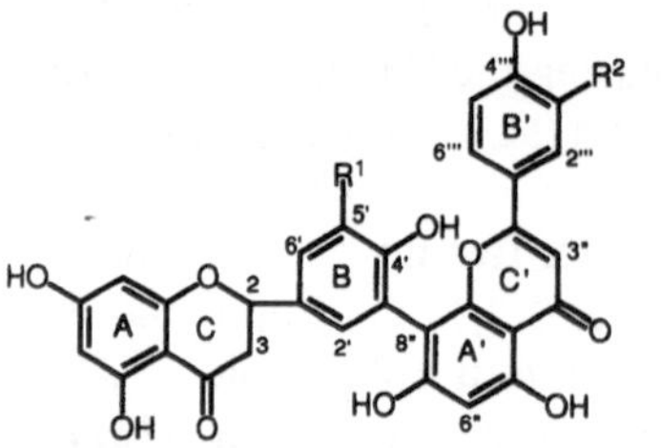

(1319) 2,3-Dihydroamentoflavone ; R¹= R²=H
(1320) 2,3-Dihydro-5'-hydroxyamentoflavone ; R¹=OH, R²=H
(1321) 2,3-Dihydro-5',3'''-dihydroxyamentoflavone ; R¹=R²=OH

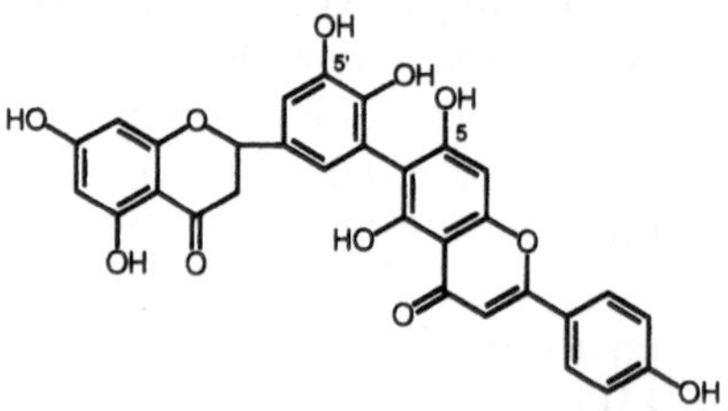

(1322) 2,3-Dihydro-5'-hydroxyrobustaflavone

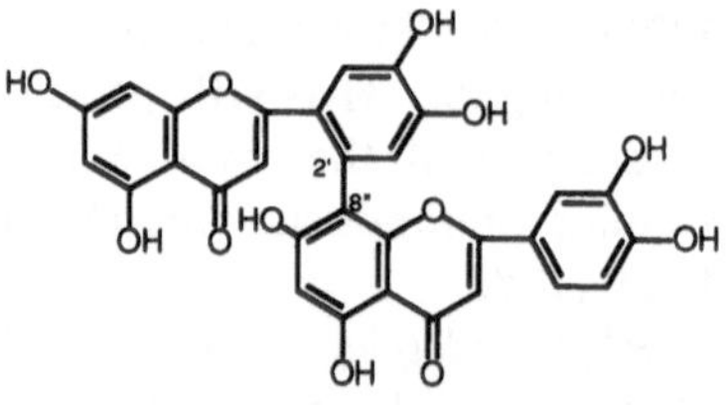

(1323) Philonotisflavone (=2',8"-Biluteolin)

Chart 77b. Biflavonoids found in the Musci and their related compounds

skeleton elongated in the 3'-position by a second dihydroflavonol, were isolated from *Hypnum cupressiforme* by means of HPLC on diol phases (*509*). The dihydroflavonol skeleton of (**1325**) and its substitution pattern were confirmed by ^{13}C NMR spectroscopy and shift calculation as well as

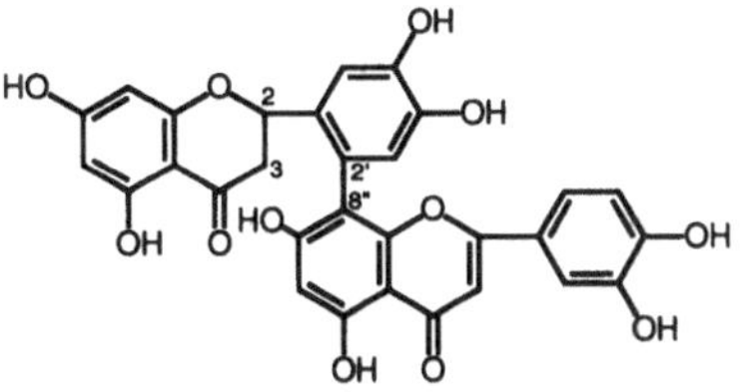

(1324) 2,3-Dihydrophilonotisflavone

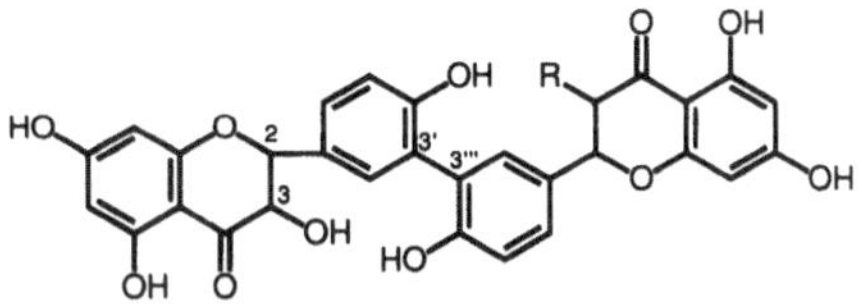

(1325) Hypnogenol A ; R=OH
(1326) Hypnogenol B ; R=H

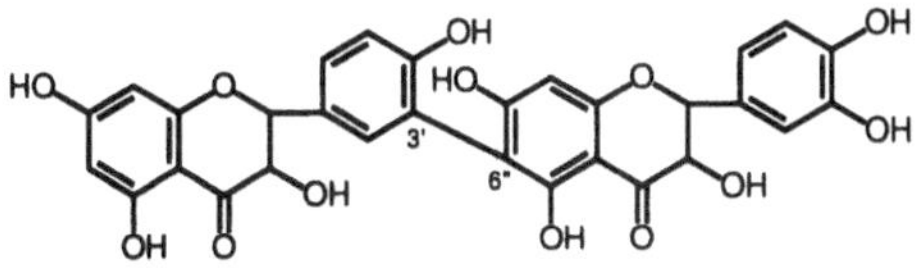

(1327) 3,5,7,4',3",5",7",3'",4'"-Nonahydroxy-3',6"-biflavanone

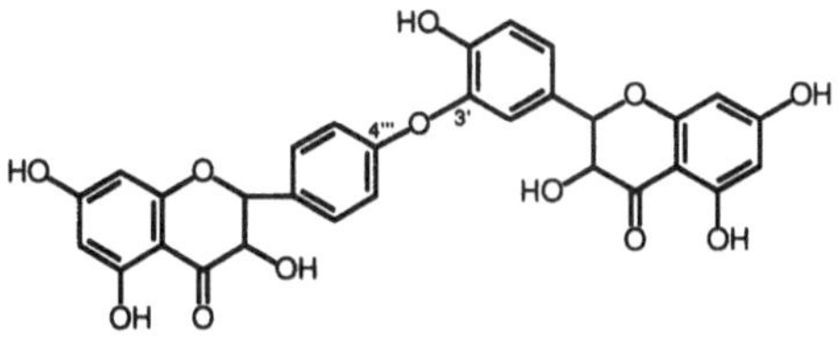

(1328) 3,5,7,4',3",5",7"-Heptahydroxy-3'-O-4'"-biflavanone

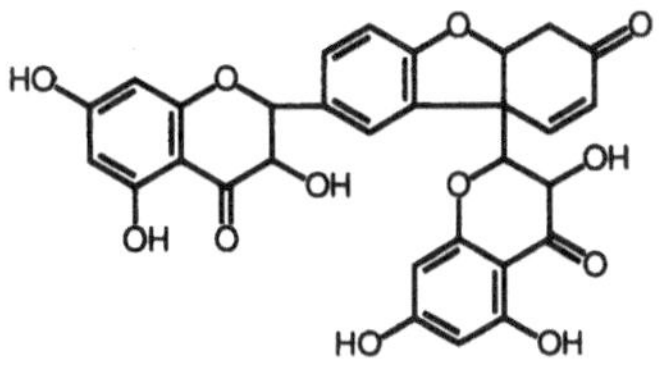

(1329) 1'",2'",3'",4'"-Tetrahydro-3,3",5,5",7,7"-hexahydroxy-4'"-keto-3',1'",4'-O-2'"-biflavanone

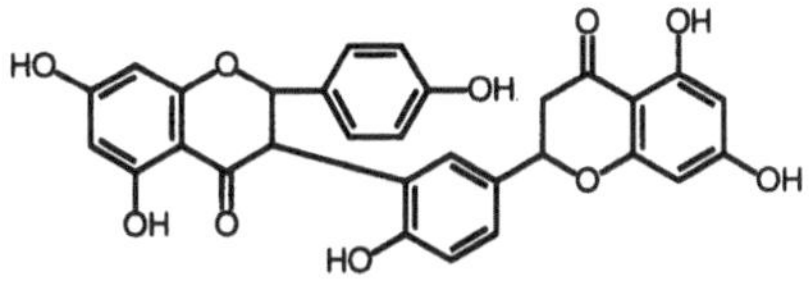

(1330) 3,3'"-Binaringenin

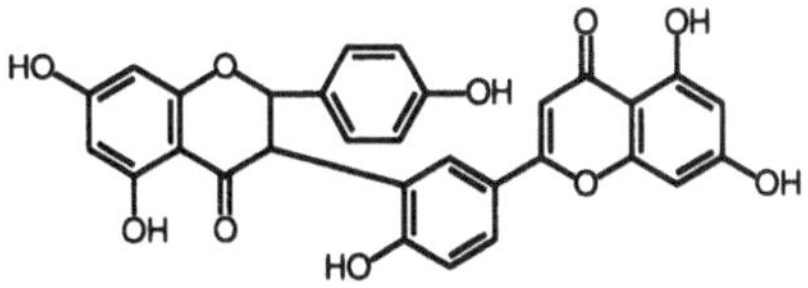

(1331) 2,3-Dihydro-3,3'"-biapigenin

Chart 77c. Biflavonoids found in the Musci

^{13}C-^{1}H-2D COSY spectrum. The 3'-3''' linkage between the subunits of the dimer was established by the presence of NOE between H-2' and H-2, and H-6' and H-2. Compound (**1327**) is the biflavonoid with 3'-6''' linked subunits derived from aromadendrin and taxifolin. The structures of the other biflavones were established by spectroscopic comparison with (**1325**). The hypnogenols (**1325, 1326**) are a new class of biflavonoids in the nature. *Homalothecium lutescens* produces two new biflavones, 3,3'''-binaringenin (**1330**) and 2,3-dihydro-3,3'''-biapigenin (**1331**) (*497a*). Their structures have been based on the ^{1}H NMR spectroscopic comparison with those of synthetic 3,3'''-biapigenin.

Campylopus holomitrium produces a unique biflavonoid, campylopus-aurone (**1332**) in which aureusidin (**1333**) is linked to eriodictyol (**1299**) by a 5'-6'' bond, along with 5',8''-biluteolin (**1309**) and 5',6''-biluteolin (**1312**) (*198a*).

3.2.4 Aurones and Anthocyanins

The sporophytes of *Funaria hygrometrica* produce bracteatin (**1335**) (*614a*) whereas some *Marchantiales* species (Hepaticae) elaborate aureusidin-6-glucuronide (**1334**) (*19*).

The 3-deoxyanthocyanins luteolinidin-5-O-glucoside (**1336**) and luteolinidin-5-O-diglucoside (**1337**) occur in *Bryum cyclophyllum* (*99*), *B. rutilans* and *B. weigelii* (*98*). Three red pigments sphagnorubins A, B

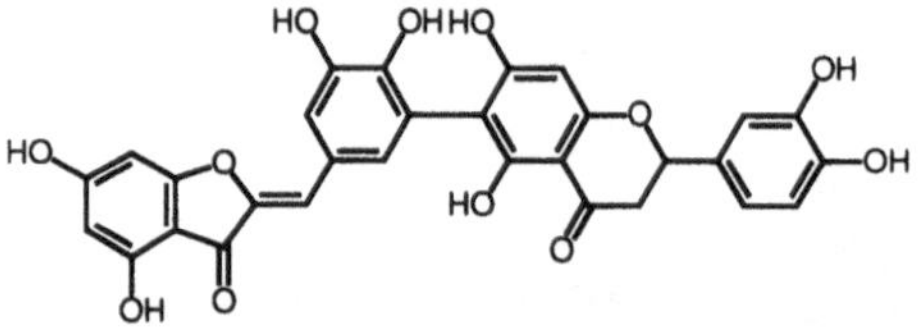

(**1332**) Campylopusaurone

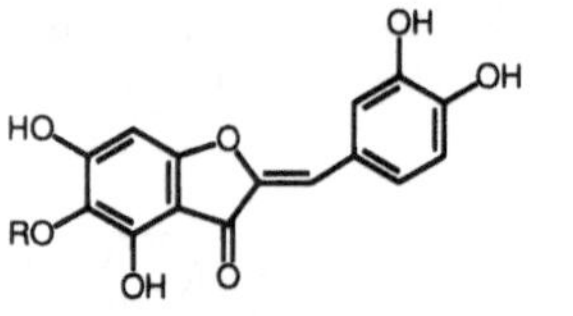

(**1333**) Aureusidin ; R=H
(**1334**) Aureusidin-6-O-glucuronide ; R=GA

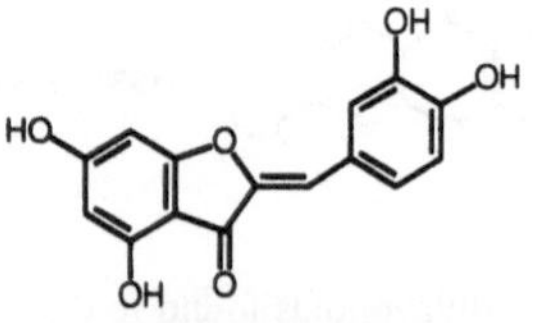

(**1335**) Bracteatin

Chart 78. Aurones found in the Musci

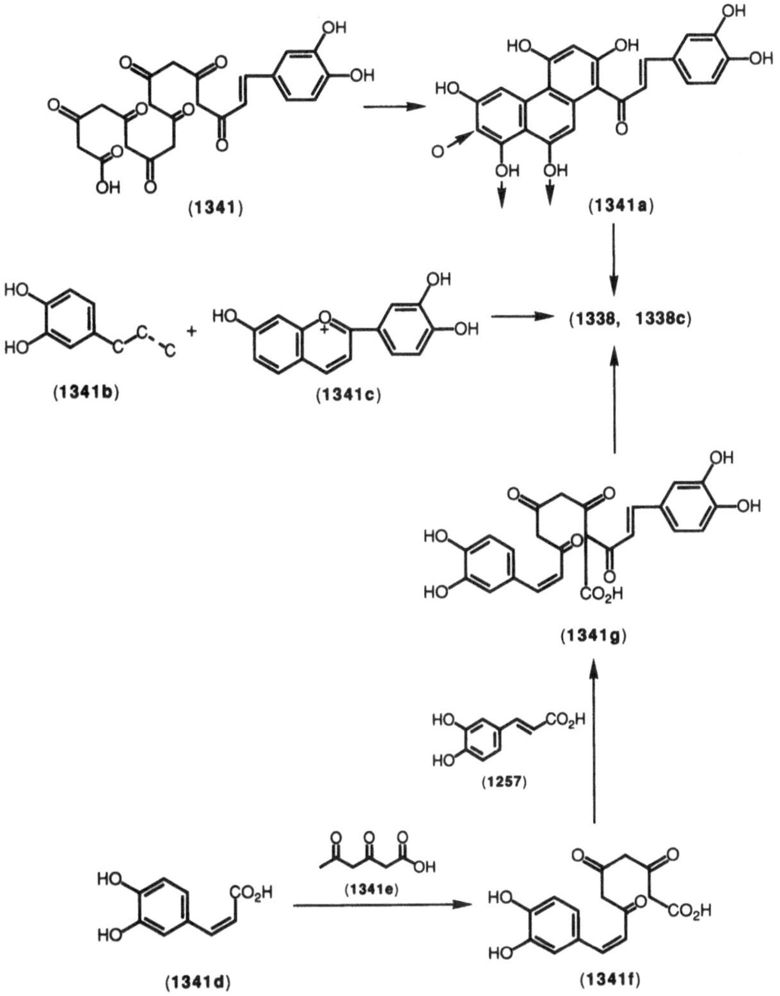

Scheme 67. Three possible biogenetic pathways for sphagnorubin A

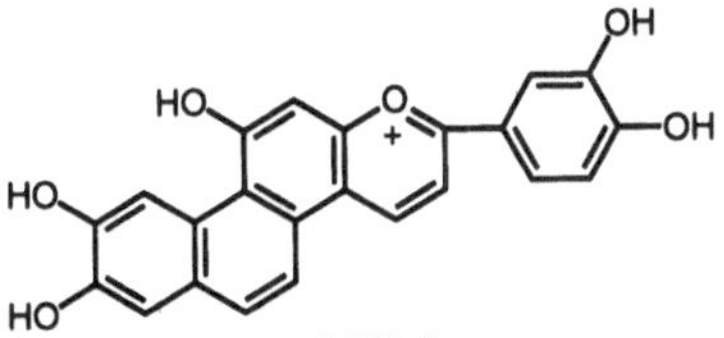

(1338c)

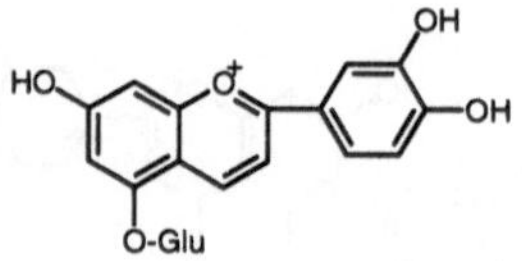

(1336) Luteolinidin-5-O-glucoside

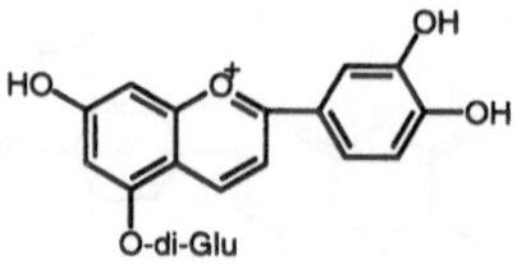

(1337) Luteolinidin-5-O-diglucoside

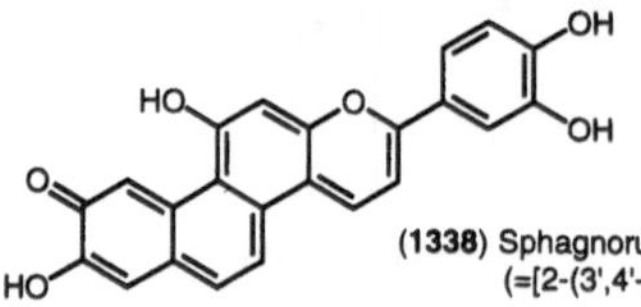

(1338) Sphagnorubin A
(=[2-(3',4'-Dihydroxyphenyl)-8,11-dihydroxy-9*H*-phenanthro [2,1-*b*] pyran-9-one)]

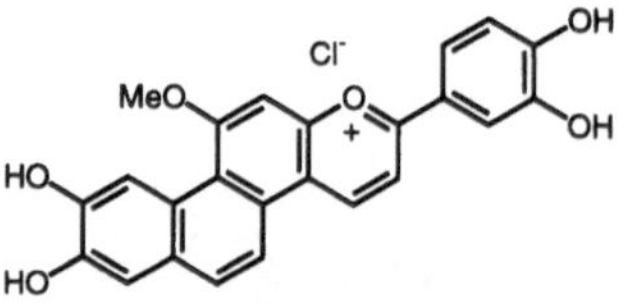

(1339) Sphagnorubin B
(=[2-(3',4'-Dihydroxyphenyl)-8,9-dihydroxy-11-methoxyphenanthro [2,1-*b*] pyryliumchloride)]

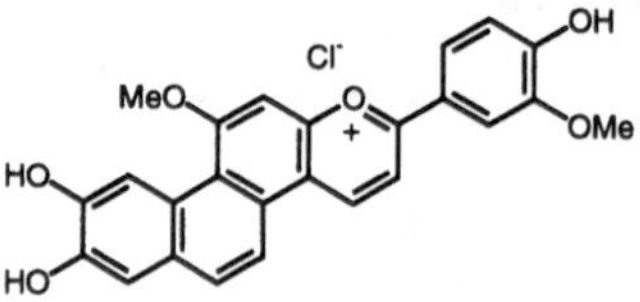

(1340) Sphagnorubin C
(=[8,9-Dihydroxy-2-(4'-hydroxy-3'-methoxyphenyl)-11-methoxyphenanthro [2,1-*b*]-pyryliumchloride)]

Chart 79. Anthocyanins found in the Musci

and C have been detected in *Sphagnum magellanicum, S. nemoreum* and *S. plum* (*476*). Sphagnorubin (= Sphagnorubin A) from *S. magellanicum* was shown to be [2-(3',4'-dihydroxyphenyl)-8,11-dihydroxy-9*H*-phenanthro[2,1-*b*]pyran-9-one] (**1338**) by chemical transformations and IR, UV, NMR and MS spectra (*610*). Three routes have been proposed for biogenesis of sphagnorubin A (**1338**) and its anhydro base (**1338c**) (Scheme 67) (*610*).

A simple screening method for separation and identification of sphagnorubins developed by MENTLEIN (*401*) led to isolation of sphagnorubin B (**1339**) and sphagnorubin C (**1340**), from the peat moss, *Sphagnum rubellum*, together with sphagnorubin A (**1338**). Structures of the new anthocyanidines were established by comparing the spectra data

(UV, IR, Mass, ^{1}H- and ^{13}C-NMR) with those of (1338) and by chemical degradation (402).

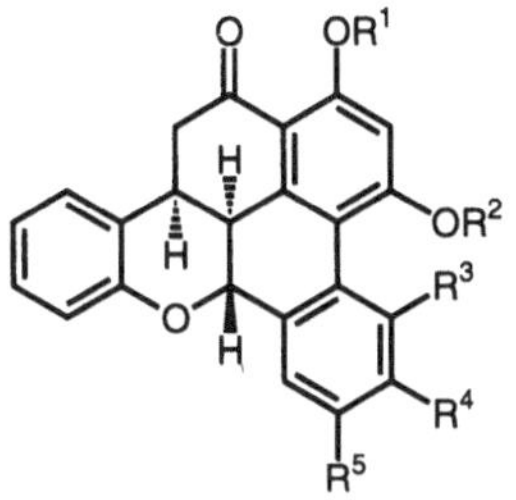

(1342) Ohioensin A ; R^1=R^2=R^3=R^4=H, R^5=OH
(1343) Ohioensin B ; R^1=R^4=R^5=H, R^2=Me, R^3=OH
(1344) Ohioensin C ; R^1=R^2=R^4=R^5=H, R^3=OH
(1345) Ohioensin D ; R^1=R^5=H, R^2=Me, R^3=R^4=OH
(1346) Ohioensin E ; R^1=R^5=H, R^2=Me, R^3=OH, R^4=OMe

Chart 80. Benzonaphthoxanthenones found in the Musci

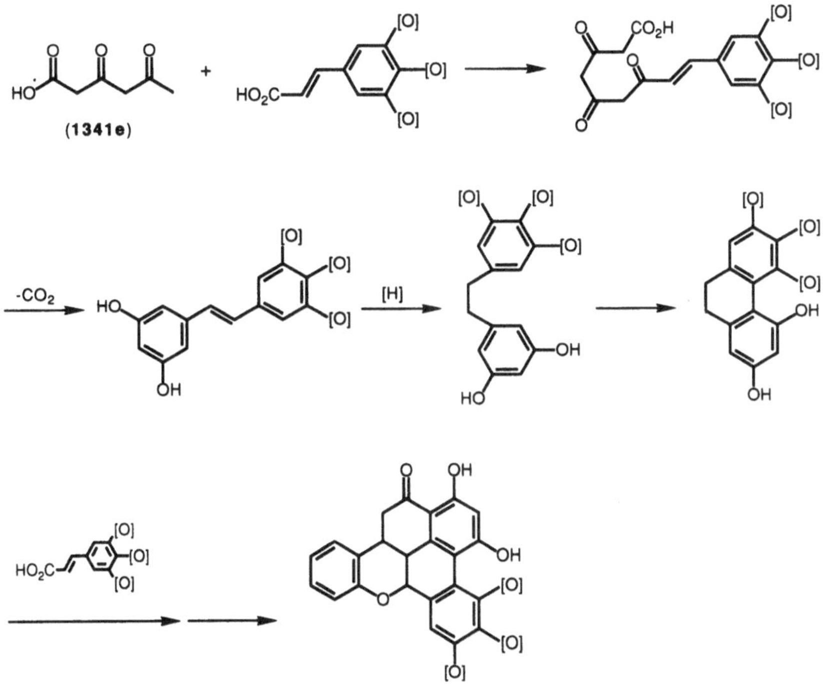

(1342 - 1346) Ohioensins

Scheme 68. Proposed biogenesis of ohioensins

3.3 Benzonaphthoxanthenones

Polytrichum species contain not only highly unsaturated fatty acids but also benzonaphthoxanthenones. From 95% ethanol extract of American *P. ohioense*, five novel benzonaphthoxanthenones, ohioensins A–E **(1342–1346)** have been isolated *(645a, 645b)*. The stereostructure of ohioensin A **(1342)** was established by a combination of spectrometry (UV, IR, MS, 2D-COSY, NOE and CD) and X-ray crystallographic analysis *(645a)*. Structures of the remaining compounds **(1343–1346)** were determined by comparing the spectral data of **(1343–1346)** with those of **(1342)** and by chemical correlation (methylation). Comparison of the CD spectra of ohioensins B–E **(1343–1346)** with that of ohioensin A **(1342)** indicated that they had the same configurations *(645b)*. Biogenesis of the ohioensins may involve condensation of *o*-hydroxycinnamate with hydroxylated phenanthrenes or 9,10-dihydrophenanthrenes as shown in Scheme 68 *(645b)*.

4. Lipids

4.1 n-Alkanes and Related Compounds

Mosses contain odd and even-membered *n*-alkanes in the range, C15–C35. The predominant *n*-alkanes of mosses are *n*-C27, 29, 31. The relative proportion of odd- and even membered *n*-alkanes of mosses is similar to that in ferns and higher plants *(271)*.

Species belonging to the Splachnaceae contain not only monoterpene hydrocarbons but also other volatile components, such as octane derivatives and lower fatty acids *(344)*. Octanal **(1347)**, 3-octanone **(1216)**, 3-octanol **(1352)**, *trans*-2-octenal **(1348)**, 1-octen-3-ol **(1213)** which is widely distributed in liverworts, 1-octanol **(1212)** and 2-octen-1-ol **(1350)** have been detected in *Splachnum luteum, S. sphaericum*. 2-Octanol **(1351)** and 2-ethylhexanal **(1349)** have also been detected in *S. sphaericum*, while *S. melanocaulon, S. vasculosum* and *Aplodon wormskioldii* produce 3-octanol. *A. wormskioldii, S. vasculosum, S. sphaericum, S. rubrum, Tayloria tenuis* and *Tetraplodon mniodes* also contain 1-octen-3-ol **(1213)**. Ethyl acetate, 2-hexanol, 6-methyl-5-hepten-2-one and 2,2-dimethyl-4-pentanol (or another isomer) and ethyl 2,3-dimethyl butylate have been detected in a few species of Splachnaceae.

Fontinalis antipyretica is a large moss growing in water which emits strong odor on being dried at room temperature *(100)*. In the ether extract, 10 compounds have been detected by GC-MS. *n*-Hexanal is the major component and is responsible for the characteristic odor. In addition, the presence of ethanal, ethyl formate, ethyl acetate, ethanol, 2-heptanone, ethyl hexanoate, hexyl acetate, octanone and ethyl heptanoate was indicated by a combination of GC-MS and GC cochromatography with authentic specimens (100). The distribution of n-alkanes, alkanals, alkanols, alkanones and alkanoates found in mosses is shown in Table IIId.

4.2 Fatty Acids

The lower fatty acids, acetic, propionic, butylic, valeric, caproic, isovaleric, phenylacetic and cyclohexylcarboxylic acids have been detected in several species of Splachnaceae as shown in Table IIId. Palmitic acid which is widespread in liverworts has been detected in *Aplodon wormskioldii*, *Splachnum sphaericum* and *S. rubrum* (344). *Fontinalis antipyretica* produces tetracosanoic acid together with volatile components (100).

Analytical methods for the lipids in bryophytes have been reviewed by BEUTELMANN et al. (105). PFAFFMANN and HARTMANN (465) reported an analytical technique for phospholipids and an assay for phospholipases from mosses. GC-MS analyses of ether-soluble carboxylic, phenolic, alkanoic, hydroxyalkanoic, hydroxyalkanedioic and dihydroxyalkanoic acids, liberated on saponification of solvent-extracted tissues of *Sphagnum palustre* and *S. cuspidatum* have been summarized by CALDICOTT and EGLINTON (128). *S. palustre* contains 16-hydroxyhexadecanoic and 10,16-dihydroxyhexadecanoic acids as the major components while 7-hydroxyhexadecanedioic and 8,16-dihydroxyhexadecanoic acids are the major fatty acids of *S. cuspidatum*.

Most mosses contain considerable amounts of long chain poly-unsaturated fatty acids, particularly acids with four and five double bonds, such as eicosatetraenoic [arachidonic (ARA)] (1223) and 5,8,11,14,17-eicosapentaenoic acids (EPA) (1224) (7, 201). Some mosses also contain acetylenic fatty acids, mainly 9,12,15-octadecatrien-6-ynoic (18:3ω3) (1222) as the major component, along with lesser amounts of 9,12-octadecadien-6-ynoic acid (18:3ω6) (1221) and 11,14-eicosadien-8-ynoic acid (1353) (7, 8). Neither ARA and EPA nor acetylenic fatty acids occur in higher plants in which fatty acids with two and three double bonds, octadecadienoic (= linoleic, 18:2) and octadecatrienoic (= linolenic, 18:3) are the predominant fatty acid constituents.

The fatty acids in *Polytrichum juniperinum*, *Hedwigia ciliata*, *Hylocomium splendens*, *Sphagnum* and *Brachithecium* + *Mnium* have been studied

Table IIId. *Chemical Constituents (Lipids and Miscellaneous) Found in the Musci*

Structure number	Name of compounds	Formula	$[\alpha]_D$	m.p.°C	Plant source	Subclass	References	Comments
(1100)	Indole acetic acid	$C_{10}H_9O_2N$			*Funaria hygrometrica*	B	(110)	
					Physcomitrella patens	B	(110)	
					Polytrichum formosum	P	(110)	
(1212)	n-Octanol	$C_8H_{18}O$			*Aplodon wormskioldii*	B	(344)	
					Splachnum luteum	B	(344)	
					S. rubrum	B	(344)	
					S. sphaericum	B	(344)	
					S. vasculosum	B	(344)	
(1213)	1-Octen-3-ol	$C_8H_{16}O_2$			*Aplodon wormskioldii*	B	(344)	
					Sphagnum luteum	S	(344)	
					S. sphaericum	S	(344)	
					S. vasculosum	S	(344)	
					Taykirua tenuis	B	(344)	
					Tetraplodon mniodes	B	(344)	
(1216)	3-Octanone	$C_8H_{16}O$			*Splachnum luteum*	B	(344)	
					S. sphaericum	B	(344)	
(1221)	9,12-Octadeca-dien-6-ynoic acid	$C_{18}H_{28}O_2$			*Amphidium mougeotii*	B	(338)	
					Ceratodon purpureus	B	(338)	
					Cymodontium polycarpum	B	(338)	
					C. strumiferum	B	(338)	
					Dicranella palustris	B	(338)	
					Dicranum fulvum	B	(338)	
					D. japonicum	B	(290, 291)	
					D. muehlembeckii	B	(338)	
					D. scoparium	B	(290, 291, 338)	
					D. spurium	B	(338)	

(1222)	9,12,15-Octadecatrien-6-ynoic acid	$C_{18}H_{26}O_2$	D. undulatum	B	(338)
			D. viride	B	(338)
			Dicranoweisia cirrata	B	(338)
			Ditrichum heteromallum	B	(338)
			Fontinalis antipyretica	B	(307)
			Orthodicranum montanum	B	(338)
			Pleuridium subulatum	B	(338)
			Rhabdiweisia crispata	B	(338)
			R. fugax	B	(338)
			Amphidium mougeotii	B	(338)
			Ceratodon purpureus	B	(8, 338)
			Cynodontium polycarpum	B	(338)
			C. strumiferum	B	(338)
			Dichodontium pellucidum	B	(338)
			Dicranella heteromalla	B	(338)
			D. jamesonii	B	(338)
			D. palustris	B	(338)
			D. schreberiana	B	(338)
			Dicranum fulbum	B	(338)
			D. fuscescens	B	(338)
			D. japonica	B	(290, 291)
			D. montanum	B	(8)
			D. muehlenbeckii	B	(338)
			D. polycetum	B	(338)
			D. scoparium	B	(290, 291, 338)
			D. spurium	B	(338)
			D. undulatum	B	(338)
			D. viride	B	(338)
			Dicranoweisia cirrata	B	(338)

Table IIId (continued)

Structure number	Name of compounds	Formula	m.p.°C	$[\alpha]_D$	Plant source	Subclass	References	Comments
					Dicranoloma cylindrothecium	B	(290)	
					Ditrichum heteromallum	B	(338)	
					D. pusillum	B	(338)	
					Fissidens areolatus	B	(290)	
					F. nobilis	B	(290)	
					Fontinalis antipyretica	B	(307)	
					Kiaeria starkei	B	(338)	
					Orthodicranum montanum	B	(338)	
					Pleuridium subulatum	B	(338)	
					Rhabdiweisia crispata	B	(338)	
					R. fugax	B	(338)	
					Dicranum scoparium	B	(290, 291)	
					Amphidium mougeotii	B	(338)	
					Atrichum undulatum	P	(500)	
					Aulacomnium androgynum	B	(500)	
					A. palustre	B	(500)	
					Bartramia pomiformis var. elongata	B	(500)	
					Blindia acuta	B	(338)	
					Brachythecium buchananii	B	(503)	
					B. reflexum	B	(500)	
					Brotherella henonii	B	(503)	
1-Monoglyceride of (**1222**)								
(**1223**)	Arachidonic acid	$C_{20}H_{32}O_2$						

Species		Reference
Calliergon cordifolium	B	(500)
Campylopus atrovirens	B	(338)
C. flexuosus	B	(338)
C. introflexus	B	(338)
C. japonicus	B	(503)
C. pyriformis	B	(338)
C. richardii	B	(503)
C. substramineus	B	(338)
Ceratodon purpureus	B	(8, 338, 500)
Climacium dendroides	B	(500)
Cratoneuron filicinum	B	(503)
Ctenidium percrassum	B	(503)
Cynodontium polycarpum	B	(338, 500)
C. strumiferum	B	(338)
Dicranella heteromalla	B	(338)
D. jamesonii	B	(338)
D. palustris	B	(338)
D. schreberiana	B	(338)
Dicranodontium denudatum	B	(338)
Dicranoloma cylindrothecium	C	(290)
Dicranum elongatum	B	(500)
D. fulvum	B	(338)
D. fuscescens	B	(338)
D. japonicum	B	(290, 291, 503)
D. montanum	B	(8)
D. muehlenbeckii	B	(338)
D. polycetum	B	(338)

Table IIId *(continued)*

Structure number	Name of compounds	Formula	m.p.°C	$[\alpha]_D$	Plant source	Subclass	References	Comments
					D. scoparium	B	(290, 291, 338, 500)	
					D. spurium	B	(338)	
					D. undulatum	B	(338)	
					D. viride	B	(338)	
					Dicranoweisia cirrata	B	(338)	
					D. crispula	B	(338)	
					Distichium capillaceum	B	(338)	
					Ditrichum heteromallum	B	(338)	
					D. pusillum	B	(338)	
					D. flexicaule	B	(338, 500)	
					D. cylindricum	B	(338)	
					Dolichomitra cymbifolia	B	(503)	
					Drepanocladus aduncus	B	(500)	
					D. tundrae	B	(500)	
					Grimmia pilifera	B	(503)	
					Encalypta streptocarpa	B	(500)	
					Fissidens adiantoides	B	(500)	
					F. areolatus	B	(290)	
					F. nobilis	B	(290, 503)	
					Fontinalis antipyretica	B	(307, 500)	
					Funaria hygrometrica	B	(500)	
					Hedwigia ciliata	B	(202, 503)	
					Herzogiella selogeri	B	(500)	
					Homalothecium sericeum	B	(500)	

Species		
Hylocomium brevirostre var. *cavifolium*	B	(503)
H. splendens	B	(202, 500)
H. splendens var. *alaskanum*	B	(500)
Hypnum cupressiforme	B	(503)
H. fujiyamae	B	(503)
H. lindbergii	B	(503)
H. oldhamii	B	(503)
H. plumaeforme	B	(503)
Leptobryum pyriforme	B	(500)
Leucobryum glaucum	B	(338, 500)
L. neilgherrense	B	(503)
L. scabrum	B	(503)
Mnium cuspidatum	B	(500)
M. hornum	B	(500)
M. medium	B	(500)
Myuroclada maximoviczii	B	(500)
Onocophorus crispofolius	B	(503)
Orthodicranum montanum	B	(338)
Palaleucobryum longifolium	B	(338)
Plagiomnium ellipticum	B	(500)
P. maximoviczii	B	(503)
Plagiothecium euryphyllum	B	(503)
P. laetum	B	(500)
P. nemorale f. *japonicum*	B	(503)
Pleuridium subulatum	B	(503)
Pleurozium schreberi	B	(500)
Pogonatum inflexum	P	(503)
P. urnigerum	P	(500)

Table IIId (continued)

Structure number	Name of compounds	Formula	m.p.°C	$[\alpha]_D$	Plant source	Subclass	References	Comments
					Pohlia nutans	B	(500)	
					Polytrichum commune	P	(500)	
					P. juniperinum	P	(202, 500)	
					Racomitrium canescens	B	(503)	
					R. heterostichum	B	(500)	
					Rhabdiweisia crispata	B	(338)	
					R. fugax	B	(338)	
					Rhizogonium spiniforme var. badakense	B	(503)	
					Rhizomnium punctatum	B	(500)	
					R. tuomikoskii	B	(503)	
					Rhodobryum roseum	B	(500)	
					Rhytidiadelphus squarrosus	B	(500)	
					Schlotheimia japonica	B	(503)	
					Scopelophila cataractae	B	(503)	
					Sphagnum angustifolium	S	(500)	
					S. cuspidatum	S	(500)	
					S. fimbriatum	S	(500)	
					S. magellanicum	S	(500)	
					S. nemoreum	S	(500)	
					S. palustre	S	(503)	
					Thamnobryum plicatulum	B	(503)	
					Thuidium glaucinum	B	(503)	
					T. recognitum	B	(500)	

(1224)	5,8,11,14,17-Eicosapentaenoic acid	$C_{20}H_{30}O_2$	
	T. recognitum var. delicatulum	B	(503)
	T. tamariscinum	B	(503)
	Tortula muralis	B	(500)
	Vesicularia ferriei	B	(503)
	Wijkia concavifolia	B	(503)
	Amphidium mougeotii	B	(338)
	Atrichum undulatum	P	(500)
	Aulacomnium androgynum	B	(500)
	A. palustre	B	(500)
	Bartramia pomiformis var. elongata	B	(503)
	Blindia acuta	B	(338)
	Brachythecium buchananii	B	(503)
	B. reflexum	B	(500)
	Brotherella henonii	B	(503)
	Calliergon cordifolium	B	(500)
	Campylopus atrovirens	B	(338)
	C. richardii	B	(503)
	C. flexuosus	B	(338)
	C. introflexus	B	(338)
	C. pyriformis	B	(338)
	C. substramineus	B	(338)
	Ceratodon purpureus	B	(338, 500)
	Climacium dendroides	B	(500)
	Cratoneuron filicinum	B	(503)
	Ctenidium percrassum	B	(503)
	Cynodontium polycarpum	B	(338, 500)

Table IIId (continued)

Structure number	Name of compounds	Formula	m.p.°C	$[\alpha]_D$	Plant source	Subclass	References	Comments
					C. strumiferum	B	(338)	
					Dichodontium pellucidum	B	(338)	
					Dicranella jamesonii	B	(338)	
					D. palustris	B	(338)	
					Dicranodontium denudatum	B	(338)	
					Dicranoloma cylindrothecium	B	(290)	
					Dicranoweisia cirrata	B	(338)	
					D. crispula	B	(338)	
					Dicranum elongatum	B	(500)	
					D. fulvum	B	(338)	
					D. japonicum	B	(290, 291, 503)	
					D. montanum	B	(8)	
					D. muehlenbeckii	B	(338)	
					D. polycetum	B	(338)	
					D. spurium	B	(338, 500)	
					D. undulatum	B	(338)	
					D. viride	B	(338)	
					Distichium capillaceum	B	(338)	
					Ditrichum heteromallum	B	(338, 500)	
					D. pusillum	B	(338)	
					D. flexicaule	B	(338)	
					D. cylindricum	B	(338)	
					Dolichomitra cymbifolia	B	(503)	
					Drepanocladus aduncus	B	(500)	

D. tundrae	B	(500)
Encalypta streptocarpa	B	(500)
Fissidens adiantoides	B	(500)
F. areolatus	B	(290)
F. nobilis	B	(290)
Fontinalis antipyretica	B	(307, 500)
Funaria hygrometrica	B	(500)
Grimmia pilifera	B	(503)
Hedwigia ciliata	B	(503)
Herzogiella selogeri	B	(500)
Homalothecium sericeum	B	(500)
Hylocomium	B	(503)
brevirostre var.		
cavifolium		
H. splendens var.	B	(500)
alaskanum		
H. suspendens	B	(500)
Hypnum cupressiforme	B	(500)
H. fujiyamae	B	(503)
H. lindbergii	B	(503)
H. oldhamii	B	(503)
H. plumaeforme	B	(503)
Kiaeria starkei	B	(338)
Leptobryum pyriforme	B	(500)
Leucobryum glaucum	B	(338, 500)
L. neilgherrense	B	(503)
L. scabrum	B	(503)
Mnium cuspidatum	B	(500)
M. hornum	B	(500)
M. medium	B	(500)
Myuroclada maximoviczii	B	(503)

Table IIId (*continued*)

Structure number	Name of compounds	Formula	[α]$_D$	m.p.°C	Plant source	Subclass	References	Comments
					Onocophorus crispofolius	B	(503)	
					Orthodicranum montanum	B	(338)	
					Palaleucobryum longifolium	B	(500)	
					Plagiomnium ellipticum	B	(500)	
					P. maximoviczii	B	(503)	
					Plagiothecium euryphyllum	B	(503)	
					P. laetum	B	(500)	
					P. nemorale f. *japonicum*	B	(503)	
					Pleuridium subulatum	B	(338)	
					Pleurozium schreberi	B	(500)	
					Pogonatum inflexum	P	(503)	
					P. urnigerum	P	(500)	
					Polytrichum commune	P	(500)	
					P. juniperinum	P	(500)	
					Racomitrium canescens	B	(503)	
					R. heterostichum	B	(500)	
					Rhabdiweisia crispata	B	(338)	
					R. fugax	B	(338)	
					Rhizogonium spiniforme var. *badakense*	B	(503)	
					Rhizomnium punctatum	B	(500)	
					R. tuomikoskii	B	(503)	
					Rhodobryum roseum	B	(500)	

No.	Compound	Formula	Species		Ref.
			Rhytidiadelphus squarrosus	B	(500)
			Scopelophila cataractae	B	(503)
			Sphagnum angustifolium	S	(500)
			S. cuspidatum	S	(500)
			S. fimbriatum	S	(500)
			S. magellanicum	S	(500)
			S. nemoreum	S	(500)
			S. palustre	S	(503)
			Thamnobryum plicatulum	B	(503)
			Thuidium glaucinum	B	(503)
			T. recognitum	B	(500)
			T. recognitum var. delicatulum	B	(503)
			T. tamariscinum	B	(503)
			Tortula muralis	B	(500)
			Vesicularia ferriei	B	(503)
			Wijkia concavifolia	B	(503)
(1347)	*n*-Octanal	$C_8H_{16}O$	*Splachnum luteum*	B	(344)
			S. sphaericum	B	(344)
(1348)	*trans*-2-Octenal	$C_8H_{14}O$	*Aplodon wormskioldii*	B	(344)
			Splachnum luteum	B	(344)
			S. sphaericum	B	(344)
(1349)	2-Ethylhexanal	$C_8H_{16}O$	*Splachnum sphaericum*	B	(344)
(1350)	2-Octen-1-ol	$C_8H_{16}O$	*Splachnum sphaericum*	B	(344)
			S. vasculosum	B	(344)
(1351)	2-Octanol	$C_8H_{18}O$	*Splachnum sphaericum*	B	(344)
(1352)	3-Octanol	$C_8H_{18}O$	*Aplodon wormskioldii*	B	(344)
			Splachnum melanocaulon	B	(344)
			S. sphaericum	B	(344)
			S. vasculosum	B	(344)

Table IIId (continued)

Structure number	Name of compounds	Formula	m.p.°C	$[\alpha]_D$	Plant source	Subclass	References	Comments
	Ethanol	C_2H_6O			*Fontinalis antipyretica*	B	(100)	
	Ethanal	C_2H_4O			*Fontinalis antipyretica*	B	(100)	
	Acetic acid	$C_2H_4O_2$			*Aplodon wormskioldii*	B	(344)	
					Splachnum luteum	B	(344)	
					S. rubrum	B	(344)	
					S. sphaericum	B	(344)	
	Propionic acid	$C_3H_6O_2$			*Aplodon wormskioldii*	B	(344)	
					Splachnum luteum	B	(344)	
					S. sphaericum	B	(344)	
	Butyric acid	$C_4H_8O_2$			*Aplodon wormskioldii*	B	(344)	
					Splachnum luteum	B	(344)	
					S. melanocaulon	B	(344)	
					S. sphaericum	B	(344)	
	Valeric acid	$C_5H_{10}O_2$			*Aplodon wormskioldii*	B	(344)	
					Splachnum luteum	B	(344)	
					S. sphaericum	B	(344)	
					Splachnum rubrum	B	(344)	
	Isovaleric acid	$C_5H_{10}O_2$			*Aplodon wormiskioldii*	B	(344)	
	Caproic acid	$C_6H_{12}O_2$			*Fontinalis antipyretica*	B	(100)	
	Ethyl acetate	$C_4H_8O_2$			*Splachnum vasculosum*	B	(344)	
	Ethyl formate	$C_3H_6O_2$			*Fontinalis antipyretica*	B	(100)	
	2-Hexanol	$C_6H_{14}O$			*Aplodon wormskioldii*	B	(344)	
					Fontinalis antipyretica	B	(100)	
					Splachnum vasculosum	B	(344)	
	6-Methyl-5-hepten-2-one	$C_8H_{14}O$			*Splachnum rubrum*	B	(344)	

No.	Compound	Formula	Species		Ref.	CD
	2,2-Dimethyl-4-pentanol	$C_7H_{16}O$	Splachnum rubrum	B	(344)	
	n-Hexanal	$C_6H_{12}O$	Fontinalis antipyretica	B	(100)	
	Hexyl acetate	$C_8H_{16}O_2$	Fontinalis antipyretica	B	(100)	
	Ethyl hexanoate	$C_8H_{16}O_2$	Fontinalis antipyretica	B	(100)	
	2-Heptenone	$C_7H_{14}O$	Fontinalis antipyretica	B	(100)	
	Ethyl heptanoate	$C_9H_{18}O_2$	Fontinalis antipyretica	B	(100)	
	Ethyl 2,3-dimethyl butylate	$C_8H_{16}O_2$	Aplodon wormskioldii	B	(344)	
(1353)	11,14-Eicosadien-8-ynoic acid	$C_{20}H_{32}O_2$	Fontinalis antipyretica	B	(307)	
(1354)	7,10,13,16,19-Docosapentaenoic acid	$C_{22}H_{34}O_2$	Campylium stellatum	B	(8)	
			Cratoneuron filicinum	B	(8)	
			Drepanocladus exannulatus	B	(8)	
(1355)	10,13-Nonadecadien-7-yn-2-one	$C_{19}H_{30}O$	Fontinalis antipyretica	B	(307)	
(1356)	10,13,16-Nonadeca-trien-7-yn-2-one	$C_{19}H_{28}O$	Fontinalis antipyretica	B	(307)	
(1357) 19.1	Dicranenone A	$C_{18}H_{24}O_3$	Dicranoloma cylindrothecium	B	(290)	CD 225 nm +24.1 CD 320 nm − 1.3
			Dicranum japonicum	B	(290, 291)	
			D. majus	B	(290)	
			D. scoparium	B	(290, 291)	
			Fissidens areolatus	B	(290)	
			Leucobryum scabrum	B	(290)	
(1358)	12-Oxo-PDA 1	$C_{18}H_{28}O_3$				

Table IIId (*continued*)

Structure number	Name of compounds	Formula	m.p.°C	$[\alpha]_D$	Plant source	Subclass	References	Comments
(1359)	13-Hydroxy-9,11,15-octadecatrien-6-ynoic acid	$C_{18}H_{26}O_3$			*Dicranum scoparium*	B	(*290, 291*)	
1,2-Diglyceride of (1359)					*Dicranum scoparium*	B	(*290, 291*)	
(1360)	Dicranenone B	$C_{18}H_{24}O_3$			*Dicranoloma cylindrothecium*	B	(*290*)	
					Dicranum majus	B	(*290*)	
					Fissidens areolatus	B	(*290*)	
(1361)	Dicranenone B1	$C_{18}H_{24}O_3$		+ 55	*Dicranoloma cylindrothecium*	B	(*290*)	
					Dicranum japonicum	B	(*290*)	
					D. majus	B	(*290*)	
(1362)	Dihydrodicranenone B	$C_{18}H_{26}O_3$			*Dicranoloma cylindrothecium*	B	(*290*)	
(1363)	12-Oxo-PDA 3	$C_{18}H_{28}O_3$			*Dicranum majus*	B	(*290*)	
					Leucobryum scabrum	B	(*290*)	
(1365)	Methyl veratrate (Methyl 3,4-dimeth-oxy benzoate)	$C_{10}H_{12}O_4$			*Dendroligotrichum dendroides*	B	(*404*)	Treated with NaOH Na$_2$S
					Dawsonia grandis	B	(*404*)	
					D. longiseta	B	(*404*)	
					D. papuana	B	(*404*)	
					D. polytrichoides	B	(*404*)	
					D. superba	B	(*404*)	

No.	Compound	Species		Ref.	Notes
(1366)	Dimethyl isohemipate $C_{12}H_{14}O_6$	Polytrichadelphus magellanicum	B	(404)	
		Polytrichum commune	P	(404)	
		Dendroligotrichum dendroides	B	(404)	Treated with NaOH Na_2S
		Dawsonia grandis	B	(404)	
		D. longiseta	B	(404)	
		D. papuana	B	(404)	
		D. polytrichoides	B	(404)	
		D. superba	B	(404)	
		Polytrichadelphus magellanicum	B	(404)	
(1367)	Dimethyl metahemipate $C_{12}H_{14}O_6$	Polytrichum commune	P	(404)	
		Dendroligotrichum dendroides	B	(404)	Treated with NaOH Na_2S
		Dawsonia grandis	B	(404)	
		D. longiseta	B	(404)	
		D. papuana	B	(404)	
		D. polytrichoides	B	(404)	
		D. superba	B	(404)	
		Polytrichadelphus magellanicum	B	(404)	
(1368)	Methyl 4,7,9-trimethoxy-2-dibenzofurancarboxylate $C_{17}H_{16}O_6$	Polytrichum commune	P	(404)	
		Dendroligotrichum dendroides	B	(404)	Treated with NaOH Na_2S
		Dawsonia grandis	B	(404)	
		D. longiseta	B	(404)	
		D. papuana	B	(404)	
		D. polytrichoides	B	(404)	
		D. superba	B	(404)	

Table IIId (*continued*)

Structure number	Name of compounds	Formula	m.p.°C	$[\alpha]_D$	Plant source	Subclass	References	Comments
					Polytrichadelphus magellanicum	B	(404)	
					Polytrichum commune	P	(404)	
(1369)	Ansamitocin P-3	$C_{32}H_{43}O_9N_2Cl$			*Isotecium subdiversiforme*	B	(479, 644)	Originated from soil microorganisms
					Thamnobryum sandei	B	(479, 644)	
(1370)	15-Methoxyansamitocin P-3	$C_{33}H_{45}O_{10}N_2Cl$			*Isotecium subdiversiforme*	B	(479, 644)	Originated from soil microorganisms
					Thamnobryum sandei	B	(479, 644)	
(1371)	Maytanbutine	$C_{36}H_{50}O_{10}N_3Cl$			*Isotecium subdiversiforme*	B	(479, 644)	Originated from microorganisms
					Thamnobryum sandei	B	(479, 644)	
(1372)	Trewiasine	$C_{37}H_{52}O_{11}N_3Cl$			*Isotecium subdiversiforme*	B	(479, 644)	Originated from soil microorganisms
					Thamnobryum sandei	B	(479, 644)	
(1373)	Deoxypodophyllotoxin	$C_{18}H_{16}O_7$			*Isotecium subdiversiforme*	B	(644)	Originated from soil microorganisms
					Thamnobryum sandei	B	(644)	
(1374)	Pheophytin a	$C_{24}H_{47}O_4N$			*Thuidium kanedae*	B	(448a)	
(1375)		$C_{55}H_{74}O_5N_4$			*Entodon rubicundus*	B	(448a)	

(1376)	13²-Hydroxy-(13²-R)-pheophytin a	$C_{55}H_{74}O_6N_4$	*Entodon rubicundus*	B	*(448a)*	
(1377)	13²-Hydroxy-(13²-S)-pheophytin a	$C_{55}H_{74}O_6N_4$	*Entodon rubicundus*	B	*(448a)*	
(1378)	13²-Hydroxy-(13²-R)-pheophytin b	$C_{55}H_{72}O_7N_4$	*Entodon rubicundus*	B	*(448a)*	
(1379)	13²-Hydroxy-(13²-S)-pheophytin b	$C_{56}H_{72}O_7N_4$	*Entodon rubicundus*	B	*(448a)*	
(1380)	13²-(MeOO)-(13²-R)-pheophytin a	$C_{56}H_{76}O_7N_4$	*Entodon rubicundus*	B	*(448a)*	
	N^6-(Δ^2-isopentenyl)-adenine (2ip)		*Funaria hygrometrica* x *Physcomitrella patens*	B	*(110, 611)*	Hybrid culture
(1396)	Acetylcholine	$C_7H_{17}O_3N$	*Funaria hygrometrica* x *Physcomitrella patens*	B	*(110, 611)*	Hybrid culture
(1401)	α-Tocoquinone	$C_{29}H_{50}O_3$	*Racomitrium japonicum*	B	*(448a)*	

B: Bryidae; P: Polytrichidae; S: Sphagnidae

by GC (*202*). The lipids obtained from the CHCl$_3$-MeOH extracts of each material were saponified with KOH. After removal of unsaponifiable components, the fatty acids were esterified with diazomethane. From 1 kg of mixed mosses consisting of *Mnium* and *Brachythecium* species, 1.3 g of the pure fatty acid methyl esters was obtained. The individual mosses contained 10% and 35% of arachidonic acid (**1223**), together with C14:0, 16:0, 16:1, 16:3, 18:0, 18:1, 18:2, 18:3, 18:4, 20:0, 20:1, 20:2, 20:3, 20:5, 22:0 and 24:0 among which C18:3 predominated.

The fatty acid composition of Polytrichaceae (4 species), Grimmi-aceae (2), Ditrichaceae (4), Dicranaceae (1), Mniaceae (4) and Amblystegi-aceae (7) has been analyzed by GC (*8*). All species contained C18:1 (oleic), 18:2 (linoleic), 18:3 (linolenic), 20:4 (arachidonic acid: ARA) (**1223**) and 20:5 (eicosapentaenoic acid: EPA) acids (**1224**). ARA (10–29%) and EPA (1–19%) as well as 18:3 (16–21%) were the predominant fatty acids in the Mniaceae. *Dicranum montanum* (Dicranaceae) and *Ceratodon purpureus* (Ditrichaceae) yielded an acetylenic fatty acid, 9,12,15-octadecatrien-6-ynoic acid (**1222**), the former in approx. 10%, the latter in 19–25% yield. *Cratoneuron filicinum*, *Campylium stellatum* and *Drepanocladus exannula-tus* (Amblystegiaceae) represented an interesting family because they produce C22:5 (7, 10, 13, 16, 19-docosapentaenoic acid) (**1354**) (*8*).

ANDERSON *et al.* (*8*) isolated 9,12,15-octadecatrien-6-ynoic acid (**1222**) from several mosses and showed that this acetylenic acid was present only in the triglyceride (TG) fraction. *Fontinalis antipyretica* also produces 9,12-octadeca-dien-6-ynoic acid (**1221**), 9,12,15-octadecatrien-6-ynoic acid (**1222**) and its C20 homologue, 11,14-eicosadien-8-ynoic acid (**1353**), as the major components of the triglycerides (*307*). These acids have been found neither in the monogalactosyl diglyceride (MGDG) nor in the digalactosyl diglyceride (DGDG) fraction. The total lipids contain satu-rated (C16:0) and monoenoic acids (C16:1-C24:1), dienoic acids (C16:ω6–C22ω6, in which 18:2ω6 is the major component), trienoic acids (C16:3ω3–20:3ω6, in which 18:3ω3 is major), tetraenoic acids (C18:4ω3–20:4ω3, in which 20:4ω5 is major), pentaenoic aids (C20:5ω3) and acetylenic acids (**1221**) (4–32%), (**1222**) (0.8–2.0%) and (**1353**) (94.5–10.1%). In addition to the above acetylenic acids, *F. antipyretica* elaborates 10,13-nonadecadien-7-yn-2-one (**1355**) and 10,13,16-nonadecatrien-7-yn-2-one (**1356**). The acetylenic fatty acid content in the triglycerides of the green and brown segments of *Dicranum elongatum* has been investigated by using capillary GC (*318*).

The fatty acids represented in the acyl lipids of 38 moss species from the four families, Ditrichaceae, Seligeriaceae, Dicranaceae and Leuco-bryaceae were studied by KOHN *et al.* (*338*). Twenty-one species of Dicranaceae and four species of Ditrichaceae contained acetylenic fatty

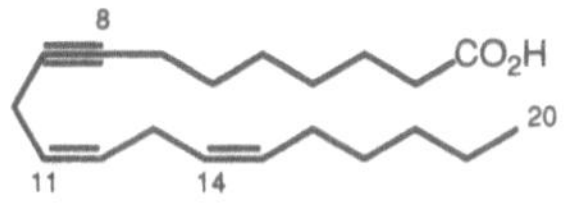

(1353) 11,14-Eicosadien-8-ynoic acid 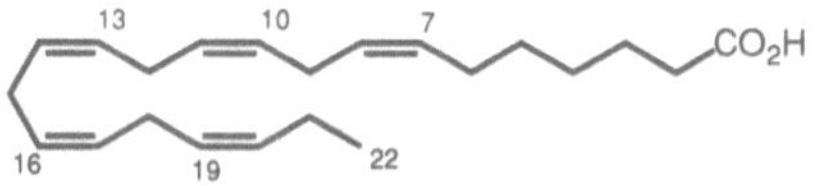   (1354) 7,10,13,16,19-Docosapentaenoic acid

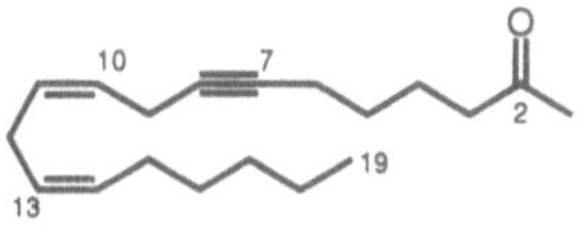

(1355) 10,13-Nonadecadien-7-yn-2-one 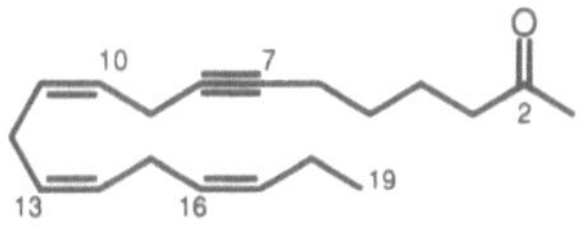 (1356) 10,13,16-Nonadecatrien-7-yn-2-one

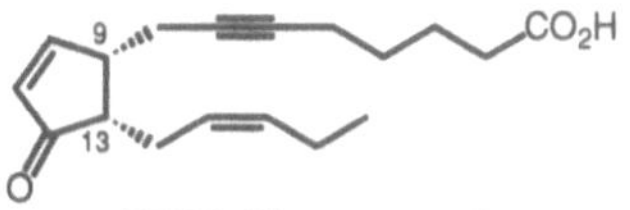

(1357) Dicranenone A 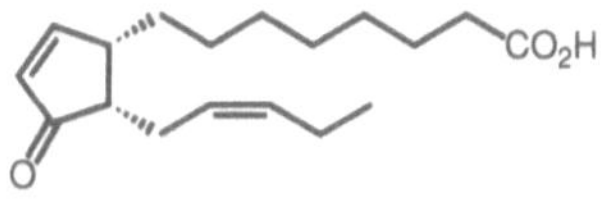   (1358) 12-Oxo-PDA 1

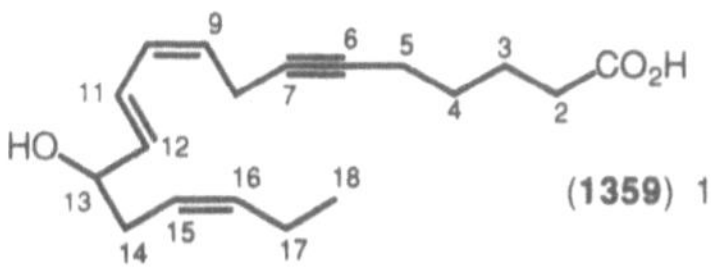

(1359) 13-Hydroxy-9,11,15-Octadecatrien-6-yonic acid

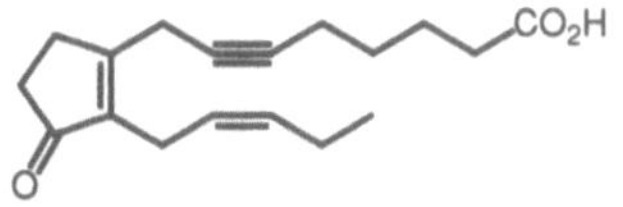

(1360) Dicranenone B 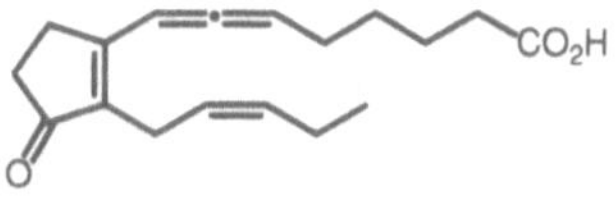   (1361) Dicranenone B1

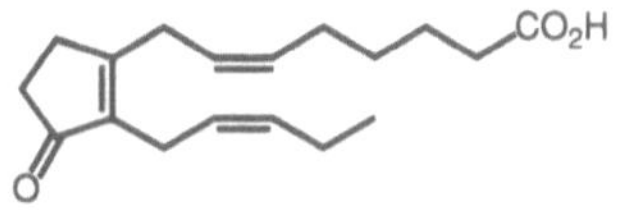

(1362) Dihydrodicranenone B 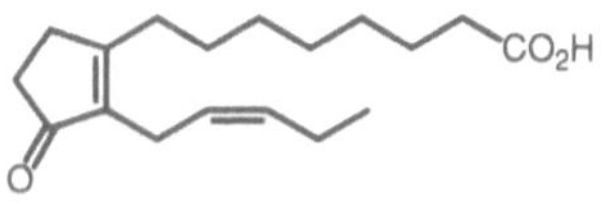   (1363) 12-Oxo-PDA 3

Chart 81. Fatty acids found in the Musci

acids as the acyl groups of the triglycerides. The major fatty acid in the triglycerides of both families was 9,12,15-octadecatrien-6-ynoic acid (1222) (10.0–88.3% mol% of the total fatty acids). 9,12-Octadecadien-6-ynoic acid (1221) was found in small amounts of less than 5 mol%. All five species of *Campylopus*, *Blindia acuta* (Seligeriaceae) and *Leucobryum glaucum* (Leucobryaceae) were free of acetylenic fatty acids.

A novel cyclopentenoyl fatty acid, dicranenone A (1357), has been isolated from the ether extract of fresh *Dicranum scoparium* together with the previously known 9Z,12Z-octadecadien-6-ynoic (**1221**), 9Z,12Z,15Z-octadecatrien-6-ynoic (**1222**) and 13-hydroxy-9Z,11E,15Z-octadeca-trien-6-ynoic acids (**1359**) and 1-monoglyceride of (**1222**) and 1,2-diglycer-ide of (**1359**) (*290, 291*). *Dicranum japonicum* furnished another new cyclopentenonic acid, dicranenone B1 (**1361**) which possesses an allenic group, together with (**1221, 1222, 1357**). The structure of (**1357**) was established by UV, IR, ^{13}C-NMR and mass spectrometry while the specific rotation and the CD spectrum suggested that the stereochemistry was the same as that of the prostaglandin A_2 (**1364**) series. Spin decoupling experiments suggested structure (**1360**) and its 10-oxo isomer as possible structures of dicranenone B1 with structure (**1360**) being preferred by analogy with (**1357**), but without specification of the stereochemistry of the allene. Acid (**1359**) corresponding to intermediate (**1222a**) has been detected as a minor constituent in *D. scoparium*. *D. majus* produces dicranenone (**1357**), dicranenone B (**1360**) and dicranenone B1 (**1361**) and 12-oxo-PDA 3 (**1363**) (*291*). *Dicranoloma cylindrothecium* also contains (**1357**), (**1360**) and (**1361**), along with dihydrodicranenone B (**1362**) and (**1222**) (*291*). An additional cyclopentenoyl fatty acid, 12-oxo-PDA 1 (**1358**) has been found in *Leucobryum scabrum*, together with 12-oxo-PDA 3 (**1363**) (*291*). *Fissidens areolatus* and *F. nobilis* elaborate 9,12,15-octadecatrien-6-ynoic acid (**1222**) and the former also produces (**1357**) and (**1360**) (*291*). It has been suggested that the acids having triple bonds are means of maximizing energy storage because acetylenic triacyglcerols are accumulated under conditions of stress (*290*).

(**1364**) PG A_2

The fatty acid composition of the lipids, steryl and wax esters of spores of the moss *Polytrichum commune* has been investigated by Karunen *et al.* (*312, 317*). The fatty acid composition of the polar lipids in *Ceratodon purpureus* and *Pleurozium schreberi* and of the mono- and diglycosyl diglyceride fractions of germinating *Polytrichum commune* spores have also been investigated using GC (*314, 316*).

The lipid content and the changes in fatty acid composition among glycolipids (MGDG and DGDG) of *Sphagnum fimbriatum* often reflect the light and temperature conditions of the habitat (*345*).

The distribution of acyl lipids and their component fatty acids in bryophytes has been reviewed by Karunen (*315*).

Polar lipids constitute the bilayer of plant membranes. Glycolipids comprise 45–50% of the monogalactosyl diacylglycerols (MGDG), 25–30% of the digalactosyl diacylglycerols (DGDG) and up to 10% of the sulfoquinovosyl diacetylglycerol (SQDG) of acyl lipids in higher plants. The major phospholipid of chloroplasts is phosphatidylglycerol (PG) and some phosphatidylcholine (PC). The extrachloroplastic lipids are PC and diphosphatidyl glycerol (DPG). These polar lipids are found universally in all plants. The mosses and liverworts differ from gymnosperms and angiosperms because they contain an additional zwitterionic lipid, diacyl-glyceryltrimethylhomoserine (DGTS). Analysis of the fatty acid moieties of the polar lipids in mosses and liverworts has shown the presence of highly unsaturated long chain fatty acids (C20–C22) in mosses and liverworts. The mosses can be classified into groups based the content of C20 polyenoates in the MGDG. Sphagnales and Polytrichales always contain a low proportion of C20 polyunsaturated fatty acids (C20:4 + C20:5) while the percentage of these acids is somewhat higher in Dicranales. Bryales (*Bryum, Rhodobryum, Aulacomnium*) also contain a low proportion of C20 polyenoates, while other species in the same order (*Herzogiella, Plagiothecium*) exhibit an extremely high level of these unsaturated acids. SEWÓN (*500*) reported the distribution of fatty acids of MGDG in 55 specimens of Bryophyta from 13 orders or suborders and found that the proportion of C20 polyenoates increased in the following order: liverworts < Sphagnales < Polytrichales < Bryales (haplolepideae lineage) < Bryales (diplolepideae pleurocarpous suborder Hypnineae). This order represents the phylogenetic branching sequence, Hypnineae being young taxa and liverworts the oldest branch (*500*).

The proportion of fatty acids in the DGDG resembles that found in the MGDG. The content of C20 polyenoates in the SQDG and the PG is very much lower than that in the MGDG and DGDG. This phenomenon is characteristic not only of bryophytes (Hepaticae and Musci) but also of ferns (Pteridophytes).

The proportion of C20 polyunsaturated fatty acids in the MGDG from some mosses changes depending on the season and the habitat. For example, the proportion of arachidonic acid is highest in summer and that of linolenic acid lowest in the MGDG from *Mnium* and *Dicranum* species, but in winter and spring, this is reversed. On the basis of the above chemical evidence it has been suggested that arachidonic acid and linolenic acid content in both mosses is controlled environmentally (*315*). The proportion of C20 polyunsaturated fatty acids may be influenced also by temperature, light intensity or length of daylight. For example, the level of C20 polyenoates in MGDG of *Polytrichum commune* protonemata is lower at higher than at lower light intensity. Darkness decreases the levels of all polyunsaturated fatty acids (16:3ω3 in the MGDG and 20:4ω6 in the PE) in *Leptobryum pyriforme* protonemata. Cold induces an increase in the proportion of all polar lipid classes and in the degree of unsaturation.

Triacylglycerols (TAG) which might be located in the globules of mosses spores and of leaf cells are reserve substances of mosses and an energy source for growth. The proportion of TAG in mosses is high in late winter or early spring when mosses are photosynthetically active. The fatty acid composition of the TAG in mosses consists of palmitic, stearic, oleic, linoleic and linolenic acids which are similar to those found in seed oils. The content of C20 polyunsaturated fatty acids is low in certain *Sphagnum*, *Polytrichum* and *Leptobryum* species, however, *Hylocomium*, *Mnium* and *Pleurozium* species include these unsaturated fatty acids in the TAG. Acetylenic fatty acids have been found in Bryaceae, Dicranaceae, Ditrichaceae, *Fontinalis antipyretica*, *F. squamosa*, *Tortula muralis* and *Schistostega pennata*.

The steryl esters of mosses are composed of C20 polyunsaturated fatty acids as well as oleic, linoleic and linolenic acids. Sphagnales contain arachidonic-rich steryl esters, however, the arachidonic acid content in the steryl ester fraction of the spores and protonemata of *Polytrichum commune* is low. This implies that the Polytrichales are highly evolved mosses.

The wax esters in mosses are comprised of fatty acids esterified with long chain aliphatic alcohols (common wax esters) or isoprenoid fatty acids (isoprenoid wax esters). The common esters are minor constituents of surface lipids whereas isoprenyl wax esters are found inside the cells. The composition of fatty acids and long chain alcohols of mosses (*Andrea*, *Dicranum*, *Pogonatum* and *Saelania*) after hydrolysis is very similar to that in higher plants, i.e. they consist of even numbered polyunsaturated acids (C16, 18, 20, 22, 24 and 26) and polysaturated alcohols (C22, 24, 26, 28 and 30). The major fatty acids and alcohols of isoprenoid wax esters found in the gametophytes of mosses (*Dicranum*, *Hygrohypnum*, *Mnium* and *Sphagnum* species) is phytenic acid (**781**) and phytol (**780**) or geranyl geraniol (**782**). It is thought that the surface lipids of mosses prevent water penetration of the leaf surfaces.

5. Miscellaneous

The endogenous plant hormone, indole acetic acid (**1100**) has been found in *Funaria hygrometrica*, *Physcomitrella patens* and *Polytrichum formosum* (*110*). Bryophytes are believed to be among the oldest of land plants and hence it is interesting to know whether they contain lignin or not (*176, 358*). The gametophytic stems of the giant mosses, *Dawsonia grandis, D. longiseta, D. papuana, D. polytrichoides, D. superba, Dendroligotrichum dendroides, Polytrichadelphus magellanicum* and *Polytrichum commune* were treated with $NaOH-Na_2S$ for the study on the presence of lignin (*404*). The solubilized material on methylation and oxidation with $KMnO_4$-$NalO_4$ in aq. NaOH, then with H_2O_2 in aq. Na_2CO_3 gave a mixture of aromatic carbocyclic acids, the major components of which were estimated as methyl esters by GC. The identified compounds were methyl 3,4-dimethoxybenzoate (= methyl veratrate) (**1365**), dimethyl isohemipate (**1366**), dimethyl metahemipate (**1367**) and methyl 4,7,9-trimethoxy-2-dibenzofurancarboxylate (**1368**). These mosses are devoid of lignin but contain another type of phenolic cell wall material. Furthermore cross polarization solid state ^{13}C-NMR spectrometry of *Dawsonia superba, Leucobryum candidum, Rhizomnium parramatense, Sphagnum cristatum* and *Thamnobryum pandum* showed that 1,3,5-trihydroxybenzene constituted the aromatic building block of these mosses. The aromatic founds in mosses can therefore not be described as lignin (*619*).

A novel 15-methoxyansamitocin P-3 (**1370**) has been isolated from two mosses, *Isothecium subdiversiforme* and *Thamnobryum sandei*, together with the known maytansinoids (**1369, 1371, 1372**) (*479*). Whole moss (20.3 kg) of *I. subdiversiforme* on extraction with ether furnished only trace amounts (estimated yield 50 µg) of (**1369–1372**). These compounds have also been obtained from *T. sandei* (6.1 kg), again only in trace amounts (in the final stage of purification, 10.1 mg of extract was subjected twice to HPLC and the eluates again purified by HPLC). The structure determination of (**1369–1372**) was carried out using ^1H-NMR spectroscopy (400 MHz, after FID accumulation of *ca.* 15,200–53,000 times). As the proportion of these macrocyclic compounds in the moss extracts was extremely low, it is possible that symbiotic microorganisms produce maytansinoids (*644*). In order to clarify this question, microorganisms from the mosses described above and from the surrounding soil were separately cultured and the chloroform extracts of the supernatants and broths checked for the presence of maytansinoids. However, the extract was inactive. On the other hand, the ether extract of the leaf mold which consisted the decayed mosses and a small amount of soil showed potent antitumor activity against P-388. Spectroscopic investigation showed the active compounds to be ansamitocin P-3 (**1369**) and deoxypodophyllotoxin (**1373**). The leaf mold originating from mosses apparently plays an important role in the regulation of the moss ecosystem since maytansinoids possess potent antimicrobial activity

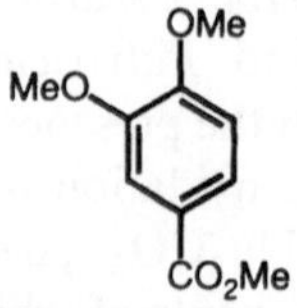

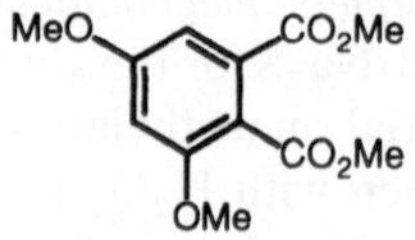

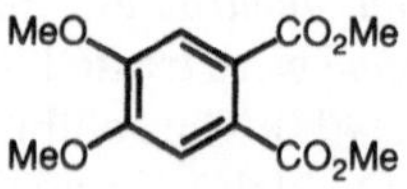

(**1365**) Methyl veratrate (**1366**) Dimethyl isohemipate (**1367**) Dimethyl metahemipate

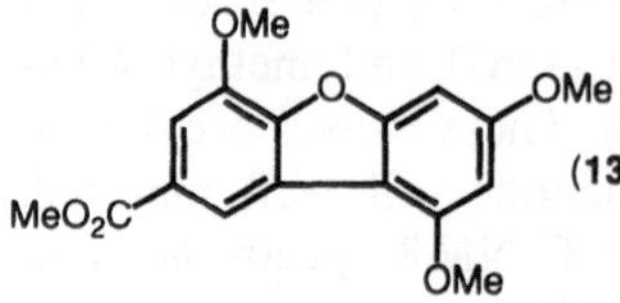

(**1368**) Methyl 4,7,9-trimethoxy-2-dibenzofurancarboxylate

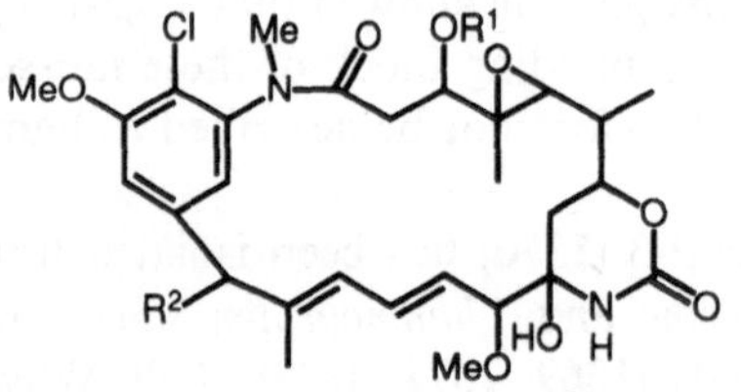

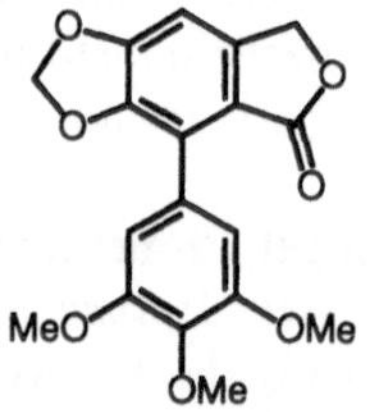

(**1373**) Deoxypodophyllotoxin

(**1369**) Ansamitocin P-3 ; R^1 = 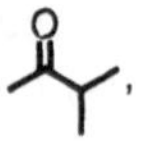 , R^2=H

(**1370**) 15-Methoxyansamitocin P-3 ; R^1 = 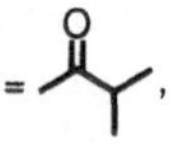 , R^2=OMe

(**1371**) Maytanbutine ; R^1 = 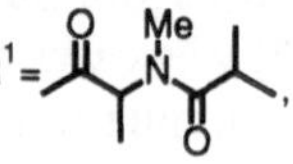 , R^2=H

(**1372**) Trewiasine ; R^1 = 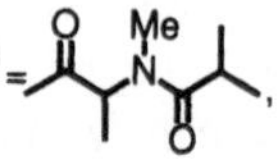 , R^2=OMe

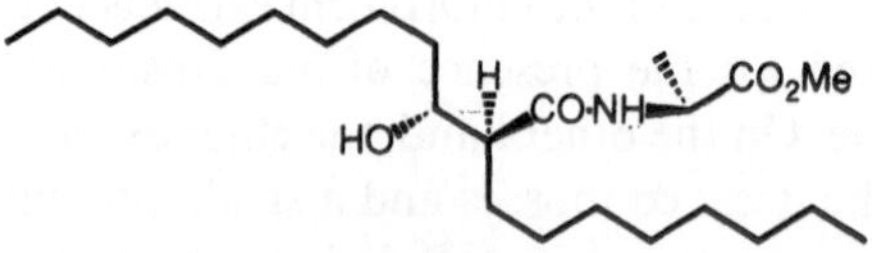

(**1374**)

Chart 82. Miscellaneous compounds found in the Musci

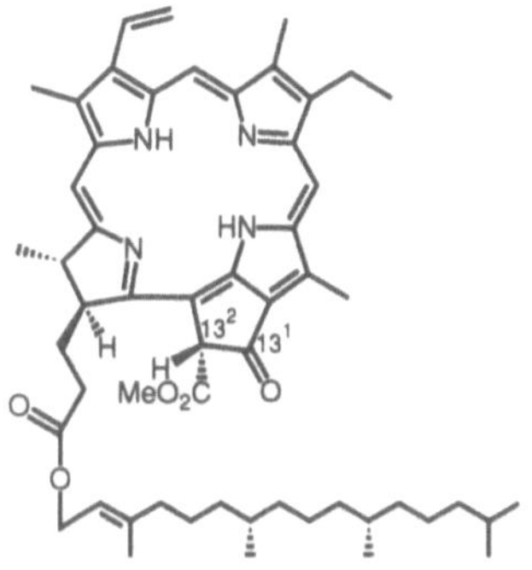

(1375) Pheophytin a

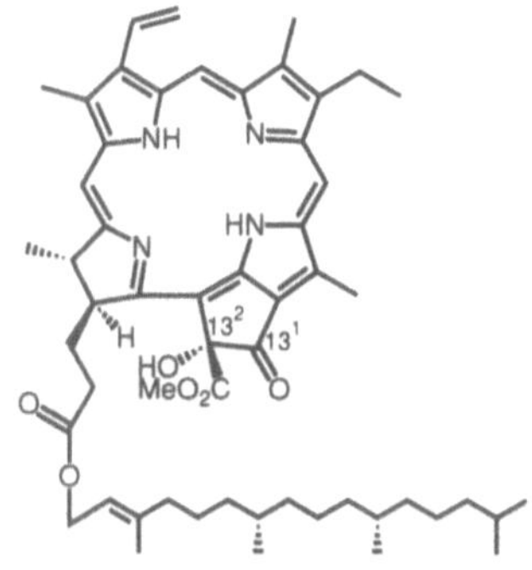

(1376) 13^2-Hydroxy-(13^2-R)pheophytin a

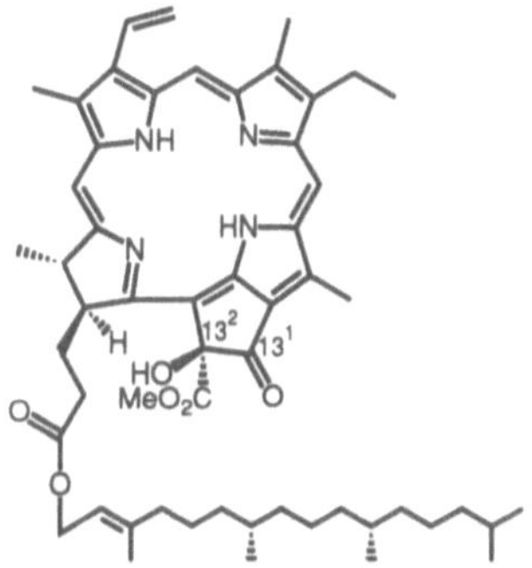

(1377) 13^2-Hydroxy-(13^2-S)pheophytin a

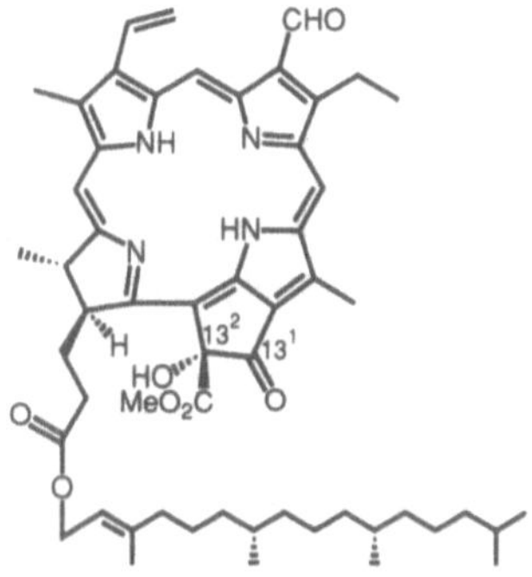

(1378) 13^2-Hydroxy-(13^2-R)pheophytin b

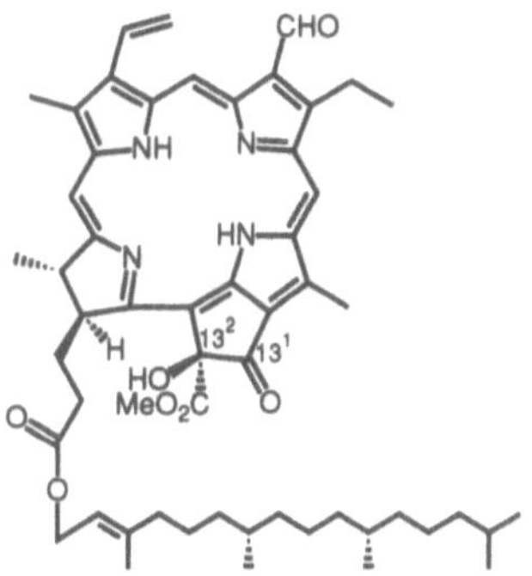

(1379) 13^2-Hydroxy-(13^2-S)pheophytin b

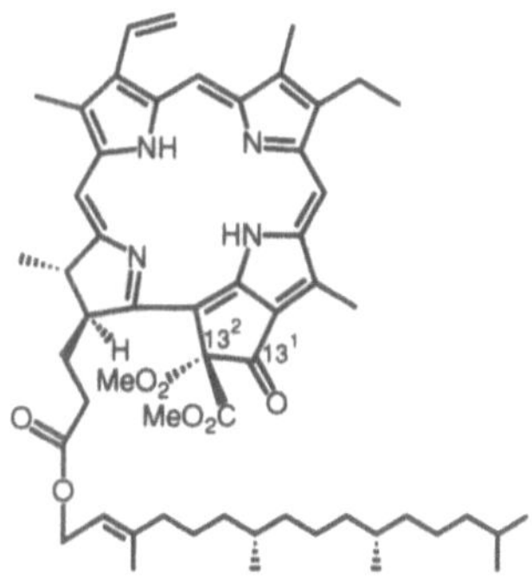

(1380) 13^2-(MeOO)-(13^2-R)pheophytin a

Chart 83. Miscellaneous compounds found in the Musci

against plant pathogenic bacteria. The presence or absence of maytan-sinoids and podophyllotoxins in *Isothecium* and *Thamnobryum* species has not been established.

Thuidium kanedae biosynthesizes not only diterpene lactone (**1233**) but also an amide (**1374**) (*448a*). *Entodon rubicundus* is a unique species because it elaborates pheophytins, pheophytin a (**1375**), 13^2-hydroxy-$(13^2$-$R)$-pheophytin a (**1376**), 13^2-hydroxy-$(13^2$-$S)$-pheophytin a (**1377**), 13^2-hydroxy-$(13^2$-$R)$-pheophytin b (**1378**) and 13^2-hydroxy-$(13^2$-$S)$-pheophytin b (**1379**) along with a novel 13^2-(MeOO)-$(13^2$-$R)$-pheophytin a (**1380**) (*448a*). These chlorophyll derivatives (**1375–1380**) have been found in silkworm excreta (*441b*).

WANG et al. (*611*) in studying cytokinins in culture media from gametophyte over-producing mutants of the moss *Physcomitrella patens* detected two cytokinins one of which was identified as N^6-$(\Delta^2$-isopentenyl)adenine (2iP).

Bryophyta, Pteridophyta and Spermatophyta have been investigated as to their ability to detoxify heavy metals like Cd^{2+} through the formation of $(\gamma$-Glu-Cys)n-Gly peptides (= phytochelatins) (*200*). *Marchantia polymorpha*, when grown in axenic culture was capable of forming 2-$(\gamma$-Glu-Cys) and 3-$(\gamma$-Glu-Cys).

IV. Chemical Constituents of Anthocerotae (Hornworts)

1. Sesquiterpenoids

Cuparane-type sesquiterpenoids such as cuparene (**284**) which are widely distributed in the Hepaticae have been detected in *Megaceros* and *Phaeoceros* species by GC-MS, in trace amounts although Anthocerotae do not contain oil bodies (*18, 68, 69*).

2. Aromatic Compounds

2.1 Cinnamic Acid Derivatives

Methanol extracts of *Anthoceros laevis* (= *Phaeoceros laevis*) and *A. punctatus* were separated into neutral and phenolic fractions by 5% NaHCO₃. The phenolic fraction, after being acidified, was chromatographed on silica gel to give methyl coumarate (**1381**) and methyl caffeate (**1382**) (*400*). *Megaceros arachnoideus* and *M. flagellaris* have been

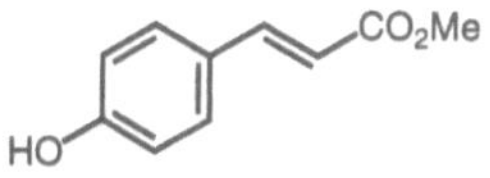

(1381) Methyl coumarate

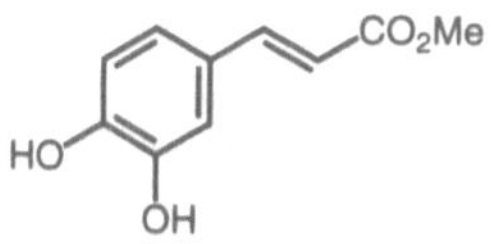

(1382) Methyl caffeate

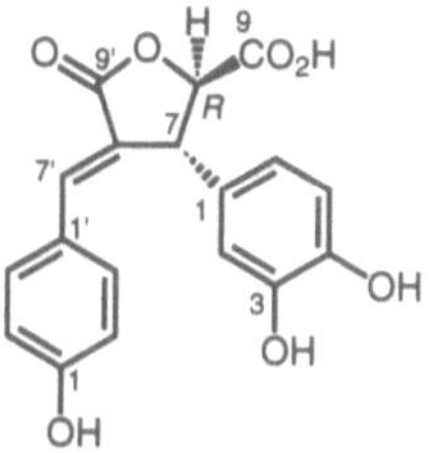

(1383) Megacerotonic acid

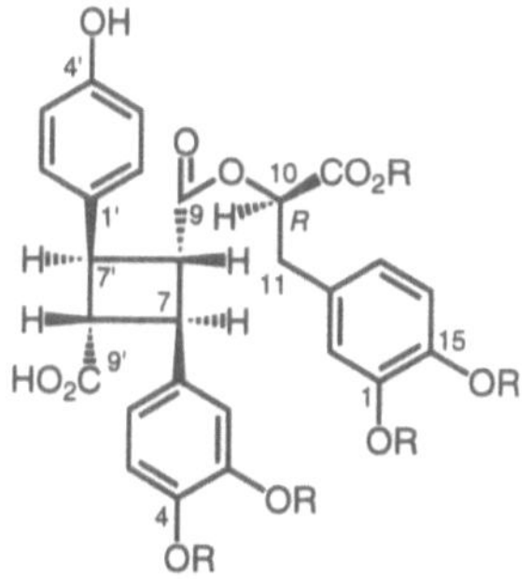

(1384a) Anthocerotonic acid ; R=H
(1384b) R=Me

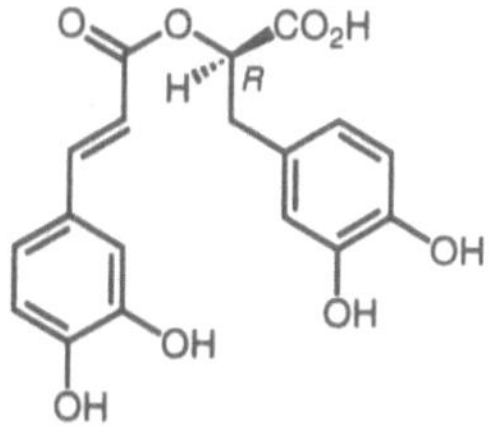

(1385) (R)-Rosmarinic acid

Chart 84. Cinnamic acid derivatives and lignans found in the Anthocerotae

investigated for the presence of flavonoids and other phenolics. A range of other phenolics was observed on 2D-TLCs, whose chromatographic behavior on cellulose TLC with H_2O and n-BuOH-2N NH_3 (1:1) as solvents suggested that they were carboxylic acids, probably cinnamic acids, with at least one free carboxyl group (416). However, flavonoids have not yet been detected in any of the Antocerotae. Caffeic acid (1257) has been isolated from *Anthoceros punctatus* (536).

2.2 Lignans

The Anthocerotae biosynthesize lignans (Table IV). 2% Acetic acid extracts of fresh *Dendroceros japonicus*, *Megaceros flagellaris*, *Notothylas*

Table IV. *Terpenoids and Aromatic Compounds Found in the Anthocerotae*

Structure number	Name of compounds	Formula	m.p. °C	$[\alpha]_D$	Plant source	Family	References	Comments
(284)	Cuparene	$C_{15}H_{22}$			*Megaceros tosanus*	A	(18, 68, 69)	
					Phaeoceros miyakeanus	A	(18, 68, 69)	
(1257)	Caffeic acid	$C_9H_8O_4$			*Anthoceros punctatus*	A	(536)	
(1381)	Methyl coumarate	$C_{10}H_{10}O_3$			*Anthoceros laevis*	A	(400)	
					(= *Phaeoceros laevis*)			
					A. punctatus	A	(400)	
(1382)	Methyl caffeate	$C_{10}H_{10}O_4$			*Anthoceros laevis*	A	(400)	
					(= *Phaeoceros laevis*)			
					A. punctatus	A	(400)	
(1383)	Megacerotonic acid	$C_{18}H_{14}O_7$		+ 233	*Anthoceros laevis*	A	(536, 537)	CD 232 nm − 11.7
					(= *Phaeoceros laevis*)			CD 305 nm + 8.5
					Dendroceros japonicus	D	(536, 537)	
					Megaceros flagellaris	A	(536, 537)	
					Notothylas temperata	N	(536, 537)	
(1384)	Anthocerotonic acid	$C_{27}H_{24}O_{11}$		+ 4.4	*Anthoceros punctatus*	A	(536, 537)	
(1385)	(R)-Rosmarinic acid	$C_{18}H_{16}O_8$			*Anthoceros punctatus*	A	(536, 537)	
					Folioceros fuciformis	A	(536, 537)	

A: Anthocerotaceae; D: Dendrocerotaceae; N: Notothyladaceae

temperata and *Anthoceros laevis* (= *Phaeoceros laevis*) were purified by HPLC to give a new lignan, megacerotonic acid (**1383**), as the major component (*536*). In addition, a second new lignan, anthocerotonic acid (**1384a**) has been isolated from fresh *Anthoceros punctatus* along with (*R*)-rosmarinic acid (**1385**) and caffeic acid (**1257**) (*536, 537*). *Folioceros fuciformis* also elaborates rosmarinic acid (*536*). The structure of (**1383**) and the absolute stereochemistry of (**1385**) were reproduced erroneously in print (*537*). The presence of *p,o*- and *p*-substituted phenyl rings in (**1383**) was established by ^{1}H- and ^{13}C-NMR spectrometry. Methylation with diazomethane gave a tetramethyl derivative which was hydrogenated or reduced with NaBH$_4$ to afford a 7′,8′-dihydro derivative or a primary alcohol resulting from reduction of the carbomethoxy group. The stereochemistry, the position of the phenyl rings and the geometry of vinyl group were established by NOESY experiments. The absolute configuration of (**1383**) is based on use of the CD chirality method (*237, 238*). The CD spectrum of (**1383**) displayed positive and negative Cotton effects at 305 nm ($\Delta\varepsilon + 8.5$) and 232 nm ($\Delta\varepsilon - 11.7$), respectively.

The structure of (**1384a**) was also established by a combination of NMR spectrometry and hydrolysis followed by methylation with diazomethane which gave a dimethyl ester (**1384b**) and secondary alcohol which was identical with methyl (3,4-dimethoxylphenyl)-lactate (**1385a**) prepared from the co-metabolite rosmarinic acid (**1385**) by alkaline hydrolysis followed by methylation (Schem 69). The *R*-configuration at C-10 of (**1384a**) was established by comparing the CD spectrum of (**1385a**) with that of (*R*)-methyl aryllactate derived from (*R*)-rosmarinic acid (**1385**). The total synthesis of rosmarinic acid has been reported by

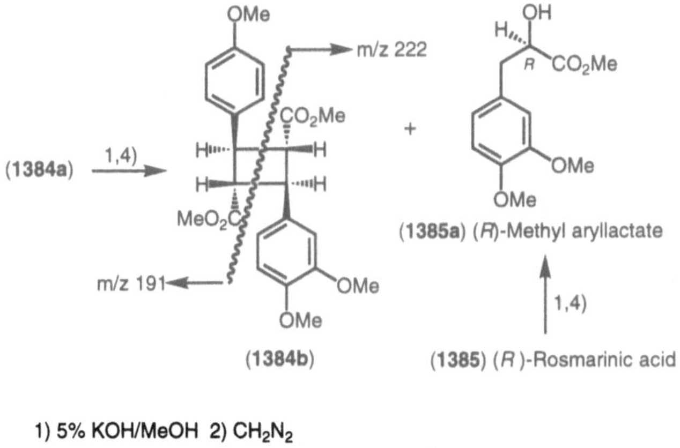

1) 5% KOH/MeOH 2) CH$_2$N$_2$

Scheme 69. Reactions of anthocerotonic and (*R*)-rosmarinic acids

ZINSMEISTER *et al.* (*647*). Anthocerotonic acid (**1384a**) may be directly derived from (*R*)-rosmarinic acid (**1385**) and *p*-coumaric acid (**1256**) in *Anthoceros punctatus* (*536, 537*). Rosmarinic acid (**1385**) has been found in the fern, *Blechnum brasiliense* (*122*), and in seed plants, especially in the Lamiaceae (*525*). This is the first record of the isolation of a lignan from non-vascular plants. Hornworts had so far been regarded as chemically simple (*24*).

V. Biologically Active Substances of Bryophytes

Generally, bryophytes are not damaged by bacteria, fungi, insects, snails and slugs. It is also known that some bryophytes contain allelo-chemicals. A number of bryophytes have been used as medicinal plants in North America (*9*), China (*165*) and Europe (*195*). More than 400 years ago, RI (*472*) stated that some *Fissidens* and *Polytrichum* species possessed diuretic activity and that these ashes promoted human head hair growth. Most bryophytes used medicinally have been applied as decoctions. Bryophytes are also crushed and the resulting powder is mixed with oil to make an ointment which reputedly cures cuts, burn and external wounds. North American Indians have used *Bryum*, *Mnium*, *Philonotis* species and *Polytrichum juniperinum* as medicinal mosses to cure burns, bruises and wounds. *Marchantia polymorpha* has been used as a diuretic in Europe. Fresh liverwort was soaked with white liquor and patients drank the resulting mixture of liquor and extracts. Some bryophytes also show antitumor activity. In addition to the biological activities described above, some bryophytes emits characteristic fragrant odors and, depending on the source, an intensely pungent, bitter and saccharine-like taste. The biological activity ascribed to the Hepaticae is due mainly to terpenoids and aromatic compounds which are constituents of oil bodies, or to hydrophilic diterpenoids.

The following bryophytes are medicinal plants and are said to possess certain physiological activity and effects:

[**Hepaticae**] *Conocephalum conicum* (antimicrobial, antifungal, anti-pyretic, antidotal activity; used to cure cuts, burns, scalds, fractures, swollen tissue, poisonous snake bites and gallstones), *Frullania tamarisci* (antiseptic activity), *Marchantia polymorpha* (antipyretic, antihepatic, antidotal, diuretic activity; used to cure cuts, fractures, poisonous snake bits, burns, scalds and open wounds), *Reboulia hemisphaerica* (for blotches, hemostasis, external wounds and bruises).

[**Musci**] *Bryum argenteum* (antidotal, antipyretic, antirhinitic activity; for bacteriosis), *Cratoneuron filicinum* (for malum cordis), *Ditrichum* *pallidum* (for convulsions, particularly in infants), *Fissidens japonicum* (diuretic activity, for growth of hair, burns and, choloplania), *Funaria hygrometrica* (for hemostasis, pulmonary tuberculosis, hematemesis, bruises, athlete's foot dermatophytosis), *Haplocladium catillatum* (antidotal, antipyretic activity; for adenopharyngitis, uropathy, mastitis, erysipelas, pneumonia, urocystitis and tympanitis), *Leptodictyum riparium* (antipyretic; for choloplania and uropathy), *Mnium cuspidatum* (for hematostasis and nosebleed), *Oreas martiana* (for anodyne, hemostasis, external wounds, epilepsy, menorrhalgia and neurasthenia), *Philonotis fontana* (antipyretic, antidotal activity; for adenopharyngitis), *Plagiopus oederi* (as a sedative, and for epilepsy, apoplexy and cardiopathy), *Polytrichum* species (diuretic acitivity; for growth of hair), *Polytrichum commune* (antipyretic, antidotal; for hemostasis, cuts, bleeding from gingivae, hematemesis and pulmonary tuberculosis), *Rhodobryum giganteum* (antipyretic, diuretic, antihypertensive; for sedation, neurasthenia, psychosis, cuts, cardiopathy and expansion of heart blood vessels), *Rhodobryum roseum* (as a sedative and for neurasthenia and cardiopathy), *Taxiphyllum taxirameum* (antiphlogistic; for hemostasis and external wounds) and *Weissia viridula* (antipyretic, antidotal; for rhinitis) (*36*).

In the earlier review (*19*) pungency and bitterness, allergenic contact dermatitis, anticancer, tumor promoting, antimicrobial and antifungal as well as antifeedant, piscicidal and schistosomisidal activities of compounds isolated from Hepaticae were discussed (*19*). Several reviews dealing with biologically active substances found in bryophytes have been published (*20–23, 29, 30, 32, 34–36*). In this section, more recently isolated biologically active compounds and their activities will be discussed.

1. Characteristic Scents

Some bryophytes emit intense mushroomy, sweet-woody, intense turpentine, sweet-mossy, fungal-like, carrot-like or seaweed-like scents when upon being crushed. Among these are *Asterella* species (resembling the higher plant, *Houttuynia cordata*), *Conocephalum conicum* (*Houttuynia cordata*-like, strong mushroomy), *Conocephalum japonicum* (also *Houttuynia cordata*-like), *Frullania* species (turpentine-like), *Funaria hygrometrica* (sea-star like), *Geocalyx graveolens*, *Jungermannia* and *Lejeunea* species (turpentine-like), *Leptolejeunea elliptica* (mixed odor of naphthalene and dried bonito), *Lophocolea heterophylla* (strongly and distinctly mossy), *L. minor* (strongly mossy), *Lophozia vicrenata* (pleasant resembling cedar oil), *Moerkia* species (intensely unpleasant), *Pellia*

endiviifolia (dried seaweed-like), *Plagiochila sciophila* (= *P. acanthophylla*
subsp. *japonica*) (sweet mossy and woody), *Plagiochila ovalifolia etc.*
(turpentine-like), *Porella cordaeana* (dried seaweed-mossy like), *Porella
vernicosa* complex (pungent and turpentine-like), *Riella* species (anise-
like), *Solenostoma obovata* (carrot-like), *Takakia lepidozioides* (mixed
odor of cinnamon and burnt wheat powder), *Trichocolea tomentella*
(sulfur-like) and *Wiesnerella denudata* (strong sweet mushroomy).

European *Porella cordaeana* emits a strong seaweed-mossy like scent.
Dimethyl sulfide (**1227**) which is one of the significant flavor chemicals of
seaweed has been found in the head space of *P. cordaeana* (*593*) which also
contains the monoterpene hydrocarbons myrcene (**1**), γ-terpinene (**11**), α-
thujene (**20**), β-sabinene (**23**), α-pinene (**27**), β-pinene (**28**), and camphene
(**37**). The characteristic odor of *P. cordaeana* may be due to the mixture of
these components (*593*). Dimethyl sulfide has also been identified in the
crude extracts of *Pellia endiviifolia* and other Metzgeriales species (*62*).

Herout (*268*) reported that the substance responsible for the mossy
scent of European *Lophocolea heterophylla* had the formula $C_{12}H_{20}O$
but the structure was not determined. Further fractionation of the ether
extract of the same species resulted in the isolation of a homomonoterpene alcohol, (−)-2-methylisoborneol (**48**), responsible for the characteristic mossy fragrance (*586*). This unusual homomonoterpenoid has also
been found among the metabolites of Actinomycetes, *Streptomyces* and
Actinomadura species (*203*). The female, male, sterile and receptacle of
Conocephalum conicum emit a strong sweet-mushroomy scent; (−)-β-
sabinene (**23**) and (+)-bornyl acetate (**39**) have been isolated as the major
components from the ether or pentane extract, together with methyl
cinnamate (**871**), (*S*)-(+)-1-octen-3-ol (**1213**) and (*R*)-(+)-1-octen-3-yl-
acetate (**1214**) which are primarily responsible for the fragrance of the
most expensive Japanese mushroom, *Tricholoma matsutake* (*524*), while
the steam distillate of *C. conicum* contained (+)-bornyl acetate (3%), 1-
octen-3-ol + β-sabinene (27%) and 1-octen-3-yl acetate (12%) (*62*). *C.
japonicum* emits a mossy scent different from *C. conicum*. The steam
distillate of *C. japonicum* contains 1-octen-3-yl acetate (**1213**) (12%),
whereas the content of 1-octen-3-ol and β-sabinene is very low (1%) (*62*).

Almost all liverworts which smell of mushrooms contain 1-octen-3-ol
and its acetate but the strong sweet-mushroomy scent of the ether extract
of *Wiesnerella denudata* is due to (+)-bornyl acetate (*39*) and mixtures of
the monoterpene hydrocarbons α-terpinene (**10**), β-phellandrene (**12b**),
terpinolene (**13**), α-pinene (**27**), β-pinene (**28**) and camphene (**37**), (*17, 58*).
The odor of the steam distillate of *W. denudata* is weaker than that of its
ether extract. The steam distillate contains nerol (**4**) (14%), neryl acetate
(**5**) (27%) and γ-terpinene (**11**) (31%), but the content of 1-octen-3-ol (7%)

and its acetate (2%) is very low (*62*). *Plagiochila ovalifolia* has a strong turpentine odor. The ether extract contains linalool (**2**), *p*-cymene (**15a**), α- and β-pinenes (**27, 28**), and camphene (**37**) whereas the essential oil of *P. ovalifolia* contained 1-octen-3-yl acetate (20%) and *trans*-nerolidol (**426**) (36%) (*62*).

The essential oil of *Plagiochila sciophila* (= *P. acanthophylla* subsp. *japonica*) which has a mild turpentine-mushroomy odor is composed of dimethyl sulfide (2%), α-pinene (3%), 3-octanone (**1216**) (2%), β-sabinene (**23**) (5%), β-barbatene (**162**) (19%) and an unidentified component ([M]$^+$ 190, base 41) (*62*).

A small French thalloid liverwort, *Targionia hypophylla*, emits a very intense and fragrant scent when crushed due primarily to *cis*- (**29**) and *trans*-pinocarveyl acetates (**30**) which have been isolated from the methanol extract (*75*).

The characteristic fragrance of a miniature liverwort, *Leptolejeunea elliptica* is due to *p*-ethylanisol (**1386**) (*442*). The intense turpentine-like odor of some species of *Frullania, Jungermannia, Lejeunea, Plagiochila* and *Porella* species which resembles that of conifer needles is caused by mixtures of simple monoterpene hydrocarbons (*36*).

The oxygenated sesquiterpene ketone, bicyclohumulenone (**479**) isolated from *Plagiochila sciophila* (= *P. acanthophylla* subsp. *japonica*) (*19*) possesses an aroma reminiscent of a variety of scents based on a strong woody note resembling the odor of patchouli, vativer, cedar wood, orris, moss and carnations. The pacifigorgiane-type sesquiterpene alcohol tamariscol (**523**) which has been isolated from European *Frullania tamarisci* subsp. *tamarisci* (*150*), Japanese *F. tamarisci* subsp. *obscura* which grows in high mountains and high latitudes, Taiwanese *F. nepalensis* and American *F. asagrayana* (*63, 511*) similarly possesses a remarkable aroma reminiscent of the woody and powdery green notes of mosses, foin, fluvia, costus, violets leaf and seaweeds. Both compounds are highly esteemed in commerce. Epoxytamariscol (**523a**) derived from tamariscol (**523**) possesses the same fragrant scent as (**523**). They are used as perfumes as such or perfume components of the powdery floral-, oriental bouquet-, fantastic chypre-, fancy violet- and white rose-types in various cosmetics (*25–28, 31*). After it had been shown that both the tertiary alcohol and the 2-methyl-1-propenyl group attached to the cyclohexane ring of tamariscol were necessary for the characteristic scent of (**523**), thirteen mini-tamariscols were synthesized by Grignard reactions of 2,7-dimethylcyclohexanone, 2-methylcyclohexanone, 4-methylcyclohexanone, cyclohexanone and cyclopentanone with vinylmagnesium bromide, 2-methyl-1-propenylmagnesium bromide and 2-methyl-2-propenylmagnesium bromide, respectively. All mini-tamariscols possessed characteristic odors but 1-hydroxy-1-(2-methyl-1-propenyl)-cyclohexane (**1387**) had a sweet mossy aroma similar to that of tamariscol itself (*511*).

A thalloid liverwort *Mannia fragrans* emits a pleasant and intense scent which can be detected at a distance of many meters. This specific fragrance is caused by a cuparane-type sesquiterpene ketone, grimaldone (**305**) (*280*). It is noteworthy that *ent*-longipinan-3,12-dione (**491f**) prepared from *ent*-12β-acetoxylongipin-2(10)-en-3-one (**491**) has a very pleasant odor resembling that of grimaldone (**305**) (*279*).

A small thalloid liverwort, an *Asterella* species grown in Pulau Dayang Bunting island, Malaysia, emits very intense unpleasant odor which is due to skatole (**1099**) (*249*).

Several species belonging to the Splachnaceae (Musci) emit volatile fecal odors to which flies are attracted. KOPONEN *et al.* (*344*) analysed the volatile components of *Aplodon*, *Splachnum*, *Tayloria* and *Tetraplodon* species by GC-MS. The identified volatiles were complex mixtures of octane derivatives (1-octen-3-ol, 3-octanone, *etc.*), fatty acids (butyric, caproic, valeric, benzoic acids, *etc.*), monoterpene hydrocarbons (α-pinene, β-pinene, *etc.*) and indole.

2. Pungency, Bitterness and Sweetness

Some genera of bryophytes produce intensely pungent, bitter or sweet substances which exhibit interesting biological activities described in the following sections. It has long been known that most bryophytes growing in North America contain unpleasantly tasting substances some of which taste like immature green pea seeds or pepper (*293*). MIZUTANI (*409*) reported that *Porella vernicosa* contained very pungent substances and that *Jamesoniella autumnalis* contained an intense bitter principle whose taste resembles that of the leaf of lilac and *Swertia japonica* or the root of *Gentiana scabra* var. *orientalis*. It is known that the sharp hot taste of *Porella* species (*P. vernicosa* complex belonging to chemo-type I) is due to polygodial (**336**) (*19*). *Porella roellii* (chemo-type I) of North America contains large amounts of polygodial (*543*). *Porella acutifolia* subsp. *tosana* which belongs to chemo-type IV (pinguisane-sacculatane/pinguisane-germacrance/guaiane-type) is also very pungent. This is due to the presence of the hydroperoxygermacranolides (**449, 450**) (*600*).

The earlier review (*19*) stated that sesquiterpene lactones, secoaromadendrane-type sesquiterpene hemiacetals and sacculatane-type diterpene dialdehyde possessing an intensely pungent taste had been isolated from some *Chiloscyphus* and *Wiesnerella*, *Plagiochila*, and *Pellia* and *Trichocoleopsis* species, respectively. The strong pungent taste of *Pallavicinia levieri* is due to sacculatal (**795**) (*251*). Plagiochiline A (**115**) from *Plagiochila* species (*19*) showed a strong pungent taste. Enzymatic treatment of (**115**) with amylase in phosphate buffer or with human saliva produced two 2,3-secoaromadendrane-type aldehydes, plagiochilal B (**126**) and furanoplagiochilal (**126d**) (*252*). The former dialdehyde has been isolated from *Plagiochila fruticosa* (*185*) and the latter monoaldehyde from *P. yokogurensis* and three other species (*19*). The dialdehyde possesses a strong pungent taste which implies that the intense pungency is not due to plagiochiline A (**115**) itself, but to plagiochilal B (**126**) whose partial structure is similar to that of the very pungent drimane dialdehyde, polygodial (**336**).

Some Lophoziaceae species are intensely bitter. Thus *Gymnocolea inflata* is amazingly and persistently bitter and induces vomiting when one chews a few leaves for several seconds. The earlier review (*19*) already mentioned that this is due to gymnocolin A (**650**), a *cis*-clerodane-type furanoditerpene lactone. The bitter tastes of *Anastrepta orcadensis*, *Barbilophozia lycopodioides* and *Scapania undulata* are due to anastreptin (**651**) (*143, 478*), barbilycopodin (**664**) (*143*) and scapanin A (**745**) (*277*), respectively. Barbilycopodin (**664**) has also been found in *B. attenuata*, *B. barbata*, *B. floerkei* (*273, 429*) and *B. hatcheri* (*429, 565*).

Jungermannia infusca has an exceedingly intense bitter taste. This is due to the presence of the infuscasides A–E (**721a, 722a, 723–725**), which are kaurane glucosides (*436*); however, the corresponding aglycones are tasteless. It is interesting that other kaurene glycosides such as stevioside from the higher plant *Stevia rebaudiana* (Compositae) are intensely sweet.

Some *Fissidens* and *Rhodobryum* species possess a sweet taste like saccharine. However, the substances responsible for the taste remain to be isolated.

3. Allergenic Contact Dermatitis

Several liverworts (*Frullania asagrayana, F. bolanderi, F. dilatata, F. eboracensis, F. franciscana, F. inflata, F. kunzei, F. nisquallensis, F. liparia, F. tamarisci, Marchantia polymorpha, Metzgeria furcata, Radula complanata* and *Schistochila appendiculata*) and mosses (*Eurhynchium organum, Isothecium stoloniferum, Leucolepis menziesii, Mnium cuspidatum, M. undulatum, Polytrichum juniperinum, Rhytidiadelphus loreus* and *Sphagnum palustre*) cause intense allergenic contact dermatitis (*36*). *Frullania* species (Hepaticae) are notable as liverworts which cause allergy. The allergy-inducing substances of *Frullania* species are sesquiterpene lactones with an α-methylene-γ-lactone group (*19*). Patients sensitive to *Frullania* react equally and strongly to (+)- (**380a**) and (−)-frullanolides (**380b**) which have been isolated from *Frullania dilatata* and *F. tamarisci*, respectively (*19*). Two groups of guinea pigs were sensitized by means of (+)- and (−)-frullanolides, respectively, and no cross-sensitization occurred (*91*). This was the first experimentally documented demonstration of stereospecificity in allergenic contact dermatitis.

There are more than 500 *Frullania* species composed of two main chemo-types. Type A contains sesquiterpene lactones with an α-methylene-γ-butyrolactone group whilst type B does not contain any sesquiterpene lactones (*19*). When patients sensitive to *Frullania* species come into contact with type A, an amazingly intense allergy is induced.

It has also been known that some *Radula* species cause allergenic contact dermatitis (*19*). *Radula complanata* is often intermingled with *F. dilatata*. In fact, (+)-frullanolide (**380a**) has been isolated from the crude extract of the mat of *R. complanata* (*66*). It has been confirmed that some mats collected in western France were contaminated by *F. dilatata* which has been detected under a binocular microscope. Thus the hapten of *R. complanata* is not due to the chemical constituents of this liverwort, but to (+)-frullanolide and other sesquiterpene lactones found in *Frullania dilatata*.

Members of the *Porella vernicosa* complex are rich sources of polygodial (**336**) which exhibits pungency and irritancy. Guinea pigs were sensitized to polygodial using intradermal injections in Freund's complete adjuvant. Cross-reaction with a mixture of racemic warburganal and (−)-warburganal (**1388**) isolated from *Polygonum hydropiper* showed a high specificity of the allergenic response because (−)-warburganal gave a stronger skin reaction than the racemic mixture (*518*).

A large beautiful liverwort *Schistochila appendiculata* collected in New Zealand causes allergenic contact dermatitis. The allergenic reaction is brought on by alkylphenols (C11, C13, C15 and C17) (**1043, 1046, 1049, 1053**), long chain alkyl salicylic acids (C11, C13, C15) (**1044, 1047, 1050**) and their potassium salts (**1045, 1048, 1051**) as well as a long chain catechol (**1052**) (*53*). Such allergenic contact dermatitis is very similar to that caused by the long chain alkylphenols of the fruit of *Ginkgo biloba* and Anacardiaceae (*162, 177*).

4. Antitumor Activity

Through the agency of the National Cancer Institute, USA, 184 species in 97 genera of mosses, 23 species in 16 genera of liverworts, and one hornwort have been screened for antitumor agents using the P388 lymphocytic leukemia test system (*512*). Ground material of each species was extracted with 90% hot EtOH in a Soxhlet apparatus for 8 hr, followed by concentration of the extract *in vacuo* and partition of the residue between chloroform or dichloromethane and water (1:1). The CHCl₃ or CH₂Cl₂ fractions were tested. Prior to 1974, the materials were extracted with 50% ethanol or with petroleum ether, ethanol and aqueous EtOH, sequentially. Extracts of 75 moss species were toxic; extracts of 43 moss species were active. Active species came from the following families; Brachytheciaceae, Dicranaceae, Hypnaceae, Mniaceae, Neckeracea, Polytrichaceae, Grimmiaceae and Thuidiaceae. The level of antitumor activity amongst the liverwort species was somewhat lower with only 4 of 23 species screened exhibiting P388 activity; these were *Bazzania trilobata* and *Porella bolanderi* (Jungermanniales) and *Conocephalum conicum* and *Dumortiera hirsuta* (Marchantiales). The only hornwort studied, *Anthoceros fusiformis*, was inactive. The substances responsible for the activities remain to be identified.

As mentioned earlier, the chemical constituents of Musci, Hepaticae and Anthocerotae differ considerably; in particular those of Hepaticae are mono-, sesqui- and diterpenoids as well as bibenzyls and cyclic bisbibenzyls and it may well be that the standard extraction procedure using 90% hot EtOH for 8 hr is too vigorous for the Hepaticae which *inter al.* contain hemiacetals, dialdehydes, a number of unstable acetylated terpenoids *etc.* Thus many biologically active constituents have been obtained by extraction of bryophytes with ether or methanol at room temperature.

As mentioned in the earlier review (*19*) a few eudesmanolides and a germacranolide possessing inhibitory activity against KB cells have been isolated from some liverworts. *Conocephalum conicum* and *Wiesnerella denudata* contain some guaianolides which exhibited anticancer activity against P-388 lymphocytic leukemia (*19*).

Riccardins A (**985a**) and B (**987a**) isolated from *Riccardia multifida* inhibited KB cells at a concentration of 10 and 12 µg/ml, respectively (*79*), as did a pinguisane-type sesquiterpene alcohol, dehydropinguisenol (**541**), at a concentration of 12.5 µg/ml (*82*). *Trocholejeunea sandvicensis* contains highly oxygenated aromatic compounds among which (**1104**) showed cytotoxic activity against KB cells (ED$_{50}$ 12.5 µg/ml) (*82*). The following compounds isolated from liverworts also showed cytotoxic activity against KB cells, values in parentheses show the concentration in µg/ml: Plagiochiline A (**115**) (0.28) from several *Plagiochila* species, eremofrullanolide (**352**) (1.70) and oxyfrullanolide (**381**) (0.80) from *F. dilatata*, epoxyfrullanolide (**383**) (2.65) prepared from (+)-frullanolide

(**380a**), 4-epiarbusculin A (**390**) (0.50) from *F. tamarisci* subsp. *obscura*, 8α-acetoxyzaluzanin D (**467**) (2.50), zaluzanin D (**468**) (1.50), and 8α-acetoxyzaluzanin C (**469**) (1.60) from *Wiesnerella denudata*, marchantin A (**993**) (8.39), B (**994**) and C (**995**) (10.0) from several *Marchantia* species and perrottetin E (**1023a**) (12.5) from *Radula perrottetii* and a few species of Marchantiales (*35, 36*). T/C (100%) values of plagiochiline A (**115**) and marchantin A (**993**) against P388 were estimated as 115 and 117%, respectively (*35, 36*). 3-Methoxy-4'-hydroxybibenzyl (**891**) from *Plagiochila* species showed cytotoxicity against monkey kidney cells (BSC) at 60 µg/well (*358a*).

Belkin et al. (*97*) reported that alcoholic or acidic extracts of *Polytrichum juniperinum* showed antitumor activity against Sarcoma 37 transferred into the muscle of CAF1 mice. However, no active compound has so far been isolated from this species. Zheng et al. (*645a, 645b*) found that the 95% ethanol extract of the moss, *Polytrichum ohioense* collected in Maryland showed cytotoxicity in the human nasopharynxcarcinoma (9KB). Fractionation of the extract resulted in the isolation of five new benzonaphthoxanthenones, ohioensins A–E (**1342–1346**) which showed cytotoxic activity not only against 9KB, but also murine P388 leukemia (9PS), human lung carcinoma (A-549), human breast adenocarcinoma (MCF-7) and human colon adenocarcinoma (HT-29): values in parentheses show the concentration in ED_{50} µg/ml: ohioensin A (**1342**) (> 10, 1.0, > 10, 9.0 and > 10 against 9KB, 9PS, A-549, MCF-7 and HT-29), ohioensin B (**1343**) (9.7, > 10, > 10, 3.4 and 4.3), ohioensin C (**1344**) (> 10, 1.0, 8.7, 6.7 and > 10), ohioensin D (**1345**) (> 10, 1.0, > 10, > 10 and > 10) and ohioensin E (**1346**) (> 10, 1.0, 6.2, > 10 and > 10). Ohioensins A, C, D and E showed moderate cytotoxicity against 9PS. Ohioensin B also showed marginal activity against HT-29 and MCF-7 cells.

Marsupellone (**489**) and acetoxymarsupellone (**490**) from *Marsupella emarginata* showed antitumor activity (IC_{50} 1 µg/ml) against P388 (*434*). Sakai et al. (*479*) found that the ether extract of *Isothecium subdiversiforme* showed intense antitumor activity (IC_{50} 3 µg/ml) against mouse lymphocytic leukemia (P388). The active fraction after being concentrated ten times gave T/C values of 149% at a dosage of 40 mg/kg, for 1–5 days. As mentioned earlier, the fresh moss (20.3 kg) gave four maytansinoids (**1369–1372**) in yields of 2.0×10^{-5} to 8.0×10^{-6} g, presumably responsible for the activity and believed to produced by microorganisms associated with the moss. 15-Methoxyansamitocin P-3 (**1370**) exhibited antitumor activity (IC_{50} 2.0×10^{-5} µg/ml) against P388 *in vitro* (*479, 643*). This activity is 100 times stronger than that of the known antitumor drug, adriamycin (IC_{50} 1.7×10^{-3} µg/ml). Compounds

(**1369, 1371, 1372**) have been isolated from the fermentation broth of *Nocardia* species (*15*), and from the higher plants, *Maytenus bunchananii* (*353*) and the Indian plant *Trewia nudiflora* (*467*). 13^2-Hydroxy-(13^2-R) pheophytin a (**1376**) and its epimer (**1377**) found in *Entodon rubicundus* have *in vitro* cytostatic activity against hepatoma tissue culture (HTC) cells (*441b*).

Some sesqui- and diterpenoids from liverworts cause irritation of the mouse ear, induction of ornithine decarboxylase (ODC), and adhesion of cultured human promyelocytic leukemia cells (HL-60), all activities associated with tumor promotion (*35, 36*) while (+)-frullanolide (**380a**) from *Frullania dilatata* inhibited tumor promotion (*455*).

5. Antimicrobial and Antifungal Activity

Several liverworts (*Bazzania* species, *Conocephalum conicum, Dumortiera hirsuta, Marchantia polymorpha, Metzgeria furcata, Pellia endiviifolia, Plagiochila* species, *Porella vernicosa* complex, *P. platyphylla, Radula* species, *etc.*) and mosses (*Atrichum antustatum, A. undulatum, Dicranum scoparium, Isothecium stoloniferum, Mnium punctatum, Orthotrichum rupestre, Polytrichum commune, P. juniperinum, Thuidium recognitum* var. *delicatulum, Tortula muralis, Sphagnum fimbriatum, S. palustre* and *S. strictum*) show antimicrobial activity (*35, 36*). Several liverworts, such as *Bazzania* species, *Conocephalum conicum, Diplophyllum albicans, Lunularia cruciata, Marchantia polymorpha, Plagiochila* species, *Porella vernicosa* complex, *Radula* species and mosses, *Atrichum undulatum, Bryum pallens, Dicranella heteromalla, D. scoparium, Microbryum delicatulum, Mnium hornum, Oligotrichum hercynicum, Plagiothecium denticulatum, Pogonatum aloides, P. urnigerum, Polytrichum commune, Pseudoscleropodium purum, Sphagnum nemoreum, S. portoricense, S. strictum* and *S. subsecundum* display antifungal activity (*35, 36*). The earlier review (*19*) mentioned isolation of three antibiotic sesquiterpenoids and four antibiotic bibenzyls from some liverworts.

The naturally-occurring prenyl bibenzyls (**918, 927**) and a chemically modified bibenzyl (**952b**) showed antimicrobial activity against *Staphylococcus aureus* (20–30 µg/ml) (*43, 66*). 3-Methoxy-4′-hydroxybibenzyl (**891**) showed antifungal activity against *Candida albicans* (MIC 125 µg/ml) and *Trichophyton mentagrophytes* (MIC 62.5 µg/ml) but not against *Escherichia coli* and *Pseudomonas aeruginosa* (*358a*).

Marchantin A (**993**) from *Marchantia chenopoda, M. polymorpha, M. paleacea* var. *diptera, M. plicata* and *M. tosana*, shows antibacterial

activity against *Acinetobacter calcoaceticus* (MIC 6.25 µg/ml), *Alcaligenes faecalis* (100), *Bacillus cereus* (12.5), *B. megaterium* (25), *B. subtilis* (25),
Cryptococcus neoformans (12.5), *Enterobacter cloacae*, *Escherichia coli*,
Proteus mirabilis, *Pseudomonas aeruginosa*, *Salmonela typhimurium* (100)
and *Staphylococcus aureus* (3.13–25) and antifungal activity against
Alternaria kikuchiana, *Aspergillus fumigatus* (MIC 100 µg/ml), *A. niger*
(25–100), *Candida albicans*, *Microsporum gypseum*, *Penicillium chryso-
enum* (100), *Piricularia oryzae* (12.5), *Rhizoctonia solani* (50), *Saccharomyces cerevisiae*, *Sporothrix schenckii* (100), *Trichophyton mentagrophytes*
(3.13) and *T. rubrum* (100) (*35, 36, 575*).

The dolabellane-type diterpenoids (**668–670**) from *Odontoschisma
denudatum* and their derivatives (**668b, 668c**) had growth inhibitory
activity against pathogenic fungi, with (**668**) of 39% against *Botrytis
cinerea*, 38% against *Rhizoctonia solani* and 22% against *Phythium
debaryanum* at a concentration of 100 ppm (*381*). Compounds (**669, 670,
668b, 668c**) showed similar activities against the three microorganisms.

Some phenolic sesquiterpenoids isolated from *Herbertus aduncus*
inhibit the growth of some plant pathogenic fungi (*394*). Thus α-
herbertenol (**308**) and β-herbertenol (**309**) exhibited 50% growth inhibition (I$_{50}$) of *Botrytis cinerea*, *Rhizoctonia solani* and *Phythium debaryanum* at 20–60 ppm while α-formylherbertenol (**310**) caused I$_{50}$ of *B.
cinerea* and *R. solani* at 8–10 ppm and β-bromoherbertenol (**309b**) (not a
natural product) produced similar results at *ca.* 15 ppm. Petroleum ether
extracts of *Fossombronia pusilla* and *F. himalayensis* had antibacterial
activity against *Escherichia coli*, *Staphylococcus aureus* and *Bacillus
subtilis etc.* at a concentration of 100 µg/ml. α-Santonin (**396**) had the
highest activity followed by 8-hydroxy-9-hydroperrottetianal (**807**), perrottetianal A (**805**) and perrottetianal B (**806**) (*491*). Tridensenal (**511**)
from *Bazzania tridens* had weak antimicrobial activity against *Staphylococcus aureus* (ATCC 3359) at 400 µg/ml (*628*). The concentration of
lunularic acid (**877**) in many liverworts may suffice to prevent fungal
attack almost completely (*497*).

(**309b**)

Dicranenone A (**1357**) from the moss *Dicranum scoparium* and dicranenone B1 (**1361**) from *D. japonicum* showed antibiotic activity against
Piricularia oryzae and *Bacillus cereus* at concentrations of 60

and 400 ppm, respectively (*290*). Similar activities were displayed by 9Z,12Z,15Z-octadecatriene-6-ynoic acid (**1222**) and 1-monoglyceride of 13-hydroxy-9Z,11Z,15Z-octadecatriene-6-ynoic acid (**1359**) (*291*). The antibiotic constituents of *Atrichum, Dicranum, Mnium, Polytrichum* and *Sphagnum* species were thought to be polyphenolic compounds (*396*).

Polar biflavonoids occur frequently in Musci and are concentrated in the hydrophilic cell walls. It has been suggested that high concentrations of biflavonoids in certain mosses cause resistance to fungal infection and deter browsing insects (*196*).

6. Insect Antifeedant and Molluscicidal Activity

As mentioned earlier (*19*) plagiochiline A (**115**), a sesquiterpene hemiacetal found in several *Plagiochila* species, is a very strong antifeedant against the African army worm (*Spodoptera exempta*). The pungent sacculatal (**795**) isolated from *Pellia endiviifolia* and *Trichocoleopsis sacculata*, two pungent eudesmanolides (**409**) and (**413**) from *Chiloscyphus polyanthos*, a germacranolide (**443**) from *Wiesnerella denudata* and the sesquiterpene lactones (**380a**) and (**380b**) from *Frullania dilatata* and *F. tamarisci*, a drimanolide (**342**) from the *Porella vernicosa* complex, and the bitter gymnocolin (**650**) from *Gymnocolea inflata* also have antifeedant activity against larvae of Japanese *Pieris* species; however, the activities are significantly less than that of plagiochiline A (**115**) (*35, 36*).

Albicanol acetate (**330**) from *Bazzania japonica* (*84*) had potent antifeedant properties against gold fish (*266*).

A series of natural drimanes and related synthetic compounds was tested for antifeedant activity against aphids. Polygodial (**336**) from the *Porella vernicosa* complex and a higher plant, *Polygonum hydropiper* and warburganal (**1388**) from the African tree *Warburgia ugandensis* were the most active substances (*41*). Natural (−)-polygodial (**336**) and the synthetic (+)-enatiomer (**1389**) showed similar levels of activity as aphid antifeedants (*41*). Polygodial (**336**) has also been found in porostome nudibranchs, *Dendrodoris* species (*456*). It was shown to inhibit feeding of fish (*166*).

The new cuparane-type sesquiterpenoids (**303, 304**) and monocyclo-farnesane-type (**506–508**) from *Ricciocarpos natans* and lunularic acid (**877**) which is widespread in liverworts and the related compounds (**1390, 1391**) have been tested against *Biomphalaria glabtata*, one of the snail vectors of schistosomiasis (Bilharzia) (*641*). Ricciocarpin A (**506**) is the most toxic compound with an LC_{100} (100% lethal concentration) of 11 ppm, ricciocarpin B (**507**) with a γ-lactone moiety causes a significant

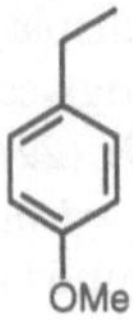

(1386) p-Ethylanisole 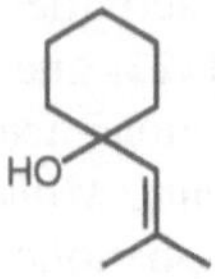   (1387) 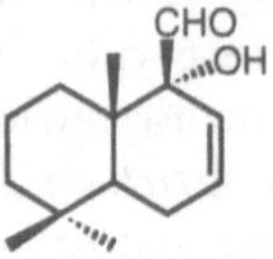   (1388) Warburganal 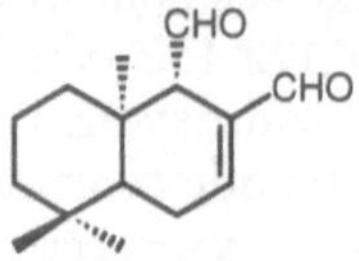   (1389) (+)-Polygodial

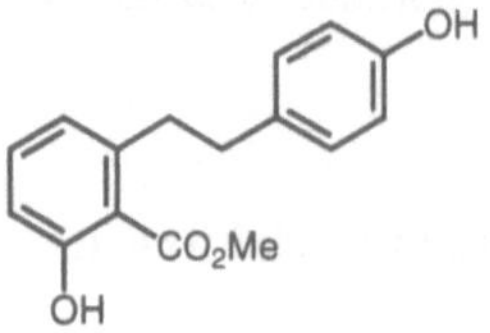

(1390) 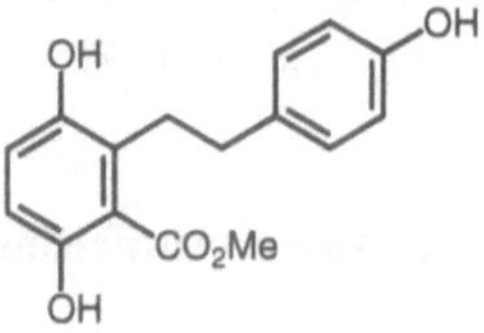   (1391)

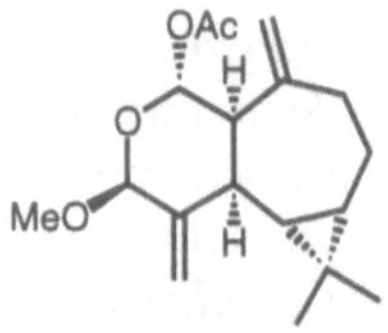

(1392) 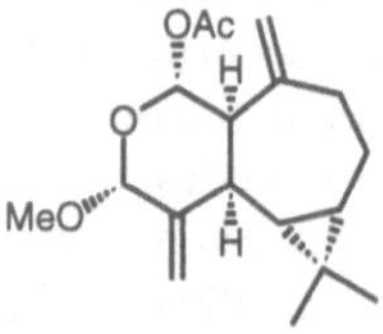   (1393)

Chart 85a. Some biologically active compounds found in the Hepaticae

reduction of the activity to an LC_{100} of 43 ppm. Cuparenolide (303) showed molluscicidal activity with LC_{100} at 32 ppm, however, cuparenolidol (304) was not active. Lunularic acid (877) and two synthetic lunularic acid methyl ester (1390) and 5-hydroxylunularic acid (1391) were active at 47, 10 and 8 ppm, respectively.

The consumption of bryophytes by animals has not been investigated in detail. It has been suggested that moss shoots are seldom eaten freely by either vertebrates or invertebrate herbivores. Moss shoots are not commonly damaged by slugs, but immature capsules are not grazed. DAVIDSON et al. (160) on investigating the phenolic constituents of the two mosses, Brachythecium rutabulum and Mnium hornum found that ferulic acid (1258) and p- (1256) and m-coumaric acids (1260) were concentrated in the wall-bound fraction of the shoot and other phenolics within the immature capsule or were common to both moss tissues. Protocatechuic (1042) and gallic acids (1252) within the shoot are responsible for the antifeedant property against slugs (160).

7. Plant Growth Regulatory Activity

It is well known that higher plants do not grow around many bryophytes. For example, *Leucobryum glaucum* and *Sphagnum* mats are almost free of other plants. HUNECK and MEINUNGER (*285*) tested 81 bryophytes (52 mosses species from 20 different families and 29 liverworts species from 14 different families) for growth regulatory activity using *Lepidium sativum* seedlings as test system. Fresh bryophytes were put into Petri dishes and moistened with distilled water. To each bryophyte seeds of *L. sativum* were added and the length of the shoots and roots measured after 10 days at room temperature under normal conditions of day and night. *Brachythecium rutabulum, Distichium capillaceum, Fissidens taxifolium* and *Hypnum cupressiforme* (Musci) and *Marchantia polymorpha* (Hepaticae) promoted the growth of shoots, *Ctenidium molluscum* and *Hypnum cupressiforme* (Musci) and *Conocephalum conicum* and *Plagiochila porelloides* (Hepaticae) accelerated the growth of roots, and *Dicranella cerviculata, Leucobryum glaucum, Orthodontium lineare*, a number of *Sphagnum* species (Musci) and *Barbilophozia floerkei, Cephalozia bicuspidata, Lepidozia reptans, Marsupella emarginata* and *Ptilidium pulcherrimum* (Hepaticae) inhibited the growth of both shoots and roots. The compounds responsible for the activities were not isolated.

Most crude extracts of bryophytes, especially those containing pungent substances, show inhibitory activity against germination, root elongation, and second coleoptile growth of lettuce, rice in husk, radishes, wheat etc. Several plant growth regulatory terpenoids have been isolated from liverworts (*19*).

The plant growth inhibitory activities of isobicyclogermacrenal (**206**) and lepidozenal (**208**) from *Lepidozia vitrea* and their hydroxy derivatives (**206a**) and (**210**) were tested against rice seedling (*383*). Compounds (**206**) and (**208**) completely inhibited the growth of leaves and roots at a concentration of 50 and 250 ppm, respectively. The 50% growth inhibition (I_{50}) of (**206**) against leaves and roots was 6 and 7 ppm. The hydroxylated product (**206a**) was less active than (**206**) while alcohol (**210**) was more active than (**208**) and inhibited the growth of leaves and roots at a concentration of 100 ppm. The enantiomeric pair, (+)-vitrenal (**594**) from *Lepidozia vitrea* and its synthetic enantiomeric (−)-vitrenal, were tested for growth regulatory activity using rice seedlings and lettuce hypocotyles (*335, 336, 383, 389*). The (+)-isomer produced strong inhibition ($I_{50} = 18$ ppm), while the (−)-isomer exhibited weak stimulatory activity ($P_{50} = 2.18 \times 10^3$ ppm) against rice seedlings and weak inhibitory activity ($I_{50} = 27.79 \times 10^3$ ppm) against lettuce hypocotyles. Some 2,3-secoaromadendrane-type sesquiterpene hemiacetals show plant growth inhibitory activity (29, 35, 36). The methoxylated plagiochilines (1392, 1393) which were derived from plagiochiline C (117) also inhibited the growth of rice seedlings (378). (−)-Polygodial (336) found in Porella species inhibits the germination and root elongation of rice in husk (19) while the synthetic (+)-polygodial (1389) exhibits a similar level of activity in phytotoxicity tests (41).

Anastrophyllum minutum contains four new sphenolobane-type diterpenoids among which 3α,4α-epoxy-5α-acetoxy-18-hydroxysphenoloba-13E,16E-diene (810) decreased shoot and root elongation of rice in husk at a concentration between 10 and 500 ppm. Compound (810) possesses low growth inhibitory activity against rice seedlings (108). It is known that lunularic acid (877) found in most of liverworts has a weak allelopathic effect (19). Drought resistance of the thallus is controlled by lunularic acid (497).

Funaria hygrometrica (Musci) elaborates ethylene which is a plant growth promoter (110).

The endogenous plant hormone, indole acetic acid (IAA) (= auxin) (1100) has been isolated from a few liverworts (9, 35, 36) and the mosses (110). The several effects of IAA on moss development include inhibition of protonema growth, stimulation of rhiziod formation, transformation of buds to filaments, torsion of young stems and complete suppression of leaves on gametophores. The hormonal system responsible for those effects includes the sequential interaction of IAA and cytokinin as a main component and perhaps cyclic AMP as an antagonistic system. Indole 3-acetonitrile (1101) has been detected in Asterella angusta and Pallavicinia canarus and might very well be the precursor of IAA (19, 35, 36). The biosynthetic pathway leading to auxin in moss protonema is shown in Scheme 70 (123).

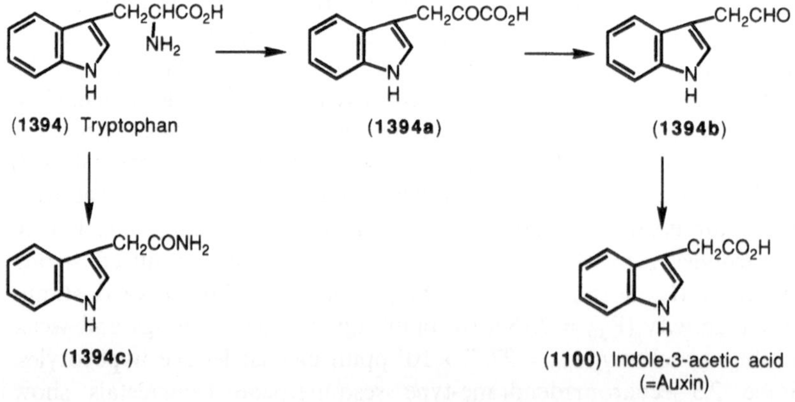

Scheme 70. Biogenetic pathway for indole-3-acetic acid (auxin) in moss protonemata

8. Superoxide Release Inhibitory Activity

Excess superoxide anion radical (O_2^-) in organisms causes various angiopathies, such as cardiac infarction, arterial sclerosis *etc.* Some flavonoids, tannins, polyphenols and carotenoids have superoxide dismutase-like activity. Clerod-3,13(16)-14-trien-17-oic acid (= infuscaic acid) (**625a**) from *Jungermannia infusca* inhibited the release of superoxide from rabbit PMN at IC_{50} 15 µg/ml and from guinea pig peritoneal macrophage induced by formyl methionyl leucyl phenylalanine (FMLP) at IC_{50} 2 µg/ml (*592*). Other clerodanes (**626, 627, 628**) show the same type of activity, however, to a lesser degree (*430*). Norpinguisone methyl ether (**567a**) from *Porella elegantula* exhibited 50% inhibition at 35 µg of the release of superoxide from the guinea pig peritoneal macrophage (*189*). The same activity (IC_{50} 7.5 µg/ml) has been found in cyclomyltaylyl-3-caffeate (**520**) from *Bazzania japonica* (*84*). Other sesquiterpenoids, plagiochilide (**121**), bicyclogermacrenal (**198**), herbertenediol (**311**), isocuparene-3,4-diol (**313**) and norpinguisone (**566**), and the diterpenoids, infuscaside A (**721a**), infuscaside B (**722a**) and perrottetianal A (**805**) from liverworts also inhibited superoxide release from guinea pig peritoneal macrophage (IC_{50} 12.5–50 µg/ml) (*35, 36*).

9. 5-Lipoxygenase, Calmodulin and Thromboxane Synthetase Inhibitory Activity

Marchantin A (**993**) from several *Marchantia* species showed 5-lipoxygenase inhibitory activity [(89% at 10^{-5} mol, 94% at 10^{-6} mol, 45% at 10^{-7} mol, 16% at 10^{-8} mol) against LTB_3 (= 5S,12R-dihydroxy-6,8,10,14-eicosatetraenoic acid)], (99% at 10^{-5} mol, 97% at 10^{-6} mol, 70% at 10^{-7} mol, 40% at 10^{-8} mol) against 5-HETE (= 5-hydroxy-6,8,11,14-eicosatetraenoic acid)] and calmodulin inhibitory activity at ID_{50} 1.85 µg/ml (*35, 36, 579*). Perrottetins A (**952a**) and D (**946a**) from *Radula perrottetii* and prenyl bibenzyls (**913, 927, 960, 964**), also from *Radula* species, riccardin A (**985a**) from *Riccardia multifida* and marchantins D (**999**) and E (**1000a**) from *Marchantia* species had calmodulin inhibitory activity (ID_{50} 2.0–95.0 µg/ml). The simple bibenzyls (**886, 892, 898, 902**) from *Radula* and *Frullania* species also showed weak calmodulin inhibitory activity (ID_{50} 100 µg/ml) as did the labdane-type diterpene diol (**735**) (ID_{50} 82 µg/ml) (*35, 36*). Perrottetin A (**952a**), prenylbibenzyls (**918, 927, 946a, 963a**), marchantins D (**999**) and E (**1000a**) and riccardin A (**985a**), also inhibited 5-lipoxygenase (76–4% at 10^{-6} mol) (*35, 36, 43, 50*).

Lunularic acid (877) which is found in almost all liverworts and has been synthesized by various methods has anti-hyaluronidase activity (IC_{50} 0.13 nM). This activity is stronger than that of tranilast (N-3',4'-dimethoxycinnamoylanthranilic acid) which is anti-allergenic agent developed in Japan for oral administration (257). Lunularic acid also inhibits thromboxane synthetase (ID_{50} 5.6×10^{-3} mol) (212).

10. Vasopressin (VP) Antagonist and Cardiotonic Activity

The prenylated bibenzyl (929) from *Radula complanata* collected in France has vasopressin antagonist activity (27 µg/ml) as do two synthetic prenyl bibenzyls (921b, 921d) which are monomethyl ethers of the naturally-occurring bibenzyls (921a, 918a) found in *Radula* species showed similar activity (57 µg/ml for 921b and 17 µg/ml for 921d) (35, 36, 43). Marchantin A (993) from *Marchantia* species increases coronary blood flow (2.5 ml/min at 0.1 mg) (35, 36). This compound may be used as a coronary vasodilator.

(921d)

11. Piscicidal Activity

Crude extracts containing pungent substances possess strong hemolytic activity. The strongest piscicides are the pungent (−)-polygodial (336) isolated from *Porella vernicosa* complex (19) and sacculatal (795) from *Pallavicinia levieri* (251), *Pellia endiviifolia* (19), *Riccardia lobata* var. *yakushimensis* (42) and *Trichocoleopsis sacculata* (19). Killie-fish (= *Oryzia latipes*) are killed within 2 hr by 0.4 ppm solution of (336) and (795), and within 20 min if the concentration is raised to 7 ppm. Killie-fish are also killed within 2 hr by a 0.4 ppm solution of synthetic very pungent (+)-polygodial (**1389**). Hence piscicidal activity is not affected by the chirality of polygodial. Polygodial is also very toxic to fresh water bitterlings which are killed within 3 min by a 0.4 ppm solution. On the other hand, isopolygodial (**337**) from cultured cells of *Porella vernicosa* and the higher plant *Polygonum hydropiper* and isosacculatal (**796**) from *Pellia*, *Riccardia* and *Trichocoleopsis* species lack piscicidal activity even at 10000 ppm. These results indicate that piscicidal activity of polygodial and sacculatal is significantly related to pungency, which in turn depends on the absolute configuration of a formyl group at C-9 position. Diplophyllin (**410**) isolated from *Chiloscyphus polyanthos* and *Diplophyllum albicans*, (+)-frullanolide (**380a**) from *Frullania dilatata* and plagiochiline A (**115**) from many *Plagiochila* species also had piscicidal activity against killie-fish (0.4 ppm–6.7 ppm/240 min (*35, 36, 60*).

12. Neuritic Sprouting Activity

Mastigophorenes A (**321**), B (**322**) and D (**324a**) from *Mastigophora diclados* exhibited neurotrophic properties at 10^{-5}–10^{-7} M, greatly accelerating neuritic sprouting and network formation in the primary neuritic cell culture derived from the fetal rat hemisphere but mastigophorene C (**322**) and the monomeric isocuparenes (**309, 311, 313**) suppressed neuritic differentiation (*186*). Plagiochilal B (**126**) from *Plagiochila fruticosa* showed not only acceleration of neurite sprouting but also enhancement of choline acetyl transferase activity in a neuronal cell culture of the fetal rat cerebral hemisphere at 10^{-5} M (*185*).

13. Muscle Relaxing Activity

Marchantin A (**993**) and related cyclic bis-bibenzyls are structurally similar to bis-bibenzylisoquinoline alkaloids such as tubocurarine (**1395**) which are pharmacologically important muscle relaxing active drugs. Surprisingly, marchantin A (**993**) and its trimethyl ether (**993a**) also had muscle relaxing activity (*255, 528*). Nicotine in frog Ringer solution effects maximum contraction of rectus abdominis in frogs (RAF) at a concentration of 10^{-6} M. After preincubation of marchantin A trimethyl ether (**993a**) (at a concentration of 2×10^{-7}–2×10^{-4} M) in Ringer solution, nicotine (10^{-8}–10^{-4} M) was added. At a concentration of 10^{-6} M, the contraction of RAF decreased by about 30%. *d*-Tubocurarine (**1395**) exhibits similar effects as does (**993a**) using acetyl choline and the same

(1395) d-Tubocurarine

results as described above obtained. Although the mechanism of action of marchantin A **(993)** and its methyl ether **(993a)** in effecting muscle relaxation is still unknown, it is interesting that these cyclic bisbibenzyls from liverworts possessing no nitrogen atoms cause concentration dependent decrease of contraction of RAF. Marchantin A and its trimethyl ether also had muscle relaxing activity *in vivo* in mice. MM2 calculations indicate that the conformation of marchantin A and its trimethyl ether and the presence of an *ortho* hydroxyl group in **(993)** and an *ortho* methoxyl group in **(993a)** contribute to the muscle relaxation activity *(528)*.

14. Miscellaneous

The liverwort *Ptychanthus striatus* and the mosses *Barbella pendula, B. enervis, Floribundaria nipponica, Hypnum plumaeforme* and *Neckeropsis nitidula* contain much vitamin B_2 *(523)*. Chickens and puppies fed a diet including the powdered bryophytes gained more weight than did control animals. The supplement did not cause any sickness or distaste *(528)*. Most mosses contain high unsaturated fatty acids. ICHIKAWA *et al.* *(290, 291)* reported the presence of prostaglandin-like fatty acids (**1357, 1358, 1360–1363**) in *Dicranum scoparium, D. japonicum* and *Leucobryum.* These and other unsaturated fatty acids are viscous liquids and it is thought that they are instrumental protecting herbivorous animals living in very cold places from cold *(468)*.

Acetylcholine **(1396)** and a cytokinin-like compound N^6-$(\Delta^2$-isopentenyl)adenine have been found in callus tissue from the hybrid of *Funaria hygrometrica* × *Physcomitrium pyriforme (110, 611)*. *Marchantia polymorpha* and *Pellia endiviifolia* (liverwort), *Atrichum undulatum* and *Mnium hornum* (moss) produce α-tocopherol (= vitamin E) **(1397)**, vitamin K **(1398)**, plastoquinone **(1399)**, plastohydroquinone **(1400)** and α-tocoquinone **(1401)** *(35, 36)*. *Racomitrium japonicum* also contains α-tocoquinone **(1401)** *(448a)*.

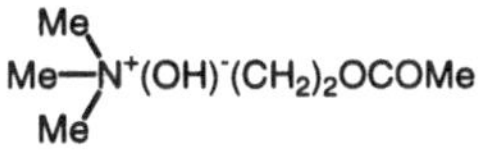

Me
Me—N⁺(OH)⁻(CH₂)₂OCOMe
Me

(1396) Acetylcholine

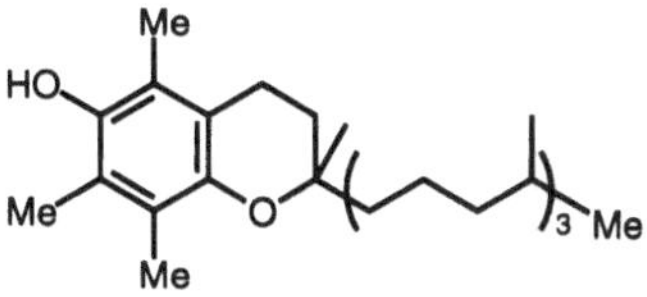

(1397) α-Tocopherol (=Vitamin E)

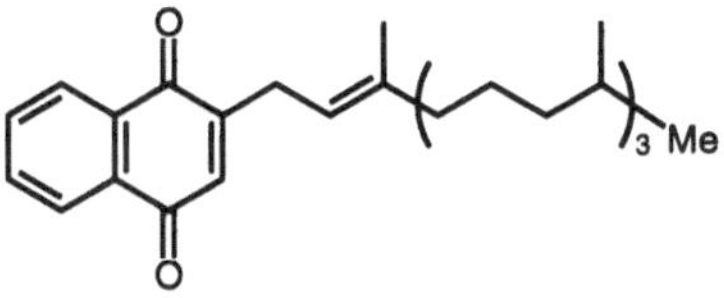

(1398) Vitamin K

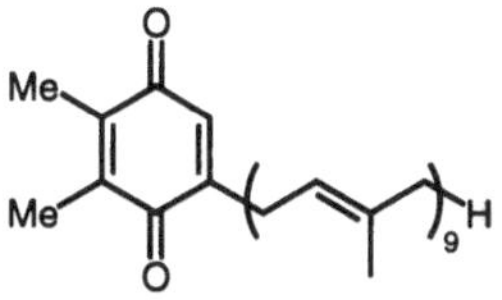

(1399) Plastoquinone

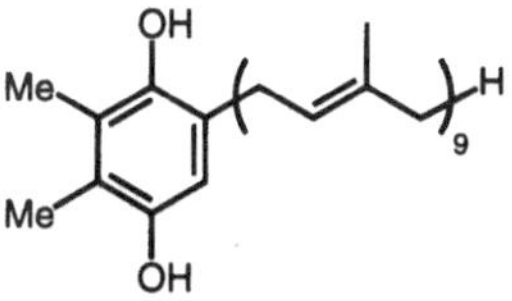

(1400) Plastohydroquinone

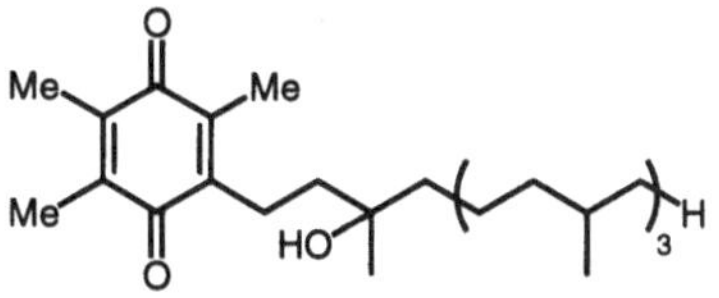

(1401) α-Tocoquinone

Chart 85b. Some biologically active compounds found in the bryophytes

VI. Chemosystematics of Hepaticae

The secondary metabolites, terpenoids and aromatic compounds such as flavonoids and bibenzyls found in bryophytes are valuable tools for studying chemosystematics. Some liverworts produce a number of complex terpenoids whereas others contain only one or two structurally simple terpenoids and aromatic compounds. If greater chemical complexity of related secondary metabolites represents an advanced character with a group of related taxa it might serve to delineate not only chemical, but evolutionary relationships within Hepaticae at the genus or family level (*33*). However, since the pattern of terpenoids and aromatic compounds often depends not only on developmental stage, season and altitudinal distribution, but also on sex, male, female and sterile forms of the same species and collections from different habitats should be examined.

In an earlier review (*19*) the chemosystematics of two subclasses, the Jungermanniidae and the Marchantiidae, and the chemical interrelationship between the Jungermanniales and the Metzgeriales were discussed briefly. In the modern classification of the Hepaticae, the Jungermanniales and Metzgeriales are united within the subclass Jungermanniidae (*493*). The chemical evidence regarding the Metzgeriales and the Jungermanniales supports Schuster's phylogenetic classification of the two orders. The significant chemical markers of 74 species, 21 genera and 33 families of the Hepaticae and Anthocerotae have been presented (*18*).

Comparative phytochemistry and taxonomy of Metzgeriales and Jungermanniales based on the occurrence of flavonoid and cinnamic acid derivatives has been discussed (*414*). In the Metzgeriales, the Metzgeriaceae, Blasiaceae and Hymenophytaceae are distinguished by their flavonoids whereas the Aneuraceae and Pelliaceae possess a characteristic pattern of cinnamic acid-type compounds. Families of the Jungermanniales can be divided into three groups: 1) those containing phenolic cinnamic acid derivatives, 2) those phenolic group with possibly flavone or flavonol derivatives and 3) those phenolic group with flavones as main components.

Basing the chemosystematics of Hepaticae on the comparative flavonoid chemistry is valid only at family, genus and species levels (*419*). The distribution of C-glycosyl flavones in 22 species of liverworts has been investigated by 2D-TLC (*413*). All compounds were isolated from an 80% aq. methanol extract. No flavones could be detected in *Herbertus*, *Mastigophora* and *Ptilidium* species. The flavonoids chemistry of *Trichocolea* and *Porella* species is not uniform. Some *Trichocolea* species produce flavonoids but others do not.

Mues *et al.* (*426*) analyzed the flavonoids of about 470 liverworts and 150 mosses by 2D-TLC and detected flavonoids. In *ca.* half of the species of each class flavonoids were detected. Liverworts are distinguished from mosses by the presence of dihydroflavones and flavonols. Mosses, on the other hand contain 3-deoxyanthocyanidins, isoflavones, biflavones and isoflavone-flavone dimers whereas liverworts do not. The predominant flavones of liverworts are of O-glucuronides, combined O-glycosides/O-glucuronides and di-C-glycosides. Flavone O-glucuronides, combined O-glycosides/O-glucuronides are predominant flavonoids of Marchantiidae (Hepaticae) and O-glycosides and di-C-glycosides are major flavonoids in Jungermanniidae. The predominant flavonoids of mosses are free aglycones and O-glycosides. Tricetin 6,8-di-C-β-D-glucopyranoside (**1199**) is considered to be a marker flavone di-C-glycoside for liverworts, since it has not been found in any other plant group. Flavone di-C-glycosides are flavonoid markers of leafy liverworts (*426*).

MARKHAM (*362, 363*) proposed the following phylogenetic relationships between liverwort orders based on flavonoid distribution shown in Fig. 1. The seven orders are related to the hypothetical ancestral stock by a process of biochemical reduction (or simplification) of flavonoids. *Takakia* has been shown to contain the key biosynthetic capabilities of all other Hepaticae orders. Within Marchantiales, Conocephalaceae and Marchantiaceae are considered to be advanced families since they elaborate basic flavone glucuronides. The Ricciaceae is a less advanced family because elaboration of basic apigenin and luteolin O-glucuronides is much less extensive. The Corsiniaceae and Sphaerocarpaceae (Sphaerocarpales) are also considered to be less evolved families on the flavonoid data. On the basis of the flavonoid features the Radulaceae, Frullaniaceae (Jungermanniales) and Metzgeriaceae (Metzgeriales) were considered as the most highly evolved families in the Jungermanniales and the Metzgeriales. However, the distribution of terpenoid and lipophilic aromatic

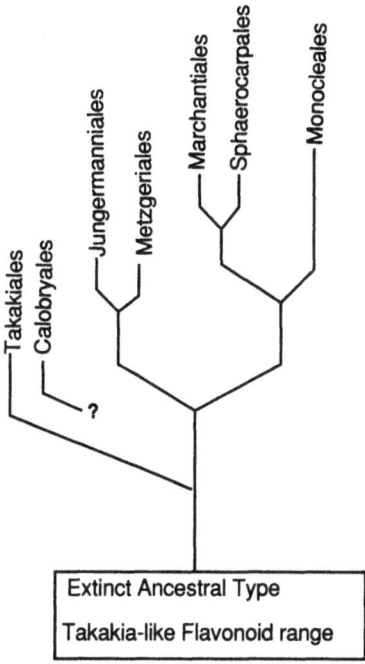

Fig. 1. Phylogenic relationships between orders of Hepaticae, based on flavonoid distribution (*363*)

compounds which have often been obtained from Hepaticae as major components of oil bodies does not accord with these suggestions. For example, neither terpenoids nor complex cyclic bis-bibenzyls which are characteristic chemical markers of Aneuraceae, Riccardiaceae and Dilaenaceae of Metzgeriales have so far been detected in any species belonging to the Metzgeriaceae. It is very dangerous to draw conclusions about the phylogeny of Hepaticae using as criteria only flavonoids or terpenoids. It is necessary to use those major metabolites which are significant endogenous characters of the Hepaticae.

Although a number of different flavonoids have been isolated from or detected in mosses, particularly Bryales, there is still no solid base for chemotaxonomic evaluations of mosses with flavonoids as marker components at the order, family or genus levels (419).

The use of electrophoresis for bryophyte taxonomy has been reviewed by KRZAKOWA (349). Several liverworts and mosses have been studied by electrophoretic methods (527). Sharp electrophoretic intraspecific polarization has not been detected in mosses and it seems that in this respect mosses are more similar to angiosperms than to liverworts.

In the following, the chemosystematics of several families of Hepaticae will be discussed in more detail.

1. Jungermanniidae

1.1 Metzgeriales

1.1.1 Metzgeriaceae

Metzgeria furcata var. *furcata* has been divided into three chemotypes on the basis of flavonoid distribution: Type I: tricetin-apigenin-type; Type II: apigenin-glycoside-type and Type III: apigenin-luteolin-type (419, 549).

1.1.2 Aneuraceae (Riccardiaceae)

Aneuraceae (= Riccardiaceae) are classified into two genera, *Aneura* and *Riccardia* on the basis of morphology. The compositon of the terpenoids and aromatic compounds of *Aneura* is significantly different from that of *Riccardia*. For example, *Riccardia multifida* produces the macrocyclic bis-bibenzyl derivatives, riccardins A (**985a**) and B (**987a**), together with a unique prenylindole (**1097**) (78, 79) which has been found in *R. chamedryfolia* (19). On the other hand, *Aneura pinguis* elaborates pinguisane- and norpinguisane-type sesquiterpenoids (*19*), which are significant chemical markers of some Jungermanniales species (*18*). There is no obvious affinity between *Riccardia lobata* var. *yakushimensis* and the other *Riccardia* species because the former species produces characteristic sacculatane- (**795**) and isosacculatane-type diterpenoids (**796**) as major components (*42*). *R. lobata* var. *yakushimensis* is morphologically similar to *Pellia endiviifolia* belonging to the Dilaenaceae and should be placed in a different genus within Riccardiaceae. The chemical results support this suggestion because both *R. lobata* var. *yakushimensis* and *P. endiviifolia* biosynthesize the sacculatal (**795**) and isosacculatal (**796**) as major components (*42*). *Riccardia crassa* is chemically quite different from the other *Riccardia* species because it biosynthesizes new type of sesquiterpene phenols, riccardiphenols A (**603a**) and B (**604**) (*583*). Electrophoretic study indicates that *Aneura pinguis* has two phenotypes. A similar intraspecific variation has been found in *Pellia endiviifolia* and *P. epiphylla* (*527*).

1.1.3 Pallaviciniaceae

Pallavicinia subciliata (= *P. longispina*) is chemically similar to *P. lyellii* because both produce aromadendrane- and bicyclogermacrane-type sesquiterpenoids as major components (*19*). *P. levieri* produces the strongly pungent diterpene dialdehyde, sacculatal (**795**), together with a chettaphanin-type diterpenoid (**777**) (*251*). Thus, *P. levieri* is chemically distinct from the other two *Pallavicinia* species. *P. levieri* is very close to *Pellia endiviifolia* [Pelliaceae (= Dilaenaceae)] because the major component of the latter species is sacculatal (*19*).

1.1.4 Blasiaceae

In Japan, only two species of Blasiaceae are known, *Blasia pusilla* and *Cavicularia densa*. The distribution of flavonoids in these two species has been investigated (*419*). Both species elaborate apigenin and crysoeriol glycosides; thus the flavonoid composition supports the notion that *Blasia* and *Cavicularia* are closely related within the family Blasiaceae. *Blasia pusilla* is a very isolated species of thalloid liverworts because it produces cyclic bis-bibenzyl dimers (**1037a–1040a**) (*262a–c*). *Blasia pusilla* is also chemically close to *Riccardia multifida* [Aneuraceae (= Riccardiaceae)] because both species produce the same riccardin-type macrocyclic bisbibenzyls, riccardins C (**988**) and F (**986**) (*249*). However, the riccardins have not been detected in *C. densa*.

1.2 Calobryales

1.2.1 Haplomitriaceae

Haplomitrium mnioides belongs to the order Calobryales which is considered to be a very primitive taxon. The Calobryales are traditionally divided into suborder Calobryineae and suborder Takakiineae (*493*) while GROLLE (*223*) classified *Takakia* and *Haplomitrium* in the independent order, Takakiales and Calobryales (Chapter I). *H. mnioides* produces complex labdane-type diterpenoids; hence, chemically speaking, it is a more advanced species than those of the Isotachidaceae and the Herbertaceae of Jungermanniales. In *Takakia lepidozioides* and *T. cerato-phylla*, neither phytanes nor labdanes have been detected, thus the genera *Haplomitrium* and *Takakia* are clearly different and the Takakiales are properly considered as an order regarded as a proper order (*77*) as mentioned by GROLLE (*223*).

1.3 Jungermanniales

1.3.1 Jungermanniaceae

The polymorphic liverwort *Jungermannia infusca* is taxonomically complex. *J. infusca* is divided into three chemotypes. Collections belonging to Type I are intensely bitter because they contain the potent bitter kaurane-type glucosides (**721a, 722a, 723–725**) (*594*). Collections belonging to Type II and III are tasteless. Type II biosynthesizes *ent*-kauranes (**693, 694, 699a, 699b, 700a, 700b, 702**), while Type III produces both clerodanes (**626–629, 631**) and (**625a, 630**) and labdanes (**738–740**). These chemical data play an important role in helping to understand the polymorphism of *J. infusca* (*594*). *Jungermannia comata* is characterized chemically since it produces only a bis-bibenzyl, perrottetin E (**1023a**) (*437*), which has also been found in leafy liverworts such as *Radula* species (*49, 50, 598*) and thalloid liverworts *Marchantia* and *Monoclea* species (*590*).

   J. truncata is chemically similar to *J. infusca* (kaurane-labdane-type). *J. vulcanicola* chemically resembles *J. infusca* (kaurane-labdane-type) although the hydroxylated positions in the kaurane skeleton are different (*437*).

   J. exsertifolia subsp. *cordifolia* produces a trachylobane-type diterpenoid (**818**) which has not been found in any other *Jungermannia* species so far (*240*). Thus, chemically, at least *J. exsertifolia* is a very isolated species of *Jungermannia*.

Nardia compressa (*19*), *N. scalaris* and *N. subclavata* (*585*) are chemically close to chemo-type II (*ent*-kaurane-type) of the *Jungermannia* because all three *Nardia* species biosynthesize *ent*-kaurane diterpenoids. However, occurrence of malonate esters of *ent*-kauranes is the most significant chemical property of the *Nardia* genus.

The genus *Demotarisia* of Jungermanniaceae resembles morphologically the genus *Jamesoniella* of Lophoziaceae. *D. linguifolia* produces clerodane-type diterpenoids as the major components (*431*) while *Jamesoniella autumnalis* produces kauranes (*431*), labdanes and clerodanes (*112*) as major components. Thus, there is no chemical affinity between *D. linguifolia* and *Jamesoniella autumnalis* except for the presence of the common clerodanes in both species.

Three chemotypes of *Mylia* species have been discovered so far. Type I (*M. taylorii*) contains aromadendrane- and secoaromadendrane-type sesquiterpenoids (*386*), type II (*M. anomala, M. verrucosa*) contains verrucosanes and neoverrucosanes (*246, 534*) and type III (*M. nuda*) contains labdanes (*624*). *Mylia anomala, M. nuda, M. taylorii* and *M. verrucosa* all produce common flavone 6,8-di-C-β-D-glucopyranosides which are widely distributed in other liverworts (*424*). The presence of the characteristic verrucosanes and common flavone-6,8-di-C-glycosides indicate that the genus *Mylia* is isolated within Jungermanniaceae intermediate between the Jungermanniaceae and the Plagiochilaceae (*419, 424*).

1.3.2 Lophoziaceae

This group has been included in Jungermanniaceae as subfamily Lophozioideae or classified as the independent family Lophoziaceae (see Table 1a).

The chemical constituents of the Lophoziaceae are very complex and some of them are structurally similar to substances found in marine organisms.

Members of Lophoziaceae are chemically divided into six chemotypes as follows: Type I, an eudesmane-type, includes *Lophozia ventricosa* and *Tritomaria quinquedentata*, type II, a cembrane-type, includes *Tetralophozia setiformis* and *Chandonanthus hirtellus*, Type III, a daucane sesquiterpenoid- and dolabellane diterpenoid-type, includes *Barbilophozia floerkei, B. hatcheri* and *B. lycopodioides*, type IV, a clerodane-type, includes *Gymnocolea inflata*, type V, a labdane-, kaurane- and clerodane-type, includes *Jamesoniella autumnalis* (*429, 565*) and type VI, a sphenolobane diterpene-type, includes *Anastrophyllum minutum* (*108*).

Jamesoniella autumnalis which has been included within Jungermanniaceae as the lone member of subfamily Jamesonielloideae or within

Lophoziaceae is chemically close to *Jungermannia truncata* and *J. infusca* (kaurane-types), which are part of the subfamily Jungermanni-oideae in Jungermanniaceae, because common *ent*-11α-hydroxykauranes have been found in *J. autumnalis* and the two *Jungermannia* species (*437*).

1.3.3 Gymnomitriaceae (Marsupellaceae)

Gymnomitrion and *Marsupella* species are common liverworts in the Gymnomitriaceae. *G. obtusum* is chemically different from *G. concinnatum* because the former contains barbatane (gymnomitrane)-type sesquiter-penoids (e.g. **162**) as major components (*19*) and the latter contains a bisabolane, *ent*-nuciferal (**218**) (*432*). *Marsupella emarginata* and *M. aquatica* are chemically close because both species elaborate longipinanes (*19, 279, 434, 434a*). *M. emarginata* var. *patens* is chemically more similar to *Gymnomitrion obtusum* than the above two *Marsupella* species since it produces barbatanes and no longipinanes (*384*).

1.3.4 Arnelliaceae

Three genera, *Arnellia*, *Gongylanthus* and *Southbya* are known in the family Arnelliaceae. The chemical constituents of only one species, *Gongylanthus ericetorum*, have been investigated and a unique cadinane-type sesquiterpene ether (**241**) has been isolated as the major component (*44*). As compound (**241**) has not been found in any liverworts so far examined, it is used as one of the chemical markers of *Gongylanthus* genus.

1.3.5 Plagiochilaceae

There are at least 3000 species of *Plagiochila*. Of these, two European (*19*), one New Zealand (*39, 188*), 12 Japanese (*19*), 13 Chilean (*46*) and 30 Peruvian *Plagiochila* species (*48*) have been investigated chemically and classified into eight chemo-types as follows: Type 1: 2,3-secoaromaden-drane-type, type II: bibenzyl-type, type III: cuparane-isocuparane-type, type IV: bibenzyl-cuparane-isocuparane-type, type V: gymnomitrane (barbatane)-bicyclogermacrane-type, type VI: bicyclogermacrane-spath-ulenol-type, type VII: pinguisane-type and type VIII: sesquiterpene lactone-type. Members of type I are further subdivided into more highly evolved or more primitive species depending the degree of oxidation and acetylation of the 2,3-secoaromadendrane-type sesquiterpenoids (*33*). *Plagiochila micropterys* which belongs to chemotype I is unique because it produces 1,4-dimethylazulene along with the 2,3-secoaromadendranes

(*433*). *P. stephensoniana* belonging to chemo-type II is also characteristic because it biosynthesizes an *epi*-verrucosane which has not been found in any *Plagiochila* species so far examined (*188*). *P. alternans* (*433*) and *P. rosariensis* which belong to chemotype VII (*48*) are chemically very distinct from the other *Plagiochila* species examined so far because they produce pinguisanes which are significant chemical markers of some species of Lejeuneaceae, Porellaceae, Trichocoleaceae and Aneuraceae (= Riccardiaceae) (*18*). Panamanian *P. moritziana* which has been placed in chemotype VIII elaborates very characteristic sesquiterpene lactone dimers (**415–419**) (*515, 516*). *P. trabeculata* which belongs to type V is quite isolated from the other *Plagiochila* species examined so far because it elaborates only barbatane-type sesquiterpenoids (*589*). *P. trabeculata* is closely related chemically to *Gymnomitrion obtusum* belonging to Marsupellaceae since both species produce the same barbatanes, although the two species are morphologically quite distinct (*19*). There is no chemical relationship between *Plagiochila rutilans* and the other *Plagiochila* species examined since the former species produces a characteristic 2,2-dimethylallylbenzene (**1096**) (*278*). European *P. spinulosa* is chemically similar to South American *P. exigua*. The terpenoids of both species are distinctly different from all other species examined (*295*). *P. sciophila* (= *P. acanthophylla* subsp. *japonica*) is one of the isolated *Plagiochila* species because it produces fusicoccane diterpenoids (**686–689**) (*261*) and unique bis-bibenzyls (**1015a, 1016a, 1017a, 1018a**) (*259*). There is some chemical affinity between *P. sciophila* and *P. corrugata* because the latter species elaborates fusicoccanes similar to those isolated from *P. sciophila* (*441a*). Japanese *P. sciophila* are divided into three forms, *P. sciophila* fo. *japonica* (Fo. 1), fo. *fragilis* (Fo. 2) and fo. *robusta* (Fo. 3), by their sesquiterpene constituents: Fo. 1 produces cyclocolorenone (**96**), β-barbatene (= gymnomitrene) (**162**), bicyclogermacrene (**196**), maalioxide (**453**) among which (**453**) is the major component, Fo. 2 contains (**162**), (**196**) and (**453**) but (**96**) is absent, while Fo. 3 only elaborates (**162**) and (**453**) (*379*).

1.3.6 Lophocoleaceae

The Lophocoleaceae are divided into five genera; *Chiloscyphus, Clasmatocolea, Heteroscyphus, Leptoscyphus* and *Lophocolea*. *Chiloscyphus polyanthos* is chemically close to *Clasmatocolea vermicularis* because two species produce the same eudesmanolide, diplophyllin (**410**), as the major secondary metabolite (*19, 60*). *Lophocolea heterophylla* is chemically unique because it contains a homomonoterpene (−)-2-methylisoborneol (**48**) which is one step more evolved than monoterpene, together with calamenene-type sesquiterpenes (**244, 248a, 253**) and eudesmanolides (**405–408**) (*586*). While *Chiloscyphus polyanthos* elaborates eudesmanolides (*19*), *C. pallescens* produces chiloscyphane- (**267–269**) and oppositane-type sesquiterpenoids (**270**) (*146, 241*), thus, there is no chemical affinity between the two *Chiloscyphus* species. *Heteroscyphus bescherellei* produces *ent*-junceic acid (**632**) as the major secondary metabolite, no other terpenoid found in species belonging to the genera *Lophocolea* and *Chiloscyphus* has been detected even by GC-MS (*83*). *H. planus* contains the highly oxygenated clerodanes (**660–663**) together with 2,3-secoaromadendranes which are significant chemical markers of Plagiochilaceae (*252*). Thus, *H. planus* is chemically rather similar to those *Plagiochila* species which belong to chemotype I (*18, 48*). The differences presented in this section indicate that there is no chemical affinity between *Chiloscyphus*, *Heteroscyphus* and *Lophocolea* although so far only six species of Lophocoleaceae have been investigated chemically.

1.3.7 Scapaniaceae

Scapania undulata is highly evolved by chemical criteria since its sesquiterpenoid features are very complex (*33*). European *S. undulata* comprises three chemical races; an (+)-*ent*-epicubenol (**233**) type, a longifolene (**481**) type and a longiborneol (**482**) type (*281, 289*). The chemical composition of *S. uliginosa* and *S. subalpina* differs considerably from that of *S. undulata* (*289*). Each of the three *Scapania* species whose segregation is based on morphological differences possesses a specific chemical profile.

Little attention has been paid to seasonal and geographical variation of secondary metabolites within the Hepaticae. HUNECK *et al.* (*282*) studied the amounts of longifolene (**481**), longiborneol (**482**) and longipinanol (**485**) in the essential oil of *Scapania undulata* in the course of one year. The yield of essential oil content reaches a maximum of 1.39% in March. The content of longifolene (~ 28%) and longiborneol (~ 34%) remains relatively constant although it reaches a distinct maximum in April, while the content of longipinanol increases slowly to 24% from January to April followed by a sharp decrease to 3% between April and May. Scottish *S. undulata* may be divided into two chemical races based on the presence or absence of specific labdane-type diterpenoids (*289*).

1.3.8 Balantiopsidaceae

Balantiopsis rosea is morphologically close to *Isotachis*, although the former is included in the Balantiopsidaceae (Jungermanniineae) and the latter in the Isotachidaceae (Herbertineae) (232). Because the chemical constituents of *B. rosea* are quite close to those of *I. japonica* except for the presence or absence of sesquiterpenoids, it has been suggested that both genera are very close and might share a common ancestor (65).

1.3.9 Adelanthaceae

The Adelanthaceae are divided into two subfamilies, Adelanthoideae and Odontoschismatoideae which contain *Odontoschisma* and *Jackiella* genera. There is no chemical affinity between *Odontoschisma denudatum* and *Jackiella javanica* because the former produces dolabellane diterpenoids (**669–672**) (381, 388) and the latter produces *ent*-verticillane diterpenoids (**837–843**) (247, 437) which have not been found in any liverworts so far examined. *Wettsteinia schusterana* (37) and *W. inversa* (623a) which belong to Adelanthoideae are chemically very isolated from the other Jungermanniales species because they biosynthesize unique naphthalene (**1057–1060**) and isocoumarin derivatives (**1061–1064**). There is no chemical affinity between *Wettsteinia* and the former two genera.

1.3.10 Schistochilaceae

The Schistochilaceae contain 54 species, five of which have been investigated chemically. So far there appear to be four chemotypes in this family, type I: a long chain alkylphenol-type (*Schistochila appendiculata*), type II: a bis-bibenzyl-type (*S. glaucescens*), type III: a 13-*epi*-neo- and 13-*epi*-homoverrucosane-type (*S. nobilis*), type IV: a 13-neoverrucosane- and 13-homoneoverrucosane-type (*S. acuminata*) and type V which contains rearranged pimaranes and clerodanes (*S. aligera*).

S. appendiculata produces long chain alkylphenols (**1043, 1047, 1049**) as major components; no terpenoids were detected even by GC-MS (53). On the other hand no alkylphenols have been detected in *S. nobilis* (54). The difference in chemistry supports the placement of the two species within *Schistochila*. *S. appendiculata* has been placed in a subgenus, *Schistochila*, within section *Schistochila* while *S. nobilis* is in subgenus *Chaetoschistochila*, section *Volantes* (496). *S. glaucescens* produces the new bis-bibenzyl derivatives (**1013a, 1014a**) (562). Similar compounds have been found not only in Jungermanniales and Metzgeriales but also in Marchantiales and Monocleales (22, 61, 71, 72, 78, 579, 590). It is noteworthy that the Taiwanese *S. acuminata* elaborates 13-neoverrucosane (**829**) and a 13-homoverrucosane (**835**) (626) while *S. nobilis* produces their 13-epimers (**833a, 834**) (54).

1.3.11 Antheliaceae

Two *Anthelia* species, *A. julacea* and *A. juratzkana*, are known in Europe. The latter species also grows in Asia. *A. julacea* produces not only *ent*-16β-hydroxykaurane (**707**), but also two unique bisabolanes, *ent*-nuciferal (**218**) and its dihydro derivative (**219**) which are significant chemical markers of this species (*435*). The major component of *A. juratzkana* is also *ent*-16β-hydroxykaurane (**707**) (*19*). The Antheliaceae are chemically similar to the Jungermanniaceae because they biosynthesize the same *ent*-kauranes as well as *ent*-nuciferal (**218**) (*434*).

1.3.12 Lepidoziaceae

There are two chemotypes of *Bazzania* species, an albicanyl caffeate-cuparane-type (type I) and a calamenane-type (type II). *B. fauriana* whose morphology is quite different from other *Bazzania* species produces bazzanenyl- (**179**), drimenyl- (**334a**) and albicanyl caffeate (**335**) as well as valencane- (**592**), barbatane- (= gymnomitrane-) (**161–163**) and eudesmane-type sesquiterpenoids (**374, 375**) (*581*). Thus, *B. fauriana* is also chemically quite distinct from the above two-types. The chemical results correlating with the morphological difference led to the proposal that *B. fauriana* represents a different chemotype III (*581*). HAYASHI and MATSUO (*264*) reported differential distribution of sesquiterpenoids in seven *Bazzania* species, *B. bidentula*, *B. japonica*, *B. pompeana*, *B. tricrenata*, *B. tridens*, *B. trilobata* and *B. yoshinagana*. The chemical markers of these *Bazzania* species are barbatanes and bazzananes. *B. pompeana* seems to be a more advanced species because of its chemical complexity. *Bazzania spiralis*, *B. harpago* and *B. praerupta* collected in East Malaysia are chemically neither closely related to each other nor related to previously examined *Bazzania* species (*19, 341*). It is interesting that *B. japonica* and *B. tridens* contain cyclomyltylane-type sesquiterpenoids (**518a, 520**) (*84, 627*). Similar cyclomyltaylanes and myltaylane (**517, 516**) have also been isolated from *Mylia taylorii* belonging to the Jungermanniaceae (*533, 535*), although there is no morphological affinity between *Bazzania* and *Mylia*. Some *Bazzania* and *Porella* species are characterized by the formation of drimanes. The former contained esterified drimanes while the latter elaborate the drimane dialdehyde, polygodial (**336**) (*33*). *Lepidozia* species produce bazzanene (**175**) as the major component. This sesquiterpene hydrocarbon is also a chemical marker of *Bazzania* species belong to the same family (*18*).

1.3.13 Calypogeiaceae

Calypogeia species characteristically produce 1,4-dimethyl azulene (**137**) and analogues which are chemical markers of this family (*18, 321, 439, 507, 538, 539*). The same substance has been isolated from South American *Macrolejeunea pallescens* (Lejeuneaceae), *Plagiochila micropterys* and *P. longispina* (Plagiochilaceae) (*433, 507*). In spite of this the Calypogeiaceae, Lejeuneaceae and Plagiochilaceae cannot be considered as related.

1.3.14 Isotachidaceae

The Isotachidaceae comprise one of the isolated families of the Jungermanniales. *Isotachis japonica* produces the unique sulfur-containing acrylates (**1228, 1229**) (*76*), simple benzoates and cinnamates which are very important chemical markers of the genus (*19*). No terpenoids have been detected in *I. japonica* even by GC-MS analysis (*76*). However, *I. humectata* and *I. haematodes* collected in South America produce mono- and/or sesquiterpenoids as well as the same benzoates and/or cinnamates found in *I. japonica* (*45, 47*). It has been suggested that *I. humectata* and *I. haematodes* are more highly evolved species than *I. japonica*.

1.3.15 Trichocoleaceae

There are three genera of Trichocoleaceae, *Trichocolea*, *Neotrichocolea* and *Trichocoleopsis*. While there is no chemical affinity among these genera (*19*), *Trichocolea pluma* is chemically quite similar to *T. tomentella* because both species produce prenylbenzoates (*51*).

1.3.16 Ptilidiaceae

The genus *Mastigophora* has been included in the Ptilidiaceae (*232*) or in the Lepicoleaceae subfamily Mastigophoroideae (*223*). The major constituents of *Ptilidium* species are pinguisane-type sesquiterpenoids (*19*) which have not been found in *Mastigophora* species. *Herbertus* species (Herbertaceae, in suborder Ptilidiineae) produce the same isocuparane (herbertane)-type sesquiterpenoids (**309, 311, 313**) (*57, 393*), as those found in *Mastigophora diclados* (*186, 191*). These results indicate that the Mastigophoroideae are almost identical chemically with the Herbertaceae and that the two families might originate from a common ancestor. CHAU's proposal *(134)* that the genus *Mastigophora* is close to *Herbertus* species is supported by the chemical results cited here.

1.3.17 Herbertaceae

The Herbertaceae which are considered to be morphologically very primitive produce only cuparanes and herbertanes (isocuparanes) *(57, 393)*. No other sesquiterpenoids have been detected in this family. The chemical data support the taxonomic position of this family. *Trichocolea tomentella* which is referred to suborder Herbertineae in the modern classification of the Hepaticae contains isoprenylbenzoates *(19)*. It is apparent that there is no chemical affinity between *Herbertus* and *Trichocolea* species.

1.3.18 Radulaceae

Radula Dum. is an isolated genus in the Jungermanniales. 61 Asian *Radula* taxa have been recognized and are taxonomically divided into three subgenera: *Radula, Cladoradula* and *Odontoradula (642)*. The chemical constituents of 13 *Radula* species have been analyzed by TLC, GC and GS-MS *(43, 49, 50, 66)*. Subgenus *Radula* produces characteristic bibenzyls with a seven-membered dihydrooxepin skeleton and subgenus *Cladoradula* biosynthesizes prenylated bibenzyls with a five-membered ring together with 2-prenyl-3,4,5-trihydroxybibenzyls *(49, 50, 66)*. *R. kojana* (subgenus *Odontoradula*) elaborates bibenzyls with a 2,2-dimethoxychromene ring skeleton along with 2-prenyl-3,5-dihydroxybibenzyls. In a study of *R. brunnea, R. constricta* and *R. okamurana* (subgenus *Radula*) and *R. chinensis* and *R. companigera* (subgenus *Cladoradula*) neither perrottetin A-type compounds (**952a, 953a, 954a**) nor the cyclized compounds (**933, 945, 946a**) could be detected by TLC, GC and GS-MS *(43)*. *R. perrottetii*, also within subgenus *Cladoradula*, is chemically different from the other *Radula* species examined so far because it elaborates 3,4,5-trihydroxy-2-(3-methyl-2-butenyl)bibenzyl (**952a**) and its cyclization products as major components *(43)*. These chemical differences among the three subgenera support the modern classification of the Radulaceae *(642)*. Almost all compounds isolated from *Radula* species are bibenzyls and/or prenyl bibenzyls and the presence of terpenoids is extremely rare. These data also support the notion that the Radulaceae comprise a quite isolated family in the Jungermanniales *(542)*.

The water soluble yellow pigments of *Radula buccinifera, R. carringtonii, R. complanata, R. grandis, R. lindenbergiana, R. nudicailis, R.*

plicata, R. tasmanica, R. uvifera and *R. wichurae* have been investigated and found to consist of different flavonoids (*415*). As mentioned earlier *Radula complanata* furnished ten flavone di-C-glycosides identified as apigenin, luteolin and tricetin di-C-glycosides, with glucose and arabinose as C-linked sugars. Tricetin-6,8-di-C-β-D-glucopyranoside (**1199**) was detected in all ten taxa and is considered as one of the chemical markers for the genus. In *R. lindenbergiana*, two different flavonoid patterns have been observed. One (the Middle Europe type) produces flavone-C-glycosides almost identical with those found in *R. complanata*. Another, the Mediterranean-type, produces the same tricetin-6,8-di-C-glycosides (**1199, 1203, 1205**), but lacks apigenin derivatives and elaborates an additional tricetin-6-C-hexoside, probably tricetin-6-C-β-glucoside (**1200**).

1.3.19 Pleuroziaceae

The Pleuroziaceae are a distincitive family of the Jungermanniales which is sometimes placed in a separate suborder Pleuroziineae because of several unique morphological features. It is considered to be closer to the Jungermanniineae than to the Radulineae and Porellineae (*422*). The chemical constituents of *Pleurozia acinosa* are similar to those of *P. gigantea*; however, the latter produces dolabellane- (**673**), fusicoccane-(**680–682**) and chettaphanin-type diterpenoids (**776**) and the distribution of sesquiterpenoids in *P. gigantea* is also different from that of *P. acinosa* (*52, 625*). In view of the greater complexity of the diterpenoids *P. gigantea* is assumed to be more advanced than *P. acinosa*. It is also interesting that fusicoccanes have been found in some *Plagiochila* species (*261, 441a*) although the Plagiochilaceae are morphologically quite different from the Pleuroziaceae.

The flavonoid pattern of ten species of Pleuroziaceae (genera *Pleurozia* and *Eopleurozia*) has been investigated using 2D-TLC (*422*). The flavone di-C-glycosides, lucenin-2 (**1164**), tricetin-6,8-di-C-glucoside (**1199**) and stellarin-2 (**1204**) were found to be marker flavonoids for *Pleurozia*, while there was no chemical affinity between *Eupleurozia* and *Pleurozia*. The species of *Pleurozia* were divided into three groups by flavonoid pattern. The first group, *P. acinosa*, *P. articulata* and *P. caledonica* produce 3'-O-glucosylated lucenin-2 (**1164**) and the marker flavonoids, lucenin-2 (**1164**), stellarin-2 (**1204**) and tricetin-6,8-di-C-glucoside (**1199**). The second group, *P. conchifolia* is characterized by the presence of the marker flavonoids as well as two further luteolin-type C-glycosides, carlinoside (**1188**) and isocarlinoside (**1189**). The third group, *P. gigantea*, *P. purpurea*, *P. giganteoides* and *P. heterophylla* only produce the marker flavonoids. These four species are suggested to be most advanced in the Pleuroziaceae because of their reduced flavone pattern (*422*). Two *Eupleurozia* sections are recognized: *E.* sec. *Eupleurozia* (*E. paradoxa*) and *E.* sec. *Ampliatae* sect. nov. (*E. simplicissima*). *Eupleurozia giganteoides* was returned to *Pleurozia* on the basis of its flavonoid pattern and morphological evidence.

1.3.20 Porellaceae

The Porellaceae are close morphologically to the Frullaniaceae; however, there is no chemical affinity at all, except for the presence of an aromadendrane-type sesquiterpene ketone, *ent*-cyclocolorenone (**96**) in both chemo-type V of *Frullania* and in the *Porella vernicosa* complex (*19*).

Fifteen *Porella* species analyzed chemically were divided into five types, a drimane-aromadendrane-type (type I), a sacculatane-type (II), a pinguisane-type (III), a pinguisane-sacculatane-type (IV), and a pseudoguaiane-sacculatane-type (V) (*18*). Type V actually produces africanes instead of pseudoguaianes (*441a, 571, 596, 597*).

Indian *Porella densifolia* subsp. *appendiculata* which belongs to chemo-type III is chemically very similar to Japanese *P. densifolia* var. *fallax*, since both species produce common pinguisanes and *ent*-kauranes (*67*). *P. cordaeana* collected in Europe and *Porella navicularis* grown in North America elaborate neither drimanes nor aromadendranes which have been found in *Porella vernicosa* complex of type I, but produce mainly striatane- and pinguisane-type sesquiterpenoids and sacculatane-type diterpenoids and therefore belong to chemo-type IV of the Porellaceae (*593, 591*).

Porella acutifolia subsp. *tosana* is chemically very similar to *P. japonica* (Type IV) because both species produce the same pinguisane-germacrane-, and guaiane-type sesquiterpenoids and the same sacculatane-type diterpenoid, perrottetianal (**805a**), except for the presence of germacra-12,8α-olide and the absence of germacra-12,6α-olide in *P. acutifolia* subsp. *tosana* (*600*). Because *P. acutifolia* subsp. *tosana* contains both 12,8α and 12,6α-olides it could be considered a more advanced species than *P. japonica* (*600*).

Porella platyphylla is considered more primitive than the other *Porella* species since it elaborates only pinguisanes as the major components (*33*). *P. platyphylla* is chemically variable at least as regards its flavonoid because the flavonoid patterns of this species collected in 16 different localities of various parts of Europe are surprisingly different (*413*). Japanese *P. caespitans* var. *setigera* is chemically very close to South American *P. swartziana* because both species produce unique africane- and secoafricane-type sesquiterpenoids (*441a, 571, 571a, 571b, 597*).

1.3.21 Frullaniaceae

The family Frullaniaceae comprises more than 500 species. Twenty-five taxa of *Frullania* have been studied chemically and divided into five chemotypes; a sesquiterpene lactone-bibenzyl-type (type I), sesquiterpene lactone-type (type II), a bibenzyl-type (type III), a monoterpene-type (type IV) and a cyclocolorenone-type (type V) (*18, 19*). *Frullania falciloba* belongs to chemotype III since it elaborates bibenzyl derivatives as major secondary metabolites and no sesquiterpene lactones (*64*). *F. serratta* is assigned to chemotype I because of the presence of eudesmane-type sesquiterpene lactones and bibenzyl derivatives (*51*). *F. hamatiloba* is chemically very distinct since the major constituents are labdanes with a sesquiterpene lactone as a minor component (*588*). *F. hamatiloba* represents as sixth chemotype, i.e. a labdane-sesquiterpene lactone-type (Type VI) (*588*). This correlates with the morphological differences between *F. hamatiloba* and other typical *Frullania* species.

F. tamarisci subsp. *obscura* grown in Asia which belongs to chemotype II, is further divided into two subtypes, Type-T and type-O (*63, 511*). Type-T produces the unusual pacifigorgiane-type sesquiterpene alcohol tamariscol (**523**) and 5α,7β(H)-eudesm-4α,6α-diol (**366**) as the major components, whereas type-O lacks these two sesquiterpenoids while eudesmane-type sesquiterpene lactones are predominant. Representatives of type-T have been found in high mountains at 1500–3000 m altitude and in the northern part of Japan (42–44 °N), while type-O occurs more frequently at lower altitudes between 32 and 40 °N. Type-T is chemically similar to the American *F. tamarisci* subsp. *asagrayana* and European *F. tamarisci* subsp. *tamarisci* which produce tamariscol (**523**), although the sesquiterpene lactones present are different. American *F. tamarisci* subsp. *nisquallensis* is chemically different from *F. tamarisci* subsp. *asagrayana* and *F. tamarisci* subsp. *tamarisci*, except for the presence of (−)-frullanolide (**380b**). Taiwanese *F. nepalensis* produces tamariscol as a minor component, but its sesquiterpene lactones differ from those in the *F. tamarisci* complex containing tamariscol.

1.3.22 Lejeuneaceae

The Lejeuneaceae are the largest family of the Hepaticae (*ca.* 80 genera and hundreds of species) and are mainly tropical in distribution. Most species are epiphytes and confined to the rain forest. About 60 species of Lejeuneaceae in 26 genera have so far been checked for the occurrence of terpenoids (*19*). It appears that most of taxa elaborate large quantities of sesquiterpenoids and/or diterpenoids, whereas only few synthesize monoterpenoids and aromatic compounds.

Subfamily Ptychanthoideae has been divided into a *Ptychanthus* complex (*Mastigolejeunea*, *Thysananthus*, *Ptychanthus* and *Tuzibeanthus*), an *Acrolejeunea* complex (*Acrolejeunea*, *Trocholejeunea* and *Frullanoides*), an *Archilejeunea* complex (*Spruceanthus* and *Archilejeunea*) and a *Lopholejeunea* complex (*Lopholejeunea* and *Marchesinia*). This classification is supported by the presence or absence of striatene- (**500, 502**), deoxopinguisone- (**536**) and pinguisanine-type sesquiterpenoids (**554**) (*18, 217, 219*). *Frullanoides densifolia* of Lejeuneaceae is chemically very close to *Porella japonica* of Porellaceae, because both species produce pinguisanes and guaianolides (*14, 554–556*). *Porella cordaeana* and *P. navicularis* also produce closely related pinguisanes, rearranged pinguisanes and monocyclofarnesanes like those found in *Frullanoides densifolia* (*591, 593*). Similar pinguisane- and the same norpinguisane-type sesquiterpenoids found in *Porella* species have been isolated from *Bryopteris filicina* belonging to the subfamily Bryopteridoideae (Lejeuneaceae) (*432*). These results further support the conclusion that the Lejeuneaceae and the Porellaceae originated from a common ancestor (*18*). *Macrolejeunea pallescens* is a very distinct species because it elaborates 1,4-dimethylazulene (**137**) which is one of the significant chemical markers of the Calypogeiaceae (*433*).

The subfamily Lejeuneoideae contains *Lejeunea*, *Leptolejeunea* and *Omphalanthus* complex, *etc*. Ten species belonging to the *Omphalanthus* complex have been investigated chemically. *Cheilolejeunea serpentina* and *C. trifaria* produce serpentiphenol (**607a**) and trifaranes (**609–613**) which have not been found in any other liverworts examined so far (*249*) while *C. excisula* and *C. imbricata* elaborate striatene-type sesquiterpenoids as the major components (*219*). Thus, there are at least two chemotypes in *Cheilolejeunea* genus. The major components of *Omphalanthus filiformis* (*441a, 571*), *O. paramicola* and *O. platycoleus* (*219*) are chamigrane-type sesquiterpenoids (**264**). Thus, the *Omphalanthus* are chemically different from the *Cheilolejeunea*. The major components of *Leucolejeunea xanthocarpa*, *L.* aff. *decurrens*, *Anoplolejeunea conferta* are striatanes indicating that these species are similar to the *Cheilolejeunea* producing such sesquiterpenes (*219*). *Leptolejeunea elliptica* belonging to the *Leptolejeunea* complex is chemically very distinct from any other species of the Lejeuneaceae so far examined because it produces a unique aromatic ether, *p*-ethylanisol (**1386**) (*219, 442*).

Nipponolejeunea is morphologically very peculiar and is nowadays placed in the monotypic subfamily Nipponolejeuneoideae. The unique presence of the monoterpenes, borneol (**38**) and bornyl acetate (**39**) (*19*) clearly underlines the patristic differences between this genus and other Lejeuneaceae (*217, 219*). Striatane-type sesquiterpenoids [striatene (**500**) and striatol (**502**)] have been detected in the subfamilies Bryopterido-ideae, Ptychanthoideae and in some members of the *Omphalanthus* complex (Lejeuneoideae), but not in Nipponolejeuneoideae, Cololejeu-neoideae or the Lejeunea complex (Lejeuneoideae). As the Bryopterioideae and the *Omphalanthus* complex are morphologically closer to Ptychan-toideae than to the other groups which have been investigated so far, the distribution of striatane- and pinguisane-type sesquiterpenoids apparent-ly corroborates morphological evidence and seems indicative of major evolutionary relationships (*219*). Five different types of flavonoid (flavon-ols, flavone di-C-glycosides, flavone C/O-glycosides, flavone O-glyco-sides and acylated flavone O-glucosides) have been observed in four *Lejeunea* species, all of which have been found in other liverworts as well (*217*). Thus, there appears to be no essential difference in flavonoid chemistry between Lejeuneaceae and the other liverworts.

2. Marchantiidae

2.1 Sphaerocarpales

The taxonomic position of *Carrpos* species is still uncertain. *Carrpos* is thought to be related to *Corsinia* and both of these genera have been united in the order Marchantiales. The phytochemical relationship of *Carrpos* to *Corsinia* has been discussed by MARKHAM (*361*). Because of the presence of flavonols, *Corsinia coriandrina* is considered to be an isolated and more primitive member of the order of the Marchantiales whilst *Carrpos* is strongly aligned with the reduced members of the "main-stream-type" Marchantialean genera, because *Carrpos sphaerocarpos* (= *Monocarpus sphaerocarpus*) elaborates biosynthetically simple fla-vone 7-O-glucuronides as well as aureusidin-6-O-glucuronide (**1334**). The phytochemical results do not support the suggested inclusion of both genera in the same isolated suborder Cordiniinae in Marchantiales. *Carrpos* is placed near Sphaerocarpos either in a separate suborder (Carripinae) or in a separate family [Carrpaceae (= Monocarpaceae)] (*361*).

2.2 Monocleales

Monoclea species are placed in a separate order Monocleales on the basis of morphology. The Monocleales differ chemically from the Metz-geriales and Marchantiales because they produce cyclic bis-bibenzyls different from those found in the other orders (*590*). *M. forsteri* grown in New Zealand produces two new acetylenic fatty acids (**1225, 1226**) as well as perrottetin- (**1023a**) and riccardin-type bis-bibenzyls (**989a, 990**) (*590*). On the other hand, *M. gottschei* biosynthesizes not only bis-bibenzyls, perrottetin E (**1023a**), marchantin C (**995**) and neomarchantin A (**1013a**), but also sesquiterpenoids. Neither riccardin-type bis-bibenzyls nor fatty acids containing an ene-yne-one partial structure have been isolated from *M. gottschei*. Reinvestigation of the constituents of the New Zealand species *M. forsteri* showed the presence of sesquiterpenoids, the content of which is considerably lower than that in *M. gottschei* and of simple composition in all habitats examined (*514*). The chemical constituents of *M. gottschei* from different geographical regions and even within the same populations vary widely. Two allopatric subspecies are recognized in *M. gottschei*, based on the morphology, *M. gottschei* subsp. *gottschei* in Chile and *M. gottschei* subsp. *elongata* in tropical America (*218*). Chemically, *M. gottschei* appears to be a more advanced species than *M. forsteri*.

It is thought that *Makinoa* species of Japan and south-east Asia might have evolved from *Monoclea via Verdoornia* growing in New Zealand (*294*). However, there is no chemical affinity between *Makinoa* and *Monoclea*.

2.3 Marchantiales

2.3.1 Targioniaceae

Targionia species are very small thalloid liverworts and produce pinane-type monoterpene acetates (**29, 30**) as major secondary meta-bolites (*75*) (Scheme 71). Apparently no chemical affinity between Tar-gioniaceae and other families belonging to Marchantiales. Recently, Whittemore (*616*) suggested from the distribution of flavonoids that the Targioniaceae are derived from ancestors similar to the modern families Exormothecaceae and Aytoniaceae.

2.3.2 Aytoniaceae (Grimaldiaceae)

The Aytoniaceae (= Grimaldiaceae) have been divided into two subfamilies, Aytonideae and Reboulioideae. *Plagiochasma rupestre* and *P. intermedium* of Aytonideae are chemically distinct, the former species producing elemane-type sesquiterpenoids as major components (*243*) and the latter the cyclic bis-bibenzyl, marchantin H (**1003**) (*550, 576*). *Reboulia hemisphaerica* and *Mannia fragrans* which belong to the second sub-

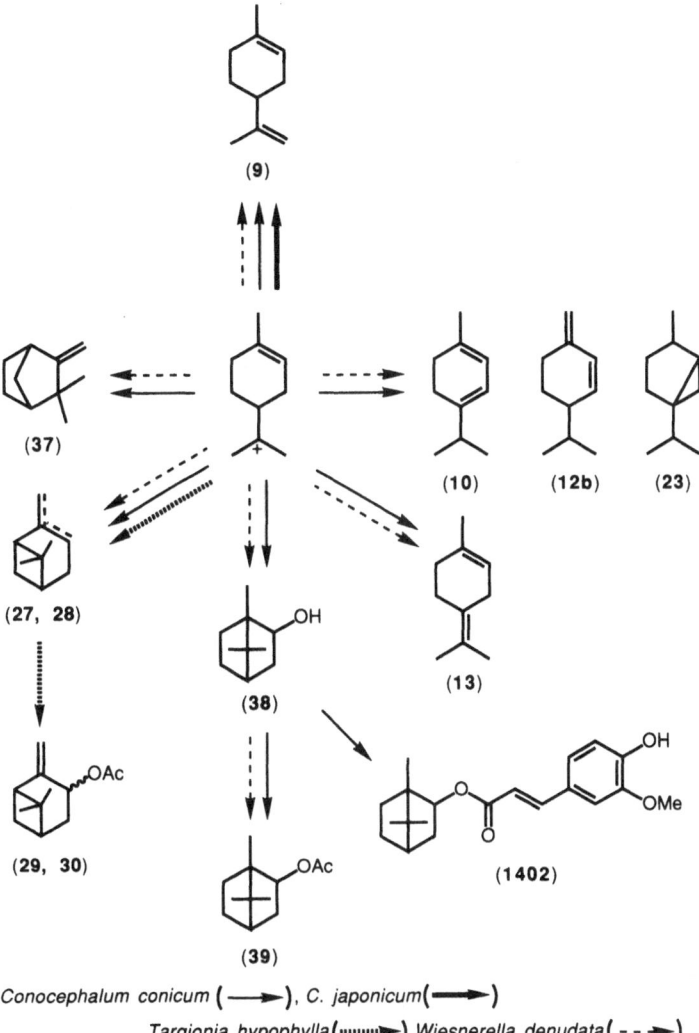

Conocephalum conicum (——►), C. japonicum(━━━►)

Targionia hypophylla(⸱⸱⸱⸱⸱►),Wiesnerella denudata(- - ►)

Scheme 71. Possible biogenetic pathways for monoterpenoids found in four liverworts belonging to the Marchantiales

family are chemically similar because they produce cuparene-type sesqui-terpenoids and macrocyclic bis-bibenzyls. There are two chemical races in *Reboulia hemisphaerica*. Material grown in Europe produces mainly cuparanes (**303, 304**) while Asiatic material elaborates aristolanes (**82, 83**) and no cuparanes (*249*). *M. fragrans* and *M. subpilosa* differ chemically, the former elaborating cuparanes and neomarchantin-type macrocyclic bis-bibenzyls (*276, 280*), and the latter cadinanes (*613*) and marchantin-type cyclic bis-bibenzyls (*612*). Aytonideae and Reboulioideae are chemically related since both subfamilies give rise to macrocyclic bis-bibenzyls. The occurrence of the common bis-bibenzyls in *Mannia*, *Plagiochasma* and *Reboulia* (Aytoniaceae) and some *Marchantia* species suggests that the Aytoniaceae have closer affinities to the Marchantiaceae (*61*).

2.3.3 Conocephalaceae

Conocephalum conicum is chemically more advanced than *C. japonicum* (= *C. supradecompositum*) since the former elaborates not only monocyclic but also bicyclic monoterpenoids and the latter produces only limonene (**9**), as a minor metabolite. *C. conicum* is also more closely related chemically to *Wiesnerella denudata* than *C. japonicum* because they produce the same monoterpenoids except for the presence or absence of bornyl ferulate (**1402**) (*19, 33*).

C. japonicum and *W. denudata* produce eudesmanolides and germacranolides. The latter species also elaborates guaianolides (**467–469**), hence *W. denudata* might be considered to be more evolved than *C. japonicum*. It has been suggested that tulipinolide (**443**) found in *Wiesnerella denudata* might be formed from costunolide (**441**) which has not been detected in Japanese *W. denudata* (*19, 33*). This suggestion was supported by the subsequent isolation of (**441**) from East Malaysian *W. denudata* (*51*).

Four enzymes, the peroxidase, glutamate-oxaloacetate transaminase, glutamic dehydrogenase and esterase of twenty-one populations of European *Conocephalum conicum* have been studied (*348*). From each population, ten thalli were investigated by means of starch gel electrophoresis. This led to conclusion that there are three chemical races in European *C. conicum*. Volatile components of ninety populations of Japanese *C. conicum* have been analyzed by GC-MS with the presence of three chemical races, α-thujene-type (77 samples), bornyl acetate-type (9 samples) and methyl cinnamate-type (4 samples) being confirmed (*580*).

Porter (*466*) analyzed the flavonoids of *Conocephalum conicum* collected at widespread sites across the northern hemisphere as well as Japanese *C. japonicum*. On the basis the flavonoids pattern, European samples of *C. conicum* were divided into four chemical races. One of the races consists of plants with less robust thalli, hence the chemical differences are correlated with a recognizable morphological truit. The North American collections of *C. conicum* showed the greatest biosynthetic diversity and share features common to both the East Asian and

European populations. This suggests that North America is the species origin. The flavonoid chemistry shows that *C. conicum* has strong affinities to that of *Preissia quadorata*. The flavonoid glycoside chemistry of Sino-Japanese collection of *C. japonicum* was clearly different from that of *C. conicum*, a distinction which is also evident from the terpenoid chemistry, hence Sino-Japanese *C. japonicum* was placed in a separate genus *Sandea* (*33, 68*).

2.3.4 Lunulariaceae

The Lunulariaceae are a chemically primitive family in suborder Marchantiineae as the genus *Lunularia*, its sole representative, produces acyclic bis-bibenzyls (**1024a, 1032**) (*255*) and their dimer (**1036**) (*249*). The family is related chemically to the Marchantiaceae rather than the Conocephalaceae.

2.3.5 Marchantiaceae

Bucegia and *Neohodgsonia* have at times been placed in their own subfamily within the Marchantiaceae. Comparative flavonoid biochemistry does not support this close relationship between the two genera because the flavonoids of *Neohodgsonia* are apigenin 7-O-glucuronide (**1403**) as well as luteolin-7-O-glucuronide (**1149**), luteolin-7,3'-di-O-glucuronide (**1276**) and luteolin-3'-O-glucuronide (**1404**) which are commonly encountered in Marchantiaceae and thus contrast significantly with those of *Bucegia* (*19, 366*). The earlier positioning of *Neohodgsonia* and *Bucegia* together as a distinct group on morphologial grounds must

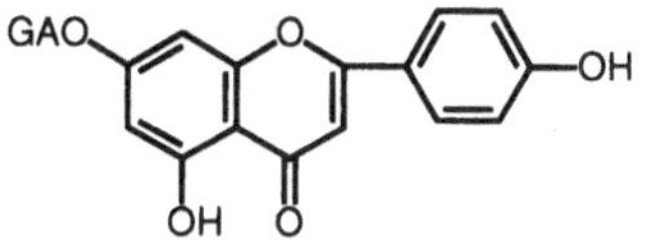

(**1403**) Apigenin-7-O-glucuronide

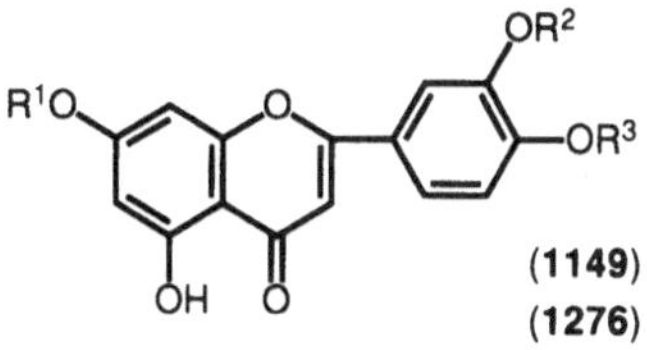

(**1149**) Luteolin-7-O-glucuronide ; R^1=GA, R^2=R^3=H
(**1276**) Luteolin-7,3'-di-O-glucuronide ; R^1=R^2=GA, R^3=H
(**1404**) Luteolin-3'-O-glucuronide ; R^1=R^3=H, R^2=GA

Chart 86. Flavonoids found in *Neohodgsonia* species

be modified such that *Bucegia* is separated from all other genera of Marchantiaceae (*366*).

Ontogenetic data have suggested that Marchantiopsidae have a close affinity with the Metzgeriales (*155*). The occurrence of common cyclic bis-bibenzyl derivatives in some species of Metzgeriales (*78, 79, 254*) and Marchantiales (*22, 61, 71, 72, 579*) may provide support for this suggestion.

Gas chromatograms of crude extracts of Japanese and French *Marchantia polymorpha* were identical. The major terpenoids of both species were (+)-β-chamigrene (**264**), (−)-cuparene (**284**) and (−)-cyclopropanecuparenol (**298**) (*74*). However, marchantin A (**993**), a cyclic bis-bibenzyl which is the most abundant constituent of Japanese *M. polymorpha*, has not been detected in French *M. polymorpha*. The latter elaborates marchantin E (**1000a**) as the major secondary metabolite. The chemical constituents of *M. paleacea* var. *diptera* resemble those of *M. polymorpha*, both species producing the same sesquiterpenoids and marchantin-type compounds. *M. tosana* is chemically different from *M. polymorpha* and *M. paleacea* var. *diptera* because the sesquiterpenoid content in *M. tosana* is very low and it produces the riccardin-type cyclic bis-bibenzyl (**986**) (*74*). Collections of Japanese *M. polymorpha*, *M. paleacea* var. *diptera* and *M. tosana* from different localities exhibited considerable degree of intraspecific uniformity both qualitatively and quantitatively when the crude extracts were checked by TLC, GC and GC-MS. The chemical constituents from either female and male thalli or capsule with pedancle were quite similar to those of sterile thalli of the same species.

Indian and French collections of *M. polymorpha* are chemically very similar as both species produce marchantin E (**1000a**) as the major secondary metabolite although marchantin A (**993**) has not been detected in the French race (*72*). *M. palmata* and *M. polymorpha* exhibit chemical affinities since both species elaborate the same cyclic bis-bibenzyl derivatives (*72*). The distribution of sesquiterpenoids and cyclic bis-bibenzyls in German *M. polymorpha* is closely related to that of Japanese *M. polymorpha*, but not to French, Indian or South African collections of the same species (*71*). New Zealand *Marchantia berteroana* is chemically similar to Japanese and French *M. polymorpha* since it produces the same cuparane-type sesquiterpenoids as those found in the latter species, although bis-bibenzyls have not been detected in *M. berteroana* (*39*). On the other hand, there is no chemical affinity between New Zealand *M. foliacea* and the four other *Marchantia* species mentioned above (*39*).

Dumortiera hirsuta and *Preissia quadorata* have been included in the Marchantiaceae. The characteristic cyclic bis-bibenzyl, marchantin C

(995) has been found in the former species (585), but not in the latter species (78). This chemical result supports the view that *Dumortiera* is allied to the *Marchantia*, but shows that there is no chemical affinity between *Preissia* and *Marchantia* (78). There are at least three chemical races of Japanese *D. hirsuta* based on elemol (190)-type, dehydronerolidol (427)-type and germacrane (437)-type sesquiterpenoids (585).

2.3.6 Ricciaceae

The Ricciaceae comprise two genera, *Ricciocarpos* and *Riccia*. *Riccia* species constitute one of the most isolated genera within the Marchantiales, since they elaborate large quantities of phytosterol mixtures (18). *Ricciocarpos natans* is chemically very different from *Riccia* because it elaborates the monocyclofarnesanes (506–508) (639, 640) and a bis-bibenzyl, riccardin C (988) as well as bibenzyl glycosides (881–883) and phenethyl glycosides (1107, 1108) (351).

VII. Chemosystematics of Musci and Anthocerotae

Although it is estimated that there are approximately 14000 species in 700 genera of the Musci, only about 320 species of 120 genera have been studied chemically (419). As mentioned earlier, the Musci are chemically simple because they do not contain oil bodies. Only flavonoids, poly-unsaturated fatty acids and steroids have been found as major secondary metabolies.

At the subclass level, the Sphagnidae and Andreaeidae are quite different from Bryidae because no flavonoids have been detected in the first two subclasses. One hundred and fifty species belonging to the Bryidae have been analyzed for flavonoid occurrence; however, it is difficult to evaluate the chemosystematics using flavonoids as chemical markers at the order, family, or genus level (419).

ANDERSON et al. (8) pointed out in 1974 that the characteristic fatty acid patterns and fatty acids markers as well as differences in the classes of lipids present in mosses could be useful in the classification of mosses. Thus, the fatty acid composition of the lipids of the Amblystegiaceae, Ditrichaceae, Grimmiaceae, Mniaceae and Polytrichaceae is quite different. Some species of Grimmiaceae stand out by their very high content of 18:3 acid while the polyunsaturated C20 acids are at an extremely low level. The major fatty acids of *Polytrichum* species are 18:1, 18:2 and

18:3, the level of 18:1 being higher than in other families Amblystegiaceae, Ditrichaceae, Grimmiaceae and Mniaceae. *Ceratodon* species which
belongs to the Ditrichaceae are distinct because they produce the
acetylenic fatty acid, 9,12,15-octadecatrien-6-ynoic acid (= 18:A3)
(**1222**). *Ceratodon purpureus* is chemically different from *Ditrichum capillaceum* and *D. inclinatum* (Ditrichaceae) because the latter two species lack
acetylenic acids. On the basis of morphology, *Dicranum* (Dicranaceae) is in a different family than *Ceratodon* (Dicranaceae), however,
Dicranum montanum is closely related to *Ceratodon* species since the fatty
acid composition of the former species is very similar to that of *Ceratodon*.
In some members of Mniaceae, the content of 20:4 arachidonic acid
(ARA) is very much higher than that in the other five families. Members of
Amblystegiaceae have in common the property that the level of 20:5 fatty
acid is higher than the level of 20:4 acid. *Cratoneuron filicinum* and
Drepanocladus exannulatus are different from all other mosses because of
their very low content of 18:2 acid and the presence of appreciable
amounts of 22:5. This could indicate that these two species are closely
related.

The fatty acids of the triglycerides of Ditrichaceae (7 species), Seligeriaceae (1), Dicranaceae (31) and Leucobryaceae (1) have been investigated
leading to an arrangement of these families into three types of fatty acids
patterns (*338*). The first type includes most of the species containing the
acetylenic acid, 9,12,15-octadecatrien-6-ynoic acid (= 18:3A) (**1222**).
Dicranaceae belong to type 1. The second is the 18:3A–18:2ω6 and
18:3ω3-type which includes *C. purpureus* and *Ditrichum pusillum*. Type 3,
the 18:3A free-linolenic acid-type, includes *Ditrichum flexicaule*, *Blindia
acuta*, *Campylopus pyriformis*, etc.

The genus *Ditrichum* is clearly divided into two groups. One includes
Ditrichum pusillum and *D. heteromallum* whose acetylenic fatty acid
pattern is similar to that of *Ceratodon purpureus*. The second group
includes *D. cylindricum* and *D. flexicaule* which lack acetylenic acids and
differ in the pattern of the other fatty acids. This difference also reflects the
differences in the morphological characters of these species. Acetylenic
acids have been detected in all of the *Dicranella* and *Dicranum* species.
The occurrence of the 18:3A fatty acid may be regarded as the genus-
specific feature of *Dicranella* and *Dicranum*. The genera *Campylopus*,
Dicranodontium, *Dicranoweisia crispula* and *Paraleucobryum* form a
group which is free of acetylenic acids. The fatty acids level confirms that
Campylopus is closely related to *Dicranodontium* and that the genera
Campylopus, *Dicranodontium* and *Paraleucobryum* can be regarded as a
separate complex within the Dicranaceae, possibly forming a link to the

Leucobryaceae. *Amphidium mougeotii* whose taxonomy is not clear might be included in the Dicranaceae on the basis of its fatty acid pattern (*338*). However, the fatty acid pattern is important only when considered together with morphological characters thus, these endogenous properties provide valuable taxonomic information for mosses.

The Anthocerotae are the smallest class of bryophytes and include about 360 species. As mentioned earlier, hornworts do not produce oil bodies. The chemical constituents of hornworts have not been studied thoroughly since, as has also been mentioned earlier, only six hornworts having been investigated chemically so far (*418*). While a few sesquiterpenoids as well as a lignan have been identified, neither flavonoids nor lunularic acid which are widely distributed in Hepaticae have been found in Anthocerotae (*426*). Thus, on the basis of limited experimental work the Anthocerotae appear to be quite different chemically from the other two classes of bryophytes. Because of the lack of the chemical information, it is not possible at this time to consider the chemosystematics of the Anthocerotae.

VIII. Chemical Relationships Between Algae, Bryophytes and Pteridophytes—Evolution of Bryophytes

The evolutionary relationships between Hepaticae, Musci and Anthocerotae (Bryophyta) and the relationship of Bryophyta to the other plant phyla are obscure. Botanists have considered the origins of the lower terrestrial plants, bryophytes and pteridophytes and their evolution and differentiation from the point of view of cytology, morphology, ecology and paleontology. Traditionally there have been two hypotheses regarding the evolution of bryophytes. The first is the progressive theory which states that bryophytes have their origin in the green algae and that the pteridophytes are in turn derived from bryophytes. The second hypothesis, the reduction theory, regards the green algae as being the ancestor of Psilophytales (*Cooksonia, Rhynia*), from which evolved both bryophytes and pteridophytes. Both hypotheses have been enthusiastically discussed and both have their proponents. It is, however, rather difficult to discuss the evolution of the lower terrestrial green plants, especially because of the absence of fossil material of bryophytes. Biochemical properties have not been considered and little attention has been paid to the investigation of the chemical interrelationships between algae, bryophytes and pteridophytes, although various characteristic secondary metabolites, such as terpenoids, flavonoids, bibenzyls, steroids and fatty acids, have been isolated from or detected in these three plant groups. European *Equisetum* fossils (2×10^8 years old) and the present-day *E. sylvaticum* (Sphenopsida) contain the same *n*-alkanes and their contents are qualitatively almost identical (*330*), indicating that these plants have preserved their biochemical properties during their long history whereas the morphological features have changed dramatically. It is hoped that a comparison of the chemical constituents of present-day algae, bryophytes and pteridophytes may provide an insight into the evolutionary relationship between these three plant groups.

On the basis of morphology bryophytes have been divided into the three classes of the Hepaticae, the Musci and the Anthocerotae. It seems that mosses are more similar to angiosperms than to the liverworts because of the absence of sharp electrophoretic intraspecific polarization in mosses (*527*). The cutin acids of *Sphagnum* moss and liverworts from the order Marchantiales have been analyzed. The relative abundances of the 8,16-, 9,16- and 10,16-dihydroxyhexadecanoic acids may be of value in the chemotaxonomy of *Sphagnum* species. From the results of their cutin acid composition, liverworts and mosses may be less closely related than generally supposed, the *Sphagnum* cutins are similar to those of higher plants (*128*). The nucleotide sequences of 5S ribosomal RNAs of four bryophytes, *Marchantia polymorpha*, *Lophocolea heterophylla* (Hepaticae), *Plagiomnium trichomanes* (Musci), *Anthoceros punctatus* (Anthocerotae) were determined by the method involving chemical degradation of $[3'-^{32}P]$RNA (*320*). The 5S rRNA sequences of *M. polymorpha* and *L. heterophylla* resembled those of *P. trichomanes* (97–99% identity). The 5S rRNA sequences of *A. punctatus* are less similar to those of the above three bryophytes (91–92% identity). These results suggest that the Hepaticae are closely related to the Musci and that both classes have evolved in different directions after emergence of the Anthocerotae. The bryophytes are less closely related to green algae plants (78–83%) than to seed plants.

Amino acid sequences of the chloroplast-type ferredoxin [2Fe-2S] of many organic photosynthetic organisms have been studied and compared with each other to establish structure-function and phylogenetic relationship (*405*). *Marchantia polymorpha* ferredoxin was purified by DE-52 and Sephadex G-75 column chromatography and the complete amino acid sequence of the carboxymethylated ferredoxin was determined. The total number of amino acid residues was 95 with tryptophan lacking; the molecular weight was calculated as 10,174, excluding the iron and sulfur atoms. *M. polymorpha* ferredoxin is placed between fern or a horsetail ferredoxin and a green algal ferredoxin.

However, as mentioned earlier, the secondary metabolites of Hepaticae, Musci and Anthocerotae are quite different and exhibit no chemical affinities except for the presence of common sterols and fatty acids.

The Hepaticae are chemically quite complex although their morphology is relatively simple. The chemical composition of the Musci is uniform, although some species are morphologically more complex than the Hepaticae. In the case of the Anthocerotae, both chemical and morphological features are simple. On the other hand, marine algae produce a large number of complex secondary metabolites, terpenoids and aromatic compounds, although their morphology is relatively simple. The pteridophytes whose morphology is quite complex also produce a number of complex secondary metabolites which are very different from those of algae and bryophytes. Considering these levels of chemical and morphological complexity, the present Hepaticae seem to be rather closer to the algae (24).

1. Similarities and Differences in Terpenoid and Steroid Content

Most liverworts examined so far elaborate a great number of terpenoids with the exception of triterpenoids and lipophilic aromatic compounds. The Musci, however, are not known to biosynthesize sesquiterpenoids or diterpenoids although a few monoterpene hydrocarbons and a kaurene-type diterpene alcohol (707) have been found in some Splachnaceae species (344) and in *Saelania glaucescens* (445), respectively. Thus, considering only lower terpenoids, it would appear that the origins of two classes are quite different. The Anthocerotae may be related distantly to the Hepaticae, but the only evidence to support this is the presence in them of trace amounts of cuparane-type sesquiterpenoids. On the whole, however, the overall chemical features of Anthocerotae are completely different from the Hepaticae and the Musci. Terpenoids and lipophilic aromatic compounds are widespread in marine algae, particularly in brown and red algae (179). Some of them are the same as, and a number of them are both structurally and chiroptically very similar to, those isolated from or detected in the Hepaticae. Table V shows the distribution of terpenoids, aromatic compounds and steroids in algae, bryophytes and pteridophytes. The Hepaticae produce a number of monoterpenoids, several of which have also been detected in steam distillates of the algae. Even in the red tide phytoplankton *Gymnodinium nagasakiense*, a monoterpene alcohol, α-terpineol (17) and sesquiterpenoids, α-cadinol (231), *epi*-cubenol (234), cubenol (235) and calamenene (244) have been identified (310).

Cadinane-, calamenane-, cubebane-, drimane-, elemane-, eudesmane-, germacrane-, muurolane- and vitrane-type sesquiterpenoids have been found in both the Phaeophyceae (brown algae) and the Hepaticae. Bisabolane-, chamigrane-, cuparane-, eudesmane- and monocyclofarnesane-type sesquiterpenoids occur in both the Rhodophyceae (red algae) and the Hepaticae. Riccardiphenol A (**603a**) (*583*) is structurally close to isocopalane-type diterpenoid (**605**) and monoterpene analogue (**606**) found in brown algae (*179, 180*). Both the Hepaticae and the Phaeophyceae biosynthesize the same skeletal sesquiterpene phenols, serpentiphenol (**607a**) (*249*) and sporochnol (**608**) (*501a*). Many soft coral species contain the same or similar sesquiterpenoids and azulene-like compounds as those found in the Hepaticae and their chiroptical properties are also the same. These terpenoids might have a dietary origin, probably from phytoplanktons or algae. Few sesquiterpenoids have been found in Chlorophyceae (green algae). Only farnesane-type sesquiterpenoids have been isolated from marine green algae (*179*).

The biogenesis of the *ent*-sesquiterpenoids found in liverworts has been discussed by Matsuo (*376*). These are biosynthesized stereospecifically by special enzymes; the enzyme-substrate complex for sesquiterpenoid synthesis in liverworts may involve a conformation inverse to that of higher plants.

Cembrane-, *cis*-clerodane-, dolabellane-, fusicoccane-, kaurane-, labdane-, pimarane-, sacculatane- and verrucosane-type and some other diterpenoids have also been isolated from the Hepaticae. As shown in Table V, labdane-type diterpenoids have been found in green and red algae; their absolute configuration is the same as that of the labdanes found in the Hepaticae. Pimarane-type diterpenoids also occur in red algae. Cembrane-type diterpenoids which have been obtained from both marine animals (*179*), *Zooxanthella* (*461, 473*) and higher plants also occur in Hepaticae (*429, 566, 567*) as do dolabellane diterpenoids which otherwise are found only in brown algae and marine animals (*179, 429, 566*). A fusicoccane diterpenoid has been found in the brown algae, *Dictyota* species (*175*). A number of skeletally different diterpenoids which are isoprenologues of eudesmane-, germacrane-, guaiane- and bourbonane-type sesquiterpenoids occur in brown algae as well as in the Hepaticae. The sacculatanes (**795, 796, 805**) from the Hepaticae and the dictyolanes (**1405**) from brown algae may be biosynthesized from the same precursor as shown in Scheme 72. Diterpenoids are relatively rare in the Chlorophyceae. Thus, the diterpenoid content of Hepaticae is closely related to that of the Phaeophyceae.

The steroidal components of the Hepaticae, the Musci, the Anthocerotae and pteridophytes are very similar. They produce campesterol

Table Va. *Distribution of Terpenoids in Algae, Bryophytes and Pteridophytes*

	Algae			Hepaticae	Anthocerotae	Musci	Pteridophytes
	Chlorophyceae	Phaeophyceae	Rhodophyceae				
Monoterpenoids	+	+	+	+		+*	+*
Sesquiterpenoids							
Bisabolanes				+			
Brasilanes		+		+			
Cadinanes		+	+	+			
Calamenenes		+		+			
Chamigranes		+	+	+			
Cubebanes		+		+			
Cuparanes			+	+			
Drimanes		+	+	+			
Elemanes		+		+			
Eudesmanes	+	+	+	+			
Farnesanes	+	+		+			
Germacranes		+		+			
Illudanes					+		
Maalianes	+		+	+			+
Monocyclofarnesanes	+			+			
Muurolanes		+		+			
Serpentiphenol		+		+			
Diterpenoids							
Geranyl linalool				+			
Abietanes				+		+	+
Clerodanes				+			+

Table Va (*continued*)

	Algae			Hepaticae	Anthocerotae	Musci	Pteridophytes
	Chlorophyceae	Phaeophyceae	Rhodophyceae				
Dolabellanes		+		+			
Dolastanes		+					
Fusicoccanes	+			+			
Geranyl geraniol (Phytol)	+	+		+		+	
Labdanes			+	+		+	+
Kauranes				+			+
Oppositanes			+	+			
Pimaranes			+	+		+	+
Sacculatanes				+			
Verrucosanes				+			
Vitranes		+		+			
Sesterterpenoids							+
Triterpenoids							
Cycloartanes						+	+
Cycloeucalenol						+	+
Cyclolaudanes							+
Dammalanes							+
Euphanes						+	
Fernanes						+	+
Friedelanes				+		+	
Hopanes				+			+

Lupanes				+	
24-Methylene-cycloartanol			+	+	
Norcycloartanes	+		+	+	+
31-Norcyclolaudenol				+	+
Obtusifoliol		+		+	+
Oleananes			+	+	+
Onoceranes					+
Serratanes					+
Squalanes	+	+	+	+	+
Taraxanes				+	+
Ursanes		+	+	+	+

* Very rare

Table Vb. *Distribution of Steroids in Algae, Bryophytes and Pteridophytes*

Steroids	Algae			Hepaticae	Anthocerotae	Musci	Pteridophytes
	Chlorophyceae	Phaeophyceae	Rhodophyceae				
Brassicasterol				+		+	
Campesterol				+	+	+	+
Cholesterol	+	+	+	+		+	+
Clerosterol	+						
Clionasterol	+					+	
Dehydrocholesterol			+				
Desmosterol			+				
Ecdysones			+				+
Ergosterol						+	
Fucosterol		+					
Isofucosterol	+						
24-Methylenecholesterol	+	+					
Siringosterol							
Sitosterol				+	+	+	+
Stigmasterol				+	+	+	+

Table Vc. *Distribution of Aromatic Compounds in Algae, Bryophytes and Pteridophytes*

	Algae			Hepaticae	Anthocerotae	Musci	Pteridophytes
	Chlorophyceae	Phaeophyceae	Rhodophyceae				
n-Alkyl phenols		+		+			
Benzoic acid deriv.				+		+	+
Benzonaphtho-xanthenones						+	
Bibenzyls	+	+		+			
Bibenzyl ethers		+	+				
Biflavones			+	+		+	+
Biphenyls		+		+			+
Biphenyl ethers		+		+			+
Chalcones	+	+		+			
Chromans		+		+			+
Chromones		+		+			+
Cinnamic acid deriv.	+	+	+	+		+	+
Diphenyl methanes	+		+				
Flavonoids	(+)*	+		+		+	+
Heptadecatrienyl resorcinol		+		+		+	
Isoprenyl hydroquinones	+	+		+			
Isoprenyl quinones		+		+			
Lignans				+			+
Styrene deriv.				+			+
Xanthones			+	+			+
Indole deriv.	+		+	+			+
Sulfur-containing compounds	+	+		+			

* Tentatively identified

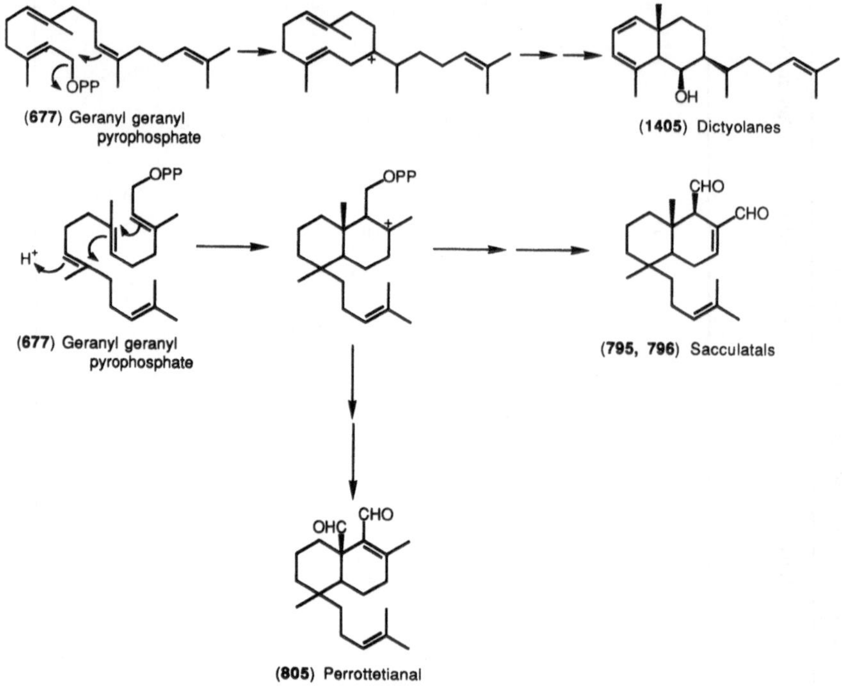

(677) Geranyl geranyl
       pyrophosphate

(1405) Dictyolanes

(677) Geranyl geranyl
       pyrophosphate

(795, 796) Sacculatals

(805) Perrottetianal

Scheme 72. Possible biogenetic pathways for dictyolanes and sacculatanes from geranyl geranyl pyrophosphate

(844a), stigmasterol (846a) and sitosterol (848a), which are also widely distributed in seed plants. In primitive Hepaticae, for example, in *Isotachis, Haplomitrium* and *Lepidozia* species, stigmasterol is the predominant sterol followed by campesterol as the second major sterol. On the other hand, the ratio of campesterol, stigmasterol and sitosterol in highly evolved liverworts like the Marchantiales is very similar to that in higher plants. Algae also produce sterols, but they are distinct from those of the Hepaticae. Green algae contain C27–C29 sterols, cholesterol (849) and clerosterol (1406). Brown algae biosynthesize fucosterol (1407) (80–90% of total content), together with 24-methylenecholesterol (1408) and cholesterol (849), while red algae also contain cholesterol, desmosterol (1409) and 22-dehydrocholesterol (1410). Recently, a sterol possessing insect moulting activity which had previously been found in pteridophytes and higher plants has been reported from red algae (*179*).

The evolutionary stage at which plants changed from producing 24β-ethylsterols to producing 24α-ethylsterols appear to involve the bryo-

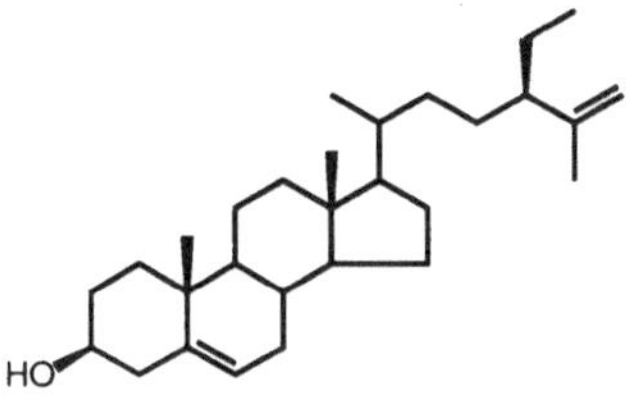

(**1406**) Clerosterol

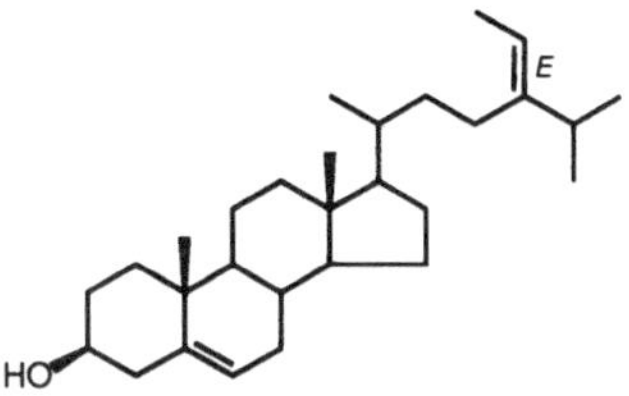

(**1407**) Fucosterol (=stigamasta-5,24(28)*E*-dien-3β-ol)

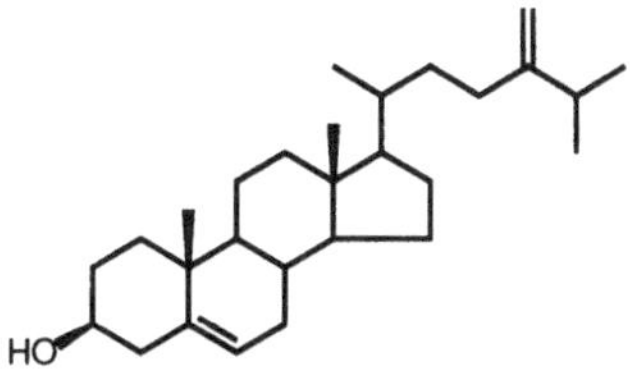

(**1408**) 24-Methylenecholesterol

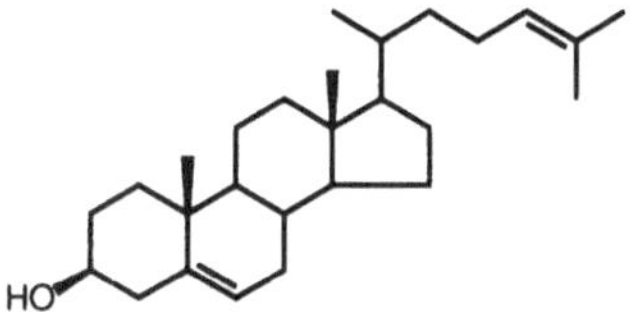

(**1409**) Desmosterol

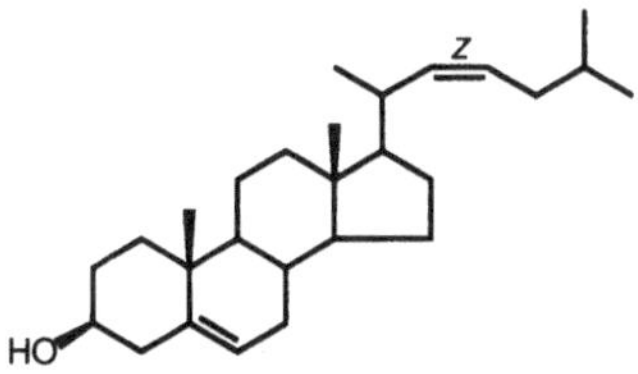

(**1410**) 22-Dehydrocholesterol
(=Cholesta-5,22*Z*-dien-3β-ol)

Chart 87. Some sterols found in algae

phyta, some species of which produce both epimers (*138, 463*). On the other hand, sitosterol in higher plants is epimerically pure 24α-ethylcholesterol (**848a**) whereas clionasterol in algae is 24β-ethylcholesterol (**848b**). The situation with respect to 24-methylcholesterol in plants is similar. Algae elaborate the 24β-methyl epimer (**844b**) exclusively, while higher plants biosynthesize epimeric mixtures of 24α- (**844a**) and 24β-methylsterols (**844b**) in which the former is predominant. All of the 24-methylcholesterol samples from bryophyta are epimeric mixtures (*e.g.* 24α/24β: 1:1 in *Marchantia polymorpha*, 1:2 in *Plagiomnium succulentum*, 1:4 in *Sphagnum palustre*). The higher level of 24β-methyl sterols and the presence epimeric mixtures of 24-ethylsterols in bryophyta is consistent with their accepted position between the thallophytes and the tracheophytes shown in Table VI (*138*).

Triterpenoids are very rare in the Hepaticae and algae. Norcycloartanes have been isolated from green algae along with squalene, three

Table VI. *Evolutionary Level of Algae, Bryophytes and Higher Plants by Fatty Acids and Sterols*

Algae——————————→Bryophytes——————————————→Higher plants		
Fatty acids		
C20–22 polyenes	C20–C24 polyenes	C16–C18
(4 to 5 double bonds)	(4 to 6 double bonds)	(3 double bonds maximum)
	longer chain and highly unsaturated	shorter chain and less unsaturated
Sterols, 24-methyl, ethyl		
α + β	α + β	α + β
(β: predominant)	(1:1)	(α: predominant)
	(1:2)	
	(1:4)	

squalene oxides and squalene alcohol. Almost all liverworts contain squalene, and traces of hopane-type triterpene hydrocarbons and alcohols have been isolated from or detected in some primitive or evolved liverworts. This phenomenon suggest that triterpenoid biosynthesis is blocked in most Hepaticae, however, hopanoids which are the significant chemical markers of pteridophytes, have often been obtained from the Musci. Thus, the distribution of common triterpenoids shows that the Musci and pteridophytes are somewhat related (*24*).

2. Similarities and Differences in Content of Aromatic Compounds

Lunularic acid (**877**), its decarboxylation product lunularin (**878**) and their related bibenzyl derivatives are widespread in the Hepaticae, but not in the Anthocerotae, the Musci or pteridophytes. However, a novel bibenzyl derivative (**909**), benzyl alcohol and some derivatives have been found in red algae (*179, 180, 211*). Thus, the Hepaticae are related to algae but not to the Musci and the Anthocerotae.

Schistochila appendiculata, Isotachis japonica and *Balantiopsis rosea* of Jungermanniales produce a large number of aromatic compounds. *S. appendiculata* biosynthesizes 3-undecyl- (**1043**), along with 3-tridecyl- (**1046**) and 3-pentadecylphenol (**1049**) as well as 6-undecyl- (**1044**), tridecyl- (**1047**) and pentadecylsalicylic acid (**1050**) and their potassium salts. 3-Tridecyl phenol (**1046**) and 6-tridecylsalicylic acid (**1047**) have also been isolated from the brown algae, *Caulocystis cephalornithos* (Sargassaceae) (322). Phloroglucinols and resorcinols which are similar to these phenolic compounds, have been reported as metabolites of brown algae (179). *Omphalanthus filiformis* elaborates 5-heptadeca-8(Z), 11(Z), 14(Z)-trienylresorcinol (1054) (441a, 571). The very similar 5-heptadecatrienyl-resorcinol (1055) has been isolated from the brown alga, *Cystophora torulosa* (179).

Some Hepaticae elaborate aromatic compounds which carry isoprenoid substituents such as 2,2-dimethylallyl- and geranylbibenzyls or 2,2-dimethylallylcatechol (278). Similar aromatic compounds are widespread in the brown algae. Some species of Jungermanniales, Marchantiales and Metzgeriales produce characteristic cyclic bis-bibenzyls possessing biphenyl and biphenyl ether linkages and many structural similar phloroglucinol derivatives having biphenyl and/or biphenyl ether linkages are present in brown algae (179, 211). No flavonoids have been detected in the brown and red algae, although a chromone derivative has been isolated from brown algae. The presence of flavonoid glycosides in Chlorophyceae has been established in only one species, *Nitella hookei* (179) while the Hepaticae, the Musci and pteridophytes produce a number of flavonoids. This is the most significant biochemical difference between algae and lower terrestrial green plants. Stilbenoids, bibenzyls and cyclized bis-bibenzyls, flavonoids and their derivatives are biosynthesized through the shikimate-malonate pathway so that in algae, the biosynthesis of flavonoids appears to be blocked.

A number of nitrogen-containing compounds has been isolated from green and red algae as well as from pteridophytes, but nitrogen-containing substances from the Hepaticae are very rare, the two prenylindoles from the *Riccardia* species of the Metzgeriales representing the only instance (79). No alkaloids have been detected in the Musci and the Anthocerotae, except for the maytansinoids found in certain mosses and or the soil on which the mosses grow (479, 644).

Two primitive liverworts, *Isotachis japonica* and *Balantiopsis rosea*, elaborate unique sulfur-containing aromatic compounds, benzyl thioacrylate (1228) and β-phenylethyl thioacrylate (1229). Some red algae produce sulfur-containing heterocyclic compounds. Similar cyclic polysulfides are present in *Dictyopteris* species (Phaeophyceae) whilst 3-methylmercaptopropylamine (1411) has been isolated from *Desmarestia* species (Phaeophyceae) (329). Dimethyl sulfide (1227) which is formed from a dimethyl β-carboxyethyl sulfonium derivative (propiothetin) (1412) by enzymatic or *in vitro* degradation is the odoriferous substance of most marine algae (Scheme 73). More than 6000 species of liverworts, in particular, the Metzgeriales, emit a strong sulfur-like odor on drying. Dimethyl sulfide has been detected in Porellaceae, Plagiochilaceae

$$MeS(CH_2)_3NH_2$$

(1411) 3-Methylmercaptopropylamine

Chart 88. A sulfur-containing compound found in algae

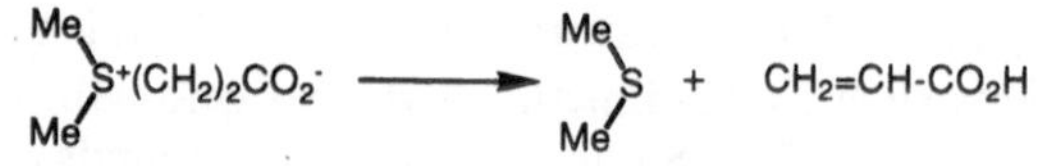

(1412) Propiothetin   (1227)   (1412a)

Scheme 73. Formation of dimethylsulfide from dimethyl β-carboxyethyl sulfonium derivative (propiothetin)

(Jungermanniales) and Dilaenaceae (= Pelliaceae) (Metzgeriales) as described in Chapter 1. On the other hand, neither dimethyl sulfide nor other sulfur containing substances have been detected in Musci and Anthocerotae. Thus, the Hepaticae are closely related to the marine algae, but not to the Musci and the Anthocerotae.

3. Similarities and Differences in Alkane and Fatty Acid Content

The distribution of n-alkanes in plants is of chemosystematic significance. In higher plants including the ferns, there is usually a maximum in the n-C27–C33 region the odd carbon number n-alkanes strongly predomination. On the other hand, the Hepaticae produce mainly n-alkanes in the C15–C20 range and only a small amount of higher homologues, while the odd/even ratio is approximately 1. The distribution of n-alkanes in the Musci is closely related to that of pteridophytes and seed plants, while the n-alkanes and the odd/even ratio in green, brown and red algae is quite similar to that of the Hepaticae. Thus, from this point of view the Hepaticae are more closely related to the algae than to other lower terrestrial green plants.

Algae—green, brown and red—contain highly unsaturated long chain fatty acids (C18:1, 18:2, 18:3, 18:4, 20:4, 20:5) as well as palmitic acid, as the major components of the fatty acid fraction (306). A large number of halogenated C15 lipids possessing an ether linkage and a triple bond have been isolated from Laurencia species which belong to the Rhodophyceae (179). The fatty acid content of the Musci is similar to that of algae with palmitic acid, C18:1, 18:2, 18:3, 18:4 and arachidonic acid (1223) as major components. Some mosses also produce other highly unsaturated fatty acids, including acetylenic acids and prostaglandine-like acids with a cyclopentenone skeleton (*8, 290, 291*). The Hepaticae also produce arachidonic and eicosapentaenoic acid (**1224**) (*503*), along with C18 acids with yne-ene and yne-en-one functional groups (*339, 340, 590*). Thus, the fatty acid profile of the Hepaticae is similar to that of the Musci, although studies of the fatty acids of the Hepaticae have so far been restricted to relatively few species (*19, 271*). A survey of distribution of the fatty acids in the plant kingdom shows that high levels of C20 and C22 unsaturated fatty acids are found in algae and decreasing amounts in bryophyta. However, some mosses growing in or near water produce an abundant C20 and C22 polyunsaturated fatty acids and it has been suggested that the C20 and C22 polyunsaturated fatty acids in such mosses may have an essential physiological function (*313*). The fatty acid profile of higher plants is different from that of the bryophytes; C16 and C18 acids rather than the C20 and C22 unsaturated fatty acids found in bryophytes with a maximum of three double bonds are predominant. The evolutionary position of bryophytes is thus reflected in their fatty acid composition as shown in Table VI. However, SEWÓN reported that bryophytes have an evolutionary pattern of their own for MGDG fatty acids and that the evolution of their fatty acid biosynthesis cannot just be described as an intermediate stage between lower and higher plants (*500*).

Botanists have suggested that bryophytes have evolved from the green algae as mentioned earlier; however, it is obvious that there is almost no chemical affinity between green algae and Hepaticae. While halogen-containing terpenoids and aromatic compounds have not been isolated from brown algae (*179, 180*), several halogenated terpenoids have been found in green algae and almost all terpenoids and aromatic compounds found in red algae possess one or two halogen atoms. This is the most characteristic chemical difference among marine algae. On the other hand, among the more than 800 species of the Hepaticae studied chemically so far, only one, *Makinoa crispata* (Metzgeriales), has yielded a halogenated terpenoid, i.e. the drimane derivative (**346**) (*258*). On this basis, the Hepaticae seem to be most closely related chemically to the Phaeophyceae among marine algae, although the Rhodophyceae contain a number of terpenoids related to those found in the Hepaticae. DAVIS (*161*) has already pointed out that the Hepaticae might be related to brown algae because of morphological and embryological similarities. It is noteworthy that the above chemical similarity between the Phaeophyceae and the Hepaticae supports this hypothesis.

On the basis of the secondary metabolites isolated from algae, bryophytes and pteridophytes, the Hepaticae and the Musci should be separated taxonomically because there are no chemical affinities between two classes. It seems that both classes might have originated from completely different ancestors. MATSUO (*376*) suggested that the Hepaticae seem to be biochemically related to the fungi rather than the algae on the basis of the presence of *ent*-sesquiterpenoids in both plant groups. This hypothesis is not tenable because as mentioned repeatedly enantiomeric terpenoids have been isolated from or detected in both marine algae and the Hepaticae.

Although no definite statement can be made at present with respect to the evolutionary processes of algae and lower terrestrial green plants, endogenous chemical characteristics may play an important role in helping to understand the evolution and differentiation of algae and lower terrestrial green plants is shown in Fig. 2 on the basis of the chemical similarities or differences described above (*24*).

CRANDAL-STOTLER (*156*) has divided bryophytes into four divisions Takakiophyta, Hepatophyta, Anthocerophyta and Bryophyta on the basis of morphological, anatomical and developmental studies. It seems that the Takakiophyta occupy either a somewhat intermediate position

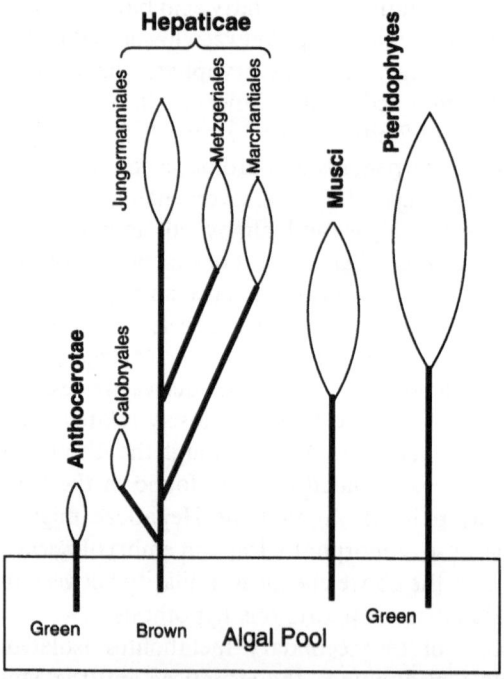

Fig. 2. Hypothetical chemical phylogeny of algae, bryophytes and pteridophytes (*24*)

between the Hepatophyta and the Bryophyta or might be included in the Hepatophyta because *Takakia* species biosynthesize characteristic sesquiterpene hydrocarbons and eudesmanolides which are also significant chemical markers of the Jungermanniales and hopane-type triterpenoids (*19*). The morphological classification of Takakiales and Calobryales is supported by their chemical properties (*77*). Recently, SMITH and DAVISON (*509a*) reclassified *Takakia ceratophylla* as a moss by the characteristics of its sporophyte and antheridia.

Acknowledgements

I am indebted to Professor WERNER HERZ (The Florida State University) whose contributions to this review went far beyond his editorial duties. Thanks are also due to Dr. FUMIHIRO NAGASHIMA (Tokushima Bunri University) for his valuable discussion and drawing the structures.

References

1. ABE, S., and Y. OHTA: Lunularic Acid in Cell Suspension Cultures of *Marchantia polymorpha.* Phytochem. **22**, 1917 (1983).

1a. ABRAHAM, W.-R., L. ERNST, L. WITTE, H.-P. HASSEN, and E. SPRECHER: New Transfused Africanols from *Leptographium lundbergii.* Tetrahedron **42**, 4475 (1986).

2. ADESOMOJU, A.A., J.I. OKOGUM, M.P. CAVA, and P.J. CARROLL: Roseadione, A Diterpene Ketone from *Hypoestes rosea.* Phytochem. **22**, 2535 (1983).

3. ADLER, J.H.: Configuration at C-24 of Sterols from the Liverwort *Pallavicinia lyellii.* Phytochem. **22**, 607 (1983).

3a. AGATA, I., T. HATANO, S. NISHIBE, and T. OKUDA: A Tetrameric Derivative of Caffeic Acid from *Rabdosia japonica.* Phytochem. **28**, 2447 (1989).

4. AIBA, C.J., R.G. CAMPOS CORREA, and O.R. GOTTLIEB: Natural Occurrence of Erdtman's Dehydrodiisoeugenol. Phytochem. **12**, 1163 (1973).

5. AKNIN, M., A. SAMB, J. MIRAILLES, V. COSTANTINO, E. FATTORUSSO, and A. MANGONI: Polysiphenol, A New Brominated 9,10-Dihydrophenanthrene from the Senegalese Red Alga *Polysiphonia ferulacea.* Tetrahedron Letters **33**, 555 (1992).

6. AMICO, V., R. CURRENTI, G. ORIENTE, M. PIATTELLI, and C. TRINGALI: 18-Hydroxy-3,7-dolabelladiene from the Brown Algae, *Dictyota dichotoma.* Phytochem. **20**, 848 (1981).

7. ANDERSON, W.H., J.L. GELLERMAN, and H. SCHLENK: Acetylenic Acids from Mosses. Lipids **10**, 501 (1975).

8. ANDERSON, W.H., J.M. HAWKINS, J.L. GELLERMAN, and H. SCHLENK: Fatty Acid Composition as Criterion in Taxonomy of Mosses. J. Hattori Bot. Lab. **38**, 99 (1974).

9. ANDO, H., and A. MATSUO: Applied Bryology. In: Advances in Bryology, Vol. 2 (W. SCHULTZE-MOTEL, ed.), p. 133. Vaduz: J. Cramer, 1984.

10. ANHUT, S., T. SEEGER, J. BIEHL, H.D. ZINSMEISTER, and H. GEIGER: Phytochemical Studies of the Moss Species *Plagiomnium elatum* and *Plagiomnium cuspidatum.* J. Hattori Bot. Lab. **67**, 377 (1989).

526 Y. ASAKAWA

11. ANHUT, S., T. SEEGER, H.D. ZINSMEISTER, and H. GEIGER: New Dihydrobiflavones from the Moss *Plagiomnium cuspidatum*. Z. Naturforsch. **44c**, 189 (1989).
12. ANHUT, S., H.D. ZINSMEISTER, R. MUES, W. BARZ, K. MACKENBROCK, J. KÖSTER, and K.R. MARKHAM: The First Identification of Isoflavones from a Bryophyte. Phytochem. **23**, 1073 (1984).
13. ANTHONSEN, T., and R. MCCRINDLE: The Constitution of Diterpenoids from *Solidago elengata* Nutt. Acta Chem. Scand. **23**, 1068 (1969).
14. ARBIYANTI, H.: Chemical Constituents of the Liverworts *Frullanoides* and *Trocholejeunea* Genera. Master Thesis, Tokushima Bunri University, 1991.
15. ASAI, M., E. MIZUTA, M. IZAWA, K. HAIBARA, and T. KISHI: Isolation, Chemical Characterization and Structure of Ansamitocin, a New Antitumor Ansamycin Antibiotic. Tetrahedron **35**, 1079 (1979).
16. ASAKAWA, Y.: Chemical Constituents of *Alnus sieboldiana* (Betulaceae), II: The Isolation and Structure of Flavonoids and Stilbenes. Bull. Chem. Soc. Jpn. **44**, 2761 (1971).
17. ASAKAWA, Y.: Comparative Study of Chemical Constituents Found in the Thalli and Female Receptacles of *Wiesnerella denudata* and *Conocephalum conicum*. J. Hattori Bot. Lab. **48**, 277 (1980).
18. ASAKAWA, Y.: Terpenoids and Aromatic Compounds as Chemosystematic Indicators in Hepaticae and Anthocerotae. J. Hattori Bot. Lab. **53**, 283 (1982).
19. ASAKAWA, Y.: Chemical Constituents of Hepaticae. In: Progress in the Chemistry of Organic Natural Products, Vol. 42 (W. HERZ, H. GRISEBACH, and G.W. KIRBY, eds.), p. 1. Wien-New York: Springer, 1982.
20. ASAKAWA, Y.: Biologically Active Substances of Hepaticae. Kagaku to Seibutsu **22**, 495 (1984).
21. ASAKAWA, Y.: Some Biologically Active Substances Isolated from Hepaticae: Terpenoids and Lipophilic Aromatic Compounds. J. Hattori Bot. Lab. **56**, 215 (1984).
22. ASAKAWA, Y.: Phytochemistry of Hepaticae: Isolation of Biologically Active Aromatic Compounds and Terpenoids. Rev. Latinoamer. Quim. **14**, 109 (1984).
23. ASAKAWA, Y.: Biologically Active Substances of Hepaticae. Seiyakukojo **5**, 284 (1985).
24. ASAKAWA, Y.: Chemical Relationships Between Algae, Bryophytes and Pteridophytes. J. Bryol. **14**, 59 (1986).
25. ASAKAWA, Y.: J.P. Patent 228409 (1985).
26. ASAKAWA, Y.: J.P. Patent 53273 (1986).
27. ASAKAWA, Y.: J.P. Patent 126013 (1986).
28. ASAKAWA, Y.: European Patent 0185921 A2 (1986).
29. ASAKAWA, Y.: Biologically Active Substances Found in Hepaticae. Farumashia **23**, 455 (1987).
30. ASAKAWA, Y.: Science Photo-Chemical Constituents of Hepaticae. Kagaku to Yakugaku no Kyoshitsu **87**, 2 (1987).
31. ASAKAWA, Y.: U.S. Patent 4659509 (1987).
32. ASAKAWA, Y.: Biologically Active Substances Found in Hepaticae. In: Studies in Natural Products Chemistry, Vol. 2. Structure Elucidation (Part A) (ATTA-UR-RAHMAN, ed.), p. 277. Amsterdam: Elsevier, 1988.
33. ASAKAWA, Y.: Chemical Evolution of Mono- and Sesquiterpenoids of Liverworts. J. Hattori Bot. Lab. **64**, 97 (1988).
34. ASAKAWA, Y.: Secondary Metabolites of Bryophytes. The Heredity **43**, 36 (1989).
35. ASAKAWA, Y.: Biologically Active Substances from Bryophytes. In: Bryophyte Development: Physiology and Biochemistry (R.N CHOPRA and S.C. BHATLA, eds.), p. 259. Boca Raton: CRC Press, 1990.

36. ASAKAWA, Y.: Terpenoids and Aromatic Compounds with Pharmacological Activity from Bryophytes. In: Bryophytes: Their Chemistry and Chemical Taxonomy (H.D. ZINSMEISTER and R. MUES, eds.), p. 369. Oxford: Oxford University Press, 1990.

36a. ASAKAWA, Y.: Biologically Active Terpenoids and Aromatic Compounds from Liverworts and the Inedible Mushroom *Cryptoporus volvatus*. In: Bioactive Natural Products: Detection, Isolation and Structural Determination (S.M. COLEGATE and R.J. MOLYNEUX, eds.), p. 319. Boca Raton: CRC Press, 1993.

37. ASAKAWA, Y.: Unpublished results.

38. ASAKAWA, Y., T. ARATANI, and G. OURISSON: Pungent Component of *Porella vernicosa* and Mossy Odor of *Frullania*. Misc. Bryol. et Lichenol. 7, 99 (1976).

39. ASAKAWA, Y., and E.O. CAMPBELL: Terpenoids and Bibenzyls from Some New Zealand Liverworts. Phytochem. 21, 2663 (1982).

40. ASAKAWA, Y., J.D. CONNOLLY, C.O. FAKUNLE, D.S. RYCROFT, and M. TOYOTA: Revised Structure of the Liverwort Sesquiterpenoid Pinguisanin. J. Chem. Res. (S) 1987, 82.

41. ASAKAWA, Y., G.W. DAWSON, D.C. GRIFFITHS, J.-Y. LALLEMAND, S.V. LEY, K. MORI, A. MUDD, M. PEZECHK-LECLAIRE, J.A. PICKETT, H. WATANABE, C.M. WOODCOCK, and Z. ZONG-NING: Activity of Drimane Antifeedants and Related Compounds Against Aphids and Comparative Biological Effects and Chemical Reactivity of (−)- and (+)-Polygodial. J. Chem. Ecol. 14, 1845 (1988).

42. ASAKAWA, Y., L.J. HARRISON, and M. TOYOTA: Occurrence of a Potent Piscicidal Diterpenedial in the Liverwort *Riccardia lobata* var. *yakushimensis*. Phytochem. 24, 261 (1985).

43. ASAKAWA, Y., T. HASHIMOTO, K. TAKIKAWA, M. TORI, and S. OGAWA: Prenyl Bibenzyls from the Liverworts *Radula perrottetii* and *Radula complanata*. Phytochem. 30, 235 (1991).

44. ASAKAWA, Y., and S. HUNECK: Unpublished results.

45. ASAKAWA, Y., and H. INOUE: Chemical Constituents of Selected Chilean Liverworts. In: Studies on Cryptogams in Southern Chile (H. INOUE, ed.), p. 109. Tokyo: Kenseisha, 1984.

46. ASAKAWA, Y., and H. INOUE: Chemical Constituents of Chilean *Plagiochila* Species. In: Studies on Cryptogams in Southern Chile (H. INOUE, ed.), p. 117. Tokyo: Kenseisha, 1984.

47. ASAKAWA, Y., and H. INOUE: Chemical Constituents of Selected Peruvian Liverworts. In: Studies on Cryptogams in Southern Peru (H. INOUE, ed.), p. 107. Tokyo: Tokai University Press, 1987.

48. ASAKAWA, Y., and H. INOUE: Chemical Constituents of Peruvian *Plagiochila* Species. In: Studies on Cryptogams in Southern Peru (H. INOUE, ed.), p. 119. Tokyo: Tokai University Press, 1987.

49. ASAKAWA, Y., K. KONDO, and M. TORI: Cyclopropanochroman Derivatives from the Liverwort *Radula javanica*. Phytochem. 30, 325 (1991).

50. ASAKAWA, Y., K. KONDO, M. TORI, T. HASHIMOTO, and S. OGAWA: Prenyl Bibenzyls from the Liverwort *Radula kojana*. Phytochem. 30, 219 (1991).

51. ASAKAWA, Y., X. LIN, K. KONDO, and Y. FUKUYAMA: Terpenoids and Aromatic Compounds from Selected East Malaysian Liverworts, Phytochem. 30, 4019 (1991).

52. ASAKAWA, Y., X. LIN, M. TORI, and K. KONDO: Fusicoccane-, Dolabellane- and Rearranged Labdane-type Diterpenoids from the Liverwort *Pleurozia gigantea*. Phytochem. 29, 2597 (1990).

528 Y. ASAKAWA

53. ASAKAWA, Y., T. MASUYA, M. TORI, and E.O. CAMPBELL: Long Chain Alkyl Phenols from the Liverwort *Schistochila appendiculata*. Phytochem. **26**, 735 (1987).
54. ASAKAWA, Y., T. MASUYA, M. TORI, and Y. FUKUYAMA: 13-Epi-*neo*- and 13-Epi-Homoverrucosane Diterpenoids from the Liverwort *Schistochila nobilis*. Phytochem. **27**, 3509 (1988).
55. ASAKAWA, Y., and R. MATSUDA: Riccardin C, A Novel Cyclic Bibenzyl Derivative from *Reboulia hemisphaerica*. Phytochem. **21**, 2143 (1982).
56. ASAKAWA, Y., R. MATSUDA, and A. CHEMINAT: Bibenzyl Derivatives from *Frullania* Species. Phytochem. **26**, 1117 (1987).
57. ASAKAWA, Y., R. MATSUDA, W.B. SCHOFIELD, and S.R. GRADSTEIN: Cuparane- and Isocuparane-Type Sesquiterpenoids in Liverworts of the Genus *Herbertus*. Phytochem. **21**, 2471 (1982).
58. ASAKAWA, Y., R. MATSUDA, and T. TAKEMOTO: Mono- and Sesquiterpenoids from *Wiesnerella denudata*. Phytochem. **18**, 1007 (1980).
59. ASAKAWA, Y., R. MATSUDA, M. TORI, and M. SONO: Efficient Preparation of Some Biologically Active Substances from Natural and Nonnatural Aromatic Compounds by *m*-Chloroperbenzoic Acid Oxidation. J. Organ. Chem. (USA) **53**, 5453 (1988).
60. ASAKAWA, Y., R. MATSUDA, M. TOYOTA, T. TAKEMOTO, J.D. CONNOLLY, and W.R. PHILLIPS: Sesquiterpenoids from *Chiloscyphus*, *Clasmatocolea* and *Frullania* Species. Phytochem. **22**, 961 (1983).
61. ASAKAWA, Y., K. OKADA, and G.W. PEROLD: Distribution of Cyclic Bis(bibenzyls) in the South African Liverwort *Marchantia polymorpha*. Phytochem. **27**, 161 (1988).
62. ASAKAWA, Y., and H. SHIOTA: Unpublished results.
63. ASAKAWA, Y., M. SONO, M. WAKAMATSU, K. KONDO, S. HATTORI, and M. MIZUTANI: Geographical Distribution of Tamariscol, A Mossy Odorous Sesquiterpene Alcohol, in the Liverwort *Frullania tamarisci* and Related Species. Phytochem. **30**, 2295 (1991).
64. ASAKAWA, Y., K. TAKIKAWA, and M. TORI: Bibenzyl Derivatives from the Australian Liverwort *Frullania falciloba*. Phytochem. **26**, 1023 (1987).
65. ASAKAWA, Y., K. TAKIKAWA, M. TORI, and E.O. CAMPBELL: Isotachin C and Balantiolide, Two Aromatic Compounds from the New Zealand Liverwort *Balantiopsis rosea*. Phytochem. **25**, 2543 (1986).
66. ASAKAWA, Y., K. TAKIKAWA, M. TOYOTA, and T. TAKEMOTO: Novel Bibenzyl Derivatives and *ent*-Cuparene-Type Sesquiterpenoids from *Radula* Species. Phytochem. **21**, 2481 (1982).
67. ASAKAWA, Y., K. TAKIKAWA, M. TOYOTA, A. UEDA, M. TORI, and S.S. KUMAR: Sesqui- and Diterpenoids from the Liverworts *Porella densifolia* subsp. *appendiculata* and *Porella densifolia* var. *fallax*. Phytochem. **26**, 1019 (1987).
68. ASAKAWA, Y., N. TOKUNAGA, T. TAKEMOTO, S. HATTORI, M. MIZUTANI, and C. SUIRE: Chemosystematics of Bryophytes, IV: The Distribution of Terpenoids and Aromatic Compounds in Hepaticae and Anthocerotae. J. Hattori Bot. Lab. **47**, 153 (1980).
69. ASAKAWA, Y., N. TOKUNAGA, M. TOYOTA, T. TAKEMOTO, S. HATTORI, M. MIZUTANI, and C. SUIRE: Chemosystematics of Bryophytes, II: The Distribution of Terpenoids in Hepaticae and Anthocerotae. J. Hattori Bot. Lab. **46**, 67 (1979).
70. ASAKAWA, Y., N. TOKUNAGA, M. TOYOTA, T. TAKEMOTO, and C. SUIRE: Chemosystematics of Bryophytes, I: The Distribution of Terpenoids in Bryophytes. J. Hattori Bot. Lab. **45**, 395 (1979).
70a. ASAKAWA, Y., and M. TORI: The Atlas of 400 MHz NMR Spectra of Natural Products. Tokyo: Hirokawa, 1993.
71. ASAKAWA, Y., M. TORI, T. MASUYA, and J.-P. FRAHM: *Ent*-Sesquiterpenoids and Cyclic Bis(bibenzyls) from the German Liverwort *Marchantia polymorpha*. Phytochem. **29**, 1577 (1990).

72. ASAKAWA, Y., M. TORI, K. TAKIKAWA, H.G. KRISHNAMURTY, and S.K. KAR: Cyclic Bis(bibenzyls) and Related Compounds from the Liverworts *Marchantia polymorpha* and *Marchantia palmata*. Phytochem. **26**, 1811 (1987).

73. ASAKAWA, Y., and M. TOYOTA: Unpublished results.

74. ASAKAWA, Y., M. TOYOTA, H. BISCHLER, E.O. CAMPBELL, and S. HATTORI: Comparative Study of Chemical Constituents of *Marchantia* Species. J. Hattori Bot. Lab. **57**, 383 (1984).

75. ASAKAWA, Y., M. TOYOTA, and A. CHEMINAT: Terpenoids from the French Liverwort *Targionia hypophylla*. Phytochem. **25**, 2555 (1986).

76. ASAKAWA, Y., M. TOYOTA, and L.J. HARRISON: Isotachin A and Isotachin B, Two Sulphur-Containing Acrylates from the Liverwort *Isotachis japonica*. Phytochem. **24**, 1505 (1985).

77. ASAKAWA, Y., M. TOYOTA, and T. MASUYA: Phytane- and Labdane-Type Diterpenoids from the Liverwort *Haplomitrium mnioides*. Phytochem. **29**, 585 (1990).

78. ASAKAWA, Y., M. TOYOTA, R. MATSUDA, K. TAKIKAWA, and T. TAKEMOTO: Distribution of Novel Cyclic Bisbibenzyls in *Marchantia* and *Riccardia* Species. Phytochem. **22**, 1413 (1983).

79. ASAKAWA, Y., M. TOYOTA, Z. TAIRA, T. TAKEMOTO, and M. KIDO: Riccardin A and Riccardin B, Two Novel Cyclic Bis(bibenzyls) Possessing Cytotoxicity from the Liverwort *Riccardia multifida* (L.) S. Gray. J. Organ. Chem. (USA) **48**, 2164 (1983).

80. ASAKAWA, Y., M. TOYOTA, R. TAKEDA, R. MATSUDA, S.R. GRADSTEIN, K. TAKIKAWA, and T. TAKEMOTO: New Diterpenoids from Lejeuneaceae, Porellaceae and Marchantiaceae. 27th Symposium on Chemistry of Terpenes, Essential Oils and Aromatics, Nagasaki, Japan. Symposium Papers, p. 58 (1983).

81. ASAKAWA, Y., M. TOYOTA, T. TAKEMOTO, and C. SUIRE: Pungent Sesquiterpene Lactones of the European Liverworts *Chiloscyphus polyanthus* and *Diplophyllum albicans*. Phytochem. **18**, 1007 (1979).

82. ASAKAWA, Y., M. TOYOTA, K. TAKIKAWA, and T. TAKEMOTO: New Normono- and Sesquiterpenoids from the Liverworts, *Lejeunea*, *Porella* and *Frullania* Species. 26th Symposium on Chemistry of Terpenes, Essential Oils and Aromatics, Yamagata, Japan. Symposium Papers, p. 172 (1982).

83. ASAKAWA, Y., M. TOYOTA, and A. UEDA: Sacculatane- and Clerodane-Type Diterpenoids from the Liverworts *Porella perrottetiana* and *Heteroscyphus bescherellei*. Phytochem. **29**, 2165 (1990).

84. ASAKAWA, Y., M. TOYOTA, A. UEDA, M. TORI, and Y. FUKAZAWA: Sesquiterpenoids from the Liverwort *Bazzania japonica*. Phytochem. **30**, 3037 (1991).

85. BAKER, R., P.H. BRINER, and D.A. EVANS: Chemical Defence in the Termite *Ancistrotermes cavithorax*: Ancistrodial and Ancistrofuran. J. Chem. Soc. Chem. Commun. **1978**, 410.

86. BAKER, R., I.F. COTTRELL, P.D. RAVENSCROFT, and C.J. SWAIN: Stereoselective Synthesis of (±)-Ancistrofuran and Its Stereoisomers. J. Chem. Soc. Perkin Trans. 1 **1985**, 2463.

87. BAKER, R., D.L. SELWOOD, C.J. SWAIN, N.M.H. WEBSTER, and J. HIRSHFIELD: Synthetic Studies Towards the Pinguisanes: Synthesis of 4-*epi*-Pinguisone. J. Chem. Soc. Perkin Trans. 1 **1988**, 471.

88. BALLIO, A., M. BRUFANI, C.G. CASINOVI, S. CERRINI, W. FEDELI, R. PELLICCIARI, B. SANTURBANO, and A. VACIAGO: The Structure of Fusicoccin A. Experientia **24**, 631 (1968).

89. BALLIO, A., C.G. CASINOVI, V. D'ALESSIO, G. GRANDOLINI, G. RANDAZZO, and C. ROSSI: 19-Deoxydideacetylfusicoccin, A Minor Metabolite of *Fusicoccum amygdali* Del. Experientia **30**, 844 (1974).

90. BANERJEE, A.K., C.A. PEÑA-MATHEUD, and M.C. DE CARRASCO: A Formal Total Synthesis of (±)-Herbertene. J. Chem. Soc. Perkin Trans. 1 **1988**, 2485.

91. BARBIER, P., and C. BENEZRA: Stereospecificity of Allergenic Contact Dermatitis (ACD) Induced by Two Natural Enantiomers, (+)- and (−)-Frullanolides in Guinea Pigs. Naturwissenschaften **69**, S 296 (1982).

92. BARROW, K.D., D.H.R. BARTON, Sir E. CHAIN, U.F.W. OHNSORGE, and R.P. SHARMA: Fusicoccin, Part III: The Structure of Fusicoccin H. J. Chem. Soc., Perkin 1 **1973**, 1590.

93. BASTARD, J., D.K. DUC, M. FÉTIZON, M.J. FRANCIS, P.K. GRANT, R.T. WEAVERS, C. KANEKO, G.V. BADDELEY, J.-M. BERNASSAU, I.R. BURFITT, P.M. WOVKULICH, and E. WENKERT: CMR Spectroscopy of Labdanic Diterpenes and Related Substances. J. Nat. Prod. **47**, 59 (1984).

94. BECKER, H.: Secondary Metabolites from *in vitro* Cultures of Liverworts. In: Bryophytes: Their Chemistry and Chemical Taxonomy (H.D. ZINSMEISTER and R. MUES, eds.), p. 339. Oxford: Oxford University Press, 1990.

95. BECKER, H., R. MUES, H.D. ZINSMEISTER, F. HERZOG, and H. GEIGER: A New Biflavone and Further Flavonoids from the Moss *Hylocomium splendens*. Z. Naturforsch. **41c**, 507 (1986).

96. BECKER, H., and G. WURZEL: Sind Lebermoose Arzneipflanzen? Pharm. in unserer Zeit **16**, 152 (1987).

97. BELKIN, M., D. FITZGERALD, and M.D. FELIX: Tumor Damaging Capacity of Plant Materials, II: Plants Used as Diuretics. J. Natl. Cancer Inst. **13**, 741 (1952–1953).

98. BENDZ, G., and O. MÅRTENSSON: Moss Pigments, II: The Anthocyanins of *Bryum rutilans* Brid. and *Bryum weigelii* Spreng. Acta Chem. Scand. **17**, 266 (1963).

99. BENDZ, G., O. MÅRTENSSON, and L. TERENIUS: Moss Pigments, I: The Anthocyanins of *Bryum cryophilum* O. Mårt. Acta Chem. Scand. **16**, 1183 (1962).

100. BENDZ, G., and L. SVENSSON: Volatile Compounds from *Fontinalis antipyretica*. Phytochem. **10**, 3283 (1971).

101. BENES, I., T. VANÉK, and M. BUDESINSKY: Nardiin, A New *ent*-Kaurane Diterpenoid from the Liverwort *Nardia scalaris*. Collect. Czech. Chem. Commun. **47**, 1873 (1982).

102. BENESOVA, V., I. BENES, H.M. CHAU, and V. HEROUT: Diterpenoids from Liverworts. Collect. Czech. Chem. Commun. **40**, 658 (1975).

103. BERNASCONI, S., M. FERRARI, P. GARIBOLDI, G. JOMMI, M. SISTI, and R. DESTRO: Synthetic Study of Pinguisane Terpenoids. J. Chem. Soc. Perkin Trans. 1 **1981**, 1994.

104. BERNASCONI, S., P. GARIBOLDI, G. JOMMI, S. MONTANARI, and M. SISTI: Total Synthesis of Pinguisone. J. Chem. Soc. Perkin Trans. 1 **1981**, 2394.

105. BEUTELMANN, P., R. EULER, G. KOHN, and E. HARTMANN: Analysis of Bryophyte Lipids. In: Methods in Bryology (J.M. GLIME, ed.), p. 173. Nichinan: J. Hattori Bot. Lab., 1988.

106. BEYER, J.: Terpene aus freilebenden Lebermoosen und deren Sterilkulturen. Ph. D Thesis, Universität Heidelberg, 1988.

107. BEYER, J., H. BECKER, and R. MARTIN: Separation of Labile Terpenoids by Low Temperature HPLC. J. Liquid Chromatog. **9**, 2433 (1986).

108. BEYER, J., H. BECKER, M. TOYOTA, and Y. ASAKAWA: Diterpenoids with a Novel Skeleton from the Liverwort *Anastrophyllum minutum*. Phytochem. **26**, 1085 (1987).

109. BHAT, S.V., B.S. BAJWA, H. DORNAUER, N.J. DE SOUZA, and H.-W. FEHLHABER: Structures and Stereochemistry of New Labdane Diterpenoids from *Coleus forskohlii* Briq. Tetrahedron Letters **1977**, 1669.

110. BHATLA, S.C., and S. DHINGRA-BABBAR: Growth Regulating Substances in Mosses. In: Bryophyte Development: Physiology and Biochemistry (R.N. CHOPRA and S.C. BHATLA, eds.), p. 79. Boca Raton: CRC Press, 1990.

111. BILLET, D., M. DURGEAT, S. HEITZ, J.P. BROUARD, and A. AHOND: Constituants D'*Evodia floribunda* Baker. II-1 ère Partie. L'Acide Floridiolique, Nouveau Diterpene de Type Clerodane. Tetrahedron Letters **1976**, 2773.

112. BLECHSCHMIDT, M., and H. BECKER: *Ent*-Labdanes and Furanoditerpenes from the Liverwort *Jamesoniella autumnalis.* J. Nat. Prod. **55**, 111 (1992).

113. BLOOR, S.J., F.J. SCHMITZ, M.B. HOSSAIN, and D. VAN DER HELM: Diterpenoids from the Gorgonian *Solenopodium stechei.* J. Organ. Chem. (USA) **57**, 1205 (1992).

114. BOHLMANN, F., A. ADLER, R.M. KING, and H. ROBINSON: *Ent*-Labdanes from *Mikania alvimii.* Phytochem. **21**, 173 (1982).

114a. BOHLMANN, F., M. GRENZ, J. JAKUPOVIC, R.M. KING, and H. ROBINSON: Four Heliangolides and Other Sesquiterpenes from *Brasilia sickii.* Phytochem. **22**, 1213 (1983).

115. BOHLMANN, F., and E. HOFFMANN: Cannabigerol-ähnliche Verbindungen aus *Helichrysum umbraculigerum.* Phytochem. **18**, 1371 (1979).

116. BOHLMANN, F., and A. SUWITA: New Sesquiterpenes from *Peteravenia schultzii.* Phytochem. **17**, 567 (1978).

117. BOHLMANN, F., A. SUWITA, and T.M. MABRY: New Labdane Derivatives and Further Constituents of *Brickellia* Species. Phytochem. **17**, 763 (1978).

118. BOHLMANN, F., and C. ZDERO: Über einen neuen Sesquiterpentyp aus *Senecio oxyriifolius.* Phytochem. **17**, 1669 (1978).

119. BOHLMANN, F., C. ZDERO, R.M. KING, and H. ROBINSON: Humulene Derivatives from *Acritopappus prunifolius.* Phytochem. **21**, 147 (1982).

120. BOHLMANN, F., C. ZDERO, R.M. KING, and H. ROBINSON: A Hydroxygermacrene and Other Constituents from *Pseudobrickellia brasiliensis.* Phytochem. **23**, 1798 (1984).

121. BOHLMANN, F., P. WAGNER, and J. JAKUPOVIC: Unusual Diterpenes and Sesquiterpene Xylosides from *Nidorella hottentotica.* Phytochem. **21**, 1109 (1982).

122. BOHM, B.A: Phenolic Compounds in Ferns, III: An Examination of Some Ferns for Caffeic Acid Derivatives. Phytochem. **7**, 1825 (1968).

123. BOPP, M.: Hormones of the Moss Protonema. In: Bryophyte Development: Physiology and Biochemistry (R.N. CHOPRA and S.C. BHATLA, eds.), p. 55. Boca Raton: CRC Press, 1990.

124. BUCKWALTER, B.L., I.R. BURFITT, H. FELIKIN, M. JOLY-GOUDKET, K. NAEMURA, M. SALOMON, E. WENKERT, and P.M. WOVKULICH: Stereoselective Conversion of Keto Groups into Methyl Vinyl Quaternary Carbon Centers. J. Am. Chem. Soc. **100**, 6445 (1978).

125. CABRERA, E., A. GARCIA-GRANADOS, A.S. DE BURUAGA, and J.M.S. DE BURUAGA: Diterpenoids from *Sideritis hirsuta* subsp. *nivalis.* Phytochem. **22**, 2779 (1983).

126. CAINE, D., and G. HASENHUETTL: Total Synthesis of *dl*-3-Oxodiplophyllin and *dl*-Yomogin. J. Organ. Chem. (USA) **45**, 3278 (1980).

127. CALDERON, J.S., L. QUIJANO, F. GOMEZ-GARIBAY, D.M. SANCHEZ, T. RIOS, and F.R. FRONCZEK: Sesquiterpene Lactones from *Stevia ovata* and Crystal Structure of 11,13-Dehydroeriolin. Phytochem. **26**, 1747 (1987).

128. CALDICOTT, A.B., and G. EGLINTON: Cutin Acids from Bryophytes: An ω-1 Hydroxyalkanolic Acid in Two Liverwort Species. Phytochem. **15**, 1139 (1976).

129. CAMPBELL, E.O., K.R. MARKHAM, and L.J. PORTER: Dendroid Liverworts of the Order Metzgeriales in New Zealand. New Zealand J. Bot. **13**, 593 (1975).

130. CANONICA, L., A. CORBELLA, P. GARIBOLDI, G. JOMMI, J. KREPINSKY, G. FERRARI, and C. CASAGRANDE: Sesquiterpenoids of *Cinnamosma fragrans* Baillon. Structure of Cinnamolide, Cinnamosmolide and Cinnamodial. Tetrahedron **25**, 3895 (1969).

532   Y. Asakawa

131. Catalan, C.A.N., I.J.S. de Fenik, P.J. de Arriazu, and W.C.M.C. Kokke: 4,5-Seco-
     African-4,5-dione from *Lippia integrifolia.* Phytochem. **31**, 4025 (1992).
132. Catalano, S., A. Marsili, I. Morelli, and M. Pacchiani: Triterpenoids and Fatty
     Acids from Mosses. Obtusifoliol from *Racomitrium lanuginosum.* Phytochem. **15**, 1178
     (1976).
133. Chang, S.-J., and C.-L. Wu: A New Labdane Alcohol from the Liverwort *Trichocolea
     pluma.* Chemistry (Chinese) **45**, 142 (1987).
134. Chau, P., and C.-L. Wu: The Polar Constituents of Two Liverwort Species *Mastigo-
     phora diclados* and *M. woodsii.* Proc. Natl. Sci. Counc. ROC (A) **11**, 124 (1987).
135. Chen, S.-M., N. Harada, and K. Nakanishi: Long Range Effect in the Exciton
     Chirality Method. J. Am. Chem. Soc. **96**, 7352 (1974).
136. Chen, J.-A., C.-D. Huang, and C.-L. Wu: Clerodane-type Diterpenoids from the
     Taiwanese Liverwort *Schistochila acuminata.* J. Chin. Chem. Soc. **39**, 263 (1992).
137. Chen, J.-A., and C.-L. Wu: Handbook of the Annual Meeting of Chin. Chem. Soc.
     Taipei, p. 283, 1989.
138. Chiu, P.-L., G.W. Patterson, and G.P. Fenner: Sterols of Bryophytes. Phytochem.
     24, 263 (1985).
139. Chopra, R.N., and S.C. Bhatla: Bryophyte Development. In: Physiology and
     Biochemistry. Boca Raton: CRC Press, 1990.
140. Clive, D.L.J., and A.C. Joussef: Synthesis of (±)-Frullanolide: An Application of
     Radical Closure. J. Organ. Chem. (USA) **55**, 1096 (1990).
141. Coates, R.M., S.K. Shah, and R.W. Mason: Stereoselective Total Synthesis of (±)-
     Gymnomitrol *via* Reduction-Alkylation of α-Cyano Ketones. J. Am. Chem. Soc. **104**,
     2198 (1982).
142. Colledge, A., W.W. Reid, and R. Russell: The Diterpenoids of *Nicotiana* Species
     and Their Potential Technological Significance. Chem. and Ind. (London) **1975**, 570.
143. Connolly, J.D.: New Terpenoids from the Hepaticae. Rev. Latinoamer. Quim. **12**,
     121 (1982).
144. Connolly, J.D.: Recent Advances in the Chemistry of the Hepaticae. In: Natural
     Products Chemistry (R.I. Zalewski and J.J. Skolik, eds.), p. 3. Amsterdam: Elsevier,
     1985.
145. Connolly, J.D.: New Compounds from the Euphorbiaceae, the Meliaceae and the
     Hepaticae. In: Studies in Natural Products Chemistry, Vol. 2. Structure Elucidation
     (Part A) (Atta-ur-Rahman, ed.), p. 261. Amsterdam: Elsevier, 1988.
146. Connolly, J.D.: Monoterpenoids and Sesquiterpenoids from the Hepaticae. In:
     Bryophytes: Their Chemistry and Chemical Taxonomy (H.D. Zinsmeister and
     R.Mues, eds.), p. 41. Oxford: Oxford University Press, 1990.
147. Connolly, J.D., L.J. Harrison, S. Huneck, and D.S. Rycroft: (+)-Eudesm-3-ene-
     6β,7α-diol from the Liverwort *Lepidozia reptans.* Phytochem. **25**, 1745 (1986).
148. Connolly, J.D., L.J. Harrison, W.R. Phillips, and D.S. Rycroft: (14R)-*Ent*-kaur-
     16-en-14-yl Hydrogen Malonate from the Liverwort *Nardia scalaris.* J. Chem. Res. (S)
     1984, 94.
149. Connolly, J.D., L.J. Harrison, and D.S. Rycroft: The Structure of Chiloscypho-
     lone, A Sesquiterpenoid with a Novel Carbon Skeleton from the Liverwort *Chiloscy-
     phus pallescens.* A Revised Structure for Chiloscyphone. J. Chem. Soc. Chem.
     Commun. **1982**, 1236.
150. Connolly, J.D., L.J. Harrison, and D.S. Rycroft: The Structure of Tamariscol, A
     New Pacifigorgiane Sesquiterpenoid Alcohol from the Liverwort *Frullania tamarisci.*
     Tetrahedron Letters **25**, 1401 (1984).
151. Connolly, J.D., and R.A. Hill: Dictionary of Terpenoids, Vol. 1: Mono- and
     Sesquiterpenoids. London: Chapman and Hall. 1991.

152. CONNOLLY, J.D., and R.A. HILL: Dictionary of Terpenoids, Vol. 2: Di- and Higher Terpenoids, p. 1382. London: Chapman and Hall, 1991.

153. CONNOLLY, J.D., W.R. PHILLIPS, and S. HUNECK: (+)-*Ent*-epicubenol from the Liverwort *Scapania undulata*. Phytochem. **21**, 233 (1982).

154. COREY, E.J., and R.D. BALANSON: A Simple Synthesis of (±)-Cedrene and (±)-Cedrol Using a Synchronous Double Annulation Process. Tetrahedron Letters **34**, 3153 (1973).

155. CRANDAL-STOTLER, B.: Morphology/Anatomy of Hepatics and Anthocerotes. In: Advances in Bryology, Vol. 1 (W. SCHULTZE-MOTEL, ed.), p. 315. Vaduz: J. Cramer, 1981.

156. CRANDAL-STOTLER, B.: Morphogenesis, Developmental Anatomy and Bryophyte Phylogenetics: Contraindications of Monophyly. J. Bryol. **14**, 1 (1986).

157. CROFT, K.D., E.L. GHISALBERTI, C.H. HOCART, P.R. JEFFRIES, C.L. RASTON, and A.H. WHITE: Absolute Configuration of (+)-Calamenene: Crystal Structure of 7-Hydroxy-calamenene. J. Chem. Soc. Perkin Trans. 1 **1978**, 1267.

158. CROMBIE, L.: Natural Products of Cannabis and Khat. Pure and Appl. Chem. **58**, 693 (1986).

159. CROMBIE, L.W., W.M.L. CROMBIE, and D.F. FIRTH: Synthesis of Bibenzyl Cannabinoids, Hybrids of Two Biogenetic Series Found in *Cannabis sativa*. J. Chem. Soc. Perkin Trans. 1 **1988**, 1263.

159a. CULLMANN, F., K.-P. ADAM, and H. BECKER: Bisbibenzyls and Lignans from *Pellia epiphylla*. Phytochem. **34**, 831 (1993).

160. DAVIDSON, A.J., J.B. HARBORNE, and R.E. LONGTON: Identification of Hydroxycinnamic Acid and Phenolic Acids in *Mnium hornum* and *Brachythecium rutablum* and Their Possible Role in Protection Against Herbivory. J. Hattori Bot. Lab. **67**, 415 (1989).

161. DAVIS, B.M.: The Origin of the *Archegonium*. Ann. Bot. **17**, 477 (1903).

162. DEVON, T.K., and A. I. SCOTT: In: Handbook of Naturally Occurring Compounds, Vol. 1. New York: Academic Press, 1975.

163. DIENES, Z., M. NÓGRÁDI, B. VERMES, and M. KAJTÁR-PEREDY: Synthese von Marchantin I, einem macrocyclischen Bis(bibenzylether) aus *Riccardia multifida*. Liebigs Ann. Chem. **1989**, 1141.

164. DILLON, J., and K. NAKANISHI: Absolute Configurational Studies of Vicinal Glycols and Amino Alcohols II with Pr(dpm)₃. J. Am. Chem. Soc. **97**, 5417 (1975).

165. DING, H.: Zhong guo Yao Yun Bao zi Zhi Wu. Shanghai: Kexue Jishu Chuban She, 1980.

166. D'ISCHIA, M., G. PROTA, and G. SODANO: Reaction of Polygodial with Primary Amines: An Alternative Explanation to the Antifeedant Activity. Tetrahedron Letters **23**, 3295 (1982).

167. DOUNIN, R.: Pigment Chlorophylliens des Bryophytes. Caroténoïdes des Andreaeales, des Sphagnales et des Hepatiques. C. R. Acad. Sci. Paris **246**, 1248 (1958).

168. EICHER, T.: Recent Results from the Synthesis of Chemical Constituents of Bryophytes. In: Bryophytes: Their Chemistry and Chemotaxonomy (H. D. ZINSMEISTER and R. MUES, eds.), p. 209. Oxford: Oxford University Press, 1990.

169. EICHER, T., K. TIEFENSEE, and R. PICK: Synthese von Bryophyten-Inhaltsstoffen, 1: Neue Synthesen der Lunular-Säure und einiger ihrer Derivate. Synthesis **1988**, 525.

170. EKMAN, R., and P. KARUNEN: Steryl and Wax Esters in *Dicranum elongatum*. Phytochem. **19**, 1243 (1980).

171. EL-FERALY, F.S., Y.-M. CHAN, E.H. FAIRCHILD, and R.W. DOSKOTCH: Peroxycostunolide and Peroxyparthenolide: Two Cytotoxic Germacranolide Hydroperoxides from

Magnolia grandiflora. The Structural Revision of Verlotorin and Artemorin. Tetrahedron Letters **1977**, 1973.

172. EL-FERALY, F.S., S.F. CHEATHAM, C.D. HUFFORD, and W.-S. LI: Optical Resolution of (±)-Dehydrodiisoeugenol: Structure Revision of Acuminatin. Phytochem. **21**, 1133 (1982).

173. ENDO, K., and H. HIKINO: Sesquiterpenoids, LIX: Absolute Configuration of Eudesma-4(14),7(11)-dien-8-one. Bull. Chem. Soc. Jpn. **52**, 2439 (1979).

174. ENDO, J., and M. NAGASAWA: Studies on Essential Oil of *Asarum caulescens*, II. Yakugaku Zasshi **94**, 1574 (1974).

175. ENOKI, N., A. FURUSAKI, K. SUEHIRO, R. ISHIDA, and T. MATSUMOTO: Epoxydictymene, A New Diterpene from the Brown Alga *Dictyota dichotoma*. Tetrahedron Letters **24**, 4341 (1983).

176. ERICKSON, M., and G. E. MIKSCHE: On the Occurrence of Lignin or Polyphenols in Some Mosses and Liverworts. Phytochem. **13**, 2295 (1974).

177. EVANS, F.J., and R.J. SCHMIDT: Plants and Plant Products That Induce Contact Dermatitis. Planta Med. **38**, 289 (1980).

178. FANG, J.-M., K.-C. HSU, and Y.-S. CHENG: Terpenoids from Leaves of *Calocedrus formosana*. Phytochem. **28**, 1173 (1989).

179. FAULKNER, D.J.: Marine Natural Products: Metabolites of Marine Algae and Herbivorous. Nat. Prod. Rep. **1984**, 251.

180. FAULKNER, D.J.: Marine Natural Products. Nat. Prod. Rep. **1992**, 321.

181. FISCHER, H.D., N.H. FISCHER, R.W. FRANEK, and E.J. OLIVIER: The Biosynthesis and Chemistry of Sesquiterpene Lactones. In: Progress in the Chemistry of Organic Natural Products, Vol. 38 (W. HERZ, H. GRISEBACH, and G.W. KIRBY, eds.), p. 138. Wien-New York: Springer, 1979.

182. FRÁTER, G.: An Accidental Synthesis of (±)-Herbertene. J. Chem. Soc. Chem. Commun. **1982**, 521.

183. FRÁTER, G., and U. MÜLLER: Mechanism of the Formation of Herbertene from *trans*-Didehydrobicyclofarnesol. J. Chem. Soc. Chem. Commun. **1988**, 1198.

183a. FRIEDEL, H.D., and R. MATUSCH: Isolierung und Strukturaufklärung epimerer 1(5),6-Guaiadiene aus Tolubalsam. Helv. Chim. Acta **70**, 1616 (1987).

184. FUJITA, T., Y. TAKEDA, and T. SHINGU: Bitter Diterpenoids from *Isodon shikokianus* var. *intermedius*. Phytochem. **18**, 299 (1979).

185. FUKUYAMA, Y., and Y. ASAKAWA: Neurotrophic Secoaromadendrane-Type Sesquiterpenes from the Liverwort *Plagiochila fruticosa*. Phytochem. **30**, 4061 (1991).

186. FUKUYAMA, Y., and Y. ASAKAWA: Novel Neurotrophic Isocuparane-Type Sesquiterpene Dimers, Mastigophorenes A, B, C and D, Isolated from the Liverwort *Mastigophora diclados*. J. Chem. Soc. Perkin Trans. 1 **1991**, 2737.

187. FUKUYAMA, Y., H. HIRONO, and M. KODAMA: Total Synthesis of (+)-Bicyclohumulenone. Chem. Letters **1992**, 167.

188. FUKUYAMA, Y., T. MASUYA, M. TORI, M. KIDO, M. WAKAMATSU, and Y. ASAKAWA: Verrucosane Diterpene from the Liverwort *Plagiochila stephensoniana*. Phytochem. **27**, 1797 (1988).

189. FUKUYAMA, Y., M. TORI, M. WAKAMATSU, and Y. ASAKAWA: Norpinguisone Methyl Ester and Norpinguisanolide, Pinguisane-Type Norsesquiterpenoids from *Porella elegantula*. Phytochem. **27**, 3557 (1988).

190. FUKUYAMA, Y., M. TOYOTA, and Y. ASAKAWA: Ent-kaurene Diterpene from the Liverwort *Plagiochila pulcherrima*. Phytochem. **27**, 1425 (1988).

191. FUKUYAMA, Y., M. TOYOTA, and Y. ASAKAWA: Mastigophorenes: Novel Dimeric Isocuparane-Type Sesquiterpenoids from the Liverwort *Mastigophora diclados*. J. Chem. Soc. Chem. Commun. **1988**, 1341.

192. GAMBACORTA, A., M. BOTTA, and S. TURCHETTA: Bicyclo[3.3.1]nonane Approach to Pinguisane Terpenoids. Total Synthesis of (±)-Pinguisone. Tetrahedron **44**, 4837 (1988).

193. GANJIAN, I., I. KUBO, and P. FLUDZINSKI: Insect Antifeedant Elemanolide Lactones from *Vernonia amygdalina*. Phytochem. **22**, 2525 (1983).

194. GARCIA-ALVAREZ, M.C., B. RODRIGUEZ, S. VALVERDE, B.M. FRAGA, and A.G. GONZÁLEZ: Carbon-13 NMR Spectra of Some *Ent*-rosane Diterpenoids. Phytochem. **20**, 167 (1981).

195. GARNIER, G., L. BEZANGER-BEAUQUENSNE, and G. DEBRAUX: Resources Medicinales de la Flore Française, Tome I. Paris: Vigot Frères, 1961.

196. GEIGER, H.: Biflavonoids in Bryophytes. In: Bryophytes: Their Chemistry and Chemical Taxonomy (H.D. ZINSMEISTER and R. MUES, eds.), p. 161. Oxford: Oxford University Press, 1990.

197. GEIGER, H., S. ANHUT, and H.D. ZINSMEISTER: Biflavones from Some Mosses. Z. Naturforsch. **43c**, 1 (1988).

198. GEIGER, H., and M. BOKEL: Die Biflavonoidausstattung von *Philonotis fontana* (Hedw.) Brid. (Bartramiaceae). Z. Naturforsch. **44c**, 559 (1989).

198a. GEIGER, H., and K.R. MARKHAM: Campylopusaurone and Auronoflavone Biflavonoid from the Moss *Campylopus clavatus* and *Campylopus holomitrium*. Phytochem. **31**, 4325 (1992).

199. GEIGER, H., W. STEIN, R. MUES, and H.D. ZINSMEISTER: Bryoflavone and Heterobryoflavone, Two New Isoflavone-Flavone Dimers from *Bryum capillare*. Z. Naturforsch. **42c**, 863 (1987).

200. GEKELER, W., E. GRILL, E.-L. WINNACKER, and M.H. ZENK: Survey of the Plant Kingdom for the Ability to Bind Heavy Metals Through Phytochelatins. Z. Naturforsch. **44c**, 361 (1989).

201. GELLERMAN, J. L., W. H. ANDERSON, and H. SCHLENK: Highly Unsaturated Lipids of *Mnium, Polytrichum, Marchanita, and Matteuccia*. The Bryologist **75**, 550 (1972).

202. GELLERMAN, J.L., and H. SCHLENK: The Occurrence of Arachidonic and Related Acids in Plants. Experientia **20**, 426 (1964).

203. GERBER, N.N.: A Volatile Metabolite of Actinomycetes, 2-Methylisoborneol. J. Antibiotics **22**, 508 (1969).

204. GERLING, K.-G., and H. WOLF: Synthese des Sesquiterpens (±)-Chiloscyphon. Tetrahedron Letters **26**, 1293 (1985).

205. GLASBY, J.S.: Encyclopaedia of the Terpenoids, p. 111. Chichester: John Wiley and Sons, 1982.

206. GLASBY, J.S.: Encyclopaedia of the Terpenoids, p. 526. Chichester: John Wiley and Sons, 1982.

206a. GLASBY, J.S.: Encyclopaedia of the Terpenoids, p. 1028. Chichester: John Wiley and Sons, 1982.

207. GLASBY, J.S.: Encyclopaedia of the Terpenoids, p. 1119. Chichester: John Wiley and Sons, 1982.

208. GLASBY, J.S.: Encyclopaedia of the Terpenoids, p. 1346. Chichester: John Wiley and Sons, 1982.

209. GLASBY, J.S.: Encyclopaedia of the Terpenoids, p. 2516. Chichester: John Wiley and Sons, 1982.

210. GLASBY, J.S.: Encyclopaedia of the Terpenoids, p. 2374. Chichester: John Wiley and Sons, 1982.

211. GLOMBITZA, K.-W.: Antibiotics from Algae. In: Marine Algae in Pharmaceutical Science (H.A. HOPPE, T. LEVRING, and Y. TANAKA, eds.), p. 303. Berlin: Walter de Gruyter, 1979.

212. Goda, Y., and U. Sankawa: Thromboxane Synthetase Inhibitors from *Allium bakeri* Regel. 105th Annual Meeting of Pharmaceutical Society of Japan, Kanazawa, Abstracts, p. 465, 1985.

213. Gonzalez, A.G., J. Darias, A. Diaz, J.D. Founeron, J.D. Martin, and C. Perez: Evidence for Biogenesis of Halogenated Chamigrenes from the Red Alga *Laurencia obtusa*. Tetrahedron Letters **1976**, 3051.

214. Gorham, J.: Phenolic Compounds, Other Than Flavonoids, from Bryophytes. In: Bryophytes: Their Chemistry and Chemical Taxonomy (H.D. Zinsmeister and R. Mues, eds.), p. 171. Oxford: Oxford University Press, 1990.

215. Gottsegen, Á., M. Nógrádi, B. Vermes, M. Kajtár-Peredy, and É. Bihátsi-Karsai: The Synthesis of Riccardin C. Tetrahedron Letters **29**, 5039 (1988).

216. Gottsegen, Á., M. Nógrádi, B. Vermes, M. Kajtár-Peredy, and É. Bihátsi-Karsai: Total Syntheses of Riccardins A, B, and C, Cytotoxic Macrocyclic Bis(bibenzyls) from Liverworts. J. Chem. Soc. Perkin Trans. 1 **1990**, 315.

217. Gradstein, S.R., Y. Asakawa, R. Mues, and R. Klein: On the Taxonomic Significance of Secondary Metabolites in the Lejeuneaceae (Hepaticae). J. Hattori Bot. Lab. **64**, 159 (1988).

218. Gradstein, S.R., R. Klein, L. Kraut, R. Mues, J. Spörle, and H. Becker: Phytochemical and Morphological Support for the Existence of Two Species in *Monoclea* (Hepaticae). Pl. Syst. Evol. **180**, 115 (1992).

219. Gradstein, S.R., R. Matsuda, and Y. Asakawa: A Chemotaxonomic Survey of Terpenoids and Aromatic Compounds in the Lejeuneaceae (Hepaticae). In: Contribution to a Monograph of the Lejeuneaceae Subfamily Ptychanthoideae (S. R. Gradstein, ed.), p. 63. Vaduz: J. Cramer, 1985.

220. Gras, J.-L.: A Direct Synthesis of α-Methylene Ketones. Tetrahedron Letters **1978**, 2111.

221. Gras, J.-L.: A Facile Entry to Vinyl Ketones. Tetrahedron Letters **1978**, 2955.

222. Gras, J.-L.: Total Synthesis of the Alleged (±)-Chiloscyphone. J. Organ. Chem. (USA) **46**, 3738 (1981).

223. Grolle, R.: Verzeichnis der Lebermoose Europas und benachbarter Gebiete. Feddes Repertorium **87**, 171 (1976).

224. Guha, P.C., and S.C. Bhattacharyya: Studies in the Santalol Series, Part II: Syntheses of *d*- and *dl*-π-Hydroxycamphor, *d*- and *dl*-Teresantalol and *d*- and *dl*-Tricyclo Ekasantalic Acid. J. Indian Chem. Soc. **21**, 271 (1944).

225. Ha, N.T.T., M. Nógrádi, J. Brlik, M. Kajtár-Peredy, and A. Wolfner: Synthesis of Marchantin H and B, Macrocyclic Bis(bibenzyls) from Liverworts. J. Chem. Res. (S) **1991**, 137.

226. Hagiwara, H., and H. Uda: A Total Synthesis of (+)-Perrottetianal A. J. Chem. Soc. Chem. Commun. **1987**, 1351.

227. Hagiwara, H., and H. Uda: Total Synthesis of (+)-Dysideapalaunic Acid. J. Chem. Soc. Chem. Commun. **1988**, 815.

228. Hagiwara, H., and H. Uda: A Total Synthesis of (−)-Sacculatal. Bull. Chem. Soc. Jpn. **62**, 624 (1989).

229. Hagiwara, H., and H. Uda: Total Synthesis of (+)-Perrottetianal A. J. Chem. Soc. Perkin Trans. 1 **1990**, 1901.

230. Hajos, Z.G., and D.R. Parrish: Synthesis and Conversion of 2-Methyl-2-(3-oxobutyl)-1,3-cyclopentanedione to the Isomeric Racemic Ketols of the [3.2.1]-Bicyclooctane and of the Perhydroindan Series. J. Organ. Chem. (USA) **39**, 1612 (1974).

231. Hajos, Z.G., and D.R. Parrish: Asymmetric Synthesis of Bicyclic Intermediates of Natural Product Chemistry. J. Organ. Chem. (USA) **39**, 1615 (1974).

232. HAMLIN, B.G.: Hepaticae of New Zealand, Parts I and II. Index of Binomials and Preliminary Checklist. Rec. Dom. Mus., Wellington **7**, 243 (1972).

233. HARADA, N., S.-M. CHEN, and K. NAKANISHI: Quantitative Definition of Exciton Chirality and the Distant Effect in the Exciton Chirality Method. J. Am. Chem. Soc. **97**, 5345 (1975).

234. HARADA, N., J. IWABUCHI, Y. YOKOTA, H. UEDA, and K. NAKANISHI: A Chiroptical Method for Determining the Absolute Configuration of Allylic Alcohols. J. Am. Chem. Soc. **103**, 5590 (1981).

235. HARADA, N., J. KOHORI, H. UDA, K. NAKANISHI, and R. TAKEDA: Absolute Stereo-chemistry of (+)-1,8a-Dihydro-3,8-dimethylazulene, a Labile Biosynthetic Inter-mediate for 1,4-Dimethylazulene. Determination by Theoretical Calculation of CD Spectra and Verification by Synthesis of Model Compounds. J. Am. Chem. Soc. **107**, 423 (1985).

236. HARADA, N., J. KOHORI, H. UDA, and K. TORIUMI: Synthesis, Absolute Stereo-chemistry, and Circular Dichroism of Chiral 1,8a-Dihydroazulene Derivatives. J. Organ. Chem. (USA) **54**, 1820 (1989).

237. HARADA, N., and K. NAKANISHI: Circular Dichroic Spectroscopy. Exciton Coupling in Organic Chemistry. Tokyo: Tokyo Kagaku Doujin, 1983.

238. HARADA, N., and K. NAKANISHI: Circular Dichroic Spectroscopy. Exciton Coupling in Organic Stereochemistry. Mill Valley: University Science Books, 1983.

239. HARADA, N., Y. YOKOTA, J. IWABUCHI, H. UDA, and M. OCHI: The Absolute Stereo-chemistry of Isoamijiol, a Dolastane Diterpene, as Determined by the C. D. Allylic Benzoate Method. J. Chem. Soc. Chem. Commun. **1984**, 1220.

239a. HARRIGAN, G.G., A. AHMAD, N.B.T.E. GLASS, A.A.L. GUNATILAKA, and D.G.I. KINGSTON: Bioactive and Other Sesquiterpenoids from *Porella cordaeana*. J. Nat. Prod. **56**, 921 (1993).

240. HARRISON, L. J., and Y. ASAKAWA: 3α,18-Dihydroxytrachyloban-19-oic Acid from the Liverwort *Jungermannia exsertifolia* subsp. *cordifolia*. Phytochem. **28**, 1533 (1989).

241. HARRISON, L. J., and Y. ASAKAWA: Oppositane and Chiloscyphane Sesquiterpenoids from the Liverwort *Chiloscyphus pallescens*. Phytochem. **30**, 3806 (1991).

242. HARRISON, L.J., and H. BECKER: A Nor-Seco-Aromadendrane from the Liverwort *Mylia taylorii*. Phytochem. **28**, 1261 (1989).

243. HARRISON, L.J., H. BECKER, J.D. CONNOLLY, and D.S. RYCROFT: Two Elemane Sesquiterpenoids from the Liverwort *Plagiochasma rupestre*. J. Chem. Res. (S) **1992**, 74.

244. HARRISON, L.J., H. BECKER, J.D. CONNOLLY, and D.S. RYCROFT: (+)-(4*S**,5*R**,7*S**,8*R**)-Eremophilia-9,11-dien-8α-ol from the Liverwort *Marsupella emar-ginata*. Phytochem. **31**, 4027 (1992).

245. HARRISON, L.J., J.D. CONNOLLY, and D.S. RYCROFT: *Ent*-Clerodane Diterpenoids from the Liverwort *Jungermannia paroica*. Phytochem. **31**, 1420 (1992).

246. HARRISON, L.J., M. TORI, and Y. ASAKAWA: (15*S*,16*S*)-2β,16-Epoxyverrucosan-16-ol, a Novel Diterpenoid from the Liverwort *Mylia taylorii*. J. Chem. Res. (S) **1986**, 212.

247. HARRISON, L.J., M. TORI, Z. TAIRA, and Y. ASAKAWA: New Verticillane-Type Diter-penoids from Adelanthaceae. 28th Symposium on Chemistry of Terpenes, Essential Oils and Aromatics, Kanazawa, Japan. Symposium Papers, p. 285, 1984.

248. HASHIMOTO, T.: Pungent Terpenoids from Liverworts and Mushrooms – Enzymatically Formed Pungent Sesquiterpenoids. 3rd Malaysian International Conference on Essential Oils and Flavour Chemicals, Langkawi, Kedah, Malaysia. Abstracts, p. 22, 1992.

249. HASHIMOTO, T., Y. ASAKAWA, K. NAKASHIMA, and M. TORI: Chemical Constituents of 25 Liverworts. J. Hattori Bot. Lab. **74**, 121 (1993).

249a. HASHIMOTO, T., K. AKAZAWA, and Y. ASAKAWA: Unpublished results.

250. Hashimoto, T., M. Horie, M. Tori, Z. Taira, and Y. Asakawa: Absolute Structures of Forskolin Related Compounds from the Liverwort *Ptychanthus striatus*. 112th Annual Meeting of Pharmaceutical Society of Japan, Fukuoka, Abstracts, p. 200, 1992.

250a. Hashimoto, T., M. Horie, M. Toyota, Z. Taira, R. Takeda, M. Tori, and Y. Asakawa: Structures of Five New Highly Oxygenated Labdane-Type Diterpenoids Ptychantins A–E, Closely Related to Forskolin from the Liverwort *Ptychanthus striatus*. Tetrahedron Letters **35**, 5457 (1994).

251. Hashimoto, T., M. Horie, A. Yasuda, M. Tori, and Y. Asakawa: New Diterpenoids from the Liverworts *Pallavicinia levieri* and *Ptychanthus striatus*. 35th Symposium on Chemistry of Terpenes, Essential Oils and Aromatics, Nagoya, Japan. Symposium Papers, p. 41, 1991.

251a. Hashimoto, T., S. Kanayama, Y. Fukuyama, S. Takaoka, M. Tori, and Y. Asakawa: Two Novel Macrocyclic Bis(bibenzyls), Isoplagiochins A and B from the Liverwort *Plagiochila fruticosa*. Tetrahedron Letters **35**, 911 (1994).

251b. Hashimoto, T., H. Koyama, S. Takaoka, Y. Kan, M. Tori, and Y. Asakawa: Chemical Constituents of Malaysian Liverworts, II: Studies on Structures of the New Skeletal Sesquiterpenoids, Trifaranes, from Cheilolejeunea. 37th Symposium on Chemistry of Terpenes, Essential Oils and Aromatics, Okinawa, Japan. Symposium Papers, p. 50, 1993.

251c. Hashimoto, T., H. Koyama, S. Takaoka, M. Tori, and Y. Asakawa: Isolation and Absolute Stereostructures of Trifarane Skeleton from the Malaysian Liverwort *Cheilolejeunea trifaria*. Tetrahedron Letters **35**, 4787 (1994).

252. Hashimoto, T., I. Nakamura, H. Tanaka, M. Tori, and Y. Asakawa: Pungent. 2,3-Secoaromadendrane-Type Sesquiterpenoids from the Liverwort *Heteroscyphus plunus*. 36th Symposium on Chemistry of Terpenes, Essential Oils and Aromatics, Nishinomiya, Japan. Symposium Papers, p. 263, 1992.

253. Hashimoto, T., I. Nakamura, M. Tori, and Y. Asakawa: Structures of New Verrucosane- and Clerodane-Type Diterpenoids from the Liverwort *Heteroscyphus planus*. 36th Symposium on Chemistry of Terpenes, Essential Oils and Aromatics, Nishinomiya, Japan. Symposium Papers, p. 260, 1992.

254. Hashimoto, T., H. Suzuki, M. Tori, and Y. Asakawa: Bis(bibenzyl) Ethers from *Pelli endiviifolia*. Phytochem. **30**, 1523 (1991).

255. Hashimoto, T., Z. Taira, M. Tori, A. Ohnishi, Y. Yamaoka, M. Takei, and Y. Asakawa: Macrocyclic Bis-bibenzyl Compounds, Marchantins and Biological Activity of Their Related Compounds. 8th Symposium on the Development and Application of Naturally Occurring Drug Materials, Tokyo, Japan. Abstracts, p. 21, 1991.

256. Hashimoto, T., M. Tori, and Y. Asakawa: Sacculaplagin, A New Highly Oxidized Sacculatane Diterpenoid Hemiacetal from the Liverwort *Plagiochila acanthophylla* subsp. *japonica*. Tetrahedron Letters **28**, 6293 (1987).

256a. Hashimoto, T., M. Tori, and Y. Asakawa: Three Dihydroisocoumarin Glucosides from *Hydrangea macrophylla* subsp. *serrata*. Phytochem. **26**, 3323 (1987).

257. Hashimoto, T., M. Tori, and Y. Asakawa: A Highly Efficient Preparation of Lunularic Acid and Some Biological Activities of Stilbene and Dihydrostilbene Derivatives. Phytochem. **27**, 109 (1988).

258. Hashimoto, T., M. Tori, and Y. Asakawa: Drimane-Type Sesquiterpenoids from the Liverwort *Makinoa crispata*. Phytochem. **28**, 3377 (1989).

259. Hashimoto, T., M. Tori, Y. Asakawa, and Y. Fukazawa: Plagiochins A, B, C, and D, New Type of Macrocyclic Bis(bibenzyls) Having a Biphenyl Linkage Between the Ortho Positions to the Benzyl Methylenes, from the Liverwort *Plagiochila acanthophylla* subsp. *japonica*. Tetrahedron Letters **28**, 6295 (1987).

260. HASHIMOTO, T., M. TORI, K. SUZUKI, and Y. ASAKAWA: Structures of New Sacculatane-type Diterpenoids from *Pellia endiviifolia*. 31st Symposium of Chemistry on Terpenes, Essential Oils and Aromatics, Kyoto, Japan. Symposium Papers, p. 239, 1987.

261. HASHIMOTO, T., M. TORI, Z. TAIRA, and Y. ASAKAWA: New Highly Oxidized Fusicoccane Diterpenoids from the Liverwort *Plagiochila acanthophylla* subsp. *japonica*. Tetrahedron Letters **26**, 6473 (1985).

262. HASHIMOTO, T., M. TOYOTA, H. TANAKA, and Y. ASAKAWA: Unpublished results.

262a. HASHIMOTO, T., T. YOSHIDA, Y. KAN, S. TAKAOKA, M. TORI, and Y. ASAKAWA: Structures of Four Novel Macrocyclic Bis(bibenzyl) Dimers, Pusilatins A–D from the Liverwort *Blasia pusilla*. Tetrahedron Letters **35**, 909 (1994).

262b. HASHIMOTO, T., T. YOSHIDA, S. KANAYAMA, S. TAKAOKA, Y. KAN, Y. FUKUYAMA, M. TORI, and Y. ASAKAWA: Structures of Novel Macrocyclic Bis(bibenzyls) Isolated from the Liverworts *Blasia pusilla* and *Plagiochila fruticosa*. 25th Symposium on the Chemistry of Natural Products, Kyoto, Japan. Symposium Papers, p. 353, 1993.

262c. HASHIMOTO, T., T. YOSHIDA, M. TORI, S. TAKAOKA, and Y. ASAKAWA; Structures of Novel Macrocyclic Bis(bibenzyls) Isolated from the Liverworts *Blasia pusilla* and *Plagichila fruticosa*. 15th International Botanical Congress, Yokohama, Japan. Abstracts, p. 370, 1993.

263. HAUSER, F.M., and S.A. POGANY: 2-Hydroxy-6-methylbenzoic Acid Derivatives. Synthesis **1980**, 814.

264. HAYASHI, S., and A. MATSUO: Differential Distribution of Sesquiterpenoids Among Some Species of the Genus *Bazzania*. Proc. Bryol. Soc. Japan **3**, 108 (1983).

265. HEGNAUER, R.: Bryophyta. In: Chemotaxonomie der Pflanzen, Bd. VII, 374. Basel: Birkhäuser, 1986.

266. HELLOU, J., R.J. ANDERSEN, and J.E. THOMPSON: Terpenoids from the Dorid Nudibranch *Cadlina luteomarginata*. Tetrahedron **38**, 1875 (1982).

267. HENDERSON, M.S., R.D.H. MURRAY, R. MCCRINDLE, and D. MCMASTER; Constituents of *Solidago* Species, Part III: The Constitution of Diterpenoids from *Solidago juncea* Ait. Can. J. Chem. **51**, 1322 (1973).

268. HEROUT, V.: An Approach to the Isolation of the Fragrant Principle of the Liverwort *Lophocolea heterophylla*. Flavour Fragr. J. **1**, 43 (1985).

268a. HIROSE, Y., S. HASEGAWA, N. OZAKI, and Y. IITAKA: Three New Terpenoid Quinone Methides from the Seed of *Chamaecypairs obtusa*. Tetrahedron Letters **24**, 1535 (1983).

269. HUANG, C.D.: Studies on the Chemical Constituents of Two Taiwanese Liverworts *Plagiochila peculiaris* and *Schistochila acuminata*. Master's Thesis, Tamkang University, Tamsui, 1991.

270. HUNECK, S.: Die Inhaltsstoffe der Laubmoose *Abietinella abietina*, *Plagiothecium undulatum* und *Tortella inclinata*. Phytochem. **10**, 3282 (1971).

271. HUNECK, S.: Chemistry and Biochemistry of Bryophytes. In: New Manual of Bryology, Vol. 1 (R. M., SCHUSTER, ed.), p. 1. Nichinan: The Hattori Botanical Laboratory, 1983.

272. HUNECK, S., Y. ASAKAWA, Z. TAIRA, A. F. CAMERON, J.D. CONNOLLY, and D.S. RYCROFT: Gymnocolin, A New *cis*-Clerodane Diterpenoid from the Liverwort *Gymnocolea inflata*. Crystal Structure Analysis. Tetrahedron Letters **24**, 115 (1983).

273. HUNECK, S., G.A. BAXTER, A.F. CAMERON, J.D. CONNOLLY, L.J. HARRISON, W.R. PHILLIPS, D.S. RYCROFT, and G.A. SIM: Dolabellane Diterpenoids from the Liverworts *Barbilophozia floerkei*, *B. lycopodioides*, *and B. attenuata*: Spectroscopic and X-Ray Studies of Structure, Stereochemistry, and Conformation. X-Ray Molecular Structure of 3S,4S;7S,8S-Diepoxy-10R, 18-Dihydroxydolabellane, 18-Acetoxy-

3S,4S;7S,8S-diepoxydolabellane, and 10R,18-Diacetoxy-3S,4S-epoxydolabell-7E-ene. J. Chem. Soc. Perkin Trans. 1 **1986**, 809.

274. HUNECK, S., G. BAXTER, A.F. CAMERON, J.D. CONNOLLY, and D.S. RYCROFT: Anadensin, A New Fusicoccane Diterpenoid from the Liverwort *Anastrepta orcadensis*. Crystal Structure Analysis. Tetrahedron Letters **24**, 3787 (1983).

275. HUNECK, S., A.F. CAMERON, J.D. CONNOLLY, M. MCLAREN, and D.S. RYCROFT: Hercynolactone, A New Carotane Sesquiterpenoid from the Liverworts *Barbilophozia lycopodioides* and *B. hatcheri*. Crystal Structure Analysis. Tetrahedron Letters **23**, 3959 (1982).

276. HUNECK, S., J.D. CONNOLLY, A.A. FREER, and D.S. RYCROFT: Grimardone, A Tricyclic Sesquiterpenoid from *Mannia fragrans*. Crystal Structure Analysis. Phytochem. **27**, 1405 (1988).

277. HUNECK, S., J.D. CONNOLLY, L.J. HARRISON, R. JOSEPH, W.R. PHILLIPS, D.S. RYCROFT, G. FERGUSON, and M. PARVEZ: New Labdane Diterpenoids from the Liverwort *Scapania undulata*. J. Chem. Res. (s) **1986**, 162.

278. HUNECK, S., J.D. CONNOLLY, L.J. HARRISON, R. JOSEPH, and T. PÓCS: 1-(3,4-Dihydroxy-5-methoxylphenyl)-3-methylbut-2-ene from the Liverwort *Plagiochila rutilans*. Phytochem. **23**, 2396 (1984).

279. HUNECK, S., J.D. CONNOLLY, D.S. RYCROFT, and A. MATSUO: (−)-*Ent*-12β-acetoxylongipin-2(10)-en-3-one, an *Ent*-Longipinane Derivative from the Liverwort *Marsupella aquatica*. Phytochem. **21**, 143 (1982).

280. HUNECK, S., J.D. CONNOLLY, D.S. RYCROFT, and N. WOODS: Pakyonol, a Macrocyclic Bisbibenzyl Diether from the Liverwort *Mannia fragrans*. J. Chem. Res. (S) **1988**, 78.

281. HUNECK, S., S. JÄNICKE, and E. HUNECK: Das Vorkommen chemischer Rassen bei *Scapania undulata* in Europa, speziell im Thüringer Wald. J. Hattori Bot. Lab. **54**, 125 (1983).

282. HUNECK, S., S. JÄNICKE, L. MEINUNGER: Seasonal Variations of the Essential Oil Components in the Liverwort *Scapania undulata*. J. Hattori Bot. Lab. **53**, 439 (1982).

283. HUNECK, S., S. JÄNICKE, L. MEINUNGER, G. SNATZKE, J.D. CONNOLLY, and Y. ASAKAWA: Seasonal Dependence of the Essential Oil from *Bazzania trilobata*. The Stereochemistry and Absolute Configuration of (−)-5-Hydroxycalamenene. J. Hattori Bot. Lab. **57**, 337 (1984).

284. HUNECK, S., S. JÄNICKE, and J. SCHMIDT: Phytosterole aus Lebermoosen. Pharmazie **39**, 784 (1984).

285. HUNECK, S., and L. MEINUNGER: Plant Growth Regulatory Activities of Bryophytes: A Contribution to the Chemical Ecology of Mosses and Liverworts. In: Bryophytes: Their Chemistry and Chemical Taxonomy (H.D. ZINSMEISTER and R. MUES, eds.), p. 289. Oxford: Oxford University Press, 1990.

286. HUNECK, S., and K. SCHREIBER: Inhaltsstoffe der Moose, XVII: Über die Inhaltsstoffe weiterer Lebermoose. J. Hattori Bot. Lab. **39**, 215 (1975).

287. HUNECK, S., K. SCHREIBER, J.D. CONNOLLY, L.J. HARRISON, and D.S. RYCROFT: Sesquiterpenoids from the Liverwort *Lophozia ventricosa*. Phytochem. **23**, 1792 (1984).

288. HUNECK, S., K. SCHREIBER, and S. JÄNICKE: Inhaltsstoffe einiger Laubmoose. Phytochem. **12**, 2533 (1973).

289. HUNECK, S., L. URBANIAK, Y. ASAKAWA, R. GROLLE, and S. JÄNICKE: The Chemistry of Three Species of *Scapania* (Hepaticae) from the Polish High Tatra Mountains. J. Hattori Bot. Lab. **60**, 187 (1986).

290. ICHIKAWA, T., M. NAMIKAWA, K. YAMADA, K. SAKAI, and K. KONDO: Novel Cyclopentenonyl Fatty Acids from Mosses, *Dicranum scoparium* and *Dicranum japonicum*. Tetrahedron Letters **24**, 3337 (1983).

291. ICHIKAWA, T., K. YAMADA, M. NAMIKAWA, K. SAKAI, and K. KONDO: New Cyclopentenonyl Fatty Acids from Japanese Mosses. J. Hattori Bot. Lab. **56**, 209 (1984).

292. INOSHIRI, S., M. SASAKI, H. KOHDA, H. OTSUKA, and K. YAMASAKI: Aromatic Glycosides from *Berchemia racemosa*. Phytochem. **26**, 2811 (1987).

293. INOUE, H.: In: Koke no Sekai, p. 162. Tokyo: Idemitsu Ltd. 1978.

294. INOUE, H.: Hepaticae Which Tell Continental Drift-Plant Research of Gondwanaland Origin. Kagakuasahi **1988**, 116.

295. INOUE, H., and Y. ASAKAWA: Phytochemical Studies on *Plagiochila spinulosa* (Dicks.) Dum. Bull. Nat. Sci. Mus. Ser. B. **14**, 143 (1988).

296. IRELAND, C., D.J. FAULKNER, J. FINER, and J. CLARDY: A Novel Diterpene from *Dollabella californica*. J. Am. Chem. Soc. **98**, 4664 (1976).

296a. IRWIN, M.A., and T.A. GEISSMAN: Sesquiterpene Lactones from *Artemisia*: Arbusculin C, Rothin-A and Rothin-B. Phytochem. **10**, 637 (1971).

297. ISHIBASHI, H., H. NAKATANI, D.J. CHOI, M. TAGUCHI, and M. IKEDA: Synthesis of (±)-β-Cuparenone Based on a Lewis Acid-Promoted [4$^+$ + 2] Polar Cycloaddition of *m*-Tolylthiomethyl Chloride. Chem. Pharm. Bull. (Tokyo) **38**, 1738 (1990).

298. ISHII, H., T. ISHIKAWA, I.-S. CHEN, and S.-T. LU: New Phenyl Propanoids Having Weak Piscicidal Action and the Related Compounds: Wutaiensol, Wutaiensal, Wutaialdehyde, and Methyl Demethoxywutaiensate. Tetrahedron Letters **23**, 4345 (1982).

299. ISHII, H., F. SEKIGUSHI, and T. ISHIKAWA: Studies on the Chemical Constituents of Rutaceous Plants, XLI: Absolute Configuration of Rutaretin Methyl Ether. Tetrahedron **37**, 285 (1981).

300. IVES, D.A.J., and A.N. O'NEILL: The Chemistry of Peat. Can. J. Chem. **36**, 926 (1958).

301. IWATSUKI, Z.: Catalog of the Mosses of Japan. Nichinan: The Hattori Botanical Laboratory, 1991.

302. IYODA, M., M. SAKAITANI, H. OTSUKA, and M. ODA: Synthesis of Riccardin B by Nickel-Catalyzed Intramolecular Cyclization. Tetrahedron Letters **26**, 4777 (1985).

302a. IZAC, R.R., M.M. BANDURRAGA, J. WASYLYK, F.W. DUNN, and W. FENICAL: Germacrene Derivatives from Diverse Marine Soft Corals (Octocorallia). Tetrahedron **38**, 301 (1982).

303. IZAC, R.R., W. FENICAL, and J.M. WRIGHT: Inflatene, an Ichthyotoxic C_{12} Hydrocarbon from the Stoloniferan Soft Coral *Clavularia inflata* var. *luzoniana*. Tetrahedron Letters **25**, 1325 (1984).

304. IZAC, R.R., S.E. POET, W. FENICAL, D.V. ENGEN, and J. CLARDY: The Structure of Pacifigorgiol, an Ichthyotoxic Sesquiterpenoid from the Pacific Gorgonian Coral *Pacifigorgia* cf. *adamsii*. Tetrahedron Letters **23**, 3743 (1982).

305. JAKUPOVIC, J., A. SCHUSTER, F. BOHLMANN, U. GANZER, R.M. KING, and H. ROBINSON: Diterpenoids and Other Constituents from Australian *Helichrysum* and Related Species. Phytochem. **28**, 543 (1989).

306. JAMIESON, G.R., and E.H. REID: The Component Fatty Acids of Some Marine Algal Lipids. Phytochem. **11**, 1423 (1972).

307. JAMIESON, G.R., and E.H. REID: Lipids of *Fontinalis antipyretica*. Phytochem. **15**, 1731 (1976).

308. JOLY, M., M. HAAG-BERRURIER, and R. ANTON: La 5′-Méthoxybilobétine, une Biflavone Extraite du *Ginkgo biloba*. Phytochem. **19**, 1999 (1980).

309. KAJIMOTO, T., M. YAMASHITA, Y. IMAMURA, K. TAKAHASHI, T. NOHARA, and M. SHIBATA: Novel Phenolics from Phytopathogenic Fungus *Helicobasidium mompa*. Chemistry Letters **1989**, 527.

542 Y. Asakawa

310. Kajiwara, T., S. Ochi, K. Kodama, K. Matsui, A. Hatanaka, T. Fujimura, and T. Ikeda: Cell-Destroying Sesquiterpenoid from Red Tide of *Gymnodinium nagasakiense*. Phytochem. **31**, 783 (1992).

311. Karlsson, B., A.-M. Pilotti, A.-C. Soderholm, T. Norin, and M. Sumimoto: The Structure and Absolute Configuration of Verticillol, a Macrocyclic Diterpene Alcohol from the Wood of *Sciadopitys verticillata* Sieb. et Zucc. (Taxodiaceae). Tetrahedron **34**, 2349 (1978).

312. Karunen, P.: Determination of Fatty Acid Composition of Spore Lipids of the Moss *Polytrichum commune* by Glass Capillary Column Gas Chromatography. Physiol. Plant. **40**, 239 (1977).

313. Karunen, P.: Fatty Acid Composition of Glycosyl Diglycerides in *Ceratodon purpureus*, *Plagiothecium laetum* and *Barbilophozia barbata*. In: Bryophytorum Bibliotheca, Vol. 13 (C. Suire, ed.), p. 365. Vaduz: J. Cramer, 1978.

314. Karunen, P.: Studies on Moss Spores, VII: Fatty Acid Composition of Mono- and Diglycosyl Diglyceride Fractions of Germinating *Polytrichum commune* Spores. The Bryologist **81**, 100 (1978).

315. Karunen, P.: The Acyl Lipids of Bryophytes. In: Bryophytes, Their Chemistry and Chemical Taxonomy (H.D. Zinsmeister and R. Mues, eds.), p. 121. Oxford: Oxford University Press, 1990.

316. Karunen, P., and E.-M. Aro: Fatty Acid Composition of Polar Lipids in *Ceratodon purpureus* and *Pleurozium schreberi*. Physiol. Plant. **45**, 265 (1979).

317. Karunen, P., and C. Liljenberg: Content and Fatty Acid Composition of Steryl and Wax Esters in Germinating Spores of *Polytrichum commune*. Physiol. Plant. **44**, 417 (1978).

318. Karunen, P., and H. Mikola: Distribution of Triglycerides in *Dicranum elongatum*. Phytochem. **19**, 319 (1980).

319. Karunen, P., H. Mikola, and R. Ekman: Occurrence of Steryl Wax Esters in *Dicranum elongatum*. Physiol. Plant. **48**, 554 (1980).

320. Katoh, K., H. Hori, and S. Osawa: The Nucleotide Sequences of 5S Ribosomal RNAs from Four Bryophyta-Species. Nucleic Acids Res. **11**, 5671 (1983).

321. Katoh, K., and R. Takeda: Growth and Production of Secondary Metabolites by Cultured Bryophyte Cells. In: Bryophytes, Their Chemistry and Chemical Taxonomy (H.D. Zinsmeister and R. Mues, eds.), p. 349. Oxford: Oxford University Press, 1990.

321a. Kato, T., C. Kabuto, N. Sasaki, M. Tsunagawa, A. Aizawa, K. Fujita, Y. Kato, and Y. Kitahara: Momilactones, Growth Inhibitors from Rice, *Oryza sativa* L. Tetrahedron Letters **1973**, 3861.

322. Kazlauskas, R., J. Mulder, P.T. Murphy, and R.J. Wells: New Metabolites from the Brown Alga *Caulocystis cephalornithos*. Aust. J. Chem. **33**, 2097 (1980).

323. Keserü, G.M., G. Mezey-Vándor, M. Nógrádi, B. Vermes, and M.K. Peredy: Total Synthesis of Plagiochins C, and D, Macrocyclic Bis(bibenzyl) Constituents of *Plagiochila acantophylla*. Tetrahedron **48**, 913 (1992).

324. Keserü, G.M., and M. Nógrádi: Biosynthesis and Molecular Strain. A Computational Study on the Conformation of Cyclic Bis(bibenzyl) Constituents of Liverwort Species. Phytochem. **31**, 1573 (1992).

325. Kesselmans, R.P.W., J.B.P.A. Wijnberg, A. de Groot, and N.K. de Vries: Synthesis of All Stereoisomers of Eudesm-11-en-4-ol. 1. Stereospecific Synthesis of the *trans*- and *cis*-Fused Octahydro-8-hydroxy-4a,8-dimethyl-2(1*H*)-naphthalenones. Conformational Analysis of the *cis*-Fused Compounds. J. Organ. Chem. (USA) **56**, 7232 (1991).

326. Kesselmans, R.P.W., J.B.P.A. Wijnberg, A.J. Minnard, R.E. Walinga, and A. de Groot: Synthesis of All Stereoisomers of Eudesm-11-en-4-ol. 2. Total Synthesis of

Selin-11-en-4α-ol, Intermedeol, Neointermedeol, and Paradisiol. First Total Synthesis of Amiteol. J. Organ. Chem. (USA) **56**, 7237 (1991).

327. KITAHARA, Y., and A. YOSHIKOSHI: The Structure of Dolabradiene. Tetrahedron Letters **26**, 1775 (1964).

328. KITAJIMA, J., T. KOMORI, and T. KAWASAKI: Studies on the Constituents of the Crude Drug "Fritillariae Bulbus", III: On the Diterpenoid Constituents of Fresh Bulbs of *Fritillaria thunbergii* Miq. Chem. Pharm. Bull. (Tokyo) **30**, 3912 (1982).

329. KNEIFEL, H.: In: Marine Algae in Pharmaceutical Science (H.A. HOPPE, T. LEVRING, and Y. TANAKA, eds.), p. 365. Berlin: Walter de Gruyter, 1979.

330. KNOCHE, H., and G. OURISSON: Organic Compounds in Fossil Plants (*Equisetum;* Horsetails). Angew. Chem. Internat. Edit. Engl. **6**, 1085 (1967).

331. KOBAYASHI, M., B. W. SON, Y. KYOGOKU, and I. KITAGAWA: Clavukerin A, a New Trinor-Guaiane Sesquiterpene from the Okinawan Soft Coral *Clavularia koellikeri.* Chem. Pharm. Bull. (Tokyo) **31**, 2160 (1983).

332. KOBAYASHI, M., B. W. SON, Y. KYOGOKU, and I. KITAGAWA: Clavukerin A, a New Trinor-Guaiane Sesquiterpene Having a Hydroperoxy Function from the Okinawan Soft Coral *Clavularia koellikeri.* Chem. Pharm. Bull. (Tokyo) **32**, 1667 (1984).

333. KODAMA, M., Y. SHIOBARA, K. MATSUMURA, and H. SUMITOMO: Total Synthesis of Marchantin A, A Cytotoxic Bis(bibenzyl) Isolated from Liverworts. Tetrahedron Letters **26**, 877 (1985).

334. KODAMA, M., Y. SHIOBARA, H. SUMITOMO, K. MATSUMURA, M. TSUKAMOTO, and C. HARADA: Total Syntheses of Marchantin A and Riccardin B, Cytotoxic Bis(bibenzyls) from Liverworts. J. Organ. Chem. (USA) **53**, 72 (1988).

335. KODAMA, M., U.S.F. TAMBUNAN, and T. TSUNODA; Total Synthesis of (−)-Vitrenal and Its Biological Activity. Tetrahedron Letters **27**, 1197 (1986).

336. KODAMA, M., U.S.F. TAMBUNAN, T. TSUNODA, and S. ITÔ: Total Synthesis of (−)-Vitrenal and Its Biological Activity. Bull. Chem. Soc. Jpn. **59**, 1897 (1986).

337. KODPINID, M., C. SADAVONGVIVAD, C. THEBTARANONTH, and Y. THEBTARANONTH: Benzyl Benzoates from the Root of *Uvaria purpurea.* Phytochem. **23**, 199 (1984).

338. KOHN, G., S. DEMMERLE, O. VANDEKERKHOVE, E. HARTMANN, and P. BEUTELMANN: Distribution and Chemotaxonomic Significance of Acetylenic Fatty Acids in Mosses of the Dicranales. Phytochem. **26**, 2271 (1987).

339. KOHN, G., O. VANDEKERKHOVE, E. HARTMANN, and P. BEUTELMANN: Acetylenic Fatty Acids in the Ricciaceae (Hepaticae). Phytochem. **27**, 1049 (1988).

340. KOHN, G., A. VIERENGEL, O. VANDEKERKHOVE, and E. HARTMANN: 9-Octadecen-6-ynoic Acid from *Riccia fluitans.* Phytochem. **26**, 2101 (1987).

341. KONDO, K., M. TOYOTA, and Y. ASAKAWA: *Ent*-eudesmane-type Sesquiterpenoids from *Bazzania* Species. Phytochem. **29**, 2197 (1990).

342. KONECNY, K., M. STREIBL, S. VASICKOVÁ, M. BUDESINSKY, D. SAMAN, K. UBIK, and V. HEROUT: Constituents of the Liverwort *Bazzania trilobata* of Czech Origin. Collect. Czech. Chem. Commun. **50**, 80 (1985).

343. KONECNY, K., K. UBIK, S. VASICKOVÁ, M. STREIBL, and V. HEROUT: Constituents of the Liverwort *Lophocolea heterophylla.* Collect. Czech. Chem. Commun. **47**, 3164 (1982).

344. KOPONEN, A., T. KOPONEN, H. PYYSALO, K. HIMBERG, and P. MANSIKKAMÄKI: Composition of Volatile Compounds in Splachnaceae. In: Bryophytes: Their Chemistry and Chemical Taxonomy (H.D. ZINSMEISTER and R. MUES, eds.), p. 437. Oxford: Oxford University Press, 1990.

345. KOSKIMIES-SOININEN, K., and H. NYBERG: Effects of Temperature and Light on the Glycolipids of *Sphagnum fimbriatum.* Phytochem. **30**, 2529 (1991).

345a. KRAUT, L., R. KLEIN, and R. MUES: Flavonoid Diversity in the Liverwort Genus *Monoclea* Hooker. Z. Naturforsch. **47c**, 794 (1992).

345b. KRAUT, L., R. MUES, and M. SIM-SIM: Acylated Flavone and Glycerol Glucosides from two *Frullania* Species. Phytochem. **34**, 211 (1993).

345c. KREISER, W.: Synthesis, Relative Absolute Configuration of Some Natural Bisabolones. In: Studies in Natural Products Chemistry, Vol. 8 (ATTA-UR-RAHMAN, ed.). Amsterdam: Elsevier, 1991.

346. KRETSCHMAR, H.C., Z.J. BARNEIS, and W.F. ERMAN: The Isolation and Synthesis of a Novel Tetracyclic Ether from East Indian Sandalwood Oil. A Facile Intramolecular Prins Reaction. Tetrahedron Letters **1970**, 37.

347. KRUIJT, R. CH., G.J. NIEMANN, C.G. DE KOSTER, and W. HEERMA: Flavonoids and Aromatic Hydroxy Acids in Lejeuneaceae Subfamily Ptychanthoideae. Cryptogamie Bryol. Lichenol. **7**, 165 (1986).

348. KRZAKOWA, M.: Isozymes as Markers of Inter- and Intraspecific Differentiation in Hepaticae. In: Bryophytorum Bibliotheca, Vol. 13 (C. SUIRE, ed.), p. 427. Vaduz: J. Cramer, 1978.

349. KRZAKOWA, M.: Usefulness of Electrophoresis for Taxonomy and Genetics of Bryophytes. In: Methods in Bryology (J.M. GLIME, ed.), p. 359. Nichinan: J. Hattori Bot. Lab., 1988.

350. KUBO, I., A. MATSUMOTO, K. HIROTSU, H. NAOKI, and W.F. WOOD: Structure and the Absolute Configuration of a New Diterpene, $(-)$-2(S),8(R)-Dihydroxyverrucosane, from the Liverwort *Gyrothyra underwoodiana* (Gyrothyraceae). J. Organ. Chem. (USA) **49**, 4644 (1984).

351. KUNZ, S., and H. BECKER: Bibenzyl Glycosides from the Liverwort *Ricciocarpos natans*. Phytochem. **31**, 3981 (1992).

352. KUO, Y.-H., and Y.-T. LIN: Two New Sesquiterpenes 3β-Hydroxycedrol and Widdringtonia Acid II–A Co-Crystal of β-Chamigrenic Acid and Hinokiic Acid. J. Chin. Chem. Soc. **27**, 15 (1980).

353. KUPCHAN, S.M., Y. KODAMA, G.J. THOMAS, and H.P.J. HINTZ: Maytanoprine and Maytanbutine, New Antileukemic Ansa Macrolides from *Maytenus buchananii*. J. Chem. Soc. Chem. Commun. **1972**, 1065.

354. KUSUMI, T., T. HAMADA, M. HARA, M.O. ISHITSUKA, H. GINDA, and H. KAKISAWA: Structure and Absolute Configuration of Isoclavukerin A, A Component from an Okinawan Soft Coral. Tetrahedron Letters **33**, 2019 (1992).

354a. LANGENBAHN, U., G. BURKHARDT, and H. BECKER: Diterpene Malonates and Other Terpenes from *Nardia succulenta* and *N. scalaris*. Phytochem. **33**, 1173 (1993).

355. LAAS, H.J., and T. EICHER: New Syntheses of Tricetin and Some of Its Analogues. J. Hattori Bot. Lab. **67**, 383 (1989).

356. LEHMANN, L., J. JAKUPOVIC, F. BOHLMANN, R.M. KING, and L. HAEGI: Azulenes, Labdanes and Furanocurcumene from *Ixiolaena leptolepis*. Phytochem. **27**, 2994 (1988).

357. LINDBERG, G., B.-G. ÖSTERDAHL, and E. NILSSON: Chemical Studies on Bryophytes, 16: 5′,8″-Biluteolin, A New Biflavone from *Dicranum scoparium*. Chem. Scripta **5**, 140 (1974).

358. LINDBERG, B., and O. THEANDER: Studies on *Sphagnum* Peat, II: Lignin in *Sphagnum*. Acta Chem. Scand. **6**, 311 (1952).

358a. LORIMER, S.D., N.B. PERRY, and R.S. TANGNEY: An Antifungal Bibenzyl from the New Zealand Liverwort, *Plagiochila stephensoniana*. Bioactivity-Directed Isolation, Synthesis, and Analysis. J. Nat. Prod. **56**, 1444 (1993).

359. MAGARI, H., H. HIROTA, and T. TAKAHASHI: A Total Synthesis of ($\pm$)-Isobicycloger-macrenal. J. Chem. Soc. Chem. Commun. **1987**, 1196.
360. MAGARI, H., H. HIROTA, T. TAKAHASHI, A. MATSUO, S. UTO, H. NOZAKI, M. NAKAYAMA, and S. HAYASHI: Synthesis of ($\pm$)-Vitrenal. Chem. Letters **1982**, 1143.
361. MARKHAM, K.R.: Phytochemical Relationships of *Carrpos* with *Corsionia* and Other *Marchantialean* Genera. Biochem. System. and Ecol. **8**, 11 (1980).
362. MARKHAM, K.R.: Distribution of Flavonoids in the Lower Plants and Its Evolutionary Significance, In: The Flavonoids Advances in Research Since 1980 (J.B. HARBORNE, ed.), p. 427. London: Chapman and Hall, 1988.
363. MARKHAM, K.R.: Bryophyte Flavonoids, Their Structures, Distribution, and Evolutionary Significance. In: Bryophytes: Their Chemistry and Chemical Taxonomy (H.D. ZINSMEISTER and R. MUES, eds.), p. 143. Oxford: Oxford University Press, 1990.
364. MARKHAM, K.R., M. ANDERSEN, and E.S. VIOTTO: Unique Biflavonoid Types from the Moss *Dicranoloma robustum*. Phytochem. **27**, 1745 (1988).
365. MARKHAM, K.R., and D.R. GIVEN: The Major Flavonoids of an Antarctic *Bryum*. Phytochem. **27**, 2843 (1988).
366. MARKHAM, K.R., and R. MUES: Taxonomically Significant 8-Hydroxyflavone Glucuronides from the Marchantialean Liverwort, *Bucegia romanica*. Phytochem. **22**, 143 (1983).
367. MARKHAM, K.R., R. MUES, M. STOLL, and H.D. ZINSMEISTER: NMR Spectra of Flavone Di-C-Glycosides from *Apometzgeria pubescens* and Detection of Rotational Isomerism in 8-C-Hexosylflavones. Z. Naturforsch. **42c**, 1039 (1987).
368. MARKHAM, K.R., C. SCHEPPARD, and H. GEIGER: ^{13}C NMR Studies on Some Naturally Occurring Amentoflavone and Hinokiflavone Biflavonoids. Phytochem. **26**, 3335 (1987).
369. MARKHAM, K.R., R. THEODOR, R. MUES, and H.D. ZINSMEISTER: 6-C-α-L-Rhamnopyranosylapigenin 7-O-β-D-glucopyranoside (Isofurcatain 7-O-β-D-glucoside), A New Flavone Glycoside from *Metzgeria furcata*. Z. Naturforsch. **37c**, 562 (1982).
370. MARSHALL, J.A., and A.R. HOCHSTETLER: The Synthesis of ($\pm$)-Geosmin and the Other 1,10-Dimethyl-9-decalol Isomers. J. Organ. Chem. (USA) **33**, 2593 (1968).
371. MARSILI, A., and I. MORELLI: Triterpenes from Mosses, I: The Occurrence of 22(29)-Hopene in *Thamnium alopecurum* (L.) Br. Eur. ssp. *eu-alopecurum* Giac. Phytochem. **7**, 1705 (1968).
371a. MARSILI, A., and I. MORELLI: Triterpenes from Mosses, II: Triterpenes from *Thuidium tamariscifolium*. Phytochem. **9**, 651 (1970).
372. MARSILI, A., and I. MORELLI, C. BERNARDINI, and M. PACCHIANI: Triterpenes from Mosses, IV: Constituents of Some Mosses. Phytochem. **11**, 2003 (1972).
373. MARSILI, A., and I. MORELLI, and A.M. IORI: Triterpenes from Mosses, III: 21-Hopene and Some Other Constituents of *Pseudoscleropodium purum*. Phytochem. **10**, 432 (1971).
374. MÅRTENSSON, O., and E. NILSSON: On the Morphological Colour of Bryophytes. Lindbergia **2**, 145 (1974).
375. MATEOS, A.F., O.F. BARRUECO, and R.R. GONZÁLEZ: Synthesis of Pinguisanes Through Furan-Terminated Cationic Cyclization. Tetrahedron Letters **31**, 4343 (1990).
376. MATSUO, A.: Selected Chemotaxonomic Characteristics of Liverwort Sesquiterpenoids. J. Hattori Bot. Lab. **53**, 295 (1982).
377. MATSUO, A., K. ATSUMI, and M. NAKAYAMA: Isolation of Seven Verrucosane Diterpenoids from the Liverwort *Scapania bolanderi*. Z. Naturforsch. **39b**, 1281 (1984).

546   Y. Asakawa

378. Matsuo, A., K. Atsumi, and M. Nakayama: Structures of *ent*-2,3-
     Secoaromadendrane Sesquiterpenoids, Which Have Plant-Growth-Inhibitor Activity
     from *Plagiochila semidecurrens* (Liverwort). J. Chem. Soc. Perkin 1 **1981**, 2816.
379. Matsuo, A., S. Hayashi, and H. Inoue: Differential Distribution of Sesquiterpenoids
     in Three Forms of *Plagiochila acanthophylla* Gott. subsp. *japonica* (Lac.) Inoue.
     Hikobia Suppl. **1**, 455 (1981).
380. Matsuo, A., O. Ishi, M. Suzuki, M. Nakayama, and S. Hayashi: Isolation of
     Enantiomeric Type Sesquiterpenoids from the Liverwort *Riccardia jackii*. Z. Natur-
     forsch. **37b**, 1636 (1982).
381. Matsuo, A., K. Kamio, K. Uohama, K. Yoshida, J.D. Connolly, and G.A. Sim:
     Dolabellane Diterpenoids from the Liverwort *Odontoschisma denudatum*. Phytochem.
     27, 1153 (1988).
382. Matsuo, A., N. Nakayama, and M. Nakayama: Enantiomeric Type Sesquiterpenoids
     of the Liverwort *Marchantia polymorpha*. Phytochem. **24**, 777 (1985).
383. Matsuo, A., H. Nozaki, N. Kubota, S. Uto, and M. Nakayama: Structures and
     Conformations of (−)-Isobicyclogermacrenal and (−)-Lepidozenal, Two Key Ses-
     quiterpenoids of the *cis*- and *trans*-10,3-Bicyclic Ring Systems, from the Liverwort
     Lepidozia vitrea: X-Ray Crystal Structure Analysis of the Hydroxy Derivative of (−)-
     Isobicyclogermacrenal. J. Chem. Soc. Perkin. Trans. 1 **1984**, 203.
384. Matsuo, A., H. Nozaki, K. Yano, S. Uto, M. Nakayama, and S. Huneck:
     Gymnomitrane Sesquiterpenoids from the Liverwort *Marsupella emarginata* var.
     patens. Phytochem. **29**, 1921 (1990).
385. Matsuo, A., S. Sato, M. Nakayama, S. Hayashi: Structure and Absolute Configu-
     ration of (−)-Taylorione, a Novel Skeletal Sesquiterpene Ketone of *ent*-1,10-Seco-
     aromadendrane Form, from *Mylia taylorii* (Liverwort). J. Chem. Soc. Perkin Trans.
     1 **1979**, 2652.
386. Matsuo, A., and D. Takaoka: Structures of New Sesquiterpenoids from the Liver-
     wort *Mylia taylorii*. In: Bryophytes: Their Chemistry and Chemotaxon-
     omy (H.D. Zinsmeister and R. Mues, eds.), p. 59. Oxford: Oxford University Press,
     1990.
387. Matsuo, A., D. Takaoka, and H. Kawahara: Soluble Carbohydrates of Liverworts.
     Phytochem. **25**, 2335 (1986).
388. Matsuo, A., K. Uohama, S. Hayashi, and J.D. Connolly: (+)-Acetoxydontoschis-
     menol, A New Dolabellane Diterpenoid from the Liverwort *Odontoschisma denu-
     datum*. Chem. Letters **1984**, 599.
389. Matsuo, A., S. Uto, H. Nozaki, and M. Nakayama: Structure and Absolute
     Configuration of (+)-Vitrenal, a Novel Carbon Skeletal Sesquiterpenoid Having
     Plant-Growth-Inhibitory Activity from the Liverwort *Lepidozia vitrea*. J. Chem. Soc.
     Perkin Trans. 1 **1984**, 215.
390. Matsuo, A., S. Uto, K. Sakuda, Y. Uchio, M. Nakayama, and S. Hayashi:
     Structures of Three New *ent*-Longipinane Type Sesquiterpenoids from the Liverwort
     Marsupella emarginata subsp. *tubulosa*. Chem. Letters **1979**, 73.
391. Matsuo, A., K. Yoshida, Y. Fukazawa, M. Nakayama, and K. Kuriyama:
     Conformational Behavior and Epoxidation of the Dolabella-3*E*,7*E*-dienoid. Chem.
     Letters **1987**, 369.
392. Matsuo, A., S. Yuki, and M. Nakayama: (−)-Herbertenediol and (−)-
     Herbertenolide, Two New Sesquiterpenoids of the *ent*-Herbertane Class from the
     Liverwort *Herberta adunca*. Chem. Letters **1983**, 1041.
393. Matsuo, A., S. Yuki, and M. Nakayama: Structures of *ent*-Herbertane Sesquiterpen-
     oids Displaying Antifungal Properties from the Liverwort *Herberta adunca*. J. Chem.
     Soc. Perkin Trans. 1 **1986**, 701.

394. MATSUO, A., S. YUKI, M. NAKAYAMA, and S. HAYASHI: (−)-Herbertene, an Aromatic Sesquiterpene with a Novel Carbon Skeleton from the Liverwort *Herberta adunca*. J. Chem. Soc. Chem. Commun. **1981**, 864.

395. MATSUO, A., S. YUKI, M. NAKAYAMA, and S. HAYASHI: Three New Sesquiterpene Phenols of the *ent*-Herbertane Class from the Liverwort *Herberta adunca*. Chem. Letters **1982**, 436.

396. MCCLEARY, J. A., and D.L. WALKINGTON: Mosses and Antibiotics. Rev. Bryol. Lichenol. **34**, 309 (1966).

397. MCCRINDLE, R., and E. NAKAMURA: Constituents of *Solidago* Species, Part VI: The Constituents of Diterpenoids from a Chemically Distinct Variety of *Solidago serotina*. Can. J. Chem. **52**, 2029 (1974).

398. MCMURRY, J. E., and G.K. BOSCH: Synthesis of Bicyclogermacrene and Lepidozene. Tetrahedron Letters **26**, 2167 (1985).

399. MEIER, B., and O. STICHER: The Use of a High Speed Spectrophotometric Detector (Diode Array) in the HPLC Analysis of Medicinal Plants. Pharm. Ind. **48**, 87 (1986).

400. MENDEZ, J., and F. SANZ-CABANILLES: Cinnamic Acid Esters in *Anthoceros* Species. Phytochem. **18**, 1409 (1979).

401. MENTLEIN, R., and E. VOWINKEL: Simple Screening Method for the Separation and Identification of Sphagnorubins, A New Class of Anthocyanidins. J. Chromatogr. **268**, 138 (1983).

402. MENTLEIN, R., and E. VOWINKEL: Die roten Wandfarbstoffe des Torfmooses *Sphagnum rubellum*. Liebigs Ann. Chem. **1984**, 1024.

403. MEZEY-VÁNDOR, G., M. NÓGRÁDI, V.P. NOVIKOV, A. WISZT, and M. KAJTÁR-PEREDY: Die Synthese von Perrottetin F und G, zwei acyclische Bis(bibenzyl)ether aus *Radula perrottetii*. Liebigs Ann. Chem. **1989**, 401.

404. MIKSCHE, G.E., and S. YASUDA: Lignin of 'Giant' Mosses and Some Related Species. Phytochem. **17**, 503 (1978).

405. MINAMI, Y., S. WAKABAYASHI, S. IMOTO, Y. OHTA, and H. MATSUBARA: Ferrodoxin from a Liverwort, *Marchantia polymorpha*. Purification and Amino Acid Sequence. J. Biochem. **98**, 649 (1985).

406. MISRA, R., R.C. PANDEY, and S. DEV: Higher Isoprenoids, IX: Diterpenoids from the Oleoresin of *Hardwickia pinnata*. Part 2: Kolavic, Kolavenic, Kolavenolic and Kolavonic Acids. Tetrahedron **35**, 979 (1979).

407. MISRA, R., R.C. PANDEY, and S. DEV: Higher Isoprenoids, X: Diterpenoids from the Oleoresin of *Hardwickia pinnata*. Part 3: Kolavenol, Kolavelool and a Nor Diterpene Hydrocarbon. Tetrahedron **35**, 985 (1979).

408. MIYASE, T., A. UENO, N. TAKIZAWA, H. KOBAYASHI, and H. KARASAWA: Studies on the Glycosides of *Epimedium grandiflorum* Morr. var. *thunbergianum* (Miq.) Nakai. I. Chem. Pharm. Bull. (Tokyo) **35**, 1109 (1987).

409. MIZUTANI, M.: On the Taste of Some Mosses. Misc. Bryol. Lichenol. **2**, 100 (1961).

410. MLOTKIEWICZ, J.A., J. MURRAY-RUST, P. MURRAY-RUST, W. PARKER, F.G. RIDDELL, J. S. ROBERTS, and A. SATTAR: A Biogenetically Significant Cyclization of Humulene-4,5-epoxide. Tetrahedron Letters **40**, 3887 (1979).

411. MORAIS, R.: In Vitro-Kultur und Sekundärstoffproduktion des Lebermooses *Reboulia hemisphaerica* (L.) Raddi. Ph.D. Thesis, Universität des Saarlandes, 1990.

412. MORAIS, R.M.S.C., L.J. HARRISON, and H. BECKER: (+)-Gymnomitr-8(12)-en-9α-ol and (*R*)-(−)-8,11-Dihydro-α-Cuparenone, Two Novel Sesquiterpenoids from the Liverwort *Reboulia hemisphaerica*. J. Chem. Res. (S) **1988**, 380.

413. MUES, R.: Occurrence and Absence of C-Glycosylflavones in Species of the Liverwort Genera *Blepharostoma*, *Herbertus*, *Mastigophora*, *Porella*, *Ptilidium* and *Trichocolea*: An Indication of Taxonomic Significance? J. Hattori Bot. Lab. **53**, 271 (1982).

414. Mues, R.: Vergleichende Untersuchungen zur Chemie und Taxonomie der Metzgeriales und Jungermanniales (Hepaticae). Ber. Deutsch. Bot. Ges. Bd. **95**, 115 (1982).

415. Mues, R.: Flavonoid Pattern of 10 *Radula* Species and Their Possible Application in
Species Differentiation. Proc. of the 3rd Meeting of the Bryologists from Central and
East Europe. Praha, Symposium Papers, p. 37 (1984).

416. Mues, R.: New Results on the Flavonoid Chemistry and Chemotaxonomy of
Hepaticae and Anthocerotae. Abstracta Botanica **9**, Suppl. 2, 171 (1985).

417. Mues, R.: Thin-Layer Chromatography (TLC) of Flavonoid Compounds from
Bryophytes. In: Methods in Bryology (J.M. Glime, ed.), p. 147. Nichinan: J. Hattori
Bot. Lab., 1988.

418. Mues, R.: Chemosystematics of Bryophytes. In: Flavonoids in Biology and Medicine,
III: Current Issues in Flavonoid Research (N.P. Das, ed.), p. 1. Singapore: National
University, 1990.

419. Mues, R.: The Significance of Flavonoids for the Classification of Bryophytes Taxa at
Different Taxonomic Rank. In: Bryophytes: Their Chemistry and Chemical Taxonomy (H.D. Zinsmeister and R. Mues, eds.), p. 421. Oxford: Oxford University Press,
1990.

420. Mues, R., S. Hattori, Y. Asakawa, and R. Grolle: Biosystematic Studies on
Frullania jackii and *F. davurica.* J. Hattori Bot. Lab. **56**, 227 (1984).

421. Mues, R., S. Huneck, J.D. Connolly, and D.S. Rycroft: Scapaniapyrone A,
A Novel Aromatic Constituent of the Liverwort *Scapania undulata.* Tetrahedron
Letters **29**, 6793 (1988).

422. Mues, R., R. Klein, and S.R. Gradstein: New Reflections on the Taxonomy of
Pleuroziaceae Supported by Flavonoid Chemistry. J. Hattori Bot. Lab. **70**, 79 (1991).

423. Mues, R., G. Leidinger, V. Lauck, H.D. Zinsmeister, T. Koponen, and K.R.
Markham: *Rhizomnium magnifolium* and *R. pseudopunctatum,* The First Mosses to
Yield Flavone Glucuronides. Z. Naturforsch. **41c**, 971 (1986).

424. Mues, R., C. Müller, U. Pröbsting, and H.D. Zinsmeister: Phytochemical Studies
on the Genus *Mylia* S. Gray (Jungermanniaceae, Hepaticae). Bull. Nat. Sci. Mus, Ser.
B. **14**, 149 (1988).

425. Mues, R., A. Strassner, and H.D. Zinsmeister: Unusual Flavonoid Glycosides for
Jungermanniales Detected in Two *Frullania* Species (Hepaticae). Cryptogamie Bryol.
Lichénol. **4**, 111 (1983).

426. Mues, R., and H.D. Zinsmeister: The Chemotaxonomy of Phenolic Compounds in
Bryophytes. J. Hattori Bot. Lab. **64**, 109 (1988).

427. Murakami, T., and N. Tanaka: Occurrence, Structure and Taxonomic Implications
of Fern Constituents. In: Progress in Organic Natural Products, Vol. 54 (W. Herz, H.
Grisebach, G.W. Kirby, and Ch. Tamm, eds.), p. 1. Wien-New York: Springer, 1988.

427a. Nabeta, K., K. Katayama, S. Nakagawara, and K. Katoh: Sesquiterpenes of
Cadinane Type from Cultured Cells of the Liverwort, *Heteroscyphus planus.*
Phytochem. **32**, 117 (1993).

428. Naegeli, P.: Spiro[4,5]decenone-Synthesis. Tetrahedron Letters **1978**, 2127.

429. Nagai, T.: Studies on Chemical Constituents of Some East European Liverworts.
Master's Thesis, Tokushima Bunri University, 1992.

430. Nagashima, F.: Phytochemical Studies on *Jungermannia, Frullania, Monoclea, Plagiochila* and *Porella* Genera (Hepaticae). Ph.D. Thesis, Tokushima Bunri University,
1991.

431. Nagashima, F., and Y. Asakawa: Rearranged Cuparane-Type Sesqui- and
Clerodane-Type Diterpenoids from the Liverwort *Demotarisia linguifolia.* Phytochem.
29, 3229 (1990).

432. Nagashima, F., and Y. Asakawa: Unpublished results.

432a. NAGASHIMA, F., H. IZUMO, M. TORI, Y. ASAKAWA, Y. KAN, and S. TAKAOKA: Terpenoids and Aromatic Compounds from Panamanian Liverwort *Bryopteris filicina* (Lejeuneaceae). 37th Symposium on Chemistry of Terpenes, Essential Oils and Aromatics. Okinawa, Japan. Symposium Papers, p. 44 (1993).

433. NAGASHIMA, F., E. NISHIOKA, K. KAMEO, C. NAKAGAWA, and Y. ASAKAWA: Terpenoids and Aromatic Compounds from Selected Ecuadorian Liverworts. Phytochem. **30**, 215 (1991).

434. NAGASHIMA, F., Y. OHI, T. NAGAI, M. TORI, Y. ASAKAWA, and S. HUNECK: Studies on the Chemical Constituents of Some East European Liverworts (3). 36th Symposium on Chemistry of Terpenes, Essential Oils and Aromatics. Nishinomiya, Japan. Symposium Papers, p. 257 (1992).

434a. NAGASHIMA, F., Y. OHI, T. NAGAI, M. TORI, Y. ASAKAWA, and S. HUNECK: Terpenoids from Some German and Russian Liverworts. Phytochem. **33**, 1445 (1993).

435. NAGASHIMA, F., Y. SARI, M. TORI, Y. ASAKAWA, and S. HUNECK: Sesquiterpenoids from Some European Liverworts. Phytochem. **34**, 1341 (1993).

435a. NAGASHIMA, F., H. TANAKA, M. TOYOTA, T. HASHIMOTO, Y. KAN, S. TAKAOKA, M. TORI, and Y. ASAKAWA: Sesqui- and Diterpenoids from the Liverworts *Plagiochila* Species. Phytochem. **36**, 1425 (1994).

436. NAGASHIMA, F., M. TOYOTA, and Y. ASAKAWA: Bitter Kaurane-Type Diterpene Glucosides from the Liverwort *Jungermannia infusca*. Phytochem. **29**, 1619 (1990).

437. NAGASHIMA, F., M. TOYOTA, and Y. ASAKAWA: Terpenoids from Some Japanese Liverworts. Phytochem. **29**, 2169 (1990).

438. NAGASHIMA, F., M. TORI, and Y. ASAKAWA: Diterpenoids from the East Malaysian Liverwort *Schistochila aligera*. Phytochem. **30**, 849 (1991).

439. NAKAGAWARA, S., K. KATOH, T. KUSUMI, H. KOMURA, K. NOMOTO, H. KONNO, S. HUNECK, and R. TAKEDA: Two Azulenes Produced by the Liverwort, *Calypogeia azurea*, During *in vitro* Culture. Phytochem. **31**, 1667 (1992).

440. NAKAMURA, H., S. TO, M. TAKAMATSU, J. KOBAYASHI, Y. OHIZUMI, K. KONDO, and Y. HIRATA: Terpenoid Constituents of Okinawan Sponge SS405. 59th Annual Meeting of Chemical Society of Japan. Yokohama, Symposium Papers, p. 1197 (1990).

441. NAKANISHI, K., and J. DILLON: A Simple Method for Determining the Chirality of Cyclic α-Glycols with Pr(DPM)₃ and Eu(DPM)₃. J. Am. Chem. Soc. **93**, 4058 (1971).

441a. NAKASHIMA, K.: Structures and Syntheses of Two Brasilane-Type Sesquiterpenoids and Chemical Constituents of Some South American Liverworts. Master Thesis, Tokushima Bunri University, 1993.

441b. NAKATANI, Y., G. OURISSON, and J.-P. BECK: Chemistry and Biochemistry of Chinese Drugs, VII: Cytostatic Pheophytins from Silkworm Excreta, and Derived Photocytotoxic Pheophorbides. Chem. Pharm. Bull. (Tokyo) **29**, 2261 (1981).

442. NAKAYAMA, M., A. MATSUO, T. KAMI, and S. HAYASHI: Volatiles from *Leptolejeunea elliptica*. Phytochem. **18**, 328 (1979).

442a. NAKAYAMA, M., T. TAKASE, and T. YOKOTA: Abscisic Acid is an Endogenous Plant Growth Regulator in Liverworts. 15th International Botanical Congress. Yokohama, Japan. Abstracts, p. 388 (1993).

443. NARAYANAN, C.R., and N.K. VENKATASUBRAMANIAN: Simple Methods to Find the Stereochemistry of the Side Chain of γ-Lactones. J. Organ. Chem. (USA) **33**, 3156 (1968).

444. NAYA, Y., Y. NAGAHAMA, and M. KOTAKE: Volatile Components of *Ledum palustre* var. *nipponicum* et *yesoense*. Heterocycles **10**, 29 (1978).

445. NILSSON, E., and O. MÅRTENSSON: Chemical Studies on Bryophytes, 11: (−)-16α-Hydroxykaurane from *Saelania glaucescens* (Hedw.) Broth. Acta Chem. Scand. **25**, 1486 (1971).

550   Y. Asakawa

446. Niwa, M., A. Nishiyama, M. Iguchi, and S. Yamamura: Sesquiterpenes from *Acorus calamus* L. Bull. Chem. Soc. Jpn. **48**, 2930 (1975).
447. Nógrádi, M., B. Vermes, and M. Kajtár-Peredy: The Constitution of Riccardin B. Tetrahedron Letters **28**, 2899 (1987).
448. Noma, M., F. Suzuki, K. Gamou, and N. Kawashima: Two Labdane Diterpenoids from *Nicotiana raimondii*. Phytochem. **21**, 395 (1982).
448a. Nozaki, H., K. Hamazaki, N. Nishimura, H. Udaka, N. Takashima, and D. Takaoka: Studies on the Constituents of *Thuidium kanedae* Sak. (Muscopsida Bryophyta). 15th International Botanical Congress. Yokohama, Japan. Abstracts, p. 370 (1993).
449. Ohta, Y., S. Abe, H. Komura, and M. Kobayashi: Prelunularic Acid, a Probable Immediate Precursor of Lunularic Acid. First Example of a "Prearomatic" Intermediate in the Phenylpropanoid-Polymalonate Pathway. J. Am. Chem. Soc. **105**, 4480 (1983).
450. Ohta, Y., S. Abe, H. Komura, and M. Kobayashi: Prelunularic Acid, a Probable Immediate Precursor of Lunularic Acid, in Suspension-Cultured Cells of *Marchantia polymorpha*. J. Hattori Bot. Lab. **56**, 249 (1984).
451. Ohta, Y., N.H. Andersen, and C.-B. Liu: Sesquiterpene Constituents of Two Liverworts of Genus *Diplophyllum*. Novel Eudesmanolides and Cytotoxicity Studies for Enantiomeric Methylene Lactones. Tetrahedron **33**, 617 (1977).
452. Ohta, Y., and Y. Hirose: The Structure of Cubenol and *Epi*-Cubenol. Tetrahedron Letters **1967**, 2073.
453. Ohta, Y., K. Katoh, and R. Takeda: Growth and Secondary Metabolites Production in Cultured Cells of Liverworts. In Bryophyte Development: Physiology and Biochemistry (R.N. Chopra and S.C. Bhatla, eds.), p. 209. Boca Raton: CRC Press, 1990.
454. Ohtani, I., T. Kusumi, Y. Kashman, and H. Kakisawa: High-Field FT NMR Application of Mosher's Method. The Absolute Configurations of Marine Terpenoids. J. Am. Chem. Soc. **113**, 4092 (1991).
455. Okamoto, H., D. Yoshida, and Y. Saito: Inhibition of 12-O-Tetradecanoylphorbol-13-acetate-induced Ornithine Decarboxylase Activity in Mouse Epidermis by Sweetening Agents and Related Compounds. Cancer Letters **21**, 29 (1983).
456. Okuda, P.K., and P.J. Scheuer: Sesquiterpenoid Constituents of Eight Porostome Nudibranchs. J. Organ. Chem. (USA) **48**, 1866 (1983).
457. Ono, K., T. Sakamoto, and Y. Asakawa: Constituents from Cell Suspension Cultures of Selected Liverworts. Phytochem. **31**, 1249 (1992).
458. Ono, K., M. Toyoto, and Y. Asakawa: Production of Terpenoids from Cultured Cells of Several Liverworts. 15th International Botanical Congress. Yokohama, Japan. Symposium Papers, p. 372 (1993).
459. Österdahl, B.-G.: Chemical Studies on Bryophytes, 23: ^{13}C NMR Analysis of a Biflavone from *Dicranum scoparium*. Acta Chem. Scand. **B37**, 69 (1983).
460. Paknikar, S.K., C.G. Naik, N.H. Anderson, and Y. Ohta: (−)-Aromadendrene (β-Diploalbicene) and (+)-*ent*-C_{10}-Epiglobulol (Diploalbicanol) from Genus *Diplophyllum*. Ind. J. Chem. **24B**, 450 (1985).
461. Papastephanou, C. and D.G. Anderson: Crassin Acetate Biosynthesis in a Cell-Free Homogenate of Zooxanthellae from *Pseudoplexaura porosa* (Houttyun); Implications to the Symbiotic Process. Comp. Biochem. Physiol. **73B**, 617 (1982).
462. Pathak, V.P., and R.N. Khanna: Synthesis of Tomentellin. Indian J. Chem. **19B**, 1077 (1980).
463. Patterson, G.W., G.R. Wolfe, T.A. Salt, and P.-L. Chiu: Sterols of Bryophytes

with Emphasis on the Configuration at C-24. In: Bryophytes: Their Chemistry and Chemical Taxonomy (H.D. ZINSMEISTER and R. MUES, eds.), p. 103. Oxford: Oxford University Press, 1990.

464. PETRAGNANI, N., H.M.C. FERRAZ, and M. YONASHIRO: A New Synthesis of ($\pm$)-Frullanolide: Application of 2-Phenylselenopropanoic Acid. Synthesis **1985**, 27.

465. PFAFFMANN, H., and E. HARTMANN: Analysis of Phospholipids and Assay for Phospholipases from Moss Tissue. In: Methods in Bryology (J.M. GLIME, ed.), p. 177. Nichinan: J. Hattori Bot. Lab., 1988.

466. PORTER, L.: Geographic Races of *Conocephalum* (Marchantiales) as Defined by Flavonoid Chemistry. Taxon **30**, 739 (1981).

467. POWELL, R.G., D. WEISLEDER, and C.R. SMITH Jr: Novel Maytansinoid Tumor Inhibitors from *Trewia nudiflora*: Trewiasine, Dehydrotrewiasine, and Demethyltrewiasine. J. Organ. Chem. (USA) **46**, 4398 (1981).

468. PRINS, H.H.TH.: Why Are Mosses Eaten in Cold Environments Only? Okios **38**, 374 (1981).

469. PYYSALO, H., A. KOPONEN, and T. KOPONEN: Studies on Entomophily in Splachnaceae (Musci), I: Volatile Compounds in the Sporophyte. Ann. Bot. Fennici. **15**, 293 (1978).

470. REFFSTRUP, T., O. HAMMERSHOY, P.M. BOLL, and H. SCHMIDT: *Philodendron scandens* Koch et Sello subsp. *oxycardium* (Schott) Bunting. A New Source of Allergenic Alkyl Resorcinols. Acta Chem. Scand. **B36**, 291 (1982).

471. REVIAL, G.: Asymmetric Michael-Type Alkylation of Chiral Imines. Enantioselective Syntheses of ($-$)-Geosmin and Two Other Related Natural Terpenes, as Well as *enant*-($+$)-Geosmin. Tetrahedron Letters **30**, 4121 (1989).

472. RI, Z.: Honzokoumoku **21**, 814 (1590).

473. RICE, J.R., C. PAPASTEPHANOU, and D.G. ANDERSON: Isolation, Localization and Biosynthesis of Crassin Acetate in *Pseudoplexaura porosa* (Houttuyn). Biol. Bull. **138**, 334 (1970).

474. RISCHMANN, M., R. MUES, H. GEIGER, H. J. LAAS, and T. EICHER: Isolation and Synthesis of 6,7-Dihydroxy-4-(3,4-dihydroxyphenyl)naphthalene-2-carboxylic Acid from *Pellia epiphylla*. Phytochem. **28**, 867 (1989).

475. RUANGRUNGSI, N., S. KASIWONG, K. LIKHITWITAYAWUID, G. L. LANGE, and C.P. DECICCO: Constituents of *Grangea maderaspatana*, a New Eudesmanolide. J. Nat. Prod. **52**, 130 (1989).

476. RUDOLPH, H.: 15 Jahre Kultur von Sphagnen unter definierten Bedingungen: Eine Übersicht über Resultate, Probleme und Perspektiven. In: Congrès International de Bryologie (C. SUIRE, éd.), Bryophytorum Bibliotheca, Bd. 13, p. 279. Vaduz: J. Cramer, 1978.

477. RUDOLPH, H., and J. SAMLAND: Occurrence and Metabolism of Sphagnum Acid in the Cell Walls of Bryophytes. Phytochem. **24**, 745 (1985).

478. RYCROFT, D.S.: Some Recent NMR Studies of Diterpenoids from the Hepaticae. In: Bryophytes: Their Chemistry and Chemical Taxonomy (H.D. ZINSMEISTER and R. MUES, eds.), p. 109. Oxford: Oxford University Press, 1990.

479. SAKAI, K., T. ICHIKAWA, K. YAMADA, M. YAMASHITA, M. TANIMOTO, A. HIKITA, Y. IJUIN, and K. KONDO: Antitumor Principles in Mosses: The First Isolation and Identification of Maytansinoids, Including a Novel 15-Methoxyansamitocin P-3. J. Nat. Prod. **51**, 845 (1988).

480. SAKAI, T., K. NISHIMURA, and Y. HIROSE: The Constituents of the Volatile Oil from the Wood of *Torreya nucifera*. Tetrahedron Letters **1963**, 1171.

481. SAMEK, Z.: The Determination of the Stereochemistry of Five-Membered α,β-Unsaturated Lactones with an Exomethylene Double Bond Based on the Allylic Long-Range Couplings of Exomethylene Protons. Tetrahedron Letters **1970**, 671.

482. SAMEK, Z.: On Terpenes, CCVIII: The Structure of Ivoxanthin, a Sesquiterpenic Lactone from *Cyclachaena xanthifolia* Fresen. Collect. Czech. Chem. Commun. **35**, 3818 (1970).

483. SAMEK, Z.: On Terpenes, CCXIX: The Structure of Grosheimin. Collect. Czech. Chem. Commun. **37**, 2611 (1972).

484. SAMEK, Z.: On the Validity of the "cis/trans" Lactone Rule for Allylic Coupling Constants of the α-Exomethylene Protons in Natural Sesquiterpenic α-Exomethylene γ-Lactones. Collect. Czech. Chem. Commun. **43**, 3210 (1978).

485. SANCHEZ-VIESCA, F., and J. ROMO: Estafiatin, a New Sesquiterpene Lactone Isolated from *Artemisia mexicana* (Willd). Tetrahedron **19**, 1285 (1963).

486. SANGAIAH, R., and G. S. K. RAO: Studies in Terpenoids, Part LV: Synthesis of 5-Hydroxycalamenene, a Phenolic Sesquiterpene from *Bazzania tricrenata*. Indian J. Chem. **21B**, 13 (1982).

487. SASA, T.: Structure of Cotylenol, the Aglycone of the Cotylenins Leaf Growth Substances. Agr. Biol. Chem. (Tokyo) **36**, 2037 (1972).

488. SASA, T., A. TAKAHAMA, and T. SHINDO: The Stereostructure of Cotylenol, the Aglycone of Cotylenins Leaf Growth Substances. Agr. Biol. Chem. (Tokyo) **39**, 1729 (1975).

489. SASA, T., and M. TOGASHI: Isolation and Structure of Cotylenin E. Agr. Biol. Chem. (Tokyo) **37**, 1505 (1973).

490. SATO, A., M. KURABAYASHI, H. NAGAHORI, A. OGISO, and H. MISHIMA: Chettaphanin-I, A Novel Furanoditerpenoid. Tetrahedron Letters **1970**, 1095.

491. SAUERWEIN, M., and H. BECKER: Growth, Terpenoid Production and Antibacterial Activity of an *in vitro* Culture of the Liverwort *Fossombronia pusilla*. Planta Med. **56**, 364 (1990).

491a. SHOENEBORN, R., and R. MUES: Flavone Di-C-glucoside from *Plagiochila jamesonii* and *Plagiochasma rupestre*. Phytochem. **34**, 1143 (1993).

492. SCHOFIELD, W.B.: In: Introduction to Bryology. New York: Macmillan, 1985.

493. SCHUSTER, R.M.: The Phylogeny of the Hepaticae. In: Bryophyte Systematics. The Systematics Association Special Volume No. 14 (G.C.S. CLARKE and J.G. DUCKETT, eds.), p. 41. London: Academic Press, 1979.

494. SCHUSTER, R.M.: Evolution, Phylogeny and Classification of the Hepaticae. In: New Manual of Bryology, Vol. 2 (R.M. SCHUSTER, ed.), p. 892. Nichinan: J. Hattori Bot. Lab., 1984.

495. SCHUSTER, R.M.: Morphology, Phylogeny and Classification of the Anthocerotae. In: New Manual of Bryology, Vol. 2 (R.M. SCHUSTER, ed.), p. 1071. Nichinan: J. Hattori Bot. Lab., 1984.

496. SCHUSTER, R., and J.J. ENGEL: Austral Hepaticae V(2). Temperate and Subantarctic Schistochilaceae of Australasia. J. Hattori Bot. Lab. **58**, 255 (1985).

497. SCHWABE, W.W.: Lunularic Acid in Growth and Dormancy of Liverworts. In: Bryophyte Development: Physiology and Biochemistry (R.N. CHOPRA and S.C. BHATLA, eds.), p. 245. Boca Raton: CRC Press, 1990.

497a. SEEGER, T., H. GEIGER, H.D. ZINSMEISTER, and W. ROZDZINSKI: Biflavonoids from the Moss *Homalothecium lutescens*. Phytochem. **34**, 295 (1993).

498. SEEGER, T., H.D. ZINSMEISTER, and H. GEIGER: The Biflavonoid Pattern of *Rhytidiadelphus squarrosus* (Hedw.) Warnst. Z. Naturforsch. **45c**, 583 (1990).

499. SEMMELHACK, M.F., and S.J. BRICKNER: Nickel-Promoted Cyclization/Carbonylation in the Preparation of α-Methylene γ-Lactones: Stereospecific Synthesis of (±)-Frullanolide. J. Am. Chem. Soc. **103**, 3945 (1981).

500. SEWÓN, P.: Fatty Acyl Composition of Monogalactosyldiacyl Glycerols in Bryophyta. Phytochem. **31**, 3461 (1992).

501. SHARMA, N.K., and V.N. SHARMA: Chemical Investigation of *Anacardium giganteum.* Indian J. Chem. **4**, 99 (1966).

501a. SHEN, Y.-C., P.I. TSAI, W. FENICAL, M.E. HAY: Secondary Metabolite Chemistry of the Caribbean Marine Alga *Sporochnus bolleanus*: A Basis for Herbivore Chemical Defence. Phytochem. **32**, 71 (1993).

502. SHIMOMURA, H., Y. SASHIDA, and T. ADACHI: Phenolic Glucosides from *Prunus grayana.* Phytochem. **26**, 249 (1987).

503. SHINMEN, Y., K. KATOH, S. SHIMIZU, S. JAREONKITMONGKOL, and H. YAMADA: Production of Arachidonic Acid and Eicosapentaenoic Acids by *Marchantia polymorpha* in Cell Culture. Phytochem. **30**, 3255 (1991).

504. SHIOBARA, Y., H. SUMITOMO, M. TSUKAMOTO, C. HAMADA, and M. KODAMA: Synthesis and Structure Confirmation of Riccardin B, A Macrocyclic Bis(bibenzyl) from the Liverwort *Riccardia multifida.* Chem. Letters **1985**, 1587.

505. SHIRAHAMA, H., K. HAYANO, Y. KANEMOTO, S. MISUMI, T. OHTSUKA, N. HASHIBA, A. FURUSAKI, S. MURATA, R. NOYORI, and T. MATSUMOTO: Conformationally Selective Transannular Cyclizations of Humulene 9,10-Epoxide. Synthesis of the Two Skeletally Different Cyclohumulanoids: DL-Bicyclohumulenone and DL-Africanol. Tetrahedron Letters **21**, 4835 (1980).

505a. SHU, Y.-F., H.-C. WEI, C.-L. WU: Novel Sesquiterpenoids from Taiwanese Liverworts *Lepidozia* Spp. 15th International Botanical Congress. Yokohama, Japan. Abstracts, p. 398 (1993).

506. SIEGEL, U., R. MUES, R. DÖNIG, and T. EICHER: A Cyclohexenone from *Plagiochila longispina.* Phytochem. **30**, 3643 (1991).

507. SIEGEL, U., R. MUES, R. DÖNIG, T. EICHER, M. BLECHSCHMIDT, and H. BECKER: Ten Azulenes from *Plagiochila longispina* and *Calypogeia azurea.* Phytochem. **31**, 1671 (1992).

508. SIEGEL, U., H.D. ZINSMEISTER, and W. STEIN: A Rapid HPLC-Fingerprint System for Flavonoids of Bryophytes. J. Hattori Bot. Lab. **67**, 389 (1989).

509. SIEVERS, H., G. BURKHARDT, H. BECKER, and H.D. ZINSMEISTER: Hypnogenols and Other Dihydroflavonols from the Moss *Hypnum cupressiforme.* Phytochem. **31**, 3233 (1992).

509a. SMITH, D.K., and P.G. DAVISON: Antheridia and Sporophytes in *Takakia ceratophylla* (Mitt.) Grolle: Evidence for Reclassification Among the Mosses. J. Hattori Bot. Lab. **73**, 263 (1993).

510. SOLAJA, B., J. HUGUET, M. KARPF, and A.S. DREIDING: The Synthesis of (±)-Isoptychanolide by Application of the α-Alkynone Cyclisation. Tetrahedron **43**, 4875 (1987).

511. SONO, M.: Odouriferant Substance of the Liverwort *Frullania tamarisci* and the Structures and Total Synthesis of the Related Sesquiterpene Alcohols of *Conocephalum conicum.* Ph.D. Thesis, Tokushima Bunri University, 1991.

512. SPJUT, R.W., M. SUFFNESS, G.M. CRAGG, and D.H. NORRIS: Mosses, Liverworts and Hornworts Screened for Antitumor Agents. Economy Botany **40**, 310 (1986).

513. SPÖRLE, J.: Phytochemische Untersuchungen an ausgewählten Panamaischen Lebermoosen. Ph.D. Thesis, Universität des Saarlandes, 1990.

514. SPÖRLE, J., H. BECKER, N.S. ALLEN, and M.P. GUPTA: Lipophilic Constituents from the Panamanian Liverwort *Monoclea gottschei* subsp. *neotropica*. J. Hattori Bot. Lab. **70**, 151 (1991).

515. SPÖRLE, J., H. BECKER, N.S. ALLEN, and M.P. GUPTA: Spiroterpenoids from *Plagiochila moritziana*. Phytochem. **30**, 3043 (1991).

516. SPÖRLE, J., H. BECKER, M.P. GUPTA, M. VEITH, and V. HUCH: Novel C-35 Terpenoids from the Panamanian Liverwort *Plagiochila moritziana*. Tetrahedron **45**, 5003 (1989).

517. STALLARD, M.O., W. FENICAL, and J.S. KITTREDGE: The Brasilenols, Rearranged Sesquiterpene Alcohols Isolated from the Marine Opisthobranch *Aplysia brasiliana*. Tetrahedron **34**, 2077 (1978).

518. STAMPF, J.-L., C. BENEZRA, and Y. ASAKAWA: Stereospecificity of Allergenic Contact Dermatitis, Part III: Experimentally Induced ACD to a Natural Sesquiterpene Dialdehyde, Polygodial in Guinea Pig. Arch. Dermatol. Res. **274**, 277 (1982).

519. STEIN, W., S. ANHUT, H.D. ZINSMEISTER, R. MUES, W. BARZ, and J. KÖSTER: New Flavone Glucoside Malonylesters from *Bryum capillare*. Z. Naturforsch. **40c**, 469 (1985).

520. STEIN, W., and H.D. ZINSMEISTER: New Flavonoids from the *Bryum pseudotriquetrum*. Z. Naturforsch. **45c**, 25 (1990).

521. STEIN, W., and H.D. ZINSMEISTER: The Occurrence of Flavonoids in the Moss Family Bryaceae. J. Hattori Bot. Lab. **69**, 195 (1991).

522. STILL, W.C., and M.J. SCHNEIDER: A Convergent Route to α-Substituted Acrylic Esters and Application to the Total Synthesis of (±)-Frullanolide. J. Am. Chem. Soc. **99**, 948 (1977).

523. SUGAWA, S.: Nutritive Values of Mosses as a Food for Domestic Animals and Fowls. Hikobia **2**, 119 (1960).

524. SUIRE, C., Y. ASAKAWA, M. TOYOTA, and T. TAKEMOTO: Chirality of Terpenoids Isolated from the Liverwort *Conocephalum conicum*. Phytochem. **21**, 349 (1982).

525. SUMARYONO, W., P. PROKSCH, T. HARTMANN, M. NIMTZ, and V. WRAY: Induction of Rosmarinic Acid Accumulation Cell Suspension Cultures of *Orthosiphon aristatus* After Treatment with Yeast Extract. Phytochem. **30**, 3267 (1991).

526. SUZUKI, H., M. NOMA, and N. KAWASHIMA: On the Structures of Two New Labdanoides Obtained from Leaf Surface of the Wild Species of Tobacco, *Nicotiana setchelli*. 26th Symposium on Chemistry of Terpenes, Essential Oils and Aromatics. Yamagata, Japan. Symposium Papers, p. 209 (1982).

527. SZWEYKOWSKI, J., and I. ODRZYKOSKI: Chemical Differentiation of *Aneura pinguis* (L.) Dum. (Hepaticeae, Aneuraceae) in Poland and Some Comments on Application of Enzymatic Markers in Bryology. In: Bryophytes: Their Chemistry and Chemical Taxonomy (H.D. ZINSMEISTER and R. MUES, eds.), p. 437. Oxford: Oxford University Press, 1990.

527a. TABACCHI, R., D. JOULAIN, S. HUNECK, and V. HEROUT: Unpublished results.

528. TAIRA, Z., T. HASHIMOTO, M. TAKEI, K. ENDO, Y. SAKIYA, and Y. ASAKAWA: Structure-Activity Relationship of Non-Nitrogen Macrocyclic Aromatic Compounds, Marchantin A and Its Analogues from Some Liverworts, Possessing *d*-Tubocurarine-Like Activity. 11th Symposium on Medicinal Chemistry. Tokushima, Japan. Symposium Papers, p. 124 (1990).

528a. TAIRA, Z., M. TAKEI, K. ENDO, T. HASHIMOTO, Y. SAKIYA, and Y. ASAKAWA: Marchantin A Trimethyl Ether: Its Molecular Structure and Tubocurarine-Like Skeletal Muscle Relaxation Activity. Chem. Pharm. Bull. (Tokyo) **42**, 52 (1994).

529. TAKAHASHI, T., Y. YAMASHITA, T. DOI, and J. TSUJI: Stereoselective Cyclopropanation of the 10-Membered Enone. Total Synthesis of Bicyclohumulenone. J. Organ. Chem. (USA) **54**, 4273 (1989).

530. TAKANO, S., M. MORIYA, and K. OGASAWARA: Enantiocontrolled Syntheses of the Cuparene Sesquiterpenes, (−)-Herbertene, (+)-β-Cuparenone, (−)-Debromoaplysin and (−)-Aplysin. Tetrahedron Letters 33, 329 (1992).

531. TAKAOKA, D., N. KOUYAMA, H. TANI, and A. MATSUO: Structures of Three Novel Dimeric Sesquiterpenoids from the Liverwort *Mylia taylorii*. J. Chem. Res. (S) 1990, 180.

532. TAKAOKA, D., A. MATSUO, and S. HAYASHI: Steryl Esters of Liverworts. Phytochem. 26, 429 (1987).

533. TAKAOKA, D., A. MATSUO, J. KURAMOTO, M. NAKAYAMA, and S. HAYASHI: (−)-Myltaylenol, a Tricyclic Sesquiterpene Alcohol with a Novel Carbon Skeleton from the Liverwort *Mylia taylorii*. J. Chem. Soc. Chem. Commun. 1985, 482.

534. TAKAOKA, D., A. MATSUO, M. NAKAYAMA, and S. HAYASHI: Three Verrucosane Diterpenoids, Verrucosane Triol and Related Compounds from the Liverwort *Mylia verrucosa*. Phytochem. 22, 1653 (1983).

535. TAKAOKA, D., H. TANI, and A. MATSUO: Cyclomyltaylenol, a Novel Tetracyclic Sesquiterpenoid from the Liverwort *Mylia taylorii*. J. Chem. Res. (S) 1988, 130.

536. TAKEDA, R., J. HASEGAWA, and M. SHINOZAKI: The First Isolation of Lignans, Megacerotonic Acid and Anthocerotonic Acid, from Non-Vasucular Plants, Anthocerotae (Hornworts). Tetrahedron Letters 31, 4159 (1990).

537. TAKEDA, R., J. HASEGAWA, and M. SHINOZAKI: Phenolic Compounds from Anthocerotae. In: Bryophytes: Their Chemistry and Chemical Taxonomy (H.D. ZINSMEISTER and R. MUES, eds.), p. 201. Oxford: Oxford University Press, 1990.

538. TAKEDA, R., and K. KATOH: 3,10-Dihydro-1,4-dimethylazulene, a Labile Biosynthetic Intermediate Isolated from Cultured Cells of Liverwort *Calypogeia granulata* Inoue. J. Am. Chem. Soc. 105, 4056 (1983).

539. TAKEDA, R., and K. KATOH: Sesquiterpenoids in Cultured Cells of Liverwort, *Calypogeia granulata* Inoue. Bull. Chem. Soc. Jpn. 56, 1265 (1983).

540. TAKEDA, R., H. NAOKI, T. IWASHITA, K. MIZUKAWA, Y. HIROSE, T. ISHIDA, and M. INOUE: Sesquiterpenoid Constituents of the Liverwort, *Ptychanthus striatus* (Lehm. et Lindenb.) Ness. Bull. Chem. Soc. Jpn. 56, 1125 (1983).

541. TAKEDA, R., Y. OHTA, and Y. HIROSE: Sesquiterpenoid Constituents of the Liverwort *Frullania brotheri* Steph. Bull. Chem. Soc. Jpn. 56, 1120 (1983).

542. TAKIKAWA, K., M. TORI, and Y. ASAKAWA: Chemical Constituents and Chemosystematics of *Radula* Species (Liverworts). J. Hattori Bot. Lab. 67, 365 (1989).

543. TAKIKAWA, K., M. TOYOTA, M. TORI, S. NOGAMI, and Y. ASAKAWA: Chemosystematics of Bryophytes (16): Terpenoids and Aromatic Compounds of *Marchantia*, *Aneura* and *Porella*. 28th Symposium on Chemistry of Terpenes, Essential Oils and Aromatics. Kanagawa, Japan. Symposium Papers, p. 102 (1984).

544. TAMAL, Y., Y. MIZUTANI, H. HAGIWARA, H. UDA, and N. HARADA: Synthesis and Absolute Stereochemistry of Optically Active Wieland-Miescher Ketone Analogues Bearing an Angular Protected Hydroxymethyl Group. J. Chem. Res. (S) 1985, 148.

545. TANAKA, N., K. NAKATANI, T. MURAKAMI, Y. SAIKI, and C.-M. CHEN: Chemische Untersuchungen der Inhaltsstoffe von *Pteris plumbaea* Christ. Chem. Pharm. Bull. (Tokyo) 26, 3260 (1978).

546. TANAKA, N., H. WADA, T. MURAKAMI, N. SAHASHI, and T. OHMOTO: Chemische und chemotaxonomische Untersuchungen der Pterophyten, LXIV: Chemische Untersuchungen der Inhaltsstoffe von *Sceptridium ternatum* var. *ternatum*. Chem. Pharm. Bull. (Tokyo) 34, 3727 (1986).

547. TAYLOR, M.D., G. MINASKANIAN, K.N. WINZENBERG, P. SANTONE, and A.B. SMITH III: Preparation, Stereochemistry, and Nuclear Magnetic Resonance Spectroscopy of

4-Hydroxy(acetoxy)biocyclo[5.1.0]octanes. Synthesis of (−)- and (±)-8,8-Dimethylbicyclo[5.1.0]oct-2-en-4-one. J. Organ. Chem. (USA) **47**, 3960 (1982).

548. Taylor, M.D., and A.B. Smith III: Total Synthesis of (+)-Hanegokedial. Tetrahedron Letters **24**, 1867 (1983).

549. Theodor, R., R. Mues, H.D. Zinsmeister, and K.R. Markham: Flavon C-Glykoside aus *Metzgeria furcata* (Hepaticae). Z. Naturforsch. **38c**, 165 (1983).

550. Tori, M.: Application of 2D NMR Techniques to Structure Determination of Natural Products. In: Studies in Natural Products Chemistry, Vol. 2. Structure Elucidation (Part A) (Atta-ur-Rahman, ed.), p. 81. Amsterdam: Elsevier, 1988.

551. Tori, M.: Synthetic Study of Biologically Active Compounds from Liverworts. In: Bryophytes: Their Chemistry and Chemical Taxonomy (H.D. Zinsmeister and R. Mues, eds.), p. 411. Oxford: Oxford University Press, 1990.

552. Tori, M.: Synthetic Studies on Liverwort Sesquiterpenoids. Rev. Latinoamer. Quim. **22**, 73 (1991).

553. Tori, M., M. Aoki, and Y. Asakawa: Chenopodene, Marchantin P and Riccardin G from the Liverwort *Marchantia chenopoda*. Phytochem. **35**, 73 (1994).

554. Tori, M., H. Arbiyanti, and Y. Asakawa: Studies on Chemical Constituents of Bolivian Liverwort *Frullanoides densifolia*. 34th Symposium on Chemistry of Terpenes, Essential Oils and Aromatics. Takamatsu, Japan. Symposium Papers, p. 4 (1990).

555. Tori, M., H. Arbiyanti, Z. Taira, and Y. Asakawa: Spirodensifolins A and B, New Rearranged Pinguisane Sesquiterpenoids, from the Liverwort *Frullanoides densifolia*. Tetrahedron Letters **33**, 4011 (1992).

556. Tori, M., H. Arbiyanti, Z. Taira, and Y. Asakawa: Terpenoids of the Liverwort *Frullanoides densifolia* and *Trocholejeunea sandvicensis*. Phytochem. **32**, 335 (1993).

557. Tori, M., T. Hasebe, and Y. Asakawa: Total Synthesis of Liverwort Sesquiterpene Ketone (±)-Chiloscyphone *via* Intramolecular Aldol Condensation. Chem. Letters **1988**, 2059.

558. Tori, M., T. Hasebe, and Y. Asakawa: Absolute Configurations of the Liverwort Sesquiterpenoids, (−)-Chiloscyphone and (+)-Chiloscypholone: Total Synthesis of Optically Active Compounds. J. Chem. Soc. Perkin Trans. 1 **1989**, 1552.

559. Tori, M., T. Hasebe, and Y. Asakawa: Total Synthesis of (−)-Chiloscyphone, Sesquiterpenoid Isolated from the Liverwort. Absolute Configuration of (−)-Chiloscyphone and (+)-Chiloscypholone. Bull. Chem. Soc. Jpn. **63**, 1706 (1990).

560. Tori, M., T. Hasebe, Y. Asakawa, K. Ogawa, and S. Yoshimura: X-Ray Crystallographic Analysis of Methyl (1S,2R,5R,6S,7R)-5,6-Dimethyl-2-(ω-camphanoyloxy)bicyclo[4.3.0]nonane-7-carboxylate. Revision of the Absolute Configurations of the Liverwort Sesquiterpenoids (−)-Chiloscyphone and (+)-Chiloscypholone. Bull. Chem. Soc. Jpn. **64**, 2303 (1991).

560a. Tori, M., K. Kosaka, and Y. Asakawa: Synthesis of Liverwort Sesquiterpene Tridensone: Correct Structure and Absolute Configuration. 37th Symposium on Chemistry of Terpenes, Essential Oils and Aromatics. Okinawa, Japan. Symposium Papers, p. 334 (1993).

561. Tori, M., T. Masuya, and Y. Asakawa: The Structures of Three New Diterpene Carboxylic Acids from the New Zealand Liverwort *Schistochila nobilis*. 32nd Symposium on Chemistry of Terpenes, Essential Oils, and Aromatics. Miyazaki, Japan. Symposium Papers, p. 197 (1988).

562. Tori, M., T. Masuya, and Y. Asakawa: Neomarchantin A and B, New Macrocyclic Bisbibenzyls from the Liverwort *Schistochila glaucescens*. J. Chem. Res. (S) **1990**, 36.

563. Tori, M., T. Masuya, and Y. Asakawa: Three Clerodane-Diterpenoids from the Liverwort *Schistochila nobilis*. Phytochem. **32**, 1229 (1993).

564. TORI, M., N. MIYAZAKI, K. KONDO, Z. TAIRA, and Y. ASAKAWA: Nepalensolide A, Novel Sesquiterpene Lactone from the Liverwort *Frullania nepalensis*. Compound Breaking the Samek Rule. A Study by NOE and X-Ray. Chem. Letters **1990**, 2115.

565. TORI, M., T. NAGAI, Y. ASAKAWA, and S. HUNECK: Studies on the Chemical Constituents of Some East European Liverworts. 35th Symposium on Chemistry of Terpenes, Essential Oils and Aromatics. Nagoya, Japan. Symposium Papers, p. 39 (1991).

566. TORI, M., T. NAGAI, Y. ASAKAWA, and S. HUNECK: Studies on the Chemical Constituents of East European Liverworts (2). 36th Symposium on Chemistry of Terpenes, Essential Oils and Aromatics. Nishinomiya, Japan. Symposium Papers, p. 255 (1992).

567. TORI, M., T. NAGAI, Y. ASAKAWA, and S. HUNECK: Setiformenol, Isolated from the Liverwort *Tetralophozia setiformis*, the First Example of Cembrane-Type Diterpene from Bryophytes. Tetrahedron Letters **34**, 643 (1993).

567a. TORI, M., T. NAGAI, Y. ASAKAWA, S. HUNECK, and K. OGAWA: Terpenoids from Six Lophoziaceae Liverworts. Phytochem. **34**, 181 (1993).

568. TORI, M., K. NAKASHIMA, and Y. ASAKAWA: Chemical Constituents of South American Liverwort *Porella swartziana*. 112th Annual Meeting of Pharmaceutical Society of Japan, Fukuoka. Abstracts, p. 198 (1992).

569. TORI, M., K. NAKASHIMA, and Y. ASAKAWA: Unpublished results.

570. TORI, M., K. NAKASHIMA, M. SONO, and Y. ASAKAWA: Total Synthesis of a Sesquiterpene Alcohol, Conocephalenol, from the Liverwort. 35th Symposium on Chemistry of Terpenes, Essential Oils and Aromatics. Nagoya, Japan. Symposium Papers, p. 214 (1991).

571. TORI, M., K. NAKASHIMA, T. TAKEDA, and Y. ASAKAWA: Studies on Chemical Constituents of South American *Porella swartziana* and *Omphalanthus filiformis*. 36th Symposium on Chemistry of Terpenes, Essential Oils and Aromatics. Nishinomiya, Japan, Symposium Papers, p. 159 (1992).

571a. TORI, M., K. NAKASHIMA, T. TAKEDA, and Y. ASAKAWA: Structure of Secoswartzianins A and B Isolated from the Liverwort *Porella swartziana*. Tetrahedron Letters **34**, 3753 (1994).

571b. TORI, M., N. NAKASHIMA, M. TOYOTA, and Y. ASAKAWA: Revised Structure of Caespitenone Isolated from the Liverwort *Porella caespitans* var. *setigera* and *Porella swartziana*. Tetrahedron Letters **34**, 3751 (1993).

572. TORI, M., M. SONO, and Y. ASAKAWA: Absolute Configuration and Synthesis of the Liverwort Sesquiterpene Alcohol Tamariscol. J. Chem. Soc. Perkin Trans. 1 **1990**, 2849.

573. TORI, M., M. SONO, K. NAKASHIMA, Y. NAKAKI, and Y. ASAKAWA: Synthesis and Relative Structure of Conocephalenol, a Sesquiterpene Alcohol Isolated from the European Liverwort *Conocephalum conicum*. J. Chem. Soc. Perkin Trans. 1 **1991**, 447.

574. TORI, M., M. SONO, Y. NISHIGAKI, K. NAKASHIMA, and Y. ASAKAWA: Studies on the Liverwort Sesquiterpene Alcohol Tamariscol. Synthesis and Absolute Configuration. J. Chem. Soc. Perkin Trans. 1 **1991**, 435.

574a. TORI, M., T. TAKEDA, K. NAKASHIMA, and Y. ASAKAWA: Chemical Constituents of the Colombian Liverwort *Porella swartziana*. 15th International Botanical Congress, Yokohama, Japan. Abstracts, p. 398 (1993).

574b. TORI, M., T. TAKEDA, K. NAKASHIMA, Y. KAN, S. TAKAOKA, and Y. ASAKAWA: Four New Sesquiterpenoids from the Colombian Liverwort *Porella swartziana*. 37th Symposium on Chemistry of Terpenes, Essential Oils and Aromatics. Okinawa, Japan. Symposium Papers, p. 53 (1993).

575. TORI, M., K. TAKIKAWA, Z. TAIRA, Y. ASAKAWA, and S. YOSHIMURA: New Sesquiterpene Lactones from the Liverwort *Frullania nepalensis*. Inapplicable Example for

Samek Rule-NOE, X-Ray and MM2. 32nd Symposium on Chemistry of Terpenes, Essential Oils, and Aromatics. Miyazaki, Japan. Symposium Papers, p. 195 (1988).

576. TORI, M., M. TOYOTA, L.J. HARRISON, K. TAKIKAWA, and Y. ASAKAWA: Total Assignment of ^{1}H and ^{13}C NMR Spectra of Marchantins Isolated from Liverworts and Its Application to Structure Determination of Two New Macrocyclic Bis(bibenzyls) from *Plagiochasma intermedium* and *Riccardia multifida*. Tetrahedron Letters **26**, 4735 (1985).

577. TORI, M., N. TSUYAMA, K. NAKASHIMA, M. SONO, and Y. ASAKAWA: Synthesis and Absolute Configuration of Valenc-1(10)-en-7α-ol, Isolated from the Liverwort *Bazzania fauriana*. J. Chem. Res. (S) **1990**, 164.

578. TORI, M., S. YOSHIMURA, C. KURODA, and Y. ASAKAWA: Revisit to the Samek Rule. A Comparative Study on Lactone Configurations by Synthesis, NMR, and MM2. Chem. Letters **1990**, 2117.

579. TOYOTA, M.: Chemical Constituents of *Marchantia polymorpha*, *Riccardia multifida* and *Plagiochila* genus (Hepaticae). Ph. D. Thesis, Tokushima Bunri University, 1987.

580. TOYOTA, M.: The Specificity and the Diversity of the Chemical Constituents of Liverworts. 36th Symposium on Chemistry of Terpenes, Essential Oils and Aromatics, Nishinomiya, Japan, Symposium Papers, I-8 (1992).

581. TOYOTA, M., and Y. ASAKAWA: Sesquiterpenoids from the Liverwort *Bazzania fauriana*. Phytochem. **27**, 2155 (1988).

582. TOYOTA, M., and Y. ASAKAWA: An Eudesmane-Type Sesquiterpene Alcohol from the Liverwort *Frullania tamarisci*. Phytochem. **29**, 3664 (1990).

583. TOYOTA, M., and Y. ASAKAWA: Sesquiterpene Derivatives and a Norsesquiterpenoid from the Liverworts, *Riccardia crassa* and *Porella caespitans* var. *setigera*. Phytochem. **32**, 137 (1993).

584. TOYOTA, M., and Y. ASAKAWA: Sesqui- and Triterpenoids of the Liverwort, *Conocephalum japonicum*. Phytochem. **32**, 1235 (1993).

585. TOYOTA, M., and Y. ASAKAWA: Diterpenoids of the Liverwort *Nardia subclavata*. Phytochem. **34**, 751 (1993).

585a. TOYOTA, M., and Y. ASAKAWA: Unpublished results.

585b. TOYOTA, M., and Y. ASAKAWA: Bibenzyl and Sesquiterpenoids from *Jubula japonica*. Phytochem. **34**, 1135 (1993).

586. TOYOTA, M., Y. ASAKAWA, and J.-P. FRAHM: Homomono- and Sesquiterpenoids from the Liverwort *Lophocolea heterophylla*. Phytochem. **29**, 2334 (1990).

587. TOYOTA, M., Y. ASAKAWA, and T. TAKEMOTO: Sesquiterpenes from Japanese Liverworts. Phytochem. **20**, 2359 (1981).

588. TOYOTA, M., F. NAGASHIMA, and Y. ASAKAWA: Labdane-Type Diterpenoids from the Liverwort *Frullania hamachiloba*. Phytochem. **27**, 1789 (1988).

589. TOYOTA, M., F. NAGASHIMA, and Y. ASAKAWA: Gymnomitrane-Type Sesquiterpenoids from the Liverwort *Plagiochila trabeculata*. Phytochem. **27**, 2161 (1988).

590. TOYOTA, M., F. NAGASHIMA, and Y. ASAKAWA: Fatty Acids and Cyclic Bis(bibenzyls) from the New Zealand Liverwort *Monoclea forsteri*. Phytochem. **27**, 2603 (1988).

591. TOYOTA, M., F. NAGASHIMA, and Y. ASAKAWA: Striatane- and Pinguisane-Type Sesquiterpenoids and Phytane-Type Diterpenoid from the Liverwort *Porella navicularis*. Phytochem. **28**, 1661 (1989).

592. TOYOTA, M., F. NAGASHIMA, and Y. ASAKAWA: A Clerodane-Type Diterpenic Acid from the Liverwort *Jungermannia infusca*. Phytochem. **28**, 2507 (1989).

593. TOYOTA, M., F. NAGASHIMA, and Y. ASAKAWA: Volatile Components and Pinguisane-Type Sesquiterpenoids from the Liverwort *Porella cordaeana*. Phytochem. **28**, 3383 (1989).

594. TOYOTA, M., F. NAGASHIMA, and Y. ASAKAWA: Clerodane, Kaurane and Labdane Diterpenoids from the Liverwort *Jungermannia infusca*. Phytochem. **28**, 3415 (1989).
595. TOYOTA, M., F. NAGASHIMA, Y. FUKUYAMA, S. HONDA, and Y. ASAKAWA: Bicyclogermacrene-Type Sesquiterpenoid from the Liverwort *Conocephalum conicum*. Phytochem. **27**, 3317 (1988).
596. TOYOTA, M., F. NAGASHIMA, K. SHIMA, and Y. ASAKAWA: Africane- and Santalane-Type Sesquiterpenoids from the Liverwort *Porella caespitans* var. *setigera*. 35th Symposium on Chemistry of Terpenes, Essential Oils and Aromatics, Nagoya, Japan. Symposium Papers, p. 28 (1991).
597. TOYOTA, M., F. NAGASHIMA, K. SHIMA, and Y. ASAKAWA: Africane- and Santalane-Type Sesquiterpenoids from the Liverwort *Porella caespitans* var. *setigera*. Phytochem. **31**, 183 (1992).
597a. TOYOTA, M., Y. OOISO, T. KUSUYAMA, and Y. ASAKAWA: Drimane-Type Sesquiterpenoids from the Liverwort *Diplophyllum serrulatum*. Phytochem. **35**, 1263 (1994).
598. TOYOTA, M., M. TORI, K. TAKIKAWA, Y. SHIOBARA, M. KODAMA, and Y. ASAKAWA: Perrottetins E, F, and G from *Radula perrottettii* (Liverwort)-Isolation, Structure Determination, and Synthesis of Perrottetin E. Tetrahedron Letters **26**, 6097 (1985).
599. TOYOTA, M., A. UEDA, and Y. ASAKAWA: Sesquiterpenoids from the Liverwort *Bazzania japonica*. 31st Symposium on Chemistry of Terpenes, Essential Oils and Aromatics. Kyoto, Japan. Symposium Papers, p. 228 (1987).
600. TOYOTA, M., A. UEDA, and Y. ASAKAWA: Sesquiterpenoids from the Liverwort *Porella acutifolia* subsp. *tosana*. Phytochem. **30**, 567 (1991).
601. TURSCH, B., J.C. BRAEKMAN, D. DALOZE, P. FRITZ, A. KELECOM, R. KARLSSON, and D. LOSMAN: Chemical Studies of Marine Invertebrates, VIII: Africanol, An Unusual Sesquiterpene from *Lemnalia africana* (Coelenterata, Octocorallia, Alcyonacea). Tetrahedron Letters **1974**, 747.
602. TUTSCHEK, R.: Isolierung und Charakterisierung der *p*-Hydroxy-β-(carboxymethyl)-zimtsäure (Sphagnumsäure) aus der Zellwand von *Sphagnum magellanicum* Brid. Z. Pflanzenphysiol. **76**, 353 (1975).
603. TUTSCHEK, R.: Untersuchungen zum Zimtsäurestoffwechsel in *Sphagnum magellanicum* Brid. In: Congrès International de Bryologie (C. SUIRE, ed.), Bryophytorum Bibliotheca Bd. 13, p. 311. Vaduz: J. Cramer, 1978.
604. TUTSCHEK, R., H. RUDOLPH, P.H. WAGNER, and R. KREHER: Struktur eines kristallinen Phenols aus der Zellwand von *Sphagnum magellanicum*. Biochem. Physiol. Pflanzen **164**, 461 (1973).
605. UYEHARA, T., Y. KABASAWA, and T. KATO: Rearrangement Approaches to Cyclic Skeletons, IV: The Total Synthesis of (±)-Pinguisone and (±)-Deoxopinguisone Based on Photochemical [1, 3] Acyl Migration of a Bicyclo[3.2.2]non-6-en-2-one. Bull. Chem. Soc. Jpn. **59**, 2521 (1986).
606. UYEHARA, T., Y. KABASAWA, T. KATO, and T. FURUTA: Photochemical Rearrangement Approach to the Total Synthesis of (±)-Pinguisone and (±)-Deoxopinguisone. Tetrahedron Letters **26**, 2343 (1985).
607. VALLE, M.G., G. APPENDINO, G.M. NANO, and V. PICCI: Prenylated Coumarins and Sesquiterpenoids from *Ferula communis*. Phytochem. **26**, 253 (1987).
608. VALTEROVA, I., C.R. UNELIUS, J. VRKOC, and T. NORIN: Enantiomeric Composition of Monoterpene Hydrocarbons from the Liverwort *Conocephalum conicum*. Phytochem. **31**, 3121 (1992).
609. VAN LIER, F.P., T.G.M. HESP, L.M. VAN DER LINDE, and A.J.A. VAN DER WEERDT: First Preparation of (+)-Spathulenol. Regio- and Stereoselective Oxidation of (+)-Aromadendrene with Ozone. Tetrahedron Letters **26**, 2109 (1985).

610. Vowinkel, E.: Torfmoosmembranochrome, 2: Die Struktur des Sphagnorubins. Chem. Ber. **108**, 1166 (1975).

611. Wang, T.L., D.J. Cove, P. Beutelmann, and E. Hartmann: Isopentenyladenine from Mutants of the Moss *Physcomitrella patens*. Phytochem. **19**, 1103 (1980).

612. Wei, H.-C., and C.-L. Wu: A New Macrocyclic Bisbibenzyl Diether, Marchantinquinone, from the Liverwort *Mannia subpilosa*. J. Chem. Res. (S) **1991**, 230.

613. Wei, H.-C., and C.-L. Wu: The Chemical Constituents of the Taiwanese Liverwort *Mannia subpilosa*. 18th IUPAC Symposium on the Chemistry of Natural Products. Strasbourg, France, Abstracts, p. 288 (1992).

614. Wei, H.-C., and C.-L. Wu: Unpublished results.

614a. Weitz, S., and R. Ikan: Bracteatin from the Moss *Funaria hygrometrica*. Phytochem. **16**, 1108 (1977).

615. Welch, S.C., and R.L. Walters: Stereoselective Total Syntheses of ($\pm$)-Longicyclene, ($\pm$)-Longicamphor, and ($\pm$)-Longiborneol. J. Organ. Chem. (USA) **39**, 2665 (1974).

616. Whittemore, A.T.: The Secondary Chemistry of the Marchantiales. In: Advances in Bryology, Vol. 4 (N.G. Miller, ed.), p. 75. Vaduz: J. Cramer, 1991.

617. Wilschke, J., E. Hoppe, and H. Rudolph: Biosynthesis of Sphagnum Acid. In: Bryophytes: Their Chemistry and Chemical Taxonomy (H.D. Zinsmeister and R. Mues, eds.), p. 253. Oxford: Oxford University Press, 1990.

618. Wilschke, J., and H. Rudolph: HPLC Analysis of Phenolics in Mosses. In: Methods in Bryology (J.M. Glime, ed.), p. 165. Nichinan: J. Hattori Bot. Lab., 1988.

619. Wilson, M.A., J. Sawyer, P.G. Hatcher, and H.E. Lerch III: 1,3,5-Hydroxybenzene Structures in Mosses. Phytochem. **28**, 1395 (1989).

620. Wright, A.D., G.M. König, and O. Sticher: New Sesquiterpenes and C_{15} Acetogenins from the Marine Red Alga *Laurencia implicata*. J. Nat. Prod. **54**, 1025 (1991).

621. Wu, C.-L.: Sesquiterpenes from Taiwanese Liverworts. J. Hattori Bot. Lab. **56**, 221 (1984).

622. Wu, C.-L.: The Chemical Constituents of Some Taiwanese Liverworts. Kor. J. Pharmacogn. **16**, 243 (1985).

623. Wu, C.-L.: Some New Findings on the Chemistry of Taiwanese Liverworts. In: Bryophytes: Their Chemistry and Chemical Taxonomy (H.D. Zinsmeister and R. Mues, eds.), p. 71. Oxford: Oxford University Press, 1990.

623a. Wu, C.-L.: Chemosystematic Correlations of Taiwanese Hepaticae. J. Chin. Chem. Soc. **39**, 655 (1992).

624. Wu, C.-L., and Y. Asakawa: The Chemical Constituents of the Liverwort *Mylia nuda*. J. Chin. Chem. Soc. **34**, 219 (1987).

625. Wu, C.-L., and Y. Asakawa: Terpenoids of *Pleurozia acinosa*. Phytochem. **27**, 940 (1988).

626. Wu, C.-L., and S.-J. Chang: Chemosystematic Features of the Constituents of Taiwanese Liverworts. J. Hattori Bot. Lab. **64**, 151 (1988).

627. Wu, C.-L., and S.-J. Chang: Cyclomyltaylane, A Tetracyclic Sesquiterpene Hydrocarbon from *Bazzania tridens*. Phytochem. **31**, 2150 (1992).

628. Wu, C.-L., S.-J. Chang, M. Tori, H. Furuta, A. Sumida, and Y. Asakawa: Tridensenal, a Sesquiterpene of Novel Skeleton from the Taiwanese Liverwort *Bazzania tridens*. J. Chin. Chem. Soc. **37**, 387 (1990).

629. Wu, C.-L., and C.-L. Chen: Oxygenated Sesquiterpenes from the Liverwort *Bazzania tridens*. Phytochem. **31**, 4213 (1992).

630. Wu, C.-L., A.-C. Chen, and C.-C. Hsieh: A GC and GC/MS Survey of the Liverwort Oil of *Schistochila rigidula*. Tamkang J. **20**, 459 (1983).

631. Wu, C.-L., and C.-H. HUANG: The Sesquiterpene Constituents of a Leafy Liverwort *Scapania robusta* Horik. Proc. Natl. Sci. Counc. B, ROC **5**, 251 (1981).

631a. Wu, C.-L., C.-H. HUANG, and T.-L. SHIH: A Sesquiterpene Oxide of a Novel Skeleton from the Liverwort *Plagiochila peculiaris.* Tetrahedron Letters **34**, 4855 (1993).

632. Wu, C.-L., and T.-J. LEE: A Spirovetivene from the Liverwort *Scapania robusta.* Proc. Natl. Sci. Counc. B, ROC **7**, 428 (1983).

633. Wu, C.-L., S.-C. LIN, and J.-A. CHEN: (+)-*Trans*-1,4a-dimethyl-1,2,3,4,4a,5,6,7-octahydronaphthalene, a Trinoreudesmene from the Taiwanese Liverwort *Bazzania fauriana.* J. Chem. Res. (S) **1991**, 50.

634. Wu, C.-L., and S. LIU: New Sesquiterpenes from Liverworts and from the Rearrangement of β-Bazzanene. Tetrahedron **39**, 2657 (1983).

635. Wu, C.-L., and W.-P. MO: The Sesquiterpene Constituents of the Liverwort *Mastigophora diclados.* Chemistry (Chinese) **43**, 108 (1985).

636. Wu, C.-L., R.-S. TSAI, and C.-M. Wu: Sesquiterpene Hydrocarbons of the Liverwort *Bazzania fauriana* (Steph) Hatt. Tamkang J. **19**, 487 (1982).

637. Wu, C.-L., F.-F. WEY, and C.-C. HSIEH: The Hydrocarbon Constituents of the Liverwort *Plagiochila kahsiana.* Chemistry (Chinese) **40**, 121 (1982).

638. Wu, C.-L., F.-F. WEY, and S.-J. HSU: Sesquiterpene Hydrocarbons of the Liverwort *Scapania ornithopodioides.* Phytochem. **21**, 2659 (1982).

639. WURZEL, G., and H. BECKER: Sesquiterpene Lactones from *in vitro* Cultures of the Liverwort *Ricciocarpos natans.* J. Hattori Bot. Lab. **67**, 373 (1989).

640. WURZEL, G., and H. BECKER: Sesquiterpenoids from the Liverwort *Ricciocarpos natans.* Phytochem. **29**, 2565 (1990).

641. WURZEL, G., H. BECKER, T. EICHER, and K. TIEFENSEE: Molluscicidal Properties of Constituents from the Liverwort *Ricciocarpos natans* and of Synthetic Lunularic Acid Derivatives. Planta Med. **56**, 444 (1990).

642. YAMADA, K., T. ICHIKAWA, M. YAMASHITA, M. TANIMOTO, A. HIKITA, K. KONDO, and K. SAKAI: Antitumor Active Substances: Isolation and Structures of a New Maytansinoid, 15-Methoxyansamitocin P-3. 31st Symposium on Chemistry of Terpenes, Essential Oils and Aromatics. Kyoto, Japan. Symposium Papers, p. 242 (1987).

644. YAMADA, K., T. ICHIKAWA, K. YAZAWA, A. MAKINO, Y. IJUIN, K. KONDO, and K. SAKAI: Antitumor Substances Containing Leaf Mold Around Mosses. 57th Annual Meeting of Chemical Society of Japan. Kyoto, Japan, Symposium Papers, p. 1240 (1989).

645. YUZAWA, Y., R. MUES, and S. HATTORI: Morphological and Chemical Studies on the Taxonomy of 14 *Frullania* Species, Subgenus *Chonanthelia.* J. Hattori Bot. Lab. **63**, 425 (1987).

645a. ZHENG, G.-Q., C.J. CHANG, T.J. STOUT, J. CLARDY, and J.M. CASSADY: Ohioensin A: A Novel Benzonaphthoxanthenone from *Polytrichum ohioense.* J. Am. Chem. Soc. **111**, 5500 (1989).

645b. ZHENG, G.-Q., C.J. CHANG, T.J. STOUT, J. CLARDY, D.K. HO, and J.M. CASSADY: Ohioensins: Novel Benzonaphthoxanthenones from *Polytrichum ohioense.* J. Organ. Chem. (USA) **58**, 366 (1993).

646. ZHENG, G.-C., A. ICHIKAWA, M.O. ISHITSUKA, T. KUSUMI, H. YAMAMOTO, and H. KAKISAWA: Cytotoxic Hydroperoxylepidozenes from the Actinia *Anthopleura pacifica* Uchida. J. Organ. Chem. (USA) **55**, 3677 (1990).

647. ZINSMEISTER, H.D., H. BECKER, and T. EICHER: Bryophytes, a Source of Biologically Active, Naturally Occurring Material? Angew. Chem. Internat. Edit. Engl. **30**, 130 (1991).

648. ZINSMEISTER, H.D., and R. MUES: The Flavonoid Chemistry of Bryophytes. Rev. Latinoamer. Quim. **11**, 23 (1980).

649. ZINSMEISTER, H.D., and R. MUES: Moose als Reservoir bemerkenswerter sekundärer Inhaltsstoffe. GIT Fachz. Lab. **31**, 499 (1987).
650. ZINSMEISTER, H.D., and R. MUES: Bryophytes as a Reservoir of Remarkable Secondary Components—A Survey. In: Plant Research and Development (Institute for Scientific Co-operation in Conjunction with the Federal Research Centre for Forestry and Forest Products and Numerous Members of German Universities, eds.), p. 12 (1988).
651. ZINSMEISTER, H.D., and R. MUES: Bryophytes: Their Chemistry and Chemical Taxonomy, p. 1. Oxford: Oxford University Press, 1990.

(Received April 21, 1994)

Author Index

Page numbers printed in *italics* refer to References

Abe, S. *525, 550*
Abraham, W.-R. *525*
Adachi, T. *553*
Adam, K.-P. *533*
Adesomoju, A.A. *525*
Adler, A. *531*
Adler, J.H. *525*
Agata, I. *525*
Ahmad, A. *537*
Ahond, A. *531*
Aiba, C.J. *525*
Aizawa, A. *542*
Akazawa, K. *537*
Aknin, M. *525*
Allen, N.S. *554*
Amico, V. *525*
Andersen, M. *545*
Andersen, N.H. *550*
Andersen, R.J. *539*
Anderson, D.G. *550, 551*
Anderson, N.H. *550*
Anderson, W.H. 452, 507, *525, 535*
Ando, H. *525*
Anhut, S. 422, *525, 526, 535, 554*
Anthonsen, T. *526*
Anton, R. *541*
Aoki, M. *556*
Appendino, G. *559*
Aratani, T. *527*
Arbiyanti, H. *526, 556*
Aro, E.-M. *542*
Asai, M. *526*
Asakawa, Y. *526–530, 534, 536–541,*
    543, 548–550, 554–560
Atsumi, K. *545, 546*

Baddeley, G.V. *530*
Bajwa, B.S. *530*
Baker, R. 201, *529*
Balanson, R.D. *533*
Ballio, A. *529*
Bandurraga, M.M. *541*
Banerjee, A.K. 147, *530*
Barbier, P. *530*
Barneis, Z.J. *544*
Barrow, K.D. *530*
Barrueco, O.F. *545*
Barton, D.H.R. *530*
Barz, W. *526, 554*
Bastard, J. *530*
Baxter, G.A. *539, 540*
Beck, J.-P. *549*
Becker, H. *530, 531, 533, 536, 537, 544, 547,*
    552–554, 561
Belkin, M. 472, *530*
Bendz, G. *530*
Benes, I. 294, *530*
Benesova, V. 277, *530*
Benezra, C. *530, 554*
Bernardini, C. *545*
Bernasconi, S. 200, *530*
Bernassau, J.-M. *530*
Beutelmann, P. 433, *530, 543, 560*
Beyer, J. *530*
Bezanger-Beauquensne, L. *535*
Bhat, S.V. *530*
Bhatla, S.C. *530, 532*
Bhattacharyya, S.C. *536*
Biehl, J. *525*
Bihátsi-Karsai, É. *536*
Billet, D. *531*

Bischler, H. *529*
Blechschmidt, M. *531, 553*
Bloor, S.J. *531*
Bohlmann, F. *531, 541, 544*
Bohm, B.A. *531*
Bokel, M. *535*
Boll, P.M. *551*
Bopp, M. *531*
Bosch, G.K. *547*
Botta, M. *535*
Braekman, J.C. *559*
Brickner, S.J. *553*
Briner, P.H. *529*
Brlik, J. *536*
Brouard, J.P. *531*
Brufani, M. *529*
Buckwalter, B.L. *531*
Budesinsky, M. *530, 543*
Burfitt, I.R. *530, 531*
Burkhardt, G. *544, 553*

Cabrera, E. *531*
Caine, D. 168, *531*
Calderon, J.S. *531*
Caldicott, A.B. 433, *531*
Cameron, A.F. *539, 540*
Campbell, E.O. *527–529, 531*
Campos Correa, R.G. *525*
Canonica, L. *531*
Carroll, P.J. *525*
Casagrande, C. *531*
Casinovi, C.G. *529*
Cassady, J.M. *561*
Catalan, C.A.N. *532*
Catalano, S. *532*
Cava, M.P. *525*
Cerrini, S. *529*
Chain, Sir E. *530*
Chan, Y.-M. *533*
Chang, C.J. *561*
Chang, S.-J. *532, 560*
Chau, H.M. *530*
Chau, P. 496, *532*
Cheatham, S.F. *534*
Cheminat, A. *528, 529*
Chen, A.-C. *560*
Chen, C.-L. *560*
Chen, C.-M. *555*
Chen, I.-S. *541*
Chen, J.-A. *532, 561*

Chen, S.-M. *532, 537*
Cheng, Y.-S. *534*
Chiu, P.-L. 280, 294, *532, 550*
Choi, D.J. *541*
Chopra, R.N. *532*
Clardy, J. *541, 561*
Clive, D.L.J. *532*
Coates, R.M. *532*
Colledge, A. *532*
Connolly, J.D. 189, 200, 357, *527, 528, 532,*
    533, 537, 539, 540, 546, 548
Corbella, A. *531*
Corey, E.J. *533*
Costantino, V. *525*
Cottrell, I.F. *529*
Cove, D.J. *560*
Cragg, G.M. *553*
Crandal-Stotler, B. 524, *533*
Croft, K.D. *533*
Crombie, L.W. 328, *533*
Crombie, W.M.L. *533*
Cullmann, F. *533*
Currenti, R. *525*

D'Alessio, V. *529*
Daloze, D. *559*
Darias, J. *536*
Davidson, A.J. 403, 476, *533*
Davis, B.M. 523, *533*
Davison, P.G. 525, *553*
Dawson, G.W. *527*
De Arriazu, P.J. *532*
Debraux, G. *535*
De Buruaga, A.S. *531*
De Buruaga, J.M.S. *531*
De Carrasco, M.C. *530*
Decicco, C.P. *551*
De Fenik, I.J.S. *532*
De Groot, A. *542*
De Koster, C.G. *544*
Demmerle, S. *543*
De Souza, N.J. *530*
Destro, R. *530*
Dev, S. *547*
Devon, T.K. *533*
De Vries, N.K. *542*
Dhingra-Babbar, S. *530*
Diaz, A. *536*
Dienes, Z. *533*
Dillon, J. *533, 549*

Ding, H. *533*
D'Ischia, M. *533*
Doi, T. *554*
Dönig, R. *553*
Dornauer, H. *530*
Doskotch, R.W. *533*
Dounin, R. *533*
Dreiding, A.S. *553*
Duc, D.K. *530*
Dunn, F.W. *541*
Durgeat, M. *531*

Eglinton, G. 433, *531*
Eicher, T. 33, 185, 322, 333, 380, *533, 544, 551, 553, 561*
Ekman, R. *533, 542*
El-Feraly, F.S. *533, 534*
Endo, J. *534*
Endo, K. *534, 554*
Engel, J.J. *552*
Engen, D.V. *541*
Enoki, N. *534*
Erickson, M. *534*
Erman, W.F. *544*
Ernst, L. *525*
Euler, R. *530*
Evans, D.A. *529*
Evans, F.J. *534*

Fairchild, E.H. *533*
Fakunle, C.O. *527*
Fang, J.-M. *534*
Fattorusso, E. *525*
Faulkner, D.J. *534, 541*
Fedeli, W. *529*
Fehlhaber, H.-W. *530*
Felikin, H. *531*
Felix, M.D. *530*
Fenical, W. *541, 553, 554*
Fenner, G.P. *532*
Ferguson, G. *540*
Ferrari, G. *531*
Ferrari, M. *530*
Ferraz, H.M.C. *551*
Fétizon, M. *530*
Finer, J. *541*
Firth, D.F. *533*
Fischer, H.D. *534*
Fischer, N.H. *534*
Fitzgerald, D. *530*

Fludzinski, P. *535*
Founeron, J.D. *536*
Fráter, G. *534*
Fraga, B.M. *535*
Frahm, J.-P. *528, 558*
Francis, M.J. *530*
Franek, R.W. *534*
Freer, A.A. *540*
Friedel, H.D. *534*
Fritz, P. *559*
Fronczek, F.R. *531*
Fujimura, T. *542*
Fujita, K. *542*
Fujita, T. *534*
Fukazawa, Y. *529, 538, 546*
Fukuyama, Y. 179, *527, 528, 534, 538, 539, 559*
Furusaki, A. *534, 553*
Furuta, H. *560*
Furuta, T. *559*

Gambacorta, A. 201, *535*
Gamou, K. *550*
Ganjian, I. *535*
Ganzer, U. *541*
Garcia-Alvarez, M.C. *535*
Garcia-Granados, A. *531*
Gariboldi, P. *530, 531*
Garnier, G. *535*
Geiger, H. *525, 526, 530, 535, 545, 551, 552*
Geissman, T.A. *541*
Gekeler, W. *535*
Gellerman, J.L. *525, 535*
Gerber, N.N. 15, *535*
Gerling, K.-G. *535*
Ghisalberti, E.L. *533*
Ginda, H. *544*
Given, D.R. *545*
Glasby, J.S. *535*
Glass, N.B.T.E. *537*
Glombitza, K.-W. *535*
Goda, Y. *536*
Gomez-Garibay, F. *531*
González, A.G. *535, 536*
González, R.R. *545*
Gorham, J. *536*
Gottlieb, O.R. *525*
Gottsegen, Á. 335, *536*
Gradstein, S.R. *528, 529, 536, 548*
Grandolini, G. *529*

Grant, P.K. *530*
Gras, J.-L. *536*
Grenz, M. *531*
Griffiths, D.C. *527*
Grill, E. *535*
Grolle, R. 488, *536, 540, 548*
Guha, P.C. *536*
Gunatilaka, A.A.L. *537*
Gupta, M.P. *554*

Ha, N.T.T. 340, *536*
Haag-Berrurier, M. *541*
Haegi, L. *544*
Hagiwara, H. 272, *536, 555*
Haibara, K. *526*
Hajos, Z.G. *536*
Hamada, C. *553*
Hamada, T. *544*
Hamazaki, K. *550*
Hamlin, B.G. *537*
Hammershoy, O. *551*
Hara, M. *544*
Harada, C. *543*
Harada, N. *532, 537, 555*
Harborne, J.B. *533*
Harrigan, G.G. *537*
Harrison, L.J. *527, 529, 532, 537, 539,*
    540, 547, 558
Hartmann, E. 433, *530, 543, 551, 560*
Hartmann, T. *554*
Hasebe, T. *556*
Hasegawa, J. *555*
Hasegawa, S. *539*
Hasenhuettl, G. *531*
Hashiba, N. *553*
Hashimoto, T. *527, 537–539, 549, 554*
Hassen, H.-P. *525*
Hatanaka, A. *542*
Hatano, T. *525*
Hatcher, P.G. *560*
Hattori, S. *528, 529, 548, 561*
Hauser, F.M. *539*
Hawkins, J.M. *525*
Hay, M.E. *553*
Hayano, K. *553*
Hayashi, S. 494, *539, 545–547, 549, 555*
Heerma, W. *544*
Hegnauer, R. *539*
Heitz, S. *531*

Hellou, J. *539*
Henderson, M.S. *539*
Herout, V. 466, *530, 539, 543, 554*
Herzog, F. *530*
Hesp, T.G.M. *559*
Hikino, H. *534*
Hikita, A. *551, 561*
Hill, R.A. *532, 533*
Himberg, K. *543*
Hintz, H.P.J. *544*
Hirata, Y. *549*
Hirono, H. *534*
Hirose, Y. *539, 550, 551, 555*
Hirota, H. *545*
Hirotsu, K. *544*
Hirshfield, J. *529*
Ho, D.K. *561*
Hocart, C.H. *533*
Hochstetler, A.R. *545*
Hoffmann, E. *531*
Honda, S. *559*
Hoppe, E. *560*
Hori, H. *542*
Horie, M. *538*
Hossain, M.B. *531*
Hsieh, C.-C. *560, 561*
Hsu, K.-C. *534*
Hsu, S.-J. *561*
Huang, C.-D. *532, 539*
Huang, C.-H. *561*
Huch, V. *554*
Hufford, C.D. *534*
Huguet, J. *553*
Huneck, E. *540*
Huneck, S. 477, 492, *527, 532, 533, 539, 540,*
    546, 548, 549, 554, 557

Ichikawa, T. 482, *540, 541, 551, 561*
Iguchi, M. *550*
Iitaka, Y. *539*
Ijuin, Y. *551, 561*
Ikan, R. *560*
Ikeda, M. *541*
Ikeda, T. *542*
Imamura, Y. *541*
Imoto, S. *547*
Inoshiri, S. *541*
Inoue, H. *527, 541, 546*
Inoue, M. *555*
Iori, A.M. *545*

Ireland, C. *541*
Irwin, M.A. *541*
Ishi, O. *546*
Ishibashi, H. 143, *541*
Ishida, R. *534*
Ishida, T. *555*
Ishii, H. *541*
Ishikawa, T. *541*
Ishitsuka, M.O. *544, 561*
Itô, S. *543*
Ives, D.A.J. *541*
Iwabuchi, J. *537*
Iwashita, T. *555*
Iwatsuki, Z. *541*
Iyoda, M. 335, *541*
Izac, R.R. *541*
Izawa, M. *526*
Izumo, H. *549*

Jakupovic, J. *531, 541, 544*
Jamieson, G.R. *541*
Jänicke, S. *540*
Jareonkitmongkol, S. *553*
Jeffries, P.R. *533*
Joly, M. *541*
Joly-Goudket, M. *531*
Jommi, G. *530, 531*
Joseph, R. *540*
Joulain, D. *554*
Joussef, A.C. *532*

Kabasawa, Y. *559*
Kabuto, C. *542*
Kajimoto, T. *541*
Kajiwara, T. *542*
Kajtár-Peredy, M. *533, 536, 547, 550*
Kakisawa, H. *544, 550, 561*
Kameo, K. *549*
Kami, T. *549*
Kamio, K. *546*
Kan, Y. *538, 539, 549, 557*
Kanayama, S. *538, 539*
Kaneko, C. *530*
Kanemoto, Y. *553*
Kar, S.K. *529*
Karasawa, H. *547*
Karlsson, B. *542*
Karlsson, R. *559*
Karpf, M. *553*
Karunen, P. 454, *533, 542*

Kashman, Y. *550*
Kasiwong, S. *551*
Katayama, K. *548*
Kato, T. *542, 559*
Kato, Y. *542*
Katoh, K. *542, 548–550, 553, 555*
Kawahara, H. *546*
Kawasaki, T. *543*
Kawashima, N. *550, 554*
Kazlauskas, R. *542*
Kelecom, A. *559*
Keserü, G.M. 345, *542*
Kesselmans, R.P.W. 159, *542*
Khanna, R.N. *550*
Kido, M. *529, 534*
King, R.M. *531, 541, 544*
Kingston, D.G.I. *537*
Kishi, T. *526*
Kitagawa, I. *543*
Kitahara, Y. *542, 543*
Kitajima, J. *543*
Kittredge, J.S. *554*
Klein, R. *536, 544, 548*
Kneifel, H. *543*
Knoche, H. *543*
Kobayashi, H. *547*
Kobayashi, J. *549*
Kobayashi, M. 24, *543, 550*
Kodama, K. *542*
Kodama, M. 207, 340, *534, 543, 553, 559*
Kodama, Y. *544*
Kodpinid, M. *543*
Kohda, H. *541*
Kohn, G. 452, *530, 543*
Kohori, J. *537*
Kokke, W.C.M.C. *532*
Komori, T. *543*
Komura, H. *549, 550*
Kondo, K. *527, 528, 540, 541, 543, 549, 551, 557, 561*
Konecny, K. *543*
König, G.M. *560*
Konno, H. *549*
Koponen, A. 468, *543, 551*
Koponen, T. *543, 548, 551*
Kosaka, K. *556*
Koskimies-Soininen, K. *543*
Köster, J. *526, 554*
Kotake, M. *549*
Kouyama, N. *555*

Koyama, H. *538*
Kraut, L. *536, 544*
Kreher, R. *559*
Kreiser, W. *544*
Krepinsky, J. *531*
Kretschmar, H.C. *544*
Krishnamurty, H.G. *529*
Kruijt, R.Ch. *544*
Krzakowa, M. 486, *544*
Kubo, I. *535, 544*
Kubota, N. *546*
Kumar, S.S. *528*
Kunz, S. *544*
Kuo, Y.-H. *544*
Kupchan, S.M. *544*
Kurabayashi, M. *552*
Kuramoto, J. *555*
Kuriyama, K. *546*
Kuroda, C. *558*
Kusumi, T. *544, 549, 550, 561*
Kusuyama, T. *559*
Kyogoku, Y. *543*

Laas, H.J. *544, 551*
Lallemand, J.-Y. *527*
Lange, G.L. *551*
Langenbahn, U. *544*
Lauck, V. *548*
Lee, T.-J. *561*
Lehmann, L. *544*
Leidinger, G. *548*
Lerch III, H.E. *560*
Ley, S.V. *527*
Li, W.-S. *534*
Likhitwitayawuid, K. *551*
Liljenberg, C. *542*
Lin, S.-C. *561*
Lin, X. *527*
Lin, Y.-T. *544*
Lindberg, G. *544*
Liu, C.-B. *550*
Liu, S. 49, *561*
Longton, R.E. *533*
Lorimer, S.D. *544*
Losman, D. *559*
Lu, S.-T. *541*

Mabry, T.M. *531*
Mackenbrock, K. *526*

Magari, H. 205, *545*
Makino, A. *561*
Mangoni, A. *525*
Mansikkamäki, P. *543*
Markham, K.R. 11, 361, 362, 408, 485, 501,
 526, 531, 535, 545, 548, 556
Marshall, J.A. *545*
Marsili, A. *532, 545*
Mårtensson, O. 403, *530, 545, 549*
Martin, J.D. *536*
Martin, R. *530*
Mason, R.W. *532*
Masuya, T. *528, 529, 534, 556*
Mateos, A.F. 201, *545*
Matsubara, H. *547*
Matsuda, R. *528, 529, 536*
Matsui, K. *542*
Matsumoto, A. *544*
Matsumoto, T. *534, 553*
Matsumura, K. *543*
Matsuo, A. 494, 512, 524, *525, 539, 540,*
 545–547, 549, 555
Matusch, R. *534*
McCleary, J.A. *547*
McCrindle, R. *526, 539, 547*
McLaren, M. *540*
McMaster, D. *539*
McMurry, J.E. *547*
Meier, B. *547*
Meinunger, L. 477, *540*
Mendez, J. *547*
Mentlein, R. 429, *547*
Mezey-Vándor, G. *542, 547*
Mikola, H. *542*
Miksche, G.E. *534, 547*
Minami, Y. *547*
Minaskanian, G. *555*
Minnard, A.J. *542*
Mirailles, J. *525*
Mishima, H. *552*
Misra, R. *547*
Misumi, S. *553*
Miyase, T. *547*
Miyazaki, N. *557*
Mizukawa, K. *555*
Mizuta, E. *526*
Mizutani, M. 468, *528, 547*
Mizutani, Y. *555*
Mlotkiewicz, J.A. *547*
Mo, W.-P. *561*

Montanari, S. *530*
Morais, R.M.S.C. *547*
Morelli, I. *532, 545*
Mori, K. *527*
Moriya, M. *555*
Mudd, A. *527*
Mues, R. 11, 484, *526, 530, 535, 536, 544, 545, 547, 548, 551–554, 556, 561, 562*
Mulder, J. *542*
Müller, C. *548*
Müller, U. *534*
Murakami, T. *548, 555*
Murata, S. *553*
Murphy, P.T. *542*
Murray, R.D.H. *539*
Murray-Rust, J. *547*
Murray-Rust, P. *547*

Nabeta, K. *548*
Naegeli, P. *548*
Naemura, K. *531*
Nagahama, Y. *549*
Nagahori, H. *552*
Nagai, T. *548, 549, 557*
Nagasawa, M. *534*
Nagashima, F. *548, 549, 558, 559*
Naik, C.G. *550*
Nakagawa, C. *549*
Nakagawara, S. 46, *548, 549*
Nakaki, Y. *557*
Nakamura, E. *547*
Nakamura, H. *549*
Nakamura, I. *538*
Nakanishi, K. *532, 533, 537, 549*
Nakashima, K. *537, 549, 557, 558*
Nakatani, H. *541*
Nakatani, K. *555*
Nakatani, Y. *549*
Nakayama, M. *545–547, 549, 555*
Nakayama, N. *546*
Namikawa, M. *540, 541*
Nano, G.M. *559*
Naoki, H. *544, 555*
Narayanan, C.R. 175, *549*
Naya, Y. *549*
Niemann, G.J. *544*
Nilsson, E. 403, *544, 545, 549*
Nimtz, M. *554*
Nishibe, S. *525*

Nishigaki, Y. *557*
Nishimura, K. *551*
Nishimura, N. *550*
Nishioka, E. *549*
Nishiyama, A. *550*
Niwa, M. *550*
Nogami, S. *555*
Nógrádi, M. *335, 533, 536, 542, 547, 550*
Nohara, T. *541*
Noma, M. *550, 554*
Nomoto, K. *549*
Norin, T. *542, 559*
Norris, D.H. *553*
Novikov, V.P. *547*
Noyori, R. *553*
Nozaki, H. *545, 546, 550*
Nyberg, H. *543*

Ochi, M. *537*
Ochi, S. *542*
Oda, M. *541*
Odrzykoski, I. *554*
Ogasawara, K. *555*
Ogawa, K. *556, 557*
Ogawa, S. *527*
Ogiso, A. *552*
Ohi, Y. *549*
Ohizumi, Y. *549*
Ohmoto, T. *555*
Ohnishi, A. *538*
Ohnsorge, U.F.W. *530*
Ohta, Y. *525, 547, 550, 555*
Ohtani, I. *550*
Ohtsuka, T. *553*
Okada, K. *528*
Okamoto, H. *550*
Okogum, J.I. *525*
Okuda, P.K. *550*
Okuda, T. *525*
Olivier, E.J. *534*
O'Neill, A.N. *541*
Ono, K. *550*
Ooiso, Y. *559*
Oriente, G. *525*
Osawa, S. *542*
Österdahl, B.-G. *544, 550*
Otsuka, H. *541*
Ourisson, G. *527, 543, 549*
Ozaki, N. *539*

Pacchiani, M. *532, 545*
Paknikar, S.K. *550*
Pandey, R.C. *547*
Papastephanou, C. *550, 551*
Parker, W. *547*
Parrish, D.R. *536*
Parvez, M. *540*
Pathak, V.P. *550*
Patterson, G.W. *532, 550*
Pellicciari, R. *529*
Peña-Matheud, C.A. *530*
Peredy, M.K. *542*
Perez, C. *536*
Perold, G.W. *528*
Perry, N.B. *544*
Petragnani, N. *551*
Pezechk-Leclaire, M. *527*
Pfaffmann, H. *433, 551*
Phillips, W.R. *528, 532, 533, 539, 540*
Piattelli, M. *525*
Picci, V. *559*
Pick, R. *533*
Pickett, J.A. *527*
Pilotti, A.-M. *542*
Pócs, T. *540*
Poet, S.E. *541*
Pogany, S.A. *539*
Porter, L. *504, 531, 551*
Powell, R.G. *551*
Prins, H.H.Th. *551*
Pröbsting, U. *548*
Proksch, P. *554*
Prota, G. *533*
Pyysalo, H. *543, 551*

Quijano, L. *531*

Randazzo, G. *529*
Rao, G.S.K. *552*
Raston, C.L. *533*
Ravenscroft, P.D. *529*
Reffstrup, T. *551*
Reid, E.H. *541*
Reid, W.W. *532*
Revial, G. *551*
Ri, Z. *464, 551*
Rice, J.R. *551*
Riddell, F.G. *547*
Rios, T. *531*
Rischmann, M. *551*

Roberts, J.S. *547*
Robinson, H. *531, 541*
Rodriguez, B. *535*
Romo, J. *552*
Rossi, C. *529*
Rozdzinski, W. *552*
Ruangrungsi, N. *551*
Rudolph, H. *403, 408, 551, 559, 560*
Russell, R. *532*
Rycroft, D.S. *527, 532, 537, 539, 540, 548, 551*

Sadavongvivad, C. *543*
Sahashi, N. *555*
Saiki, Y. *555*
Saito, Y. *550*
Sakai, K. *472, 540, 541, 551, 561*
Sakaitani, M. *541*
Sakamoto, T. *550*
Sakiya, Y. *554*
Sakuda, K. *546*
Salomon, M. *531*
Salt, T.A. *550*
Saman, D. *543*
Samb, A. *525*
Samek, Z. *552*
Samland, J. *408, 551*
Sanchez, D.M. *531*
Sanchez-Viesca, F. *552*
Sangaiah, R. *552*
Sankawa, U. *536*
Santone, P. *555*
Santurbano, B. *529*
Sanz-Cabanilles, F. *547*
Sari, Y. *549*
Sasa, T. *552*
Sasaki, M. *541*
Sasaki, N. *542*
Sashida, Y. *553*
Sato, A. *552*
Sato, S. *546*
Sattar, A. *547*
Sauerwein, M. *552*
Sawyer, J. *560*
Scheppard, C. *545*
Scheuer, P.J. *550*
Schlenk, H. *525, 535*
Schmidt, H. *551*
Schmidt, J. *540*
Schmidt, R.J. *534*

Schmitz, F.J. *531*
Schneider, M.J. *554*
Schofield, W.B. *528, 552*
Schreiber, K. *540*
Schuster, A. *541*
Schuster, R.M. *552*
Schwabe, W.W. *552*
Scott, A.I. *533*
Seeger, T. *525, 526, 552*
Sekigushi, F. *541*
Selwood, D.L. *529*
Semmelhack, M.F. *553*
Sewón, P. *455, 523, 553*
Shah, S.K. *532*
Sharma, N.K. *553*
Sharma, R.P. *530*
Sharma, V.N. *553*
Shen, Y.-C. *553*
Shibata, M. *541*
Shih, T.-L. *561*
Shima, K. *559*
Shimizu, S. *553*
Shimomura, H. *553*
Shindo, T. *552*
Shingu, T. *534*
Shinmen, Y. 382, *553*
Shinozaki, M. *555*
Shiobara, Y. *543, 553, 559*
Shiota, H. *528*
Shirahama, H. *553*
Shoeneborn, R. *552*
Shu, Y.-F. *553*
Siegel, U. 45, 322, *553*
Sievers, H. *553*
Sim, G.A. *539, 546*
Sim-Sim, M. *544*
Sisti, M. *530*
Smith, D.K. *525, 553*
Smith III, A.B. *555, 556*
Smith Jr., C.R. *551*
Snatzke, G. *540*
Sodano, G. *533*
Soderholm, A.-C. *542*
Solaja, B. 202, *553*
Son, B.W. *543*
Sono, M. *528, 553, 557, 558*
Spjut, R.W. *553*
Spörle, J. *536, 553, 554*
Sprecher, E. *525*
Stallard, M.O. *554*

Stampf, J.-L. *554*
Stein, W. *535, 553, 554*
Sticher, O. *547, 560*
Still, W.C. *554*
Stoll, M. *545*
Stout, T.J. *561*
Strassner, A. *548*
Streibl, M. *543*
Suehiro, K. *534*
Suffness, M. *553*
Sugawa, S. *554*
Suire, C. *528, 529, 554*
Sumaryono, W. *554*
Sumida, A. *560*
Sumimoto, M. *542*
Sumitomo, H. *543, 553*
Suwita, A. *531*
Suzuki, F. *550*
Suzuki, H. *538, 554*
Suzuki, K. *539*
Suzuki, M. *546*
Svensson, L. *530*
Swain, C.J. *529*
Szweykowski, J. *554*

Tabacchi, R. *554*
Taguchi, M. *541*
Taira, Z. *529, 537–539, 554, 556, 557*
Takahama, A. *552*
Takahashi, K. *541*
Takahashi, T. 178, *545, 554*
Takamatsu, M. *549*
Takano, S. 143, 147, *555*
Takaoka, D. 187, *546, 550, 555*
Takaoka, S. *538, 539, 549, 557*
Takase, T. *549*
Takashima, N. *550*
Takeda, R. 200, *529, 537, 538, 542, 549, 550,*
   555
Takeda, T. *557*
Takeda, Y. *534*
Takei, M. *538, 554*
Takemoto, T. *528, 529, 554, 558*
Takikawa, K. *527–529, 555, 557–559*
Takizawa, N. *547*
Tamal, Y. *555*
Tambunan, U.S.F. *543*
Tanaka, H. *538, 539, 549*
Tanaka, N. *548, 555*
Tangney, R.S. *544*

Tani, H. 555
Tanimoto, M. 551, 561
Taylor, M.D. 40, 555, 556
Terenius, L. 530
Theander, O. 544
Thebtaranonth, C. 543
Thebtaranonth, Y. 543
Theodor, R. 545, 556
Thomas, G.J. 544
Thompson, J.E. 539
Tiefensee, K. 533, 561
To, S. 549
Togashi, M. 552
Tokunaga, N. 528
Tori, M. 59, 186, 527–529, 534, 537–539, 549, 555–560
Toriumi, K. 537
Toyota, M. 527–530, 534, 538, 539, 543, 549, 550, 554, 555, 557–559
Tringali, C. 525
Tsai, P.I. 553
Tsai, R.-S. 561
Tsuji, J. 554
Tsukamoto, M. 543, 553
Tsunagawa, M. 542
Tsunoda, T. 543
Tsuyama, N. 558
Turchetta, S. 535
Tursch, B. 559
Tutschek, R. 559

Ubik, K. 543
Uchio, Y. 546
Uda, H. 536, 537, 555
Udaka, H. 550
Ueda, A. 528, 529, 559
Ueda, H. 537
Ueno, A. 547
Unelius, C.R. 559
Uohama, K. 546
Urbaniak, L. 540
Uto, S. 545, 546
Uyehara, T. 201, 559

Vaciago, A. 529
Valle, M.G. 559
Valterova, I. 559
Valverde, S. 535
Vandekerkhove, O. 543
Van der Helm, D. 531

Van der Linde, L.M. 559
Van der Weerdt, A.J.A. 559
Vanék, T. 530
Van Lier, F.P. 559
Vasicková, S. 543
Veith, M. 554
Venkatasubramanian, N.K. 549
Vermes, B. 533, 536, 542, 550
Vierengel, A. 543
Viotto, E.S. 545
Vowinkel, E. 547, 560
Vrkoc, J. 559

Wada, H. 555
Wagner, P. 531, 559
Wakabayashi, S. 547
Wakamatsu, M. 528, 534
Walinga, R.E. 542
Walkington, D.L. 547
Walters, R.L. 560
Wang, T.L. 460, 560
Wasylyk, J. 541
Watanabe, H. 527
Weavers, R.T. 530
Webster, N.M.H. 529
Wei, H.-C. 553, 560
Weisleder, D. 551
Weitz, S. 560
Welch, S.C. 179, 560
Wells, R.J. 542
Wenkert, E. 530, 531
Wey, F.-F. 561
White, A.H. 533
Whittemore, A.T. 502, 560
Wijnberg, J.B.P.A. 542
Wilschke, J. 403, 408, 560
Wilson, M.A. 560
Winnacker, E.-L. 535
Winzenberg, K.N. 555
Wiszt, A. 547
Witte, L. 525
Wolf, H. 535
Wolfe, G.R. 550
Wolfner, A. 536
Wood, W.F. 544
Woodcock, C.M. 527
Woods, N. 540
Wovkulich, P.M. 530, 531
Wray, V. 554
Wright, A.D. 560

Wright, J.M. *541*
Wu, C.-L. 49, 141, 295, *532, 553, 560, 561*
Wu, C.-M. *561*
Wurzel, G. *530, 561*

Yamada, H. *553*
Yamada, K. *540, 541, 551, 561*
Yamamoto, H. *561*
Yamamura, S. *550*
Yamaoka, Y. *538*
Yamasaki, K. *541*
Yamashita, M. *541, 551, 561*
Yamashita, Y. *554*
Yano, K. *546*
Yasuda, A. *538*
Yasuda, S. *547*
Yazawa, K. *561*
Yokota, T. *549*

Yokota, Y. *537*
Yonashiro, M. *551*
Yoshida, D. *550*
Yoshida, K. *546*
Yoshida, T. *539*
Yoshikoshi, A. *543*
Yoshimura, S. *556–558*
Yuki, S. *546, 547*
Yuzawa, Y. *561*

Zdero, C. *531*
Zenk, M.H. *535*
Zheng, G.-C. *561*
Zheng, G.-Q. 472, *561*
Zinsmeister, H.D. 328, 330, 333, 464, *525, 526, 530, 535, 545, 548, 552–554, 556, 561, 562*
Zong-Ning, Z. *527*

Subject Index

Abietanes 211, 513

Abietatriene 211

ar-Abietatriene 211, 247

Abietinella abietina 395

Abietinella sp. 392

Abscisic acid 130, 185, 186

Acacetin 374, 378

Acacetin-7-O-digalacturonide 364, 374

Acacetin-7-O-diglucuronide 364, 374

Acacetin-7-O-rhamnoarabinosyl
  galacturonide 364, 374

Acacetin-7-O-rhamnoglucosyl
  glucuronide 364, 374

Acacetin-7-O-rhamnoxylosyl
  galacturonide 364, 374

(U-^{14}C)-Acetate 408

Acetic acid 433, 446

3α-Acetoxybicyclogermacrene 54, 55, 93

14-Acetoxycadina-4,11-diene 61, 62, 97

8α-Acetoxy-β-cyclocostunolide 118, 162,
  163

18-Acetoxy-3S,4S,7S,8S-
  diepoxydolabellane 220, 250

6-Acetoxy-12,16-dihydroxydolabella-
  3E,7E-diene 220, 251

17-Acetoxy-1β,12-dihydroxy-15,16-epoxy-
  cis-ent-clerod-3,13(16),14-trien-6α,
  18-olide 218, 249

9α-Acetoxy-2β,13β-
  dihydroxyverrucosane 265, 275

19α-Acetoxy-*ent*-labda-8(17),12E,14-
  triene 238, 243, 257

19α-Acetoxy-*ent*-labda-8(17),12E,
  14-trien-3α-ol 242, 259

19α-Acetoxy-*ent*-labda-8(17),12E,14-
  trien-3β-ol 242, 259

6-Acetoxy-3,4-epoxy-12-hydroxy-
  dolabella-7E-en-16-al 221, 251

6β-(6'-Acetoxy)-β-glucopyranosyl-

ent-15α,20-dihydroxykaur-16-ene 235,
  255

20-(2'-Acetoxy)-β-glucopyranosyl-*ent*-15α-
  hydroxykaur-16-ene 235, 256

20-(2'-Acetoxy)-β-glucopyranosyl-*ent*-
  6-keto-15α-hydroxykaur-16-ene 235, 256

2-Acetoxy-3-hydroxybicyclogermacrene 54,
  55, 94

3-Acetoxy-2-hydroxy-
  bicyclogermacrene 54, 55, 94

7-Acetoxy-8-hydroxycalamenene 61, 64,
  100

9α-Acetoxy-2β-hydroxy-
  verrucosane 265, 275

2β-Acetoxy-11α-hydroxy-
  verrucosane 265, 275

Acetoxyisoplagiochilide 37, 40, 77

2-Acetoxy-8-ketogermacrene 122, 172

12β-Acetoxylongipin-2(10)-en-3-one 181

ent-12β-Acetoxylongipin-2(10)-en-3-one
  127, 180, 468

(−)-*ent*-12β-Acetoxylongipin-2(10)-
  en-3-one 181

Acetoxymarsupellone 126, 179, 180, 472

Acetoxyodontoschismenol 220, 250

(+)-Acetoxyodontoschismenol 223

9β-Acetoxy-10α-ovalifolianal 39, 78

9α-Acetoxy-10β-ovalifolianal 39, 78

9α-Acetoxyovalifoliene 37, 77

(−)-3-Acetoxytaylorione 39, 41, 78

6β-Acetoxyvitranoxide 116, 160, 161

8α-Acetoxyzaluzanin C 178, 472

8α-Acetoxyzaluzanin D 125, 162, 176, 178,
  472

Acetylcholine 451, 481, 482, 483

Acidic methanol 160

Acinetobacter calcoaceticus 474

Acoradiene 26, 66

Acoranes 26, 200, 202

Acritopappus prunifolius 178
Acrolejeunea mariana 86, 281
Acrolejeunea pusilla 89, 125, 131, 133, 134, 281, 284, 288
Acrolejeunea pycnoclada 86, 113, 125, 133, 134
Acrolejeunea torulosa 16, 19, 20, 86, 96, 120, 133, 134, 281, 284, 288
Actinomadura sp. 466
Actinomyces sp. 25
Actinomycetes 466
Adelanthaceae 493
Adelanthoideae 493
Adelanthus bisetulus 89
Adelanthus lindenbergianus 80, 84, 89, 120, 281, 284, 291, 295
Adenopharyngitis 465
Adriamycin 472
Africanes 27, 28, 30
Africanol 30
Agarospirane 203, 204
Agarospirene 203, 204
Agarospirol 203, 204
Aglycones 375
Ajanol 158
Albicanal 111, 153, 154
Albicanic acid 111, 153, 154
Albicanol 111, 153, 154
Albicanol acetate 475
Albicanyl acetate 111, 153, 154
Albicanyl caffeate 112, 152, 153, 154, 494
Alcaligenes faecalis 474
Algae 5, 60, 513–517
Aligerol 262, 268
n-Alkanes 380, 432, 433
Alkanoic acid 433
n-Alkyl phenols 517
Allergenic contact dermatitis 337, 465, 469, 470
Alloaromadendrene 33, 34, 68
Alnus sieboldiana 328
Alternaria kikuchiana 474
Amblystegiaceae 452, 507, 508
Amentoflavone 424, 425
Amorphanes 60, 62
α-Amorphene 60, 62, 97
Amphidium mougeotii 434, 435, 436, 441, 509
Amylase 38, 41
α-Amyrin 392, 397, 402, 403

β-Amyrin 392, 397, 402
Anacardiaceae 351, 470
Anacardium gigantheum 351
Anadensin 225, 226, 227, 251
Anagigantic acid 351
Anastrepta orcadensis 217, 226, 249, 251, 281, 284, 288, 469
Anastreptene 24, 33, 34, 43, 44, 46, 69, 170
Anastreptin 217, 249, 469
(−)-9-Anastreptone 33
Anastrophyllum minutum 69, 71, 80, 87, 89, 99, 263, 264, 272, 478, 489
Ancistrofuran 185
Ancistrotermes cavithorax 185
Andrea rupestris 396
Andrea sp. 456
Andreaeidae 362, 401, 507
Aneura latissima 284, 288
Aneura pinguis 190, 486, 487
Aneura sp. 486
Aneuraceae 190, 484, 486, 487, 491
Angustifolene 108, 141, 142
Annonaceae 296
Anodyne 465
Anoplolejeunea conferta 500
Ansamitocin P-3 450, 457, 458
Anthelia julacea 58, 96, 254, 391, 494
Anthelia juratzkana 391, 494
Anthelia sp. 229, 494
Antheliaceae 494
Antheraxanthin 397, 403
Anthoceros fusiformis 471
Anthoceros laevis 460, 462, 463
Anthoceros punctatus 408, 460, 461, 462, 463, 464, 510
Anthocerotaceae 462
Anthocerotae 5, 11, 362, 423, 460, 461, 471, 509, 510, 511, 512, 513, 514, 515, 516, 517, 520, 521, 522, 524
Anthocerotales 362
Anthocerotonic acid 461, 462, 463, 464
Anthocyanins 428, 430
Anthopleura pacifica 56
Antibacterial activity 473, 474
Antibiotic activity 474
Anticancer activity 465, 471
Antidotal activity 464, 465
Antifeedant activity 465, 475, 476
Antifungal activity 464, 465, 473, 474
Antihepatic activity 464

Anti-hyaluronidase activity 480
Antihypertensive activity 465
Antimicrobial activity 457, 464, 465, 473, 474
Antiphlogistic activity 465
Antipyretic activity 464, 465
Antirhinitic activity 464
Antiseptic activity 464
Antitrichia curtipipendula 415, 423
Antitumor activity 457, 464, 470, 471, 472
Apigenin 362, 363, 374, 375, 378, 409, 410, 424, 425
Apigenin-7,4'-di-O-galacturonide 363, 374
Apigenin-6,8-di-C-glucoside 364, 374, 410
Apigenin-7,4'-di-O-glucuronide 363, 374
Apigenin-7,4'-dimethyl ether 363, 374, 380
Apigenin-6-C-β-D-glucopyranoside-7-O-β-D-glucopyranoside 379
Apigenin-4'-O-glucoside 363, 374
Apigenin-7-O-glucoside 364, 374, 410, 421
Apigenin-7-O-glucoside-4'-O-galacturonide 363, 374
Apigenin-6-C-glucoside-7-O-glucoside 376
Apigenin-7-O-glucoside-6''-malonate 412, 420, 421
Apigenin-7-O-glucuronide 505
Apigenin-7-O-glucuronide-4'-O-rhamnosylglucuronide 363, 374
Apigenin-7-O-neohesperidoside 364, 374, 410
Apigenin-7-O-neohesperidoside-6''-malonate 411, 420
Apigenin-7-O-neohesperidoside-6''-malonylester 409
Apigenin-6-C-rhamnoside 379
Apigenin-6-C-α-L-rhamnoside-7-O-β-D-glucopyranoside 378
Apigenin-6-C-rhamnoside-7-O-glucoside 365, 376
Apigenin-7-O-rhamnosylglucuronide-4'-O-glucuronide 363, 374
Apigenin-6,7,4'-trimethyl ether 365, 376
Aplodon sp. 403, 468
Aplodon wormskioldii 404, 406, 432, 433, 434, 445, 446, 447
Apometzgerin 379, 380
Apometzgerin-7-O-glucuronide 409, 413, 421
Apoplexy 465
Arabinose 374, 387

Araceae 351
Arachidonic acid 382, 383, 386, 433, 436, 452, 455, 456, 508, 522, 523
Arbusculin A 117, 162, 163
Arbusculin B 162
(+)-Arbusculin B 117, 162, 163
(−)-Arbusculin B 117, 162, 163
Archilejeunea mariana 80, 84, 93, 128, 284, 288
Archilejeunea olivacea 71, 99, 100, 281, 284, 288
Archilejeunea parviflora 16, 20, 86, 100, 281, 284
Aristolanes 32
ent-Aristolan-10β-ol 32, 67
Aristol-1,8-diene 67
Aristol-1(10),8-diene 32
ent-Aristol-9-en-12β-al 32, 67
(+)-Aristol-9-ene 32, 67
ent-Aristol-9-en-12β-oic acid 32, 67
ent-Aristolone 32, 67
Arnellia sp. 490
Arnelliaceae 490
Aromadendranes 33
(+)-Aromadendrene 33, 34, 67
(−)-Aromadendrene 33
Aromadendrin 428
Aromatic compounds 298
Artemisia mexicana 176
Artemisia sp. 162
Arterial sclerosis 479
Asarum caulescens 160
Ascaridole 15
Aspergillus fumigatus 474
Aspergillus niger 474
Asteraceae 243
Asterella angusta 478
Asterella sp. 317, 318, 358, 465, 468
Athlete's foot dermatophytosis 465
Atractylodes japonica 160
Atrichum antustatum 473
Atrichum sp. 475
Atrichum undulatum 394, 436, 441, 473, 482
Aulacomnium androgynum 436, 441
Aulacomnium palustre 398, 399, 400, 401, 436, 441
Aulacomnium sp. 455
Aureusidin 428
Aureusidin-6-O-glucuronide 428, 501
Aurones 362, 409, 428

Auroxanthin 398, 403
Auxin 478
Aytoniaceae 61, 188, 338, 343, 380, 502
Aytonideae 502, 504
Azulene-1,4-carboxylic acid 45
Azulenes 42, 43
Azulenoids 54

Bacillus cereus 474
Bacillus megaterium 474
Bacillus subtilis 474
Bacteriosis 464
Balantiopsidaceae 492
Balantiopsis cancellata 89, 96, 296, 298
Balantiopsis erinacea 96, 281, 284, 296, 298, 299
Balantiopsis rosea 71, 89, 100, 114, 296, 298, 299, 316, 358, 384, 390, 492, 493, 520, 521
Balantiopsis sp. 159
Barbatanes 47, 48
α-Barbatene 47, 48, 51, 80
β-Barbatene 33, 47, 48, 51, 80, 467, 491
(−)-β-Barbatene 49
Barbatenes 47, 51
Barbella enervis 482
Barbella pendula 482
Barbifusicoccine A 227, 252
Barbifusicoccine B 227, 252
Barbilophozia attenuata 219, 250, 469
Barbilophozia barbata 111, 152, 223, 250, 386, 469
Barbilophozia floerkei 71, 111, 115, 124, 152, 160, 176, 219, 220, 227, 250, 252, 469, 477, 489
Barbilophozia hatcheri 71, 111, 152, 223, 250, 273, 281, 284, 288, 469, 489
Barbilophozia lycopodioides 111, 152, 219, 250, 273, 469, 489
Barbilophozia sp. 219
Barbilycopodin 219, 220, 222, 223, 250, 469
Bartramia pomiformis var. elongata 436, 441
Bartramiaceae 423
Batatifolin-7-O-sophoroside 371, 377
Bayer-Villiger oxidation 196
Bazzananes 49, 50
Bazzanene 494
α-Bazzanene 49, 50, 51, 84

β-Bazzanene 49, 50, 51, 52, 84
Bazzanenes 47
Bazzanenol 49, 50, 51, 85
(+)-Bazzanenol 49
Bazzanenyl caffeate 50, 51, 85, 494
Bazzanetin 50, 51, 85
Bazzania angustifolia 22, 24, 49, 80, 84, 85, 102, 104, 108, 135
Bazzania bidentula 494
Bazzania fauriana 22, 24, 25, 47, 49, 50, 69, 80, 83, 84, 85, 89, 95, 98, 99, 102, 104, 107, 108, 111, 112, 114, 116, 131, 135, 152, 153, 160, 190, 204, 494
Bazzania harpago 16, 81, 494
Bazzania japonica 54, 94, 107, 111, 112, 130, 153, 187, 291, 295, 475, 479, 494
Bazzania pompeana 51, 85, 494
Bazzania praerupta 16, 69, 71, 89, 104, 111, 113, 121, 171, 494
Bazzania sp. 49, 140, 151, 152, 157, 158, 188, 204, 281, 283, 284, 288, 294, 295, 473, 494
Bazzania spiralis 113, 116, 160, 252, 281, 284, 288, 494
Bazzania stolonifera 69, 87, 102, 105, 135
Bazzania tricrenata 141, 494
Bazzania tridens 32, 57, 67, 69, 71, 80, 81, 84, 85, 89, 95, 105, 108, 111, 115, 116, 130, 159, 185, 187, 474, 494
Bazzania trilobata 47, 61, 64, 70, 75, 83, 98, 99, 100, 105, 111, 141, 471, 494
Bazzania yoshinagana 494
Benzoic acid 296, 403, 404, 407
Benzoic acid derivatives 517
Benzonaphthoxanthenones 431, 432, 517
Benzophenone 403, 406, 408
Benzoyl chloride 145, 154
Benzoyl peroxide 145
Benzyl alcohol 403, 404
Benzyl benzoate 296, 297, 298
Benzyl cis-cinnamate 296, 297, 298
Benzyl trans-cinnamate 296, 297, 298
Benzyl trans-β-methyl thioacrylate 384, 390
p-Benzyloxybenzaldehyde 347
Benzyl thioacrylate 521
Benzyltriphenylphosphonium bromide 340
Berchemia racemosa 361
Bergamotanes 51, 52

trans-α-Bergamotene 51, 52
trans-β-Bergamotene 51, 52, 85
Betulaceae 328
3,3‴-Biapigenin 428
Bibenzyl ethers 517
Bibenzyls 296, 320, 324, 326, 327, 329, 331, 334, 517
Bicycloelemanes 51, 53
Bicycloelemene 51, 53, 85
(−)-Bicyclo[5.1.0]enone 40
Bicyclogermacranes 54, 55
Bicyclogermacrenal 479
(−)-Bicyclogermacrenal 54, 55, 93
Bicyclogermacrene 25, 43, 46, 56, 57, 491
(−)-Bicyclogermacrene 38
ent-Bicyclogermacrene 54, 55, 57, 89
Bicyclogermacrene-14-al 54, 55, 93
Bicyclohumulenone 467
(+)-Bicyclohumulenone 126, 178, 179
(±)-Bicyclohumulenone 178
Bi-dihydroflavonols 409, 422
Biflavones 362, 409, 422, 517
Biflavonoids 374, 409, 424, 426, 427
(*R*)-3,3′-Biherbertenediol 150
(*S*)-3,3′-Biherbertenediol 150
2′,6″-Biluteolin 416, 423, 424
2′,8″-Biluteolin 417, 425, 426
5′,6″-Biluteolin 415, 423, 424, 425, 428
5′,8″-Biluteolin 415, 422, 423, 424, 425, 428
3,3‴-Binaringenin 417, 427, 428
Biological activity 321, 464
Biomphalaria glabtata 475
Biphenyl ethers 517
Biphenyls 517
Birch reduction 340, 347
Bisabolanes 57, 59, 200, 202, 513
Bisabolatriene 58, 59, 95
α-Bisabolene 57, 95
(*S*)-(+)-*Z*-α-Bisabolene 57, 59, 95
β-Bisabolene 57, 59, 95
(6*R*,7*R*)-(+)-α-Bisabolol 57, 59, 95
(6*S*,7*S*)-(−)-α-Bisabolol 59, 95
Bis-bibenzyls 335, 336, 342, 344, 346, 350
Bis-(14*R*)-*ent*-kaur-16-en-14-yl malonate 233, 255
Bis-*ent*-kaur-16-en-15β-yl-malonate 234, 255
Bis-*ent*-labda-8(17),13-dien-15-yl malonate 245, 260
(1,2)-Bis-norphytone 246, 262, 267
Bitaylorione 35, 36, 76
Blasia pusilla 308, 313, 317, 335, 337, 350, 358, 363, 364, 366, 367, 368, 487
Blasiaceae 484, 487
Blechnum brasiliense 464
Blepharolejeunea densifolia 281, 284, 288
Blepharolejeunea incongrua 120, 261, 281, 284, 288
Blindia acuta 436, 441, 453, 508
Borneol 14, 21, 500
(−)-L-Bornesitol 384, 387
Bornyl acetate 14, 21, 500
(+)-Bornyl acetate 466
Bornyl ferulate 504
Boron tribromide 145
Botrytis cinerea 474
Bourbonanes 57
β-Bourbonene 59, 96
Brachiolejeunea laxifolia 367, 375
Brachiolejeunea phyllorhiza 366, 367, 375
Brachytheciaceae 471
Brachythecium buchananii 436, 441
Brachythecium reflexum 436, 441
Brachythecium rivulare 392, 395, 396, 397
Brachythecium rutabulum 403, 404, 405, 407, 476, 477
Brachythecium sp. 392, 433, 452
Bracteatin 417, 428
Brasilanes 57, 59, 60, 513
Brasilenol 60
Brasilenol acetate 60
Brasilia sickii 35
Brassicasterol 280, 290, 293, 294, 392, 395, 516
Breast adenocarcinoma (MCF-7) 472
Brickellia californica 171
Brittonin B 302, 325, 326
p-Bromobenzoyl chloride 27, 29, 227
β-Bromoherbertenol 474
(1*S*,3*SR*,4*S*,7*R*,8*SR*)-(+)-7-Bromo-2,3,3a,4,5,7,8,8-octahydro-1,4-dimethoxy-8a-methyl-6(1H)-azulene-6-ethylene acetal 45
(+)-Brothenolide 117, 162, 163, 166
Brotherella henonii 436, 441
Brotherella recurvans 394
Bryaceae 423, 456

Bryales 362, 374, 455, 486
Bryidae 401, 406, 408, 419, 451, 507
Bryoflavone 416, 423, 424
Bryophytes 5, 44, 464
Bryopteridoideae 500, 501
Bryopterin A 132, 194, 200
Bryopterin B 132, 194, 200
Bryopterin C 132, 194, 200
Bryopterin D 134, 196, 199
Bryopteris diffusa 128, 261, 281, 284, 288
Bryopteris filicina 71, 95, 122, 124, 132, 134, 173, 176, 196, 200, 226, 251, 284, 310, 341, 500
Bryopteris trinitensis 284
Bryum argenteum 409, 410, 411, 412, 413, 464
Bryum capillare 409, 410, 411, 412, 414, 415, 416, 421, 422, 423
Bryum cryophilum 418
Bryum cyclophyllum 428
Bryum pallens 473
Bryum pallescens 410, 411, 413, 421
Bryum pseudotriquetrum 409, 410, 411, 412, 413
Bryum rutilans 418, 428
Bryum schleicheri 410, 411, 412, 414, 415, 416, 421, 422
Bryum sp. 422, 455, 464
Bryum ventricisum 398, 399, 400, 401
Bryum weigelii 418, 428
Bucegia sp. 505, 506
Butylic acid 433, 446
Butyllithium-nitrobenzene 144

Cadalene 61, 64, 100
Cadalenes 61, 64
Cadina-4,11-dien-14-al 61, 62, 97
Cadina-4,11-dien-14-oic-acid 97
(+)-Cadina-4,11-dien-14-oic acid 62
Cadina-4,11-dien-14-ol 61, 62, 97
Cadinanes 60, 62, 513
γ-Cadinene 60, 62, 96
δ-Cadinene 60, 62, 96
α-Cadinol 60, 62, 97, 511
T-Cadinol 60, 62, 97
Cadlina luteomarginata 153
Caespitenone 27, 28, 30, 66
Caffeic acid 403, 405, 407, 461, 462, 463
Caffeoyl diglycoside 403
1-O-β-D-(3'-Caffeoyl)-glucopyranosyl

glycerol 319, 361
1-O-β-D-(4'-Caffeoyl)-glucopyranosyl glycerol 319, 361
1-O-β-D-(6'-Caffeoyl)-glucopyranosyl glycerol 319, 361
5-O-Caffeoyl shikimate 355
Calacoranes 61, 63
α-Calacorene 61, 99
β-Calacorene 61, 99
Calamenanes 61, 63, 64
Calamenene 61, 65, 511
cis-Calamenene 64, 65
trans-Calamenene 65
(1*S*,4*S*)-Calamenene 98
Calamenenes 47, 513
Calliergon cordifolium 437, 441
Calmodulin inhibitory activity 479
Calobryales 266, 485, 488, 524, 525
Calobryineae 488
Calypogeia azulea 42, 45, 46, 69, 78, 79, 89, 98
Calypogeia granulata 22, 24, 42, 54, 68, 69, 70, 78, 79, 80, 85, 86, 89, 93, 94, 113
Calypogeia muelleriana 54, 69, 70, 85, 86, 89, 93, 94, 113
Calypogeia neesiana subsp. *subalpina* 387, 388, 389
Calypogeia peruviana 54, 69, 70, 78, 79, 85, 86, 89, 93, 94, 113
Calypogeia sp. 42, 46, 54, 157, 294, 495
Calypogeia tosana 54, 69, 71, 78, 79, 85, 86, 90, 93, 94, 113, 297, 299
Calypogeia trichomanis 42, 54, 69, 71, 79, 85, 86, 90, 93, 94, 113
Calypogeiaceae 42, 46, 495, 500
Campesterol 280, 281, 293, 294, 392, 394, 512, 516, 518
Campesteryl behenate 280, 293
(−)-Camphanic acid 144
(1*S*)-(−)-Camphanic chloride 138, 139
Camphene 12, 14, 466, 467
(−)-Camphene 21
(−)-Camphonyl bromide 144
(+)-Camphor 14, 21
D-(+)-Camphor 15
Camphorenaldehyde 14, 21
Campylium stellatum 447, 452
Campylopus atrovirens 437, 441
Campylopus clavatus 415, 417
Campylopus flexuosus 437, 441

Campylopus holomitrium 415, 417, 428

Campylopus introflexus 392, 395, 396, 397, 437, 441

Campylopus japonicus 437

Campylopus pyriformis 437, 441, 508

Campylopus richardii 437, 441

Campylopus sp. 392, 453, 508

Campylopus substramineus 437, 441

Campylopusaurone 417, 428

Candida albicans 473, 474

(±)-*o*-Cannabichromene 328

o-Cannabigerol 328

p-Cannabigerol 328

Cannabis sativa 328

Caproic acid 433, 446

Carbohydrates 384

1-Carbomethoxy-2,3-dihydroxy-4'-methoxybibenzyl 302, 325, 326

2'-Carboxy-4,3'-dihydroxybibenzyl-3-O-β-D-glucopyranoside 299, 320, 321

Carboxylic acid 197, 433

4-Carboxy-1-methoxycarbonylazulene 43, 79

Cardiac infarction 479

Cardiopathy 465

Cardiotonic activity 480

3-Carene 391

Δ³-Carene 14, 19, 391, 393

2-Caren-4β-ol 207

Carlinoside 372, 377, 380, 497

Carotanes 152

α-Carotene 398, 403

β-Carotene 398, 403

β,β-Carotene-4-one 403

Carripinae 501

Carrpaceae 501

Carrpos sp. 501

Carrpos sphaerocarpos 501

(−)-Carvone 188

Caryophyllanes 65

β-Caryophyllene 65, 100, 189, 190

β-Caryophyllene oxide 65, 101

(−)-β-Caryophyllene oxide 65

Caulocystis cephalornithos 520

Caulocystis sp. 351

Cavicularia densa 364, 368, 369, 487

Cedar wood oil 204

Cedranes 65

α-Cedrene 65, 102

β-Cedrene 65, 102

Cembranes 211, 212

Cephalozia bicuspidata 477

Ceratodon purpureus 434, 435, 437, 441, 452, 454, 508

Ceratodon sp. 508

Chalcones 517

Chamaecydin 391, 396

Chamaecyparis obtusa 391

Chamberlainia acuminata 398, 399, 400, 401

Chamigranes 65, 136, 513

α-Chamigrene 65, 102, 136

β-Chamigrene 65, 102, 136, 188, 209

(+)-β-Chamigrene 506

ent-β-Chamigrene 65

ent-Chamigrenic acid 65, 103, 136

Chandonanthone 211, 212, 247

Chandonanthus hirtellus 16, 69, 95, 211, 247, 489

Chandonanthus setiformis 211, 223, 247, 250

Cheilolejeunea excisula 102, 128, 129, 500

Cheilolejeunea imbricata 128, 500

Cheilolejeunea serpentina 135, 209, 500

Cheilolejeunea trifaria 129, 135, 183, 209, 500

Chenopodene 135, 210

Chettaphanins 236, 245

Chiloscyphanes 136, 137

Chiloscypholone 103, 136, 137

(+)-Chiloscypholone 138

Chiloscyphone 103, 136, 137, 138

(+)-Chiloscyphone 138

(−)-Chiloscyphone 137, 138, 139

(±)-Chiloscyphone 138

Chiloscyphus hookeri 66, 71, 81, 83, 90

Chiloscyphus pallescens 103, 115, 126, 136, 160, 179, 492

Chiloscyphus pallido-virens 68, 81, 104

Chiloscyphus polyanthos 81, 99, 102, 119, 136, 168, 475, 481, 491, 492

Chiloscyphus sp. 140, 468, 491, 492

Chloroform 199, 471

7α-Chloro-6β-hydroxyconfertifolin 112, 153, 156

m-Chloroperbenzoic acid 141

Chlorophyceae 512, 513, 514, 515, 516, 517, 521

Chlorophyll 246

Cholesta-5,22Z-dien-3β-ol 519

Cholesterol 280, 290, 293, 294, 392, 395, 516, 518

Choline acetyl transferase activity 481

Choloplania 465

Chonanthelia sp. 375

Chromans 517

Chromic acid 272

Chromic oxide 141

Chromones 517

1,8-Cineole 14, 19

trans-Cinnamic acid 403, 404, 407

Cinnamic acid derivatives 460, 461, 517

Cinnamolide 112, 152, 156

Cinnamosma fragrans 153

9α-Cinnamoxygymnomitrol 47

Cirsilol 370, 377

Citral 328

Cladocarpicae 375

Cladoradula sp. 496

Clasmatocolea humilis 51, 85, 86, 89, 90, 122, 173, 281, 284, 288

Clasmatocolea sp. 491

Clasmatocolea vermicularis 69, 72, 81, 87, 90, 95, 99, 102, 116, 117, 119, 168, 284, 491

Clavukerin A 24, 25

Clavukerin B 24

ent-Cleroda-13,14-dien-13ζ-ol 247

Clerodanes 212, 213, 215, 216, 217, 218, 513

Cleroda-3,12(*E*),14-trien-11ζ-ol 219, 249

ent-Cleroda-3,12(*E*),14-trien-11ζ-ol 217

Clerod-3,13*E*-dien-15-al-17-oic acid 213, 247

Clerod-3,13*Z*-dien-15-al-17-oic acid 213, 247

Clerod-3,13*E*-dien-15,17-dial 213, 248

Clerod-3,13*Z*-dien-15,17-dial 213, 247

Clerod-3,13(14)-dien-15-ol-17-al 215, 248

cis-Clerod-3,12*E*,14-trien-18-al 216, 249

Clerod-3,13(16),14-trien-17-al 215, 248

cis-Clerod-3,13(16),14-trien-18-al 216, 249

cis-Clerod-3,12*E*,14-trien-18-oic acid 216, 248

cis-Clerod-3,12*Z*,14-trien-18-oic acid 216, 249

Clerod-3,13(16),14-trien-17-oic acid 212, 213, 247, 479

cis-Clerod-3,13(16),14-trien-18-oic acid 215, 248

cis-Clerod-3,12*E*,14-trien-18-ol 216, 249

cis-Clerod-3,13(16),14-trien-18-ol 216, 249

Clerosterol 516, 518, 519

Climacium dendroides 395, 437, 441

Climacium sp. 392

Clionasterol 290, 293, 294, 395, 516, 519

Colartin 117, 162, 163

Coleus forskohlii 243

Collins reagent 204

Colon adenocarcinoma (HT-29) 472

(−)-*ent-trans*-Communic acid 238, 243, 257

trans-Communol 243

ent-trans-Communol acetate 238, 243, 257

Compositae 31, 46, 53, 162, 171, 173, 176, 178, 212, 214, 240, 328, 469

Conicumol 135, 208

Conocephalaceae 338, 485, 504

Conocephalenol 59, 60, 96

(+)-Conocephalenol 59

(−)-Conocephalenol 61

Conocephalum conicum 12, 16, 17, 18, 19, 20, 21, 54, 59, 60, 86, 87, 93, 96, 97, 100, 102, 104, 115, 121, 122, 130, 135, 140, 159, 171, 173, 186, 208, 281, 283, 284, 288, 291, 294, 295, 296, 298, 317, 318, 358, 385, 464, 465, 466, 471, 473, 477, 503, 504, 505

Conocephalum japonicum 16, 54, 86, 90, 94, 117, 162, 171, 292, 295, 381, 385, 386, 465, 466, 503, 504, 505

Conocephalum supradecompositum 54, 504

Cooksonia sp. 509

Copaanes 136, 140

α-Copaene 104, 140

Cope rearrangement 207

Cordiniinae 501

Corsinia coriandrina 501

Corsinia sp. 382, 501

Corsiniaceae 485

Costunolide 122, 159, 162, 167, 173, 174, 504

(+)-Costunolide 167, 173

Cotton effect 35, 49, 63, 138, 150, 153, 154, 159, 160, 166, 183, 186, 187, 188, 195, 196, 203, 219, 221, 224, 228, 229, 231, 234, 244, 269, 270, 272, 277, 330, 332, 333, 334, 354, 463

m-Coumaric acid 403, 405, 407, 476

o-Coumaric acid 403, 405, 407

p-Coumaric acid 403, 404, 407, 464, 476

Cratoneuron filicinum 437, 441, 447, 452, 464, 508
o-Cresol isobutyrate 64
Crotonaldehyde 322
Cruciatin 313, 346, 347, 349
Cryptococcus neoformans 474
Cryptoxanthin 399, 403
Crysoeriol-7,4'-di-O-glucoside 369, 376
Crysoeriol-7-O-glucoside 368, 376
Crysoeriol-7-O-glucuronide 369, 376, 409, 411
Crysoeriol-7-O-neohesperidoside 369, 376
Ctenidium molluscum 395, 397, 477
Ctenidium percrassum 437, 441
Ctenidium sp. 392
Cubebanes 136, 140, 513
α-Cubebene 104, 140
(−)-α-Cubebene 140
β-Cubebene 104, 140
(−)-β-Cubebene 140
(+)-*ent*-Cubebene 60
Cubeb oil 60
Cubebol 104, 140
Cubenol 61, 62, 511
Cuparanes 140, 142, 144, 147, 513
Cuparene 50, 104, 141, 142, 143, 147, 148, 460, 462
(−)-Cuparene 140, 141, 146, 506
(±)-Cuparene 143
α-Cuparenol 108, 140, 142
Cuparenolide 109, 141, 144, 146, 476
Cuparenolidol 109, 141, 144, 476
α-Cuparenone 140, 146
(*R*)-(−) α-Cuparenone 108, 141, 142
(+)-β-Cuparenone 143
(±)-β-Cuparenone 143
Cupareno-quinone 140
α-Cuprenene 108, 142
ent-α-Cuprenene 141
β-Cuprenene 108, 141, 142
γ-Cuprenene 108, 141, 142
δ-Cuprenene 141
(−)-δ-Cuprenene 108, 141, 142
ε-Cuprenene 141
(+)-ε-Cuprenene 109, 142
(+)-Cuprenenol 109, 141, 144
ent-Cuprenenol 109, 144
(−)-*ent*-Cuprenenol 141
Cupric oxide 338
Curcuma zedoaria 173

ar-Curcumene 57, 58
(−)-*ar*-Curcumene 59, 95
Cycloartanes 514
Cycloart-23-en-3β,25-diol 291, 293, 295
Cycloartenol 291, 293, 295, 392, 395, 403
Cyclobazzanene 50
Cyclocitral 14, 21, 23
Cyclocolorenone 491
(+)-Cyclocolorenone 33
ent-Cyclocolorenone 34, 71, 498
α-Cyclocostunolide 118, 164, 167
(+)-α-Cyclocostunolide 167
β-Cyclocostunolide 117, 159, 162, 163, 166
ent-β-Cyclocostunolide 117, 163
(−)-*ent*-β-Cyclocostunolide 162
γ-Cyclocostunolide 162, 163
β-Cyclodextrin 12
γ-Cyclodextrin 12
Cycloeucalenol 280, 392, 397, 402, 514
α-Cyclogermacrone 116, 159, 160, 161
β-Cyclogermacrone 115, 160, 161
ent-β-Cyclogermacrone 160
Cyclohexanone 467
Cyclohexylcarboxylic acid 433
Cyclolaudanes 514
Cyclolaudenol 280, 392, 396, 402
Cyclomyltaylane 130, 187
Cyclomyltaylane-3-ol 130, 187
Cyclomyltaylane-5-ol 130, 187
Cyclomyltaylanes 186, 187
Cyclomyltaylenol 130, 186, 187
Cyclomyltaylyl-3-caffeate 130, 187, 479
Cyclopentanone 196, 197, 467
(−)-Cyclopropanecuparenol 109, 141, 142, 506
p-Cymene 12, 13, 18, 467
Cynodontium polycarpum 434, 435, 437, 441
Cynodontium strumiferum 434, 435, 437, 442
ent-Cyperone 160
(+)-α-Cyperone 160
ent-α-Cyperone 116, 161
Cystophora torulosa 351, 521
Cytokinin 478
Cytokinins 460
Cytostatic activity 473
Cytotoxic activity 471, 472

Dammalanes 514
Daucanes 152

Dawsonia grandis 448, 449, 457
Dawsonia longiseta 448, 449, 457
Dawsonia papuana 448, 449, 457
Dawsonia polytrichoides 448, 449, 457
Dawsonia superba 398, 399, 400, 401, 448, 449, 457
10-Deacetoxybarbilycopodin 220, 221, 250
Deacetoxyinfuscaside A 236
n-Decanal 381, 386
Dehydroabietic acid 211, 247
Dehydrocholesterol 516
22-Dehydrocholesterol 518, 519
Dehydroconfertifolin 112, 152, 154
Dehydrodeoxopinguisone 131, 191
(+)-4(15)-Dehydroglobulol 33, 34, 75
Dehydroisoporelladiolide 125, 176, 177
(+)-4(15)-Dehydroledol 33, 34, 75
4,5-Dehydronerolidol 121, 171
7′,8′-Dehydroperrottetin F 312, 346, 347
Dehydropinguisanin 133, 197, 198
Dehydropinguisenol 131, 191, 197, 471
Dehydropinguisenol methyl ether 131, 191
Dehydropinguisone 131, 191, 192
Dehydrosaussurea lactone 51, 53, 89
11,12-Dehydroursolic acid 397
11,12-Dehydroursolic acid lactone 392, 402, 403
Demotarisia linguifolia 109, 141, 217, 249, 489
Demotarisia sp. 489
Demotarisiol 109, 144
Dendroceros japonicus 461, 462
Dendrocerotaceae 462
Dendrodoris sp. 475
Dendroligotrichum dendroides 448, 449, 457
12-Deoxo-1β,11α-dihydroxy-sacculatanolide 263, 270
Deoxopinguisone 131, 191, 193, 197, 200, 201
(±)-Deoxopinguisone 201
Deoxopinguisone-12,15-dimethyl ester 132, 194
Deoxopinguisone methyl ester 132, 194, 200
Deoxyanthocyanins 362, 409
Deoxypodophyllotoxin 450, 457, 458
Desmarestia sp. 521
Desmosterol 516, 518, 519
10*R*,18-Diacetoxy-3*S*,4*S*,7*S*,8*S*-

diepoxydolabellane 220, 250
10*R*,18-Diacetoxy-3*S*,4*S*-epoxydolabell-7*E*-ene 220, 221, 250
3α,14-Diacetoxy-2-hydroxybicyclo-germacrene 54, 55, 94
6,16-Diacetoxy-12-hydroxydolabella-3*E*,7*E*-diene 221, 251
(3*S*,6*R*)-Diacetoxy-(7*R*)-hydroxymanoyloxide 242, 259
(3*S*,6*R*)-Diacetoxy-(7*S*)-hydroxymanoyloxide 245, 260
(3*S*,7*S*)-Diacetoxy-(6*R*)-hydroxy-manoyloxide 245, 260
ent-9,14-Diacetoxylongipin-2(10)-en-3-one 127, 180
9,14-Diacetoxymarsupellone 127, 180, 181
(3*S*,6*R*)-Diacetoxy-7-oxomanoyloxide 245, 260
Diacyl-glyceryltrimethylhomoserine 455
Diazomethane 65, 321, 463
Di-*p*-bromobenzoate 205
3,3′-Dibromo-4,4′,5,5′-tetrahydroxybibenzyl 326
1,4-Dicarboxyazulene 43, 79
(1*R*,2*S*)-2,3-Dicarboxy-6,7-dihydroxy-1-(3′,4′-dihydroxy)-phenyl-1,2-dihydronaphthalene 315, 354
2,3-Dicarboxy-6,7-dihydroxy-1-(3′,4′-dihydroxy)-phenyl-1,2-dihydro-naphthalene-10-methyl ester 315, 354
2,3-Dicarboxy-6,7-dihydroxy-1-(3′,4′-dihydroxy)-phenyl-1,2-dihydro-naphthalene-9,5″-shikimic acid ester 315, 354
2,3-Dichloro-5,6-dicyano-1,4-benzoquinone 141
Dichloromethane 471
Dichodontium pellucidum 435, 442
Dicranaceae 382, 423, 452, 456, 471, 508, 509
Dicranales 455
Dicranella cerviculata 477
Dicranella heteromalla 435, 437, 473
Dicranella jamesonii 435, 437, 442
Dicranella palustris 434, 435, 437, 442
Dicranella schreberiana 435, 437
Dicranella scoparium 473
Dicranella sp. 508
Dicranenone A 447, 453, 454, 474

Dicranenone B 448, 453, 454
Dicranenone B1 448, 453, 454, 474
Dicranodontium denudatum 437, 442
Dicranodontium sp. 508
Dicranoloma cylindrothecium 436, 437, 442, 447, 448, 454
Dicranoloma robustum 415, 416, 417, 425
Dicranoloma scoparium 415
Dicranolomin 416, 423, 424, 425
Dicranoweisia cirrata 435, 438, 442
Dicranoweisia crispula 438, 442, 508
Dicranum elongatum 392, 393, 395, 397, 437, 442, 452
Dicranum flagilifolium 394
Dicranum fulvum 434, 435, 437, 442
Dicranum fuscescens 435, 437
Dicranum japonicum 434, 435, 437, 442, 447, 448, 454, 474, 482
Dicranum majus 447, 448, 454
Dicranum montanum 435, 437, 442, 452, 508
Dicranum muehlenbeckii 434, 435, 437, 442
Dicranum polycetum 394, 435, 437, 442
Dicranum robustum 410, 423
Dicranum scoparium 394, 422, 423, 434, 435, 436, 438, 447, 448, 453, 454, 473, 474, 482
Dicranum sp. 455, 456, 475, 508
Dicranum spurium 434, 435, 438, 442
Dicranum undulatum 435, 438, 442
Dicranum viride 435, 438, 442
Dictyolanes 518
Dictyopteris sp. 61, 140, 173, 521
Dictyota sp. 512
trans-Didehydrobicyclofarnesol 147, 148, 149
[1,9-^{13}C]-*trans*-Didehydro-bicyclofarnesol 147
[4a-^{13}C]-*trans*-Didehydro-bicyclofarnesol 147
1α,3β-Di(3,4-dihydroxyphenyl)-2α,4β-dibaz-zanenylcyclobutylate 85
Diels-Alder reaction 36
Diethyl[4-[2-methoxy-5-(1,3-dioxan-2-yl)-phenoxy]benzyl] phosphate 335
Digalactosyl diacylglycerols 455
Dihydroagarofuran 119, 165, 170
2,3-Dihydroamentoflavone 425, 426
(11S)-Dihydroarbusculin A 117, 162, 163
(8aR)-1,8a-Dihydroazulene 44

2,3-Dihydro-3,3'''-biapigenin 417, 427, 428
2,3-Dihydro-2',6''-biluteolin 416, 423, 426
2'',3''-Dihydro-5',6''-biluteolin 416, 423, 426
Dihydrobrassicasterol 280, 283, 293, 394
22-Dihydrobrassicasterol 294
Dihydrochalcones 362
Dihydrocostunolide 122, 173, 174
(R)-(−)-8,11-Dihydro-α-cuparenone 109, 141, 142
Dihydro-β-cyclocostunolide 163
(11S)-Dihydro-β-cyclocostunolide 117, 162
Dihydrodicranenone B 448, 453, 454
2,3-Dihydrodicranolomin 416, 423, 426
2,3-Dihydro-5',3'''-dihydroxy-amentoflavone 416, 425, 426
3,10-Dihydro-1,4-dimethylazulene 43, 46, 80
1,8a-Dihydro-3,8-dimethylazulene 44
(+)-1,8a-Dihydro-3,8-dimethylazulene 42, 43, 80
(8aS)-(+)-1,8a-Dihydro-6,8a-dimethylazulene 44
ent-Dihydrodiplophyllolide 119, 165, 168
11(13)-Dihydro-4α,5β-epoxy-8-epi-inunolide 123, 173, 174
Dihydroeremofrullanolide 157
Dihydroestafiatin 125, 176, 178
Dihydroflavones 362
Dihydroflavonol 409
Dihydrofrullanolide 117, 161, 163
(−)-Dihydrofrullanolide 167
Dihydro-β-frullanolide 117, 162, 163
2,3-Dihydro-5'-hydroxy-amentoflavone 416, 425, 426
2,5-Dihydro-5-hydroxy-4-(4'-hydroxy-phenyl)-2-furanone 403, 406, 407
2,3-Dihydro-5'-hydroxy-robustaflavone 416, 425, 426
Dihydroinversin 314, 352, 353
Dihydroisoalantolactone 168
ent-Dihydroisoalantolactone 119, 165
Dihydrolinguifolide 217, 249
(1S,8aS)-(+)-1,8a-Dihydro-1-methoxy-6,8a-dimethylazulene 44
(1S,8aS)-(+)-1,8a-Dihydro-1-methoxy-8a-methylazulene 44
(8aS)-(+)-1,8a-Dihydro-8a-methylazulene 44

Dihydromylione 33, 34, 41, 75
(−)-Dihydromylione 35
Dihydrooxyfrullanolide 117, 163
2,3-Dihydrophilonotisflavone 417, 425,
    427
Dihydropinosylvin 330
14,15-Dihydroscapanin A 240
(−)-Dihydrotaylorione 35
(11R)-Dihydrotulipinolide 123, 174
(11S)-Dihydrotulipinolide 123, 173, 174
11,13-Dihydrovernodalin 53
3α,4α-Dihydroxy-african-2(6)-en-5-one 27,
    28, 66
Dihydroxyalkanoic acid 433
5′-Dihydroxyamentoflavone 409
5′,3‴-Dihydroxyamentoflavone 415, 422,
    424, 425
ent-4β,10α-Dihydroxyaromadendrane 34,
    35, 75
3,5-Dihydroxybibenzyl 328
5,4′-Dihydroxybibenzyl-2-O-β-D-
    glucopyranoside 299, 320, 321
5,8-Dihydroxycalamenene 61, 63, 64, 100
3,5-Dihydroxy-6-carbomethoxy-2-(3-
    methyl-2-butenyl)bibenzyl 306, 331
6β,7α-Dihydroxyconfertifolin 112, 153, 156
2,3-Dihydroxycuparene 107, 140, 142, 146
6,7-Dihydroxy-4-(3,4-dihydroxyphenyl)-
    naphthalene-2-carboxylic acid 314, 354
3,4-Dihydroxy-5,3′-dimethoxybibenzyl
    302, 325, 326
3,3′-Dihydroxy-4,5;4′,5′-dimethylenedioxy-
    bibenzyl 302, 325, 326
3,5-Dihydroxy-4-(3,7-dimethyl-2,6-
    octadienyl)bibenzyl 327
3,5-Dihydroxy-4(3,7-dimethyl-2,7-
    octadienyl)bibenzyl 325, 327
6,12-Dihydroxydolabella-3E,7E-diene 220,
    251
5S,12R-Dihydroxy-6,8,10,14-
    eicosatetraenoic acid 479
3,5-Dihydroxy-2-(2,3-epoxy-3-
    methylbutyl)bibenzyl 304, 329
3,5-Dihydroxy-4-(2,3-epoxy-3-
    methylbutyl)bibenzyl 302, 326, 333
5α,8β-Dihydroxyeudesm-4(15),7(11)-dien-
    12,8-olide 118, 164
2,5-Dihydroxy-4-formyl-6-methoxy-
    acetophenone 318, 360
3,5-Dihydroxy-4-geranylbibenzyl 303,

    325, 327, 328
8,16-Dihydroxyhexadecanoic acid 433
10,16-Dihydroxyhexadecanoic acid 433
6α,22-Dihydroxyhopane 295
8,9-Dihydroxy-2-(4′-hydroxy-3′-methoxy-
    phenyl)-11-methoxyphenanthro[2,1-b]-
    pyryliumchloride 418, 430
4,5-Dihydroxyisocuparene 143
1-(3,4-Dihydroxy-5-methoxybenzyl)-3-
    methylbut-2-ene 317, 358, 359
3,5-Dihydroxy-2-(3-methyl-2-
    butenyl)bibenzyl 303, 327, 332
3,4′-Dihydroxy-4-(3-methyl-2-
    butenyl)bibenzyl 303, 325, 327
3,5-Dihydroxy-4-(3-methyl-2-butenyl)
    bibenzyl 302, 326, 330
2,5-Dihydroxy-4-methyl-6-methoxy-
    acetophenone 318, 360
2β,13β-Dihydroxy-9-oxoverrucosane 265,
    275
14,14′-Dihydroxyperrottetin E 312, 346, 347
2-(3′,4′-Dihydroxyphenyl)-8,9-dihydroxy-
    11-methoxyphenanthro[2,1-b]-
    pyryliumchloride 418, 430
2-(3′,4′-Dihydroxyphenyl)-8,11-dihydroxy-
    9H-phenanthro[2,1-b]pyran-9-
    one 418, 430
β-(3,4-Dihydroxyphenyl)-ethyl-O-β-D-
    glucoside 319, 360, 361
5′,3‴-Dihydroxyrobustaflavone 415, 423,
    424, 425
1β,11β-Dihydroxysacculatanolide 263, 270
1β,11α-Dihydroxysacculatenolide 263, 271
3,5-Dihydroxystilbene 328
3α,18-Dihydroxytrachyloban-19-oic
    acid 264, 273, 274
2β,8β-Dihydroxyverrucosane 265, 275
2β,9α-Dihydroxyverrucosane 265, 275, 277
2β,9α-Dihydroxyverrucos-13-ene 265, 275
1,6-Diketogermacrene 121, 172
Dilaenaceae 486, 487, 522
3,4-Dimethoxybenzaldehyde 347
3-(3′,4′-Dimethoxybenzyl)-7-hydroxy-5-
    methoxyphthalide 316, 357, 358
3,4-Dimethoxybenzyltriphenylphos-
    phonium bromide 325
3,4′-Dimethoxybibenzyl 300, 323, 324
3,5-Dimethoxybibenzyl 300, 323, 324, 333
3,5-Dimethoxy-6-carbomethoxy-2-(3-
    methyl-2-butenyl)bibenzyl 331

N-3',4'-Dimethoxycinnamoylanthranilic acid 480

3,4-Dimethoxycinnamyl alcohol 51, 152

3,4'-Dimethoxy-4-hydroxybibenzyl 301, 323, 324

3,4-Dimethoxy-5-hydroxy-9,10-dihydrophenanthrene 315, 355, 356

2,5-Dimethoxy-3-hydroxy-phenanthrene 316, 355, 356

3',4'-Dimethoxyluteolin-7-O-rhamnoarabinosyl galacturonide 369, 376

3',4'-Dimethoxyluteolin-7-O-rhamnoxylosyl galacturonide 369, 376

3,5-Dimethoxy-2-(3-methyl-2-butenyl)bibenzyl 327

3,5-Dimethoxy-4-(3-methyl-2-butenyl)bibenzyl 326

3,3'-Dimethoxy-4,5-methylene-dioxybibenzyl 301, 324, 325

3,3'-Dimethoxy-4,5-methylenedioxy-4'-hydroxybibenzyl 301, 324, 325

6α,11α-Dimethoxypinguis-5(10)-ene 132, 194

6α,11β-Dimethoxypinguis-5(10)-ene 132, 198

2,2-Dimethoxypropane 195

3,4-Dimethoxystyrene 317, 358, 359

2,2-Dimethylallylbibenzyl 521

2,2-Dimethylallylbromide 330

2,2-Dimethylallylcatechol 521

p-Dimethylaminobenzoyl chloride 321

Dimethylaminopyridine 154

1,4-Dimethylazulene 42, 43, 44, 45, 46, 47, 78, 490, 495, 500

2,6-Dimethylcyclohexanone 25

2,7-Dimethylcyclohexanone 467

2,2-Dimethyl-7,8-dihydroxy-5-(2-phenylethyl)chromene 305, 329, 330

2,2-Dimethyl-7,8-dimethoxy-5-(2-phenylethyl)chromene 329

3,4;3',4'-Dimethylenedioxybibenzyl 301, 325, 326

Dimethylformamide 154

1,4-Dimethyl-3-formylazulene 43, 45, 79

(8S, 8aS)-(−)-3,8-Dimethyl-1,2,6,7,8,8a-hexahydroazulene 24

2,2-Dimethyl-5-hydroxy-6-carboxy-7-(2-phenylethyl)chromene 305, 329

2,2-Dimethyl-5-hydroxy-7-(2-phenylethyl)chromene 305, 329

2,2-Dimethyl-7-hydroxy-5-(2-phenylethyl)chromene 304, 329

3,7-Dimethylindene-5-carbaldehyde 43, 79

3,7-Dimethylindene-5-carboxaldehyde 42, 44

Dimethyl isohemipate 449, 457, 458

Dimethyl isophthalate 143

Dimethyl kolavate 212, 213

Dimethyl metahemipate 449, 457, 458

3,7-Dimethyl-5-methoxy-carbonylindene 42, 43, 79

2,2-Dimethyl-5-methoxy-7-(2-phenylethyl)chromene 305, 329

2,2-Dimethyl-7-methoxy-5-(2-phenylethyl)chromene 304, 329

(4aS,5S,8aS)-(−)-5β,8aβ-Dimethyl-5α-(4-methyl-3-pentenyl)-3,4,4a,5,6,7,8,8a-octahydro-1(1H)-naphthalenone 269

2-(3,7-Dimethyl-2,7-octadienyl)-3,5-dihydroxybibenzyl 303, 325, 327

trans-1,4a-Dimethyl-(1,2,3,4,4a,5,6,7)-octahydronaphthalene 22, 24

2,2-Dimethyl-4-pentanol 432, 447

Dimethyl sulfide 384, 390, 466, 467, 521, 522

Dimethylsulfoxide 338

Diosmetin-7-O-glucoside 412, 420

Diosmetin-7-O-glucoside-6''-malonate 409, 412, 420

(−)-3,10-Dioxotaylori-4-ene 39, 41, 78

Diphenyl methanes 517

Diphosphatidyl glycerol 455

Diplasiolejeunea patelligera 71, 72, 100, 128, 292, 295

β-Diploalbicene 33

Diploalbicanol 33

Diplophyllin 168, 169, 170, 481, 491

ent-Diplophyllin 119, 165, 168

Diplophyllolide 168, 169

ent-Diplophyllolide 119, 165, 168

Diplophyllum albicans 33, 67, 75, 119, 153, 168, 473, 481

Diplophyllum serrulatum 69, 72, 111, 129, 153, 155, 183

Diplophyllum sp. 152

Diplophyllum taxifolium 33, 67, 75

Diploptene 291, 293, 295, 392, 395, 403

Diplopterol 291, 293, 295

Distichium capillaceum 438, 442, 477

Diterpenoids 27, 247, 391
Ditrichaceae 382, 452, 456, 507, 508
Ditrichum capillaceum 508
Ditrichum cylindricum 438, 442, 508
Ditrichum flexicaule 438, 442, 508
Ditrichum heteromallum 435, 436, 438, 442, 508
Ditrichum inclinatum 508
Ditrichum lineare 398, 399, 400, 401
Ditrichum pallidum 464, 465
Ditrichum pusillum 436, 438, 442, 508
Ditrichum tundrae 442, 443
Diuretic activity 464, 465
Docosanol 380
Docosanyl hexadecanoate 380
7,10,13,16,19-Docosapentaenoic acid 447, 452, 453
Dolabella californica 225
Dolabellanes 219, 220, 221, 225, 227, 514
Dolabellanoids 224, 227
Dolabellatriene 223
Dolabradiene 268
Dolastanes 514
Dolichomitra cymbifolia 438, 442
Drepanocladus aduncus 438, 442
Drepanocladus exannulatus 447, 452, 508
Drepanocladus tundrae 438
Drimanes 152, 154, 156, 513
Drimenin 112, 152, 156
Drimeninol 112, 152, 154
Drimenol 12, 111, 152, 153, 154
Drimenyl caffeate 112, 153, 154, 494
Dumortiera hirsuta 121, 122, 171, 173, 281, 284, 288, 299, 309, 320, 338, 471, 473, 506, 507
Dumortiera sp. 59, 507

Ecdysones 516
Echinenone 403
11,14-Eicosadien-8-ynoic acid 433, 447, 452, 453
Eicosapentaenoic acid 382, 383, 386, 452, 523
5Z,8Z,11Z,14Z,17Z-Eicosapentaenoic acid 382, 383, 386, 433, 441
Eicosatetraenoic acid 433
5Z,8Z,11Z,14Z-Eicosatetraenoic acid 382, 383, 386
Elatin 409, 412, 420

Elemanes 51, 53, 513
Elema-1,4(15),11-trien-3-al 53, 89
Elema-1,4(15),11-trien-3,14-olide 53, 89
α-Elemene 51, 53, 86
β-Elemene 51, 53, 86
γ-Elemene 51, 53, 87
δ-Elemene 51, 53, 87
Elemol 51, 53, 89
Ellagic acid 317, 358, 359
Encalypta streptocarpa 438, 443
Enterobacter cloacae 474
Entodon rubicundus 450, 451, 460, 473
Entodon seductrix 398, 399, 400, 401
4-Epiarbusculin A 117, 163, 472
Epibrasilenol 60
Epicubenol 60, 61, 511
(−)-Epicubenol 60, 62
(+)-*ent*-Epicubenol 60, 62, 97
(−)-Epicyclopropanecuparenol 109, 141, 142
ent-10-Epiglobulol 33
(+)-C-10-Epiglobulol 33, 34, 75
10-Epi-iso-α-gurjunene B 177
Epilepsy 465
3-Epi-myliol 35
(−)-3-Epi-myliol 33, 34, 75
3α,4α-Epoxy-5α-acetoxy-18-hydroxy-sphenoloba-13E,16E-diene 263, 264, 273, 478
3α,4α-Epoxy-5α-acetoxy-18-hydroxysphenoloba-13Z,16E-diene 264, 273
3α,4α-Epoxy-5α-acetoxysphenoloba-13E,16E,18-triene 264, 273
3α,4α-Epoxy-5α-acetoxysphenoloba-13Z,16E,18-triene 264, 273
11ζ,12-Epoxychiloscypholone 103, 136, 137
ent-3β,4β-Epoxyclerod-13E-en-15-al 215, 248
ent-3β,4β-Epoxyclerod-13Z-en-15-al 215, 248
ent-3β,4β-Epoxyclerod-14-en-13ζ-ol 215, 248
6β,7β-Epoxyconfertifolin 112, 153, 156
(11S,12Z)-8α,12-Epoxy-5α,11α-dihydroxy-labda-12,14-dien-1-one 239, 258
3α,4α-Epoxy-5α,18-dihydroxysphenoloba-13E,16E-diene 264, 273
4α,5β-Epoxy-8-epi-inunolide 123, 173, 174, 175

Epoxyfrullanolide 471
4α,5β-Epoxy-7α,8β,11α-H-germacra-1(10)-en-12,8α-olide 123, 173, 174
3α,4α-Epoxy-5α-hydroxysphenoloba-13Z,16E,18-triene 264, 273
Epoxylutein 399, 403
Epoxytamariscol 467
2β,16-Epoxyverrucosane-16-ol 278
(15S,16S)-2β,16-Epoxyverrucosan-16-ol 265, 279
Equisetum sp. 510
Equisetum sylvaticum 510
Eremofrullanolide 157, 471
(+)-(4S*,5R*,7S*,8R*)-Eremophila-9,11-dien-8α-ol 113, 157
Eremophilane 137
Eremophilanes 157
Eremophilene 113, 157
Ergosterol 516
Eriodictyol 421, 425, 428
Erysipelas 465
Escherichia coli 473, 474
Estafiatin 125, 176, 178
Ethanal 433, 446
Ethanethiol 145
Ethanol 221, 433, 446, 471
Ethyl acetate 27, 432, 433, 446
Ethyl acetoacetate 322
p-Ethylanisol 467, 476, 500
24-Ethyl-5,22-cholestadienol 392
24-Ethylcholest-5,22-dien-3-ol 294
24α-Ethylcholest-5,22-dien-3β-ol 280
24-Ethylcholest-5-en-3-ol 294
24α-Ethylcholest-5-en-3β-ol 280
24α-Ethylcholesterol 280, 519
24β-Ethylcholesterol 280, 519
Ethyl 2,3-dimethyl butylate 432, 447
Ethyl formate 433, 446
Ethyl heptanoate 433, 447
2-Ethylhexanal 432, 445
Ethyl hexanoate 433, 447
ent-4(15),7(11)-Eudesmadien-8-one 115, 160, 161
(−)-*ent*-4(15),7(11)-Eudesmadien-8-one 160
ent-4(15),11(12)-Eudesmadien-8-one 116, 161
Eudesmanal 115, 158, 159
Eudesmanes 15, 157, 158, 161, 163, 164, 165, 513

Eudesmanolides 157, 159, 162, 167, 168
12,6-Eudesmanolides 161
5α,7β(H)-Eudesm-4α,6α-diol 159, 499
(+)-5α,7β(H)-Eudesm-4α,6α-diol 115, 158
Eudesm-3-en-6α-acetoxy-7α-ol 115, 159, 161
Eudesm-3-en-6β,7α-diol 115, 159, 161
(+)-Eudesm-3-en-6β,7α-diol 160
ent-Eudesm-4(15)-en-6α,7α-diol 115, 158
Eudesm-4(15)-en-6β,7α-diol 115, 158, 159
Eudesm-4(15)-en-6β,7β-diol 160
Eudesm-3-en-7α-ol 115, 158, 159
α-Eudesmol 25, 114, 158
β-Eudesmol 25, 114, 158
Eupatilin 370, 377
Euphanes 514
Eupleurozia giganteoides 498
Eupleurozia paradoxa 366, 368, 369, 380, 498
Eupleurozia simplicissima 364, 369, 372, 373, 380, 498
Eupleurozia sp. 380, 497, 498
Eurhynchium organum 469
Euriccia sp. 381
Exo-3-methyl-3-borneol 15, 22, 23
Exo-2-methylfenchol 15, 22, 23

Farnesanes 171, 513
α-Farnesene 119, 171
trans-β-Farnesene 120, 171
trans-Farnesol 171
trans,cis-Farnesol 121, 171, 200
Farnesyl pyrophosphate 25, 44, 152
trans,cis-Farnesyl pyrophosphate 188, 202
Fatty acids 381, 383, 433, 453
Fenchone 14, 21
Fernanes 514
Fern-7-ene 392, 397, 402
Fern-9(11)-ene 392, 397, 402
Ferredoxin 510
Ferula communis 173
(+)-α-Ferulene 32, 67
Ferulic acid 403, 405, 407, 476
Fissidens adiantoides 438, 443
Fissidens areolatus 436, 438, 443, 447, 448, 454
Fissidens japonicum 465
Fissidens nobilis 436, 438, 443, 454
Fissidens sp. 464, 469
Fissidens taxifolium 477

Flavanols 421
Flavanones 421
Flavone O-glucuronides 362, 374
Flavone C-glycosides 362, 374, 378, 409
Flavone O-glycosides 362, 374, 409
Flavones 374, 376, 377, 379, 408, 420, 421
Flavonoids 361, 362, 363, 375, 377, 408, 409, 505, 517
Flavonols 362, 379, 421
Floribundaria nipponica 482
Fo.-1 491
Fo.-2 491
Fo.-3 491
Folioceros fuciformis 462, 463
Fontinalis antipyretica 398, 399, 401, 432, 433, 435, 436, 438, 443, 446, 447, 452, 456
Fontinalis squamosa 456
Formic acid 49, 147, 158
1-Formyl-2,3-dihydroxycuparene 108, 140, 142
α-Formylherbertenol 146, 474
(−)-α-Formylherbertenol 110, 143, 144
2-Formyl-5-hydroxyisocuparene 143
Formyl methionyl leucyl phenylalanine 479
4-Formyl-1-methoxycarbonylazulene 43, 46, 79
Forskolin 242, 243
Fossombronia himalayensis 474
Fossombronia pusilla 19, 20, 22, 100, 114, 118, 119, 162, 170, 263, 271, 474
Fossombronia sp. 158
(+)-Fragrolide 153
Friedelanes 514
Friedelin 291, 293, 295, 395, 403
3α-Friedelinol 392, 397, 402, 403
Fructose 387
Frullania africana 366, 367, 369
Frullania apiculata 116
Frullania arecae 364, 366, 367, 368, 369
Frullania asagrayana 116, 117, 467, 469
Frullania bicornistipula 69, 86, 100, 116, 117, 124, 127, 162
Frullania bolanderi 469
Frullania bonincola 301, 325
Frullania brachyclada 363, 364, 366, 367, 368, 369
Frullania brasiliensis 57, 86, 90, 95, 116, 117, 162, 281, 284, 288

Frullania brotheri 117
Frullania californica 122, 173
Frullania clavata 81, 90, 111
Frullania confertiloba 364, 366, 367, 368, 369
Frullania davurica 84, 87, 90, 96, 100, 104, 114, 281, 285, 288, 300, 301, 323, 363, 365, 367, 370, 371, 373, 375, 381
Frullania dilatata 116, 375, 469, 470, 471, 473, 475, 481
Frullania eboracensis 469
Frullania ecklonii 363, 364, 366, 367, 368, 369, 370, 375
Frullania ericoides 301, 325
Frullania falciloba 19, 71, 81, 84, 90, 101, 281, 285, 288, 300, 301, 316, 323, 357, 499
Frullania franciscana 469
Frullania gaudichaudii 71, 81, 105
Frullania gibbosa 366, 367, 370
Frullania gradsteinii 366, 367, 370
Frullania hamatiloba 117, 162, 244, 259, 260, 499
Frullania inflata 469
Frullania jackii 68, 84, 87, 90, 96, 101, 104, 281, 285, 288, 300, 301, 363, 365, 367, 370, 371, 373, 375
Frullania kunzei 469
Frullania laxiflora 367, 370
Frullania liparia 469
Frullania muscicola 319, 361, 365, 367, 368, 370, 371
Frullania nepalensis 118, 130, 162, 188, 467, 499
Frullania nisquallensis 469
Frullania obscura 366, 367, 369
Frullania ovistipula 367
Frullania parvistipula 301, 302, 325
Frullania pluricarinata 366, 367, 370
Frullania polysticta 371, 372, 373
Frullania riojaneirensis 363, 364, 366, 367, 369
Frullania serratta 86, 113, 116, 117, 118, 122, 123, 157, 162, 173, 302, 325, 499
Frullania sp. 118, 126, 140, 152, 161, 162, 176, 178, 182, 292, 295, 323, 375, 465, 467, 469, 470, 479, 498, 499
Frullania sphaerocephala 87, 101, 116, 117, 364, 366, 367, 368, 369

Frullania tamarisci 118, 188, 464, 469, 470, 475
Frullania tamarisci subsp. *asagrayana* 130, 188, 499
Frullania tamarisci subsp. *nisquallensis* 117, 122, 173, 499
Frullania tamarisci subsp. *obscura* 113, 115, 117, 118, 122, 130, 159, 188, 467, 472, 499
Frullania tamarisci subsp. *tamarisci* 130, 167, 188, 467, 499
Frullania ternatensis 116
Frullania tolimana 366, 367, 370
Frullania usamiensis 117, 162
Frullania vethii 363
Frullania wallichiana 364, 366, 367, 368, 369
Frullania yunnanensis 167
Frullaniaceae 485, 499
Frullanoides densifolia 90, 101, 102, 125, 131, 133, 134, 176, 199, 200, 231, 254, 372, 377, 500
Frullanolide 166
(+)-Frullanolide 116, 161, 470, 471, 473, 481
(−)-Frullanolide 116, 161, 162, 166, 167, 470, 499
(±)-Frullanolide 162
(+)-β-Frullanolide   117,   162,   163, 166
Fucose 387
Fucosterol 516, 518, 519
Funaria hygrometrica 417, 418, 428, 434, 438, 443, 451, 456, 465, 478, 482
Furanodiene 122, 173, 174
Furanoeudesma-1,3-diene 116, 160, 161
Furanogermacra-1(10),4-diene 122, 173, 174
Furanopinguisanol 131, 191, 197
Furanoplagiochilal 38, 41, 469
Fusicoccadiene 226, 227, 251
Fusicoccanes 225, 226, 227, 228, 514
Fusicogigantepoxide 226, 227, 251
Fusicogigantone A 226, 227, 251
Fusicogigantone B 226, 227, 251
Fusicoplagin A 227, 228, 252
Fusicoplagin B 227, 228, 252
Fusicoplagin C 227, 228, 252
Fusicoplagin D 227, 228, 252
Fusicorrugatol 226, 227, 252

Gackstroemia magellanica 19, 89, 96, 101, 121, 134, 171, 203, 281, 285, 288
D-Galactose 387
Gallic acid 403, 404, 407, 476
Gentiana scabra var. *orientalis* 468
Gentiana sp. 409
Gentianaceae 409
Geocalyx graveolens 465
Geosmin 22, 24, 25
Geranial 13, 16, 214
Geraniol 13, 16
Geranyl acetate 13, 14, 16, 296
Geranylacetone 56
Geranylbibenzyl 521
2-Geranyl-3,5-dihydroxybibenzyl 303, 325, 327, 328, 333
2-Geranyl-3,5-dimethoxybibenzyl 327
Geranyl geraniol 267, 393, 456, 514
Geranyl geranyl pyrophosphate 225, 229, 272, 274, 518
Geranyl linalool 513
2-Geranyl-3,5,4′-trihydroxybibenzyl 303, 327, 328
Germacra-1(10),5-dien-4,11-diol 122, 172, 173
ent-1(10)*E*,5*E*-Germacradien-11-ol 122, 172, 173
Germacranes 171, 172, 174, 513
Germacranolides 162, 175, 176
Germacra-12,6α-olide 498
Germacra-12,8α-olide 498
ent-Germacra-4(15),5,10(14)-trien-1α-ol 121, 171, 172
Germacra-4(15),5,10(14)-trien-1β-ol 172
ent-Germacra-4(15),5,10(14)-trien-1β-ol 121, 171, 172
Germacra-4(15),5,10(14)-trien-1β-yl acetate 171
Germacrene-B 121, 171, 172
Germacrene-D 121, 171, 172
Germacrone 160
Ginkgo biloba 351, 425, 470
ent-Globulol 33, 34, 75
(+)-*ent*-Globulol 33, 35
C-10-*epi*-Globulol 33
6β-β-Glucopyranosyl-*ent*-15α,20-dihydroxykaur-16-ene 235, 256
20-β-Glucopyranosyl-*ent*-6-keto-15α-hydroxykaur-16-ene 235, 256
β-D-Glucose 321

Glucose 374
(+)-Glucose 391
(+)-D-Glucose 388
3'-O-Glucosylated lucenin-2 497
Glycosides 375
C-Glycosylflavones 362
Gomeraldehyde 237, 238, 257
epi-Gomeraldehyde 237, 239, 257
Gongylanthus ericetorum 61, 97, 490
Gongylanthus sp. 323, 490
Gongylantoxide 62, 97
Gorgonanes 176
Grangea maderaspatana 162
Grignard reaction 467
Grimaldiaceae 61, 343, 502
Grimaldone 109, 143, 144, 468
Grimmia pilifera 438, 443
Grimmiaceae 423, 452, 471, 507, 508
Growth inhibitory activity 474, 477
Guaianes 176, 177, 178
Guaiazulene 43, 46, 47, 79
(1S,10R)-Guai-4,6-diene 124, 176, 177
Guai-4,11(12)-diene 124, 176, 177
α-Guaiene 176, 177
β-Guaiene 124, 176, 177
(1R*,5S*,7S*,10R*)-Guai-3-en-6-one-1-
    ol 124, 177
Guai-4(15)-en-6-one-1-ol 176
(1R*,5S*,7S*,10R*)-Guai-4(15)-en-6-one-1-
    ol 124, 177
Guai-1(10),3,11(13)-trien-14,2β,12,6α-
    diolide 125, 177
α-Gurjunene 33, 34, 68, 124
(−)-α-Gurjunene 33
β-Gurjunene 33, 34, 69
γ-Gurjunene 124, 176, 177
Gymnocolea inflata 26, 66, 68, 96, 102, 104,
    105, 120, 121, 171, 216, 249, 469, 475,
    489
Gymnocolea sp. 140, 171
Gymnocolin 216, 217, 219, 249, 469, 475
Gymnodinium nagasakiense 511
Gymnomitranes 47, 48
(8R)-(+)-Gymnomitran-9-one 48, 49, 83
(−)-Gymnomitr-8(12)-en-15-al 48, 49, 83
Gymnomitrene 47, 491
(−)-Gymnomitr-8(12)-en-15-oic acid 48,
    49, 83
(+)-Gymnomitr-8(12)-en-9-one 48, 49, 83
(+)-Gymnomitr-8(12)-en-9α-ol 48, 83

(−)-Gymnomitr-8(12)-en-15-ol 48, 49, 83
Gymnomitriaceae 490
Gymnomitrion concinnatum 58, 96, 107, 490
Gymnomitrion obtusum 490, 491
Gymnomitrion sp. 490
Gymnomitrol 47, 48, 49, 83
(±)-Gymnomitrol 47
Gyrothyra underwoodiana 265, 277

Halichondria panicea 272, 278
Hamatilobene A 242, 244, 259
Hamatilobene B 244, 245, 260
Hamatilobene C 244, 245, 260
Hamatilobene D 244, 245, 260
Hamatilobene E 244, 245, 260
Hanegokedial 37, 77
ent-Hanegokedial 38
(+)-ent-Hanegokedial 40
Haplocladium catillatum 465
Haplomitrenolide A 237, 239, 257
Haplomitrenolide B 237, 239, 258
Haplomitrenolide C 237, 239, 258
Haplomitrenone 246, 262, 267
Haplomitriaceae 488
Haplomitrium mnioides 237, 246, 257, 258,
    262, 488
Haplomitrium sp. 488, 518
Hardwickia pinnata 212
Hedwigia ciliata 433, 438, 443
Helichrysum ambiguum subsp.
    ambiguum 46
Helichrysum umbraculigerum 328
Helicobasidium mompa 146
Hematemesis 465
Hematostasis 465
Hemolytic activity 480
Hemostasis 465
Hepaticae 5, 6, 12, 13, 14, 16, 23, 24, 26, 28,
    31, 32, 33, 34, 36, 37, 39, 43, 48, 50, 52,
    53, 54, 55, 57, 58, 59, 60, 62, 63, 64, 65,
    66, 136, 137, 140, 142, 144, 150, 153, 154,
    156, 157, 158, 161, 163, 164, 165, 171,
    172, 174, 176, 177, 178, 179, 180, 182,
    185, 187, 188, 190, 191, 194, 198, 199,
    203, 205, 206, 207, 208, 210, 211, 212,
    213, 215, 216, 217, 218, 220, 221, 227,
    228, 230, 232, 233, 234, 235, 238, 239,
    242, 243, 244, 245, 247, 267, 268, 270,
    271, 273, 274, 275, 279, 280, 293, 294,
    295, 297, 298, 320, 324, 326, 327, 329,

331, 334, 335, 336, 342, 344, 346, 350, 352, 353, 354, 356, 357, 359, 360, 361, 362, 374, 376, 377, 379, 383, 384, 385, 391, 403, 408, 422, 423, 428, 455, 464, 465, 469, 471, 476, 483, 484, 485, 486, 492, 496, 499, 509, 510, 511, 512, 513, 514, 515, 516, 517, 519, 520, 521, 522, 523, 524

Heptadecatrienylresorcinol 517

5-Heptadecatrienylresorcinol 521

5-Heptadeca-8Z, 11Z, 14Z-trienylresorcinol 351, 521

5-Heptadeca-8Z,11Z,14Z-trienyl-resorcinol monomethylether 314, 352

3-Heptadecenylphenol 313, 351, 352

3,5,7,4′,3″,5″,7″-Heptahydroxy-3′-O-4‴-biflavone 417, 427

n-Heptanal 381, 385

2-Heptanone 433

2-Heptenone 447

Herbertaceae 143, 488, 495, 496

Herbertanes 140, 150

Herbertene 109, 144, 147, 148, 149

(±)-Herbertene 147

(−)-Herbertene 143, 147

(R)-(+)-Herbertene 149

Herbertenediol 145, 146, 147, 149, 150, 151, 479

(−)-Herbertenediol 110, 143, 144, 145, 147, 151

α-Herbertenol 110, 144, 145, 146, 147, 474

(−)-α-Herbertenol 143, 144, 145

β-Herbertenol 110, 147, 474

(−)-β-Herbertenol 143, 144, 146

(−)-Herbertenolide 111, 143, 144

Herbertineae 493, 496

Herbertus acanthelius 110, 146

Herbertus aduncus 105, 107, 108, 109, 110, 111, 140, 143, 146, 474

Herbertus divergens 90, 104, 105, 107, 110, 146

Herbertus sakuraii 107, 109, 110, 111, 143, 146

Herbertus sp. 140, 143, 362, 484, 495, 496

Herbertus subdentatus 105, 107, 109, 110, 140, 146

Hercynolactone 111, 152, 153, 272

Herzogiella selogeri 438, 443

Herzogiella sp. 455

Heterobryoflavone 416, 423, 424

Heteroscyphol 219, 249

Heteroscyphone A 218, 219, 250

Heteroscyphone B 218, 219, 250

Heteroscyphone C 218, 219, 250

Heteroscyphone D 218, 219, 250

Heteroscyphus aselliformis 69, 72, 102, 105

Heteroscyphus bescherellei 214, 248, 492

Heteroscyphus planus 19, 39, 58, 60, 64, 66, 76, 78, 86, 95, 96, 97, 98, 100, 101, 102, 104, 105, 135, 140, 204, 219, 249, 250, 265, 266, 278, 492

Heteroscyphus sp. 491, 492

Hexamethylphosphoric acid 145

n-Hexanal 381, 385, 432, 447

n-Hexane 42

2-Hexanol 432, 446

n-Hexanol 381, 385

Hexyl acetate 433, 447

Himachalanes 177, 178

α-Himachalene 125, 177, 178

β-Himachalene 125, 177, 178

γ-Himachalene 177, 178

Homalothecium lutescens 417, 428

Homalothecium sericeum 438, 443

Homomonoterpenoids 15, 16, 23

Homoverrucosan-5β-ol 266, 279

(+)-Homoverrucosan-5β-ol 278

13-epi-Homoverrucosan-5β-ol 265, 279

Hopanes 514

Hop-6α,22-diol 292

Hop-12β,22-diol 292, 295

Hop-22,23-diol 292, 295

Hop-22(29)-ene 291, 293, 295, 392, 395, 403

Hop-22-ol 291, 293, 295

Hop-6α,11α,22-triol 292, 295

Houttuynia cordata 465

Humulanes 177, 179

Humulene 31, 60, 61

α-Humulene 61, 125, 178, 179

Humulene epoxide 178

Humulenyl acetate 126, 178, 179

Hydrangea macrophylla var. otaksa 322

Hydrangea sp. 321

Hydrangenol 320, 321, 322

Hydrangenol-8-O-β-glucoside 320, 322, 323

1α-Hydroperoxy-4α,5β-epoxygermacra-10(14),11(13)-dien-12,8α-olide 123, 173, 174

1β-Hydroperoxy-4α,5β-epoxygermacra-
    10(14),11(13)-dien-12,8α-olide 123, 173,
    174
5α-Hydroperoxylepidozenolide 62, 98
3α-Hydroxy-5α-acetoxyafrican-2(6)-en-4-
    one 27, 28, 66
(3S)-Hydroxy-(7R)-
    acetoxymanoyloxide 245, 260
Hydroxyalkanedioic acid 433
Hydroxyalkanoic acid 433
5'-Hydroxyamentoflavone 415, 424, 425
6-Hydroxyapigenin 365, 374, 375, 376
6-Hydroxyapigenin-7-O-diglucoside 365,
    376
ent-8β-Hydroxyaristolene 32, 67
m-Hydroxybenzaldehyde 317, 340, 358,
    359
p-Hydroxybenzaldehyde 317, 358, 359,
    403, 404
p-Hydroxybenzoic acid 403, 404, 407, 408
3-Hydroxybibenzyl 300, 323, 324
3α-Hydroxybicyclogermacrene 54, 55, 94
5-Hydroxycalamenene 61, 64
(1S,4S)-(−)-5-Hydroxycalamenene 99
7-Hydroxycalamenene 61, 63, 64
(1R,4R)-7-Hydroxycalamenene 63
(1S,4R)-7-Hydroxycalamenene 64, 65, 100
(1S,4S)-7-Hydroxycalamenene 63, 100
8-Hydroxycalamenene 61, 63, 64, 100
p-Hydroxy-β-(carboxymethyl)-cinnamic
    acid 407
o-Hydroxycinnamate 432
2-Hydroxycuparene 107, 140, 142, 143,
    144, 146
3-Hydroxycuparene 107, 140, 142
7α-Hydroxydeoxopinguisone 131, 191, 197
4-Hydroxydeoxopinguisone-12,15-
    dimethyl ester 132, 194
4-Hydroxy-3,5-dimethoxyallylbenzene
    317, 359
3-Hydroxy-4,5-dimethoxyallylbenzene
    317, 359
2-Hydroxy-3,4'-dimethoxybibenzyl 301,
    324
3-Hydroxy-4,3'-dimethoxybibenzyl 301,
    323, 324
8-Hydroxy-6,7-dimethoxy-3-methyl-
    isocoumarin 314, 352, 353
2-Hydroxy-3,6-dimethoxy-
    phenanthrene 316, 355, 356

2-Hydroxy-3,7-dimethoxy-
    phenanthrene 316, 356
3-Hydroxy-4,5-
    dimethylmethoxybibenzyl 330
ent-7α-Hydroxydiplophyllolide 119, 165,
    168
18-Hydroxy-4,8-dolabelladiene 221, 251
18-Hydroxydolabell-7E-en-3-one 220, 222,
    250
5-Hydroxy-6,8,11,14-eicosatetraenoic
    acid 479
ent-8β-Hydroxyeudesm-3,11-diene 116,
    160, 161
6β-Hydroxyeudesm-3-ene 116, 160, 161
5α-Hydroxyeudesm-4(15),7(11),8(9)-trien-
    12,8-olide 118, 164
4β-Hydroxygermacra-1(10),5-diene 122,
    172, 173
(+)-8β-Hydroxygymnomitran-9-one 48,
    49, 83
9α-Hydroxygymnomitryl acetate 47, 48,
    83
9α-Hydroxygymnomitryl cinnamate 47,
    48, 83
7-Hydroxyhexadecanedioic acid 433
16-Hydroxyhexadecanoic acid 433
8-Hydroxy-9-hydroperrottetianal 263,
    271, 474
5-Hydroxyisobicyclohumulenone 126, 178,
    179
5-Hydroxyisocuparene 143
6α-Hydroxyisodrimeninol 112, 152, 154
1β-Hydroxyisosacculatal 262, 270
ent-16β-Hydroxykaurane 231, 232, 254,
    391, 393, 494
ent-16β-Hydroxykauran-3-one 231, 232,
    254
(16R)-ent-11α-Hydroxykauran-15-one 229,
    230, 253
(16S)-ent-11α-Hydroxykauran-15-one 231,
    232, 254
(16R)-ent-18-Hydroxykauran-15-one 229,
    230
ent-14α-Hydroxykaurene 235, 257
15-Hydroxykaurene 229
ent-15α-Hydroxykaurene 230, 236, 253
ent-3β-Hydroxykauren-15-one 231, 232,
    254
ent-11α-Hydroxykauren-15-one 229, 230,
    253

ent-18-Hydroxykauren-15-one 229, 230, 253

ent-15α-Hydroxykauren-3β-yl acetate 231, 232, 254

5-Hydroxylunularic acid 476

6-Hydroxyluteolin 370, 375, 377

6-Hydroxyluteolin-7-O-acylglucoside 375

6-Hydroxyluteolin-6-O-diglucoside 371, 377

6-Hydroxyluteolin-7,3'-dimethyl ether 370, 377

6-Hydroxyluteolin-7-O-glucoside 370, 377, 411

6-Hydroxyluteolin-7-O-glucoside-3''-hydroxy-3-methylglutarate 371, 377

6-Hydroxyluteolin-7-O-glucoside-4''-hydroxy-3-methylglutarate 371, 377

6-Hydroxyluteolin-7-O-glucoside-6''-hydroxy-3-methylglutarate 371, 377

6-Hydroxyluteolin-7-O-glucoside-6''-malonate 409, 412, 420

6-Hydroxyluteolin-6-O-glucoside-7-O-xyloglucoside 371, 377

6-Hydroxyluteolin-3'-methylether 371, 375, 377

6-Hydroxyluteolin-7-O-rhamnoside 371, 377

6-Hydroxyluteolin-7-O-sophoroside 370, 375, 377

6-Hydroxyluteolin-6-O-xyloside-7-O-glucoside 370, 377

3-Hydroxy-4-methoxybibenzyl 300, 323, 324

3-Hydroxy-4'-methoxybibenzyl 301, 323, 324

2-Hydroxy-7-methoxydeoxopinguisone 132, 194

3-Hydroxy-5-methoxy-2-(3-hydroxy-3-methylbutyl)bibenzyl 304, 329

3-Hydroxy-5-methoxy-2-(3-methyl-2-butenyl)bibenzyl 304, 327

3-Hydroxy-5-methoxy-4-(3-methyl-2-butenyl)bibenzyl 302, 325, 326

2-Hydroxy-11ζ-methoxypinguis-5(10),6-diene 132, 194

3-Hydroxy-5-methoxyprenylbibenzyl 328

3-Hydroxy-5-methoxystilbene 328, 330

3-Hydroxy-4-(3-methyl-2-butenyl)bibenzyl 302, 325, 326

3-Hydroxy-4-(3-methyl-2-butenyl)-4'-methoxybibenzyl 303, 325, 327

4-Hydroxy-4-methyl-cyclohex-2-en-1-one 22, 23

3-Hydroxy-4,5-methylene-dioxybibenzyl 300, 323, 324

4-Hydroxymethyl-1-methoxycarbonylazulene 43, 79

1-Hydroxy-1-(2-methyl-1-propenyl)-cyclohexane 468

13-Hydroxy-9,11,15-octadecatrien-6-ynoic acid 448, 453, 454, 475

10-Hydroxy-8*E*-octadecen-6-ynoic acid 383, 387

10-Hydroxyoctadec-6-yn-8*E*-enoic acid 382

ent-(5*R*,6*S*,9*R*)-4α-Hydroxyoppositan-10-one 103, 136, 137

2β-Hydroxy-9-oxoverrucosane 265, 275

14-Hydroxyperrottetin E 312, 346, 347

14'-Hydroxyperrottetin E 312, 346, 347

14-Hydroxyperrottetin E-11'-methyl ether 312, 345, 346, 347

14-Hydroxyperrottetin F 347

6-Hydroxy-4-(2-phenylethyl)-benzofuran 305, 329

13²-Hydroxy-(13²-*R*)-pheophytin a 451, 459, 460, 473

13²-Hydroxy-(13²-*S*)-pheophytin a 451, 459, 460

13²-Hydroxy-(13²-*R*)-pheophytin b 451, 459, 460

13²-Hydroxy-(13²-*S*)-pheophytin b 451, 459, 460

2-Hydroxypinguisenene 132, 194

4'-Hydroxyradulanin H 307, 332, 333, 334

5'-Hydroxyrobustaflavone 415, 424, 425

11α-Hydroxysacculatanolide 263, 270

1β-Hydroxysacculatanolide 262, 270

2-Hydroxy-3,4,7-trimethoxy-9,10-dihydrophenanthrene 315, 355, 356

5-Hydroxy-7,8,4'-trimethoxyflavone 366, 376, 380

7α-Hydroxyvalenc-1(10)-ene 135, 204, 206

2β-Hydroxyverrucosane 265, 275

Hygrohypnum luridum 393

Hygrohypnum smithii 398, 399, 400, 401

Hygrohypnum sp. 456

Hylocomiaceae 423, 425

Hylocomium brevirostre var. *cavifolium* 439, 443

Hylocomium sp. 456
Hylocomium splendens 415, 423, 433, 439
Hylocomium splendens var. *alaskanum* 393,
    439, 443
Hymenophytaceae 378, 484
Hymenophyton flavellatum 378
Hymenophyton leptopodum 378
Hypnaceae 471
Hypnineae 455
Hypnobryales 409, 422
Hypnogenol A 417, 425, 427
Hypnogenol B 417, 425, 427
Hypnum cupressiforme 414, 417, 422, 426,
    439, 443, 477
Hypnum cupressiforme subsp.
    imponens 395, 396, 397
Hypnum fujiyamae 439, 443
Hypnum lindbergii 439, 443
Hypnum oldhamii 439, 443
Hypnum plumaeforme 391, 396, 439,
    443, 482
Hypnum sp. 392
Hypolaetin 420, 421
Hypolaetin-7-O-glucoside 409, 413, 420

Illudanes 513
Indenes 42, 43
Indole 403, 406, 408
Indole acetic acid 318, 360, 361, 434, 456,
    478
Indole-3-acetic acid 478
Indole 3-acetonitrile 360, 478
Indole derivatives 517
Inflatene 24
Inflatenone 26, 66
Infuscaic acid 213, 247, 479
Infuscaside A 235, 236, 255, 469, 479
Infuscaside B 235, 236, 256, 469, 479
Infuscaside C 235, 236, 256, 469
Infuscaside D 235, 236, 256, 469
Infuscaside E 235, 236, 256, 469
Inhibitory activity 337, 471, 477
Insect moulting activity 518
Intermediol 160
Inversin 314, 352, 353
(±)-Ionone 183
β-Ionone 22, 24, 25
Isoabienol 239, 257
(+)-Isoabienol 237
Isoaffinetin 373, 379

Isoafricanol 28, 30, 67
ent-Isoalantolactone 118, 165, 168
Isoalbicanal 111, 153, 154
Isobazzanene 49, 50, 85
Isobicyclogermacranes 54, 55
Isobicyclogermacrenal 56, 477
(−)-Isobicyclogermacrenal 54, 55, 57, 94
(±)-Isobicyclogermacrenal 57
Isobicyclogermacrene 54, 56
(+)-Isobicyclogermacrenol 54
Isocarlinoside 372, 377, 380, 497
Isochiloscyphone 136, 137
Isoclavukerin A 24
Isocoumarins 352, 353
Isocuparanes 140, 144, 150
Isocuparene 109, 143
Isocuparene-3,4-diol 111, 144, 147,
    150, 479
Isocyclobazzanene 50
Isodolabradiene 268
Isodrimeninol 112, 152, 154
Isoflavones 362, 409, 422
Isoflavonoids 422
Isofucosterol 516
Isofurcatain-7-O-β-D-glucopyranoside 378
Isofurcatain-7-O-glucoside 365, 376
Iso-α-gurjunene 176
Iso-α-gurjunene B 124, 176, 177
Isolepidozene 56
Isolongifolene 126, 179, 180
Isomarchantin C 310, 341, 342
Isonaviculol 131, 191, 199
Isoorientin 409, 413, 420
Isoorientin-7-O-glucoside 372, 375, 377
N^6-(Δ^2-Isopentenyl)adenine 451, 460, 482
Isoperrottetin A 313, 346, 347
Isopinguisanin 192, 193
Isopinguisanolide 134, 193, 198
Isopinocamphone 14, 21
Isoplagiochin A 311, 344, 345
Isoplagiochin B 311, 344, 345
Isoplagiochin C 311, 344, 345
Isoplagiochin D 311, 344, 345, 347
Isopolygodial 112, 152, 154, 481
Isoporelladiolide 125, 176, 177
Isoprenyl hydroquinones 517
Isoprenyl quinones 517
2(R)-Isopropenyl-6-hydroxy-4-(2-
    phenylethyl)dihydrobenzofuran 306,
    330, 331

(±)-Isoptychanolide 201
Isoriccardin C 308, 336, 341
Isosacculatal 262, 269, 270, 481, 487
Isoscoparin-7-O-glucoside 409, 413, 420
Isoscutellarein 420, 421
Isoscutellarein-7-O-glucoside 409, 413, 420
Isoshaftoside 412, 420
Isotachidaceae 488, 493, 495
Isotachin A 384, 390
Isotachin B 384, 390
Isotachin C 384, 390
Isotachioside 318, 360
Isotachis haematodes 69, 72, 90, 261, 281,
    285, 296, 298, 495
Isotachis humectata 19, 21, 72, 90, 96, 114,
    157, 252, 281, 285, 296, 298, 299, 495
Isotachis japonica 287, 294, 296, 318, 361,
    384, 390, 493, 495, 520, 521
Isotachis sp. 492, 518
Isothecium sp. 460
Isothecium stoloniferum 469, 473
Isothecium subdiversiforme 450, 457, 472
Isotomentellin 296, 297, 298
Isovaleric acid 433, 446
Isoverticillene 266, 279, 280
ent-Isoverticillenol 266, 279, 280
Ixiolaena leptolepis 46

Jackiella javanica 72, 90, 121, 171, 266, 278,
    279, 315, 355, 493
Jackiella sp. 493
Jamesoniella autumnalis 218, 219, 231, 243,
    249, 250, 253, 257, 259, 468, 489
Jamesoniella colorata 66, 81, 90
Jamesoniella sp. 489
Jamesoniellide A 218, 219, 250
Jamesoniellide B 218, 219, 250
Jamesonielloideae 489
Jensenia connivens 378
Jensenia erythropus 81, 90, 120, 252
Jones oxidation 153, 229
Jones reagent 182, 220, 221
Jubula japonica 71, 81, 124, 126, 176, 178,
    312, 347
Julaceal 58, 59, 96
Junceic acid 215
(−)-Junceic acid 214
ent-Junceic acid 215, 248, 492
Jungermannia autumnalis 490
Jungermannia colorata 71

Jungermannia comata 311, 345, 488
Jungermannia exsertifolia
    subsp. *cordifolia* 264, 273, 281, 285,
    288, 488
Jungermannia infusca 140, 212, 214, 231,
    236, 237, 247, 248, 253, 254, 255, 256,
    257, 287, 294, 387, 388, 389, 469, 479,
    488, 490
Jungermannia obovata 15, 22
Jungermannia paroica 214, 247, 248
Jungermannia rosulans 109, 141
Jungermannia sp. 212, 229, 231, 465, 467,
    488, 489, 490
Jungermannia subulata 252, 261, 285, 297,
    299
Jungermannia thermarum 262, 267
Jungermannia torticalyx 387, 388, 389
Jungermannia truncata 231, 253, 488, 490
Jungermannia vulcanicola 12, 18, 57, 95,
    231, 254, 488
Jungermanniaceae 188, 217, 488, 489,
    490, 494
Jungermanniales 22, 26, 33, 51, 54, 60, 65,
    135, 157, 173, 178, 182, 190, 229, 236,
    237, 266, 268, 271, 275, 292, 297, 319,
    323, 338, 341, 347, 351, 362, 373, 374,
    375, 378, 389, 390, 471, 484, 485, 487,
    488, 493, 495, 496, 497, 520, 521, 522,
    524, 525
Jungermanniidae 47, 484
Jungermanniineae 492, 497
Jungermannioideae 490
Jungermanool 244
Juniperus squamata 65

Kaempferol 378, 379
Kaempferol-3,4'-di-O-glucoside 413, 421
Kaempferol-3-O-galactoside 413, 421
Kaempferol-3-O-galactoside-4'-
    O-glucoside 409, 413, 421
Kaempferol-6-C-glucoside 372, 378, 379
Kaempferol-3-O-glucoside 413, 421
Kaempferol-3-O-glucoside-6''-
    malonate 413, 421
Kaempferol-6-C-glucoside-3-O-
    glucoside 372, 378, 379
Kaempferol-3-methyl ether 372, 377, 379,
    411
Kaempferol-3-O-neohesperidoside 413,
    421

Kaempferol-3-O-rhamnosyl-
  glucoside 413, 421
ent-Kauran-3β,16β-diol 231, 232, 254
Kauranes 229, 230, 232, 233, 234, 235, 514
(16R)-*ent*-Kauran-15-one 229, 232, 254
ent-Kauren-3β,15α-diol 231, 232, 254
ent-Kaur-16-en-7α,15β-diol 229, 232, 253
Kaurene 229
ent-Kaurene 230, 252
(14R)-*ent*-Kaur-16-en-15β-hydrogen
  malonate 233, 255
ent-Kauren-18-oic acid 229, 230, 253
Kauren-15-one 229
ent-Kauren-15-one 230, 253
ent-Kauren-15-one-18-oic acid 229, 230,
  253
ent-Kaur-16-en-15β-yl *ent*-labdan-8(17),
  13-dien-15-yl malonate 234, 236, 255
(14R)-*ent*-Kaur-16-en-14-yl epi-bornyl
  malonate 233, 255
ent-Kaur-16-en-15β-yl epi-bornyl malonate 234, 255
(14R)-*ent*-Kaur-16-en-14-yl fenchyl
  malonate 233, 255
(14R)-*ent*-Kaur-16-en-14-yl-hydrogen
  malonate 232, 254
(14R)-*ent*-Kaur-16-en-14-yl malonate 232
Kaurenyl malonates 229
(14R)-*ent*-Kaur-16-en-14-yl phytyl
  malonate 233, 255
ent-Kaur-16-en-15β-yl phytyl
  malonate 234, 236, 255
7-Keto-8-carbomethoxypinguisenol 131,
  191, 197
5-Keto-7α,8β,11α-H-germacra-1(10)-en-
  12,8α-olide 123, 173, 174
7-Ketoisodrimenin 112, 152, 156
7-Ketoisodrimenin-5-ene 112, 152, 156
2-Keto-norpinguisone methyl ester 196
10-Keto-8E-octadecen-6-ynoic acid 382,
  383, 387
Kiaeria starkei 436, 443
(−)-Kolavelool 212, 213, 214, 237,
  244, 247
Kolavenic acid 212, 213

Labda-12,14-dien-7,8-diol 238, 244, 257
Labda-12E,14-dien-8α,11ζ-diol 238, 240,
  257
Labda-7,13E-dien-15-ol 238, 244, 257

ent-Labda-7,13E-dien-15-ol 237
Labda-7,14-dien-13-ol 238, 257
(+)-Labda-7,14-dien-13-ol 237
Labda-12,14-dien-8α-ol 237, 238, 257
ent-Labda-8(17),13-dien-15-yl hydrogen
  malonate 245, 260
Labdanes 236, 238, 239, 242, 243, 245, 514
ent-Labda-8(17),12E,14-trien-3β-ol 242,
  259
Labiatae 243
Lactone dimer A 118
Lactone dimer B 118
Lamiaceae 464
Laurencia sp. 522
Laurencia subopposita 138
(−)-Ledene 33
ent-(−)-Ledene 34, 70
Leguminosae 422
Lejeunea albescens 72, 80, 81, 84, 86, 87, 90,
  121, 282, 285
Lejeunea cavifolia 372, 377
Lejeunea discrenata 288
Lejeunea discreta 71, 72, 90, 121, 129, 134,
  282, 285
Lejeunea flava 112, 152
Lejeunea glaucescens 16, 72, 282, 285, 288
Lejeunea lumbricoides 66, 68, 72, 86, 121,
  128
Lejeunea sp. 81, 84, 86, 90, 171, 178, 183,
  192, 282, 285, 288, 465, 467, 501
Lejeuneaceae 47, 182, 183, 190, 375, 377,
  491, 495, 499, 500
Lejeuneoideae 500, 501
Lepicoleaceae 143, 495
Lepidium sativum 477
Lepidolejeunea ornata 16, 21, 72, 252, 285
Lepidozanes 54, 57
Lepidozenal 57, 95, 477
(−)-Lepidozenal 57
Lepidozene 56
Lepidozenol 57
(4S*,5S*,6R*,7R*)-1(10)E-Lepidozen-5-
  ol 56, 57, 95
Lepidozenolide 62, 97
(−)-Lepidozenolide 61
Lepidozia borneensis 84, 87, 102, 105
Lepidozia concinna 108, 140
Lepidozia fauriana 61, 97, 98, 116, 160
Lepidozia reptans 81, 90, 105, 115, 160, 282,
  285, 288, 477

Lepidozia sp. 494, 518
Lepidozia vitrea 54, 57, 61, 94, 95, 98, 115, 116, 128, 135, 159, 160, 182, 205, 303, 328, 477
Lepidoziaceae 152, 188, 494
Leptobryum pyriforme 439, 443, 456
Leptobryum sp. 456
Leptodictyum riparium 465
Leptolejeunea elliptica 465, 467, 500
Leptoscyphus liebmanianus 69, 72, 81, 90, 120, 290
Leptoscyphus sp. 491
Lethocolea glossophylla 72, 90
Leucobryaceae 452, 453, 508, 509
Leucobryum candidum 457
Leucobryum glaucum 392, 397, 439, 443, 453, 477
Leucobryum neilgherrense 439, 443
Leucobryum scabrum 439, 443, 447, 448, 454
Leucobryum sp. 482
Leucodontaceae 423
Leucolejeunea aff. *decurrens* 71, 102, 128, 129, 282, 285, 289, 500
Leucolejeunea sp. 323
Leucolejeunea xanthocarpa 204, 500
Leucolepis menziesii 469
Levierol 245, 260
(+)-Licarin A 355
(−)-Licarin A 315, 355, 356
Lignans 461, 517
Lignin 457
Limonene 12, 13, 15, 16, 504
Linalool 13, 16, 246, 467
R-Linalool 183
Linalyl acetate 13, 16
Linguifolide 217, 249
Linoleic acid 294, 383, 433, 452, 456
α-Linoleic acid 383
γ,γ-Linoleic acid 383
Linolenic acid 294, 381, 383, 433, 452, 455, 456
5-Lipoxygenase inhibitory activity 479
Lippia integrifolia 30
Lithium diisopropylamide 226
Longibornanes 179, 180
Longiborneol 126, 136, 179, 180, 492
ent-Longiborneol 179
Longicyclanes 179, 180
Longicyclene 179, 180

ent-Longicyclene 179
Longifolanes 179, 180
Longifolene 126, 179, 180, 492
ent-Longifolene 179
ent-Longipinan-3,12-dione 468
(−)-*ent*-Longipinane 181
Longipinanes 179, 180
Longipinanol 126, 179, 180, 492
α-Longipinene 126, 180
ent-α-Longipinene 179
β-Longipinene 126, 180
ent-β-Longipinene 179
Longipinenes 181
Lophocolea bidentata 15, 18, 19, 20, 21, 22, 24, 25, 60, 61, 69, 72, 75, 80, 81, 87, 96, 97, 102, 104, 105, 113, 119, 121, 168, 171, 261, 317, 358, 380, 381, 385, 386
Lophocolea coadunata 81, 87, 99, 104, 118, 119, 168, 285
Lophocolea heterophylla 15, 16, 17, 18, 19, 20, 21, 22, 25, 58, 60, 61, 64, 69, 72, 75, 80, 81, 87, 95, 96, 97, 98, 99, 100, 102, 104, 105, 113, 115, 116, 118, 120, 121, 122, 125, 160, 168, 171, 173, 178, 297, 299, 317, 358, 380, 381, 385, 386, 465, 466, 491, 510
Lophocolea minor 465
Lophocolea sp. 140, 171, 491, 492
Lophocoleaceae 491, 492
Lopholejeunea eulopa 68, 99, 101, 105
Lopholejeunea howei 17, 99, 105, 282, 285
Lopholejeunea subfusca 68, 69, 72, 99, 121, 285
Lophozia ventricosa 22, 24, 87, 90, 101, 102, 105, 108, 115, 116, 124, 125, 160, 176, 178, 218, 249, 282, 285, 289, 291, 295, 489
Lophozia vicrenata 465
Lophoziaceae 223, 226, 469, 489, 490
Lophozioideae 489
Lucenin-2 369, 377, 378, 380, 411, 497
Lucenin-2 3′-glucoside 372, 377, 380
Lung carcinoma (A-549) 472
Lunularia cruciata 312, 313, 345, 347, 473
Lunularia sp. 505
Lunulariaceae 505
Lunularic acid 297, 299, 320, 321, 322, 323, 337, 350, 474, 475, 476, 478, 480, 509, 520
Lunularic acid methyl ester 476

600 Subject Index

Lunularin 299, 320, 325, 337, 350, 520
Lupanes 515
Lupeol 392, 397, 402, 403
Lutein 399, 403
Luteolin 362, 366, 375, 376, 380, 409, 410, 422, 423, 424, 425
Luteolin-6-C-arabinoside-8-C-glucoside 372, 377
Luteolin-6,8-di-C-arabinoside 370, 377, 380
Luteolin-6,8-di-C-glucoside 369, 374, 377, 411
Luteolin-7,3'-di-O-glucoside 368, 376
Luteolin-6,8-di-C-glucoside-3'-glucoside 372, 377
Luteolin-7,3'-di-O-glucuronide 409, 412, 420, 505
Luteolin-7,4'-di-O-glucuronide 368, 376
Luteolin-3',4'-dimethyl ether 367, 375, 376
Luteolin-6,7-dimethyl ether 370, 377
Luteolin-7,3'-dimethyl ether 367, 376
Luteolin-7-O-gentiobiose 368, 376
Luteolin-6-C-β-D-glucopyranoside-8-C-α-L-rhamnoside 409
Luteolin-3'-O-glucoside 368, 376
Luteolin-4'-O-glucoside 368, 376
Luteolin-6-C-glucoside 413, 420
Luteolin-6-β-D-glucoside 409
Luteolin-7-O-glucoside 367, 375, 376, 411
Luteolin-8-C-glucoside 413, 420
Luteolin-8-C-β-D-glucoside 409
Luteolin-6-C-glucoside-8-C-arabinoside 372, 377
Luteolin-7-O-glucoside-6''-hydroxy-3-methylglutarate 368, 376
Luteolin-7-O-glucoside-6''-malonate 368, 376, 409, 411
Luteolin-6-C-glucoside-8-C-rhamnoside 412, 420
Luteolin-3'-O-glucuronide 505
Luteolin-4'-O-glucuronide 368, 376, 380
Luteolin-7-O-glucuronide 368, 376, 409, 411, 505
Luteolinidin-5-O-diglucoside 418, 428, 430
Luteolinidin-5-O-glucoside 418, 428, 430
Luteolin-7-O-neohesperidoside 412, 420
Luteolin-7-O-neohesperidoside-6''-malonate 412, 420
Luteolin-7-O-neohesperidoside-6''-malonyl ester 409
Luteolin-4'-O-rhamnosyl-β-D-galacturonide 368, 376

Luteolin-6,3,4'-trimethyl ether 370, 377
Luteolin-7,3',4'-trimethyl ether 367, 375, 376
Lutonarin 372, 375, 377, 409, 411
Lycopodiaceae 295
Lymphocytic leukemia (P388) 471, 472

Maalianes 182, 513
Maalian-5-ol 128, 182
ent-Maalian-5-ol 182
β-Maaliene 127, 182
γ-Maaliene 128, 182
ent-γ-Maaliene 182
ent-Maali-4(15)-en-1β-ol 183
(+)-ent-Maali-4(15)-en-1β-ol 128, 182
Maaliol 128, 182
Maalioxide 491
ent-Maalioxide 124, 176
Macrolejeunea pallescens 47, 78, 247, 495, 500
Macrolejeunea sp. 212
Magnolia kachirachirai 355
Makinoa crispata 72, 81, 91, 107, 112, 153, 156, 261, 282, 285, 289, 290, 523
Makinoa sp. 502
Malum cordis 464
Mannia fragrans 109, 143, 282, 285, 289, 311, 343, 468, 502, 503
Mannia sp. 188, 343
Mannia subpilosa 61, 97, 130, 187, 292, 295, 310, 343, 503
Mannitol 388
(−)-D-Mannitol 391
(−)-Manool 244
Manoyl oxide 238, 257
(+)-Manoyl oxide 237
Marchantia berteroana 81, 91, 101, 105, 107, 282, 285, 289, 299, 320, 506
Marchantia chenopoda 58, 95, 135, 210, 299, 308, 309, 320, 337, 338, 473
Marchantia diptera 282, 283, 285, 289, 294, 310, 340
Marchantia foliacea 96, 97, 104, 121, 171, 282, 285, 289, 506
Marchantia paleacea var. diptera 17, 19, 20, 65, 101, 102, 105, 107, 109, 113, 141, 237, 257, 261, 299, 308, 309, 310, 313, 316, 317, 320, 338, 347, 355, 358, 366, 380, 473, 506
Marchantia palmata 103, 114, 261, 282, 285, 289, 308, 309, 310, 335, 341, 506

Marchantia plicata 87, 103, 104, 122, 173, 282, 285, 290, 308, 338, 473

Marchantia polymorpha 65, 80, 81, 85, 87, 95, 101, 102, 103, 105, 107, 108, 109, 110, 114, 116, 130, 135, 141, 143, 146, 160, 186, 204, 207, 237, 257, 261, 280, 282, 284, 285, 289, 290, 297, 299, 308, 309, 310, 311, 316, 317, 318, 320, 321, 335, 337, 338, 340, 341, 345, 356, 358, 361, 366, 382, 383, 386, 460, 464, 469, 473, 477, 482, 506, 510, 519

Marchantia tosana 17, 72, 85, 91, 97, 101, 308, 309, 316, 337, 338, 355, 473, 506

Marchantia sp. 140, 356, 382, 472, 479, 480, 488, 506, 507

Marchantiaceae 337, 338, 381, 485, 505, 506

Marchantiales 22, 51, 60, 65, 135, 157, 173, 181, 237, 266, 292, 297, 319, 335, 337, 338, 340, 343, 347, 362, 373, 374, 381, 389, 428, 471, 472, 485, 493, 501, 502, 503, 506, 510, 518, 521, 524

Marchantiidae 47, 362, 484, 501

Marchantiineae 505

Marchantin A 308, 335, 336, 337, 338, 339, 340, 341, 472, 473, 479, 480, 481, 482, 506

Marchantin A trimethyl ether 341, 481

Marchantin B 308, 335, 336, 337, 338, 340, 341, 472

Marchantin C 309, 336, 337, 338, 340, 341, 343, 472, 502, 506

Marchantin C dimethyl ether 309, 336, 338

Marchantin C monomethyl ether 309, 336, 338

Marchantin D 309, 336, 337, 340, 341, 479

Marchantin E 309, 336, 337, 340, 341, 347, 479, 506

Marchantin E trimethyl ether 341

Marchantin F 310, 336, 337, 341

Marchantin G 310, 336, 337, 340, 341

Marchantin H 310, 340, 341, 342, 343, 502

Marchantin I 310, 340, 341, 342

Marchantin J 310, 341, 342

Marchantin K 310, 341, 342

Marchantin L 310, 341, 342

Marchantin M 310, 342, 343

Marchantin N 310, 342, 343

Marchantin O 309, 336, 338

Marchantin P 309, 336, 338

Marchantinquinone 310, 342, 343

Marchantiopsidae 506

Marchesinia brachiata 101, 282, 285, 289, 317, 358

Marsupella aquatica 49, 127, 179, 490

Marsupella emarginata 49, 113, 126, 127, 157, 179, 472, 477, 490

Marsupella emarginata subsp. *tubulosa* 179

Marsupella emarginata var. *patens* 49, 58, 81, 83, 95, 387, 388, 389, 490

Marsupella sp. 490

Marsupellaceae 179, 490, 491

Marsupellol 126, 179, 180

Marsupellone 126, 179, 180, 472

(−)-Marsupellone 179

ent-Marsupellone 181

Mastigolejeunea humilis 68, 128, 282, 285, 289

Mastigolejeunea undulata 99, 285

Mastigophora diclados 47, 69, 79, 81, 84, 91, 98, 99, 105, 108, 109, 110, 111, 126, 130, 135, 147, 152, 179, 190, 246, 262, 282, 284, 286, 289, 290, 291, 294, 295, 481, 495

Mastigophora sp. 143, 204, 362, 484, 495, 496

Mastigophora undulata 282, 289

Mastigophora woodsii 109, 110, 147

Mastigophorene A 111, 147, 150, 151, 481

Mastigophorene B 111, 147, 150, 151, 481

Mastigophorene C 111, 147, 150, 151, 481

Mastigophorene D 111, 147, 150, 151, 152, 481

Mastigophoroideae 495

Mastitis 465

Maytanbutine 450, 458

Maytansinoids 457, 460, 472

Maytenus bunchananii 473

(+)-Mayurone 204, 205

McLafferty-type cleavage 186

Megaceros arachnoideus 460

Megaceros flagellaris 460, 461, 462

Megaceros sp. 460

Megaceros tosanus 462

Megacerotonic acid 461, 462, 463

Menorrhalgia 465

13^2-(MeOO)-(13^2-R)-pheophytin a 451, 459, 460

Methanol 41, 193, 195, 199, 231, 471

5′-Methoxyamentoflavone-4′-methyl ether 426

15-Methoxyansamitocin P-3 450, 457, 458, 472
m-Methoxybenzaldehyde 325, 340
2-Methoxybenzyl benzoate 296, 297, 298, 384
3-(4'-Methoxybenzyl)-5,6-dimethoxyphthalide 316, 357
2-Methoxybenzyl trans-β-methyl-thioacrylate 384, 390
3-Methoxybibenzyl 300, 323, 324, 333
5-Methoxybicyclogermacrene 54, 55, 94
14-Methoxybicyclogermacrene 54, 55, 94
5'-Methoxybilovetin 425, 426
7-Methoxycadalene 64, 65, 100
(1S,4R)-7-Methoxycalamenene 64, 100
7-Methoxydehydropinguisenene 131, 191
(1S)-7-Methoxy-1,2-dihydrocadalene 64, 100
4β-Methoxyeudesmanal 115, 158
Methoxyfrullanolide 118, 164, 167
3-Methoxy-4-hydroxybibenzyl 300, 323, 324
3-Methoxy-4'-hydroxybibenzyl 300, 323, 324, 472, 473
3-Methoxy-5-hydroxy-2-(3-methyl-2-butenyl)bibenzyl 304, 327
2-Methoxy-4-hydroxyphenyl-1β-O-glucoside 318, 360
p-Methoxyacetophenone 189
3-Methoxy-4-(3-methyl-2-butenyl)-4'-hydroxybibenzyl 303, 325, 327
4-Methoxymethyl-1-carboxyazulene 43, 47, 79
3-(3'-Methoxy-4',5'-methylenedioxy-benzyl)-5,7-dimethoxyphthalide 316, 357
3-Methoxy-4,5-methylenedioxy-4'-hydroxybibenzyl 301, 325, 326
6-Methoxy-7,8-methylenedioxy-3-methylisocoumarin 352
2-Methoxy-5-methyl-1,4-quinone 343
4-p-(Methoxyphenylethyl)-cyclohex-2-en-1-one 299, 322, 324
4'-Methoxyscutellarein-7-O-glucronylrhamnoside 366, 376
(+)-21α-Methoxyserrat-14-en-3-one 291, 293, 294
(R)-Methyl aryllactate 463
4-Methylazulene 45
4-Methylazulene-1-carbaldehyde 43, 45, 46, 47, 79

4-Methylazulene-1-carboxylic acid 47
(R)-(+)-Methylbenzylamine 25
2-Methylbornanes 23
2-Methyl-2-bornene 15, 22, 23
2-Methylbutanol 381, 386
6-(3-Methyl-2-butenyl) indole 318, 358, 360
7-(3-Methyl-2-butenyl) indole 318, 358, 360
Methyl caffeate 460, 461, 462
4-Methyl-1-carboxyazulene 43, 45, 79
24-Methyl-5,22-cholestadienol 392, 395
24-Methyl-5,7,22-cholestatrienol 392, 397, 402
24-Methylcholesterol 294, 519
24α-Methylcholesterol 280
24β-Methylcholesterol 280
Methyl cinnamate 296, 297, 298, 466
Methyl cis-clerod-3,13(16),14-trien-18-oate 215, 248
Methyl coumarate 460, 461, 462
2-Methylcyclohexanone 467
4-Methylcyclohexanone 467
Methyl 3α,18-diacetoxytrachyloban-19-oate 264, 274
Methyl 3,4-dimethoxybenzoate 296, 297, 298, 448, 457
Methyl (3,4-dimethoxylphenyl)-lactate 463
3(R)-Methyl-5,6-dimethoxy-7,8-methylene-dioxydihydroisocoumarin 353
2-Methylenebornane 15, 22, 23
24-Methylenecholesterol 516, 518, 519
24-Methylenecycloartanol 280, 392, 397, 402, 515
3,4-Methylenedioxy-3'-methoxybibenzyl 301, 323, 324
Methylenetriphenylphosphorane 33
4-Methyl-1-formylazulene 43, 45, 79
6-Methyl-5-hepten-2-one 432, 446
3-Methyl-6-hydroxy-7-carboxy-8-(2-phenylethyl)-3,5-cyclopropano-chroman 307, 334
(−)-Methyl 13-hydroxy-cleroda-3,14-dien-18-carboxylate 212, 213, 247
Methyl 3α-hydroxy-18-oleanen-28-oate 292, 295
3-Methyl-6-hydroxy-8-(2-phenylethyl)-3,5-cyclopropanochroman 307, 334
Methyl iodide 51, 152
2-Methylisoborneol 15
(−)-2-Methylisoborneol 15, 22, 23, 466, 491
1-Methyl-3-isopropylbenzene 13, 14, 18

3-Methylmercaptopropylamine 521, 522
4-Methyl-1-methoxycarbonylazulene 42, 43, 46, 47, 79
Methyl-3-methoxy(3-formylphenoxy) benzoate 335
3-Methyl-6-methoxy-8-(2-phenylethyl)-3,5-cyclopropanochroman 307, 334
Methyl 2-methyl-3,4-methylenedioxy-6-methoxybenzoate 317, 358, 359
2(S)-2-Methyl-2-(4-methyl-3-pentenyl)-6-carboxy-7-hydroxy-5-(2-phenylethyl)chromene 305, 329
2(S)-2-Methyl-2-(4-methyl-3-pentenyl)-7-hydroxy-5-(2-phenylethyl) chromene 305, 329
2(S)-2-Methyl-2-(4-methyl-3-pentenyl)-7-methoxy-5-(2-phenylethyl) chromene 329
Methyl myrtenate 14, 21
Methyl plagiochilate 39, 78
2-Methyl-1-propenylmagnesium bromide 467
2-Methyl-2-propenylmagnesium bromide 467
2-Methyl-1,4-quinone 343
6-Methylsalicylic acid ethyl ester 322
Methyl trachyloban-19-oate 274
Methyl 4,7,9-trimethoxy-2-dibenzofurancarboxylate 449, 457, 458
Methyl veratrate 448, 457, 458
Metzgeria albinea 286, 289
Metzgeria furcata 373, 378, 469, 473
Metzgeria furcata var. *furcata* 486
Metzgeria furcata var. *uvula* 365, 373, 378
Metzgeria sp. 372, 378
Metzgeriaceae 484, 485, 486
Metzgeriales 22, 26, 135, 153, 157, 190, 266, 269, 271, 292, 319, 335, 340, 347, 362, 373, 374, 375, 378, 390, 466, 484, 485, 486, 493, 506, 521, 522, 523, 524
Mevalonic acid 46
Microbryum delicatulum 473
Microsporum gypseum 474
Mikania alvimii 243
Mineral acid 193
Minitamariscols 467
Mniaceae 423, 452, 471, 507, 508
Mnium cuspidatum 393, 398, 399, 400, 439, 443, 465, 469
Mnium hornum 403, 404, 405, 407, 439, 443, 473, 476, 482

Mnium medium 393, 439, 443
Mnium punctatum 473
Mnium sp. 433, 452, 455, 456, 464, 475
Mnium undulatum 469
Moerkia sp. 465
Molluscicidal activity 476
Momilactone A 391, 396
Momilactone B 391, 396
Monocarpaceae 501
Monocarpus sphaerocarpus 501
Monoclea forsteri 51, 68, 69, 71, 72, 85, 88, 95, 97, 98, 101, 103, 104, 105, 113, 120, 124, 127, 157, 176, 286, 308, 309, 311, 335, 337, 338, 343, 345, 372, 382, 386, 387, 502
Monoclea gottschei 502
Monoclea gottschei subsp. *elongata* 502
Monoclea gottschei subsp. *gottschei* 502
Monoclea gottschei subsp. *neotropica* 68, 73, 101, 103, 114, 124, 128, 157, 176, 246, 262, 286, 309, 311, 338, 343, 345, 363, 364, 365, 366, 367, 368, 369, 370, 371
Monoclea sp. 140, 157, 171, 182, 488, 501, 502
Monocleaceae 382
Monocleales 60, 65, 135, 266, 292, 319, 347, 373, 374, 389, 485, 493, 501
Monocleic acid 383, 387
Monocleolic acid 383, 387
Monocyclofarnesanes 182, 185, 513
γ-Monocyclofarnesol 185
trans-γ-Monocyclofarnesol 129, 183
Monocyclonerolidol 183
β-Monocyclonerolidol 129, 182, 185
Monoethoxymarchantin A 341
Monogalactosyl diacylglycerols 455
Monoterpenoids 12, 13, 14, 15, 16, 23, 503
Murashige-Skoog-2 medium 297, 382
Murashige-Skoog-3 medium 64
Murashige-Skoog-12 medium 382
Musci 5, 8, 267, 362, 374, 382, 391, 393, 402, 407, 408, 420, 421, 422, 423, 424, 426, 427, 428, 430, 431, 453, 455, 458, 459, 464, 468, 471, 507, 509, 510, 511, 512, 513, 514, 515, 516, 517, 520, 521, 522, 523, 524
Muscle relaxing activity 481, 482
Muurolanes 60, 62, 513
α-Muurolene 60, 62, 96
γ-Muurolene 60, 62, 96

Mylia anomala 265, 277, 364, 369, 373, 378, 489
Mylia nuda 17, 19, 20, 21, 59, 71, 84, 88, 96, 98, 99, 105, 108, 114, 124, 157, 176, 237, 257, 310, 341, 364, 369, 378, 489
Mylia sp. 186, 187, 275, 489
Mylia taylorii 33, 35, 40, 41, 75, 76, 78, 128, 130, 182, 186, 265, 277, 282, 286, 289, 290, 291, 294, 295, 364, 369, 373, 378, 383, 387, 388, 389, 391, 489, 494
Mylia verrucosa 264, 265, 275, 282, 286, 289, 290, 294, 364, 369, 373, 378, 383, 387, 388, 489
(+)-Myli-4(15)-en-9-one 33, 34, 75
Myliol 33, 34, 35, 41, 75, 277
Mylione 35
Myltaylanes 186, 187, 494
Myltaylenol 130, 186, 187
Myltayl-4(12)-en-15-ol 130, 186
Myltaylorione A 35, 36, 75
Myltaylorione B 35, 36, 75
Myo-inositol 388
Myrcene 12, 13, 16, 246, 466
Myrtenol 14, 21
Myuroclada maximoviczii 439, 443

n-Nananal 381
Naphthalene 314, 352, 353
Naphthalenes 352, 353, 354
Nardia compressa 489
Nardia scalaris 19, 30, 67, 70, 81, 105, 109, 143, 229, 231, 234, 246, 253, 254, 255, 257, 262, 291, 294, 489
Nardia sp. 229, 489
Nardia subclavata 81, 212, 234, 247, 254, 255, 257, 311, 345, 489
Nardia succulenta 229, 236, 244, 253, 254, 255, 260
Nardiin 231, 232, 233, 254
Nasopharynxcarcinoma (9KB) 472
Naviculide 246, 262, 267
Naviculol 131, 191, 196, 199
Neckera crispa 395, 396
Neckera sp. 392
Neckeraceae 471
Neckeropsis nitidula 482
Neo-β-carotene U 399, 403
Neodrimanoxide 208
Neohodgsonia sp. 505

Neointermediol 115, 158
Neolignans 355
Neomarchantin A 311, 342, 343, 502
Neomarchantin B 311, 342, 343
Neotrichocolea bissetii 91, 133, 193
Neotrichocolea sp. 495
13-*epi*-Neoverrucosan-5β,20-diol 266, 279
Neoverrucosan-5β-ol 265, 275, 278
13-*epi*-Neoverrucosan-5β-ol 265, 278, 279
Neoxanthin 399, 403
Neoxanthin neo A 400, 403
Nepalensolide A 118, 162, 164, 166, 167
(+)-Nepalensolide A 166
Nepalensolide B 118, 162, 164, 167
Nepalensolide C 118, 162, 164, 167
Nepalensolide D 118, 162, 164
Neral 13, 14, 16, 212
Nerol 13, 14, 16, 466
Nerolidol 246
(+)-Nerolidol 171
trans-Nerolidol 467
(+)-*trans*-Nerolidol 121, 171
Neryl acetate 13, 14, 16, 466
Neteroclada confluens 91, 95, 121, 171, 282, 286, 289
Neurasthenia 465
Neuritic sprouting activity 481
Nicotiana setchelli 237
Nicotiana tabacum 237
Nicotine 481
Nidorella hottentotica 240
Nidorella lactone 240
Nipponolejeuneoideae 500
Nitella hookei 521
Nitric acid 143
Nocardia sp. 473
Nodifloretin-7-O-glucoside-6″-hydroxy-3-methylglutarate 371, 377
Nodifloretin-6-O-glucoside-7-O-xyloside 371, 377
Nodifloretin-7-monomethyl ether 370, 377
Nodifloretin-7-O-rhamnoside-4′-O-glucuronide 371, 377
Nodifloretin-7-O-sophoroside 371, 377
Nodifloretin-7-O-xyloglucoside 371, 377
Nodifloretin-6-O-xyloside-7-O-glucoside 371, 377
Nonacosan-10-ol 380, 385
Nonacosan-10-one 385
n-Nonacosan-10-one 380

10,13-Nonadecadien-7-yn-2-one 447, 452, 453
10,13,16-Nonadecatrien-7-yn-2-one 447, 452, 453
3,5,7,4′,3″,5″,7″,3‴,4‴-Nonahydroxy-3′,6″-biflavone 417, 427
Nonanal 386
n-Nonanal 381
(+)-Nootkatone 205
Noracoranes 26
Norafricanes 27, 28
Norafricanone 27, 28, 67
Norcycloartanes 515
31-Norcyclolaudenol 280, 392, 396, 402, 515
2-Nor-1,3-epoxy-1,10-secoaromadendra-1(5),3-dien-10-one 39, 78
Norpinguisanes 190, 199
Norpinguisanolide 134, 195, 196, 199
Norpinguisone 134, 196, 199, 200, 479
Norpinguisone methyl ester 134, 193, 195, 196, 197, 199, 200, 479
Norsecoaromadendranes 33, 39
Norswartzianin 28, 30, 67
Notholanic acid 325, 326
Notothyladaceae 462
Notothylas temperata 461, 462
(−)-Nuciferal 59, 96
ent-Nuciferal 490, 494
(−)-*ent*-Nuciferal 58

Obtusifoliol 280, 392, 397, 402, 515
Octadecadienoic acid 433
9,12-Octadecadien-6-ynoic acid 381, 383, 386, 433, 434, 452, 453, 454
Octadecatrienoic acid 433
9,12,15-Octadecatrien-6-ynoic acid 381, 383, 386, 433, 435, 452, 453, 454, 475, 508
9-Octadecen-6-ynoic acid 381, 382, 383, 386
Octanal 381, 432
n-Octanal 445
Octanol 381
1-Octanol 381, 432
3-Octanol 432, 445
2-Octanol 432, 445
n-Octanol 385, 434
Octanone 381, 433
3-Octanone 381, 385, 432, 434, 467

Octan-3-yl acetate 381, 385
trans-2-Octenal 432, 445
2-Octen-1-ol 432, 445
1-Octen-3-ol 381, 432, 434, 466
(S)-(+)-1-Octen-3-ol 381, 385, 466
1-Octen-3-yl acetate 381
(R)-(+)-1-Octen-3-yl acetate 381, 385, 466
Octenol 381
Odontoradula sp. 496
Odontoschisma denudatum 223, 250, 251, 474, 493
Odontoschisma sp. 493
Odontoschismatoideae 493
Ohioensin A 419, 431, 432, 472
Ohioensin B 419, 431, 432, 472
Ohioensin C 419, 431, 432, 472
Ohioensin D 419, 431, 432, 472
Ohioensin E 419, 431, 432, 472
Oleananes 515
Oleic acid 294, 383, 452, 456
Oligotrichum hercynicum 473
Omphalane 209
Omphalanthus filiformis 19, 65, 84, 91, 101, 103, 135, 208, 314, 351, 500, 521
Omphalanthus paramicola 500
Omphalanthus platycoleus 207, 500
Omphalic acid 135, 208
Onoceranes 515
Onocophorus crispofolius 439, 444
Onopordin-7,4′-di-O-polysaccharide 372, 377
Oppositanes 136, 514
Oppositol 138
Optical activity 333
Orcadensin 217, 249
Oreas martiana 465
Orellinic acid methyl ester 317, 358, 359
Orientin 409, 413, 420
Ornithine decarboxylase 473
Orobol 414, 422, 423, 428
Orobol-7-O-diglucoside 414, 422
Orobol-7-O-glucoside 414, 422
Orobol-7-O-glucoside-6″-malonate 414, 422
Orthodicranum montanum 435, 436, 439, 444
Orthodontium lineare 477
Orthotrichum rupestre 473
Oryza sativa 391
Oryzia latipes 480

Ovalifolienal 37, 39, 77
Oximitra sp. 382
ent-9-Oxo-α-chamigrene 103, 136
(−)-*ent*-9-Oxo-α-chamigrene 65, 136
(±)-3-Oxodiplophyllin 168
ent-3-Oxodiplophyllin 119, 165
19α-Oxo-*ent*-labda-8(17),12*E*,14-
      trien-3β-ol 242, 259
9-Oxogymnomitryl acetate 47, 48, 83
3-Oxonorpinguisone 201
12-Oxo-PDA 1 447, 453, 454
12-Oxo-PDA 3 448, 453, 454
Oxyfrullanolide 116, 163, 471

Pacifigorgia cf. *adamsii* 190
Pacifigorgiane 190
Pacifigorgianes 188
Pacifigorgiol 188, 190
Pakyonol 311, 342, 343
Palaleucobryum longifolium 439, 444
Paleatin A 313, 346, 347, 348
Paleatin B 313, 346, 347, 348
Pallavicinia canarus 478
Pallavicinia levieri 246, 260, 262, 269, 468,
      469, 480, 487
Pallavicinia longispina 487
Pallavicinia lyellii 280, 282, 284, 286, 289,
      290, 487
Pallavicinia sp. 487
Pallavicinia subciliata 240, 258, 487
Pallaviciniaceae 487
Pallavicinin 258
(+)-Pallavicinin 239, 240
Pallavicinioideae 378
Palmitic acid 380, 383, 433, 456, 522
Palmitoleic acid 383
Paraleucobryum longifolium 393
Paraleucobryum sp. 508
Patchoulanes 188, 190
α-Patchoulene 130, 190
β-Patchoulene 131, 190
Pectolinarigenin-7-O-galacturonide 365,
      376
Pectolinarigenin-7-O-glucuronide 365, 376
Peculiaroxide 135, 208
Pedalitin 370, 377
Pellepiphyllin 301, 323, 324
Pellia endiviifolia 104, 262, 263, 268, 269,
      312, 345, 384, 390, 466, 473, 475, 480,
      482, 487

Pellia epiphylla 66, 70, 101, 113, 176, 312,
      314, 315, 347, 354, 487
Pellia neesiana 262, 269
Pellia sp. 157, 468, 481
Pelliaceae 484, 487, 522
Penicillium chrysogenum 474
6-Pentadecenylsalicylic acid 351
3-Pentadecylphenol 313, 352, 520
Pentadecylsalicylic acid 520
6-Pentadecylsalicylic   acid   313,   351,
      352
3,4,5,3′,4′-Pentamethoxybibenzyl 302, 325,
      326
n-Pentane 33
Perchloric acid 147
3,6-Peroxocupar-1-ene 109, 144
Perrottetianal 272, 498, 518
(+)-Perrottetianal 272
Perrottetianal A 263, 271, 474, 479
Perrottetianal B 263, 271, 474
Perrottetianals 162
Perrottetin A 306, 330, 331, 479
Perrottetin B 306, 330, 331
Perrottetin C 306, 330, 331
Perrottetin D 306, 330, 331, 332, 479
Perrottetin E 311, 328, 330, 337, 341, 343,
      344, 345, 347, 348, 349, 351, 472, 488,
      502
Perrottetin E-11′-methyl ether 312, 344,
      345
Perrottetin F 312, 328, 330, 344, 345, 347,
      348
Perrottetin G 312, 330, 344, 345
Perrottetin H 312, 346, 347
Perrottetinene 307, 334
Peteravenia schultzii 160
Petroleum ether 471
Phaeoceros laevis 460, 462, 463
Phaeoceros miyakeanus 462
Phaeoceros sp. 460
Phaeophyceae 512, 513, 514, 515, 516, 517,
      521, 523
Phellandrene 393
α-Phellandrene 12, 13, 17, 391
β-Phellandrene 12, 13, 17, 391, 466
(+)-β-Phellandrene 12
Phenanthrenes 355, 356
Phenol 403, 406, 408
Phenolic acid 433
Phenylacetic acid 403, 406, 408, 433

Phenylacetoaldehyde 317, 358, 359, 403, 404
Phenylacetylene 403, 406, 408
L(U-¹⁴C)-Phenylalanine 408
L-(2,3,4,5,6-³H)-Phenylalanine 408
Phenyl cinnamate 296, 297, 298
Phenylethyl alcohol 403, 404
β-Phenylethyl benzoate 296, 297, 298
β-Phenylethyl cinnamate 296
β-Phenylethyl *trans*-cinnamate 297, 299
β-Phenylethyl dihydrocinnamate 296, 297, 299
β-Phenylethyl *trans*-β-methyl thioacrylate 384, 390
β-Phenylethyl thioacrylate 521
Pheophytin a 450, 459, 460
Philodendron scandens subsp. *oxycardium* 351
Philonotis fontana 398, 399, 400, 401, 415, 416, 417, 425, 465
Philonotis sp. 464
Philonotisflavone 417, 425, 426
Phosphatidylcholine 455
Phosphatidylglycerol 455
N-(Phosphonomethyl)glycine 408
Phthalides 355, 357
Physcomitrella patens 434, 451, 456, 460
Physcomitrium pyriforme 482
Physiological activity 464
Phytadienes 246, 261, 267
Phytanes 246, 267
Phytenic acid 246, 267, 393, 456
Phythium debaryanum 474
Phytochelatins 460
Phytol 246, 261, 267, 393, 456, 514
Phytyl phytenate 246, 262, 267
Picea ajamensis 159
Piceaceae 159
Pieris sp. 475
ent-Pimara-8(14),15-dien-19-oic acid 246, 262, 268
ent-Pimara-8(14),15-dien-19-ol 262, 268
Pimaranes 246, 268, 514
α-Pinene 12, 14, 15, 391, 393, 466, 467
(−)-α-Pinene 19
β-Pinene 12, 14, 15, 466, 467
(−)-β-Pinene 20
Pinguisanes 190, 191, 194, 197, 198, 201, 202
Pinguisanin 133, 192, 193, 197, 198

4-*epi*-Pinguisanol 201
Pinguisanolide 133, 193, 197, 198
Pinguisenal 133, 197, 198
Pinguisenene 131, 191, 200
α-Pinguisene 131, 191, 193
Pinguisenene methyl ester 132, 194
Pinguisenol 131, 191, 197
Pinguisone 191, 193, 200
(±)-Pinguisone 201
Pinguisone methyl ester 191
cis-Pinocarveol 12
trans-Pinocarveol 12, 14, 21
cis-Pinocarveyl acetate 12, 14, 21, 467
trans-Pinocarveyl acetate 12, 14, 21, 467
Pinocarvone 14, 21
Pinosylvin 328, 334
Pinosylvin methylether 334
Pinosylvin monomethyl ether 328
Piperitone 57
Piricularia oryzae 474
Piscicidal activity 465, 480, 481
Plagiochasma intermedium 310, 340, 502
Plagiochasma rupestre 51, 53, 89, 126, 181, 292, 295, 363, 370, 380, 502
Plagiochasma sp. 343
Plagiochila acanthoda 73, 91
Plagiochila acanthophylla subsp. *japonica* 19, 20, 82, 126, 176, 178, 228, 252, 263, 272, 309, 310, 311, 312, 338, 340, 343, 345, 381, 384, 385, 390, 466, 467, 491
Plagiochila adiantoides 73, 76, 87, 88, 91, 105
Plagiochila alternans 131, 491
Plagiochila amazonica 70, 91
Plagiochila asplenioides 73, 76, 81, 286, 317, 358
Plagiochila beskeana 73, 77, 81, 84, 87, 88, 91, 290
Plagiochila bifaria 88, 91
Plagiochila bispinosa 17, 91, 97, 98, 99, 105, 107, 108, 110, 140, 143, 146, 291, 295
Plagiochila bursata 73, 82
Plagiochila chacabucensis 99, 105, 125, 289, 301, 325
Plagiochila cipaconensis 70, 73, 76, 88, 91, 286, 290
Plagiochila corniculata 73, 290
Plagiochila corrugata 226, 237, 251, 252, 257, 491

Plagiochila cristatissima 73, 76, 91, 290
Plagiochila cucullata 70, 73, 76, 88, 91, 98, 106, 286, 290
Plagiochila dichotoma 70, 73, 88, 91, 106, 286, 290
Plagiochila dilatata 70, 76, 82, 88, 91, 286, 290
Plagiochila dura 68, 76, 77, 82, 106, 107, 128, 182, 252, 282, 286, 289
Plagiochila duricaulis 91, 98, 99, 106, 107, 108, 125
Plagiochila elata 19, 20, 73, 97, 91, 99, 125, 291, 295
Plagiochila engelii 80, 82, 88, 91, 125, 261, 283, 286, 289
Plagiochila excisa 84, 88, 91, 290
Plagiochila exigua 491
Plagiochila falcata 70, 73, 76, 77, 91, 98, 106, 283, 286, 290
Plagiochila friabilis 70, 73, 82, 91, 283, 286, 290
Plagiochila fruticosa 37, 38, 54, 76, 77, 94, 311, 345, 469, 481
Plagiochila fuegiensis 68, 113, 125, 283, 286, 301, 323
Plagiochila gayana 73, 76, 77, 87, 88, 91, 101, 120, 283, 286
Plagiochila geniculata 17, 68, 73, 128, 226, 251
Plagiochila goebeliana 73, 76, 88, 91
Plagiochila guayrapurinensis 70, 76, 88, 91, 106, 286, 290
Plagiochila guilleminiana 76, 77, 82, 88, 91
Plagiochila hondurensis 51, 73, 82, 84, 87, 88, 89, 91, 122, 173, 263, 271
Plagiochila hookeriana 73, 77, 91
Plagiochila jamesonii 372, 373, 378
Plagiochila kroneana 70, 73, 84, 88, 91, 106, 290
Plagiochila lecheri 68, 77, 97, 99, 120, 125, 128, 182, 291, 295
Plagiochila longispina 46, 47, 78, 79, 299, 322, 495
Plagiochila micropterys 47, 76, 77, 78, 490, 495
Plagiochila moritziana 66, 69, 70, 73, 77, 84, 88, 91, 93, 97, 103, 119, 126, 128, 168, 169, 179, 182, 226, 251, 491
Plagiochila neesiana 66, 75, 80, 82, 85, 88, 92, 283, 286

Plagiochila oresitropha 17, 19, 21, 71, 92, 99, 290
Plagiochila ovalifolia 35, 40, 73, 75, 77, 78, 92, 121, 128, 171, 182, 226, 251, 380, 385, 466, 467
Plagiochila oxyphylla 73, 92, 98, 290
Plagiochila pachyloma 19, 73, 87, 88, 92
Plagiochila panamensis 87, 92, 130, 190, 226, 251
Plagiochila parvidens 66, 71, 73, 80, 82, 86, 88, 92, 98, 120, 125, 283, 286, 289
Plagiochila parvitexta 73, 98, 106, 291
Plagiochila peculiaris 39, 67, 70, 77, 78, 82, 85, 92, 98, 99, 103, 106, 108, 113, 114, 135, 208, 211, 247, 291, 295
Plagiochila pittieri 73, 76, 77, 92, 106
Plagiochila porelloides 477
Plagiochila pulcherrima 73, 76, 80, 88, 92, 229, 252, 253, 286
Plagiochila retrospectans 131, 192
Plagiochila rosariensis 82, 88, 92, 131, 133, 286, 491
Plagiochila rutilans 317, 358, 491
Plagiochila sciophila 19, 20, 82, 126, 176, 178, 228, 252, 263, 272, 309, 310, 311, 312, 338, 340, 343, 345, 381, 384, 385, 390, 466, 467, 491
Plagiochila sciophila fo. *fragilis* 491
Plagiochila sciophila fo. *japonica* 491
Plagiochila sciophila fo. *robusta* 491
Plagiochila scopulosa 73, 77, 82, 88, 92, 120
Plagiochila sp. 33, 37, 38, 140, 157, 161, 177, 182, 192, 204, 272, 323, 467, 468, 469, 471, 472, 473, 475, 481, 490, 491, 492, 497
Plagiochila spinulosa 19, 92, 228, 252, 302, 315, 317, 325, 355, 358, 491
Plagiochila squamurifera 19
Plagiochila stephensoniana 19, 20, 68, 73, 92, 265, 277, 300, 491
Plagiochila subdura 16, 17, 19, 66, 68, 97, 106, 107, 108, 110, 140, 143, 146, 300, 301, 314, 352
Plagiochila tambillensis 20, 74, 92, 291
Plagiochila tenerrima 70, 74, 77, 82, 88, 92, 117, 287, 291
Plagiochila trabeculata 17, 47, 70, 74, 82, 83, 92, 93, 291, 491
Plagiochila verruculosa 20, 92
Plagiochila yokogurensis 203, 469

Plagiochilaceae 47, 54, 192, 489, 490, 492, 495, 497, 521
Plagiochilal A 37, 40, 77
Plagiochilal B 37, 38, 40, 41, 77, 469, 481
Plagiochilic acid 39, 78
Plagiochilide 37, 38, 77, 479
Plagiochiline A 37, 38, 40, 41, 76, 469, 471, 472, 475, 481
Plagiochiline B 37, 76
Plagiochiline C 37, 39, 76, 478
Plagiochiline D 37, 77
Plagiochiline E 37, 77
Plagiochiline H 37, 77
Plagiochiline J 37, 38, 39, 40, 77
Plagiochiline K 37, 38, 39, 40, 77
Plagiochiline L 39, 78
Plagiochiline M 39, 78
Plagiochiline N 39, 40, 78
Plagiochin A 311, 343, 344
Plagiochin B 311, 343, 344
Plagiochin C 311, 343, 344, 345
Plagiochin D 311, 343, 344, 345
Plagiodiila acanthophylla 466
Plagiomnium cuspidatum 409, 410, 411, 413, 416, 425
Plagiomnium elatum 409, 410, 411, 412, 413, 415, 424
Plagiomnium ellipticum 439, 444
Plagiomnium maximoviczii 439, 444
Plagiomnium succulentum 392, 394, 395, 519
Plagiomnium trichomanes 510
Plagiopus oederi 465
Plagiospirolide A 119, 165, 168, 169, 170
Plagiospirolide B 119, 165, 168, 169, 170
Plagiospirolide C 119, 165, 169, 170
Plagiospirolide D 119, 165, 169, 170
Plagiospirolide E 119, 165, 169, 170
Plagiothecium denticulatum 473
Plagiothecium euryphyllum 439, 444
Plagiothecium laetum 439, 444
Plagiothecium nemorale f. *japonicum* 439, 444
Plagiothecium sp. 455
Plant growth regulatory activity 477, 478
Plastohydroquinone 482, 483
Plastoquinone 482, 483
Platinum oxide 338
Pleuridium subulatum 435, 436, 439, 444
Pleurodiol 244, 245, 260

Pleurozia acinosa 20, 32, 67, 86, 87, 88, 89, 92, 103, 237, 247, 257, 369, 372, 373, 380, 497
Pleurozia articulata 369, 372, 373, 380, 497
Pleurozia caledonica 369, 372, 373, 380, 497
Pleurozia conchifolia 369, 372, 373, 380, 497
Pleurozia gigantea 74, 224, 226, 237, 244, 247, 251, 257, 260, 283, 287, 289, 380, 497
Pleurozia giganteoides 380, 497
Pleurozia heterophylla 380, 497
Pleurozia purpurea 497
Pleurozia sp. 212, 380, 497, 498
Pleuroziaceae 380, 497, 498
Pleuroziineae 497
Pleurozium schreberi 393, 394, 395, 439, 444, 454
Pleurozium sp. 456
Pneumonia 465
Podophyllotoxins 460
Pogonatum aloides 473
Pogonatum inflexum 439, 444
Pogonatum sp. 456
Pogonatum urnigerum 439, 444, 473
Pohlia longicollis 398, 399, 400, 401
Pohlia nutans 440
Pohlia wahlenbergii 403
Polygodial 112, 152, 154, 194, 468, 469, 470, 475, 494
(+)-Polygodial 476, 478, 481
(−)-Polygodial 475, 478, 480
Polygonum hydropiper 470, 475, 481
Polypodiaceae 295
Polysyphonia ferulacea 356
Polytrichaceae 452, 471, 507
Polytrichadelphus magellanicum 449, 450, 457
Polytrichales 362, 455, 456
Polytrichidae 401, 419, 451
Polytrichum commune 393, 394, 395, 397, 398, 399, 400, 401, 440, 444, 449, 450, 454, 455, 456, 457, 465, 473
Polytrichum formosum 434, 456
Polytrichum formosum subsp. *euformosum* var. *typicum* 396
Polytrichum juniperinum 433, 440, 444, 464, 469, 472, 473
Polytrichum ohioense 419, 432, 472
Polytrichum sp. 432, 456, 464, 465, 475, 507

Porella acutifolia subsp. *tosana* 123, 125,
    131, 173, 176, 197, 263, 271, 468, 498
Porella bolanderi 471
Porella caespitans var. *setigera* 27, 32, 66,
    67, 134, 202, 263, 271, 498
Porella cordaeana 16, 18, 19, 20, 21, 32, 67,
    112, 114, 129, 131, 133, 134, 152, 183,
    196, 197, 263, 271, 291, 381, 384, 386,
    390, 466, 498, 500
Porella densifolia 197
Porella densifolia subsp. *appendiculata* 20,
    21, 74, 86, 92, 101, 103, 129, 131, 134,
    183, 229, 253, 261, 283, 287, 289, 498
Porella densifolia var. *fallax* 74, 86, 92, 101,
    103, 129, 183, 229, 253, 498
Porella elegantula 92, 131, 134, 193, 195,
    261, 263, 271, 283, 287, 289, 479
Porella fauriei 152
Porella japonica 176, 498, 500
Porella navicularis 129, 131, 134, 183, 196,
    199, 246, 262, 263, 271, 498, 500
Porella perrottetiana 237, 257, 261, 263, 272
Porella platyphylla 132, 133, 134, 193, 283,
    287, 289, 362, 473, 498
Porella roellii 32, 67, 112, 152, 211, 247, 468
Porella sp. 192, 229, 362, 467, 468, 478, 484,
    494, 498, 500
Porella squamurifera 87, 92, 283, 287
Porella swartziana 27, 30, 56, 66, 67, 70, 92,
    95, 101, 121, 122, 124, 173, 176, 283, 287,
    291, 498
Porella vernicosa 111, 112, 131, 132, 134,
    152, 197, 200, 466, 468, 470, 473, 475,
    480, 481, 498
Porellaceae 190, 271, 491, 498, 500, 521
Porelladiolide 125, 176, 178
Porellapinguisanolide 133, 197, 198
Porellapinguisenone 131, 191, 197
Potassium 6-pentadecyl salicylate 313, 352
Potassium 6-tridecyl salicylate 313, 352
Potassium 6-undecyl salicylate 313, 352
Pratensein 414, 422
Pratensein-7-O-glucoside 414, 422
Pratensein-7-O-glucoside-6″-
    malonate 415, 422
Preissia quadorata 308, 311, 335, 343, 505,
    506
Preissia sp. 507
Prelunularic acid 299, 320, 321, 322
Propionic acid 433, 446

Propiothetin 521, 522
Prostaglandin A_2 454
Proteus mirabilis 474
Protocatechuic acid 403, 404, 476
Ps-1 182
Ps-2 182
Ps-2′ 182
Ps-3 240, 242, 259
Ps-4 240, 242, 259
Ps-5 240, 242, 259
Ps-7 240, 242, 259
Ps-m-3 196
Pseudobrickellia brasiliensis 173
Pseudomonas aeruginosa 473, 474
Pseudoscleropodium purum 395, 396, 397,
    473
Pseudoscleropodium sp. 392
Psilophytales 509
Psychosis 465
Pteridophytes 5, 455, 513, 514, 515, 516,
    517, 520, 521, 524
Ptilidiaceae 190, 495
Ptilidiineae 495
Ptilidium ciliare 386
Ptilidium pulcherrimum 477
Ptilidium sp. 192, 362, 484, 495
Ptychanolactone 133, 198, 199
Ptychanolide 133, 193, 196, 197, 198, 199,
    200, 201
Ptychanthoideae 377, 500, 501
Ptychanthus striatus 122, 128, 129, 131,
    133, 173, 182, 199, 200, 237, 240, 241,
    243, 244, 259, 482
Ptychantin A 240, 242, 259
Ptychantin B 240, 242, 243, 259
Ptychantin C 240, 242, 259
Ptychantin D 240, 242, 259
Ptychantin E 241, 242, 259
Ptychantin F 242, 259
Ptychantin G 242, 259
Ptychantin H 242, 259
Pulegone 13, 14, 18
Pulmonary tuberculosis 465
Pusilatin A 313, 350
Pusilatin B 313, 350
Pusilatin C 313, 350
Pusilatin D 313, 350
Pyridine 60, 145, 154, 168, 267
Pyridinium chlorochromate 12, 187
Pyrrolidine 403, 406, 408

Rabdosia japonica 355

Rabdosiin 355

Racemic warburganal 470

Racomitrium canescens 440, 444

Racomitrium heterostichum 440, 444

Racomitrium japonicum 392, 395, 397, 403, 404, 451, 482

Racomitrium lanuginosum 392, 396, 397, 415, 416, 423

Radula boryana 18, 51, 69, 85, 89, 92, 106, 120, 124, 125, 128, 178

Radula brunnea 496

Radula buccinifera 106, 300, 301, 302, 306, 323, 325, 496

Radula carringtonii 496

Radula chinensis 496

Radula companigera 496

Radula complanata 106, 301, 302, 303, 306, 307, 316, 328, 332, 333, 358, 469, 470, 480, 496, 497

Radula constricta 496

Radula frondescens 80, 114, 120, 300, 301, 303, 323, 328

Radula grandis 496

Radula javanica 108, 140, 300, 303, 306, 307, 323, 328, 333, 334

Radula kojana 107, 120, 283, 287, 290, 303, 304, 305, 312, 328, 345, 496

Radula lindenbergiana 496, 497

Radula nudicailis 496

Radula okamurana 496

Radula oyamensis 106, 302, 303, 325, 328

Radula perrottetii 108, 261, 304, 305, 306, 307, 312, 313, 318, 328, 330, 332, 334, 345, 347, 361, 472, 479, 496

Radula plicata 497

Radula sp. 140, 158, 171, 176, 182, 303, 323, 325, 328, 345, 470, 473, 479, 480, 488, 496

Radula tasmanica 497

Radula tokiensis 303, 328

Radula uvifera 497

Radula variabilis 108, 140, 300, 303, 306, 307, 323, 328, 333

Radula voluta 120, 303, 304, 306, 325, 328, 332

Radula wichurae 497

Radulaceae 485, 496

Radulanin A 306, 325, 333, 334

Radulanin B 333, 334

Radulanin C 306, 325, 334

Radulanin H 307, 333, 334

Radulanin I 307, 333, 334

Radulanin J 307, 333, 334

Radulanin K 307, 333, 334

Radulanin L 307, 332, 333, 334

Radulanolide 316, 357, 358

Reboulia hemisphaerica 32, 48, 49, 61, 67, 80, 82, 83, 85, 89, 93, 97, 106, 108, 109, 130, 135, 141, 187, 210, 287, 292, 295, 308, 309, 335, 338, 363, 366, 380, 464, 502, 503

Reboulia sp. 188, 343

Rebouliadienol 135, 210

Reboulioideae 502, 504

Retro-Diels-Alder reaction 169

Rhabdiweisia crispata 435, 436, 440, 444

Rhabdiweisia fugax 435, 436, 440, 444

Rhamnose 374, 388

Rhamnosides 375

Rhinitis 465

Rhizoctonia solani 474

Rhizogonium spiniforme var. *badakense* 440, 444

Rhizomnium magnifolium 409, 411, 412, 413, 414

Rhizomnium parramatense 457

Rhizomnium pseudopunctatum 409, 411, 412, 413, 414

Rhizomnium punctatum 409, 440, 444

Rhizomnium tuomikoskii 440, 444

Rhodobryum giganteum 465

Rhodobryum roseum 440, 444, 465

Rhodobryum sp. 455, 469

Rhodophyceae 512, 513, 514, 515, 516, 517, 522, 523

Rhynia sp. 509

Rhytidiadelphus loreus 469

Rhytidiadelphus sp. 392

Rhytidiadelphus squarrosus 415, 416, 425, 440, 445

Rhytidiadelphus squarrosus subsp. *eusquarrosus* 396

Rhytidiadelphus squarrosus subsp. *squarrosus* 395

Rhytidiadelphus triquetrus 395

Rhytidiadelphus triquetrus var. *typicus* 392, 395, 396, 397

Riccardia andina 66, 74, 75, 82, 93, 95, 106, 226, 251

Riccardia chamedryfolia 128, 182, 318, 358, 486
Riccardia crassa 114, 135, 209, 487
Riccardia jackii 32, 67, 71, 74, 80, 84, 85, 93, 106, 113, 114, 115, 159, 261, 315, 355
Riccardia lobata var. *yakushimensis* 17, 262, 269, 480, 487
Riccardia multifida 307, 308, 310, 318, 335, 340, 358, 471, 479, 486, 487
Riccardia prehensilis 114, 287, 289
Riccardia sp. 89, 158, 261, 323, 481, 486, 487, 521
Riccardiaceae 190, 486, 487, 491
Riccardin A 307, 335, 336, 337, 471, 479, 486
Riccardin B 308, 335, 336, 351, 471, 486
Riccardin C 308, 321, 335, 336, 337, 341, 487, 507
Riccardin D 308, 336, 337
Riccardin E 308, 336, 337, 487
Riccardin F 308, 336, 337, 487
Riccardin G 308, 336, 337
Riccardiphenol A 135, 208, 209, 487, 512
Riccardiphenol B 135, 208, 209, 487
Riccia duplex 386
Riccia fluitans 381, 386
Riccia sp. 381, 507
Ricciaceae 141, 143, 184, 381, 485, 507
Ricciella sp. 381
Ricciocarpin A 129, 184, 185, 475
(±)-Ricciocarpin A 185
Ricciocarpin B 129, 184, 185, 475
Ricciocarpos natans 17, 106, 109, 129, 141, 184, 261, 297, 299, 308, 319, 320, 321, 335, 361, 381, 386, 475, 507
Ricciocarpos sp. 381, 507
Ricciofuranol 129, 184, 185
Riella sp. 466
Robustaflavone 423, 424
Roivainenia jacquinotii 69, 82, 93, 107, 114, 120, 287
ent-Rosa-5,15-diene 268
(*R*)-Rosmarinic acid 461, 462, 463, 464
Rothin A 162
Rothin A acetate 118, 162, 164

β-Sabinene 12, 14, 466, 467
(−)-β-Sabinene 18, 466
Sabinene hydrate 14, 19
(+)-D-Saccharofructose 388

Saccharomyces cerevisiae 474
Saccogyna viticulosa 135, 207
Saccogynol 135, 207
Sacculaplagin 263, 271, 272
Sacculaporellin 263, 271, 272
Sacculatal 262, 268, 269, 270, 469, 475, 480, 487
(−)-Sacculatal 269
Sacculatals 518
Sacculatanes 268, 270, 271, 514
Sacculatanolide 262, 269, 270
Saelania glaucescens 391, 393, 511
Saelenia sp. 456
Salidroside 319, 360, 361
Saliva 38, 41
Salmonela typhimurium 474
Samek rule 162, 165, 166
(−)-Sandaracopimaric acid 246, 262, 268
Sandea sp. 505
α-Santalan-12(*R*),13-diol 134, 203
Santalanes 202, 203
β-Santalene 134, 203
α-Santalol 202, 203
α-Santonin 118, 164, 474
(−)-α-Santonin 162
Saponarin 376, 409, 410
Sarcoma-37 472
Sargassaceae 520
Saussurea lactone 51, 53, 89
Saxifragaceae 322
Scapania bolanderi 212, 247, 265, 277
Scapania crasiretis 57, 74, 95
Scapania glaucoviridis 89
Scapania javanica 74, 82
Scapania maxima 135, 203
Scapania ornithopodioides 67, 69, 70, 82, 85, 86, 87, 93, 95, 103, 106, 109, 113, 114, 119, 120, 128, 130, 157, 171, 190, 246, 261
Scapania ornithopodioides var. *trifidum* 99
Scapania parvidens 387, 388, 389
Scapania reflexa 107
Scapania robusta 68, 69, 70, 85, 97, 98, 106, 107, 109, 113, 114, 125, 126, 135, 157, 178, 179, 203
Scapania sp. 59, 60, 65, 157, 171, 182, 212, 492
Scapania stephanii 388, 389
Scapania subalpina 93, 98, 120, 126, 179, 492

Scapania uliginosa 60, 74, 97, 126, 179, 492
Scapania undulata 60, 97, 126, 179, 240, 257, 258, 314, 354, 469, 492
Scapaniaceae 153, 492
Scapaniapyrone A 314, 353, 354
Scapanin 239, 240
Scapanin A 239, 240, 258, 469
Scapanin B 239, 240, 258
Scapanin B-type labdane 258
Scapanin G 239, 240, 258
Scapanins 241
Sceptridium japonicum 329
Sceptridium ternatum var. *ternatum* 328
Schiffneria hyalina 71, 246, 261, 283, 287, 290
Schiffneriolejeunea nymannii 103, 129, 283, 287
Schiffneriolejeunea omphalanthoides 68, 71, 74, 129, 283, 287
Schistochila acuminata 68, 70, 74, 82, 85, 93, 103, 106, 109, 135, 214, 248, 249, 265, 266, 278, 493
Schistochila aligera 17, 74, 89, 93, 96, 98, 214, 248, 262, 267, 493
Schistochila appendiculata 313, 351, 469, 470, 493, 520
Schistochila glaucescens 309, 311, 338, 343, 493
Schistochila glaucoviridis 85, 93, 98, 106
Schistochila laminigera 66, 74, 82, 93, 106, 261
Schistochila nobilis 82, 214, 249, 265, 278, 493
Schistochila reflexa 74, 85, 106
Schistochila rigidula 68, 278
Schistochila sp. 204
Schistochilaceae 493
Schistochilic acid A 214, 216, 249
Schistochilic acid B 214, 216, 249
Schistochilic acid C 214, 216, 249
Schistosomiasis 475
Schistosomisidal activity 465
Schistostega pennata 456
Schlotheimia japonica 440
Schuster's phylogenetic classification 484
Sciadopitys verticillata 279
Sclareol 237, 238
8-*epi*-Sclareol 237, 238, 257
13-*epi*-Sclareol 237, 238
Scleropodium touretii 392, 396, 397

Scopelophila cataractae 440, 445
Scutellarein 365, 374, 375
Scutellarein-7-O-diglucoside 365, 376
Scutellarein-7-O-glucoxylglucoside 365, 376
Scutellarein-7-O-glucoside-6″-hydroxy-3-methylglutarate 365, 376
Scutellarein-7-O-glucoside-6″-malonate 365, 376, 410
Scutellarein-7-O-glucoside-6″-malonyl ester 409
Scutellarein-6-O-glucoside-7-O-rhamnoglucoside 365, 376
Scutellarein-7-O-xyloside 365, 376
Scutellarein-6-O-xyloside-7-O-glucoside 365, 376
Secoafricanes 27, 28
Secoaromadendranes 33, 39
2,3-Secoaromadendranes 345
Secoclerodanes 212
Secoswartzianin A 27, 28, 30, 67
Secoswartzianin B 28, 30, 67
Seligeriaceae 452, 453, 508
Selgin 379, 380
Selgin-6,8-di-C-glucoside 373, 378, 379, 411
Selgin-7,5′-di-O-glucuronide 409, 414, 421
Selina-4,11-diene 114, 157, 158
Selina-11-en-4-ol 115, 158
ent-Selina-11-en-4-ol 159
α-Selinene 15, 157
ent-α-Selinene 113, 158, 160
(−)-α-Selinene 158
β-Selinene 157, 162
ent-β-Selinene 113, 158
γ-Selinene 114, 158
δ-Selinene 114, 157, 158
(−)-δ-Selinene 158, 203
Senecio oxyriifolius 31
Sephadex G-75 510
Sephadex LH-20 54, 209, 236, 337
Serpentiphenol 135, 209, 210, 500, 512, 513
Serratanes 515
β-Sesquiphellandrene 58, 59, 95
Sesquiterpenoids 5, 25, 26, 27, 66
Setiformenol 211, 212, 247
Shaftoside 412, 420
Shikimate 408
Shikimic acid 317, 355, 358, 359
Sideritis sp. 237

Silica gel 33, 42, 138, 193, 209, 219, 236, 337, 460
Silica gel-Lobor 12
Siringosterol 516
Sitosterol 280, 288, 293, 294, 392, 394, 516, 518, 519
Skatole 318, 358, 360, 468
Sodium 335
Sodium cyanoborohydride 145
Sodium hydride 154
Solenopodium stochei 211
Solenostoma obovata 466
Solenostoma sp. 229
Solidago elongata 212
Solidago juncea 214
Solidago serotina 214
Solidago sp. 212
Solvent shift method 175
Southbya sp. 490
β-Spathulene 33, 34, 71
Spathulenol 33, 35
(+)-Spathulenol 33
ent-Spathulenol 33, 34, 71
Sphaerocarpaceae 485
Sphaerocarpales 362, 374, 485, 501
Sphagnales 362, 408, 455, 456
Sphagnic acid 403, 405, 407, 408
Sphagnidae 401, 406, 419, 451, 507
Sphagnorubin 430
Sphagnorubin A 418, 428, 429, 430
Sphagnorubin B 418, 428, 430
Sphagnorubin C 418, 430
Sphagnum acid 403, 405, 407
Sphagnum angustifolium 440, 445
Sphagnum aongstroemii 405
Sphagnum balticum 405
Sphagnum centrale 405
Sphagnum compactum 405
Sphagnum contortum 405
Sphagnum cristatum 457
Sphagnum cuspidatum 405, 433, 440, 445
Sphagnum fallax 405
Sphagnum fimbriatum 405, 440, 445, 454, 473
Sphagnum fuscum 393, 405
Sphagnum imbricatum 405
Sphagnum jensenii 405
Sphagnum lindbergii 405
Sphagnum magellanicum 405, 406, 407, 418, 430, 440, 445

Sphagnum majus 405
Sphagnum molle 405
Sphagnum nemoreum 405, 418, 430, 440, 445, 473
Sphagnum obtusum 405
Sphagnum palustre 392, 394, 395, 405, 433, 440, 445, 469, 473, 519
Sphagnum papillosum 405
Sphagnum plum 418, 430
Sphagnum portoricense 473
Sphagnum quinquefarium 405
Sphagnum riparium 405
Sphagnum rubellum 405, 418, 430
Sphagnum sp. 392, 397, 398, 399, 400, 401, 403, 404, 405, 408, 433, 456, 475, 477, 510
Sphagnum squarrosum 406
Sphagnum strictum 473
Sphagnum subfulvum 406
Sphagnum subsecundum 406, 473
Sphagnum temellum 406
Sphagnum teres 392, 396, 397, 406
Sphagnum urussowii 405
Sphagnum warnstorfii 406
Sphenolobanes 272, 273, 274
Sphenolobus minutus 70, 83
Sphenopsidae 510
Spinuloplagin A 228, 252
Spinuloplagin B 228, 252
Spiroclerodanes 212, 218
Spirodensifolin A 134, 198, 199
Spirodensifolin B 134, 198, 199, 200
Spiropinguisanes 201
Spiropinguisanin 134, 197, 198
Spirovetivanes 202, 203, 204
Spirovetivene 203, 204
α-Spirovetivene 135, 203
β-Spirovetivene 135, 203
Splachnaceae 391, 432, 433, 468, 511
Splachnum luteum 391, 393, 406, 432, 434, 445, 446
Splachnum melanocaulon 432, 445, 446
Splachnum rubrum 391, 393, 404, 406, 432, 433, 434, 446, 447
Splachnum sp. 391, 403, 468
Splachnum sphaericum 404, 406, 432, 433, 434, 445, 446
Splachnum vasculosum 404, 406, 432, 434, 445, 446
Spodoptera exempta 475

Sporochnol 210, 512
Sporochnus bolleanus 209
Sporothrix schenckii 474
Spruceanthus polymorphus 129, 183
Squalanes 515
Squalene 290, 293, 294, 395, 403, 519, 520
Staphylococcus aureus 473, 474
Stearic acid 380, 383, 456
Stellarin-2 373, 378, 379, 380, 411, 497
Stephaniella paraphyllina 74, 93
Steroids 293, 402, 511, 516
Sterols 519
Stevia rebaudiana 469
Stevioside 469
Stictolejeunea balfourii var. *bekkei* 17, 18, 74, 125, 287, 302, 325
Stictolejeunea squamata 74, 261, 283, 287, 372, 377
Stigamasta-5,24(28)E-dien-3β-ol 519
Stigmasterol 280, 284, 293, 294, 392, 394, 516, 518
Stigmasteryl glucoside 287, 293
Stigmasteryl-3β-glucoside 294
Streptomyces sp. 466
Striatene 128, 182, 183, 184, 185, 500
Striatenic acid 129, 185
Striatenone 129, 183, 185
Striatol 129, 182, 183, 185, 501
Stypodium zonale 209
Styrene derivatives 517
(+)-Sucrose 391
Sulfoquinovosyl diacetylglycerols 455
Sulfur-containing compounds 384, 517
Sulfuric acid 278
Superoxide dismutase-like activity 479
Superoxide release inhibitory activity 479
Swartzianin A 28, 30, 66
Swartzianin B 28, 30, 66
Swartzianin C 28, 30, 67
Swartzianin D 28, 30, 67
Swertia japonica 468
Symbiezidium barbiflorum 20, 287
Symbiezidium transversale var. *hookeriana* 283, 287
Symphyogyna brasiliensis 89, 119, 170, 226, 251
Symphyogyna brongniartii 22, 25, 83, 93, 104, 114, 158, 263, 271
Symphyogyna podophylla 378
Symphyogyna sp. 140

Symphyogynoideae 378
Syzygiella anomala 70, 74, 120, 287, 291

Takakia ceratophylla 488, 525
Takakia lepidozioides 466, 488
Takakia sp. 485, 488, 525
Takakiales 374, 485, 488, 525
Takakiineae 488
Tamariscol 130, 188, 189, 190, 467, 468, 499
(+)-Tamariscol 188
(−)-Tamariscol 188
(±)-Tamariscol 189
Tamariscol MW-222 188
Taraxanes 515
Taraxerol 392, 397, 402, 403
Taraxerone 392, 397, 402
Targionia hypophylla 12, 17, 20, 21, 107, 111, 237, 257, 467, 503
Targionia sp. 152, 502
Targioniaceae 12, 502
Taxifolin 428
Taxiphyllum taxirameum 465
Taxodiaceae 279
Tayloria sp. 468
Tayloria tenuis 432, 434
Taylorione 35, 36, 40, 41, 76, 277
Termarol 262, 267, 268
Terpenoids 5, 511, 513
α-Terpinene 12, 13, 17, 466
γ-Terpinene 12, 13, 17, 466
Terpinene-4-ol 13, 18
α-Terpineol 12, 13, 18, 511
Terpinolene 12, 13, 18, 466
α-Terpinyl acetate 13, 14, 18
Tetracosanoic acid 433
Tetrahydrofuran 226, 321
1‴,2‴,3‴,4‴-Tetrahydro-3,3″,5,5″,7,7″-hexahydroxy-4‴-keto-3′,4′-O-2‴-biflavone 417, 427
Tetrahydroscapanin A 240
3,5,7,4′-Tetrahydroxy-3′-(3″-formyl-6″-hydroxyphenyl)flavanone 414, 421
5,7,3′,4′-Tetrahydroxyisoflavone-7-O-glucoside 422
Tetralophozia setiformis 74, 211, 223, 247, 489
Tetraphenylethene 335
Tetraphidales 362, 421
Tetraphis pellucida 421

Tetraplodon mniodes 432, 434
Tetraplodon sp. 468
Thamnium alopecurum subsp.
    eualopecurum 395
Thamnium sp. 392
Thamnobryum pandum 457
Thamnobryum plicatulum 440, 445
Thamnobryum sandei 450, 457
Thamnobryum sp. 460
Thuidiaceae 471
Thuidiopsis furfurosa 398, 399, 400, 401
Thuidium glaucinum 440, 445
Thuidium kanedae 391, 392, 395, 396, 397,
    450, 460
Thuidium recognitum 398, 399, 400, 401,
    440, 445
Thuidium recognitum var. *delicatulum* 441,
    445, 473
Thuidium sp. 392
Thuidium tamariscifolium 392, 397
Thuidium tamariscinum 441, 445
(+)-Thujanol 12
(−)-Thujanol 12, 14, 18
1-*epi*-Thujanol 14, 18
α-Thujene 12, 13, 466
(+)-α-Thujene 12, 18
(−)-α-Thujene 12
Thujopsanes 202, 205
ent-Thujopsan-7β-ol 135, 205
Thujopsene 204
(+)-Thujopsene 204
(−)-Thujopsene 204, 205
ent-Thujopsene 135, 205
(+)-*ent*-Thujopsene 204
(+)-Thujopsenone 204, 205
ent-Thujopsenone 135, 204, 205
(−)-*ent*-Thujopsenone 204
Thymol 13, 18
Thysananthus amazonicus 17, 21, 70, 74, 87,
    121, 129, 261, 287, 291
Thysananthus convolutus 68, 75, 87, 89, 93,
    101, 113, 114, 121, 283, 287, 291
Thysananthus fruticosus 101, 129, 131, 283,
    287
Thysananthus mollis 68, 75, 101, 113, 283,
    287, 291
Thysananthus pterobryoides 71, 75, 85, 93,
    103, 107
Thysananthus sp. 158
α-Tocopherol 482, 483

δ-Tocopherol 318, 360, 361
α-Tocoquinone 451, 482, 483
p-Toluenesulfonic acid 141, 145
Tomentellin 296, 297, 298
Tomentellol 296, 297, 298
Torreya nucifera 58
Torreyol 60, 62, 97
Tortula muralis 441, 445, 456, 473
Tosylhydrazine 33
Tracheophytes 65
Trachylobanes 273, 274
Tranilast 480
(+)-Trehalose 389, 391
Trewia nudiflora 473
Trewiasine 450, 458
ent-9,11α,14-Triacetoxylongipin-2(10)-en-
    3-one 127, 180
ent-9,11β,14-Triacetoxylongipin-2(10)-en-
    3-one 127, 180
9,11α,14-Triacetoxymarsupellone A 127,
    179, 180
9,11β,14-Triacetoxymarsupellone B 127,
    179, 180
Triacylglycerols 456
Triandrophyllum sp. 178
Triandrophyllum subtrifidum 283, 352
Triandrophyllum subtrifidum var.
    trifidum 93, 98, 99, 102, 107, 110, 125,
    146, 283, 287, 291, 314
Tricetin 379, 380
Tricetin-6-C-arabinopyranoside-8-C-
    glucoside 378
Tricetin-6-C-arabinoside-8-C-
    glucoside 373, 379
Tricetin-6,8-di-C-arabinoside 373,
    379
Tricetin-6,8-di-C-α-L-arabinoside 378
Tricetin-6,8-di-C-glucopyranoside 378
Tricetin-6,8-di-C-β-D-
    glucopyranoside 377, 378, 484, 497
Tricetin-6,8-di-C-glucoside 372, 379, 380,
    497
Tricetin-7,3′-di-O-glucuronide 409, 414,
    421
Tricetin-3′,4′-dimethyl ether-7-O-
    monoglucuronide 409, 413, 421
Tricetin-6-C-β-D-glucopyranoside-8-C-α-
    L-arabinopyranoside 379
Tricetin-6-C-glucoside 373, 379
Tricetin-6-C-β-glucoside 497

Tricetin-6-C-glucoside-8-C-arabinoside 373, 379
Tricetin-7-O-glucoside-3'-O-glucoside-6''-hydroxy-3-methyl-glutarate 373, 379
Tricetin-6-C-β-D-glucosyl-8-C-α-L-arabinoside 378
Trichocolea mollissima 296, 298
Trichocolea pluma 17, 18, 20, 68, 69, 96, 104, 111, 243, 246, 259, 261, 296, 298, 495
Trichocolea sp. 140, 152, 362, 484, 495, 496
Trichocolea tomentella 243, 261, 296, 298, 466, 495, 496
Trichocoleaceae 190, 491, 495
Trichocolein 296, 297, 298
Trichocoleopsis sacculata 193, 268, 475, 480
Trichocoleopsis sp. 468, 481, 495
Tricholoma matsutake 466
Trichophyton mentagrophytes 473, 474
Trichophyton rubrum 474
Tricin 379, 380
Tricin-6,8-di-C-glucoside 373, 379, 380
α-Tricodiene 49, 51
Tricyclene 202, 203
3-Tridecylphenol 313, 351, 352, 520
Tridecylsalicylic acid 520
6-Tridecylsalicylic acid 313, 351, 352
Tridensenal 186, 474
(−)-Tridensenal 130, 185
Tridensene 130, 187
(+)-Tridensone 186
(−)-Tridensone 130, 185, 186
Triethylamine 321
Trifarienol A 135, 209, 210
Trifarienol B 135, 209, 210
Trifarienol C 135, 209, 210
Trifarienol D 135, 209, 210
Trifarienol E 135, 209, 210
1,3,5-Trihydroxybenzene 457
5,3',4'-Trihydroxybibenzyl-2-O-β-D-glucopyranoside 299, 320, 321
5,7,3'-Trihydroxy-4'-methoxyisoflavone 422
3,4,5-Trihydroxy-2-(3-methyl-2-butenyl)bibenzyl 496
2β,9α,13β-Trihydroxyverrucosane 264, 275, 276
2,4,5-Trimethoxyallylbenzene 317, 359
3-(3',4',5'-Trimethoxybenzyl)-5,7-dimethoxyphthalide 316, 357

3,4,5-Trimethoxybenzyltriphenylphosphonium bromide 340
3,4,7-Trimethoxy-9,10-dihydrophenanthrene 315, 355, 356
3,4,4'-Trimethoxy-2-hydroxybibenzyl 301, 324
2,4,7-Trimethoxynaphthalene 314, 352, 353
2,4,5-Trimethoxystyrene 317, 359
(−)-2-(1,2,2-Trimethyl)-cyclopentyl-6-methyl-1,4-quinone 108, 140, 142
(1S)-1,2,2-Trimethyl-m-tolyl-cyclopentane 109, 143, 144
Trinoranastreptene 22, 24, 25, 43, 44, 46
Triphenylphosphitemethiodide 145
Tris-normonoterpenoids 15, 16, 23
Tris-norsesquiterpenoids 24, 66
Triterpenoids 293, 294, 295, 402
Tritomaria quinquedentata 75, 119, 168, 287, 386, 489
Trocholejeunea sandvicensis 56, 95, 131, 132, 133, 197, 301, 316, 318, 325, 358, 361, 471
Trocholejeunin 318, 360
Tryptophan 478
Tubocurarine 481
d-Tubocurarine 481, 482
Tulipinolide 122, 162, 173, 174, 176, 504
Tumor promoting activity 465
Tylimanthin A 306, 331, 332
Tylimanthin B 306, 331, 332
Tylimanthin C 306, 331, 332
Tylimanthus urvilleanus 120, 283, 287, 306, 332
Tympanitis 465

Ullmann coupling 335
Umbelliferae 173
6-Undecylcatechol 313, 351, 352
3-Undecylphenol 313, 351, 352, 520
3-Undecylsalicylic acid 351
6-Undecylsalicylic acid 313, 351, 352, 520
Urbanodendron verrucosum 355
Urocystitis 465
Uropathy 465
Ursanes 515
Ursolic acid 392, 397, 402
Uvaria purpurea 296

Valencanes 202, 206

Valencene 204, 206
Valeric acid 433, 446
Vanillic acid 403, 404, 407
Vanillin 403, 404, 407
Vasopressin antagonist activity 480
Velutin 367, 376
Ventricosenediolide 217, 218, 249
Ventricosin A 115, 160, 161
Ventricosin B 124, 176
Vernonia sp. 53
Verrucosanes 273, 275, 277, 278, 279, 514
ent-Verticillanediol 266, 279, 280
ent-13-epi-Verticillanediol 266, 279, 280
Verticillanes 278, 280
Verticillene 266, 279, 280
(+)-Verticillol 279
ent-Verticillol 266, 278, 280
ent-13-epi-Verticillol 266, 279, 280
Vesicularia ferriei 441, 445
Vicenin-2 364, 374, 378, 380, 410
4-Vinylguaiacol 317, 358, 359
Vinylmagnesium bromide 467
Violaxanthin 400, 403
Viridiflorol 34, 75
(−)-Viridiflorol 33
Vitamin B₂ 482
Vitamin E 482, 483
Vitamin K 482, 483
Vitexin-2″-rhamnoside 412, 420
Vitranes 205, 206, 514
Vitrenal 135, 206, 207
(+)-Vitrenal 205, 206, 477
(−)-Vitrenal 205, 477
(1R,6R,7S,10R)-Vitr-4-en-14-al 205
Volatile terpenoids 64

Warburganal 475, 476
(−)-Warburganal 470
Warburgia ugandensis 475
Weissia viridula 465
Wettstein A 314, 353
Wettstein B 314, 353
Wettstein C 314, 353
Wettsteinia inversa 314, 352, 493
Wettsteinia schusterana 314, 353, 493
Wettsteinia sp. 493
Wettsteinolide 314, 353
Widdranes 205, 207
Widdrene 207
(−)-Widdrol 135, 207
Wiesnerella denudata 16, 17, 18, 19, 21, 60,
    97, 101, 118, 121, 122, 123, 125, 162, 171,
    173, 176, 261, 291, 308, 309, 317, 338,
    358, 380, 385, 466, 471, 472, 475, 503,
    504
Wiesnerella sp. 468
Wijkia concavifolia 441, 445
Wittig reaction 33, 335, 340, 347
Wolff-Kishner reduction 231, 277

Xanthones 517
Xylose 374, 389
Xylosides 375

Zaluzanin D 178, 472
Zeaxanthin 401, 403
α-Zeorin 292, 293, 295
Zieranes 205
Zierone 207
Zooxanthella sp. 512

SpringerChemistry

Fortschritte der Chemie organischer Naturstoffe

Progress in the Chemistry

of Organic Natural Products

Founded by L. Zechmeister
Edited by W. Herz, G. W. Kirby, R. E. Moore, W. Steglich,
and Ch. Tamm

Volume 64

1995. 22 partly coloured figures. VII, 216 pages. Cloth DM 250,–, öS 1750,–
Subscription price: Cloth DM 225,–, öS 1575,–. ISBN 3-211-82533-9

Contents:
A. G. González and J. Bermejo Barrera: Chemistry and Sources
of Mono- and Bicyclic Sesquiterpenes from Ferula Species.
G. Prota: The Chemistry of Melanins and Melanogenesis.
H. J. M. Gijsen, J. B. P. A. Wijnberg, and Ae. de Groot: Structure,
Occurrence, Biosynthesis, Biological Activity, Synthesis, and
Chemistry of Aromadendrane Sesquiterpenoids.

Volume 63

1994. VII, 216 pages. Cloth DM 220,–, öS 1540,–
Subscription price: Cloth DM 198,–, öS 1386,–. ISBN 3-211-82443-X

Contents:
A. B. Ray and M. Gupta: Withasteroids, a Growing Group of
Naturally Occurring Steroidal Lactones.
L. Rodríguez-Hahn, B. Esquivel, and J. Cárdenas: Clerodane
Diterpenes in Labiatae.

 SpringerWienNewYork

P.O. Box 89, A-1201 Wien • New York, NY 10010, 175 Fifth Avenue
Heidelberger Platz 3, D-14197 Berlin • Tokyo 113, 3-13, Hongo 3-chome, Bunkyo-ku

SpringerChemistry

Fortschritte der Chemie organischer Naturstoffe

Progress in the Chemistry

of Organic Natural Products

Founded by L. Zechmeister
Edited by W. Herz, G. W. Kirby, R. E. Moore, W. Steglich,
and Ch. Tamm

Volume 64

1995. 22 partly coloured figures. VII, 216 pages. Cloth DM 250,–, öS 1750,–
Subscription price: Cloth DM 225,–, öS 1575,–. ISBN 3-211-82533-9

Contents:
A. G. González and J. Bermejo Barrera: Chemistry and Sources
of Mono- and Bicyclic Sesquiterpenes from Ferula Species.
G. Prota: The Chemistry of Melanins and Melanogenesis.
H. J. M. Gijsen, J. B. P. A. Wijnberg, and Ae. de Groot: Structure,
Occurrence, Biosynthesis, Biological Activity, Synthesis, and
Chemistry of Aromadendrane Sesquiterpenoids.

Volume 63

1994. VII, 216 pages. Cloth DM 220,–, öS 1540,–
Subscription price: Cloth DM 198,–, öS 1386,–. ISBN 3-211-82443-X

Contents:
A. B. Ray and M. Gupta: Withasteroids, a Growing Group of
Naturally Occurring Steroidal Lactones.
L. Rodríguez-Hahn, B. Esquivel, and J. Cárdenas: Clerodane
Diterpenes in Labiatae.

 SpringerWienNewYork

P.O.Box 89, A-1201 Wien • New York, NY 10010, 175 Fifth Avenue
Heidelberger Platz 3, D-14197 Berlin • Tokyo 113, 3-13, Hongo 3-chome, Bunkyo-ku

SpringerChemistry

Fortschritte der Chemie organischer Naturstoffe

Progress in the Chemistry

of Organic Natural Products

Founded by L. Zechmeister
Edited by W. Herz, G. W. Kirby, R. E. Moore, W. Steglich,
and Ch. Tamm

Volume 62

1993. 52 figures. VIII, 330 pages. Cloth DM 280,–, öS 1960,–
Subscription price: Cloth DM 252,–, öS 1764,–. ISBN 3-211-82402-2

Contents:
Sujata V. Bhat: Forskolin and Congeners.
L. Minale, R. Riccio and F. Zollo: Steroidal Oligoglycosides and
Polyhydroxysteroids from Echinoderms.

Volume 61

1993. 4 figures. IX, 206 pages. Cloth DM 220,–, öS 1540,–
Subscription price: Cloth DM 198,–, öS 1386,–. ISBN 3-211-82388-3

Contents:
D. G. I. Kingston, A. A. Molinero, and J. M. Rimoldi: The Taxane
Diterpenoids.

 SpringerWienNewYork

P.O.Box 89, A-1201 Wien • New York, NY 10010, 175 Fifth Avenue
Heidelberger Platz 3, D-14197 Berlin • Tokyo 113, 3-13, Hongo 3-chome, Bunkyo-ku